Lecture Notes in Computer Science 12747

More information about this subseries at http://www.springer.com/series/7407

Maciej Paszynski · Dieter Kranzlmüller ·
Valeria V. Krzhizhanovskaya ·
Jack J. Dongarra · Peter M. A. Sloot (Eds.)

Computational Science – ICCS 2021

21st International Conference
Krakow, Poland, June 16–18, 2021
Proceedings, Part VI

 Springer

Editors
Maciej Paszynski (iD)
AGH University of Science and Technology
Krakow, Poland

Valeria V. Krzhizhanovskaya (iD)
University of Amsterdam
Amsterdam, The Netherlands

Peter M. A. Sloot (iD)
University of Amsterdam
Amsterdam, The Netherlands

ITMO University
St. Petersburg, Russia

Nanyang Technological University
Singapore, Singapore

Dieter Kranzlmüller (iD)
Ludwig-Maximilians-Universität München
Munich, Germany

Leibniz Supercomputing Center (LRZ)
Garching bei München, Germany

Jack J. Dongarra (iD)
University of Tennessee at Knoxville
Knoxville, TN, USA

ISSN 0302-9743   ISSN 1611-3349   (electronic)
Lecture Notes in Computer Science
ISBN 978-3-030-77979-5   ISBN 978-3-030-77980-1   (eBook)
https://doi.org/10.1007/978-3-030-77980-1

LNCS Sublibrary: SL1 – Theoretical Computer Science and General Issues

This Springer imprint is published by the registered company Springer Nature Switzerland AG
The registered company address is: Gewerbestrasse 11, 6330 Cham, Switzerland

Preface

Welcome to the proceedings of the 21st annual International Conference on Computational Science (ICCS 2021 - https://www.iccs-meeting.org/iccs2021/).

In preparing this edition, we had high hopes that the ongoing COVID-19 pandemic would fade away and allow us to meet this June in the beautiful city of Kraków, Poland. Unfortunately, this is not yet the case, as the world struggles to adapt to the many profound changes brought about by this crisis. ICCS 2021 has had to adapt too and is thus being held entirely online, for the first time in its history.

These challenges notwithstanding, we have tried our best to keep the ICCS community as dynamic and productive as always. We are proud to present the proceedings you are reading as a result of that.

ICCS 2021 was jointly organized by the AGH University of Science and Technology, the University of Amsterdam, NTU Singapore, and the University of Tennessee.

The International Conference on Computational Science is an annual conference that brings together researchers and scientists from mathematics and computer science as basic computing disciplines, as well as researchers from various application areas who are pioneering computational methods in sciences such as physics, chemistry, life sciences, engineering, arts, and humanitarian fields, to discuss problems and solutions in the area, identify new issues, and shape future directions for research.

Since its inception in 2001, ICCS has attracted an increasing number of attendees and higher quality papers, and this year is not an exception, with over 350 registered participants. The proceedings have become a primary intellectual resource for computational science researchers, defining and advancing the state of the art in this field.

The theme for 2021, "**Computational Science for a Better Future**," highlights the role of computational science in tackling the current challenges of our fast-changing world. This conference was a unique event focusing on recent developments in scalable scientific algorithms, advanced software tools, computational grids, advanced numerical methods, and novel application areas. These innovative models, algorithms, and tools drive new science through efficient application in physical systems, computational and systems biology, environmental systems, finance, and other areas.

ICCS is well known for its excellent lineup of keynote speakers. The keynotes for 2021 were given by

- **Maciej Besta**, ETH Zürich, Switzerland
- **Marian Bubak**, AGH University of Science and Technology, Poland | Sano Centre for Computational Medicine, Poland
- **Anne Gelb**, Dartmouth College, USA
- **Georgiy Stenchikov**, King Abdullah University of Science and Technology, Saudi Arabia
- **Marco Viceconti**, University of Bologna, Italy

- **Krzysztof Walczak**, Poznan University of Economics and Business, Poland
- **Jessica Zhang**, Carnegie Mellon University, USA

This year we had 635 submissions (156 submissions to the main track and 479 to the thematic tracks). In the main track, 48 full papers were accepted (31%); in the thematic tracks, 212 full papers were accepted (44%). A high acceptance rate in the thematic tracks is explained by the nature of these tracks, where organisers personally invite many experts in a particular field to participate in their sessions.

ICCS relies strongly on our thematic track organizers' vital contributions to attract high-quality papers in many subject areas. We would like to thank all committee members from the main and thematic tracks for their contribution to ensure a high standard for the accepted papers. We would also like to thank *Springer, Elsevier,* and *Intellegibilis* for their support. Finally, we appreciate all the local organizing committee members for their hard work to prepare for this conference.

We are proud to note that ICCS is an A-rank conference in the CORE classification.

We wish you good health in these troubled times and look forward to meeting you at the conference.

June 2021

Maciej Paszynski
Dieter Kranzlmüller
Valeria V. Krzhizhanovskaya
Jack J. Dongarra
Peter M. A. Sloot

Organization

Local Organizing Committee at AGH University of Science and Technology

Chairs

Maciej Paszynski
Aleksander Byrski

Members

Marcin Łos
Maciej Woźniak
Leszek Siwik
Magdalena Suchoń

Thematic Tracks and Organizers

Advances in High-Performance Computational Earth Sciences: Applications and Frameworks – IHPCES

Takashi Shimokawabe
Kohei Fujita
Dominik Bartuschat

Applications of Computational Methods in Artificial Intelligence and Machine Learning – ACMAIML

Kourosh Modarresi
Paul Hofmann
Raja Velu
Peter Woehrmann

Artificial Intelligence and High-Performance Computing for Advanced Simulations – AIHPC4AS

Maciej Paszynski
Robert Schaefer
David Pardo
Victor Calo

Biomedical and Bioinformatics Challenges for Computer Science – BBC

Mario Cannataro
Giuseppe Agapito

Mauro Castelli
Riccardo Dondi
Italo Zoppis

Classifier Learning from Difficult Data – CLD²

Michał Woźniak
Bartosz Krawczyk

Computational Analysis of Complex Social Systems – CSOC

Debraj Roy

Computational Collective Intelligence – CCI

Marcin Maleszka
Ngoc Thanh Nguyen
Marcin Hernes
Sinh Van Nguyen

Computational Health – CompHealth

Sergey Kovalchuk
Georgiy Bobashev
Stefan Thurner

Computational Methods for Emerging Problems in (dis-)Information Analysis – DisA

Michal Choras
Robert Burduk
Konstantinos Demestichas

Computational Methods in Smart Agriculture – CMSA

Andrew Lewis

Computational Optimization, Modelling, and Simulation – COMS

Xin-She Yang
Leifur Leifsson
Slawomir Koziel

Computational Science in IoT and Smart Systems – IoTSS

Vaidy Sunderam
Dariusz Mrozek

Computer Graphics, Image Processing and Artificial Intelligence – CGIPAI

Andres Iglesias
Lihua You
Alexander Malyshev
Hassan Ugail

Data-Driven Computational Sciences – DDCS

Craig Douglas

Machine Learning and Data Assimilation for Dynamical Systems – MLDADS

Rossella Arcucci

**MeshFree Methods and Radial Basis Functions in Computational
Sciences – MESHFREE**

Vaclav Skala
Marco-Evangelos Biancolini
Samsul Ariffin Abdul Karim
Rongjiang Pan
Fernando-César Meira-Menandro

Multiscale Modelling and Simulation – MMS

Derek Groen
Diana Suleimenova
Stefano Casarin
Bartosz Bosak
Wouter Edeling

Quantum Computing Workshop – QCW

Katarzyna Rycerz
Marian Bubak

**Simulations of Flow and Transport: Modeling, Algorithms
and Computation – SOFTMAC**

Shuyu Sun
Jingfa Li
James Liu

**Smart Systems: Bringing Together Computer Vision, Sensor Networks
and Machine Learning – SmartSys**

Pedro Cardoso
Roberto Lam

João Rodrigues
Jânio Monteiro

Software Engineering for Computational Science – SE4Science

Jeffrey Carver
Neil Chue Hong
Anna-Lena Lamprecht

Solving Problems with Uncertainty – SPU

Vassil Alexandrov
Aneta Karaivanova

Teaching Computational Science – WTCS

Angela Shiflet
Nia Alexandrov
Alfredo Tirado-Ramos

Uncertainty Quantification for Computational Models – UNEQUIvOCAL

Wouter Edeling
Anna Nikishova

Reviewers

Ahmad Abdelfattah
Samsul Ariffin Abdul
   Karim
Tesfamariam Mulugeta
   Abuhay
Giuseppe Agapito
Elisabete Alberdi
Luis Alexandre
Vassil Alexandrov
Nia Alexandrov
Julen Alvarez-Aramberri
Sergey Alyaev
Tomasz Andrysiak
Samuel Aning
Michael Antolovich
Hideo Aochi
Hamid Arabnejad
Rossella Arcucci
Costin Badica
Marina Balakhontceva

Bartosz Balis
Krzysztof Banas
Dariusz Barbucha
Valeria Bartsch
Dominik Bartuschat
Pouria Behnodfaur
Joern Behrens
Adrian Bekasiewicz
Gebrail Bekdas
Mehmet Belen
Stefano Beretta
Benjamin Berkels
Daniel Berrar
Sanjukta Bhowmick
Georgiy Bobashev
Bartosz Bosak
Isabel Sofia Brito
Marc Brittain
Jérémy Buisson
Robert Burduk

Michael Burkhart
Allah Bux
Krisztian Buza
Aleksander Byrski
Cristiano Cabrita
Xing Cai
Barbara Calabrese
Jose Camata
Almudena Campuzano
Mario Cannataro
Alberto Cano
Pedro Cardoso
Alberto Carrassi
Alfonso Carriazo
Jeffrey Carver
Manuel Castañón-Puga
Mauro Castelli
Eduardo Cesar
Nicholas Chancellor
Patrikakis Charalampos

Henri-Pierre Charles
Ehtzaz Chaudhry
Long Chen
Sibo Cheng
Siew Ann Cheong
Lock-Yue Chew
Marta Chinnici
Sung-Bae Cho
Michal Choras
Neil Chue Hong
Svetlana Chuprina
Paola Cinnella
Noélia Correia
Adriano Cortes
Ana Cortes
Enrique
  Costa-Montenegro
David Coster
Carlos Cotta
Helene Coullon
Daan Crommelin
Attila Csikasz-Nagy
Loïc Cudennec
Javier Cuenca
António Cunha
Boguslaw Cyganek
Ireneusz Czarnowski
Pawel Czarnul
Lisandro Dalcin
Bhaskar Dasgupta
Konstantinos Demestichas
Quanling Deng
Tiziana Di Matteo
Eric Dignum
Jamie Diner
Riccardo Dondi
Craig Douglas
Li Douglas
Rafal Drezewski
Vitor Duarte
Thomas Dufaud
Wouter Edeling
Nasir Eisty
Kareem El-Safty
Amgad Elsayed
Nahid Emad

Christian Engelmann
Roberto R. Expósito
Fangxin Fang
Antonino Fiannaca
Christos
  Filelis-Papadopoulos
Martin Frank
Alberto Freitas
Ruy Freitas Reis
Karl Frinkle
Kohei Fujita
Hiroshi Fujiwara
Takeshi Fukaya
Wlodzimierz Funika
Takashi Furumura
Ernst Fusch
David Gal
Teresa Galvão
Akemi Galvez-Tomida
Ford Lumban Gaol
Luis Emilio
  Garcia-Castillo
Frédéric Gava
Piotr Gawron
Alex Gerbessiotis
Agata Gielczyk
Adam Glos
Sergiy Gogolenko
Jorge
  González-Domínguez
Yuriy Gorbachev
Pawel Gorecki
Michael Gowanlock
Ewa Grabska
Manuel Graña
Derek Groen
Joanna Grzyb
Pedro Guerreiro
Tobias Guggemos
Federica Gugole
Bogdan Gulowaty
Shihui Guo
Xiaohu Guo
Manish Gupta
Piotr Gurgul
Filip Guzy

Pietro Hiram Guzzi
Zulfiqar Habib
Panagiotis Hadjidoukas
Susanne Halstead
Feilin Han
Masatoshi Hanai
Habibollah Haron
Ali Hashemian
Carina Haupt
Claire Heaney
Alexander Heinecke
Marcin Hernes
Bogumila Hnatkowska
Maximilian Höb
Jori Hoencamp
Paul Hofmann
Claudio Iacopino
Andres Iglesias
Takeshi Iwashita
Alireza Jahani
Momin Jamil
Peter Janku
Jiri Jaros
Caroline Jay
Fabienne Jezequel
Shalu Jhanwar
Tao Jiang
Chao Jin
Zhong Jin
David Johnson
Guido Juckeland
George Kampis
Aneta Karaivanova
Takahiro Katagiri
Timo Kehrer
Christoph Kessler
Jakub Klikowski
Alexandra Klimova
Harald Koestler
Ivana Kolingerova
Georgy Kopanitsa
Sotiris Kotsiantis
Sergey Kovalchuk
Michal Koziarski
Slawomir Koziel
Rafal Kozik

Bartosz Krawczyk
Dariusz Krol
Valeria Krzhizhanovskaya
Adam Krzyzak
Pawel Ksieniewicz
Marek Kubalcík
Sebastian Kuckuk
Eileen Kuehn
Michael Kuhn
Michal Kulczewski
Julian Martin Kunkel
Krzysztof Kurowski
Marcin Kuta
Bogdan Kwolek
Panagiotis Kyziropoulos
Massimo La Rosa
Roberto Lam
Anna-Lena Lamprecht
Rubin Landau
Johannes Langguth
Shin-Jye Lee
Mike Lees
Leifur Leifsson
Kenneth Leiter
Florin Leon
Vasiliy Leonenko
Roy Lettieri
Jake Lever
Andrew Lewis
Jingfa Li
Hui Liang
James Liu
Yen-Chen Liu
Zhao Liu
Hui Liu
Pengcheng Liu
Hong Liu
Marcelo Lobosco
Robert Lodder
Chu Kiong Loo
Marcin Los
Stephane Louise
Frederic Loulergue
Hatem Ltaief
Paul Lu
Stefan Luding

Laura Lyman
Scott MacLachlan
Lukasz Madej
Lech Madeyski
Luca Magri
Imran Mahmood
Peyman Mahouti
Marcin Maleszka
Alexander Malyshev
Livia Marcellino
Tomas Margalef
Tiziana Margaria
Osni Marques
M. Carmen Márquez
    García
Paula Martins
Jaime Afonso Martins
Pawel Matuszyk
Valerie Maxville
Pedro Medeiros
Fernando-César
    Meira-Menandro
Roderick Melnik
Valentin Melnikov
Ivan Merelli
Marianna Milano
Leandro Minku
Jaroslaw Miszczak
Kourosh Modarresi
Jânio Monteiro
Fernando Monteiro
James Montgomery
Dariusz Mrozek
Peter Mueller
Ignacio Muga
Judit Munoz-Matute
Philip Nadler
Hiromichi Nagao
Jethro Nagawkar
Kengo Nakajima
Grzegorz J. Nalepa
I. Michael Navon
Philipp Neumann
Du Nguyen
Ngoc Thanh Nguyen
Quang-Vu Nguyen

Sinh Van Nguyen
Nancy Nichols
Anna Nikishova
Hitoshi Nishizawa
Algirdas Noreika
Manuel Núñez
Krzysztof Okarma
Pablo Oliveira
Javier Omella
Kenji Ono
Eneko Osaba
Aziz Ouaarab
Raymond Padmos
Marek Palicki
Junjun Pan
Rongjiang Pan
Nikela Papadopoulou
Marcin Paprzycki
David Pardo
Anna Paszynska
Maciej Paszynski
Abani Patra
Dana Petcu
Serge Petiton
Bernhard Pfahringer
Toby Phillips
Frank Phillipson
Juan C. Pichel
Anna
    Pietrenko-Dabrowska
Laércio L. Pilla
Yuri Pirola
Nadia Pisanti
Sabri Pllana
Mihail Popov
Simon Portegies Zwart
Roland Potthast
Malgorzata
    Przybyla-Kasperek
Ela Pustulka-Hunt
Alexander Pyayt
Kun Qian
Yipeng Qin
Rick Quax
Cesar Quilodran Casas
Enrique S. Quintana-Orti

Ewaryst Rafajlowicz
Ajaykumar Rajasekharan
Raul Ramirez
Célia Ramos
Marcus Randall
Lukasz Rauch
Vishal Raul
Robin Richardson
Sophie Robert
João Rodrigues
Daniel Rodriguez
Albert Romkes
Debraj Roy
Jerzy Rozenblit
Konstantin Ryabinin
Katarzyna Rycerz
Khalid Saeed
Ozlem Salehi
Alberto Sanchez
Aysin Sanci
Gabriele Santin
Rodrigo Santos
Robert Schaefer
Karin Schiller
Ulf D. Schiller
Bertil Schmidt
Martin Schreiber
Gabriela Schütz
Christoph Schweimer
Marinella Sciortino
Diego Sevilla
Mostafa Shahriari
Abolfazi
  Shahzadeh-Fazeli
Vivek Sheraton
Angela Shiflet
Takashi Shimokawabe
Alexander Shukhman
Marcin Sieniek
Nazareen
  Sikkandar Basha
Anna Sikora
Diana Sima
Robert Sinkovits
Haozhen Situ
Leszek Siwik

Vaclav Skala
Ewa
  Skubalska-Rafajlowicz
Peter Sloot
Renata Slota
Oskar Slowik
Grazyna Slusarczyk
Sucha Smanchat
Maciej Smolka
Thiago Sobral
Robert Speck
Katarzyna Stapor
Robert Staszewski
Steve Stevenson
Tomasz Stopa
Achim Streit
Barbara Strug
Patricia Suarez Valero
Vishwas Hebbur Venkata
Subba Rao
Bongwon Suh
Diana Suleimenova
Shuyu Sun
Ray Sun
Vaidy Sunderam
Martin Swain
Jerzy Swiatek
Piotr Szczepaniak
Tadeusz Szuba
Ryszard Tadeusiewicz
Daisuke Takahashi
Zaid Tashman
Osamu Tatebe
Carlos Tavares Calafate
Andrei Tchernykh
Kasim Tersic
Jannis Teunissen
Nestor Tiglao
Alfredo Tirado-Ramos
Zainab Titus
Pawel Topa
Mariusz Topolski
Pawel Trajdos
Bogdan Trawinski
Jan Treur
Leonardo Trujillo

Paolo Trunfio
Ka-Wai Tsang
Hassan Ugail
Eirik Valseth
Ben van Werkhoven
Vítor Vasconcelos
Alexandra Vatyan
Raja Velu
Colin Venters
Milana Vuckovic
Jianwu Wang
Meili Wang
Peng Wang
Jaroslaw Watróbski
Holger Wendland
Lars Wienbrandt
Izabela Wierzbowska
Peter Woehrmann
Szymon Wojciechowski
Michal Wozniak
Maciej Wozniak
Dunhui Xiao
Huilin Xing
Wei Xue
Abuzer Yakaryilmaz
Yoshifumi Yamamoto
Xin-She Yang
Dongwei Ye
Hujun Yin
Lihua You
Han Yu
Drago Žagar
Michal Zak
Gabor Závodszky
Yao Zhang
Wenshu Zhang
Wenbin Zhang
Jian-Jun Zhang
Jinghui Zhong
Sotirios Ziavras
Zoltan Zimboras
Italo Zoppis
Chiara Zucco
Pavel Zun
Pawel Zyblewski
Karol Zyczkowski

Contents – Part VI

Simulations of Flow and Transport: Modeling, Algorithms and Computation

Smart Systems: Bringing Together Computer Vision, Sensor Networks and Machine Learning

Software Engineering for Computational Science

Solving Problems with Uncertainty

Quantum Computing Workshop

Implementing Quantum Finite Automata Algorithms on Noisy Devices

Utku Birkan[1,2](✉), Özlem Salehi[3,7], Viktor Olejar[4], Cem Nurlu[5], and Abuzer Yakaryılmaz[6,7]

[1] Department of Computer Engineering, Middle East Technical University, Ankara, Turkey
utku.birkan@metu.edu.tr
[2] Department of Physics, Middle East Technical University, Ankara, Turkey
[3] Department of Computer Science, Özyeğin University, İstanbul, Turkey
[4] Institute of Mathematics, P.J. Šafárik University in Košice, Košice, Slovakia
viktor.olejar@student.upjs.sk
[5] Department of Physics, Boğaziçi University, İstanbul, Turkey
[6] Center for Quantum Computer Science, University of Latvia, Rīga, Latvia
abuzer@lu.lv
[7] QWorld Association, Tallinn, Estonia
ozlem.salehi@qworld.net
https://qworld.net

Abstract. Quantum finite automata (QFAs) literature offers an alternative mathematical model for studying quantum systems with finite memory. As a superiority of quantum computing, QFAs have been shown exponentially more succinct on certain problems such as recognizing the language $\mathtt{MOD_p} = \{a^j \mid j \equiv 0 \mod p\}$ with bounded error, where p is a prime number. In this paper we present improved circuit based implementations for QFA algorithms recognizing the $\mathtt{MOD_p}$ problem using the Qiskit framework. We focus on the case $p = 11$ and provide a 3 qubit implementation for the $\mathtt{MOD_{11}}$ problem reducing the total number of required gates using alternative approaches. We run the circuits on real IBM quantum devices but due to the limitation of the real quantum devices in the NISQ era, the results are heavily affected by the noise. This limitation reveals once again the need for algorithms using less amount of resources. Consequently, we consider an alternative 3 qubit implementation which works better in practice and obtain promising results even for the problem $\mathtt{MOD_{31}}$.

Keywords: Quantum finite automata · Quantum circuit · Rotation gate · Quantum algorithms

1 Introduction

Quantum finite automata literature offers an alternative mathematical model for studying quantum systems with finite memory. Many different models have

M. Paszynski et al. (Eds.): ICCS 2021, LNCS 12747, pp. 3–16, 2021.
https://doi.org/10.1007/978-3-030-77980-1_1

been proposed with varying computational powers [5]. Moore-Crutchfield quantum finite automaton (MCQFA) [9] is one of the earliest proposed models which is obtained by replacing the transition matrices of the classical finite automata by unitary operators. Despite the fact that they are weaker than their classical counterparts in terms of their language recognition power, for certain languages MCQFAs have been shown to be more succinct. One such example is the language $\texttt{MOD}_p = \{a^j \mid j \equiv 0 \mod p\}$, where p is a prime number: MCQFAs were shown to be exponentially more space-efficient than their classical counterparts [3].

Experimental demonstration of quantum finite automata has recently gained popularity. In [12], the authors implement an optical quantum finite automaton for solving promise problems. A photonic implementation for $\texttt{MOD}_p$ problem is presented in [8]. MCQFA for the $\texttt{MOD}_p$ problem has been also implemented using a circuit based approach within Qiskit and Rigetti frameworks by Kālis in his Master's Thesis [7].

As a continuation of [7], in this paper we present improved circuit based implementations for MCQFA recognizing the $\texttt{MOD}_p$ problem using the Qiskit framework. We start with the naive implementation proposed in [7] and provide a new implementation which reduces both the number of qubits and the number of required basis gates, due to an improved implementation of the multi-controlled rotation gate around y-axis and the order in which the gates are applied. We demonstrate the results of the experiments carried out by IBMQ backends for both the improved naive implementation and the optimized implementation of [7]. Regarding the optimized implementation, we experimentally look for the parameters which would minimize the maximum error rate.

We also propose a 3-qubit parallel implementation which works better in practice for the $\texttt{MOD}_{11}$ and $\texttt{MOD}_{31}$ problems. The choice of parameters for this implementation heavily influences the outcomes unlike the optimized implementation where this choice does not have a huge impact on the acceptance probabilities.

We conclude by suggesting a new implementation for the rotation gate around y-axis, taking into account the new *basis gates*, the gates that are implemented at the hardware level–that have been recently started to be used by IBM. This new proposal lays the foundations for future work on the subject.

The source code of our quantum circuits can be accessed from https://gitlab.com/qworld/qresearch/research-projects/qfa-implementation/-/tree/iccs-2021.

2   Background

We assume that the reader is familiar with the basic concepts and terminology in automata theory and quantum computation. We refer the reader to [5,10,11] for details.

Throughout the paper, Σ denotes the finite input alphabet, not containing the left and right-end markers (¢ and \$, respectively), and $\tilde{\Sigma}$ denotes $\Sigma \cup \{¢, \$\}$. For a string $w \in \Sigma^*$, its length is denoted by $|w|$ and, if $|w| > 0$, w_i denotes the ith symbol of w. For any given input string w, an automaton processes string $\tilde{w} = ¢w\$$ by reading it symbol by symbol and from left to right.

There are several models of quantum finite automata (QFAs) in the literature with different computational powers [5]. In this paper, we focus on the known most restricted model called as *Moore-Crutchfield quantum finite automaton* (MCQFA) model [9].

Formally, a d-state MCQFA is a 5-tuple

$$M = (\Sigma, Q, \{U_\sigma \mid \sigma \in \tilde{\Sigma}\}, q_s, Q_A),$$

where $Q = \{q_1, \ldots, q_d\}$ is the finite *set of states*, U_σ is the *unitary operator* for symbol $\sigma \in \tilde{\Sigma}$, $q_s \in Q$ is the *start state*, and $Q_A \subseteq Q$ is the *set of accepting states*.

The computation of M is traced by a d-dimensional vector, called the state vector, where j^{th} entry corresponds to state q_j. At the beginning of computation, M is in quantum state $|q_s\rangle$, a zero vector except its sth entry, which is 1. For each scanned symbol, say σ, M applies the unitary operator U_σ to the state vector. After reading symbol \$, the state vector is measured in the computational basis. If an accepting state is observed, the input is accepted. Otherwise, the input is rejected.

For a given input $w \in \Sigma^*$, the final state vector $|v_f\rangle$ is calculated as

$$|v_f\rangle = U_\$ U_{w_{|w|}} \cdots U_{w_1} U_\mathptext{¢} |q_s\rangle.$$

Let $|v_f\rangle = (\alpha_1 \ \alpha_2 \ \cdots \ \alpha_d)^T$. Then, the probability of observing the state q_j is $|\alpha_j|^2$, and so, the accepting probability of M on w is $\sum_{q_j \in Q_A} |\alpha_j|^2$.

3 MOD$_p$ Problem and QFA Algorithms

For any prime number p, we define language

$$\text{MOD}_p = \{a^j \mid j \equiv 0 \mod p\}.$$

Ambainis and Freivalds [3] showed that MCQFAs are exponentially more succinct than their classical counterparts, i.e., MOD$_p$ can be recognized by an MCQFA with $\mathcal{O}(\log p)$ states with bounded error, while any probabilistic finite automaton requires at least p states to recognize the same language with bounded error. The MCQFA constructions given in [3] were improved later by Ambainis and Nahimovs [4].

3.1 2-State QFA

We start with giving the description of a 2-state MCQFA that accepts each member of MOD$_p$ with probability 1 and rejects each nonmember with a nonzero probability.

Let M_p be an MCQFA with the set of states $Q = \{q_1, q_2\}$, where q_1 is the starting state and the only accepting state. The identity operator is applied when reading ¢ or \$. Let $\Sigma = \{a\}$, which is often denoted as a unary alphabet. For each symbol a, the counter-clockwise rotation with angle $2\pi/p$ on the unit circle is applied:

$$U_a = \begin{pmatrix} \cos\left(2\pi/p\right) & -\sin\left(2\pi/p\right) \\ \sin\left(2\pi/p\right) & \cos\left(2\pi/p\right) \end{pmatrix}.$$

The minimal rejecting probability of a non-member string w by the automaton M_p is $\sin^2\left(|w| \cdot 2\pi/p\right)$, which gets closer to 0 when $|w|$ approaches an integer multiple of p. One may notice that instead of $2\pi/p$, it's also possible to use the rotation angle $k \cdot 2\pi/p$ for some $k \in \{1,\ldots,p-1\}$. It is easy to see that the rejecting probability of each non-member differs for different values of k, but the minimal rejecting probability will not be changed when considering all non-members.

On the other hand, to obtain a fixed error bound, we can execute more than one 2-state MCQFA in parallel, each of which uses a different rotation angle.

3.2 $\mathcal{O}(\log p)$-State QFAs

Here we explain how to combine 2-state MCQFAs with different rotation angles to obtain a fixed error bound.

First, we define the 2-state MCQFA M_p^k as same as M_p except for the rotation angle for the symbol a, which is now $k \cdot 2\pi/p$ where $k \in \{1,\ldots,p-1\}$.

Then we define the 2d-state MCQFA M_p^K, where K is a set formed by d many k values: $K = \{k_1,\ldots,k_d\}$ and each $k_j \in K$ is an integer between 1 and $p-1$. The MCQFA M_p^K executes d 2-state MCQFAs $\{M_p^{k_1},\ldots,M_p^{k_d}\}$ in parallel. The state set of M_p^K is formed by d pairs of $\{q_1,q_2\}$:

$$\{q_1^1, q_2^1, q_1^2, q_2^2, \ldots, q_1^d, q_2^d\}.$$

The state q_1^1 is the starting state and the only accepting state. At the beginning of the computation, M_p^K applies a unitary operator $U_{\cent}$ when reading the symbol $\cent$ and enters the following superposition:

$$|q_1^1\rangle \xrightarrow{U_{\cent}} \frac{1}{\sqrt{d}}|q_1^1\rangle + \frac{1}{\sqrt{d}}|q_1^2\rangle + \cdots + \frac{1}{\sqrt{d}}|q_1^d\rangle.$$

In other words, we can say that M_p^K enters an equal superposition of 2-state MCQFAs $M_p^{k_1}, M_p^{k_2}, \ldots, M_p^{k_d}$. Until reading the right end-marker, M_p^K executes each 2-state sub-automaton, $M_p^{k_j}$, in parallel, where $M_p^{k_j}$ rotates with angle $2\pi k_j/p$. Thus, the overall unitary matrix of M_p^K for symbol a is

$$U_a = \bigoplus_{j=1}^{d} R_j = \begin{pmatrix} R_1 & 0 & \cdots & 0 \\ 0 & R_2 & \cdots & 0 \\ \vdots & \vdots & \ddots & \vdots \\ 0 & 0 & \cdots & R_d \end{pmatrix},$$

where

$$R_j = \begin{pmatrix} \cos\left(2\pi k_j/p\right) & -\sin\left(2\pi k_j/p\right) \\ \sin\left(2\pi k_j/p\right) & \cos\left(2\pi k_j/p\right) \end{pmatrix}.$$

After reading the symbol $, we apply the unitary operator $U_\$ = U_\mathbb{C}^{-1}$. This overall algorithm gives us an exponential advantage of quantum computation over classical computation for some suitable values for K, for each p. It was shown [3] that, for each p, there exists a set of K with $d = \mathcal{O}(\log p)$ elements such that M_p^K recognizes MOD$_p$ with a fixed error bound.

4   MOD$_p$ Implementations

In this section, we present our implementation schema in Qiskit and results on simulators and real machines.

4.1   Single Qubit Implementation

We start with a single qubit implementation. An example implementation of 2-state MCQFA M_7 for MOD$_7$ is given in Fig. 1 where the input string is aaa:

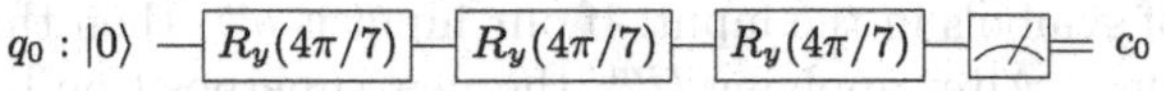

$$q_0 : |0\rangle \quad \boxed{R_y(4\pi/7)} \boxed{R_y(4\pi/7)} \boxed{R_y(4\pi/7)} \boxed{\measuredangle} = c_0$$

Fig. 1. Single qubit MOD$_7$ implementation

This circuit has one qubit (q_0) and one bit (c_0). There are different rotation operators (gates) in Qiskit. Here we use R_y gate, which is defined on the Bloch sphere and takes the twice of the rotation angle as its parameter to implement the rotation on the unit circle on the $|0\rangle - |1\rangle$ plane. The outcome of the measurement at the end is written to the classical bit c_0.

4.2   Three-Qubit Implementations of MOD$_p$

A Naive Implementation. To implement the unitary operator given in Subsect. 3.2, we use controlled gates, the conditional statements of the circuits. The implementation cost of the controlled gates are expensive and unfortunately, the straightforward implementation of the above algorithm is costly.

Kālis [7] gave a four-qubit implementation of the above algorithm for the problem MOD$_{11}$, where three qubits are used to simulate four sub-automata and one ancilla qubit is used to implement the controlled operators.

Here we present our implementation schema by using only 3 qubits. All diagrams are obtained by using Qiskit [1]. We use three qubits called q_2, q_1, q_0. We implement $U_\mathbb{C}$ operator by applying Hadamard gates to q_2 and q_1. The initial state is $|000\rangle$. After applying Hadamard operators, we will have the following superposition, in which we represent the state of q_0 separately:

$$|v_\mathbb{C}\rangle = \frac{1}{2}|00\rangle \otimes |0\rangle + \frac{1}{2}|01\rangle \otimes |0\rangle + \frac{1}{2}|10\rangle \otimes |0\rangle + \frac{1}{2}|11\rangle \otimes |0\rangle$$

The unitary matrix for symbol a is represented as follows:

$$U_a = \begin{pmatrix} R_1 & 0 & 0 & 0 \\ 0 & R_2 & 0 & 0 \\ 0 & 0 & R_3 & 0 \\ 0 & 0 & 0 & R_4 \end{pmatrix}$$

In order to implement U_a, we apply R_1 when $q_2 \otimes q_1$ is in state $|00\rangle$, R_2 when $q_2 \otimes q_1$ is in state $|01\rangle$, R_3 when $q_2 \otimes q_1$ is in state $|10\rangle$, and R_4 when $q_2 \otimes q_1$ is in state $|11\rangle$.

We pick $K = \{1, 2, 4, 8\}$. Then, after applying U_a, the new superposition $(U_a U_\mathbb{C} |000\rangle)$ becomes

$$|v_1\rangle = \frac{1}{2}|00\rangle \otimes R_y\left(2\pi/11\right)|0\rangle + \frac{1}{2}|01\rangle \otimes R_y\left(4\pi/11\right)|0\rangle$$
$$+ \frac{1}{2}|10\rangle \otimes R_y\left(8\pi/11\right)|0\rangle + \frac{1}{2}|11\rangle \otimes R_y\left(16\pi/11\right)|0\rangle.$$

Once we have a block for U_a, then we can repeat it in the circuit as many times as the number of symbols in the input. If our input is a^m, then the block for U_a is repeated m times. After applying U_a^m, the new superposition becomes

$$|v_m\rangle = \frac{1}{2}|00\rangle \otimes R_y^m\left(2\pi/11\right)|0\rangle + \frac{1}{2}|01\rangle \otimes R_y^m\left(4\pi/11\right)|0\rangle$$
$$+ \frac{1}{2}|10\rangle \otimes R_y^m\left(8\pi/11\right)|0\rangle + \frac{1}{2}|11\rangle \otimes R_y^m\left(16\pi/11\right)|0\rangle.$$

After reading the whole input, before the measurement, we apply the Hadamard gates which correspond to the operator for symbol $.

There is no single-gate solution we can use to implement all of these operators though. Besides, the controlled operators are activated only when all of the control qubits are in state $|1\rangle$. This is why X gates are used; to activate the control qubits when they are in state $|0\rangle$. Figure 2 depicts a circuit implementing U_a.

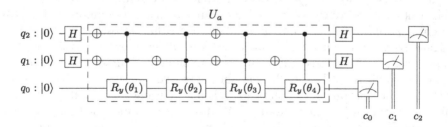

Fig. 2. A naive MOD$_p$ circuit implementation

Initially, two X gates (represented as $\oplus$ in the circuit diagrams) are applied and R_1 is implemented as the controlled rotation gate will be activated only when the control qubits are initially in state $|00\rangle$. Next, we apply X gate to q_2 and so the controlled qubits are activated when in state $|01\rangle$ and we implement R_2. Similarly, we implement R_3 by applying X gates to both qubits so that the controlled qubits are activated when in state $|10\rangle$, and finally we implement R_4 so that the control qubits are activated in state $|11\rangle$.

Next, we discuss how to reduce the number of X gates and an improved implementation is given in Fig. 3. Initially, R_4 is implemented as the controlled operators will be activated only in state $|11\rangle$. Next, we apply X gate to q_2 and so the controlled qubits are activated when in state $|10\rangle$, and so, we implement R_3. We apply X gate to q_1 and similarly the controlled qubits are activated when in state $|11\rangle$ so that we implement R_1. Finally, we apply X gate to q_2 again to implement R_2. Note, that we apply one more X gate at the end so that the initial value of q_2 is restored.

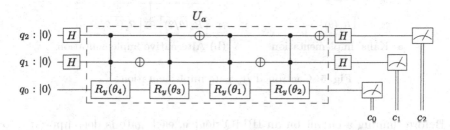

Fig. 3. An improved (but still naive) MOD_p circuit implementation

The number of X gates can be reduced further by omitting the last X gate in the above diagram and changing the order of R_y gates in the next round so that they follow the values of the controlled qubits. For each scanned a, either one of the blocks is applied, alternating between the two, starting with the first block. Overall, three X gates are used instead of four X gates for a single a. Note that when the input length is odd, we should always use an extra X gate before the final pair of Hadamard gates. The blocks are depicted below (Fig. 4).

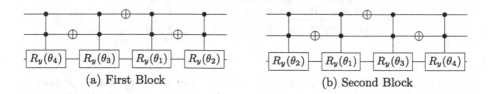

(a) First Block   (b) Second Block

Fig. 4. Reducing the NOT gates using two blocks to implement U_a

When we implement the circuit in Qiskit, there are different approaches to implement the multi-controlled rotation gate. One option is Kālis' implementation [7] that takes advantage of Toffoli gates and controlled rotations as seen in Fig. 5a. Another possibility is to use the built-in RYGate class in Qiskit. An alternative implementation of controlled R_y gate presented in [6] is given in Fig. 5b. In this implementation, the rotation gates applied on the target qubit cancel each other unless the control qubits are in state $|11\rangle$ which is checked by the Toffoli gate, thus yielding the same effect as an R_y gate controlled by two qubits. As a further improvement, the Toffoli gate can be replaced with the simplified Toffoli gate, (also referred to as Margolus gate in Qiskit) which has a reduced cost compared to Toffoli gate. This replacement does not affect the overall algorithm as only the states corresponding to first and second sub-automata and that of third and fourth sub-automata are swapped.

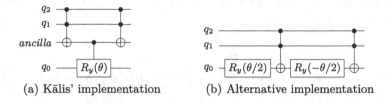

(a) Kālis' implementation   (b) Alternative implementation

Fig. 5. Controlled R_y gate implementations

Before running a circuit on an IBMQ device, each gate is decomposed into basis gates U_1, U_2, U_3 and CX^1 and this decomposition also depends on the backend on which the circuit is run and the physical qubits used on the machine. In the table given below, the number of basis gates required to implement the rotation gate with two controls using the two approaches is given for the IBMQ Santiago and IBMQ Yorktown machines with the default optimization level (Table 1).

Table 1. Number of basis gates required by the controlled rotation gate implementations

	RYGate				Alternative			
	U_1	U_2	U_3	CX	U_1	U_2	U_3	CX
Santiago	0	0	6	11	1	2	1	9
Yorktown	0	0	6	8	4	1	2	6

Next, we present some experimental results about MOD_{11} problem comparing the performance of the 4 qubit implementation which was originally proposed in [7] and our improved version with 3 qubits where the number of X gates are reduced and the controlled rotation gates are implemented using the alternative approach. The acceptance probability of each word is the number of times the states $|000\rangle$ and $|0000\rangle$ are observed divided by the number of shots (which was

---

[1] The set of basis gates was changed to $CX, I, R_z, \sqrt{X}, X$ in January 2021.

taken as 1000 for the experiments) for 3 qubit and 4 qubit implementations, respectively. The results do not look promising as the acceptance probabilities are around 0.125 and 0.0625 for the 3 qubit and 4 qubit implementations, which are the probabilities of observing a random result. Ideally, the acceptance probabilities for the word lengths 11 and 22 would have been close to 1 (Fig. 6).

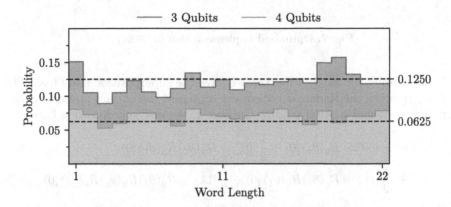

Fig. 6. Acceptance probabilities for $\texttt{MOD}_{11}$ naive implementation

There are three main sources of error in IBMQ machines: number of qubits, circuit depth and the number of CX gates. Even though our 3-qubits naive implementation reduced the number of qubits and CX gates, this improvement was not enough to have any meaningful result. Each Margolous gate still requires 3 CX gates which is better compared to the Toffoli gate which requires 6 but it is still not enough [1]. In addition, the connectivity of the underlying hardware requires some additional CX gates when the 4-qubits circuit is transpiled. Nevertheless, our implementation provides a significant improvement in the number of basis gates required for the implementation of the algorithm. In Table 2, we list the number of basis gates required by both implementations for word length 11 using the default optimization level by IBMQ Santiago backend.

An Optimized Implementation. The circuit construction above can be improved by sacrificing some freedom in the selection of rotation angles as proposed in [7]. In the circuit diagram given in Fig. 7, only the controlled rotation operators are used, where the unitary matrix for symbol a is as follows:

$$\tilde{U}_a = \begin{pmatrix} R_1 & 0 & 0 & 0 \\ 0 & R_2R_1 & 0 & 0 \\ 0 & 0 & R_3R_1 & 0 \\ 0 & 0 & 0 & R_3R_2R_1 \end{pmatrix}$$

Thus, each sub-automaton applies a combination of rotations among three rotations.

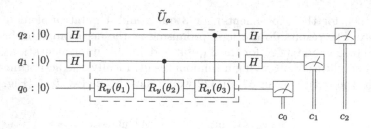

Fig. 7. Optimized implementation for MOD$_{11}$

Upon reading the left end-marker, we have the superposition state $|v_\phi\rangle$. The block $\tilde{U}_a$ between the Hadamard operators represent the unitary operator corresponding to symbol a. After reading the first a, the new superposition becomes

$$
\begin{aligned}
|\tilde{v}_1\rangle =\ & \frac{1}{2}|00\rangle \otimes R_y(\theta_1)|0\rangle + \frac{1}{2}|01\rangle \otimes R_y(\theta_2)R_y(\theta_1)|0\rangle \\
& + \frac{1}{2}|10\rangle \otimes R_y(\theta_3)R_y(\theta_1)|0\rangle + \frac{1}{2}|11\rangle \otimes R_y(\theta_3)R_y(\theta_2)R_y(\theta_1)|0\rangle.
\end{aligned}
$$

By letting $\phi_1 = \theta_1$, $\phi_2 = \theta_1 + \theta_2$, $\phi_3 = \theta_1 + \theta_3$, and $\phi_4 = \theta_1 + \theta_2 + \theta_3$, the state $|\tilde{v}_1\rangle$ can be equivalently expressed as

$$
\begin{aligned}
|\tilde{v}_1\rangle =\ & \frac{1}{2}|00\rangle \otimes R_y(\phi_1)|0\rangle + \frac{1}{2}|01\rangle \otimes R_y(\phi_2)|0\rangle \\
& + \frac{1}{2}|10\rangle \otimes R_y(\phi_3)|0\rangle + \frac{1}{2}|11\rangle \otimes R_y(\phi_4)|0\rangle.
\end{aligned}
$$

Compared to the 3 qubit implementation presented in the previous subsection, this implementation uses less number of gates and especially the number of CX gates is reduced. Furthermore, as no multi-controlled gates are required, the number of CX gates used by the IBMQ Santiago machine even reduces when the default optimization is used. The number of required basis gates for various implementations is given in Table 2 for word length 11.

Table 2. Number of basis gates required by naive and optimized implementations ran on IBMQ Santiago

Implementation	Basis gates			
	U_1	U_2	U_3	CX
3 Qubits Naive	270	15	121	270
4 Qubits Naive	561	154	114	1471
Optimized	0	4	44	55

We conducted experiments on IBMQ Santiago machine with two different set of values of k, $\{2, 4, 8\}$ and $\{4, 9, 10\}$. A discussion about the choice of the value of k is presented in the next subsection. The results are still far from the ideal as it can be seen in Fig. 8. We also plotted the ideal results we got from the simulator. When compared with the naive implementation, we observe that the acceptance probabilities fluctuate until a certain word length but after some point they tend to converge to $\frac{1}{8}$, the probability of selecting a random state.

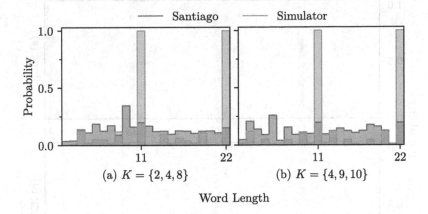

(a) $K = \{2, 4, 8\}$ (b) $K = \{4, 9, 10\}$

Word Length

Fig. 8. Acceptance probabilities for $\mathtt{MOD}_{11}$ optimized implementation

A Parallel Implementation. Recall that in the single qubit implementation, we have a single automaton, but the problem is, that we get arbitrarily large error for nonmember strings and we try to reduce it by running multiple sub-automata in parallel. In this section we provide some experimental results about $\mathtt{MOD}_p$ problem where we simply run each sub-automaton using a single qubit. Although theoretically this approach has no memory advantage, it works better in real devices as no controlled gates are used while still providing a space advantage in terms of number of bits in practice.

Consider the circuit diagram given in Fig. 9. Using three qubits, we run three automata in parallel with three different rotation angles.

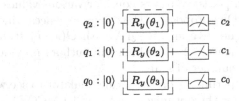

Fig. 9. Parallel $\mathtt{MOD}_{11}$ circuit where each qubit implements an MCQFA

In this implementation, the unitary operators corresponding to ¢ and \$ are identity operators. Upon reading the first a, the new state becomes

$$|v_1'\rangle = R_y(\theta_1) |0\rangle \otimes R_y(\theta_2) |0\rangle \otimes R_y(\theta_3) |0\rangle .$$

We conducted experiments on IBMQ Santiago machine for MOD_{11} problem with $K = \{1, 2, 4\}$ and for MOD_{31} problem with $K = \{8, 12, 26\}$. The results are summarized in Fig. 10.

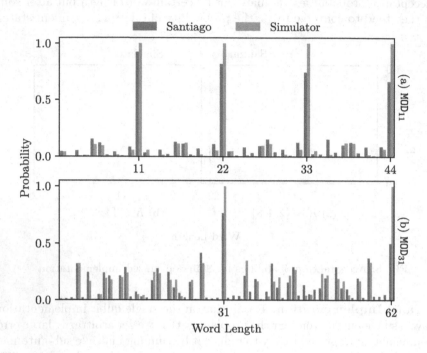

Fig. 10. Acceptance probabilities for MOD_{31} and MOD_{11} parallel implementations

From the graphs above we can see that the experimental results on real machines coincide with the simulator outcome especially for small word lengths. The number of required basis gates to implement the parallel implementation is simply three times the length of the word.

Choosing Values of k. In [4], the authors consider various values of k for the optimized implementation. One proposal is the cyclic sequences which work well in numerical experiments and give an MCQFA with $\mathcal{O}(\log p)$ states. Another proposal is the AIKPS sequences [2] for which the authors provide a rigorous proof but it requires larger number of states.

We conducted several experiments on the local simulator to see which values of k produce better results for the optimized and parallel MCQFA implementations. We define the maximum error as the highest acceptance probability for a nonmember string and we investigated for which values of k, the maximum error is minimized. Experimental findings are listed below, in Table 3.

Table 3. Standard deviation and mean values for compared circuits

	Min	Max	Mean	Std. Dev.
MOD_{11} optimized	0.010	0.080	0.034	0.018
MOD_{11} parallel	0.109	0.611	0.270	0.148
MOD_{31} parallel	0.319	0.974	0.615	0.167

In the optimized implementation, the maximum error ranges between 0.01 and 0.08 with a standard deviation of 0.018 so it can be concluded that the choice of different values of k does not have a significant impact on the success probabilities, for this specific case. When we move on to parallel implementation, we observe that the maximum error probabilities vary heavily depending on the choice of the value of k. Interquartile range is much larger this time. Furthermore, MOD_{31} results include error values as high as 0.97 in the third quartile. MOD_{11} also has some outliers that show far greater error than we would like to work with. With this insight, we picked the values of k accordingly in the parallel implementation, which had an impact on the quality of the results we obtained.

5   Conclusion and Future Work

The goal of this study was to investigate circuit implementations for quantum automata algorithms solving MOD_p problem. As a way of dealing with the limitations of NISQ devices, we considered different implementation ideas that reduce the number of gates and qubits used. Our findings contribute to the growing field of research on efficient implementations of quantum algorithms using limited memory.

Recently, the basis gates of IBMQ backends were reconfigured as CX, I, R_z, $\sqrt{X}$, and X. As a result, a new methodology should be developed in order to reduce the number of required gates. For instance, the circuit with two consecutive R_y gates that is transpiled by Qiskit is given in Fig. 11a. Instead of this design, we propose the implementation given in Fig. 11b, which would reduce the number of required basis gates. We use the fact that $R_y(\theta) = \sqrt{X} \cdot R_z(\theta) \cdot \sqrt{X} \cdot X$, hence multiple rotations can be expressed as $R_y^n(\theta) = \sqrt{X} \cdot R_z^n(\theta) \cdot \sqrt{X} \cdot X$.

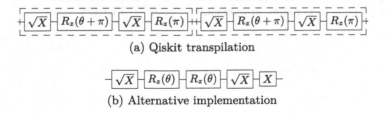

(a) Qiskit transpilation

(b) Alternative implementation

Fig. 11. R_y gate implementations with the new basis set

Acknowledgements. This research project started during QIntern2020 program of QWorld (July–August 2020) and later continued under the QResearch Department of QWorld.

We thank to anonymous reviewers for their helpful comments and Mārtiņš Kālis for giving a presentation during QIntern2020 program on the details of his master thesis and kindly answering our questions.

Birkan and Nurlu were partially supported by TÜBİTAK scholarship "2205 – Undergraduate Scholarship Program". Yakaryılmaz was partially supported by the ERDF project Nr. 1.1.1.5/19/A/005 "Quantum computers with constant memory".

References

1. Abraham, H., et al.: Qiskit: an open-source framework for quantum computing (2019). https://doi.org/10.5281/zenodo.2562110
2. Ajtai, M., Iwaniec, H., Komlós, J., Pintz, J., Szemerédi, E.: Construction of a thin set with small Fourier coefficients. Bull. Lond. Math. Soc. **22**(6), 583–590 (1990)
3. Ambainis, A., Freivalds, R.: 1-way quantum finite automata: strengths, weaknesses and generalizations. In: 39th Annual Symposium on Foundations of Computer Science, FOCS 1998, pp. 332–341. IEEE Computer Society (1998)
4. Ambainis, A., Nahimovs, N.: Improved constructions of quantum automata. Theor. Comput. Sci. **410**(20), 1916–1922 (2009)
5. Ambainis, A., Yakaryılmaz, A.: Automata and quantum computing. CoRR abs/1507.01988 (2015). http://arxiv.org/abs/1507.01988
6. Barenco, A., et al.: Elementary gates for quantum computation. Phys. Rev. A **52**(5), 3457–3467 (1995)
7. Kālis, M.: Kvantu Algoritmu Realizācija Fiziskā Kvantu Datorā (Quantum Algorithm Implementation on a Physical Quantum Computer). Master's thesis, University of Latvia (2018)
8. Mereghetti, C., Palano, B., Cialdi, S., Vento, V., Paris, M.G.A., Olivares, S.: Photonic realization of a quantum finite automaton. Phys. Rev. Res. **2**(1), 013089–013103 (2020)
9. Moore, C., Crutchfield, J.P.: Quantum automata and quantum grammars. Theor. Comput. Sci. **237**(1–2), 275–306 (2000)
10. Nielsen, M.A., Chuang, I.L.: Quantum Computation and Quantum Information: 10th Anniversary Edition, 10th edn. Cambridge University Press, USA (2011)
11. Sipser, M.: Introduction to the Theory of Computation, 1st edn. International Thomson Publishing (1996)
12. Tian, Y., Feng, T., Luo, M., Zheng, S., Zhou, X.: Experimental demonstration of quantum finite automaton. npj Quant. Inf. **5**(56), 1–5 (2019)

OnCall Operator Scheduling for Satellites with Grover's Algorithm

Antonius Scherer[2], Tobias Guggemos[1]([✉]), Sophia Grundner-Culemann[1],
Nikolas Pomplun[2], Sven Prüfer[2], and Andreas Spörl[2]

[1] MNM-Team, Ludwig-Maximilian Universität (LMU) München, Munich, Germany
{guggemos,grundner-culemann}@nm.ifi.lmu.de
[2] German Space Operation Center, German Aerospace Center, Wessling, Germany
{antonius.scherer,nikolas.pomplun,sven.pruefer,andreas.spoerl}@dlr.de

Abstract. The application of quantum algorithms on some problems in NP promises a significant reduction of time complexity. This work uses Grover's Algorithm, designed to search an unstructured database with quadratic speedup, to find valid a solution for an instance of the on-call operator scheduling problem at the German Space Operation Center. We explore new approaches in encoding the problem and construct the Grover oracle automatically from the given constraints and independent of the problem size. Our solution is not designed for currently available quantum chips but aims to scale with their growth in the next years.

Keywords: Grover's algorithm · On-call scheduling · Satellites

1 Introduction

Grover's algorithm [15] is one of the best-known algorithms offering significant speed-up for computational problems on a quantum computer. It features a quadratic speed-up for every problem which can be described as a search in an unsorted database. Hence, it allows significant improvements for problems where finding solutions requires searching through a large part of the input space. The algorithm consists of three major parts (see Fig. 1):

State preparation creates a superposition of the search space.
Grover Iteration is repeated $1/k \cdot \sqrt{N}$ times to solve a search problem for an input space with N elements and $k \geq 1$ valid solutions. It has two parts:
    Oracle A search function $f(x)$ is applied on the input state, which outputs 1 for the searched values $\hat{x}$.
    Diffusion The amplitude (and therefore the measurement probability) of the searched input states is increased.
Measurement outputs a binary string that represents $\hat{x}$ with high probability.

Satisfiability (SAT) problems are problems where a solution for a set of *variables* has to be found satisfying a given set of *constraints*. Although there is a variety

© Springer Nature Switzerland AG 2021
M. Paszynski et al. (Eds.): ICCS 2021, LNCS 12747, pp. 17–29, 2021.
https://doi.org/10.1007/978-3-030-77980-1_2

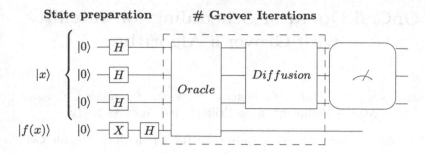

Fig. 1. Major steps of Grover's algorithm.

of classical SAT-solvers that allow finding solutions under a large number of constraints, the general satisfaction problem for an arbitrary number of constraints is believed to be NP-hard [5,8]. Grover's algorithm certainly does not change such problems' complexity, however, once we see scalable quantum computers beyond the NISQ-era, using them may offer significant improvements.

The German Space Operation Center (GSOC) hosts multiple applications where a SAT-solver is used to find a solution for a given problem. One of them is "Spacecraft On-Call Scheduling" (SOCS), where the operators' time-table for various spacecraft missions needs to be determined. Currently, it deals with 24 h shifts for ~50 operators and 20 positions (missions) over a period of 180 days.

Scope of this Work

This work aims to solve the aforementioned problem with a quantum computer using Grover's algorithm. However, a quantum device that is able to leverage the potentials of Grover's algorithm is far beyond the horizon. Hence, we present a method for encoding the variables of the SOCS problem into a quantum state which can be used for Grover's algorithm once such a device is available. Additionally, we develop an algorithm for representing constraints as a circuit, which is then used as the *Oracle*. To ensure correctness, we evaluate the approach with *Qiskit* for a reduced problem size that fits the currently available IBMq simulator.

2   Spacecraft On-Call Scheduling

The GSOC as a part of the German Aerospace Center (DLR) operates a variety of spacecraft missions. As depicted in Fig. 2, some of those missions require the constant presence of one or more *operators* per subsystem – called *position* – and *time period*. The operators are organized through an on-call spacecraft operator scheduling which is conducted multiple times per year to incorporate updated unavailability times or new personnel. A valid schedule has to allocate approximately 50 operators on 20 positions over a period of 180 days and may need to be updated if changes in the assumptions arise later on. On-call shifts are scheduled in whole days and every operator can cover a certain subset of positions

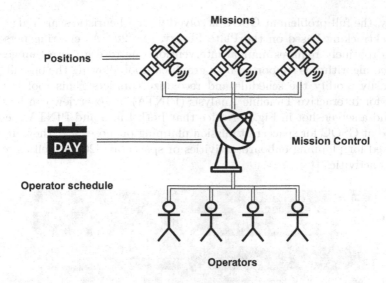

Fig. 2. Overview of the use case "Spacecraft On-Call Scheduling"

depending on their training and responsibilities. A valid schedule depends on various inputs from the spacecraft missions as well as the operators themselves and must satisfy the following constraints:

   (i) Operators can only work on some given positions.
  (ii) Per tuple of day and position at least one operator needs to be assigned.
 (iii) An operator can be assigned to at most one position a day.
 (iv) Operators can specify days in advance when they are unavailable.
  (v) A partial on-call schedule may be supplied and needs to be obeyed.[1]
 (vi) Operators can work at most two out of any three consecutive weeks.
(vii) Operators can work at most 35 days out of any 105 consecutive days.
(viii) Operators shall work preferably whole weeks.
 (ix) All operators shall work a similar amount of days.

Notice that **(viii)** and **(ix)** are not constraints but rather optimization goals. As implementing all these constraints efficiently in full generality is rather complicated and is certainly not feasible on todays quantum computers, we adapt the constraints **(ii)**, **(iii)** and **(vi)** and restrict ourselves to them. This means that we consider the following constraints in this paper:

A Per tuple of day and position exactly one operator needs to be assigned.
B Out of three consecutive days, an operator is only allowed to work two.
C An operator can be assigned to at most one position a day.

---

[1] This is needed e. g., for updating an on-call schedule during the year when operators have updated their vacations, for example. In this case only the future will be replanned, but the past may influence the applied constraints in the future.

Currently, the full problem at GSOC is solved with a heuristic search algorithm using backtracking based on the Plato library, see [20]. As creating personnel schedules routinely requires manual intervention due to e. g., late changes, the automated algorithm is supported by a graphical tool allowing the on-call planner to easily modify the schedule and recognize conflicts. This tool is called Program for Interactive Timeline Analysis (PINTA), an overview can be found in [21] and a screenshot in Fig. 3. Notice that both Plato and PINTA are tools developed at GSOC for spacecraft mission planning purposes, i. e. they are commonly used for planning onboard activities of spacecraft [26] as well as related onground activities [16].

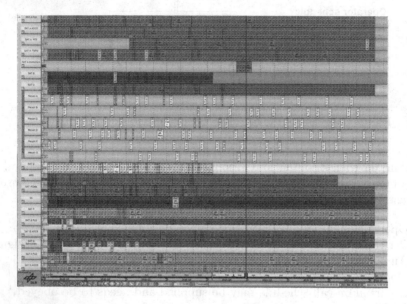

Fig. 3. A screenshot of an on-call schedule from PINTA

3   Related Work

Spacecraft On-Call Scheduling is closely related to the nurse scheduling problem, an especially hard version of the personnel scheduling problem [8]. Linear programming [1,19], simulated annealing [2,11], and tabu search [3,4] are classical approaches for calculating solutions. An overview of the complexity of various personnel scheduling variants and classical solvers is provided in [5].

It is well known that Grover's algorithm [15] can be used to solve such optimization problems with a single solution. It can further be extended to work without knowledge of the exact or multiple solutions [7] or for highly structured combinatorial search problems [17]. How to exploit the structure of NP-complete problems to perform a nested quantum search was shown in [9]. This is quadratically faster than classical nested search and exponentially faster than unstructured quantum search. Iterative application of Grover's algorithm allows searching the optimum of an objective function [13], and can be used to solve constrained polynomial binary optimization problems. Due to their complexity and ubiquity, NP problems are also the focus of various other quantum algorithms. Especially for combinatorial optimization problems, the Quantum Approximate Optimization Algorithm (QAOA) [12] and the Variational Quantum Eigensolver (VQE) [22] play an important role for NISQ era devices [23]. For example, the graph coloring problem, which is similar to how we encode the scheduling problem, and the Traveling Salesman Problem have been approached with space efficient QAOA methods [14,24]. There are also approaches using quantum annealing for the nurse scheduling problem [18].

4   Encoding of Variables and Constraints

This section presents a method for encoding the SOCS problem's variables as qubits on which we apply Grover's algorithm. In contrast to a naive approach, we reduce the number of qubits by a factor of ~10. Additionally, we present a way to encode the three constraints into an oracle function.

4.1   Variables

As a first step, we encode the variables into Grover's input register. After the Grover iterations, the input register is measured, which results in a binary representation of a valid schedule. A naive approach represents all combinations of a day d, position p and operator o in the input state as depicted in Fig. 4a:

$$|\psi\rangle_{d,p,o} = \begin{cases} |0\rangle, & \text{operator } o \text{ is not assigned to position } p \text{ at day } d \\ |1\rangle, & \text{operator } o \text{ is to position } p \text{ at day } d \end{cases}$$

This requires a quantum register with $d \cdot p \cdot o$ qubits to encode the search space. For the use case presented in Sect. 2, the problem volume amounts to $o \cdot p \cdot d = 50 \cdot 20 \cdot 180 = 180,000$ qubits, each of which is a binary representation of whether an operator is scheduled for a certain day and position (so-called *time-position*) or not.

Encoding values in binary can help to reduce the contribution of one variable from linear to logarithmic. As the variable *days* has the biggest impact on the number of qubits in our case with a value of 180, one may try to apply this to days, i. e. counting the days starting from one and assigning them this number in a binary representation. This way one could encode allocations of a day-position-operator combination by assigning binary encoded day values to position–operator combinations. This would reduce the number of qubits to $o \cdot p \cdot \log_2 d$ – in our use case 7,492 qubits when encoded, *e.g.,* for 4 days, as

$$|\psi_1\psi_0\rangle_{p,o} = \begin{cases} |00\rangle, & \text{operator } o \text{ is assigned to position } p \text{ at day } \mathbf{0} \\ \cdots \\ |11\rangle, & \text{operator } o \text{ is assigned to position } p \text{ at day } \mathbf{3} \end{cases} \tag{1}$$

and correspondingly for larger numbers of days (powers of 2, for simplicity).

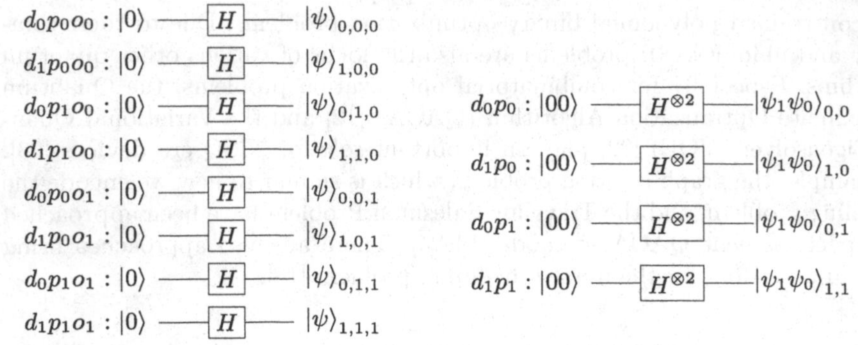

$d_0 p_0 o_0 : |0\rangle$ —[H]— $|\psi\rangle_{0,0,0}$

$d_1 p_0 o_0 : |0\rangle$ —[H]— $|\psi\rangle_{1,0,0}$

$d_0 p_1 o_0 : |0\rangle$ —[H]— $|\psi\rangle_{0,1,0}$ $d_0 p_0 : |00\rangle$ —[$H^{\otimes 2}$]— $|\psi_1 \psi_0\rangle_{0,0}$

$d_1 p_1 o_0 : |0\rangle$ —[H]— $|\psi\rangle_{1,1,0}$ $d_1 p_0 : |00\rangle$ —[$H^{\otimes 2}$]— $|\psi_1 \psi_0\rangle_{1,0}$

$d_0 p_0 o_1 : |0\rangle$ —[H]— $|\psi\rangle_{0,0,1}$

$d_1 p_0 o_1 : |0\rangle$ —[H]— $|\psi\rangle_{1,0,1}$ $d_0 p_1 : |00\rangle$ —[$H^{\otimes 2}$]— $|\psi_1 \psi_0\rangle_{0,1}$

$d_0 p_1 o_1 : |0\rangle$ —[H]— $|\psi\rangle_{0,1,1}$ $d_1 p_1 : |00\rangle$ —[$H^{\otimes 2}$]— $|\psi_1 \psi_0\rangle_{1,1}$

$d_1 p_1 o_1 : |0\rangle$ —[H]— $|\psi\rangle_{1,1,1}$

(a) Naive, 1-hot approach to encoding (b) Improved approach to encoding
(8 qubits, 2 operators)   (8 qubits, 4 operators)

Fig. 4. Input state for four time-positions and (a) two resp. (b) four operators.

Notice, however, that one needs to be careful with such efficient encodings as they reduce the size of the representable state space and may in fact rule out valid solutions. The encoding in Eq. 1 forces every operator to work every position exactly once within the given number of days. It is easy to construct examples in which *all* solutions to the problem are removed via this encoding, e. g., two operators, one position and three days. As the number of days is larger than the available operators, one operator would need to work twice, which cannot be represented in this encoding. This problem cannot be easily fixed as there are 2^d possible subsets of all available days when an operator may work at a particular position, thus parametrizing all of them eliminates the logarithmic advantage that one gained. Notice also that parametrizing sets of scheduled days makes expressing day-wise constraints rather cumbersome.

However, in our case, constraint **A** actually says that for every day–position combination, *exactly* one operator needs to be scheduled. Therefore, encoding the *operators* instead of *days* in the above binary fashion is actually possible and only reduces the state space in a way that only invalid solutions are removed. Although the number of required qubits is not decreased as much, the circuit size can be decreased drastically, since we implement constraint **A** directly in the encoding, see Sect. 4.2 for details. In Fig. 4b it can be seen that with the same available amount of qubits, e. g., eight, twice as many operators, i. e. four operators instead of two, can be assigned to four time-positions. The required number of qubits for the full problem is $d \cdot p \cdot \log_2 o$ – in our case $20,318$ – and the assignment looks as follows for four operators:

$$|\psi_1 \psi_0\rangle_{d,p} = \begin{cases} |00\rangle, & \text{operator } \mathbf{0} \text{ is assigned to position } p \text{ at day } d \\ \cdots \\ |11\rangle, & \text{operator } \mathbf{3} \text{ is assigned to position } p \text{ at day } d \end{cases} \tag{2}$$

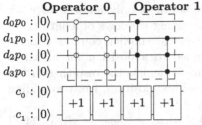
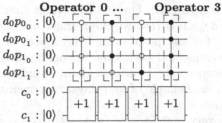
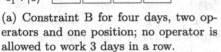

(a) Constraint B for four days, two operators and one position; no operator is allowed to work 3 days in a row.

(b) Constraint C for four Operators, one day and two positions; no operator works the same day at two different positions.

Fig. 5. Exemplary circuits Constraint **B** and **C** that increment the counter $|c_1 c_0\rangle$ if one of the constraints is violated.

Notice that this encoding still has some peculiarities. Imagine, for example, that the total number of operators is not a power of two. In that case there will be states that schedule a non-existent operator and thus do not solve the initial problem although they may satisfy all constraints. There are various ways how to deal with this, e.g., by preparing initial states such that the corresponding amplitudes are always zero or by considering them as operators that have an outage during every timeslot. For the current paper we will restrict to numbers of operators which are powers of two.

4.2   Constraints

A schedule is valid if none of the constraints defined in Sect. 2 are violated. Hence, we encode *conflicts* in the oracle function and sum up the number of conflicts for a state. Only if the amount of conflicts is zero, a state is considered a valid schedule and the Oracle function returns one.

For counting the conflicts, we use a controllable *Increment*-gate – depicted as $\boxed{+1}$ in Fig. 5. The constraints themselves are encoded as follows:

Constraint A: Assigning multiple operators to the same position at one day is impossible by construction: Since every *time-position* is encoded in an individual set of $\log_2 o$ qubits, this satisfies the "One operator per time-position" constraint automatically.

Constraint B: For the constraint that an operator is only allowed to work two days in a row, we check this constraint for gliding windows of length three days for any position and any operator. The global conflict counter register is incremented each time an operator is assigned to a position for all three consecutive days during a gliding window. Figure 5a depicts the resulting circuit with an example of four days, two operators and one position, i. e. two gliding windows.

Every state where $(d_0 p_0 = d_1 p_0 = d_2 p_0) \vee (d_1 p_0 = d_2 p_0 = d_3 p_0)$, shall increment the global conflict counter register. $\boxed{+1}$ gates are activated for all configurations of the control qubits that represent invalid *time-positions*.

This results in $p \cdot o \cdot (d - 2)$ incrementors, however one can see that at most $p \cdot (d - 2)$ can be activated at the same time as it is not possible to have two operators both working the same position three out of three given days due to constraint **A**.

Constraint C: Similarly, the constraint that an operator can only be assigned to at most one position per day is implemented. The idea here is that for any day and any operator we can verify that the operator is not scheduled for two positions. The global counter register is thus incremented each time an operator is assigned to more than one position per day. Figure 5b depicts the resulting circuit with an example of one day, four operators and two positions. Each of the four operators 0–3 is represented by one configuration of the control-qubits activating the $\boxed{+1}$ gate, which is activated if $d_0 p_{00} = d_0 p_{10} \wedge d_0 p_{01} = d_0 p_{11}$. Notice that this results in $d \cdot o \cdot \binom{p}{2}$ incrementors as we need to check on every day that each pair of distinct positions is not occupied by the same operator. However, again due to constraint **A**, for every time-position there is exactly one operator scheduled, meaning that at most $d \cdot \binom{p}{2}$ incrementers can register a constraint violation at the same time.

We can use these calculations to estimate the number of required counter qubits to avoid overflows of the counter register. Notice that such an overflow would effectively mean that the Grover oracle would mark states with particular numbers of constraint violations as valid and amplify their state correspondingly. This would mean that our amplified end results could actually be invalid if the counter register is not large enough. Combining **B** and **C**, we see that we have at most $p(d-2)+d\binom{p}{2}$ constraint violations and, since a count of zero constraint violations needs to be represented, we thus need at least $\lceil \log_2 \left(p(d-2) + d\binom{p}{2} + 1 \right) \rceil$ qubits to represent this number in binary. However, depending on the constraints and the parameters it might be possible that this estimate is too crude and that it is not possible that all of these constraint violations can be achieved at the same time, so a smaller number may be sufficient.

5   Evaluation

We evaluate the solutions with Qiskit's QASM simulator provided by IBM, which simulates a noiseless quantum computer with up to 32 qubits. Due to the

Table 1. Simulation results for different problem sizes used for the evaluation.

Case	I	II	III	IV	V	X
Operators	4	4	4	4	8	4
Postions	2	2	2	2	2	3
Days	3	4	5	6	3	3
Used qubits	15	21	25	29	21	23
Used counter qubits	2(3)	4(4)	4(4)	4(4)	2(3)	4(4)
Percentage solutions	0.223	0.107	0.048	0.022	0.587	0
Grover iterations	1	2	3	5	2	–
Success rate	0.99	0.99	0.99	0.96	0.88	0

limitation of available qubits and a maximum simulation time, only reduced problem sizes are feasible. Table 1 summarizes our simulation of six different configurations of operators, positions and days. There are five valid (**Case I–V**) and one invalid (**Case X**) configurations, all of which are problems with $\log_2 o \cdot p \cdot d \leq 30$.

As an example, Case I reads as follows: Given 4 operators, 2 positions and 3 days, the number of used qubits[2] to run the algorithm is 15. The minimal number of counter qubits was empirically evaluated and maybe less than the upper bounds given in Sect. 4.2. The latter is given in paranthesis, e. g., for Case I $\lceil \log_2 \left(2 \cdot (3 - 2) + 3\binom{2}{2} + 1 \right) \rceil$ is 3.

Further, the percentage of valid solutions in the overall search space is given. With an input size of 12 qubits, the number of possible schedules is $2^{12} = 4096$ of which 912 ($= 22.27\%$) are valid schedules. The number of Grover iterations to reach the first maximum of amplitude amplification can be approximated with $\left\lfloor \frac{\pi}{4} \sqrt{\frac{1}{s}} \right\rfloor$ [6], where s is the percentage of valid solutions as given in Table 1.[3] The success rates are calculated by running the algorithm 8000 times and counting measured results that are valid schedules (see e. g., Fig. 6).

Notice that all but one of the shown cases have 2 positions, whereas one can easily see that Case X with 3 positions has no valid solution at all. The quantum algorithm verifies this true negative result, by amplifying non of the input states.

Case I to V in Table 1 show that with an increasing number of days, the number of required qubits increases. This is mostly due to constraint **B** which introduces a greater amount of conflicts if the number of days is bigger than three. The effect is also clearly visible in Fig. 7, where the success probability (y-axis) of the configuration is plotted with respect to the size of the counter

---

[2] The number of used qubits is given by the sum of counts of problem-encoding qubits, necessary global conflict counter register qubits and one Grover phase-flip qubit, so e. g., $\log_2 4 \cdot 2 \cdot 3 + 2 + 1 = 15$.

[3] Except for Case V, where the approximations are not valid, but the exact formula reproduces the numbers given in the table.

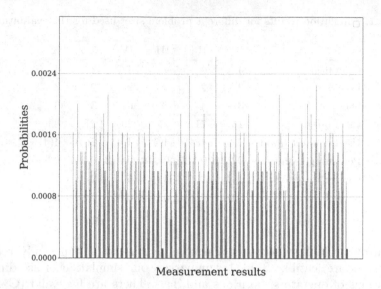

Fig. 6. Statistic for 8000 shots resulting (in)valid schedules in (red)blue for Case I. (Color figure online)

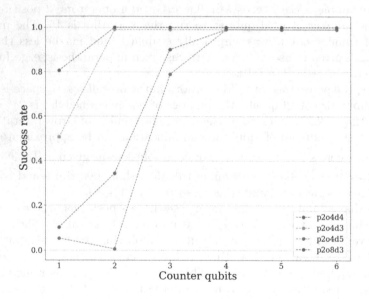

Fig. 7. Heuristical evaluation of minimal number of required counter qubits.

register. Configurations where constraint **B** comes into play require a larger counter register to achieve a high success probability.

The correctness of the implementation is further validated by plotting success rates against a larger number of Grover iteration. The expected sinusoidal-squared behavior is clearly visible in Fig. 8.

Unfortunately, we were not able to test the overall circuit on IBM's real quantum devices, since the circuit depth exceeded their limits. To receive a comparable result, we introduced an error rate on the used gates which imitates those of current IBM quantum hardware to the smallest configuration. The error decreases the success probability to $\sim$33%, which is only slightly higher than the random distribution of $\sim$22%. Therefore, states with correct solutions are no longer distinguishable from a random distribution in this case. This indicates that current error rates are too high to deliver results with our approach even if we would have enough qubits to fit our problem size.

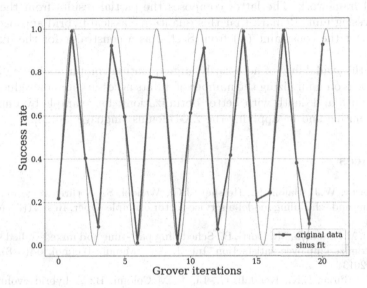

Fig. 8. Fit for expected $\sin^2$-behavior of success rate as a function of Grover iterations in Case I. Fitting parameters: $p = 1.002 \cdot \sin^2(\pi(0.313r + 1.155)) + 0.000$

6   Conclusion and Future Work

This paper successfully presents how to find a valid schedule for the "on-call spacecraft operator scheduling"-problem at the German Space Operating Center by using a Grover's quantum search algorithm. The presented encoding of variables and constraints is transferable to similar nurse-scheduling problems. Our algorithm creates a Qiskit circuit for *Grover's Oracle* with three variables of arbitrary size and complies with three selected constraints[4]. Due to the absence of a sufficiently large and stable quantum computer, we validated our circuits with affordable small input sizes on a quantum simulator capable of processing up to 32 qubits, by various means.

---

[4] Including qubit initialisation and measurement operations.

Future Work

What we showed are just the first steps towards solving a classical problem on a real quantum computer. To execute the necessary real quantum operations of the algorithms on a quantum computer, these need an improvement in qubit stability and quality of their control – or the availability of fully error-corrected qubits.

Quantum computers are currently way too small for real life problems. But the problem can be divided in parts, *e.g.*, in fragments of days or weeks. This approach allows to run the quantum algorithm as a quantum subroutine within a classical framework. The latter composes the partial results from the quantum processing unit. Remark that this quantum-classical-hybrid approach asks to implement the constraint (**v**) from Sect. 2 as a constraint for the quantum algorithm.

While the availability of necessarily large quantum computers is pending, we will also focus on minimizing the number of qubits necessary and on reducing the circuit's width and depth with better optimizations, for example by improving conflict counting and by applying the ZX-calculus language [10,25].

References

1. Abernathy, W.J., Baloff, N., Hershey, J.C., Wandel, S.: A three-stage manpower planning and scheduling model-a service-sector example. Oper. Res. **21**(3), 693–711 (1973)
2. Akbari, M., Zandieh, M., Dorri, B.: Scheduling part-time and mixed-skilled workers to maximize employee satisfaction. Int. J. Adv. Manuf. Technol. **64**(5–8), 1017–1027 (2013)
3. Bai, R., Burke, E.K., Kendall, G., Li, J., McCollum, B.: A hybrid evolutionary approach to the nurse rostering problem. IEEE Trans. Evol. Comput. **14**(4), 580–590 (2010)
4. Bard, J.F., Wan, L.: The task assignment problem for unrestricted movement between workstation groups. J. Sched. **9**(4), 315–341 (2006)
5. Van den Bergh, J., Beliën, J., De Bruecker, P., Demeulemeester, E., De Boeck, L.: Personnel scheduling: a literature review. Eur. J. Oper. Res. **226**(3), 367–385 (2013)
6. Boyer, M., Brassard, G., Høyer, P., Tapp, A.: Tight bounds on quantum searching. Fortschr. Phys. **46**(4–5), 493–505 (1998). https://doi.org/10.1002/(SICI)1521-3978(199806)46:4/5⟨493::AID-PROP493⟩3.0.CO;2-P
7. Brassard, G., Hoyer, P., Mosca, M., Tapp, A.: Quantum amplitude amplification and estimation. Contemp. Math. **305**, 53–74 (2002)
8. Burke, E.K., De Causmaecker, P., Berghe, G.V., Van Landeghem, H.: The state of the art of nurse rostering. J. Sched. **7**(6), 441–499 (2004)
9. Cerf, N.J., Grover, L.K., Williams, C.P.: Nested quantum search and NP-complete problems. arXiv preprint arXiv:quant-ph/9806078 (1998)
10. Coecke, B., Horsman, D., Kissinger, A., Wang, Q.: Kindergarden quantum mechanics graduates (... or how I learned to stop gluing LEGO together and love the ZX-calculus). CoRR abs/2102.10984 (2021). https://arxiv.org/abs/2102.10984.pdf

11. Cordeau, J.F., Laporte, G., Pasin, F., Ropke, S.: Scheduling technicians and tasks in a telecommunications company. J. Sched. **13**(4), 393–409 (2010)
12. Farhi, E., Goldstone, J., Gutmann, S.: A quantum approximate optimization algorithm. arXiv preprint arXiv:1411.4028 (2014)
13. Gilliam, A., Woerner, S., Gonciulea, C.: Grover adaptive search for constrained polynomial binary optimization. arXiv preprint arXiv:1912.04088 (2019)
14. Glos, A., Krawiec, A., Zimborós, Z.: Space-efficient binary optimization for variational computing (2020)
15. Grover, L.K.: A fast quantum mechanical algorithm for database search. In: Miller, G.L. (ed.) Proceedings of the 28th Annual ACM Symposium on Theory of Computing, pp. 212–219. ACM, New York (1996)
16. Hartung, J., Nibler, R., Peat, C., Spörl, A.K., Wörle, M.T., Lenzen, C.: GSOC SoE-Editor 2.0 - A Generic Sequence of Events Tool (2016). https://doi.org/10.2514/6.2016-2568
17. Hogg, T.: Highly structured searches with quantum computers. Phys. Rev. Lett. **80**(11), 2473 (1998)
18. Ikeda, K., Nakamura, Y., Humble, T.S.: Application of quantum annealing to nurse scheduling problem. Sci. Rep. **9**(1), 1–10 (2019)
19. Jaumard, B., Semet, F., Vovor, T.: A generalized linear programming model for nurse scheduling. Eur. J. Oper. Res. **107**(1), 1–18 (1998)
20. Lenzen, C., Wörle, M.T., Mrowka, F., Spörl, A., Klaehn, R.: The algorithm assembly set of Plato. In: 12th International Conference on Space Operations, SpaceOps 2012 (2012). ePoster. https://elib.dlr.de/75694/
21. Nibler, R., Mrowka, F., Wörle, M.T., Hartung, J., Lenzen, C., Peat, C.: PINTA and TimOnWeb - (more than) generic user interfaces for various planning problems. In: 10th International Workshop on Planning and Scheduling for Space, IWPSS 2017 (July 2017). https://elib.dlr.de/114095/
22. Peruzzo, A., et al.: A variational eigenvalue solver on a quantum processor. eprint. arXiv preprint arXiv:1304.3061 (2013)
23. Sanders, Y.R., et al.: Compilation of fault-tolerant quantum heuristics for combinatorial optimization. PRX Quantum **1**(2), 020312 (2020)
24. Tabi, Z., et al.: Quantum optimization for the graph coloring problem with space-efficient embedding. In: 2020 IEEE International Conference on Quantum Computing and Engineering (QCE), pp. 56–62 (2020). https://doi.org/10.1109/QCE49297.2020.00018
25. van de Wetering, J.: ZX-calculus for the working quantum computer scientist. arXiv preprint arXiv:2012.13966 (2020). https://arxiv.org/pdf/2012.13966.pdf
26. Wörle, M.T., Spörl, A., Hartung, J., Lenzen, C., Mrowka, F.: The mission planning system for the firebird spacecraft constellation. In: Proceedings of the 14th International Conference on Space Operations, Daejeon, Republic of Korea (2016)

Multimodal Container Planning: A QUBO Formulation and Implementation on a Quantum Annealer

F. Phillipson[✉][iD] and I. Chiscop

TNO, PO Box 96800, 2509 JE The Hague, The Netherlands
frank.phillipson@tno.nl

Abstract. Quantum computing is developing fast. Real world applications are within reach in the coming years. One of the most promising areas is combinatorial optimisation, where the Quadratic Unconstrained Binary Optimisation (QUBO) problem formulation is used to get good approximate solutions. Both the universal quantum computer as well as the quantum annealer can handle this kind of problems well. In this paper, we present an application on multimodal container planning. We show how to map this problem to a QUBO problem formulation and how the practical implementation can be done on the quantum annealer produced by D-Wave Systems.

Keywords: Multimodal container planning · Quantum computing · Quantum annealing · QUBO modelling

1 Introduction

In container logistics various transportation concepts are known [28]. Unimodal transport usually refers to loading cargo onto a single transportation mode, usually a truck, and driving it from origin to destination. Multimodal transport refers to transporting cargo from origin to destination by more than one transportation mode. In intermodal transport, again more than one transportation mode may be used, but the containment unit (container) must always be of standardised size. Co-modal transport looks like multimodal transport, but requires a consortium of shippers and has a focus on exploiting the benefits of each transportation mode in a smart way. Finally, synchromodal transport is a version of intermodal transport that focuses on real-time planning flexibility and coordination between different shippers, both using large amounts of real-time data.

When looking at the planning of container flows, it is often categorised on three levels of problems: on strategic, tactical and operational level [27]. Strategic problems concern long-term investments in the transportation network, for example, where to build new terminals. Tactical problems may concern service design, for example, determining how many times in a month a barge should make a round-trip. Operational problems revolve around using a current network in an optimal way for problems occurring in the present. Little attention

© Springer Nature Switzerland AG 2021
M. Paszynski et al. (Eds.): ICCS 2021, LNCS 12747, pp. 30–44, 2021.
https://doi.org/10.1007/978-3-030-77980-1_3

has been given to operational problems [22]. Operational problems can be divided into various classes: (a) problems in which containers are assigned to existing barge services, (b) problems in which the routes of barges are determined for a given demand, and (c) problems in which both the assignment of containers to barges and the routes of barges are decided upon [29].

In an operational, dynamic, environment with a lot of uncertainty, or in a synchromodal environment where real time system characteristics are used in the planning, short computation times are crucial, since these make it possible repeat the planning process with a high frequency. To this end, efficient algorithms are proposed [28] or techniques are developed to reduce the complexity of the problem [19]. Next to software and algorithm enhancements to speed up the computation, also hardware developments can play a role. A promising direction here is quantum computing. Note that this will not be a solution for the short term, where the current generation of quantum computers is far too small to compete with classical computers and their advanced solvers and algorithms. This work is meant to sketch a new direction for medium and long term solutions.

The past decade has seen the rapid development of the two paradigms of quantum computing, quantum annealing and gate-based quantum computing. In 2011 D-Wave Systems announced the release of the world's first commercial quantum annealer[1] operating on a 128-qubit architecture, which has since been continually extended up to the 2048-qubit version, available from 2017[2]. D-Wave announced in 2019 a (more than) 5000 qubit system available mid-2020, using their new Pegasus-technology-based chip with 15 connections per qubit[3]. Even more recently, quantum supremacy is claimed to have been achieved by Google's 54-qubit Sycamore gate-based computer [2]. These technological advances have led to a renewed interest in finding classical intractable problems suited for quantum computing.

A particular category of problems that are expected to fit well on quantum devices are (combinatorial) optimisation problems. An important trail in implementing optimisation problems on quantum devices is the Quadratic Unconstrained Binary Optimisation (QUBO) problem. These QUBO problems can be solved by a universal (gate-based) quantum computer using a QAOA (Quantum Approximate Optimisation Algorithm) implementation [11] or by a quantum annealer, such as D-Wave [18]. For already a large number of combinatorial optimisation problems the QUBO representation is known [15,20]. Well-known examples that have been implemented on one of D-Wave's quantum processors include maximum clique [5], capacitated vehicle routing [13], minimum vertex cover [24], set cover with pairs [4], Multi-Service Location Set Covering Problem [6], traffic flow optimisation [23] and integer factorisation [17]. These studies have

---

[1] https://www.dwavesys.com/news/d-wave-systems-sells-its-first-quantum-computing-system-lockheed-martin-corporation.

[2] https://www.dwavesys.com/press-releases/d-wave%C2%A0announces%C2%A0d-wave-2000q-quantum-computer-and-first-system-order.

[3] https://www.dwavesys.com/press-releases/d-wave-previews-next-generation-quantum-computing-platform.

shown that although the current generation of D-Wave annealers may not yet have sufficient scale, precision and connectivity to allow faster or higher quality solutions, they have the suitable infrastructure for modelling real-world instances of these problems, effectively decomposing these into smaller sub-problems and solving these on a real Quantum Processing Unit (QPU). The QUBO problems can also be solved by simulated (quantum) annealing on a conventional computer. Then, however, no quantum advantage can be expected.

In this paper we propose a QUBO formulation for a container assignment problem and show the implementation on both simulated annealing as well as on the D-Wave QPU. The size of the current generation of quantum computers limits us to small problems, with a limited number of decision variables. Therefore, we propose first a QUBO-formulation that only uses two potential paths for each container, and expand this to four alternatives per container and a generalisation to 2^q paths.

2   Mathematical Framework

In this work a deterministic container-to-mode problem is studied. To give a formal definition the framework of [8] is used. The following $\bar{R}|\bar{D}, [D2R]|social(1)$-problem is meant: to assign freight containers to transportation modes, such that the containers reach their destinations before a deadline against minimum total cost, given that the transport modes have fixed given schedules and all features of the problem are deterministic [16]. These problems are often solved using minimum cost multi-commodity flow problems on time-dependent graphs and space-time networks [1,9]. The non-negative two-commodity flow problem on a directed graph, however, is proven to be NP-complete [10]. Finding a flow with minimum cost and with at least two commodities must be at least as difficult, however, the LP-relaxation of the studied problems almost always has an integral optimum [16]. To study the potential of quantum computing for this problem we propose a QUBO formulation.

2.1   Approach

Given the state of the quantum devices, which are quite small at this moment, we will limit the size of the problem by giving each container a limited number of paths through the transportation network, here either 2 or 4. For this we propose the following approach:

1. Select a set (2 or 4) of alternative routes per container; In the case of 2 alternatives we take the best multimodal path and the uni-modal (trucking) path. In the case of 4 alternative routes, we take again the uni-modal path and the 3 maximum dissimilar paths.
2. Define the QUBO-problem.
3. Solve the QUBO-problem.
4. Evaluate the solution and go back to step 1 if necessary to select different sets of alternatives.

2.2   QUBO Formulation

We now give the definition of the QUBO-problem [15] and propose the formulation of the QUBO for the two problems. The QUBO is expressed by the optimisation problem:

$$\text{QUBO: min/max } y = x^t Q x, \tag{1}$$

where $x \in \{0,1\}^n$, the decision variables and Q is a $n \times n$ coefficient matrix. QUBO problems belong to the class of NP-hard problems. Another formulation of the problem, often used, equals

$$\text{QUBO: min/max } H = x^t q + x^t Q x, \tag{2}$$

or a combination of multiple of these terms

$$\text{QUBO: min/max } H = A \cdot H_A + B \cdot H_B + \cdots, \tag{3}$$

where $A, B, \ldots$ are weights that can used to tune the problem and include constraints into the QUBO. We will use this representation in the next sections, where we will present the QUBO formulation for the 2 and 4 alternative route problems.

Creating a QUBO. For already a large number of combinatorial optimisation problems the QUBO representation is known [15,20]. Many constrained integer programming problems can be transformed easily to a QUBO representation. Assume that we have the problem

$$\min y = c^t x, \text{ subject to } Ax = b, \tag{4}$$

then we can bring the constraints to the objective value, using a penalty factor λ for a quadratic penalty:

$$\min y = c^t x + \lambda (Ax - b)^t (Ax - b). \tag{5}$$

Using $P = Ic$, the matrix with the values of vector c on its diagonal, we get

$$\min y = x^t P x + \lambda (Ax - b)^t (Ax - b) = x^t P x + x^t R x + d = x^t Q x, \tag{6}$$

where matrix R and the constant d follow from the matrix multiplication and the constant d can be neglected, where it does not influence the optimisation problem.

2-Alternative Route QUBO. We look at a situation where containers have to be shipped in an intermodal or synchromodal network. To limit the number of options, we give every container two possible paths through the network, one using a truck only, the other a multimodal or multitrack path through the network. Each track or modality in this network has a specific capacity.

Table 1. Parameters and decision variables.

Indices	
i	Containers, $i \in \{1, \dots, n\}$
j	Tracks, $j \in \{1, \dots, m\}$

Parameters	
c_i^b	Costs for using multimodal route container i
c_i^t	Costs for using truck by container i
V_j	Capacity of track j
r_i	Route for container i
$r_{ij} = \begin{cases} 1 & \text{if route } i \text{ contains track } j \\ 0 & \text{otherwise} \end{cases}$	

Decision variables	
$x_i = \begin{cases} 1 & \text{if container } i \text{ is transported by truck} \\ 0 & \text{if container } i \text{ is transported by barge route } r_i \end{cases}$	

We define a QUBO representation for this problem. We assume that all routes are feasible, with regard to due dates etc. (Table 1).

The binary integer linear program for the problem is formulated as follows:

$$\min \sum_{i=1}^{n} c_i^b + (c_i^t - c_i^b)x_i, \tag{7}$$

$$\text{such that} \quad \sum_{i=1}^{n} (1 - x_i)r_{ij} \leq V_j \quad \forall j \in \{1, \dots m\}. \tag{8}$$

Here for every container i it has to be decided whether it is sent using a truck ($x_i = 1$) or using the multitrack multimodal path ($x_i = 0$). Equation 7 minimises the costs of the choices, where per container i the choice $x_i = 0$ leads to costs c_i^b and $x_i = 1$ to $c_i^b + (c_i^t - c_i^b) = c_i^b - c_i^b + c_i^t = c_i^t$. Equation 8 makes sure that the capacity constraints of the modalities or tracks are met.

Based on the model in Eq. (7–8), the QUBO formulation of the problem can also be derived. The overall solution for this QUBO is given by:

$$\min \quad A \cdot H_A + B \cdot H_B, \tag{9}$$

$$\text{with } H_A = \sum_{i=1}^{n} c_i^b + (c_i^t - c_i^b)x_i, \tag{10}$$

$$H_B = \sum_{j=1}^{m} \left(\sum_{i=1}^{n} (1 - x_i)r_{ij} + \sum_{k=0}^{K} 2^k y_{jk} - V_j \right)^2, \tag{11}$$

where A and B denote penalty coefficients to be applied such that the constraints will be satisfied and y_{ik} denotes additional slack variables. These $K \cdot m$

binary slack variables are necessary to remodel Eq. 8 into equality constraints, as they are required in a QUBO-formulation. The parameter K follows from $K = \max_j \left(\log_2(V_j) \right)$. When determining the penalty coefficients, we can set $A = 1$ and look for a good value for B. By rule of thumb, the gain of violating a constraint must be lower than the costs. This means in this problem that $B > \max_i c_i^b$.

Note that if capacity is bounding for all tracks, Eq. 8 changes to an equality, which removes the need for the slack variables in Eq. 11. If the capacity is not bounding for all tracks, i.e., the capacity is very high, Eq. 8 and Eq. 11 disappear from the problem for certain values of j, instead of allowing the slack variables for very high values.

4-Alternative Route QUBO. If we give the problem, next to trucking, three multitrack and multimodal options to choose from, we get the following model. Now we use an extra binary variable to represent this choice: {11} stands for trucking, {00} for route 0, {10} for route 1 and {01} for route 2 (Table 2).

Table 2. Parameters and decision variables.

Parameters	
c_i^k	Costs for using multimodal route k for container i
c_i^t	Costs for using truck by container i
V_j	Capacity of track j
r_{ik}	Route $k = 0, 1, 2$ for container i
r_{ikj}	$= \begin{cases} 1 & \text{if route } k \text{ for container } i \text{ contains track } j \\ 0 & \text{otherwise} \end{cases}$
Decision variables	
$\{x_{2i-1} \quad x_{2i}\} =$	$\begin{cases} \{1\ 1\} & \text{if container } i \text{ is transported by truck} \\ \{0\ 1\} & \text{if container } i \text{ is transported by route } r_{i2} \\ \{1\ 0\} & \text{if container } i \text{ is transported by route } r_{i1} \\ \{0\ 0\} & \text{if container } i \text{ is transported by route } r_{i0} \end{cases}$

The binary integer linear program for the problem is formulated as follows:

$$\min \sum_{i=1}^{n} c_i^0 + (c_i^1 - c_i^0)x_{2i-1} + (c_i^2 - c_i^0)x_{2i} + (c_i^t + c_i^0 - c_i^1 - c_i^2)x_{2i-1}x_{2i},$$

$$(12)$$

such that $\sum_{i=1}^{n}(1 - x_{2i-1})(1 - x_{2i})r_{i0j} + x_{2i-1}(1 - x_{2i})r_{i1j}$

$$+ x_{2i}(1 - x_{2i-1})r_{i2j} \leq V_j \quad \forall j \in \{1, \ldots m\}. \qquad (13)$$

Here again Eq. 12 minimises the costs of the choices and Eq. 13 makes sure that the capacity constraints of the modalities or tracks are met.

Based on the model in Eq. (12–13), the QUBO formulation of the problem can also be derived. The overall formulation is given by:

$$\min \quad H = A \cdot H_A + B \cdot H_B \tag{14}$$

$$\text{with } H_A = \sum_{i=1}^{n} c_i^0 + (c_i^1 - c_i^0)x_{2i-1} + (c_i^2 - c_i^0)x_{2i} + (c_i^t + c_i^0 - c_i^1 - c_i^2)x_{2i-1}x_{2i},$$
$$\tag{15}$$

$$H_B = \sum_{j=1}^{m} \left(\sum_{i=1}^{n} (1 - x_{2i-1})(1 - x_{2i})r_{i0j} + x_{2i-1}(1 - x_{2i})r_{i1j} \right.$$

$$\left. + x_{2i}(1 - x_{2i-1})r_{i2j} + \sum_{k=0}^{K} 2^k y_{jk} - V_j \right)^2, \tag{16}$$

where y_{ik} denotes additional slack variables. These binary slack variables are necessary to remodel Eq. 13 into equality constraints, as they are required in a QUBO-formulation. Again, A, B and K denote penalty coefficients and the number of slack variables per track. They will be set conform the rules explained earlier. Note that H_B contains higher order terms that has to be reduced to quadratic terms to obtain a QUBO. This can be done using the D-Wave functionality dimod.higherorder.utils.make_quadratic.

Table 3. Parameters and decision variables.

Parameters	
c_i^k	Costs for using multimodal route $k = 0, ..., 2^q - 2$ for container i
$c_i^{2^q-1}$	Costs for using truck by container i
V_j	Capacity of track j
r_{ik}	Route $k = 0, 1, ..., 2^q - 2$ for container i
r_{ikj}	$= \begin{cases} 1 & \text{if route } k \text{ for container } i \text{ contains track } j \\ 0 & \text{otherwise} \end{cases}$
Decision variables	
$\{x_{q \cdot i-q+1} \cdots x_{q \cdot i}\}$	$= \begin{cases} 1 = \{1 \ldots 1\} & \text{if container } i \text{ is transported by truck} \\ B = \{b_1 \ldots b_q\} & \text{if container } i \text{ is transported by route } r_{i\theta}, \theta = \sum_{j=1}^{q} 2^{j-1}b_j, \\ b_i & \text{binary variables} \end{cases}$

Generic 2^q-alternative Route QUBO. We can generalise this formulation to 2^q alternative routes if we use q qubits per container (Table 3).

The binary integer linear program for the problem is formulated as follows:

$$\min \sum_{i=1}^{n} \sum_{b_1=0}^{1} \cdots \sum_{b_q=0}^{1} t_{i,b} x_{q \cdot i-q+1}^{b_1} \cdot \cdots \cdot x_{q \cdot i}^{b_q} \tag{17}$$

such that

$$\sum_{i=1}^{n} \sum_{b \in B_1} (1 - x_{q \cdot i - q + 1})^{1-b_1} \cdot \ldots \cdot (1 - x_{qi})^{1-b_q} \cdot x_{q \cdot i - q + 1}^{b_1} \cdot \ldots \cdot x_{q \cdot i}^{b_q} r_{i\theta_b j} \leq V_j$$

$$\forall j \in \{1, \ldots m\} \tag{18}$$

where a recursive relation is defined for the parameter

$$t_{i,b} = c_i^{\theta_b} - \sum_{b' \in B_b} t_{i,b'}, \tag{19}$$

and defining

$$\theta_b = \sum_{j=1}^{q} 2^{j-1} b_j, \quad B_b = \{b' = \{0,1\}^q | b' \preceq b, b' \neq b\}. \tag{20}$$

Here $b' \preceq b$ says that no single element of b' is greater than the corresponding element of b. Based on this model, the generalised QUBO formulation of the problem can be derived. The overall formulation is given by:

$$\min \quad H = A \cdot H_A + B \cdot H_B \tag{21}$$

with

$$H_A = \sum_{i=1}^{n} \sum_{b_1=0}^{1} \cdots \sum_{b_q=0}^{1} t_{i,b} x_{q \cdot i - q + 1}^{b_1} \cdot \ldots \cdot x_{q \cdot i}^{b_q}, \tag{22}$$

$$H_B = \sum_{j=1}^{m} \Big(\sum_{i=1}^{n} \sum_{b \in B_1} (1 - x_{q \cdot i - q + 1})^{1-b_1} \cdot \ldots \cdot (1 - x_{qi})^{1-b_q} \cdot$$

$$x_{q \cdot i - q + 1}^{b_1} \cdot \ldots \cdot x_{q \cdot i}^{b_q} r_{i\theta_b j} + \sum_{k=0}^{K} 2^k y_{jk} - V_j \Big)^2, \tag{23}$$

where y_{ik} denotes additional slack variables. Again, A, B and K denote penalty coefficients and the number of slack variables per track. They will be set conform the rules explained earlier. Again, H_B has to be reduced to quadratic terms only.

2.3 Solving the QUBO Problem

For solving the QUBO problem we use the D-Wave system. Implementing problems on a quantum device asks for specific Quantum Software Engineering [25]. We will sketch a number of specific implementation issues for this problem.

Quantum Annealing. The quantum devices produced by D-Wave Systems are practical implementations of quantum computation by adiabatic evolution [12]. The evolution of a quantum state on D-Wave's QPU is described by a time-dependent Hamiltonian, composed of initial Hamiltonian H_0, whose ground state is easy to create, and final Hamiltonian H_1, whose ground state encodes the solution of the problem at hand. The system is initialised in the ground state of the initial Hamiltonian, i.e., $H(0) = H_0$. The adiabatic theorem states that if the system evolves according to the Schrödinger equation, and the minimum spectral gap of $H(t)$ is not zero, then for time T large enough, $H(T)$ will converge to the ground state of H_1, which encodes the solution of the problem. This process is known as quantum annealing. Goal is to find the eigenstates of H, where the eigenvalues are the energy values. The eigenstates with the lowest eigenvalue are the ground states of the system. Although here we are not concerned with the technical details, it is worthwhile to mention that it is not possible to estimate an annealing time T to ensure that the system always evolves to the desired state. Since there is no estimation of the annealing time, there is also no optimality guarantee.

The D-Wave quantum annealer can accept a problem formulated as an Ising Hamiltonian or rewritten as its binary equivalent, in QUBO formulation. Next, this formulation needs to be embedded on the hardware. In the most developed D-Wave 2000Q version of the system, the 2048 qubits are placed in a Chimera architecture [21]: a 16×16 matrix of unit cells consisting of 8 qubits. This allows every qubit to be connected to at most 5 or 6 other qubits. With this limited hardware structure and connectivity, fully embedding a problem on the QPU can sometimes be difficult or simply not possible. In such cases, the D-Wave system employs built-in routines to decompose the problem into smaller sub-problems that are sent to the QPU, and in the end reconstructs the complete solution vector from all sub-sample solutions. The first decomposition algorithm introduced by D-Wave was *qbsolv* [3], which gave a first possibility to solve larger scale problems on the QPU. Although *qbsolv* was the main decomposition approach on the D-Wave system, it did not enable customisations, and therefore is not particularly suited for all kinds of problems. The new decomposition approaches D-Wave offers are D-Wave Hybrid and the Hybrid Solver Service, offering more customisations.

Next to the QPU also a CPU can be used within the programming environment of D-Wave, using for example simulated annealing (SA), a conventional meta-heuristic.

2.4   Embedding

Some restrictions are introduced by the hardware design of the D-Wave hardware. The current implementation has a qubit connectivity based on a Chimera structure. The QUBO problem at hand has to be transformed to this structure. Because of the limited chip sizes we have currently, a compact formulation of the QUBO is important, but also a compact transformation to the chip design. This problem is known as minor-embedding. Here the vertices correspondent to problem variables and edges exist if $Q_{ij} \neq 0$, where Q represents the qubo-matrix. Because of the limited connectivity of the chip, a problem variable has to be duplicated to multiple (connected) qubits. Those qubits should have the same value, meaning the weight of their connection should be such that it holds in the optimisation process. All these qubits representing the same variables are part of a so-called chain, and their edge weights is called the chain strength (λ), which is an important value in the optimisation process. In [7], Coffrin gives a thorough analysis. He indicates that if λ is sufficiently large, optimal solutions will match $\lambda \geq \sum_{ij} |Q_{ij}|$. However, the goal is to find the smallest possible value of λ, to avoid re-scaling of the problem. Coffrin also indicates that finding the smallest possible setting of λ can be NP-hard. Also other research has been performed on selecting an optimal choice for this chain strength [26], but at the moment there is no solid criterion for choosing a value. A rule of thumb that is suggested[4] is $\lambda = \max_{ij} Q_{ij}$. It may be necessary to use the quantum annealer with multiple values of the chain strength in order to determine which value for λ is optimal for a given problem [14].

Table 4. Parameters used in the 2-alternative route case study.

$$c_b = (2, 7, 1, 6, 2, 4, 8, 7, 7, 10)$$
$$c_t = (23, 25, 23, 17, 24, 22, 19, 16, 21, 17)$$
$$v = (5, 5, 5, 5, 5, 5, 5, 5, 5, 5, 5, 5)$$

$r_1 = (1, 0, 1, 0, 0, 1, 0, 1, 0, 0, 0, 0)$	$r_2 = (1, 0, 1, 0, 1, 0, 1, 0, 0, 0, 0, 0)$
$r_3 = (1, 0, 1, 0, 0, 1, 0, 1, 0, 0, 0, 0)$	$r_4 = (1, 0, 0, 0, 0, 0, 1, 0, 1, 1, 0, 0)$
$r_5 = (1, 0, 1, 0, 1, 0, 1, 0, 0, 0, 0, 0)$	$r_6 = (0, 1, 0, 1, 0, 1, 0, 1, 0, 0, 0, 0)$
$r_7 = (1, 0, 1, 0, 1, 0, 1, 0, 0, 0, 0, 0)$	$r_8 = (1, 0, 1, 0, 1, 0, 1, 0, 0, 0, 0, 0)$
$r_0 = (1, 0, 1, 0, 1, 0, 1, 0, 0, 0, 0, 0)$	$r_{10} = (0, 1, 0, 1, 0, 1, 0, 1, 0, 0, 0, 0)$

3   Numerical Results

To give some idea about the approach and performance we present a case study for the 2-alternative route approach.

3.1   2-Alternative Route Case Study

We use a case consisting of 10 containers ($n = 10$) and 12 tracks ($m = 12$). The parameters are given in Table 4. In the optimal solution, containers 4, 7 and 8 are transported by trucks and the others by barges, resulting in a total costs of 85, using the standard Matlab solver. As explained, this size of problems poses no challenge for classical solvers. The goal of this exercise is to show th potential of quantum computing when the quantum computers have more mature sizes. The QUBO following from Eq. 9 leads to a 46×46 matrix.

---

[4] https://support.dwavesys.com/hc/en-us/community/posts/360034852633-High-Chain-Break-Fractions.

3.2   Simulated Annealing

The annealing processes (both simulated as quantum) are stochastic processes. This means that if we do a high number of queries we get a probability distribution for the resulting solution. The first important parameter to fix is this number of queries. We chose 500 samples. Next the values for A and B are important. We can set $A = 1$ and look for a good value for B. Rule of thumb is that the gain of violating a constraint must be higher than the consequent reduction in cost. This means in this problem that $B > c_b$, so $B > 10$. In Fig. 1 the probability distribution of four values for B. We see that choosing B too low ($B = 6$) gives all solutions that are better than the optimal solution, but they all violate one or more constraints. Choosing B too high gives solutions that are feasible, but they are (far) away of the optimal solution. Choosing $B = 12$ gives a range of solutions that include the optimal value, which will be found in most runs. The computation time for approach is around one minute, much more than conventional solvers, that ask for less than one second.

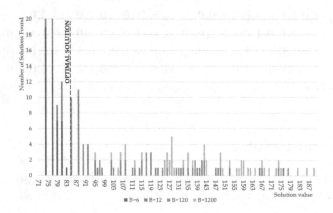

Fig. 1. Results for four values of B for the simulated annealing approach.

3.3   Quantum Annealing

To use D-Wave's QPU[5] now two parameters have to be determined. Again the value for A and B are important. By setting $A = 1$ we can choose a good value for B. Here we expect that the same value as established for simulated annealing will work. Here, we also have to determine a suitable value for the chain strength λ. We start with the rule of thumb indicated in the previous section, $\lambda = \max_{ij} Q_{ij}$. Note that Eq. 1–3 imply that λ is dependent on the value of B, because B is part of the Q matrix. If we choose the values 3, 6 and 12 for B, then the values 120, 240 and 480 follow for λ. If we use $\lambda \geq \sum_{ij} |Q_{ij}|$, the values $4,673$, $9,341$ and

---

[5] We used the D-Wave 2000 system. Only recently, D-Wave presented the 5000 qubit version.

18, 687 follow. In Table 5 the results from the annealing are depicted. For each combination $\{\lambda, B\}$ we present the minimum value, the average value and the percentage of feasible solutions found using 1000 readouts. The minimum value comes from the solution with the lowest energy that does not violate any of the constraints. The average value is the average objective value of all solutions that do not violate any of the constraints.

Table 5. Results for Quantum Annealing for various combinations of the chain strength and the penalty value B. Each result shows the minimum value, average value, and percentage of allowed solutions.

Chain strength	B								
	3			6			12		
	Min.	Avg.	Perc.	Min.	Avg.	Perc.	Min.	Avg.	Perc.
1	85	111	52%	92	135	96%	128	167	100%
2	85	103	5%	96	129	50%	134	139	97%
5	99	110	2%	98	120	11%	104	124	72%
10	98	98	1%	100	117	18%	129	120	40%
120	92	118	58%	104	128	87%	92	126	93%
240	90	145	93%	115	130	83%	106	124	80%
480	96	147	95%	131	135	93%	95	146	95%
4, 673	110	149	85%						
9, 341				142	153	98%			
18, 687							158	160	99%

Note in Table 5 that the best (the optimal) solution is found in the left upper corner, for low values for both λ and B. The suggested value by the rule of thumb does not give good solutions in this example nor do the upper bound values. We also see that the best average values appear at a value $\lambda = 10$ for all values for B. This value for λ also comes with the lowest percentage feasible solutions, indicating that the probability distribution, like we see in Fig. 1, shifts to the left for these values of λ. For two values ($\lambda = 10$ and $\lambda = 240$ and $B = 6$ the distribution is plotted in Fig. 2, confirming this idea. Note that for $\lambda = 240$ the best solution found, based on minimum energy, has objective value 115. In Fig. 2 we see that much better feasible solutions were found (even the optimal solution). These solutions follow from solution that have a chain break(s) and have a higher energy value. Same holds for $\lambda = 10$. Here the solution with objective value 90 is found, where 100 is reported as best solution. Calculation time of this approach is around $10\,\mu s$ (pure QPU time) for each readout, leading to 10 ms for 1000 readouts. This is more than conventional solvers ask for this size of problems, however, the idea is that Quantum Annealing will scale better for bigger problems, when the hardware is ready for that size of problems.

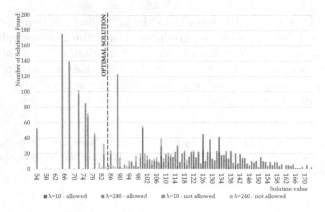

Fig. 2. Results for $B = 6$ and two values for λ for the quantum annealing approach.

4   Conclusions

The rise of quantum computing and the promising application to combinatorial optimisation ask for re-formulation of many real-world problems to match the form these quantum computer can handle, the QUBO formulation, where QUBO stands for Quadratic Unconstrained Binary Optimisation. In this paper we proposed two QUBO-formulations for multimodal container planning, where sets containing a limited number of paths are used. For each container an assignment is found that minimises the overall costs of transporting all containers to their destination. These formulations are proposed to sketch to potential of quantum computing for this kind of problems when the quantum computer will be a more matured technology. When implementing these problem formulations on D-Wave's quantum annealer, there are a number of implementation issues such as finding the right embedding, defining the chain strength and finding the right penalty functions. We showed this is addressed by doing a parameter grid search. For further research a more general idea for finding these parameters is recommended. Also the extension of the set with alternative paths is recommended and a integrated way to include promising paths to this set. Here we think of introducing a quantum variant of the known classical technique of column generation, Quantum Column Generation.

References

1. Andersen, J., Crainic, T.G., Christiansen, M.: Service network design with management and coordination of multiple fleets. Eur. J. Oper. Res. **193**(2), 377–389 (2009)
2. Arute, F., et al.: Quantum supremacy using a programmable superconducting processor. Nature **574**(7779), 505–510 (2019)
3. Booth, M., Reinhardt, S.P., Roy, A.: Partitioning optimization problems for hybrid classical/quantum execution. Technical report, D-Wave Systems (September 2017)

4. Cao, Y., Jiang, S., Perouli, D., Kais, S.: Solving set cover with pairs problem using quantum annealing. Sci. Rep. **6**(1), 33957 (2016)
5. Chapuis, G., Djidjev, H., Hahn, G., Rizk, G.: Finding maximum cliques on the d-wave quantum annealer. J. Sig. Process. Syst. **91**(3–4), 363–377 (2018)
6. Chiscop, I., Nauta, J., Veerman, B., Phillipson, F.: A hybrid solution method for the multi-service location set covering problem. In: International Conference On Computational Science (ICCS) (2020)
7. Coffrin, C.J.: Challenges with chains: Testing the limits of a d-wave quantum annealer for discrete optimization. Technical report, Los Alamos National Laboratory, U.S. (2019)
8. De Juncker, M.A.M., Huizing, D., del Vecchyo, M.R.O., Phillipson, F., Sangers, A.: Framework of synchromodal transportation problems. ICCL 2017. LNCS, vol. 10572, pp. 383–403. Springer, Cham (2017). https://doi.org/10.1007/978-3-319-68496-3_26
9. Ding, B., Yu, J.X., Qin, L.: Finding time-dependent shortest paths over large graphs. In: Proceedings of the 11th International Conference on Extending Database Technology: Advances in Database Technology, pp. 205–216. ACM (2008)
10. Even, S., Itai, A., Shamir, A.: On the complexity of time table and multi-commodity flow problems. In: 16th Annual Symposium on Foundations of Computer Science, SFCS 1975, pp. 184–193. IEEE (1975)
11. Farhi, E., Goldstone, J., Gutmann, S.: A quantum approximate optimization algorithm. arXiv preprint arXiv:1411.4028 (2014)
12. Farhi, E., Goldstone, J., Gutmann, S., Sipser, M.: Quantum computation by adiabatic evolution. arXiv:quant-ph/0001106v1 (2000)
13. Feld, S., et al.: A hybrid solution method for the capacitated vehicle routing problem using a quantum annealer. Front. ICT **6**, 13 (2019)
14. Foster, R.C., Weaver, B., Gattiker, J.: Applications of quantum annealing in statistics. arXiv preprint arXiv:1904.06819 (2019)
15. Glover, F., Kochenberger, G., Du, Y.: A tutorial on formulating and using QUBO models. arXiv preprint arXiv:1811.11538 (2018)
16. Huizing, D.: General methods for synchromodal planning of freight containers and transports. Master's thesis, Delft University of Technology, The Netherlands (2017)
17. Jiang, S., Britt, K.A., McCaskey, A.J., Humble, T.S., Kais, S.: Quantum annealing for prime factorization. Sci. Rep. **8**(1), 17667 (2018)
18. Johnson, M.W., et al.: Quantum annealing with manufactured spins. Nature **473**(7346), 194–198 (2011)
19. Kalicharan, K., Phillipson, F., Sangers, A., De Juncker, M.: Reduction of variables for solving logistic flow problems. In: 2019 6th International Physical Internet Conference (IPIC) (2019)
20. Lucas, A.: Ising formulations of many NP problems. Front. Phys. **2**, 5 (2014)
21. McGeoch, C.C.: Adiabatic Quantum Computation and Quantum Annealing: Theory and Practice. Synthesis Lectures on Quantum Computing, vol. 5, no. 2, pp. 1–93 (2014)
22. Mes, M.R.K., Iacob, M.E.: Synchromodal transport planning at a logistics service provider. In: Zijm, H., Klumpp, M., Clausen, U., Hompel, M. (eds.) Logistics and Supply Chain Innovation. Lecture Notes in Logistics. Springer, Cham (2016). https://doi.org/10.1007/978-3-319-22288-2_2
23. Neukart, F., Compostella, G., Seidel, C., Dollen, D.V., Yarkoni, S., Parney, B.: Traffic flow optimization using a quantum annealer. Front. ICT **4**, 29 (2017)

24. Pelofske, E., Hahn, G., Djidjev, H.: Solving large minimum vertex cover problems on a quantum annealer. In: Proceedings of the 16th ACM International Conference on Computing Frontiers, CF 2019, pp. 76–84. ACM, New York (2019)
25. Piattini, M., et al.: The Talavera manifesto for quantum software engineering and programming. In: Proceedings of the 1st QANSWER Workshop (2020)
26. Rieffel, E.G., Venturelli, D., O'Gorman, B., Do, M.B., Prystay, E.M., Smelyan-skiy, V.N.: A case study in programming a quantum annealer for hard operational planning problems. Quantum Inf. Process. **14**(1), 1–36 (2014). https://doi.org/10.1007/s11128-014-0892-x
27. van Riessen, B., Negenborn, R.R., Dekker, R.: Synchromodal container transporta-tion: an overview of current topics and research opportunities. In: Corman, F., Voß, S., Negenborn, R.R. (eds.) ICCL 2015. LNCS, vol. 9335, pp. 386–397. Springer, Cham (2015). https://doi.org/10.1007/978-3-319-24264-4_27
28. SteadieSeifi, M., Dellaert, N.P., Nuijten, W., Van Woensel, T., Raoufi, R.: Mul-timodal freight transportation planning: a literature review. Eur. J. Oper. Res. **233**(1), 1–15 (2014)
29. Zweers, B.G., Bhulai, S., van der Mei, R.D.: Optimizing barge utilization in hinter-land container transportation. Nav. Res. Logistics (NRL) **66**(3), 253–271 (2019)

Portfolio Optimisation Using the D-Wave Quantum Annealer

Frank Phillipson[1]([⊠]) and Harshil Singh Bhatia[2]

[1] TNO, The Netherlands Organisation for Applied Scientific Research,
The Hague, The Netherlands
frank.phillipson@tno.nl
[2] Department of Computer Science and Engineering, Indian Institute of Technology,
Jodhpur, India

Abstract. The first quantum computers are expected to perform well on quadratic optimisation problems. In this paper a quadratic problem in finance is taken, the Portfolio Optimisation problem. Here, a set of assets is chosen for investment, such that the total risk is minimised, a minimum return is realised and a budget constraint is met. This problem is solved for several instances in two main indices, the Nikkei225 and the S&P500 index, using the state-of-the-art implementation of D-Wave's quantum annealer and its hybrid solvers. The results are benchmarked against conventional, state-of-the-art, commercially available tooling. Results show that for problems of the size of the used instances, the D-Wave solution, in its current, still limited size, comes already close to the performance of commercial solvers.

Keywords: Quantum portfolio optimisation · Quadratic
unconstrained binary optimisation · Quantum annealing · Genetic
algorithm

1 Introduction

Portfolio management is the problem of selecting assets (bonds, stocks, commodities) or projects in an optimal way. Classical portfolio management, as introduced by Markowitz, focuses on efficient (expected) mean-variance combinations [20], and has led to a broad spectrum of optimisation problems: single and multi-objective [29,35], single and multi-period [18,33], without and with [10,18] transaction costs, deterministic or stochastic [26,27] in all possible combinations. One of the basic problems is the single objective, maximising expected return, under budget and risk constraints. Risk is expressed by the covariance matrix of all assets. In this paper we consider a variant minimising the risk, under budget and return constraints. This is applied in family trust and pension funds, where a specific return is needed for future liabilities and a low risk is desirable. This leads to a quadratic optimisation problem with continuous or

© Springer Nature Switzerland AG 2021
M. Paszynski et al. (Eds.): ICCS 2021, LNCS 12747, pp. 45–59, 2021.
https://doi.org/10.1007/978-3-030-77980-1_4

binary variables. The variables are continuous if a fraction of the budget is allocated to an asset. Binary variables can be found when the choice is whether or not to invest in a specific asset or project.

Solving this kind of problems is not trivial. Integer quadratic programming (IQP) problems are NP-hard, the decision version of IQP is NP-complete [6]. Binary quadratic programming problems are also NP-hard in general [12], however, specific cases are polynomially solvable [17]. Classical solvers, such as IBM-CPLEX, Gurobi and Localsolver, are still getting better in solving these problems for bigger instances. Next to these classical solvers, heuristic approaches exist, based on meta-heuristics like Particle Swarms, Genetic Algorithms, Ant Colony and Simulated Annealing. A recent overview of these approaches for Portfolio Optimisation can be found in [36].

Quadratic optimisation problems with binary decision variables are expected to be a sweet-spot for near future quantum computing [32], using Quantum Annealing [22] or the Quantum Approximate Optimisation Algorithm (QAOA) on a gate model quantum computer [9]. Quantum computing is the technique of using quantum mechanical phenomena such as superposition, entanglement and interference for doing computational operations. The type of devices that are capable of doing such quantum operations are still actively being developed and are called quantum computers. We distinguish between two paradigms of quantum computing devices: gate model based (or gate based) and quantum annealers. A practically usable quantum computer is expected to be developed in the next few years. In less than ten years quantum computers are expected to outperform everyday computers, leading to breakthroughs in artificial intelligence [24], the discovery of new pharmaceuticals and beyond [13,23]. Currently, various parties, such as Google, IBM, Intel, Rigetti, QuTech, D-Wave and IonQ, are developing quantum chips, which are the basis of the quantum computer [31]. The size of these computers is limited, with the state-of-the-art being around 70 qubits for gate-based quantum computers and 5000 qubits for quantum annealers. In the meantime, progress is being made on algorithms that can be executed on those quantum computers and on the software (stack) to enable the execution of quantum algorithms on quantum hardware [28].

This also means that Portfolio Optimisation is one of the promising applications of quantum computing in finance [25]. The work of [8] presents an implementation of Markowitz's portfolio selection on D-Wave, where the expected return is maximised whilst minimising the covariance (risk) of the portfolio under a budget constraint. They formulate this as an Ising problem and solve it on the D-Wave One, having 128 qubits, for 63 potential investments within $20\,\mu s$ on the quantum processor. They indicate that the solution depends on the weights added to each of the objectives and constraints. The same is done in [34], where reverse quantum annealing is used to optimise the risk-adjusted returns by the use of the metrics of Sharpe ratio. In [21], the stock returns, variances and covariances are modelled in the graph-theoretic maximum independent set (MIS) and weighted maximum independent set (WMIS) structures under combinatorial optimisation. These structures are mapped into the Ising physics model representation of the underlying D-Wave One system. This is benchmarked against the MATLAB standard function quadprog.

A newer version of the D-Wave hardware is used in [4]. They also perform a stock selection out of a universe of U.S. listed, liquid equities based on the Markowitz formulation and the Sharpe ratio. They approach this rst classically, then by an approach also using the D-Wave 2000Q. The results show that practitioners can use a D-Wave system to select attractive portfolios out of 40 U.S. liquid equities. The research has been extended to 60 U.S. liquid equities [5].

Algorithms have also been created for the gate model quantum computer. In [30] an algorithm is given for Portfolio Optimisation that runs in poly($log(N)$), where N is the size of the historical return data set. The number of required qubits here is also N. An alternative method for a gate model quantum computer is given by [16]. In [11] a quantum-enhanced simulation algorithm is used to approximate a computationally expensive objective function. Quantum Amplitude Estimation (QAE) provides a quadratic speed-up over classical Monte Carlo simulation. Combining QAE with quantum optimisation can be used for discrete optimisation problems like Portfolio Optimisation.

With the introduction of the 5000 qubit Advantage quantum system by D-Wave and the new hybrid solver tooling a new heuristic approach is operational for in-production quantum computing applications[1]. In this paper we use the new functionality of D-Wave for the portfolio management problem and compare its performance with other state-of-the-art, commercially available tooling.

There exists other work that compares D-Wave's quantum annealer to other optimisation tooling on other application areas. We name a few examples. A QUBO formulation for the railway dispatching problem has been proposed in [7]. They further benchmark their solutions for the polish railway network on the state of the art D-Wave hardware. In [2], the authors have benchmarked the 5000 Advantage system for the garden optimisation problem class. Their results demonstrate that the Advantage system and the new hybrid solvers can solve large instances in less time than its predecessor.

In the remainder of this paper first the Portfolio Optimisation problem is formulated. This is a quadratic, constrained, binary optimisation problem that can in principle be solved by commercial solvers. The solvers we use in this paper as benchmark for the quantum approach are presented in Sect. 3. In Sect. 4 the implementation of the problem on the D-Wave quantum annealer is shown. The results of solving instances of two main stock indices, the Nikkei225 and the S&P500, on the quantum annealer and the comparison with the benchmark approaches, will be presented in Sect. 5. We end with some conclusions and recommendations.

2 Problem Description

We look at a Portfolio Optimisation problem, where we have N assets to invest in, $P_1, ..., P_N$. The expected return of assets i equals μ_i and the risk of the asset, denoted by the standard deviation, equals σ_i. The returns of the assets

---

[1] https://www.D-Wavesys.com/press-releases/d-wave-announces-general-availability-first-quantum-computer-built-business.

are correlated, expressed by the correlation ρ_{ij} for the correlation between assets i and j. We have now the return vector $\mu = \{\mu_i\}$ and the risk matrix $\Sigma = \{\sigma_{ij}\}$ where $\sigma_{ij} = \sigma_i^2$ if $i = j$ and $\sigma_{ij} = \rho_{ij}\sigma_i\sigma_j$ if $i \neq j$.

Assume that we have a budget to select n assets out of N. We want the return to be higher than a certain value R^* and are searching for that set of n assets that realise the target return against minimal risk. We therefore define $x_i = 1$ if asset i is selected and $x_i = 0$ otherwise. This gives the following optimisation problem:

$$\min x^T \Sigma x, \tag{1}$$

$$\text{s.t.} \sum_{i=1}^{N} x_i = n, \tag{2}$$

$$\mu^T x \geq R^*, \tag{3}$$

which is a quadratic, constrained, binary optimisation problem.

3 Benchmarks

We want to compare the performance of the D-Wave hardware with other state of the art, commercially available tooling. For this we selected two solvers and two meta-heuristics.

The first solver is LocalSolver [1], which is a black-box local-search solver of 0–1 programming with non-linear constraints and objectives. A local-search heuristic is designed according to the following three layers: Search Strategy, Moves and Evolution Machinery. LocalSolver performs structured Moves tending to maintain the feasibility of solutions at each iteration whose evaluation is accelerated by exploiting invariants induced by the structure of the model. Unlike other math optimisation software, LocalSolver hybridises different optimisation techniques dynamically. We used an academic licensed local version.

The second solver is Gurobi [15], from Gurobi Optimisation, Inc., which is a powerful optimiser designed to run in multi-core with capability of running in parallel mode. Gurobi uses the Branch and Bound Algorithm to solve Mixed-Integer Programming (MIP) models. It is based on four basic principles: presolve, cutting planes, heuristics, and parallelism. Each node in the Branch and Bound search tree is a new MIP. For our current MIQP (Mixed Integer Quadratic Programming), a Simplex Algorithm is used to solve the root node. We used an academic licensed local version of Gurobi.

The third benchmark is a standard MATLAB implementation of Genetic Algorithms, which is a method for solving both constrained and unconstrained optimisation problems based on a natural selection process that mimics biological evolution. The algorithm repeatedly modifies a population of individual solutions. It is a stochastic, population-based algorithm that searches randomly by mutation and crossover among population members.

The last benchmark is the Simulated Annealing approach as implemented by D-Wave in their Ocean environment. Their sampler implements the simulated annealing algorithm, based on the technique of cooling metal from a high temperature to improve its structure (annealing). This algorithm often finds good solutions to hard optimisation problems.

4   Implementation

In this paper we use the newly introduced (2020) functionality of D-Wave for the portfolio management problem and compare its performance with the commercial solvers described in Sect. 3. In this section we describe how the problem can be implemented on the D-Wave hardware. For this, the QUBO representation will be derived, a method to find the parameters is given and D-Wave's quantum and hybrid algorithms are explained.

4.1   QUBO Representation

The D-Wave hardware solves Ising or QUBO problems. The QUBO [14] is expressed by the optimisation problem:

$$\text{QUBO: min/max } y = x^t Q x, \tag{4}$$

where $x \in \{0,1\}^n$ are the decision variables and Q is a $n \times n$ coefficient matrix. Another formulation of the problem, often used, equals

$$\text{QUBO: min/max } H = x^t q + x^t Q x, \tag{5}$$

or a combination of multiple of these terms

$$\text{QUBO: min/max } H = \lambda_1 \cdot H_1 + \lambda_2 \cdot H_2 + \cdots, \tag{6}$$

where $\lambda_1, \lambda_2, \ldots$ are weights that can used to tune the problem and include constraints into the QUBO. For already a large number of combinatorial optimisation problems the QUBO representation is known [14,19]. Many constrained binary programming problems can be transformed easily to a QUBO representation. Assume that we have the problem

$$\min y = c^t x, \text{ subject to } Ax = b, \tag{7}$$

then we can bring the constraints to the objective value, using a penalty factor λ for a quadratic penalty:

$$\min y = c^t x + \lambda (Ax - b)^t (Ax - b). \tag{8}$$

Using $P = Ic$, the matrix with the values of vector c on its diagonal, we get

$$\min y = x^t P x + \lambda (Ax - b)^t (Ax - b) = x^t P x + x^t R x + d = x^t Q x, \tag{9}$$

where matrix R and the constant d follow from the matrix multiplication and the constant d can be neglected, as it does not influence the optimisation problem.

A QUBO problem can be easily translated into a corresponding Ising problem of N variables s_i $(i = 1, .., N)$ with $s_i \in \{-1, 1\}$ given by :

$$\min y = \sum_{i=1}^{N} h_i s_i + \sum_{i=1}^{N} \sum_{j=i+1}^{N} J_{ij} s_i s_j. \tag{10}$$

The Ising model and QUBO model are related by $s_i = 2x_i - 1$.

4.2   QUBO Formulation

We now create a QUBO formulation for the problem given in Eq. (1)–(3). In case Eq. (3) is an equality, this problem can be translated easily to a QUBO formulation:

$$\min \left(\lambda_0 x^T \Sigma x + \lambda_1 \left(\sum_{i=1}^{N} x_i - n \right)^2 + \lambda_2 (\mu^T x - R^*)^2 \right). \tag{11}$$

In the case of inequalities in Eq. (3) we have to add additional K slack variables y_k $(k = 1, \ldots, K)$, where $K = \lfloor \log_2(\sum_{i=1}^{N}(\mu_i)) \rfloor$. Scaling the μ values (in thousands, millions, etc.) will help reduce the number of variables. This leads to

$$\min \left(\lambda_0 x^T \Sigma x + \lambda_1 \left(\sum_{i=1}^{N} x_i - n \right)^2 + \lambda_2 \left(\mu^T x - R^* - \sum_{k=1}^{K} 2^k y_k \right)^2 \right). \tag{12}$$

4.3   Finding QUBO Parameters

It is known that the performance of the D-Wave hardware is depending strongly on the choice of the penalty coefficient λ. When determining the penalty coefficients, we can set $\lambda_0 = 1$ and look for good values for λ_1 and λ_2. In Fig. 1 it is shown that the choice of λ_1 greatly influences the best found solution. Values for λ_1 lower that 500 gives an invalid solution, i.e., the solutions do not meet the constraints given in the original problem. Values for λ_1 higher than 500 give allowed solutions. A value of 500 or just above would be optimal. Rule of thumb is that the gain of violating a constraint must be lower than the costs. For λ_1 this means that violating the associated constraint in the optimal solution x^* for stock i gives a benefit of $\sum_{j=1}^{N} \sigma_{ij} x_j^*$. Around the optimal value, the biggest benefit would be for stock i for which the sum of the smallest n values of σ_{ij} is the largest, meaning

$$\widehat{\lambda_1} = \max_i \sum_{j=1}^{n} \sigma_{i\{j\}},$$

where $\sigma_{i\{j\}}$ represents the j-th smallest covariance value for asset i.

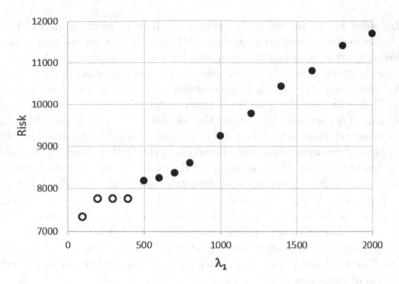

Fig. 1. Relation between Risk and setting of λ. Open dots are violating the original constraints, while closed dots are valid solutions.

This is more complicated for λ_2. Again, we have to look at the optimal solution x^* and exchange a zero and one in this solution, such that n stocks are chosen, leading to x'. This is done such that $(x^* - x')^T \Sigma (x^* - x')/\mu^T (x^* - x')$ is maximised. The procedure we used is as follows:

1. $A1 =$ average difference between smallest n sums $S_i = \sum_{j=1}^{n} \sigma_{i\{j\}}$,
2. $A2 =$ average positive difference in μ_i between these n stocks,
3. $\widehat{\lambda_2} = A1/A2$.

Note that these values are a first estimation and can be the starting point for an eventual grid search of the optimal parameters. In Table 1 we show the results of the grid search and the comparison with the estimated values. It is shown that the estimation is accurate (enough), such that the grid search can be skipped in practice.

4.4 Solving QUBO

The most recent hardware, D-Wave Advantage, features a qubit connectivity based on Pegasus topology. The QUBO problem has to be transformed to this structure. Due to the current limitation of the chip size, a compact formulation of the QUBO and an efficient mapping to the graph is required. This problem is known as Minor Embedding. Minor Embedding is NP-Hard and can be handled by D-Wave's System automatically, hence we do not attempt to optimise it. While some qubits in the chip are connected using external couplers, the D-Wave QPU (Quantum Processing Unit) is not fully connected. Hence a problem variable has to be duplicated to multiple connected qubits. Those qubits should have the same value, meaning the weight of their connection should be such that it holds in the optimisation process. All these qubits representing the same variables are part of a so-called chain, and their edge weights is called the chain strength (γ), which is an important value in the optimisation process. In [3], it has been indicated that if γ is large enough, the optimal solutions will match $\gamma \geq \Sigma_{ij}|Q_{ij}|$. However the goal is to find the smallest value to avoid rescaling the problem. The problem of finding the smallest γ is NP-Hard. On a higher level, D-Wave offers hybrid solvers. These solvers implement state of the art classical algorithms with intelligent allocation of the QPU to parts of the problem where it benefits most. These solvers are designed to accommodate even very large problems. This means that also most parameters of the embedding are set automatically. By default, samples are iterated over four parallel solvers. The top branch implements a classical Tabu Search that runs on the entire problem until interrupted by another branch completing. The other three branches use different decomposers to sample out a part of the current sample set to different samplers. The D-Wave's System sampler uses the energy impact as the criteria for selection of variables.

5   Results

We solved the Portfolio Optimisation problem for two indices, the Nikkei225 and the S&P500. We solved several instances for both indices on the D-Wave annealer, using the implementation described in Sect. 4, and all four benchmark approaches described in Sect. 3. The instances were run on a Intel(R) Core(TM) i7-8565U CPU@1.80 Ghz 8 GB RAM personal computer. We used the newest version of the D-Wave Leap environment solvers, Hybrid Binary Quadratic Model Version 2, for binary problems. The solver is, due to the underlying QPU, a stochastic solver. For this, we ran the hybrid solver five times for each problem instance and performed a parameter grid search for λ_1 and λ_2 for each problem instance.

We used the implementation first to select a number of stocks of the Nikkei225 index. The Nikkei225 is a stock market index for the Tokyo Stock Exchange. It has been calculated daily by the Nihon Keizai Shimbun (The Nikkei) newspaper since 1950. It is a price-weighted index, operating in the Japanese Yen, and its components are reviewed once a year. The Nikkei measures the performance of 225 large, publicly owned companies in Japan from a wide array of industry sectors. We took quarterly data of the Nikkei225 index of the last five year. The used return (μ) of each stock is the five year return, and the covariance σ_{ij} the calculated covariance over all quarters of the last 5 years. From this index, we took a subset N of which n are selected to minimise the risk under a budget and return (R^*) constraint. Note that (R^*) and n are related here: $R^* = 3000$ and $n = 20$ means an average 5-year return of the portfolio of $3000/20 = 150$ percent.

In Table 1 we show the 20 test instances we created on the Nikkei225, each having a different combination of (N, n, R^*). For each instance we show the size of the coefficient matrix of the QUBO (Size(Q)), the calculated (λ) and optimal $\widehat{\lambda_2}$ value of the parameters and the resulting value of the objective function of the best solution found by the hybrid solver. Also the best known solution is shown.

Table 1. 20 instances for the Nikkei225 index with their parameters and HQPU solutions. For the best solution, values marked with * have been proven optimal

N	n	R^*	Size(Q)	$\widehat{\lambda_1}$	$\widehat{\lambda_2}$	λ_1	λ_2	HQPU	Best
50	10	0	60	76	0	70	0	256	256*
50	25	0	60	320	0	365	0	3,035	3,035*
50	10	1,200	60	76	0.04	76	0.1	321	321*
50	25	1,200	60	320	0.10	365	0.1	3,058	3,058*
100	20	0	111	137	0	100	0	809	809*
100	50	0	111	607	0	700	0	10,999	10,999*
100	20	2,500	111	137	0.05	100	0.1	1,126	1,126*
100	50	2,500	111	607	0.09	700	0.1	11,065	11,065*
150	20	0	161	116	0	100	0	628	628*
150	50	0	161	502	0	500	0	7,993	7,993*
150	20	3,000	161	116	0.04	100	0.1	1,271	1,256*
150	50	3,000	161	502	0.10	500	0.1	8,315	8,270*
200	20	0	211	100	0	100	0	530	529
200	50	0	211	428	0	500	0	6,000	5,950
200	20	3,250	211	100	0.04	100	0.1	1,308	1,262
200	50	3,250	211	428	0.08	500	0.1	6,859	6,500
225	20	0	236	88	0	88	0	484	482
225	50	0	236	385	0	385	0	5,468	5,379
225	20	3,500	236	88	0.05	100	0.1	1,504	1,455
225	50	3,500	236	385	0.07	385	0.1	6,610	6,131

In Table 2 the results of all benchmark tooling per test instance is given. In the table the results are shown of the hybrid solver (HQPU), Simulated Annealing (SA), Genetic Algorithm (GA), the Gurobi solver (GB) and LocalSolver (LS). We see that LocalSolver finds for all instances the best solution. Gurobi also finds this solution, however, it runs out of memory (locally) for instances bigger than $N = 150$. For the instances it solves, it proves optimality by closing the optimality gap to 0%. LocalSolver is not able to close this optimality gap within reasonable time for most larger instances. The HQPU gives reasonable results, optimal for the smaller instances and well within 5% of the optimal solution for the larger cases, with exception for the last instance. The Simulated Annealing

Table 2. Results of all the methods. Values marked with * are proven optimal. Values marked with † for LocalSolver are not proven optimal, solver stops before GAP = 0. Bold values in HQPU, SA, GA columns represent that they achieved optimal solutions while italicised values represent solutions within 5% of the solution. All solutions in GB and LS are optimal in nature.

N	n	R^*	HQPU	SA	GA	GB	LS	Best solution
50	10	0	**256**	**256**	**256**	256	256	256*
50	25	0	**3,035**	**3,035**	3,236	3,035	3,035†	3,035*
50	10	1,200	**321**	**321**	**321**	321	321	321*
50	25	1,200	**3,058**	**3,058**	3,236	3,058	3,058	3,058*
100	20	0	**809**	**809**	1,102	809	809†	809*
100	50	0	**10,999**	**10,999**	12,318	10,999	10,999	10,999*
100	20	2,500	**1,126**	*1,135*	1,450	1,126	1,126	1,126*
100	50	2,500	**11,065**	**11,065**	12,397	11,065	11,065	11,065*
150	20	0	**628**	**628**	1,047	628	628†	628*
150	50	0	**7,993**	*8,321*	10,561	7,993	7,993†	7,993*
150	20	3,000	*1,271*	*1,260*	2,937	1,256	1,256†	1,256*
150	50	3,000	*8,315*	**8,270**	12,659	8,270	8,270†	8,270*
200	20	0	*530*	*534*	1,151	-	529†	529
200	50	0	*6,000*	*6,030*	9,397	-	5,950†	5,950
200	20	3,250	*1,308*	*1,312*	2,525	-	1,262†	1,262
200	50	3,250	6,859	6,983	11,804	-	6,500†	6,500
225	20	0	*484*	*489*	1,143	-	482†	482
225	50	0	*5,468*	*5,472*	9,438	-	5,379†	5,379
225	20	3,500	*1,504*	1,754	4,276	-	1,455†	1,455
225	50	3,500	6,610	6,835	11,802	-	6,131†	6,131

stays close to the HQPU solution and the Genetic Algorithm implementation underperforms in comparison with the other solvers.

The performance is also depending on the computation time. In Table 3 the calculation times are listed. Here only the time the solver requires are mentioned, the time to build the problem is out of scope. Best solving times are realised by Gurobi. LocalSolver performs well regarding to finding a good solution. However, for most cases the optimality gap is not closed to 0% which means the solver keeps running until the maximum admitted time is achieved. Simulated Annealing and Genetic algorithms have increasing calculation times for the instances. The HQPU is quite fast and independent from the instance size here. The times displayed in the table correspond to one call to the hybrid solver, where we used 5 calls per instance. D-Wave allows the user to set a time limit. When the user does not specify this, a minimum time limit is calculated and used. For all instances here we did not adjust the time limit.

Table 3. Pure solver times in seconds of all the methods. For LocalSolver the time to find the best binary solution is given and between brackets the time to prove optimality. The HQPU time is per query, in the analysis we used 5 queries per problem and a parameter grid search was performed prior to these runs. Pure solver means time required just for solving the problem and doesn't include the initialisation time or the time needed to send the problem to the Leap network.

N	n	R^*	HQPU	SA	GA	GB	LS
50	10	0	1.0	8	58	<1	2 (210)
50	25	0	1.3	7	55	<1	2 (>600)
50	10	1,200	1.0	2	53	<1	2 (35)
50	25	1,200	0.9	2	61	<1	2 (20)
100	20	0	1.1	9	89	<1	3 (>600)
100	50	0	1.1	20	88	<1	2 (76)
100	20	2,500	1.0	9	82	<1	2 (556)
100	50	2,500	1.1	9	86	<1	3 (43)
150	20	0	1.3	37	128	<1	2 (>600)
150	50	0	1.4	51	130	<1	4 (>600)
150	20	3,000	1.4	2	71	<1	3 (>600)
150	50	3,000	1.2	7	74	<1	16 (>600)
200	20	0	1.2	26	157	-	3 (>600)
200	50	0	1.1	33	161	-	4 (>600)
200	20	3,250	1.1	28	106	-	21 (>600)
200	50	3,250	1.1	32	128	-	7 (>600)
225	20	0	1.2	31	169	-	3 (>600)
225	50	0	1.1	43	168	-	12 (>600)
225	20	3,500	1.3	32	131	-	165 (>600)
225	50	3,500	1.2	37	140	-	207 (>600)

The second exercise we perform on the bigger S&P500. The S&P500 is a stock market index that measures the stock performance of 500 large companies listed on stock exchanges in the United States. It is one of the most commonly followed equity indices. The S&P500 index is a capitalisation-weighted index and the 10 largest companies in the index account for 26% of the market capitalisation of the index: Apple Inc., Microsoft, Amazon.com, Alphabet Inc., Facebook, Johnson & Johnson, Berkshire Hathaway, Visa Inc., Procter & Gamble and JPMorgan Chase. For this index we created five instances, which are depicted with the results of the solvers in Table 4 and Table 5.

Again we see that Gurobi is not able to run instances bigger than $N = 150$. LocalSolver finds the best solutions but is not able to close the optimality gap. In the three largest instances the gap stayed 100% within the allowed 600 s. Genetic algorithms performed poorly again. The hybrid approach and Simu-

Table 4. Results of all the methods. Values marked with * are proven optimal. Values marked with † for LocalSolver are not proven optimal, solver stops before GAP = 0.

N	n	R^*	Size(Q)	HQPU	SA	GA	GB	LS	Best
100	50	3500	112	5518	5518	6959	5518	5518†	5518*
200	50	3500	213	3121	3141	5896	-	3121†	3121
300	50	3500	314	2492	2525	6825	-	2485†	2485
400	50	3500	414	1537	1677	6846	-	1504†	1504
500	50	3500	515	1287	1570	9014	-	1286†	1286

Table 5. Pure solver times in seconds of all the methods. For LocalSolver the time to find the best binary solution is given and between brackets the time to prove optimality. The HQPU time is per query, in the analysis we used 5 queries per problem and a parameter grid search was perfumed prior to these runs.

N	n	R^*	HQPU	SA	GA	GB	LS
100	50	3500	1.0	8	61	<1	2 (>600)
200	50	3500	1.3	28	92	-	3 (>600)
300	50	3500	1.6	52	124	-	7 (>600)
400	50	3500	1.9	92	155	-	18 (>600)
500	50	3500	5.8	135	221	-	16 (>600)

lated Annealing were able to stay close to the LocalSolver solutions, within 3% except for the last instance. Although we gave the HQPU more time to solve, we adjusted the time limit by hand, it stayed on 18% above the best solution found by LocalSolver, as the Simulated Annealing solver did. The solving times of the HQPU stay quite reasonable as compared to the other solvers.

6   Conclusions and Further Research

Quantum computing is still in its early stage. However, the moment of quantum computers outperforming classical computers is coming closer. In this phase, hybrid solvers, using classical methods combined with quantum computing, are promising. The quantum paradigm that is already maturing is quantum annealing combined with the Hybrid Solvers that D-Wave is offering. In this paper we showed the performance of this hybrid approach, applied to a quadratic optimisation problem occurring in finance, Portfolio Optimisation. Here a portfolio of assets is selected, minimised the total risk of the portfolio. This portfolio has to meet some return and budget constraints.

We suggested an implementation of this problem on the newest D-Wave quantum annealer using the hybrid solver. The performance on a problem where 50 stocks needs to be selected from an existing index, the Nikkei225 or the S&P500,

was already reasonable, in comparison with the performance of other commercially available solvers. Of these solvers, LocalSolver performed the best, finding the optimal solution in all cases very quickly. However, proving optimality, and thus finishing the optimisation task, was not realised within reasonable time. The hybrid quantum solver was able to find a solution within 3% of the optimal solution for the S&P500 index up to 400 stocks. Further improvement of the hybrid solvers and the enlargement of the QPU in the coming years lead to the expectation that this computing paradigm is close to real business applications.

For further research we recommend to dive into the hybrid approach and look for improvements. The used method is a standard method, whereas D-Wave offers the opportunity to tailor the methodology in the Leap environment. Also other quantum optimisation algorithms could be used, like the Quantum Approximate Optimization Algorithm on a gate model quantum computer.

Acknowledgments. The authors want to thank Hans Groenewegen for providing the financial data underlying this research and Irina Chiscop for her valuable review and comments. This work was supported by the Dutch Ministry of Economic Affairs and Climate Policy (EZK), as part of the Quantum Delta NL programme.

References

1. Benoiot, T., Estellon, B., Gardi, F., Megel, R., Nouioua, K.. Localsolver 1.x: a black-box local-search solver for 0–1 programming. 4OR Q. J. Belgian **9**, 299–316 (2011). French and Italian Operations Research Societies
2. Calaza, C.D.G., Willsch, D., Michielsen, K.: Garden optimization problems for benchmarking quantum annealers (2021)
3. Coffrin, C.J.: Challenges with chains: testing the limits of a d-wave quantum annealer for discrete optimization. Technical report, Los Alamos National Lab. (LANL), Los Alamos, NM (United States) (2019)
4. Cohen, J., Khan, A., Alexander, C.: Portfolio optimization of 40 stocks using the dwave quantum annealer. arXiv preprint arXiv:2007.01430 (2020)
5. Cohen, J., Khan, A., Alexander, C.: Portfolio optimization of 60 stocks using classical and quantum algorithms. arXiv preprint arXiv:2007.08669 (2020)
6. Del Pia, A., Dey, S.S., Molinaro, M.: Mixed-integer quadratic programming is in NP. Math. Program. **162**(1–2), 225–240 (2017)
7. Domino, K., Koniorczyk, M., Krawiec, K., Jałowiecki, K., Gardas, B.: Quantum computing approach to railway dispatching and conflict management optimization on single-track railway lines (2021)
8. Elsokkary, N., Khan, F.S., La Torre, D., Humble, T.S., Gottlieb, J.: Financial portfolio management using d-wave quantum optimizer: the case of Abu Dhabi securities exchange. Technical report, Oak Ridge National Lab. (ORNL), Oak Ridge, TN (United States) (2017)
9. Farhi, E., Harrow, A.W.: Quantum supremacy through the quantum approximate optimization algorithm. arXiv preprint arXiv:1602.07674 (2016)
10. Fulga, C., Stanojević, B.: Single period portfolio optimization with fuzzy transaction costs. In: 20th International Conference EURO Mini Conference on Continuous Optimization and Knowledge-Based Technologies, EurOPT 2008. Vilnius Gediminas Technical University (2008)

11. Gacon, J., Zoufal, C., Woerner, S.: Quantum-enhanced simulation-based optimization. arXiv preprint arXiv:2005.10780 (2020)
12. Garey, M.R., Johnson, D.S.: Computers and Intractability, vol. 174. Freeman San Francisco (1979)
13. Gemeinhardt, F.G.: Quantum Computing: A Foresight on Applications, Impacts and Opportunities of Strategic Relevance. Ph.D. thesis, Universität Linz (2020)
14. Glover, F., Kochenberger, G., Du, Y.: A tutorial on formulating and using QUBO models. arXiv preprint arXiv:1811.11538 (2018)
15. Gurobi Optimization, L.: Gurobi optimizer reference manual (2020). http://www.gurobi.com
16. Kerenidis, I., Prakash, A., Szilágyi, D.: Quantum algorithms for portfolio optimization. In: Proceedings of the 1st ACM Conference on Advances in Financial Technologies, pp. 147–155 (2019)
17. Li, D., Sun, X., Gu, S., Gao, J., Liu, C.: Polynomially solvable cases of binary quadratic programs. In: Chinchuluun, A., Pardalos, P., Enkhbat, R., Tseveendorj, I. (eds) Optimization and Optimal Control. Springer Optimization and Its Applications, vol 39. Springer, New York (2010). https://doi.org/10.1007/978-0-387-89496-6_11
18. Liagkouras, K., Metaxiotis, K.: Multi-period mean-variance fuzzy portfolio optimization model with transaction costs. Eng. Appl. Artif. Intell. **67**, 260–269 (2018)
19. Lucas, A.: Ising formulations of many NP problems. Front. Phys. **2**, 5 (2014)
20. Markowitz, H.: Harry Markowitz: Selected Works, vol. 1. World Scientific (2009)
21. Marzec, M.: Portfolio optimization: applications in quantum computing. In: Handbook of High-Frequency Trading and Modeling in Finance, pp. 73–106 (2016)
22. McGeoch, C.C.: Adiabatic Quantum Computation and Quantum Annealing: Theory and Practice. Synthesis Lectures on Quantum Computing, vol. 5, no. 2, pp. 1–93 (2014)
23. Möller, M., Vuik, C.: On the impact of quantum computing technology on future developments in high-performance scientific computing. Ethics Inf. Technol. **19**(4), 253–269 (2017)
24. Neumann, N., Phillipson, F., Versluis, R.: Machine learning in the quantum era. Digitale Welt **3**(2), 24–29 (2019)
25. Orus, R., Mugel, S., Lizaso, E.: Quantum computing for finance: overview and prospects. Rev. Phys. **4**, 100028 (2019)
26. Pang, T., Hussain, A.: A stochastic portfolio optimization model with complete memory. Stoch. Anal. Appl. **35**(4), 742–766 (2017)
27. Pang, T., Varga, K.: Portfolio optimization for assets with stochastic yields and stochastic volatility. J. Optim. Theory Appl. **182**(2), 691–729 (2019)
28. Piattini, M., et al.: The Talavera manifesto for quantum software engineering and programming. In: QANSWER, pp. 1–5 (2020)
29. Radulescu, M., Radulescu, C.Z.: A multi-objective approach to multi-period: portfolio optimization with transaction costs. In: Masri, H., Pérez-Gladish, B., Zopounidis, C. (eds.) Financial Decision Aid Using Multiple Criteria. MCDM, pp. 93–112. Springer, Cham (2018). https://doi.org/10.1007/978-3-319-68876-3_4
30. Rebentrost, P., Lloyd, S.: Quantum computational finance: quantum algorithm for portfolio optimization. arXiv preprint arXiv:1811.03975 (2018)
31. Resch, S., Karpuzcu, U.R.: Quantum computing: an overview across the system stack. arXiv preprint arXiv:1905.07240 (2019)
32. Ronagh, P., Woods, B., Iranmanesh, E.: Solving constrained quadratic binary problems via quantum adiabatic evolution. Quantum Inf. Comput. **16**(11–12), 1029–1047 (2016)

33. Skaf, J., Boyd, S.: Multi-period portfolio optimization with constraints and transaction costs. In: Working Manuscript. Citeseer (2009)

34. Venturelli, D., Kondratyev, A.: Reverse quantum annealing approach to portfolio optimization problems. Quantum Mach. Intell. **1**(1–2), 17–30 (2019)

35. Xidonas, P., Mavrotas, G., Hassapis, C., Zopounidis, C.: Robust multiobjective portfolio optimization: a minimax regret approach. Eur. J. Oper. Res. **262**(1), 299–305 (2017)

36. Zanjirdar, M.: Overview of portfolio optimization models. Adv. Math. Finan. Appl. **5**(4), 1–16 (2020)

Cross Entropy Optimization
of Constrained Problem Hamiltonians
for Quantum Annealing

Christoph Roch[✉], Alexander Impertro, and Claudia Linnhoff-Popien

LMU Munich, Munich, Germany
`christoph.roch@ifi.lmu.de`

Abstract. This paper proposes a Cross Entropy approach to shape constrained Hamiltonians by optimizing their energy penalty values. The results show a significantly improved solution quality when run on D-Wave's quantum annealing hardware and the numerical computation of the eigenspectrum reveals that the solution quality is correlated with a larger minimum spectral gap. The experiments were conducted based on the Knapsack-, Minimum Exact Cover- and Set Packing Problem. For all three constrained optimization problems we could show a remarkably better solution quality compared to the conventional approach, where the energy penalty values have to be guessed.

Keywords: Quantum annealing · Cross entropy method · Optimization · Minimum spectral gap · Constrained Hamiltonian · D-Wave Systems

1   Introduction

Experimental quantum computing has gained a lot of attention in the last decade, since quantum computers have been commercialized by their companies [2,3]. There exist different types of quantum computing hardware. The more known one is the quantum gate-model computer. It is the quantum pendant to our classical computers, which work with classical logical gates. The other type of quantum computers are quantum annealers, which are particularly designed for solving or finding good approximations to optimization problems. D-Wave Systems is the first company, which has made their quantum annealing hardware available for public and a lot of research has been done since then [1,5,13,23].

D-Wave's Quantum Annealing (QA) algorithm, which is implemented in hardware, is based on the adiabatic quantum computing principle [20]. First, one has to map the corresponding optimization problem to a so called Ising Hamiltonian in order to execute it on the hardware. The fundamental process of QA then is to physically interpolate between an initial Hamiltonian, whose minimal energy configuration (or ground state) is easy to prepare, and a problem Hamiltonian, whose minimal energy configuration, that corresponds to the best

© Springer Nature Switzerland AG 2021
M. Paszynski et al. (Eds.): ICCS 2021, LNCS 12747, pp. 60–73, 2021.
https://doi.org/10.1007/978-3-030-77980-1_5

solution of the defined Ising problem Hamiltonian, is sought. According to the adiabatic theorem, if this process is executed slowly enough, and the coherence domain is sufficiently large, the probability to stay in the ground state and thus in the minimal energy configuration of the problem Hamiltonian is close to one [6]. However, due to thermal fluctuations or a non-adiabatic anneal process the system can leap from the ground to an excited state. The minimum distance between the ground state and the first excited state throughout any point in the anneal process is called the minimum spectral gap. Since it is physically hard to ensure long coherence times in quantum systems, one can not increase the anneal time arbitrarily in order to avoid computational errors by jumping to excited states.

That is why we address this problem experimentally, not by adjusting the anneal time, but by applying a Cross-Entropy (CE) method to optimize the hyperparameters of the solution landscape, which are represented by the energy penalty values of the constrained Hamiltonian of the optimization problem. We can show that by optimizing the penalty values, the minimum spectral gap of the problem Hamiltonian is scaled and shifted. In effect, D-Wave's QA algorithm has a higher chance of remaining in the ground state, which results in a significant increase in solution quality as compared to the conventional approach without CE optimization. For our experiments we used the Knapsack Problem (KP), Minimum Exact Cover (MEC) Problem and the Set Packing (SP) Problem to verify our approach. We experimentally reveal the linear correlation between the size of the minimum spectral gap and the corresponding approximation ratio of the D-Wave annealer to the best known solution (BKS) of the problem instance.

2 Background

2.1 Quantum Annealing Algorithm

Quantum annealing is a metaheuristic for solving complex optimization and decision problems [15]. D-Wave's quantum annealing algorithm is implemented in hardware, and it is designed to find the lowest energy state of a spin glass. Such a system can be described by an Ising Hamiltonian of the form

$$\mathcal{H}(s) = \sum_i h_i s_i + \sum_{i<j} J_{ij} s_i s_j \tag{1}$$

where h_i is the on-site energy of qubit i, J_{ij} are the interaction energies of two qubits i and j, and s_i represents the spin $(-1, +1)$ of the i-th qubit. The basic process of quantum annealing is to physically interpolate between an initial Hamiltonian $\mathcal{H}_I$ with an easy to prepare minimal energy configuration (or ground state), and a problem Hamiltonian $\mathcal{H}_P$, whose minimal energy configuration, that corresponds to the best solution of the defined problem, is sought (see Eq. (2)). The physical principle on which the D-Wave computation process is based on can be described by a time-dependent Hamiltonian as stated in Eq. (2).

$$\mathcal{H}(t) = A(t)\mathcal{H}_I + B(t)\mathcal{H}_P \tag{2}$$

The anneal functions $A(t)$ and $B(t)$ must satisfy $B(t=0)=0$ and $A(t=\tau)=0$, with τ being the total evolution time. When the state evolution changes from $t=0$ to $t=\tau$, the annealing process, described by $\mathcal{H}(t)$, leads to the final form of the Hamiltonian corresponding to the objective Ising problem that needs to be minimized. Therefore, the ground state of the initial Hamiltonian $\mathcal{H}(0)=\mathcal{H}_I$ evolves to the ground state of the problem Hamiltonian $\mathcal{H}(\tau)=\mathcal{H}_P$. The measurements performed at time τ deliver low energy states of the Ising Hamiltonian as stated in Eq. (1). According to the adiabatic theorem, if this transition is executed sufficiently slowly (i.e., τ is large enough), and the coherence domain is large enough, the probability to stay in the ground state of the problem Hamiltonian is close to one [6].

However, due to a non-adiabatic anneal process the system can jump from the ground to an excited state. The minimum distance between the ground state and the first excited state the—one with the lowest energy apart from the ground state—throughout any point in the anneal process is called the minimum spectral gap $g_{\min}$ of $\mathcal{H}(t)$, and is defined as

$$g_{\min} = \min_{0 \le t \le T} \min_{j \ne 0} [E_j(t) - E_0(t)] \tag{3}$$

where $E_j(t)$ is any higher lying energy state and $E_0(t)$ the ground state [14]. The adiabatic theorem states that staying in the ground state is enforced by setting the change rate of the time-dependent Hamiltonian $\mathcal{H}(t)$ proportional to $1/g_{\min}^\delta$, with δ depending on the distribution of eigenvalues at higher energy levels. δ may range from one to even three in some circumstances [12,17,26]. To understand the efficiency of adiabatic quantum computing, we need to analyze $g_{\min}$, but in practice, this is a difficult task [7], which we try to approach experimentally in this work.

For the sake of completeness, note that there exists an alternative and often used formulation to the Ising spin glass system. The so called Quadratic Unconstrained Binary Optimization (QUBO) formulation [8], which is mathematically equivalent and uses 0 and 1 for the spin variables [27]. The quantum annealer is also able to minimize the functional form of the QUBO formulation $x^T Q x$, with x being a vector of binary variables $\{0,1\}$ of size n, and Q being an $n \times n$ real-valued matrix describing the relationship between the variables. Given the matrix $Q : n \times n$, the annealing process tries to find binary variable assignments $x \in \{0,1\}^n$ to minimize the objective function.

2.2 Knapsack Problem

In the NP-Complete Knapsack Problem [16], n items are given, each having a certain weight w_α and a certain value c_α. The items must be picked in a way that the total weight of the items is less than or equal to the knapsack capacity W, i.e., $\sum_{\alpha=1}^{n} w_\alpha x_\alpha \le W$, and the total sum of the item values $\sum_{\alpha=1}^{n} c_\alpha x_\alpha$ is maximized. Variable x_α is set 1 if the item is packed in the knapsack and 0 otherwise [19]. In order to implement the KP on D-Wave's quantum computer using QA, we need to encode the objective function of the KP into a Hamiltonian which is diagonal in the computational basis.

The weight constraint can be encoded in the following quadratic Hamiltonian, as stated in [18]:

$$\mathcal{H}_1 = A \left(1 - \sum_{n=1}^{W} y_n\right)^2 + A \left(\sum_{n=1}^{W} n y_n - \sum_{\alpha=1}^{N} w_\alpha x_\alpha\right)^2 \tag{4}$$

while the objective function is straightforwardly

$$\mathcal{H}_2 = -B \sum_{\alpha=1}^{N} c_\alpha x_\alpha. \tag{5}$$

Here, y_n for $1 \leq n \leq W$ is a binary variable, which is set to 1, if the final weight of the knapsack is n and 0 otherwise. $\mathcal{H}_1$ enforces that the weight can only take exactly one value and that the weight of the items in the knapsack equals the value we claimed it did. The parameters A, B are chosen according to

$$0 < B \cdot \max(c_\alpha) < A \tag{6}$$

in order to penalize violations of the weight constraint. Note that one can reduce the number of binary variables using the so called log trick to $N + \lfloor 1 + logW \rfloor$ [18].

2.3   Minimum Exact Cover Problem

The Minimum Exact Cover Problem is an NP-Hard constrained optimization problem, where a set $U = \{1, \ldots, n\}$, and subsets $V_i \subseteq U (i = 1, \ldots, N)$ are given, such that $U = \bigcup_i V_i$.

The task is to find the minimum number of sets V_i with the elements of those sets being disjoint, and the union of the sets is U. The Hamiltonians $\mathcal{H}_3 = \mathcal{H}_1 + \mathcal{H}_2$ are stated in [18]:

$$\mathcal{H}_1 = A \sum_{\alpha}^{n} \left(1 - \sum_{i:\alpha \in V_i} x_i\right)^2 \tag{7}$$

$$\mathcal{H}_2 = B \sum_i x_i \tag{8}$$

In Eq. (7) α denotes the elements of U, while i denotes the subsets V_i. $\mathcal{H}_1 = 0$, if every element is included exactly one time, which implies that the unions of the subsets are disjoint. With the additional Hamiltonian $\mathcal{H}_2$ the smallest number of subsets is sought. The ground state of this Hamiltonian will be $m * B$, where m is the smallest number of subsets required for the complete union. The ratio of the penalty values A and B can be determined by regarding the worst case scenario which is that there are a very small number of subsets with a single common element, whose union is U. To ensure this does not happen, one can set

$$A > n \cdot B. \tag{9}$$

2.4 Set Packing Problem

The Set Packing Problem is also an NP-hard constrained optimization problem. Given a setup as in Sect. 2.3, its difficulty lies in finding the maximum number of subsets V_i which are all disjoint. In [18] the following Hamiltonians $\mathcal{H}_3 = \mathcal{H}_1 + \mathcal{H}_2$ are given:

$$\mathcal{H}_1 = A \sum_{i,j:V_i \cap V_j \neq \emptyset} x_i x_j \tag{10}$$

$$\mathcal{H}_2 = -B \sum_i x_i \tag{11}$$

$\mathcal{H}_1$ is minimized only when all subsets are disjoint, while $\mathcal{H}_2$ simply counts the number of included sets. Choosing the penalty values

$$B < A \tag{12}$$

ensures that it is never favorable to violate the constraint $\mathcal{H}_1$. Considering there will always be a penalty of at least A per extra set included. Just as the Minimum Exact Cover Problem the Set Packing Problem requires N spins.

2.5 Cross-Entropy Method

The Cross-Entropy method is a Monte Carlo method for importance sampling and optimization, and is known to perform well on combinatorial optimization problem with noisy objective functions [24,25].

We experimentally implement the method using a common CE algorithm (see Algorithm 1), as stated for example in [28]. In each step of the iterative optimization, a set of points $a_1...a_n$ from the distribution p is sampled, based on its current parameterization Φ_{g-1} (line 3). To each point $a_1...a_n$, the objective function f of the optimization problem assigns values $v_1...v_n$ (line 4). Finally, a fraction ρ of elite samples is chosen based on a selection routine (line 5 and 6) and used to compute a new parameterization of p, Φ_g.

Algorithm 1. Cross-Entropy

1: **function** OPTIMIZE(p, Φ_0, f, ρ, n, G)
2:     **for** $g = 1 \to G$ **do**
3:         $a_1...a_n \sim p(\cdot|\Phi_{g-1}), \mathbf{a} \leftarrow a_1...a_n$
4:         $v_1...v_n \sim f(a_1)...f(a_n), \mathbf{v} \leftarrow v_1...v_n$
5:         sort $\mathbf{a}$ according to $\mathbf{v}$
6:         $\Phi_g \leftarrow \text{argmax}_\Phi \prod_{i=1}^{\lceil n\rho \rceil} p(a_i|\Phi)$
        return a_1

The convergence rate of the algorithm critically depends on the distribution p, as it determines the new sample points $\mathbf{a}$ for each generation. The initial distribution $p(\cdot|\Phi_0)$ should be chosen such that it reproduces optimal samples as closely as possible. However, when this is not possible, a generally applicable approach is to choose a distribution which covers the entire sample space. This increases the probability for the algorithm to evolve towards a good solution already in early generations.

After each iteration, a maximum likelihood estimate of the currently chosen elite fraction is done to update the parameterization Φ_g according to the following rule $\mu_g = \frac{\sum_{i=1}^{\lceil n\rho \rceil} X_i}{\lceil n\rho \rceil}$, $\sigma_g^2 = \frac{\sum_{i=1}^{\lceil n\rho \rceil} (X_i - \mu_g)^T (X_i - \mu_g)}{\lceil n\rho \rceil}$ and $\Phi_g = \langle \mu_g, \sigma_g^2 \rangle$, which is valid for a multivariate Gaussian distribution. Other critical parameters are the selected fraction ρ of elite samples, the population size n and the number of generations G, which must be carefully adjusted for a given problem in order to maximize the likelihood of finding a good solution. In Sect. 4, we explain how the CE method is adapted for our specific task.

3   Related Work

In 2017, Mark W. Coffey studied the Knapsack Problem within an Adiabatic Quantum Computing (AQC) framework. He mapped the optimization problem to an Ising model and used small problem instances to evaluate his approach. He points the relevance of theoretical and numerical investigations regarding the minimum spectral gap and its location in the anneal path out, in order to improve AQC [11].

More insights brought Choi in 2019. She theoretically showed, that adjusting the energy penalty of the Ising Maximum weighted Independent Set Problem, one may change the quantum evolution from one that has an anti-crossing to one that does not have, or the other way around, and thus drastically change the minimum spectral gap [10]. Following this insight, we propose an adapted CE method to automatically adjust the penalty values of constrained Hamiltonians to influence the quantum evolution of D-Wave's QA hardware, or more precisely the size and location of the corresponding minimum spectral gap, so that D-Wave's quantum annealer has a higher chance of remaining in the ground state.

In previous work, we already applied CE in the field of gate based quantum computers [22]. Similarly to this work, we shaped the solution landscape of the Knapsack problem Hamiltonian, which allowed the classical optimizer of the hybrid Quantum Approximate Optimization Algorithm (QAOA) to find better gate parameter and hence resulted in an improved performance.

4   QA with Cross-Entropy

In our approach we use an adapted CE method to optimize the penalty values A and B of our problem Hamiltonians. Varying those values significantly changes the energy landscape of the corresponding constrained optimization problem, and therefore also influences the pathway of the Quantum Annealing algorithm. By adjusting those penalty values it is possible to scale and shift the minimum spectral gap g_{min} and hence improve the probability of staying in the ground state throughout the anneal process.

An example can be seen in Fig. 1. In Fig. 1a the time-dependent eigenspectrum of a Minimum Exact Cover Hamiltonian with randomly selected penalty values is visualized. The same setup, but with optimized penalty values, found by CE, can be seen in Fig. 1b. The histograms in Fig. 1c and d show the corresponding solution qualities w.r.t the best known solution (BKS). The solution quality is associated with the approximation ratio, which is calculated as follows: $Approx.\ ratio = \frac{\#BKS}{\#Measurements}$ with #BKS being the number of measuring the BKS and #Measurements being the total number of measurements (default 100).

In Fig. 1a the minimum spectral gap g_{min} is ≈ 1 and located in the middle of the anneal process, while in Fig. 1b the gap g_{min} is ≈ 3 and shifted to the beginning of the anneal process. The impact of scaling and shifting g_{min} also reflects in an improved quantum annealing solution quality in Fig. 1d. Therefore, optimizing the penalty values of the problem Hamiltonian, in a way that g_{min} increases, decreases the likelihood of (thermal) excitations out of the instantaneous ground state, and consequently allows the quantum annealer to reach a better solution quality.

Regarding the problem Hamiltonians, the choice of penalty values are restricted by Eq. (6), (9) and (12), respectively for the Knapsack, Minimum Exact Cover and Set Packing Problem [18]. To satisfy those constraints, we use a modified CE optimization scheme (see Algorithm 2), in which the penalty values A, B are sampled from truncated normal distributions p. Since the allowed values for A depend on the choice of B, we first draw a value for B (line 3) with an appropriately chosen sampling range Γ_B. Afterwards, the value for A is drawn over a sampling range $\Gamma_A(B)$, such that the penalization constraint of the corresponding optimization problem is satisfied (line 5). This is done for n samples. For each sample, we construct the corresponding Hamiltonian, as described in [18] and run D-Wave's QA heuristic to assign a value $v_1...v_n$ corresponding to the approximation ratio of the best found solution for each A_i, B_i-pair (line 7). This is done iteratively for a specified number of generations G. In Fig. 2, the process of CE with QA and the fitness of each generation for the MEC problem Hamiltonian $\mathcal{A}$ is shown.

5   Evaluation

5.1   Experimental Setup

For the experimental evaluation, we used D-Wave's 2000Q quantum annealer (solver name: DW_2000Q_6 (lower noise)). We used the standard anneal schedule with $20\,\mu s$ and the standard energy scales $A(t)$ and $B(t)$, as stated in [4], which are required to compute the energy of a problem at a specific point in the annealing process. For embedding the problem instances on the D-Wave QPU, we used the minorminer library [9], which tries to find an efficient embedding of the logical problem graph to the physical architecture, called chimera graph,

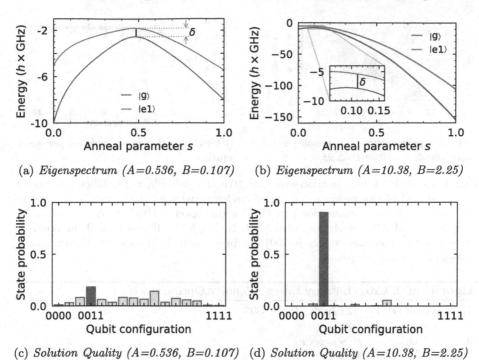

(a) *Eigenspectrum (A=0.536, B=0.107)*   (b) *Eigenspectrum (A=10.38, B=2.25)*

(c) *Solution Quality (A=0.536, B=0.107)*   (d) *Solution Quality (A=10.38, B=2.25)*

Fig. 1. Eigenspectra and solution qualities for MEC problem instance $\mathcal{A}$. Problem $\mathcal{A}$ consists of four subsets with numbers ranging from 1 to 4, $[(1,2),(1,3),(1,2,4),(3)]$. The BKS (marked in dark blue) is 0011, i.e., sets with index 2 and 3 are in the solution. The $g_{min} = \delta$ marks the minimum spectral gap, which is ≈ 1 for random penalty values A and B and ≈ 3 for optimized ones, see Figs. a and b, respectively. Figures c and d show the corresponding solution qualities, i.e., the state probability of the BKS compared to the other solutions. (Color figure online)

of the D-Wave quantum annealer. Since not every problem graph fits directly to the architecture, due to the sparse connectivity of the qubits on the QPU, physical ancillary qubits need to be used to represent one logical qubit. Additionally, it is known that not every qubit of the D-Wave QPU has the same physical quality. Thus, the embedding has an influence on the solution quality of the problem, as stated in [21]. That means, the same graph structure embedded on different physical qubits of the QPU leads to different solution qualities. In Fig. 3 one graph structure of the MEC problem instance $\mathcal{A}$ is embedded ten times, each time on different qubits of the QPU. For every embedding we used the same 100 individuals (penalty value pairs), their fitness was averaged over three runs per individual, to draw a fair comparison. One can see, that in the best embedding (the third one) the mean solution quality is around 71%, while in the worst embedding (the second one) the mean solution quality is around 50%. Consequently, the embedding can not be influenced directly and thus may also impact the overall solution quality of the CE optimization.

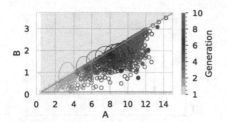

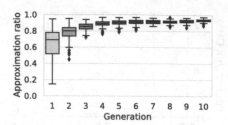

(a) *CE optimization (best penalty value pair found: A=10.38,B=2.25)*

(b) *Fitness of each individuum per generation*

Fig. 2. Example CE optimization with $G = 10$ is shown in Fig. a. The ellipses represent the μ_g and σ_g^2 of generation g. The filled circles correspond to the best ρ fraction of individuals. The best values found by CE, for this specific MEC problem instance $\mathcal{A}$, were 10.38 and 2.25 for A, respectively B. In Fig. b the fitness of each individuum per generation (population size is 100) computed with D-Wave's 2000Q annealer is represented by the boxplots.

Algorithm 2. Cross-Entropy Energy Penalty Optimization

1: **function** OPTIMIZE(p, Φ_0, f, ρ, n, G)
2:     **for** $g = 1 \to G$ **do**
3:         $B_1...B_n \sim p(\cdot|\Phi_{g-1}, \Gamma_B)$
4:         $\mathbf{B} \leftarrow B_1...B_n$
5:         $A_1...A_n \sim p(\cdot|\Phi_{g-1}, \Gamma_A(\mathbf{B}))$
6:         $\mathbf{A} \leftarrow A_1...A_n$
7:         $v_1...v_n \sim QA(A_1, B_1)...QA(A_n, B_n)$
8:         $\mathbf{v} \leftarrow v_1...v_n$
9:         sort $\mathbf{A}, \mathbf{B}$ according to $\mathbf{v}$
10:         $\Phi_g \leftarrow \text{argmax}_\Phi \prod_{i=1}^{\lceil n\rho \rceil} p(A_i, B_i|\Phi)$
        return A_1, B_1

In Table 1 the parameter settings of the CE method are listed. Since we are using a truncated normal distribution to sample from, we need to specify additional clipping parameters for both penalty values A and B. The sampling range of A is computed according to the corresponding problem constraints.

We used four problem instances for each optimization problem to test our CE approach with D-Wave's QA algorithm. The instances per optimization problem, named $\mathcal{A}$, $\mathcal{B}$, $\mathcal{C}$ and $\mathcal{D}$ range from 4 to 7 logical qubits, respectively. Since D-Wave's quantum annealing hardware is still in its infancy regarding the number of qubits and their connectivity, those logical problems already led to physical ancillary qubits to enable a valid embedding. Those ancillary qubits are coupled by a chain strength parameter to ensure that by measuring they collapse to the same basis state. However, this additional parameter also influences the overall solution quality and makes it harder for CE to optimize large problem instances. That is why we rather used small instances, to demonstrate the effect of CE. However, our approach is theoretically also applicable to larger problems.

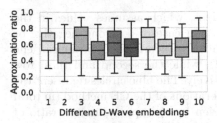

Fig. 3. Different D-Wave embeddings for MEC problem instance $\mathcal{A}$.

Table 1. Cross-entropy parameter settings

CE attributes							
G	n	ρ	γ^*	Min σ^2	σ_0^2	μ_0	B sample range
10	100	0.1	0.5	0.1	1.0	0.0	$[0.1, 10.0]$

*The learning rate specifies the amount of changes from Φ_{g-1} to Φ_g.

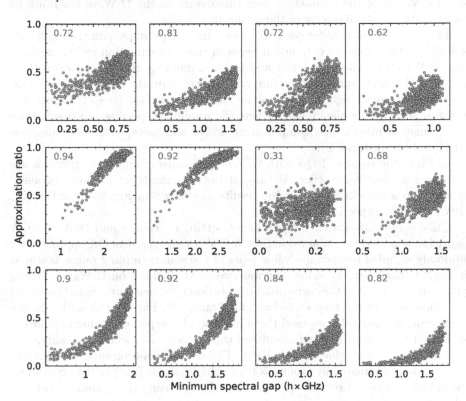

Fig. 4. Correlation plots of the approximation ratio to the BKS and the minimum spectral gap. The blue circles represent the individuals (penalty value pairs) over ten generations of the CE method (in total 1000 individuals). The first row represents the KP instances $\mathcal{A}$–$\mathcal{D}$, while the second and third row represents the MEC and SP instances, respectively. The red number is the correlation coefficient. (Color figure online)

5.2 Results and Discussion

In Fig. 4 the linear correlation of the approximation ratio to the BKS and the minimum spectral gap is plotted. The number in the upper left corner is the

Pearson product-moment correlation coefficient, which assumes values in the range of -1 to $+1$, where the extrema occur in case of a strong negative or positive correlation, while a value of 0 indicates uncorrelated variables. The first row represents the KP instances $\mathcal{A} - \mathcal{D}$, while the second and third row represents the MEC and SP instances, respectively. The blue circles represent the individuals (penalty value pairs) over ten generations of the CE method (in total 1000 individuals). The solution quality is given by the approximation ratio of the BKS and can be calculated by dividing the BKS counts by the number of measurements (default 100). Note that we averaged the approximation ratio for each individual of the population over three runs on the D-Wave hardware to compensate the stochasticity of the quantum system.

The results show that for each problem instance (perhaps with the exception of MEC $\mathcal{C}$), there exists a correlation between the approximation ratio computed with D-Wave's quantum annealer and the minimum spectral gap. Notice that, within the broadest range of the minimum spectral gap, the approximation ratio is in general much higher and in some cases near to 1.0 (see MEC problem instance $\mathcal{A}$ and $\mathcal{B}$), while in cases, where the minimum spectral gap is small the approximation ratio is also comparatively lower and more variance occurs (see MEC problem instance $\mathcal{C}$).

In Fig. 5 the results of QA with CE and the conventional QA approach are compared against each other. We tested both methods on different problem instances as stated in Sect. 5.1. The results are shown in Fig. 5a–c for the KP, MEC, and SP, respectively.

CE was initialized with the parameter setting of Table 1 and D-Wave's QA algorithm was used as explained in Sect. 5.1. W.r.t the classical QA approach, we randomly sampled five penalty value pairs of the same sampling range as stated in Table 1. Each penalty value pair was executed 10 times on D-Wave's 2000Q quantum annealer and the corresponding solution qualities (approximation ratio) are represented in the respective box plot "random". For the QA with CE box plots, named "optimal", we used the ρ fraction of the penalty value population of the last CE generation and calculated their fitness, i.e., solution quality.

The results show, that for the KP in Fig. 5a, the solution quality, w.r.t the mean could be increased by around 500% in the best case (see problem $\mathcal{B}$) and by around 85% in the worst case (see problem $\mathcal{D}$), by using the optimized penalty values.

Also for the MEC problem in Fig. 5b, the solution qualities were increased by around 170% in problem $\mathcal{C}$) in the best case and by 30% in the worst case in problem $\mathcal{A}$. Furthermore note, that in problem $\mathcal{A}$ and $\mathcal{B}$ an approximation ratio of nearly 100% could be reached with the optimized penalty values.

In Fig. 5c a significant growth in the quality of the solution can be seen, too, for the SP problem. In the best cases (problem $\mathcal{B}$ and $\mathcal{D}$) an increase of around 600% could be achieved while in the worst case (problem $\mathcal{A}$) the optimized penalty values still led to an 80% increase in solution quality.

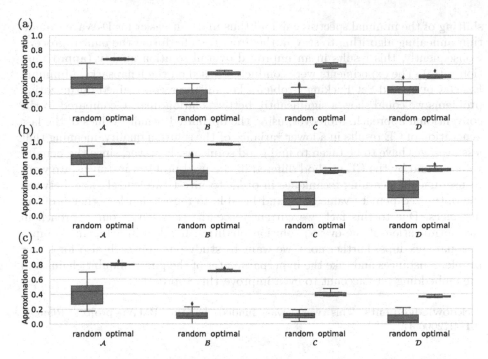

Fig. 5. Solution qualities for the four problem instances $\mathcal{A}$–$\mathcal{D}$ of the KP, MEC and SP problem instances are represented in Fig. a–c, respectively. The "random" boxplots represent the approximation ratio of five randomly sampled penalty value pairs (each run 10 times), while the "optimal" boxplots represent the approximation ratio of the ρ fraction of the penalty value pair population of the 10th generation of the CE method.

The overall solution quality in general decreases with the size of the problem instances. However, this is obvious since the number of possible solution and therefore the whole solution space increases.

Another feature is the comparatively small variance of the approximation ratio of the optimized penalty values, over all problem instances, which can be seen in Fig. 5. That is due to the values being picked from the best ρ fraction of individuals of the last CE generation, while the randomly sampled energy penalty value pairs may contain disadvantageously ones. However, this does not detract from the fact that our CE approach is quite stable w.rt. the stochastic quantum system.

6   Conclusion

In this paper we have presented a Cross Entropy approach to shape constrained Hamiltonians by optimizing their penalty values. We showed by the numerical computation of the eigenspectrum that this optimization leads to a scaling and shifting of the minimal spectral gap and thus makes it easier for D-Wave's quantum annealing algorithm to stay in the ground state during the anneal process. Consequently, this results in an improved overall solution quality (approximation ratio). The experiments were conducted based on the Knapsack-, Minimum Exact Cover- and Set Packing Problem. For all three constrained optimization problems we could show a significantly better solution quality compared to the conventional approach. Moreover, using the optimized penalty values of the last generation of CE results in a lower variance of the solution quality, meaning that less averages have to be taken to find good solutions. However, since the penalty values found by the CE method differed in the used problem instances, we want to use machine learning techniques in order to investigate correlations between the optimized penalty value pairs and be able to reuse them for other problem instances. Due to this fact, we currently see the strength of our approach in the optimization of the overall solution qualities, but less in achieving computational speedups. Furthermore, we want to study our approach also for larger problem instances and take the hyperparameter of the physical qubit chains and the embedding into account to even improve this approach.

Acknowledgements. This work was funded by the BMWi project PlanQK (01MK20005I).

References

1. D-wave launches leap, the first real-time quantum application environment. https://www.dwavesys.com/press-releases/d-wave-launches-leap-first-real-time-quantum-application-environment. Accessed 10 Jan 2021
2. D-wave systems sells its first quantum computing system to lockheed martin corporation. https://www.dwavesys.com/news/d-wave-systems-sells-its-first-quantum-computing-system-lockheed-martin-corporation. Accessed 10 Jan 2021
3. IBM makes quantum computing available on IBM cloud to accelerate innovation. https://www-03.ibm.com/press/us/en/pressrelease/49661.wss. Accessed 10 Jan 2021
4. QPU-specific anneal schedules. https://support.dwavesys.com/hc/en-us/articles/360005267253-What-are-the-Anneal-Schedule-Details-for-this-QPU-. Accessed 10 Jan 2021
5. Adachi, S.H., Henderson, M.P.: Application of quantum annealing to training of deep neural networks. arXiv preprint arXiv:1510.06356 (2015)
6. Albash, T., Lidar, D.A.: Adiabatic quantum computation. Rev. Mod. Phys. **90**(1), 015002 (2018)
7. Amin, M.H., Choi, V.: First-order quantum phase transition in adiabatic quantum computation. Phys. Rev. A **80**(6), 062326 (2009)
8. Boros, E., Hammer, P.L., Tavares, G.: Local search heuristics for quadratic unconstrained binary optimization (QUBO). J. Heuristics **13**(2), 99–132 (2007)
9. Cai, J., Macready, W.G., Roy, A.: A practical heuristic for finding graph minors. arXiv preprint arXiv:1406.2741 (2014)
10. Choi, V.: The effects of the problem Hamiltonian parameters on the minimum spectral gap in adiabatic quantum optimization. Quantum Inf. Process. **19**(3), 90 (2020)

11. Coffey, M.W.: Adiabatic quantum computing solution of the knapsack problem. arXiv preprint arXiv:1701.05584 (2017)
12. Farhi, E., Goldstone, J., Gutmann, S., Sipser, M.: Quantum computation by adiabatic evolution. arXiv preprint arXiv:quant-ph/0001106 (2000)
13. Feld, S., et al.: A hybrid solution method for the capacitated vehicle routing problem using a quantum annealer. Front. ICT **6**, 13 (2019)
14. Grant, E.K., Humble, T.S.: Adiabatic Quantum Computing and Quantum Annealing (July 2020)
15. Kadowaki, T., Nishimori, H.: Quantum annealing in the transverse Ising model. Phys. Rev. E **58**, 5355–5363 (1998)
16. Karp, R.M.: Reducibility among combinatorial problems. In: Miller, R.E., Thatcher, J.W., Bohlinger, J.D. (eds.) Complexity of Computer Computations. The IBM Research Symposia Series. Springer, Boston (1972). https://doi.org/10.1007/978-1-4684-2001-2_9
17. Lidar, D.A., Rezakhani, A.T., Hamma, A.: Adiabatic approximation with exponential accuracy for many-body systems and quantum computation. J. Math. Phys. **50**(10), 102106 (2009)
18. Lucas, A.: Ising formulations of many NP problems. Front. Phys. **2**, 5 (2014)
19. Martello, S.: Knapsack Problems: Algorithms and Computer Implementations. Wiley-Interscience Series in Discrete Mathematics and Optimization (1990)
20. McGeoch, C.C.: Adiabatic Quantum Computation and Quantum Annealing: Theory and Practice. Synthesis Lectures on Quantum Computing, vol. 5(2), pp. 1–93 (2014)
21. Padilha, D.: Solving NP-hard Problems on an Adiabatic Quantum Computer. UNSW School of Mechanical and Manufacturing Engineering (2014)
22. Roch, C., Impertro, A., Phan, T., Gabor, T., Feld, S., Linnhoff-Popien, C.: Cross entropy hyperparameter optimization for constrained problem Hamiltonians applied to QAOA. arXiv preprint arXiv:2003.05292 (2020)
23. Roch, C., et al.: A quantum annealing algorithm for finding pure Nash equilibria in graphical games. In: Krzhizhanovskaya, V.V., et al. (eds.) ICCS 2020. LNCS, vol. 12142, pp. 488–501. Springer, Cham (2020). https://doi.org/10.1007/978-3-030-50433-5_38
24. Rubinstein, R.: The cross-entropy method for combinatorial and continuous optimization. Methodol. Comput. Appl. Probab. **1**(2), 127–190 (1999)
25. Rubinstein, R.Y., Kroese, D.P.: The Cross-Entropy Method: A Unified Approach to Monte Carlo Simulation, Randomized Optimization and Machine Learning. Information Science & Statistics. Springer, New York (2004)
26. Schaller, G., Mostame, S., Schützhold, R.: General error estimate for adiabatic quantum computing. Phys. Rev. A **73**(6), 062307 (2006)
27. Su, J., Tu, T., He, L.: A quantum annealing approach for Boolean satisfiability problem. In: Proceedings of the 53rd Annual Design Automation Conference, p. 148. ACM (2016)
28. Weinstein, A., Littman, M.L.: Open-loop planning in large-scale stochastic domains. In: 27th AAAI Conference on Artificial Intelligence (2013)

Classification Using a Two-Qubit Quantum Chip

Niels M. P. Neumann[⊠]

TNO, P.O. Box 96800, 2509JE The Hague, The Netherlands
niels.neumann@tno.nl

Abstract. Quantum computing has great potential for advancing machine learning algorithms beyond classical reach. Even though full-fledged universal quantum computers do not exist yet, its expected benefits for machine learning can already be shown using simulators and already available quantum hardware. In this work, we consider a distance-based classification algorithm and modify it to be run on actual early stage quantum hardware. We extend upon earlier work and present a non-trivial reduction using only two qubits. The algorithm is consequently run on a two-qubit silicon spin quantum computer. We show that the results obtained using the two-qubit silicon spin quantum computer are similar to the theoretically expected results.

Keywords: Classification · Machine learning · Quantum computing · Hardware implementations

1 Introduction

Whether we are aware of it or not, machine learning has taken a prominent role in our lives. For example, various algorithms are used to process text [1,2] and speech [3,4]. Furthermore, machine learning algorithms exist to recognise patterns [5] and also more specifically focused on recognising faces [6]. These are just a few of the many applications of machine learning.

In general, however, we can distinguish between three different types of machine learning: supervised, unsupervised and reinforced machine learning. In supervised machine learning, the machine is given annotated data, which is then used to train a model to annotate unseen data. Examples of supervised machine learning are decision trees, support vector machines and neural networks. Unsupervised machine learning algorithms use data without annotation and instead assign the labels themselves. Examples of unsupervised machine learning are clustering algorithms, such as k-means clustering. The third and last type is reinforced machine learning, or reinforcement learning, where a reward function is used that quantifies the 'goodness' of a solution. Based on the reward function, model parameters are adjusted. A prominent example of reinforcement learning applied in practice is AlphaGo, the first algorithm to beat a human at the game of Go [7]. As annotated data is in general expensive to gather, often a fourth

© Springer Nature Switzerland AG 2021
M. Paszynski et al. (Eds.): ICCS 2021, LNCS 12747, pp. 74–83, 2021.
https://doi.org/10.1007/978-3-030-77980-1_6

type is considered: semi-supervised learning. Here a model is initially trained with a small set of annotated data, after which the rest of the learning is run unsupervised with a larger set of unlabeled data.

Each of these types has its own challenges. Two common challenges across the four different types are however lack of data and intractability of the training phase, meaning that training the model is too complex. The intractability is often overcome by running the algorithms on (a clusters of) computers with more computational power. Instead, we may also opt for a completely different way of computing: doing computations using quantum computers. In quantum computers, computations are done using quantum bits (qubits) instead of classical bits. Classical bits are always in one of two possible definite states, zero or one. Qubits on the other hand can be in a superposition of the different computational basis states. A superposition is a complex linear combination of these states. Upon measurement a label corresponding to only one of the computational basis states is found. The probability to find each of the labels corresponds to the absolute value squared of the corresponding amplitudes. After the measurement, the state is projected onto a space corresponding to the found label and information on the initial superposition before the measurement is lost. Another key property of quantum states is that two qubits or, more general, multiple quantum states can be entangled. Entanglement implies correlations between two systems beyond what is possible classically. A more elaborated introduction to quantum computing is given in [8].

These quantum mechanical properties can be used to enhance classical computing and machine learning specifically [9,10]. Quantum algorithms can for instance replace computationally expensive (sub)routines in classical algorithms, thereby enhancing the algorithm as a whole. Examples of computationally complex subroutines where quantum computing can offer benefits are sampling from probability distributions [11] and inverting matrices [12].

Another example of where quantum computing will provide improvements over classical algorithms is classification. In [13] for example, two quantum computing methods are proposed to classify data. The first method is a variational quantum classifier [14], similar to classical support vector machines (SVMs). In the second method, a kernel function is estimated and optimised directly. Instead of explicitly and iteratively training a machine learning model, a classifier can also be implemented directly from the data points, as in [15]. There, a controlled-SWAP-gate between the training data and a test data point is applied and two measurements are applied to classify the test data point. An example of distance-based classification is given in [16]. Here, a label is assigned based on a distance measure evaluated on the training points and a new test point. The classical complexity of this algorithm is $\mathcal{O}(NM)$, with N the number of data points and M the number of features. The time complexity of this quantum algorithm is constant, given efficient state preparation. The number of qubits n required is logarithmic in the number of data points, i.e., $n = \mathcal{O}(\log N)$.

For this constant complexity to hold, efficient state-preparation is a necessity. This can however pose challenges [17], and, if the state is prepared explicitly, even can result in an exponential overhead in the number of qubits. In general, an $m \times m$-qubit unitary gate can be decomposed exactly in $\mathcal{O}(m^3 4^m)$ single qubit and CNOT-gates [18], which was later reduced to $\mathcal{O}(2^m)$ quantum gates [19,20]. Instead of explicit state preparation, the state can also be obtained from the output of other quantum processes or by using a quantum RAM [21], a register or quantum computer that stores specific states. Note that for the latter, the challenge shifts from efficiently preparing a quantum state to efficiently preparing and storing the states in the quantum RAM. An approach proposed in [22] uses a divide-and-conquer algorithm for quantum state preparation, with a poly-logarithmic complexity in N, the number of data points, compared to a classical complexity of $\mathcal{O}(N)$. The complexity of the quantum distance-based classifier of [16] thus depends on the complexity of state-preparation.

Especially for near-term devices with limited resources, state preparation can limit the applicability of this algorithm. A reduction to overcome this limitation is presented in [23]. By formulating the algorithm as a quantum channel and considering a single data point at a time, similar performance is reached as with the original algorithm. Conditional on the measurement of the ancilla qubit either a new data point is chosen uniformly at random or the label is assigned to the test point. We consider a different non-trivial reduction of the distance-based classifier to be run on near-term hardware, classifying data points in one of two classes. Our algorithm can be used as benchmarking algorithm for comparison with classical devices and to compare different quantum chips. In this work, we compare the classical theoretical results with the results obtained through a decoherence-free quantum simulation and by running the algorithm on a two-qubit silicon spin quantum chip developed by QuTech. In Sect. 2 we briefly explain the distance-based classifier presented in [16]. In Sect. 3 we present a non-trivial reduction of the algorithm to a two-qubit version. The results of running the algorithm on the quantum hardware and on the simulator are presented and compared to the theoretically expected results in Sect. 4. Conclusions are given in Sect. 5.

2   Distance-Based Classifier

In this section we explain the quantum distance-based classifier proposed in [16]. This algorithm classifies a test point, based on its distances to data points in a data set. The algorithm returns a binary variable representing the label of the test point.

Consider a data set $\mathcal{D} = \{\mathbf{x}^i, y^i\}_{i=0}^{N-1}$ with data points $\mathbf{x}^i \in \mathbb{R}^M$ and labels $y^i \in \{\pm 1\}$. Let $\tilde{\mathbf{x}} \in \mathbb{R}^M$ be an unlabeled data point, the goal is to assign the label $\tilde{y}$ to this data point $\tilde{\mathbf{x}}$. The algorithm presented in [16], implements the threshold function

$$\tilde{y} = \text{sgn} \left(\sum_{i=0}^{N-1} y^i \left[1 - \frac{1}{4N} \left\| \tilde{\mathbf{x}} - \mathbf{x}^i \right\|^2 \right] \right), \tag{1}$$

where $\text{sgn} \colon \mathbb{R} \to \{-1, 1\}$ is the signum function and $\kappa(\tilde{\mathbf{x}}, \mathbf{x}) = 1 - \frac{1}{4N} \left\| \tilde{\mathbf{x}} - \mathbf{x} \right\|^2$ is the similarity function or kernel.

Without loss of generality we can assume that the data points in $\mathcal{D}$ are normalised. The data points can then be encoded in qubits:

$$\mathbf{x} = (x_0, \ldots, x_{M-1})^T \mapsto |\mathbf{x}\rangle = \sum_{j=0}^{M-1} x_j |j\rangle ,$$

with x_j the j-th coefficient of $\mathbf{x}$ and $|j\rangle$ the j-th computational basis state. Let us consider the quantum state

$$|\mathcal{D}\rangle = \frac{1}{\sqrt{2N}} \sum_{i=0}^{N-1} |i\rangle \left(|0\rangle |\tilde{\mathbf{x}}\rangle + |1\rangle |\mathbf{x}^i\rangle\right) |y^i\rangle . \tag{2}$$

Here, the first register $|i\rangle$ is an index register, indexing the data points. The second register is an ancilla qubit entangled with the test point and the i-th data point. The fourth register encodes the label y^i. In case of only two classes, the fourth register is only a single qubit. Note that binary labels y and labels $s \in \{\pm 1\}$ are directly related via $y = (s+1)/2$.

The algorithm starts from the quantum state of Eq. (2) and consists of a Hadamard operation on the ancilla qubit, a measurement of that qubit and a measurement of the fourth register. Due to the probabilistic nature of quantum algorithms, multiple measurement rounds should be used. The label of the test point is assigned based on the measurement of the fourth register, conditional on the first measurement giving a 0. Results where the first measurement gives a 1 should be neglected.

After the Hadamard gate we are left with

$$|\mathcal{D}\rangle = \frac{1}{2\sqrt{N}} \sum_{i=0}^{N-1} |i\rangle \left(|0\rangle \left(|\tilde{\mathbf{x}}\rangle + |\mathbf{x}^i\rangle\right) + |1\rangle \left(|\tilde{\mathbf{x}}\rangle - |\mathbf{x}^i\rangle\right)\right) |y^i\rangle . \tag{3}$$

Measuring the ancilla qubit and only continuing with the algorithm if the $|0\rangle$-state is measured, leaves us with

$$|\mathcal{D}\rangle = \frac{1}{2\sqrt{Np_{acc}}} \sum_{i=0}^{N-1} |i\rangle |0\rangle \left(|\tilde{\mathbf{x}}\rangle + |\mathbf{x}^i\rangle\right) |y^i\rangle . \tag{4}$$

Here, p_{acc} is the probability of measuring 0, given by

$$p_{acc} = \frac{1}{4N} \sum_i \left\| \tilde{\mathbf{x}} + \mathbf{x}^i \right\|^2 . \tag{5}$$

If instead the $|1\rangle$-state is measured, the algorithm should be aborted and run again, which can also be taken care of in a post-processing step. The probability of obtaining a label $\tilde{y} = 1$ is given by

$$\mathbb{P}(\tilde{y} = 1) = \frac{1}{4Np_{acc}} \sum_{i|y^i=1} \left\| \tilde{\mathbf{x}} + \mathbf{x}^i \right\|^2 . \tag{6}$$

If both classes have the same number of data points and the data points are normalised, we have

$$\frac{1}{4N} \sum_i \left\| \tilde{\mathbf{x}} + \mathbf{x}^i \right\|^2 = 1 - \frac{1}{4N} \sum_i \left\| \tilde{\mathbf{x}} - \mathbf{x}^i \right\|^2 .$$

Therefore, the algorithm implements the classifier of Eq. (1). By evaluating the algorithm multiple times, the most likely class is obtained.

Note that, as discussed in the introduction, the constant complexity of this algorithm is under the assumption of efficient state preparation, for instance using a quantum RAM. Extensions on this work and relaxation of assumptions in the original work are given in [24]. One of these assumptions is that all classes contain the same number of data points. In the next section we will reduce this distance-based classifier to a two qubit version.

3   Reduction to a Two-Qubit Version

In this section we present a non-trivial reduction of the distance-based classifier to a two qubit version to be run on few qubit quantum hardware. This reduced algorithm can consequently be used for comparing the fidelity of pairs of qubits due to the simple nature of this algorithm. The algorithm proposed in this section produces the same probability distribution for the measured labels as the original distance-based classifier for a given data set. In our approach, we use the same qubit for both encoding the data points as well as encoding the labels.

For a two-qubit version of the algorithm, we consider a training set $\mathcal{D} = \{(\mathbf{x}^0, -1), (\mathbf{x}^1, 1)\}$ and a test point $\tilde{\mathbf{x}}$, with each data point having two features. We can encode the data points as

$$\left|\mathbf{x}^0\right\rangle = \cos(\theta/2)\left|0\right\rangle - \sin(\theta/2)\left|1\right\rangle$$
$$\left|\mathbf{x}^1\right\rangle = \cos(\phi/2)\left|0\right\rangle - \sin(\phi/2)\left|1\right\rangle$$
$$\left|\tilde{\mathbf{x}}\right\rangle = \cos(\omega/2)\left|0\right\rangle - \sin(\omega/2)\left|1\right\rangle,$$

such that $R_y(\theta)\left|0\right\rangle = \left|\mathbf{x}^0\right\rangle$. Without loss of generality we may assume $\theta = 0$. Furthermore, note that for two data points, the index register and the label register have the same value. Hence, the two can be combined and the initial state is given by

$$\frac{1}{2}\left|0\right\rangle\left(\left|0\right\rangle\left|\tilde{\mathbf{x}}\right\rangle + \left|1\right\rangle\left|0\right\rangle\right)$$
$$+ \frac{1}{2}\left|1\right\rangle\left(\left|0\right\rangle\left|\tilde{\mathbf{x}}\right\rangle + \left|1\right\rangle\left|\mathbf{x}^1\right\rangle\right).$$

The ratio of the probabilities when measuring the first register is then given by

$$\frac{\mathbb{P}(\left|y^i\right\rangle = \left|0\right\rangle)}{\mathbb{P}(\left|y^i\right\rangle = \left|1\right\rangle)} = \frac{\cos^2\left(\frac{\omega}{4}\right)}{\cos^2\left(\frac{\omega-\phi}{4}\right)}, \tag{7}$$

with ω and ϕ depending on the data points. For a further reduction to only two qubits we set $t = \cos^2\left(\frac{\omega}{4}\right) / \cos^2\left(\frac{\omega-\phi}{4}\right)$, and define

$$\omega' = \begin{cases} 4\arctan\left(\frac{1-\sqrt{t}}{1+\sqrt{t}}\right) & \text{if } t \neq 1 \\ 0 & \text{else} \end{cases}. \tag{8}$$

For $t = 1$ both classes are equally likely. We propose the quantum circuit shown in Fig. 1 for classification. The resulting probability distribution is equal to that of the original classifier.

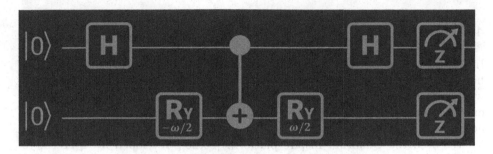

Fig. 1. A two qubit classification quantum circuit. The operations are a Hadamard gate (H), rotations around the Y-axis (R_y), a controlled-NOT operation ($CNOT$) and two measurements. The used angle depends on the data points.

The quantum circuit in Fig. 1 produces the desired probability distributions. The first gates prepare the desired quantum state and the last Hadamard-gate is similar to the operation in the original algorithm. The initial state is given by

$$\frac{1}{\sqrt{2}} \left(\cos(\omega'/2) \left|00\right\rangle + \sin(\omega'/2) \left|01\right\rangle + \left|11\right\rangle \right) \tag{9}$$

and the quantum state before the measurements is

$$\frac{1}{2} \left(\cos(\omega'/2) \left|00\right\rangle + \cos(\omega'/2) \left|10\right\rangle \right.$$
$$\left. + (1 + \sin(\omega'/2)) \left|01\right\rangle + (1 - \sin(\omega'/2)) \left|11\right\rangle \right).$$

When measuring the first (left-most) qubit and only continuing if the $\left|0\right\rangle$-state is measured, we have

$$\frac{1}{2\sqrt{p'_{acc}}} \left(\cos(\omega'/2) \left|00\right\rangle + (1 + \sin(\omega'/2)) \left|01\right\rangle \right), \tag{10}$$

with p'_{acc} the acceptance probability given by

$$p'_{acc} = \frac{1 + \sin(\omega'/2)}{2}. \tag{11}$$

Note that this acceptance probability differs from the one given in Eq. (5), however the conditional probabilities of measuring the labels does match that of the original algorithm. This acceptance probability only depends on the distribution of the considered data points on the unit circle.

The method above can be generalised to arbitrary data sets with two classes. Instead of a single data point, a representative of each data set is used in the classification. This follows as for normalised data, only the angle of the data point is considered and each angle has an equal contribution. Hence, the mean of the angles can be used for the representative data point. This step requires the computation of two means, both of $N/2$ angles.

4   Results

In this section we present the results when running the algorithm on quantum hardware and on a quantum simulator. These results are compared to the results we expect based on a theoretical evaluation of the quantum operations. These expected probabilities are obtained from the quantum state in Eq. (10) and the acceptance probability from Eq. (11).

We used the Quantum Inspire platform [25, 26], developed by QuTech, to obtain our results. This online platform hosts two quantum chips: a 2-qubit silicon spin chip and a 5-qubit transmon chip. Furthermore, a quantum simulator based on the QX programming language is available [27]. We used the publicly available 2-qubit silicon spin chip for our results.

For the experiments we use the Iris flower data set [28]. This data set contains 150 data points, equally distributed over three classes. Each data point has four features. We only consider the Setosa and Versicolor class and the first two features of the data points: the width and length of the sepal leaves. We standardise and normalise the data points and then randomly sample data points to form the data sets to run the algorithm with. Additionally, we randomly sample another data point, not used yet, together with its corresponding label. This data point is used as test point. For the first data set, we sample the test point from the Setosa class, for the second data set we sample the test point from the Versicolor class. The three data points of each data set can now be written as

$$|\mathbf{x}^0\rangle = |0\rangle$$
$$|\mathbf{x}^1\rangle = \cos(\phi/2)\,|0\rangle - \sin(\phi/2)\,|1\rangle$$
$$|\tilde{\mathbf{x}}\rangle = \cos(\omega/2)\,|0\rangle - \sin(\omega/2)\,|1\rangle,$$

with appropriate angles ϕ and ω. We identify label 0 with the Setosa class and label 1 with the Versicolor class.

For the first data set we have $\mathbf{x}^0 = (1,0)$, $\mathbf{x}^1 = (-0.9929, 0.1191)$ and $\tilde{\mathbf{x}} = (0.9939, 0.1103)$, which correspond with Iris samples 34, 75 and 13, respectively. The corresponding angles are $\phi \approx -6.0445$ and $\omega \approx -0.2210$. For the second data set we randomly chose Iris samples 21, 58 and 82. Hence, the data points are given by $\mathbf{x}^0 = (1,0)$, $\mathbf{x}^1 = (-0.1983, 0.9802)$ and $\tilde{\mathbf{x}} = (0.5545, 0.8322)$. The corresponding angles are $\phi \approx -3.5407$ and $\omega \approx -1.9662$. Both data sets are shown in Fig. 2. Based on a visual inspection, the first data set should be easier to correctly classify than the second.

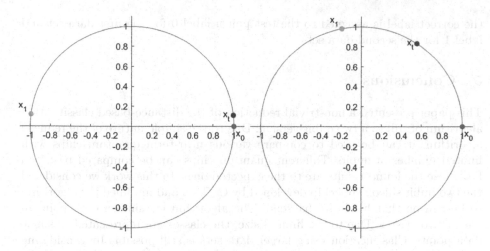

Fig. 2. Data set 1 (left) and data set 2 (right) used for the experiments. The label indicates the data points. The test point is given by x_t.

For both data sets, we determine t and ω' and consequently run the circuit as shown in Fig. 1 on the quantum simulator and on the 2-qubit silicon spin chip with 2,048 circuit evaluations in total. Additionally, we present the theoretically expected probabilities, which correspond with an infinite number of evaluations. The found probabilities are shown in Table 1 and the different between the simulation results and the theoretical results are due to the finite number of shots for the simulations. This table also shows the acceptance probabilities. The shown probabilities for both labels are conditional on the ancilla qubit being in the $|0\rangle$-state.

We see a significant different between the acceptance probability for both data sets. This results from the initial distribution of the data points on the unit circle. Different acceptance probabilities are expected for different initial distributions. For the second data set, the probabilities corresponding to both classes matches well with the expected probabilities. For the first data set, the probabilities differ more, likely due to decoherence effects. Note that in all cases,

Table 1. Shown are the results for classifying $\bar{x}$. Hardware and simulation results are shown as well as the theoretical values. The results hardware and simulation results are taken from 2048 measurement rounds.

		p_{acc}	$\mathbb{P}(y^m) = -1$	$\mathbb{P}(y^m) = 1$
Data set 1	Hardware	0.83544	0.7744	0.2256
	Simulation	0.9893	0.9877	0.0123
	Theoretical	0.9870	0.9870	0.0130
Data set 2	Hardware	0.3755	0.4655	0.5345
	Simulation	0.4863	0.4719	0.5281
	Theoretical	0.5232	0.4768	0.5232

the correct label is assigned to the test point: label 0 for the first data set and label 1 for the second data set.

5 Conclusions

This paper presented a non-trivial reduction of the distance-based classification algorithm of [16] to a two qubit version. Due to the simple nature of the reduced algorithm, it can be used to compare various near-term quantum chips with limited number of qubits. Different quantum chips can be compared based on how close the found results are to the expected ones. In this work we considered the two qubit silicon spin chip developed by QuTech and modified the algorithm to be run on that hardware backend. The algorithm classifies a data point in one of two classes. Due to the limited size, the classes are represented as single data points. Classification using larger data sets is still possible by considering the single data point used in the algorithm as being a representative of the entire class. Thereby, arbitrarily sized data sets can be used, after a suitable preprocessing step.

The obtained probability distributions are similar to the theoretically expected results, however with the differences resulting from decoherence and different rotation angles. For small rotation angles, the probability of introducing errors is also smaller. We found that the same probability distribution is produced as one would obtain with the original multi-qubit approach, however, less qubits are used. We tested the algorithm with two random data sets and in both cases, the correct label was assigned to the test point. The hardware used to produce these results was a two-qubit silicon spin quantum chip, developed by QuTech and hosted publicly on the Quantum Inspire platform.

Acknowledgements. This work is the result of the quantum technology project of TNO. There are no conflict of interests for the authors.

References

1. Hogervorst, A.C.R., van Dijk, M.K., Verbakel, P.C.M., Krijgsman, C.: Handwritten character recognition using neural networks. In: Kappen, B., Gielen, S. (eds.) Neural Networks: Artificial Intelligence and Industrial Applications, pp. 337–343. Springer, London (1995). https://doi.org/10.1007/978-1-4471-3087-1_62
2. Kim, K.I., Jung, K., Park, S.H., Kim, H.J.: Support vector machines for texture classification. IEEE Trans. Patt. Anal. Mach. Intell. **24**(11), 1542–1550 (2002)
3. Graves, A., Mohamed, A., Hinton, G.: Speech recognition with deep recurrent neural networks. In: 2013 IEEE International Conference on Acoustics, Speech and Signal Processing, pp. 6645–6649 (2013)
4. Abdel-Hamid, O., Mohamed, A., Jiang, H., Deng, L., Penn, G., Yu, D.: Convolutional neural networks for speech recognition. IEEE/ACM Trans. Audio Speech Lang. Process. **22**(10), 1533–1545 (2014)
5. Bishop, C.M.: Pattern Recognition and Machine Learning. Springer, New York (2006)

6. Lawrence, S., Giles, C. L., Tsoi, A.C., Back, A.D.: Face recognition: a convolutional neural-network approach. In: IEEE Transactions on Neural Networks, vol. 8, no. 1, pp. 98–113 (1997)
7. Silver, D., et al.: Mastering the game of go with deep neural networks and tree search. Nature **529**, 484–489 (2016)
8. Nielsen, M.A., Chuang, I.L.: Quantum Computation and Quantum Information: 10th, Anniversary edn. Cambridge University Press (2010)
9. Schuld, M., Sinayskiy, I., Petruccione, F.: An introduction to quantum machine learning. Contemp. Phys. **56**(2), 172–185 (2015)
10. Neumann, N.M.P., Phillipson, F., Versluis, R.: Machine learning in the quantum era. Digitale Welt **3**(2), 24–29 (2019)
11. Perdomo-Ortiz, A., Benedetti, M., Realpe-Gómez, J., Biswas, R.: Opportunities and challenges for quantum-assisted machine learning in near-term quantum computers. Quantum Sci. Technol. **3**(3), 030502 (2018)
12. Harrow, A.W., Hassidim, A., Lloyd, S.: Quantum algorithm for linear systems of equations. Phys. Rev. Lett. **103**(15), 150502 (2009)
13. Havlíček, V., et al.: Supervised learning with quantum-enhanced feature spaces. Nature **567**, 209–212 (2019)
14. Mitarai, K., Negoro, K., Kitagawa, M., Fujii, K.: Quantum circuit learning. Phys. Rev. A **98**(3), 032309 (2018)
15. Park, D.K., Blank, C., Petruccione, F.: The theory of the quantum kernel-based binary classifier. Phys. Lett. A **384**(21), 126422 (2020)
16. Schuld, M., Fingerhuth, M., Petruccione, F.: Implementing a distance-based classifier with a quantum interference circuit. EPL (Europhysics Letters), vol. 119, p. 60002, Sep 2017. arXiv: 1703.10793
17. Aaronson, S.: Read the fine print. Nature Phys. **11**, 291–293 (2015)
18. Barenco, A., et al.: Elementary gates for quantum computation. Phys. Rev. A **52**, 3457–3467 (1995)
19. Möttönen, M., Vartiainen, J.J., Bergholm, V., Salomaa, M.M.: Transformation of quantum states using uniformly controlled rotations. Quantum Inf. Comput. **5**(6), 467–473 (2005)
20. Shende, V.V., Bullock, S.S., Markov, I.L.: Synthesis of quantum-logic circuits. IEEE Trans. Comput.-Aided Des. Integr. Circ. Syst. **25**(6), 1000–1010 (2006)
21. Giovannetti, V., Lloyd, S., Maccone, L.: Quantum random access memory. Phys. Rev. Lett. **100** (2008)
22. Araujo, I.F., Park, D.K., Petruccione, F., da Silva, A.J.: A divide-and-conquer algorithm for quantum state preparation. Sci. Rep. **11**(1), 1–12 (2021)
23. Sadowski, P.: Quantum distance-based classifier with distributed knowledge and state recycling. Int. J. Quantum Inf. **16**, 1840013 (2018)
24. Wezeman, R., Neumann, N., Phillipson, F.: Distance-based classifier on the quantum inspire. Digitale Welt **4**, 85–91 (2020)
25. QuTech. Quantum inspire. https://www.quantum-inspire.com (2020)
26. Last, T., et al.: Quantum inspire: quTech's platform for co-development and collaboration in quantum computing. In: Sanchez, M.I., Panning, E.M. (eds.) Novel Patterning Technologies for Semiconductors, MEMS/NEMS and MOEMS 2020, vol. 11324, pp. 49–59. International Society for Optics and Photonics, SPIE (2020)
27. Khammassi, N., Ashraf, I., Fu, X., Almudever, C.G., Bertels, K.: QX: A high-performance quantum computer simulation platform. In: Design Automation Test in Europe Conference Exhibition (DATE), vol. 2017, pp. 464–469 (2017)
28. Fisher, R.A.: The use of multiple measurements in taxonomic problems. Annu. Eugenics **7**, 179–188 (1936)

Performance Analysis of Support Vector Machine Implementations on the D-Wave Quantum Annealer

Harshil Singh Bhatia[1] and Frank Phillipson[2(✉)]

[1] Department of Computer Science and Engineering, Indian Institute of Technology, Jodhpur, India

[2] TNO, The Netherlands Organisation for Applied Scientific Research, The Hague, The Netherlands
frank.phillipson@tno.nl

Abstract. In this paper a classical classification model, Kernel-Support Vector machine, is implemented as a Quadratic Unconstrained Binary Optimisation problem. Here, data points are classified by a separating hyperplane while maximizing the function margin. The problem is solved for a public Banknote Authentication dataset and the well-known Iris Dataset using a classical approach, simulated annealing, direct embedding on the Quantum Processing Unit and a hybrid solver. The hybrid solver and Simulated Annealing algorithm outperform the classical implementation on various occasions but show high sensitivity to a small variation in training data.

Keywords: Quadratic Unconstrained Binary Optimisation · Quantum annealing · Support vector machine · Performance analysis

1 Introduction

With the advancement in quantum computing and machine learning in recent times, the combination of both fields has emerged, termed *Quantum Machine Learning* [20]. Quantum computing is a technique of using quantum mechanical phenomena, such as superposition and entanglement, for solving computation problems. The current technology has established two paradigms: gate-based quantum computers and quantum annealers. The size of these computers is still limited, with the state of the art gated model-based quantum computer having 72 qubits (Google's Bristlecone) and a quantum annealer having 5000 qubits (D-Wave's Advantage [10]). Also various other firms, such as IBM, Rigetti, Qutech and IonQ, are working towards a practically usable quantum computer. To this end, we need to work on quantum algorithms and quantum software engineering skills [24].

Support Vector Machines (SVMs) [26] are supervised learning models that are used for classification and regression analysis. SVMs are known for their

© Springer Nature Switzerland AG 2021
M. Paszynski et al. (Eds.): ICCS 2021, LNCS 12747, pp. 84–97, 2021.
https://doi.org/10.1007/978-3-030-77980-1_7

stability, i.e., they don't produce largely different classifications with a small variation in training data. SVMs are preferred over Deep Learning algorithms when a relatively small training data set is available.

The SVM can be implemented on a gated model-based quantum computer with time complexity logarithmic in the size of vectors and training samples. This idea was proposed by Rebentrost et al. [25]. In [18], a quantum support vector machine was experimentally realised on a four-qubit NMR (nuclear magnetic resonance) test bench. The work of [17] establishes quantum annealing as an effective method for classification of certain simplified computational biology problems. The quantum annealer showed a slight advantage in classification performance and nearly equalled the ranking performance of the other state of the art implementations, for fairly small datasets.

In [27], Willsch et al. use a non-linear kernel to train a model on the D-Wave 2000 Quantum Annealer. For small test cases, the quantum annealer gave an ensemble of solutions that often generalise better for the classification of unseen data, than a single global minimum provided, which is given by a classically implemented SVM. In [23] this approach is used, together with two other machine learning approaches, to classify mobile indoor/outdoor locations. One of their findings is that the quantum annealing results show a solution with more confidence than the simulated annealing solution.

One of the significant limitations of classical algorithms using a non-linear kernel is that the kernel function has to be evaluated for all pairs of input feature vectors which can be of high dimensions. In [4], Chatterjee et al. propose using generalised coherent states as a calculation tool for quantum SVM and the realization of generalised coherent states via experiments in quantum optics indicate the near term feasibility of this approach.

In this paper, we use the formulation proposed by Willsch et al. [27] for the SVM modeled as a Quadratic Unconstrained Binary Optimisation (QUBO) problem, and look at its implementation and performance on the 5000 qubits [10] quantum annealer manufactured by D-Wave Systems, both directly on the chip and using the hybrid solver offered by D-Wave Systems. The aim of the paper is to study the difference in behaviour of the quantum and classical algorithm for fixed hyperparamaters. The code for the implementation using DWave Samplers is available at [1].

This paper is structured as follows. In Sect. 2, we give a short introduction on quantum annealers. The problem description and its formulation are given in Sect. 3. Section 4 deals with the various implementations of the Support Vector Machine. The classification results are presented in Sect. 5. We end with some conclusions and future research prospects.

2   Annealing Based Quantum Computing

Quantum annealing is based on the work of of Kadowaki and Nishimori [15]. The idea of quantum annealing is to create an equal superposition over all possible states. Then, by slowly turning on a problem-specific magnetic field, the qubits interact with each other and move towards the state with the lowest energy.

The challenging task of quantum annealing is to formulate the desired problem in such terms that it corresponds to finding a global minimum, such that it can also be implemented on the hardware of the quantum device. The most advanced implementation of this paradigm is the D-Wave quantum annealer. This machine accepts a problem formulated as an Ising Hamiltonian, or rewritten as its binary equivalent, in QUBO formulation. The QUBO, Quadratic Unconstrained Binary Optimisation problem [12], is expressed by the optimisation problem:

$$\text{QUBO:} \quad \min_{x \in \{0,1\}^n} x^t Q x, \tag{1}$$

where $x \in \{0,1\}^n$ are the decision variables and Q is a $n \times n$ coefficient matrix. QUBO problems belong to the class of NP-hard problems [19]. Many constrained integer programming problems can easily be transformed to a QUBO representation. For a large number of combinatorial optimisation problems the QUBO representation is known [12,19].

Next, this formulation needs to be embedded on the hardware. In the most advanced D-Wave Advantage version of the system, the 5000 qubits are placed in a Pegasus architecture [10] containing $\mathcal{K}_4$ and $\mathcal{K}_{6,6}$ sub-graphs. Pegasus qubits have a degree 15, i.e., they are externally coupled to 15 different qubits. The QUBO problem has to be transformed to this structure. Due to the current limitation of the chip size, a compact formulation of the QUBO and an efficient mapping to the graph is required. This problem is known as Minor Embedding. Minor Embedding is an NP-Hard problem and is automatically handled by the D-Wave's system [5]. Sometimes, fully embedding a problem on the Quantum Processing Unit (QPU) is difficult or simply not possible. In such cases, the D-Wave system employs built-in routines to decompose the problem into smaller sub-problems that are sent to the QPU, and in the end reconstructs the complete solution vector from all sub-sample solutions. The first decomposition algorithm introduced by D-Wave was *qbsolv* [2], which gave a possibility to solve problems of a large scale on the QPU. Although *qbsolv* was the main decomposition approach on the D-Wave system, it did not enable customisations of the workflow, and therefore is not particularly suited for all kinds of problems. The new decomposition approaches that D-Wave offers are D-Wave Hybrid [10] and the Hybrid Solver Service [8], offering more customization options of the workflow.

3   Problem Description

Support Vector Machines are supervised learning models that are used for classification and regression analysis. The training set is given as $\{x_1, x_2, \ldots x_n\}$ that are d-dimensional vectors in a some space $\chi \in \mathcal{R}^d$, where d is the dimension of each vector, i.e., the number of attributes of the training data. We are also given their labels $\{y_1, y_2, \ldots, y_n\}$, where $y_i \in \{1, -1\}$. SVMs use hyperplanes that separate the training data by a maximal margin. In general, SVMs allow one to project the training data in the space χ to a higher dimensional feature space

$\mathcal{F}$ by $z = \varPhi(x)$, where $z \in \mathcal{F}$. To find the optimal separating hyperplane having the maximum margin, the algorithm minimizes the following equation:

$$\min \frac{1}{2}||w||^2, \tag{2}$$

$$\text{subject to:} \quad y_i(w^T\varPhi(x_i) + b) \geq 1, \quad \forall i = 1, ..., n, \tag{3}$$

where w is the normal vector for the separating hyperplane given by the equation, $w = \sum_i \alpha_i y_i \mathcal{K}(x_i, x)$, which can be transferred into its dual form by maximising its primal Lagrangian. This is further formulated as a quadratic programming problem:

$$\text{minimise} \left(\frac{1}{2} \sum_{i=1}^{n} \sum_{j=1}^{n} \alpha_i \alpha_j y_i y_j \mathcal{K}(x_i, x_j) - \sum_{i=1}^{n} \alpha_i \right), \tag{4}$$

$$\text{subject to } 0 \leq \alpha_i \leq C \quad \forall i = 1, ..., n, \tag{5}$$

$$\sum_{i=1}^{n} \alpha_i y_i = 0, \tag{6}$$

where α_i is the weight assigned to the training sample x_i. If $\alpha_i > 0$, then x_i is a support vector. C is a regularisation parameter that controls the trade-off between achieving a low training error and a low testing error such that a generalization can be obtained for unseen data. The function $\mathcal{K}(x_i, x_j) = \varPhi(x_i)^T \varPhi(x_j)$ is Mercer's kernel function which allows us to calculate the dot product in high-dimensional space without explicitly knowing the non-linear Mapping. There are different forms of kernel functions, however, the SVM with a Gaussian Kernel (or RBF-kernel) has been popular because of its ability to handle cases with non-linear relation between classes and features and doing so while having less parameters. The Gaussian Kernel is defined as

$$\mathcal{K}(x_i, x_j) = e^{-\gamma||x_i - x_j||^2} \tag{7}$$

where $\gamma > 0$ is the hyperparameter.

The coefficients define a decision boundary that separates the vector space in two regions, corresponding to the predicted class labels. The decision function is fully specified by the support vectors and is used to predict samples around the optimal hyperplane. It is formulated as follows:

$$f(x) = w^T\varPhi(x) + b = \sum_{i=1}^{n} \varPhi(x_i)\varPhi(x) + b, \tag{8}$$

$$f(x) = \sum_{i=1}^{n} \alpha_i \mathcal{K}(x_i, x) + b. \tag{9}$$

4   Implementation

In this section we describe the used data sets, the QUBO formulation and the solvers that are used for benchmarking.

4.1   Data

Two public datasets have been used. First, a Standard Banknote Authentication[1] dataset has been used. We only considered two attributes: variance of the wavelet function and skewness of the Wavelet function. We also used the well known Iris[2] dataset, while only considering Iris-sentosa and Iris-versicolor for binary classification. The following 2 attributes have been used for classification here: sepal length and sepal width.

The data has been randomised before training, and each datapoint was scaled according to the following:

$$x_{new} = \frac{x - x_{min}}{x_{max} - x_{min}}. \tag{10}$$

In our experiment, subsets of various sizes are taken for comparison of the scalability of the implementations. We use two thirds of the data points as training data and the remaining one third as our validation set.

4.2   QUBO Formulation

To translate the quadratic programming formulation of Eqs. (4)–(6) to a QUBO formulation there are two main steps. First, the input has to be translated to binary input, using the encoding:

$$\alpha_n = \sum_{k=0}^{K-1} B^k a_{K_{n+k}}, \tag{11}$$

with $a_{K_{n+k}} \in \{0, 1\}$ binary variables, K the number of binary variables to encode α_n and B the base used for the encoding, usually $B = 2$. More details about the choice for K can be found in [27]. The second step is to translate the constraints (Eq. (5)–(6)) to the objective function (Eq. 4), using a penalty factor ξ for a quadratic penalty. The value of ξ is determined before the training phase (if no particular value of ξ is known, a good strategy is to try exponentially growing sequences, $\xi = \{..., 10^{-4}, 10^{-3}, 10^{-2}, ...\}$) (We have optimized the value of ξ). The resulting objective function then becomes:

$$\frac{1}{2} \sum_{n,m,k,j} a_{K_{n+k}} a_{K_{m+j}} B^{k+j} y_n y_m \mathcal{K}(x_n, x_m)$$
$$- \sum_{n,k} B^k a_{K_{n+k}} + \xi \left(\sum_{n,k} B^k a_{K_{n+k}} y_n \right)^2. \tag{12}$$

4.3   Solvers

Four main implementations are used. A short description of all four are given here.

---

[1] http://archive.ics.uci.edu/ml/datasets/banknote+authentication.
[2] http://archive.ics.uci.edu/ml/datasets/Iris/.

QPU Implementation. The first implementation is directly on the D-Wave Quantum Processor Unit (QPU). First, a minor embedding has to be created, as described in Sect. 2. Next, due to the architecture of the quantum chip, there are limitations to the number of direct connections between qubits. While some qubits in the chip are connected using external couplers, the D-Wave QPU is not fully connected. Hence a problem variable has to be duplicated to multiple connected qubits. All these qubits representing the same variables are part of a so-called chain. For this, a penalty value (edge weights of the chain), called chain strength (λ), has to be found. In [6], Coffrin et al. provide a thorough analysis for its selection procedure. The DwaveSampler [10], automatically determines an initial value based on the QUBO at hand.

Hybrid Solvers. For our second implementation, we use the Hybrid solvers offered by D-Wave Advantage [10], which implement state of the art classical algorithms with intelligent allocation of the QPU to parts of the problem where it benefits most, i.e., the sampler uses the energy impact as the criteria for selection of variables. These solvers are designed to accommodate even very large problems. This means that most parameters of the embedding are set automatically. By default, samples are iterated over four parallel solvers. The top branch implements a classical Tabu Search [13] that runs on the entire problem until interrupted by another branch completing. The other three branches use different decomposers to sample out parts of the current sample set and send it to different samplers. HQPU acts as a black box solver and the specific part of the problem which gets embedded on the QPU is unknown.

Simulated Annealing. Simulated Annealing, is implemented using the SimulatedAnnealerSampler() sampler from the D-Wave Ocean Software Development Kit [9]. Simulated Annealing is a probabilistic method proposed by Kirkpatric, Gelett and Vecchi in [16] for finding the global minimum of a cost function. It is a meta-heuristic to approximate global optimisation in a large search space for optimisation problems. It works by emulating the physical process of first heating and then slowly cooling the system to decrease defects, thus minimizing energy. In each iteration the Simulated Annealing heuristic considers a neighbouring state s^* of the current state s and probabilistically decides whether to move the system to state s^* or not. Here, the probability distribution is based on a scale proportional to temperature. The heuristic accepts points that lower the objective, but also accepts points that raise the objective with a certain probability, hence avoiding being trapped in a local minimum. To converge the algorithm an annealing schedule is decided, which decreases the temperature as the heuristic proceeds. Lowering the temperature reduces the error probability, and hence decreasing the extent of the search, which in turn leads the heuristic to converge to a global minimum.

Classical Implementation. Finally, to implement the SVM's classically, we have used the scikit-learn [21] Python library. Scikit-learn uses LIBSVM [3] for implementing a support vector machine. The Quadratic formulation, Eq. (4), is formulated in its dual form:

$$\text{minimise } \frac{1}{2}\alpha^T Q\alpha - e^T\alpha, \tag{13}$$

subject to Eq. (5)-(6). Here $Q_{ij} = y_i y_j \mathcal{K}(x_i, x_j)$ and $e = [1, .., 1]$. Q is a dense positive semi-definite matrix and might become too large to store. To tackle this a decomposition method is used that modifies only a subset of Q. This subset B leads to a smaller optimisation problem. LIBSVM uses Sequential Minimal Optimisation [11], which restricts B to only two elements. Hence, a simple two variable problem is solved at each iteration without the need of any quadratic programming optimisation software like CPLEX [7] or Gurobi [14].

5   Results

We ran the simulations for the Standard Banknote Authentication dataset and the well known Iris Dataset, using the implementations discussed in Subsect. 4.3. The simulations were run using the following parameter values: $K = 2$, $B = 2$, $C = 3$, $\xi = 0.001$, $\gamma = 16$. We do not intend to optimize C, γ as our aim is to study the behaviour for same hyper parameters. However ξ has been optimized.

We use Key Performing Indicators (KPI), which are measurable values that depicts how effectively a classification model has performed. We have used Accuracy, $F1$-score, Precision and Recall as our KPI's. Accuracy is the fraction of samples that have been classified correctly, Precision is the proportion of correct positive identifications over all positive identifications, Recall is the proportion of correct positive identifications over all actual positives:

$$\text{Accuracy} = \frac{tp + tn}{tp + tn + fp + fn}, \quad \text{Recall} = \frac{tp}{tp + fn}, \quad \text{Precision} = \frac{tp}{tp + fp},$$

where tp is true positive, fp is false positive, tn is true negative and fn is false negative. The $F1$-score is a way of combining the precision and recall of the model, and it is defined as the harmonic mean of the model's precision and recall:

$$F1 = \frac{tp}{tp + \frac{1}{2}(fp + fn)}.$$

The probability or the certainty with which a class is predicted by the model, is defined as the confidence of the classifier. The higher the absolute value of the decision function, as shown in Eq. (9), for a given data point, the more probable it is that the data point belongs to a particular class. Figures 1-6 represent the contour plot of the decision function, with the horizontal and vertical axis representing the data points and the decision function for the corresponding points being represented by the colour gradient.

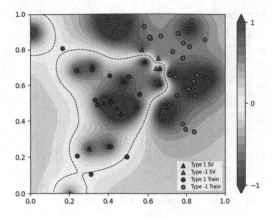

Fig. 1. Contour plot of decision function, support vectors and input data points implemented using the QPU on the Standard Banknote Authentication dataset with 60 samples.

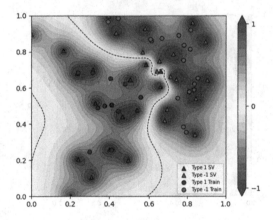

Fig. 2. Contour plot of decision function, support vectors and input data points implemented using the *HQPU and SA* solvers on the Standard Banknote Authentication dataset with 60 samples.

In Table 1, we benchmark the results of the previously discussed implementations from Sect. 4.3 on two different randomized versions of the well known Iris Dataset while varying the size of the input dataset (N). For each instance we show the accuracy, $F1$-score and Lagrangian value(-Energy) for the best solution found by the solvers. We see that the hybrid solver (HQPU) and Simulated Annealing (SA) produce the same classifiers for all instances of the data. In our previous research we have seen that the hybrid solver outperforms SA while scaling for larger datasets [22].

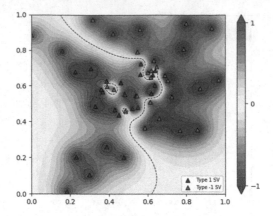

Fig. 3. Contour plot of decision function and support vectors implemented using the *HQPU and SA* solvers on the Standard Banknote Authentication dataset with 60 samples.

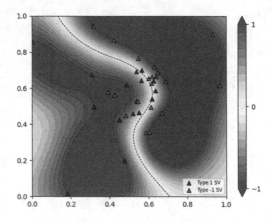

Fig. 4. Contour plot of decision function and support vectors implemented using the scikit-learn library on the Standard Banknote Authentication dataset with 150 samples.

In Table 2, we show the results of the previously discussed implementations on the Standard Banknote Authentication dataset. Similar to Table 1, the HQPU and SA produced identical classifiers. The HQPU and SA give reasonable results in comparison to the scikit-learn (Classical Implementation). We observe that the QPU is not able to find embeddings for instances larger that 90. QPU produces higher energy, inefficient solutions in comparison to HQPU and SA. From Figs. 2 and 3, we observe that the most efficient solutions in terms of energy (produced by the QUBO formulation), have sparsely situated dark regions instead of a smoother dark region. These dark regions are concentrated around the support vectors. A smoother plot is obtained for the QPU simulation (Fig. 1), which is credited to the inability of the solver to find the minimum energy solution. The

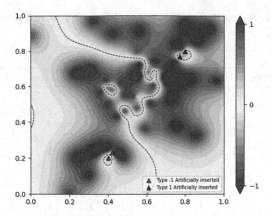

Fig. 5. Contour plot of decision function and artificially inserted data points using the HQPU and SA solvers on the Standard Banknote Authentication dataset with 150 samples.

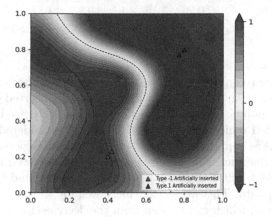

Fig. 6. Contour plot of decision function and artificially inserted data points using the scikit-learn library on the Standard Banknote Authentication dataset with 150 samples.

solution obtained by the QPU has already been disregarded and improved by the HQPU and SA solvers. Unlike SA and HQPU, scikit-learn (Fig. 4) produces smoother plots, i.e., plots with large dense regions. Hence, the classifier has a higher confidence when predicting data. Even though the hyperparamters are same, scikit-learn has a softer margin, i.e., it allows the SVM to make some mistakes while keeping the margin at the maximum so that other points can be classified correctly.

Table 1. Results for the iris dataset. Data 1 and Data 2 denote 2 different randomised version of the same.

	Data 1					Data 2				
Size	$F1$	Prec	Rec	Acc	Lagr	$F1$	Prec	Rec	Acc	Lagr
HQPU										
30	1	1	1	1	6.61	0.88	0.8	1	0.9	5.44
60	1	1	1	1	7.36	1	1	1	1	8.5
90	1	1	1	1	8.75	1	1	1	1	9.33
SA										
30	1	1	1	1	6.61	0.88	1	1	0.9	5.44
60	1	1	1	1	7.36	1	1	1	1	8.5
90	1	1	1	1	8.75	1	1	1	1	9.33
Scikit-learn										
30	1	1	1	1	4.32	0.88	1	1	0.9	4.03
60	0.96	0.9	1	0.96	3.16	1	0.9	1	1	2.17
90	1	1	1	1	3.13	1	1	1	1	2.06

To test the error tolerance of the classifiers, we ran another simulation of the Banknote Authentication dataset, but with 4 artificially inserted data points. We inserted 2 Type 1 points in dark blue region and similarly, we inserted 2 Type -1 points in dark red region. The SA/HQPU (Fig. 5) considered the artificially inserted datapoints as support vectors and created a small region around it giving a completely new classifier. The accuracy and $F1$ score showed a variation, and decreased from (0.94, 0.94) to (0.88, 0.89). The classifier ended up overfitting (hard margin) the data. This works extremely well for the training set, but not for the validation set, which can be seen from the decrease in the value of KPI's.

Although scikit-learn (Fig. 6), did consider the points as support vectors, it doesn't create a new region for it, hence demonstrating a softer margin. The KPI's for this classifier also remained unaffected by the insertion of the new data.

Table 2. Results for the two randomized versions (denoted as Data 1, Data 2) of the Standard Banknote Authentication dataset

Size	Data 1					Data 2				
	$F1$	Prec	Rec	Acc	Lagr	$F1$	Prec	Rec	Acc	Lagr
HQPU										
30	1	1	1	1	7.12	0.66	1	0.5	0.8	7.44
60	0.93	0.88	1	0.95	13.81	0.88	0.8	1	0.9	11.45
90	0.83	1	0.7	0.83	15.98	0.86	0.81	0.92	0.86	16.79
120	0.91	0.95	0.88	0.90	31.04	0.875	0.875	0.875	0.9	21.30
150	0.95	1	0.89	0.94	38.72	0.84	0.88	0.88	0.88	31.85
SA										
30	1	1	1	1	7.13	0.66	1	0.5	0.8	7.45
60	0.93	0.88	1	0.95	13.81	0.88	0.8	1	0.9	11.45
90	0.83	1	0.7	0.83	15.98	0.93	0.81	0.92	0.95	16.8
120	0.91	0.95	0.88	0.90	31.04	0.87	0.875	0.875	0.9	21.30
150	0.95	1	0.89	0.94	38.72	0.84	0.88	0.8	0.88	31.85
Scikit-Learn										
30	1	1	1	1	2.49	0.57	0.4	1	0.7	−4.45
60	1	1	1	1	5.99	0.88	1	0.8	0.9	−15.92
90	0.77	0.71	0.83	0.80	8.30	0.94	0.94	0.94	0.93	−21.18
120	0.91	0.9	0.9	0.9	13.43	0.85	0.82	0.88	0.88	−18.57
150	0.92	0.92	0.92	0.92	4.63	0.85	0.77	0.94	0.88	−34.23
QPU										
30	1	1	1	1	12	0.8	1	0.5	0.9	5.9
60	0.93	0.88	1	0.95	96.6	0.75	0.9	1	0.8	8.4
90	0.77	1	1	0.76	294	0.8	0.75	8	0.8	6.53

6 Conclusion

Quantum machine learning is still in its early stages and the implementation of machine learning paradigms on a quantum computer has shown immense capability. Quantum computing will provide the computational power needed when classical computers are reaching the upper cap of their performance, In this phase, hybrid solvers produce promising results by combining state of the art classical algorithms with a quantum annealer. In this paper, we compare the performance of hybrid solver, simulated annealing and direct QPU embedding for implementing the support vector machine. While comparing the sensitivity of the QUBO formulation in all the implementation to variation in input data.

   With the limited development of the current hardware, the Quantum Processing Unit (QPU) is only able to solve small instances and is unable to find a suitable embedding for larger instances (> 90), for larger instances decomposition algorithms are required. Inefficient classifications are obtained by the QPU (for the small instances), which are outperformed by other implementations.

HQPU and SA have produced the same classifiers for all the input data, but from previous research we know that HQPU scales better. In Table 2, HQPU and SA have been able to outperform scikit-learn's implementation.

The scikit-learn implementation has produced higher confidence plots, while HQPU and SA have produced highly sparse plots centered around support vectors, which isn't desirable for the generalisation of a classifier. The QUBO formulations have been highly sensitive to small variations showing a hard margin for the same input parameters. A slight variation (2 misclassified in a data set consisting of 100 datapoints) lead to different classifiers. HQPU and SA were found to overfit the data for the same parameters as scikit-learn. The classical implementation on the other hand wasn't affected by a slight variation in the input dataset.

Future research comprises understanding some of the results further. First there is the sensitivity to small variations of the data. We want to investigate whether this comes from possible overfitting by the QUBO-based implementations, which follow from the various plots. Second, we saw a significant difference in Lagrangian value between the QUBO-based and the Scikit-Learn approaches. There might be other objective functions suitable for the QUBO formulation than currently used.

Acknowledgments. The authors thank Irina Chiscop and Tariq Bontekoe for their valuable review and comments.

References

1. Bhatia, H.: Support vector machine implementation on d-wave quantum annealer. https://github.com/HarshilBhatia/-Support-Vector-Machine-Implementation-on-D-Wave-Quantum-Annealer
2. Booth, M., Reinhardt, S.P., Roy, A.: Partitioning Optimization Problems for Hybrid Classical/Quantum Execution. Technical report, D-Wave Systems (2017)
3. Chang, C.C., Lin, C.J.: LIBSVM: a library for support vector machines. ACM Trans. Intell. Syst. Technol. **2**, 27:1–27:27. software available at http://www.csie.ntu.edu.tw/~cjlin/libsvm (2011)
4. Chatterjee, R., Yu, T.: Generalized coherent states, reproducing kernels, and quantum support vector machines. Quantum Inf. Comput. **17**(15–16), 1292–1306 (2017)
5. Choi, V.: Minor-embedding in adiabatic quantum computation: I. the parameter setting problem. Quantum Inf. Process. **7**(5), 193–209 (2008)
6. Coffrin, C.J.: Challenges with chains: testing the limits of a d-wave quantum annealer for discrete optimization. Technical report, Los Alamos National Lab. (LANL), Los Alamos, NM (United States) (2019)
7. Cplex, I.I.: V12. 1: User's manual for CPLEX. International Business Machines Corporation. **46**(53), 157 (2009)

8. D-Wave-Systems: D-wave hybrid solver service: an overview. https://www. dwavesys.com/sites/default/files/14-1039A-A_D-Wave_Hybrid_Solver_Service_An_ Overview.pdf
9. D-Wave-Systems: D-wave ocean sdk. https://github.com/dwavesystems/dwave-ocean-sdk
10. D-Wave-Systems: The d-wave advantage system: an overview (2020). https:// www.dwavesys.com/sites/default/files/14-1049A-A_The_D-Wave_Advantage_ System_An_Overview_0.pdf (2020)
11. Fan, R.E., Chen, P., Lin, C.: Working set selection using second order information for training support vector machines. J. Mach. Learn. Res. **6**, 1889–1918 (2005)
12. Glover, F., Kochenberger, G., Du, Y.: A tutorial on formulating and using QUBO models. arXiv preprint arXiv:1811.11538 (2018)
13. Glover, F., Laguna, M.: Tabu search. In: Du, D.Z., Pardalos, P.M. (eds.) Handbook of Combinatorial Optimization. Springer, Boston (1998). https://doi.org/10.1007/ 978-1-4613-0303-9_33
14. Gurobi Optimization, L.: Gurobi optimizer reference manual (2020). http://www. gurobi.com (2020)
15. Kadowaki, T., Nishimori, H.: Quantum annealing in the transverse ising model. Phys. Rev. E **58**(5), 5355 (1998)
16. Kirkpatrick, S., Gelatt, C.D., Vecchi, M.P.: Optimization by simulated annealing. Sci. **220**(4598), 671–680 (1983)
17. Li, R.Y., Di Felice, R., Rohs, R., Lidar, D.A.: Quantum annealing versus classical machine learning applied to a simplified computational biology problem. npj Quantum Inf. **4**(1), 1–10 (2018)
18. Li, Z., Liu, X., Xu, N., Du, J.: Experimental realization of a quantum support vector machine. Phys. Rev. Lett. **114**(14), 140504 (2015)
19. Lucas, A.: Ising formulations of many np problems. Front. Phys. **2**, 5 (2014)
20. Neumann, N., Phillipson, F., Versluis, R.: Machine learning in the quantum era. Digitale Welt **3**(2), 24–29 (2019)
21. Pedregosa, F., et al.: Scikit-learn: machine learning in python. J. Mach. Learn. Res. **12**, 2825–2830 (2011)
22. Phillipson, F., Bhatia, H.S.: Portfolio optimisation using the d-wave quantum annealer. In: International Conference on Computational Science, pp. 1–14. Springer (2021)
23. Phillipson, F., Wezeman, R.S., Chiscop, I.: Three quantum machine learning approaches or mobile user indoor-outdoor detection. In: 3rd International Conference on Machine Learning for Networking (2020)
24. Piattini, M., et al.: The talavera manifesto for quantum software engineering and programming. In: QANSWER, pp. 1–5 (2020)
25. Rebentrost, P., Mohseni, M., Lloyd, S.: Quantum support vector machine for big data classification. Phys. Rev. Lett. **113**(13) (2014). https://doi.org/10.1103/ physrevlett.113.130503
26. Wang, L.: Support Vector Machines: Theory and Applications, vol. 177. Springer Science & Business Media (2005)
27. Willsch, D., Willsch, M., De Raedt, H., Michielsen, K.: Support vector machines on the d-wave quantum annealer. Comput. Phys. Commun. **248**, 107006 (2020)

Adiabatic Quantum Feature Selection for Sparse Linear Regression

Surya Sai Teja Desu[1(✉)], P. K. Srijith[1(✉)], M. V. Panduranga Rao[1(✉)], and Naveen Sivadasan[2(✉)]

[1] Indian Institue of Technology Hyderabad, Hyderabad, India
cs17b21m000002@iith.ac.in, {srijith,mvp}@cse.iith.ac.in
[2] TCS Research, Hyderabad, India
naveen.sivadasan@tcs.com

Abstract. Linear regression is a popular machine learning approach to learn and predict real valued outputs or dependent variables from independent variables or features. In many real world problems, its beneficial to perform sparse linear regression to identify important features helpful in predicting the dependent variable. It not only helps in getting interpretable results but also avoids overfitting when the number of features is large, and the amount of data is small. The most natural way to achieve this is by using 'best subset selection' which penalizes non-zero model parameters by adding ℓ_0 norm over parameters to the least squares loss. However, this makes the objective function non-convex and intractable even for a small number of features. This paper aims to address the intractability of sparse linear regression with ℓ_0 norm using adiabatic quantum computing, a quantum computing paradigm that is particularly useful for solving optimization problems faster. We formulate the ℓ_0 optimization problem as a Quadratic Unconstrained Binary Optimization (QUBO) problem and solve it using the D-Wave adiabatic quantum computer. We study and compare the quality of QUBO solution on synthetic and real world datasets. The results demonstrate the effectiveness of the proposed adiabatic quantum computing approach in finding the optimal solution. The QUBO solution matches the optimal solution for a wide range of sparsity penalty values across the datasets.

Keywords: Adiabatic quantum computing · Sparse linear regression · Feature selection

1 Introduction

Most of the real world application of machine learning arising from various domains such as web, business, economics, astronomy and science involve solving regression problems [17,20,26,27]. Linear regression is a popular machine learning technique to solve the regression problems where a real valued scalar response variable (dependent variable or output) is predicted using explanatory variables (independent variables or input features) [1,12]. It assumes the

© Springer Nature Switzerland AG 2021
M. Paszynski et al. (Eds.): ICCS 2021, LNCS 12747, pp. 98–112, 2021.
https://doi.org/10.1007/978-3-030-77980-1_8

output variable to be an affine function of the input variables and learns the model parameters (weight coefficient parameters) from the data by minimizing the least squares error. In many practical applications, we are not only interested in learning such functions but also in understanding the importance of the input features in determining the output. Identifying relevant features helps in interpretability, which is important for many real-world applications. Moreover, these problems are often associated with a large number of input variables and consequently large number weight parameters. When the data is limited, least squares approach to learn the parameters will result in over-fitting and poor predictive performance [1].

Sparse linear regression has been proposed to address the above limitations that are associated with least squares regression. It encourages sparse weight vectors by adding to the loss function, appropriate norms over the weight vectors. Using ℓ_0 norm over the weight vector, which is known as the *best subset selection method* [12,13], is an effective approach to identify features useful for linear regression. ℓ_0 norm counts the number of non-zero components in the weight vector and favours solutions with smaller number of non-zero elements. Consequently, one can identify the important features as the ones with non-zero weight vectors. Best subset selection has been theoretically shown to achieve optimal risk inflation [10]. However, solving an optimization problem involving ℓ_0 norm is non-convex and intractable as the number of possible choices of non-zero elements in weight vectors is exponentially large. In fact, this is an NP-hard problem and can become intractable even for small values of data dimension and subset size [21]. Approaches such as greedy algorithm based *Forward- and Backward-Stepwise Selection* [12,28] and heuristics-based integer programming [16,19] were proposed to select subset of features for linear regression. However, they are sub-optimal and are not close to the optimal solution of best subset selection. Instead of modelling sparsity exactly through ℓ_0 norm, a common practice is to use its convex approximation using ℓ_1 norm, known as Lasso regression [23]. However, it will lead to sub-optimal selection of features, and incorrect models as the shrinkage property can result in a weight vector with many elements zero. It can also lead to a biased model as it heavily penalizes the weight coefficients, even the ones corresponding to the relevant and active features [11,18].

In this paper, we propose to use adiabatic quantum computing technique for training sparse linear regression with ℓ_0 norm. Given the NP-hardness of the problem, a tractable solution is not expected even with quantum computers. However, quantum computers have been found to be effective in speeding up training of machine learning algorithms. In particular, adiabatic quantum computing is found to be excellent in solving hard optimization problems [24]. Adiabatic quantum computers like D-Wave 2000Q (from D-Wave) can efficiently solve quadratic unconstrained binary optimization (QUBO) problems. Machine learning algorithms relying on optimization techniques for parameter estimation can benefit from (adiabatic) quantum computing. It has been effectively used to train machine learning models like deep belief networks (DBN) [5] and least squares regression [2,4,6,8,25]. While much of the previous work focuses only on least squares regression, in this paper we solve the ℓ_0 regularized least squares regression problem. We derive a Quadratic Unconstrained Binary Optimization (QUBO) problem for the same, and solve it using the D-Wave adiabatic quantum computer. We report an extensive evaluation of the performance of the QUBO approach on synthetic and real-world data. Our experimental results indicate that the QUBO based approach for solving large scale sparse linear regression problems shows promise. Our main contributions are summarized as follows.

1. We propose an adiabatic quantum computing approach for the best subset selection based sparse linear regression problem.
2. We formulate the sparse linear regression problem as a Quadratic Unconstrained Binary Optimization (QUBO) problem. Our approach uses QUBO to solve best subset selection optimally and computes the corresponding regression coefficients using the standard least squares regression.
3. We conduct experiments on the DWave quantum computer using synthetic and real-world datasets to demonstrate the performance of the proposed approach. For a wide range of sparsity penalty values, the QUBO solution matches the optimal solution across the datasets.

2   Background

In this section, we give a brief overview of the adiabatic quantum computing approach for solving optimization problems. We refer the interested reader to the excellent survey by Venegas-Andraca et al. [24].

2.1   Optimization Problems

Given a set S of n elements, denote by $\mathcal{P}(S)$ its power set. Consider a function $f : \mathcal{P}(S) \to \mathbb{R}$. The *combinatorial optimization* problem (we consider, without loss of generality, the minimization problem) is to find $P \in \mathcal{P}(S)$ such that $f(P) = \min_{P_i \in \mathcal{P}(S)}(f(P_i))$. In the absence of any structure, these optimization problems are NP-hard.

2.2   Three Formulations of Optimization

It is possible to formulate these (NP-hard) minimization problems as

- Minimization of pseudo-Boolean functions:
  Consider a function $f : \{0,1\}^n \to \mathbb{R}$ such that

$$f(\mathbf{x}) = \sum_{I \subseteq [n]} \gamma_I \prod_{x_j \in I} x_j,$$

where $\gamma_I \in \mathbb{R}$. The degree of this multilinear polynomial is $max\{I\}$. The problem is to find the $\mathbf{x}$ for which $f(\mathbf{x})$ is minimum.

- Quadratic Unconstrained pseudo-Boolean Optimization (QUBO):
  Consider a function

$$f_B(\mathbf{x}) = \sum_{i=1}^{n} \sum_{j=1}^{i} \beta_{ij} x_i x_j,$$

  where $x_i \in \{0,1\}$ and $\beta_{ij} \in \mathbb{R}$. Notice that this is a degree two polynomial. Again, the problem in this case is to find $\mathbf{x}$ for which $f(\mathbf{x})$ is minimum.
- The Ising Model:
  Given (i) a graph $G = (V, E)$, real valued weights h_v assigned to each vertex $v \in V$, and real valued weights J_{uv} assigned to each edge $(u, v) \in E$, (ii) A set of Boolean variables called *spins*: $S = \{s_1, \ldots, s_n\}$ where $s_u \in \{-1, 1\}$, the spin s_v corresponding to the vertex v and (iii) an *energy function*

$$E(S) = \sum_{v \in V} h_v s_v + \sum_{(u,v) \in E} J_{uv} s_u s_v.$$

The problem here is to find an assignment to the spins that minimizes the energy function.

It is well known that any pseudo-Boolean function minimization problem can be reformulated as a QUBO minimization in polynomial time [14,24]. Further, a QUBO problem can be very easily converted to an optimization problem in the Ising Model in a straightforward manner, namely by using $s_i = 2x_i - 1$.

2.3   Adiabatic Quantum Computation

We now briefly discuss the adiabatic quantum computing paradigm. The interested reader is referred to [9]. In what follows, we assume a basic knowledge of quantum mechanics, in particular, the notion of a Hamiltonian operator.

To solve an optimization problem, the system is initially prepared in the ground state of an initial Hamiltonian H_I. Naturally, this ground state should be "easy to prepare". The ground state of a "final" Hamiltonian H_F encodes the solution to the optimization problem. Let the system, prepared in the ground state of H_I, evolve as per the following *interpolating* Hamiltonian for a time duration T.

$$H\left(\frac{t}{T}\right) = \left(1 - \frac{t}{T}\right) H_I + \frac{t}{T} H_F.$$

The adiabatic gap theorem says that if the $T \geq O(\frac{1}{g_{min}^2})$, where g_{min} is the spectral gap, the system will stay in the ground state of the Hamiltonian with high probability. Therefore, with a high probability, the system will end up in the ground state of the final Hamiltonian, the solution of the optimization that we are searching for. This evolution is analogous to classical simulated annealing; the difference being that due to phenomena like superposition and entanglement that are peculiar to quantum mechanics, the system *tunnels* through local peaks instead of going over them. In what follows, we will use the *quantum annealing* interchangeably with adiabatic quantum computing.

2.4 Ising Model Implementation

Quantum computers like the Quantum Processing Unit (QPU) of D-Wave, that are based on the adiabatic quantum computing, typically implement an Ising model [15]. Let G_{QC} be the graph with qubits as vertices and connectors as edges. For a vertex i in such a graph, one can talk of h_i the magnetic field on qubit i and J_{ij} the so-called coupling strength between qubits i and j. Further, let σ_i^z be the Pauli z matrix acting on qubit i. Then,

The problem Hamiltonian H_F has the Ising formulation:

$$H_F = \sum_i h_i \sigma_i^z + \sum_{i<j} J_{ij} \sigma_i^z \sigma_j^z. \tag{1}$$

While the graph structure in the Ising optimization formulation can be generic, it may not be isomorphic to the G_{QC}, which for D-wave QPU's current implementation is a chimera graph. However, it is possible to obtain a minor embedding in a subgraph of G_{QC} corresponding to G [24]. Then, it is possible to obtain h_i and J_{ij} from the Ising formulation of the minimization problem. The ground state of the Hamiltonian H_F then corresponds the spin assignment to S that minimizes the function in Eq. (1).

This yields the following approach for using quantum annealing to solve a minimization problem:

1. Pose the minimization problem as a pseudo-Boolean function minimization problem
2. Convert the pseudo-Boolean function to a QUBO formulation
3. The QUBO formulation is then converted to an Ising model formulation on the native QPU

The minimization is achieved through adiabatic evolution.

2.5 Linear Regression

We consider a regression problem with input-output pair $(\mathbf{x}, y)$, where $\mathbf{x} \in \mathcal{R}^d$ and $y \in \mathcal{R}$ (for e.g. house price prediction, grade prediction). We assume the training dataset to be $\mathcal{D} = \{\mathbf{x}_i, y_i\}_{i=1}^N$. The goal is to learn a function $f : \mathcal{R}^d \to \mathcal{R}$ with good generalization performance from the training dataset. Linear regression is one of the widely used approach for regression problems which assumes y is a linear function of $\mathbf{x}$. Non-linear regression can also be modelled through linear regression framework by considering higher order powers of $\mathbf{x}$. In either case, the functional form is linear in terms of the parameters $\mathbf{w} \in \mathcal{R}^d$ such that $y = \mathbf{w}^T \mathbf{x}$.

In practice, one cannot find a function which passes through all the data points. Hence, the function is learnt to be as close as possible to the observations. Training in linear regression involves learning the parameters $\mathbf{w}$ such that squared error between actual values and the values given by the function f is minimum. The resulting optimization problem would be to find $\mathbf{w}$ such that it minimizes the following squared error loss [1]:

$$\min_{\mathbf{w}} \sum_{i=1}^{N} (y_i - \mathbf{w}^T \mathbf{x}_i)^2. \tag{2}$$

The solution to the above problem can be computed in closed form as

$$\mathbf{w} = (X^T X)^{-1} X^T \mathbf{y}, \tag{3}$$

where we assume X is a $N \times d$ matrix formed by stacking the input data as N row vectors, $\mathbf{y}$ is an N dimensional column vector of the training data outputs. We assume that the matrix X is column normalized where each column of X is divided by its ℓ_2 norm.

However, on datasets with large dimensions and limited data, learning the parameters by minimizing the squared loss alone can result in learning a complex model which will overfit on the data. Under these circumstances, one could achieve a very low training error but a high test error and consequently poor generalization (predictive) performance. For instance, in predicting tumour based on genetic information, the input dimension is in the order of thousands while the number of samples is often in the order of hundreds. In such problems, only a few dimensions are relevant and contribute to the output prediction. Instead of solving the regression problem using all the input features, solving using the best subset of features could lead to functions with better generalization performance. Moreover, this can help in identifying features which are relevant to the regression problem and this aids in interpret ability. This is ideally achieved by adding a ℓ_0 regularization term over the parameters to the least squares loss. ℓ_0 norm regularizer is defined as

$$\|\mathbf{w}\|_0 = \sum_{j=1}^{d} \mathrm{I}\{\mathbf{w}_j \neq 0\}.$$

This counts the number of non-zero elements in the weight vector and consequently it determines the dimensions which are active and relevant. The resulting regularized loss function to estimate $\mathbf{w}$ is given as

$$\min_{\mathbf{w}} \sum_{i=1}^{N} (y_i - \mathbf{w}^T \mathbf{x}_i)^2 + \lambda \|\mathbf{w}\|_0. \tag{4}$$

However, ℓ_0 norm is discontinuous, and the above optimization problem is known to be NP-Hard. However, it gives the best possible solution to sparse linear regression [12].

3 Proposed Methodology

Adiabatic quantum computing system in D-Wave requires the problem to be specified as a QUBO expression [24]. The QUBO expression is then internally converted to an Ising model and is fed to the D-Wave quantum computing system. We propose an approach to express the regularized least squares regression problem with ℓ_0 norm given by (4) as a QUBO problem. Existing approaches (eg [6]) for the related problem of standard least square regression assumes a bounded word length k for the elements of $\mathbf{w}$ in order to construct a QUBO formulation and the number of logical qubits needed depends on k as well as the dimension d. We attempt to construct a QUBO formulation to solve the subset selection optimally. We then solve the values of the selected elements of $\mathbf{w}$ classically. In this way, we avoid any assumption on the word length for the elements of $\mathbf{w}$. Furthermore, the number of logical qubits used in our case is only a function of the dimension d. We introduce a binary vector (selection vector) $\mathbf{z} \in \{0, 1\}^d$, with z_j determining whether dimension j is selected. The best subset selection problem can be solved as an optimization problem involving both the selection vector $\mathbf{z}$ and the weight vector $\mathbf{w}$. We first rewrite the original formulation in (4) using the selection vector as follows

$$\min_{\mathbf{w}, \mathbf{z}} \sum_{i=1}^{N} (y_i - \mathbf{w}^T \mathbf{x'}_i)^2 + \lambda \|\mathbf{z}\|_0, \tag{5}$$

where vector $\mathbf{x'}_i = \mathbf{x}_i \circ \mathbf{z}$, is the element-wise product between the input vector $\mathbf{x}_i$ and the selection vector $\mathbf{z}$. For any arbitrarily fixed $\mathbf{z}$, solution for $\mathbf{w}$ can be obtained by first solving

$$\mathbf{w'} = (X'^T X')^{-1} X'^T y, \tag{6}$$

where X' is the $N \times d'$ submatrix of the $N \times d$ matrix X, where X' is obtained by retaining only the $\|z\|_0 = d'$ columns of X corresponding to non-zeros in the vector $\mathbf{z}$. Vector $\mathbf{w'}$ is simply a projection of the d' components of the target vector $\mathbf{w}$ corresponding to the d' non-zeros in $\mathbf{z}$. Since, the remaining components of $\mathbf{w}$ are all zeros, $\mathbf{w}$ is trivially obtained from $\mathbf{w'}$.

We use Lemma 1 (Sect. 3.2), to obtain the following first order approximation $(X'^T X')^{-1} \approx \alpha(2I - \alpha X'^T X')$, where $\alpha = 2/(d+1)$. Substituting this in (6) yields $\mathbf{w'} = \alpha(2I - \alpha X'^T X')X'^T y$. We can also view X' as the $N \times d$ matrix $X \times \mathbf{diag}(\mathbf{z})$, where the $\mathbf{diag}(\mathbf{z})$ is a diagonal matrix whose diagonal corresponds to the vector $\mathbf{z}$. Clearly, in this X', columns corresponding to the zeros in $\mathbf{z}$ are all zeros. With this definition of X', it is straightforward to verify that, we can directly write the expression for $\mathbf{w}$ in place of $\mathbf{w'}$ as

$$\mathbf{w} = \alpha(2I - \alpha X'^T X')X'^T y. \tag{7}$$

3.1   QUBO Formulation

We derive an equivalent QUBO formulation for (4) using (5) and the expression for $\mathbf{w}$ from (7). Using (7), we obtain ith component w_i of $\mathbf{w}$ as

$$w_i = \alpha \left(\left(2 - \alpha z_i^2 \sum_{p=1}^{N} x_{pi}^2 \right) \sum_{k=1}^{N} x_{ki} z_i y_k - \sum_{j=1, j \neq i}^{d} \sum_{p=1}^{N} x_{pi} z_i . x_{pj} z_j \sum_{k=1}^{N} x_{kj} z_j y_k \right).$$

Recalling that for each column i of X, $\sum_{p=1}^{N} x_{pi}^2 = 1$, we expand the above expression to obtain

$$
w_i = \alpha(2 - \alpha z_i^2) \sum_{k=1}^{N} x_{ki} z_i y_k \quad - \quad \alpha^2 \sum_{j=1, j \neq i}^{d} \sum_{p=1}^{N} x_{pi} z_i . x_{pj} z_j \sum_{k=1}^{N} x_{kj} z_j y_k
$$

$$
= z_i \left(\alpha(2 - \alpha) \sum_{k=1}^{N} x_{ki} y_k \quad - \quad \alpha^2 \sum_{j=1, j \neq i}^{d} z_j \sum_{p=1}^{N} x_{pi} x_{pj} \sum_{k=1}^{N} x_{kj} y_k \right). \tag{8}
$$

The last step follows because $z_i^c = z_i$ for binary z_i, for all positive integer c. Substituting the above expression for w_i, we can rewrite (5) as

$$
\min_{\mathbf{z}} \sum_{t=1}^{N} \left(y_t - \sum_{i=1}^{d} w_i x_{ti} z_i \right)^2 + \lambda \sum_{r=1}^{d} z_r
$$

$$
= \min_{\mathbf{z}} \sum_{t=1}^{N} \left(y_t - \sum_{i=1}^{d} \left(\alpha(2-\alpha) \sum_{k=1}^{N} x_{ki} y_k - \alpha^2 \sum_{j=1, j \neq i}^{d} z_j \sum_{p=1}^{N} x_{pi} x_{pj} \sum_{k=1}^{N} x_{kj} y_k \right) x_{ti} z_i \right)^2 + \lambda \sum_{r=1}^{d} z_r.
$$

Denoting the elements of $X^T X$ by p_{ij} and the elements of $X^T y$ by q_i, the above minimization can be reformulated as

$$
\min_{\mathbf{z}} \sum_{t=1}^{N} \left(y_t - \sum_{i=1}^{d} \left(\alpha(2-\alpha)q_i - \alpha^2 \sum_{j=1, j \neq i}^{d} z_j p_{ij} q_j \right) x_{ti} z_i \right)^2 + \lambda \sum_{r=1}^{d} z_r
$$

$$
= \min_{\mathbf{z}} \sum_{t=1}^{N} \left(y_t - \sum_{i=1}^{d} \left(\alpha(2-\alpha)q_i x_{ti} z_i - \sum_{j=1, j \neq i}^{d} \alpha^2 p_{ij} q_j x_{ti} z_i z_j \right) \right)^2 + \lambda \sum_{r=1}^{d} z_r.
$$

Letting $b(t)_{ij} = \alpha^2 (p_{ij} q_j x_{ti} + p_{ji} q_i x_{tj})$ and $q_i' = \alpha(2-\alpha)q_i$, the above minimization becomes

$$
\min_{\mathbf{z}} \sum_{t=1}^{N} \left(y_t - \sum_{i=1}^{d} q_i' x_{ti} z_i + \sum_{1 \leq i < j \leq d} b(t)_{ij} z_i z_j \right)^2 + \lambda \sum_{r=1}^{d} z_r. \tag{9}
$$

Expanding (9) further and again simplifying using $z_i^c = z_i$, we finally obtain

$$
\min_{\mathbf{z}} \left(y + \sum_{i=1}^{d} (e_i + \lambda) z_i + \sum_{1 \leq i < j \leq d} f_{ij} z_i z_j + \sum_{1 \leq i < j < k \leq d} g_{ijk} z_i z_j z_k + \sum_{1 \leq i < j < k < l \leq d} h_{ijkl} z_i z_j z_k z_l \right),
$$

where

$$e_i = \sum_{t=1}^{N} \left(q_i'^2 x_{ti}^2 - 2q_i' x_{ti} y_t \right),$$

$$f_{ij} = \sum_{t=1}^{N} \left(b(t)_{ij}^2 + 2\left(q_i' q_j' x_{ti} x_{tj} - (q_i' x_{ti} + q_j' x_{tj} - y_t) b(t)_{ij} \right) \right),$$

$$g_{ijk} = \sum_{t=1}^{N} \left(2\left(b(t)_{ij} b(t)_{ik} + b(t)_{ij} b(t)_{jk} + b(t)_{ik} b(t)_{jk} \right) \right.$$
$$\left. - q_k' x_{tk} b(t)_{ij} - q_i' x_{ti} b(t)_{jk} - q_j' x_{tj} b(t)_{ik} \right),$$

$$h_{ijkl} = \sum_{t=1}^{N} 2\left(b(t)_{ij} b(t)_{kl} + b(t)_{ik} b(t)_{jl} + b(t)_{il} b(t)_{jk} \right), \quad \text{and} \quad y = \sum_{t=1}^{N} y_t^2.$$

We make use of the in-built method 'make_quadratic()' in Ocean SDK provided by D-Wave [3] to obtain a quadratic equivalent (QUBO expression) of the above binary optimization problem. D-wave uses efficient approaches [24] to turn higher degree terms to quadratic terms by introducing auxiliary binary variables. In our case, it follows that the final QUBO formulation would involve $O(d^4)$ logical qubits.

The QPU finds an optimal solution for $\mathbf{z}$. An optimal solution for $\mathbf{z}$ corresponds to an optimal feature selection. Once an optimal choice of features is known, the problem of solving $\mathbf{w}$ reduces to the standard least squares regression problem (Eq. (3)), which can be solved efficiently using any of the existing methods.

3.2   Approximation of $(X^T X)^{-1}$

We derive a bound on $(X^T X)^{-1}$, which is a direct adaptation of the well-known Neumann series [22]. However, we include the whole proof here for completeness. We recall that every column of the $N \times d$ matrix X is ℓ_2 normalized. We assume that $X^T X$ is full rank. Clearly, following Lemma also holds true for any $(X'^T X')^{-1}$, where X' is a submatrix of X obtained by retaining some $d' \leq d$ columns of X.

Lemma 1. $(X^T X)^{-1} = \alpha \lim_{n \to \infty} \sum_{i=0}^{n} (I - \alpha X^T X)^i$ *for positive* $\alpha \leq \frac{2}{d+1}$.

From Lemma 1, a kth order approximation of $(X^T X)^{-1}$ is given by $(X^T X)^{-1} \approx \alpha \sum_{i=0}^{k} (I - \alpha X^T X)^i$, where $\alpha = 2/(d+1)$. In order to prove the Lemma, we first show the following Claim.

Claim. Eigenvalues of $X^T X$ are in the range $[0, d]$.

Proof. Since $X^T X$ is a $d \times d$ Gram matrix, $X^T X$ is a symmetric positive semidefinite matrix. Hence, all eigenvalues of $X^T X$ are non-negative. Let $U[\lambda]U^T$ be the eigendecomposition of $X^T X$ where $[\lambda]$ is the diagonal matrix of eigenvalues of $X^T X$. We note that U is an orthonormal $d \times d$ matrix where $UU^T = I$. Thus, for any vector $\mathbf{a}$,

$$\left\| \mathbf{a} U^T \right\|_2^2 = \left\| \mathbf{a} \right\|_2^2.$$

Let $U = [\mathbf{u}_1, \ldots, \mathbf{u}_d]$, where $\mathbf{u}_i$s are column vectors. Similarly, let $U^T = [\mathbf{v}_1, \ldots, \mathbf{v}_d]$. Let $X = AU^T$, where $A = XU$. Let the $N \times d$ matrix A be viewed as stacking of N row vectors $\mathbf{a}_1, \ldots, \mathbf{a}_N$ in the same order.

Since, columns of X are normalized and since $X = AU^T$, it is easy to verify that for any $j \in \{1, \ldots, d\}$, $\sum_{i=1}^N \langle \mathbf{a}_i, \mathbf{v}_j \rangle^2 = 1$. It follows that,

$$\sum_{j=1}^d \sum_{i=1}^N \langle \mathbf{a}_i, \mathbf{v}_j \rangle^2 = d.$$

Interchanging the summations, we obtain $\sum_{i=1}^N \sum_{j=1}^d \langle \mathbf{a}_i, \mathbf{v}_j \rangle^2 = d$.

The inner summation is $\left\| \mathbf{a}_i U^T \right\|_2^2$. Recalling that $\left\| \mathbf{a}_i U^T \right\|_2^2 = \left\| \mathbf{a}_i \right\|_2^2$, it follows that $\sum_{i=1}^N \left\| \mathbf{a}_i \right\|_2^2 = d$. In other words, $\sum_{i=1}^N \sum_{j=1}^d a_{ij}^2 = d$, where a_{ij} is the entry at ith row and jth column of matrix A, which is also the jth component of the vector $\mathbf{a}_i$.

Recalling that $X = AU^T$ and that $X^T X = U[\lambda]U^T$, we have

$$X^T X = U A^T A U^T = U[\lambda]U^T.$$

In other words, $A^T A = [\lambda]$. The jth diagonal entry of $A^T A$ is given by $\sum_{i=1}^N a_{ij}^2$. However, recalling that $\sum_{i=1}^N \sum_{j=1}^d a_{ij}^2 = d$, it follows that jth diagonal entry of $A^T A$ satisfies

$$\sum_{i=1}^N a_{ij}^2 \leq \sum_{i=1}^N \sum_{j=1}^d a_{ij}^2 = d.$$

It follows that every entry of $[\lambda]$ is in the range $[0, d]$. $\square$

Proof (Lemma 1). Let the eigendecomposition of $X^T X = U[\lambda]U^T$. It follows that the eigendecomposition of

$$\alpha X^T X = U[\alpha\lambda]U^T.$$

where $[\alpha\lambda]$ is the scalar multiplication of α with the eigenvalue diagonal matrix $[\lambda]$.

Let $Y = \alpha X^T X$. From Claim 3.2, it follows that all eigenvalues of a full rank $X^T X$ are in $(0, d]$. Consequently, all eigenvalues of Y are in $(0, 2)$ for $\alpha \leq 2/(d+1)$.

Let $B = I - Y$. Since eigenvalues of Y are in $(0, 2)$, it follows that the eigenvalues of B are in $(-1, 1)$. Since all eigenvalues of B are strictly less than 1 in absolute value, it follows that $\lim_{n\to\infty} B^n = 0$. In other words, $\lim_{n\to\infty} \sum_{i=0}^{n} B^i$ is a convergent series. Following Neumann series gives

$$\lim_{n\to\infty} (I - B) \sum_{i=0}^{n} B^i = \lim_{n\to\infty} \left(\sum_{i=0}^{n} B^i - \sum_{i=0}^{n} B^{i+1} \right) = \lim_{n\to\infty} (I - B^{n+1}) = I.$$

Thus, $Y^{-1} = (I - B)^{-1} = \lim_{n\to\infty} \sum_{i=0}^{n} B^i$ Recalling $Y = \alpha X^T X$, where α is a positive scalar, and $B = I - Y$, we obtain

$$\alpha^{-1}(X^T X)^{-1} = \lim_{n\to\infty} \sum_{i=0}^{n} (I - Y)^i.$$

The result follows by observing that, $(X^T X)^{-1} = \alpha \lim_{n\to\infty} \sum_{i=0}^{n} (I - Y)^i$.

4   Experimental Results

For experimental evaluation of our approach, we consider both synthetic datasets as well as real world data. The QUBO formulation is run on D-Wave 2000Q quantum computer which has 2048 qubits. For a given dataset, we run the corresponding QUBO instance on D-Wave for multiple runs. Each run outputs a feature selection $\mathbf{z}$. The corresponding $\mathbf{w}$ and the objective function value is obtained from Eq. (6) and Eq. (4) respectively. The final solution is chosen as the one that minimizes the objective function. We compare the quality of QUBO solution with optimal solution for this NP-hard problem computed by exhaustive search in the classical setting. This exhaustive search is performed over all possible feature selections to find the solution that minimizes Eq. (4). Here again, for a given feature selection, $\mathbf{w}$ is obtained from Eq. (6).

Synthetic datasets were generated using randomly chosen X and a randomly chosen sparse vector $\mathbf{w}$. The outputs are obtained as $\mathbf{y} = X^\top \mathbf{w}$. We generate a separate synthetic input data for input dimensions in the range $5, \ldots, 10$. For each of these dimensions, the number of samples N in the input data is fixed as 3000. For each of the resulting six input datasets, we use QUBO to solve Eq. (4) for 5 different λ values. Table 1 summarises QUBO results. For each input data and λ combination, the table gives the ℓ_0 norm of $\mathbf{w}$, which corresponds to the cardinality of the selected features, computed using QUBO and the optimal $\mathbf{w}$ computed classically by exhaustive search. Columns 6 and 7 of the table give the values of the objective function Eq. (4) corresponding to $\mathbf{w}$ obtained through the classical solution and QUBO respectively. A set of $\lambda \times d$ values were used so as to reduce the sparsity penalty λ for increasing d. This balances the contributions of the regression gap and the sparsity penalty in the objective function. The last five rows of Table 1 show results for 500 runs, while the rest are for 100 runs. We can observe that the cardinality of $\mathbf{w}$, i.e. the number of features selected, is the same for the QUBO solution and classical solution for most values of λ across all the synthetic datasets. Similarly, the objective function values are also very close for the QUBO and classical solution. For data with input dimension 10, we also experimented with 500 QUBO runs. These are included as the last 5 rows in the table. As shown in the table, increasing the number of runs further improved the performance of the QUBO solver in finding optimal solutions. For higher values of λ, we observe an increased gap between the QUBO solution and the optimal solution. This is partly due to the fact that higher values of λ induces sparser $\mathbf{z}$. In such situations, choosing α closer to $2/(d' + 1)$ in Eq. (7), where d' is the cardinality of the optimal feature set, can yield a better inverse approximation compared to the existing $2/(d+1)$ value. However, the value of d' is not known *a priori*. Hence, using a higher order inverse approximation or running QUBO with different permissible values of α are potential ways to mitigate this issue.

We also measured the mean squared error (MSE) for the optimal classical and the QUBO settings on the train and test data. The test error was measured using a held-out test data of 1000 points generated separately for each d, using the same distribution as the training data. In particular, we measured the absolute gap between the test errors of classical and QUBO as well as between the train

Table 1. Experimental results on synthetic datasets for different d values.

Number of points (N)	Number of features (d)	$\lambda \times d$	$\|\mathbf{w}\|_0$ classical	$\|\mathbf{w}\|_0$ QUBO	Objective value classical	Objective value QUBO	Preprocessing time (sec)	Processing time (sec)
3000	5	10	1	2	2.43	4	0.3802	1.7997
3000	5	1	2	2	0.4	0.4	0.3015	1.463
3000	5	0.1	2	2	0.04	0.04	0.2921	1.8042
3000	5	0.01	2	2	0.004	0.004	0.3087	1.5339
3000	5	0.001	2	2	0.0004	0.0004	0.3158	1.7676
3000	6	10	1	3	3.1993	5	0.5311	1.6666
3000	6	1	2	3	0.4542	0.5	0.5421	1.6226
3000	6	0.1	3	3	0.05	0.05	0.5286	1.6272
3000	6	0.01	3	3	0.005	0.005	0.5807	1.5697
3000	6	0.001	3	3	0.0005	0.0005	0.5341	1.6663
3000	7	10	2	3	2.9767	4.2858	0.8584	1.9038
3000	7	1	2	3	0.4053	0.4286	0.985	1.8351
3000	7	0.1	3	3	0.0429	0.0429	0.9155	1.8839
3000	7	0.01	3	3	0.0043	0.0043	0.8463	1.8471
3000	7	0.001	3	3	0.0005	0.0005	0.8983	1.8971
3000	9	10	2	4	2.3813	4.4573	2.0121	3.7735
3000	9	1	3	4	0.3468	0.4445	1.9516	4.4245
3000	9	0.1	4	4	0.0445	0.0445	2.0444	4.4954
3000	9	0.01	4	4	0.0045	0.0045	1.9706	4.5724
3000	9	0.001	4	4	0.0005	0.0005	1.9625	4.5866
3000	10	10	3	5	3.1369	5.0124	3.8412	7.2634
3000	10	1	4	5	0.4128	0.5	3.8206	7.5119
3000	10	0.1	5	6	0.05	0.1573	3.6595	7.2924
3000	10	0.01	5	5	0.005	0.0174	3.6433	7.3122
3000	10	0.001	5	6	0.0005	0.0127	3.6286	7.3667
3000	10	10	3	5	3.1369	5	3.0189	6.6345
3000	10	1	4	5	0.4128	0.5	3.1579	7.244
3000	10	0.1	5	5	0.05	0.05	3.0992	7.0419
3000	10	0.01	5	5	0.005	0.005	2.8799	7.3774
3000	10	0.001	5	5	0.0005	0.0005	2.8831	7.1387

Table 2. Results on the real-world Diabetes dataset for 100 QUBO runs.

Number of points (N)	Number of features (d)	λ	$\|w\|_0$ classical	$\|w\|_0$ QUBO	Objective value classical	Objective value QUBO	Preprocessing time (sec)	Processing time (sec)
442	10	10000	6	6	11561403.16	11615190.26	0.52024	7.71478
442	10	1000	8	8	11502623.87	11519510.11	0.52523	8.0403
442	10	100	9	7	11494877.38	11554655.73	0.51418	7.75863
442	10	10	10	7	11493995.03	11511792.48	0.52244	7.51821
442	10	1	10	8	11493905.03	11511518.11	0.51872	7.47994

errors. For values of $d \times \lambda \leq 1$, the gap was less than 10^{-6}. For $d \times \lambda = 10$ case, the gap was of the order of 10^{-3}.

The preprocessing time is the time taken to create the QUBO formulation using the D-wave libraries. The processing time includes the network time to upload the problem instance to D-wave, network time to download the solution and the execution time on the D-Wave quantum computer. We used the default annealing time ($20\,\mu s$) provided in the D-Wave Sampler API (SAPI) servers [3] and with this, the execution time for each run was around 11 milli seconds. The reported processing time is almost entirely the data upload and download overhead. Processing time is averaged over the number of QUBO runs.

Table 2 shows the QUBO performance for the real-world Diabetes dataset [7]. The number of QUBO runs here is 100. For this dataset, smaller λ values did not induce any sparsity even for the classical optimal solution, due to high values of regression error term in the objective function. Hence, we chose larger λ values to induce sparsity. Here again, we observe that the objective function values are similar for the QUBO solution and the classical solution, and increasing the number of QUBO iterations can possibly narrow the gap further.

5   Future Directions

We believe that this work is an important step towards applying Adiabatic Quantum Computing for efficiently solving practical and large scale sparse linear regression problems. This paper leaves open several future directions to explore. Our experiments were limited by a restricted access to the D-wave infrastructure. Experiments involving larger problem instances needs to be done to measure the quantum advantage. Moreover, experimentation with different annealing schedules and durations are likely to yield faster convergence to the optimal solution. Different ways of reducing the effect of inverse approximation can be explored. Examples are higher order approximation and approximations using multiple feasible values of α. A practical direction would be to apply the QUBO based regression model on real world data and compare with other models to measure generalization error.

References

1. Bishop, C.M.: Pattern Recognition and Machine Learning. Springer, New York (2006)
2. Borle, A., Lomonaco, S.J.: Analyzing the quantum annealing approach for solving linear least squares problems. In: Das, G.K., Mandal, P.S., Mukhopadhyaya, K., Nakano, S. (eds.) WALCOM 2019. LNCS, vol. 11355, pp. 289–301. Springer, Cham (2019). https://doi.org/10.1007/978-3-030-10564-8_23
3. D-Wave: D-wave's ocean software. https://docs.ocean.dwavesys.com/
4. Date, P., Potok, T.: Adiabatic quantum linear regression. arXiv:2008.02355 (2020)
5. Date, P., Schuman, C., Patton, R., Potok, T.: A classical-quantum hybrid approach for unsupervised probabilistic machine learning. In: Arai, K., Bhatia, R. (eds.) FICC 2019, Volume 2. LNNS, vol. 70, pp. 98–117. Springer, Cham (2020). https://doi.org/10.1007/978-3-030-12385-7_9
6. Date, P., Arthur, D., Pusey-Nazzaro, L.: Qubo formulations for training machine learning models. arXiv:2008.02369 (2020)
7. Diabetes dataset: https://www4.stat.ncsu.edu/boos/var.select/diabetes.html
8. El-Mahalawy, A.M., El-Safty, K.H.: Classical and quantum regression analysis for the optoelectronic performance of NTCDA/p-Si UV photodiode. arXiv:2004.01257 (2020)
9. Farhi, E., Goldstone, J., Gutmann, S., Sipser, M.: Quantum computation by adiabatic evolution (2000)
10. Foster, D.P., George, E.I.: The risk inflation criterion for multiple regression. The Annals of Statistics, pp. 1947–1975 (1994)
11. Friedman, J.H.: Fast sparse regression and classification. Int. J. Forecast. **28**(3), 722–738 (2012)
12. Hastie, T., Tibshirani, R., Friedman, J.: The Elements of Statistical Learning. SSS, Springer, New York (2009). https://doi.org/10.1007/978-0-387-84858-7
13. Hocking, R.R., Leslie, R.N.: Selection of the best subset in regression analysis. Technometrics **9**(4), 531–540 (1967)
14. Ishikawa, H.: Transformation of general binary MRF minimization to the first-order case. IEEE Trans. Pattern Anal. Mach. Intell. **33**(6), 1234–1249 (2011)
15. Kadowaki, T., Nishimori, H.: Quantum annealing in the transverse Ising model. Phys. Rev. E **58**, 5355–5363 (1998)
16. Konno, H., Yamamoto, R.: Choosing the best set of variables in regression analysis using integer programming. J. Global Optim. **44**, 273–282 (2009)
17. Leatherbarrow, R.: Using linear and non-linear regression to fit biochemical data. Trends Biochem. Sci. **15**, 455–458 (1990)
18. Mazumder, R., Friedman, J.H., Hastie, T.: SparseNet: coordinate descent with non-convex penalties. J. Am. Stat. Assoc. **106**(495), 1125–1138 (2011)
19. Miyashiro, R., Takano, Y.: Subset selection by Mallows' Cp: a mixed integer programming approach. Expert Syst. Appl. **42**, 325–331 (2015)
20. Montgomery, D., Peck, E.A., Vining, G.G.: Introduction to Linear Regression Analysis. Wiley, Hoboken (2001)
21. Natarajan, B.K.: Sparse approximate solutions to linear systems. SIAM J. Comput. **24**, 227–234 (1995)
22. Riesz, F., Sz. Nagy, B.: Functional Analysis. Frederick Ungar Publishing Company, New York (1955)
23. Tibshirani, R.: Regression shrinkage and selection via the lasso. J. Roy. Stat. Soc. Ser. B (Methodol.) **58**, 267–288 (1996)

24. Venegas-Andraca, S.E., Cruz-Santos, W., McGeoch, C., Lanzagorta, M.: A cross-disciplinary introduction to quantum annealing-based algorithms. Contemp. Phys. **59**, 174–197 (2018)
25. Wang, G.: Quantum algorithm for linear regression. Phys. Rev. A **96**(1), 012335 (2017)
26. Wu, B., Tseng, N.: A new approach to fuzzy regression models with application to business cycle analysis. Fuzzy Sets Syst. **130**, 33–42 (2002)
27. Yatchew, A.: Nonparametric regression techniques in economics. J. Econ. Lit. **36**, 669–721 (1998)
28. Zhang, T.: Adaptive forward-backward greedy algorithm for sparse learning with linear models. Adv. Neural Inf. Process. Syst. **21**, 1921–1928 (2009)

EntDetector: Entanglement Detecting Toolbox for Bipartite Quantum States

Roman Gielerak[1] (ID), Marek Sawerwain[1](✉) (ID), Joanna Wiśniewska[2] (ID),
and Marek Wróblewski[1] (ID)

[1] Institute of Control and Computation Engineering, University of Zielona Góra,
Licealna 9, 65-417 Zielona Góra, Poland
{R.Gielerak,M.Sawerwain,M.Wroblewski}@issi.uz.zgora.pl
[2] Institute of Information Systems, Faculty of Cybernetics, Military University of
Technology, Gen. S. Kaliskiego 2, 00-908 Warsaw, Poland
JWisniewska@wat.edu.pl

Abstract. Quantum entanglement is an extremely important phenomenon in the field of quantum computing. It is the basis of many communication protocols, cryptography and other quantum algorithms. On the other hand, however, it is still an unresolved problem, especially in the area of entanglement detection methods. In this article, we present a computational toolbox which offers a set of currently known methods for detecting entanglement, as well as proposals for new tools operating on two-partite quantum systems. We propose to use the concept of combined Schmidt and spectral decomposition as well as the concept of Gramian operators to examine a structure of analysed quantum states. The presented here computational toolbox was implemented by the use of Python language. Due to popularity of Python language, and its ease of use, a proposed set of methods can be directly utilised with other packages devoted to quantum computing simulations. Our toolbox can also be easily extended.

Keywords: Quantum entanglement · Quantum software · Numerical computations

1 Introduction

Quantum entanglement [2,15,20] is the physical phenomenon, already described in [5] by Einstein, Podolsky, Rosen. Currently, quantum entanglement for quantum states [18] is the basis of many quantum information processing protocols such as teleportation [4], communication [10] and cryptographic protocols as well [3].

On the other hand, the quantum entanglement phenomenon is still not a well-understood problem. One of the main issues is the criterion for detecting entanglement [7] in quantum pure and mixed states. In the case of bipartite pure states, the Schmidt criterion can be successfully applied – it gives an unambiguous answer whether we are dealing with an entangled state. However, in the

© Springer Nature Switzerland AG 2021
M. Paszynski et al. (Eds.): ICCS 2021, LNCS 12747, pp. 113–126, 2021.
https://doi.org/10.1007/978-3-030-77980-1_9

case of quantum states described by a density matrix, so-called mixed states, the problem of detecting entanglement is still not solved in computationally effective way. The general criterion of entanglement detection for density matrix cases is related to the entanglement witness theory [14], and so far, there is no simple and fast (especially computationally) entanglement verification criterion for such type of states. In fact due to the proven NP-hardness [6,12] of the problem "entangled or separable?", it is hardly to expect that such efficient algorithms do exist on the classical side of computational technologies.

In this paper, we present a selected set of computational methods devoted to numerical studies of bipartite entanglement in quantum states. The aim is to develop a publicly available set of functions in Python, which will allow utilising the proposed package of computational methods also in combination with other packages related to quantum computing, such as QuTiP [16] and Qiskit [1].

In addition to the implementation of the basic generally known methods, our EntDetector package also offers an access to methods of entanglement testing using so-called Gramian matrices [8], as an additional tool for checking the entanglement level between individual qubits in a given pure bipartite state. The Gramian calculation techniques also allow us to provide a new method of easy calculation of the density matrix in the case of a bipartite system.

We use Python language in our package, because it is currently very popular in the field of quantum computing simulation. It seems that available Python software do not offers enough tools in area of entanglement detection although, of course, it is necessary to indicate the existence of the packages like Qubit4Matlab [23] or QETLAB [17], which offer a support in this field. However, they require a Matlab software [19].

The article is organized as follows: Sect. 2 introduces the basic concepts and briefly describes the basic math engine that is used in the EntDetector package. The technical aspects of the package, including examples of usage, are presented in Sect. 3. A summary is provided in Sect. 4. This paper ends with acknowledgements and bibliography sections.

2   Mathematical Framework

2.1   Spectral and Schmidt Decomposition

Let $\mathcal{H} = \mathbb{C}^{d^2} = \mathbb{C}^d \otimes \mathbb{C}^d$ be a bipartite (A and B) finite dimensional Hilbert space and let $Q \in E(\mathcal{H})$ be a given quantum state. It is well known that the spectrum of Q (counting multiplicity) $\sigma(Q) = (\lambda_1, \lambda_2, \ldots, \lambda_{d^2})$, is purely discrete (it is important to point out that the list forming spectrum is ordered in non-increasing way) and the following spectral decomposition is valid:

$$Q = \sum_{i=1}^{d^2} \lambda_i |\Psi_i\rangle\langle\Psi_i| \tag{1}$$

where the orthogonal (and normalised) system of eigenfunctions $|\Psi_i\rangle$ of Q forms a complete orthonormal system of $\dim \mathcal{H} = d^2$.

It is important to point out in this moment that the material presented in this report is valid in the current form only under the assumption that the spectrum of Q is simple which means that all of the corresponding eigenvalues are non-degenerated. The general case is technically more involved and will be presented (due to the limited space here) in an another paper [9].

Each eigenfunction $|\Psi_i\rangle$ can be expanded by the use of the Schmidt decomposition [11]:

$$|\Psi_i\rangle = \sum_{j=1}^{d} \tau_j^i |\psi_j^i\rangle \otimes |\vartheta_j^i\rangle, \tag{2}$$

where $\tau_j^i \geq 0$, $\sum_{j=1}^{d}(\tau_j^i)^2 = 1$ and systems $\{\psi_j^i\}$ form a complete orthonormal systems in part A, and resp. $\{\vartheta_j^i\}$ in B.

Let us define the following matrix SaSD(Q) (named Schmidt and Spectral Data) connected to the analysed state Q: it is $(d^2, (d+1))$ matrix in which we collect all the appearing Schmidt coefficients τ_i^n in the (d^2, d) block building from the first d^2 rows and d columns of SaSD(Q), and in the last column we localize the eigenvalues of Q. Graphically the map SaSD defined on $E(\mathcal{H})$ looks like:

$$\text{SaSD}(Q) = \begin{pmatrix} \tau_1^1 & \cdots & \tau_d^1 & \lambda_1 \\ \vdots & \ddots & \vdots & \vdots \\ \tau_1^{d^2} & \cdots & \tau_d^{d^2} & \lambda_{d^2} \end{pmatrix}, \tag{3}$$

or in coordinates:

$$\text{SaSD}(Q)_{j,i} = \tau_j^i \quad \text{for } j = 1 : d^2, i = 1 : d, \text{ and}$$
$$\text{SaSD}(Q)_{j,d+1} = \lambda_j \quad \text{for } j = 1 : d^2.$$

The map SaSD is uniquely defined (all eigenvalues are different) for each Q. However:

Proposition 1. *Let U_1, U_2 be a pair of unitary maps in $\mathbb{C}^d$ i.e. U are $SU(d)$ group elements, then we define its action on Q as*

$$Q \to (U_1^\dagger \otimes U_2^\dagger)(Q)(U_1 \otimes U_2), \tag{4}$$

then

$$\text{SaSD}\left((U_1^\dagger \otimes U_2^\dagger)(Q)(U_1 \otimes U_2)\right) = \text{SaSD}(Q). \tag{5}$$

This means that on each $2(d^2 - 1)$ – dimensional orbit of the local action of the group $SU(d)$ in the space $E(\mathcal{H})$ of dimension (real) $d^4 - 1$, the $(d^2(d+1) - (d^2+1))$-dimensional table SaSD do not change its value. For the pair of qudits of an arbitrary dimensions from Eq. (5) it follows that the matrix SaSD is locally unitary invariant. However, still the complete knowledge of SaSD(Q) is not sufficient for the complete (modulo $SU(d) \otimes SU(d)$) determination of Q as there is still deficit in dimensions at least as large as $\delta = d^4 - d^3 - 2d^2 + 2$, which for the case of two qubits gives $\delta = 2$.

Example 1. Let us assume that we have states Q_p, Q_s with the corresponding SaSD tables:

$$
\text{SaSD}(Q_p) = \begin{pmatrix} \tau_1^1 \cdots \tau_n^1 & 1 \\ 1 \cdots 0 & 0 \\ \vdots \ddots \vdots & \vdots \\ 1 \cdots 0 & 0 \end{pmatrix}, \quad \text{SaSD}(Q_s) = \begin{pmatrix} 1 \cdots 0 & \lambda_1 \\ \vdots \ddots \vdots & \vdots \\ 1 \cdots 0 & \lambda_{d^2} \end{pmatrix}. \tag{6}
$$

Then Q_p is a pure state and Q_s is a separable state.

Remark 1. It should be noted, once again, that in this part of our paper we only discussed case when the spectrum of Q is simple. In general case, the situation is more complicated and will be discussed in an another paper [9].

Any scalar function defined on the basis of $\text{SaSD}(Q)$ will be locally $SU(d) \otimes SU(d)$ invariant scalar. In particular, the function that is named von Neumann entropy (vNEN) of the SaSD, which is defined below:

$$
\text{vNEN}(Q) = \sum_{i=1}^{d^2} \sum_{j=1}^{d} \left(-(\tau_j^i)^2 \log\left((\tau_j^i)^2\right) \right) - \sum_{i=1}^{d^2} \lambda_i \log(\lambda_i), \tag{7}
$$

is monotone non-increasing under the action of any local quantum operations [9] and invariant under the action of local unitary operations, where the standard convention $0 \cdot \log 0 = 0$ is being used.

Proposition 2. *The function* vNEN, *defined on* $E(\mathcal{H})$, *is continuous in* $\| \cdot \|_1$ *topology and is* $SU(d) \otimes SU(d)$ *invariant.*

Remark 2. An extensive study of the introduced von Neumann entropy and other locally unitary invariant functions build on SaSD, and, what is even more important, the study of functions which are monotonic (in the sense of majorization theory) and also in the sense of (S)LOCC semi-order relation will be presented elsewhere [9].

Remark 3. Some computational examples of the use of SaSD are presented at point Sect. 3.3 of this paper.

From the very definition of vNEN it follows that

$$
\sup_{Q \in E(\mathcal{H})} \text{vNEN}(Q) = (d^2 + 2) \log d, \tag{8}
$$

It easy to observe that the SaSD table on which the introduced global entropy vNEN attains its maximal value looks like :

$$
\text{SaSD}(Q) = \begin{pmatrix} \frac{1}{\sqrt{d}} \cdots \frac{1}{\sqrt{d}} & \frac{1}{d^2} \\ \vdots \ddots \vdots & \vdots \\ \frac{1}{\sqrt{d}} \cdots \frac{1}{\sqrt{d}} & \frac{1}{d^2} \end{pmatrix}, \tag{9}
$$

and corresponds to the $(d^2 - 1)^2$ – dimensional manifold of maximally mixed states in $E(\mathcal{H})$.

The extended SaSD table, denoted as exSaSD, is obtained from SaSD by adding to SaSD, defined in Eq. 3, two additional columns obtained from Schmidt decompositions (Eq. 2):

$$
\text{exSaSD}(Q) = \left(
\begin{array}{ccc|c|ccc|ccc}
\tau_1^1 & \cdots & \tau_d^1 & \lambda_1 & \psi_1^1 & \cdots & \psi_d^1 & \vartheta_1^1 & \cdots & \vartheta_d^1 \\
\vdots & \ddots & \vdots & \vdots & \vdots & \ddots & \vdots & \vdots & \ddots & \vdots \\
\tau_1^{d^2} & \cdots & \tau_d^{d^2} & \lambda_{d^2} & \psi_1^{d^2} & \cdots & \psi_d^{d^2} & \vartheta_1^{d^2} & \cdots & \vartheta_d^{d^2}
\end{array}
\right). \tag{10}
$$

Locally available information on state Q, as is well known, is contained in the corresponding density matrices. Owing to the obtained below decomposition, the computation of the corresponding reduced density matrices is very easy now. Taking Eq. 2 into account, and after few lines of calculations, we obtain:

$$
Q^B = \text{Tr}_A (Q) = \text{Tr}_A \left(\sum_{i=1}^{d^2} \lambda_i |\Psi_i\rangle\langle\Psi_i| \right) = \sum_{i=1}^{d^2} \lambda_i Q_i^B, \tag{11}
$$

where the operators $Q_i^B = \sum_{j=1}^{d} |\tau_j^i|^2 |\vartheta_j^i\rangle\langle\vartheta_j^i|$ are states on the subsystem B.

Similarly, for the reduced density matrix connected to the observer relating to the subsystem A:

$$
Q^A = \text{Tr}_B (Q) = \text{Tr}_B \left(\sum_{n=1}^{d^2} \lambda_n |\Psi_n\rangle\langle\Psi_n| \right) = \sum_{n=1}^{d^2} \lambda_n Q_n^A, \tag{12}
$$

where $Q_n^A = \sum_{i=1}^{d} |\tau_i^n|^2 |\psi_i^n\rangle\langle\psi_i^n|$ are states, this time, on the subsystem A.

The obtained systems of operators $\{Q_n^A\}$ and $\{Q_n^B\}$ are consisting of non-negative, and therefore Hermitian, operators and are locally measurable, each of them. In particular, the squares of the Schmidt coefficients τ_i^n in the Schmidt decompositions of the parent state Q eigenfunctions are observable (measurable) quantities.

Proposition 3. *Let $\mathcal{H} = \mathbb{C}^{d^2} = \mathbb{C}^d \otimes \mathbb{C}^d$ be a bipartite Hilbert space and let $Q \in E(\mathcal{H})$. Let (Q^A, Q^B) be the pair of corresponding reduced density matrices and let*

$$
Q_n^A = \sum_{i=1}^{d} |\tau_i^n|^2 |\psi_i^n\rangle\langle\psi_i^n| \quad \text{and corr.} \quad Q_n^B = \sum_{i=1}^{d} |\tau_i^n|^2 |\vartheta_i^n\rangle\langle\vartheta_i^n|, \tag{13}
$$

for $n = 1 : d^2$ be the corresponding operators as defined by the use of exSaSD(Q). Then $Q^A = \sum_{n=1}^{d^2} \lambda_n Q_n^A$ and $Q^B = \sum_{n=1}^{d^2} \lambda_n Q_n^B$.

As the local information is invariant under local unitary action, we can define, in according to the ideas presented in [8], the following interesting invariant and monotonic functions.

Definition 1. *Gramian functions of Q:*

$$G(Q_n^A) = \det\left(1_{\mathbb{C}^d} + Q_n^A\right) = \prod_{j=1}^{d}\left(1 + (\tau_j^n)^2\right), \tag{14}$$

for $n = 1:d^2$ and $1_{\mathbb{C}^d}$ represents an identity operator in space of the subsystem A. Identical definitions are also valid in part B of the analysed system.

Proposition 4. *For each n the value $G(Q_n^A)$ is invariant under the action of local unitary group, for any unitary map U acting on $\mathbb{C}^d$:*

$$G(UQ_n^A U^\dagger) = G(Q_n^A) \quad \text{and resp.} \quad G(UQ_n^B U^\dagger) = G(Q_n^B). \tag{15}$$

Proof. Obvious. □

Lemma 1. *For each $n = 1:d^2$,*

$$\sup_{Q\in E(\mathbb{C}^{d^2})} G(Q_n^A) = \sup_{Q\in E(\mathbb{C}^{d^2})} G(Q_n^B) = \prod_{j=1}^{d}\left(1 + \frac{1}{d}\right), \tag{16}$$

is true.

Let us define also logarithmic volumes of Q_n^A (resp. Q_n^B) as:

$$g^A(n) = \log G(Q_n^A) = \log G(Q_n^B) = g^B(n), \tag{17}$$

and also the complete Gram volumes of Q as

$$\mathfrak{g}^A(Q) = \prod_{k=1}^{d^2} G(Q_k^A) = \mathfrak{g}^B(Q) = \prod_{k=1}^{d^2} G(Q_k^B). \tag{18}$$

Together with its logarithmic counterparts :

$$\mathfrak{lg}^A(Q) = \log \prod_{k=1}^{d^2} G(Q_k^A) = \mathfrak{lg}^B(Q) = \log \prod_{k=1}^{d^2} G(Q_k^B). \tag{19}$$

Proposition 5. *The complete Gram volume $\mathfrak{g}^A(Q)$ of a given $Q \in E(\mathbb{C}^{d^2})$ and its logarithmic counterpart are locally $SU(d) \otimes SU(d)$ invariants of Q. The same is true for part B of the analysed system. Additionally:*

$$\sup_{Q\in E(\mathbb{C}^{d^2})} \mathfrak{g}^A(Q) = \sup_{Q\in E(\mathbb{C}^{d^2})} \mathfrak{g}^B(Q) = \left(\prod_{j=1}^{d}\left(1 + \frac{1}{d}\right)\right)^{d^2}, \tag{20}$$

and

$$\sup_{Q\in E(\mathbb{C}^{d^2})} \mathfrak{lg}^A(Q) = \sup_{Q\in E(\mathbb{C}^{d^2})} \mathfrak{lg}^B(Q) = d^2\left(\sum_{j=1}^{d}\log\left(1 + \frac{1}{d}\right)\right). \tag{21}$$

Another approach to certain aspects of reduced density matrices structure is based on the use of the Schmidt decomposition method in the Hilbert-Schmidt space of operators build on the space $\mathbb{C}^d \otimes \mathbb{C}^d$. This will be discussed in a separate paper [9].

2.2 Some Remarks on the Majorization Theory

For a given finite sequence $\underline{a} = (a_1, \ldots, a_n)$ where $a_i \in \mathbb{R}$, we apply the operation of ordering in non-increasing order and denote the result as $\underline{a}^{\geq}$. Of particular interest, is the image of this operation when applied point-wise to the finite dimensional simplex $C_+^n(1) := \{\underline{a} = (a_1, \ldots, a_n), a_i \in \mathbb{R}, a_i \geq 0, \sum_{i=1}^n a_i = 1\}$. This will be denoted as $C^{\geq}$.

Let us recall the standard definition of majorizations. Let $a, b \in C^{\geq}$. Then we say that $\underline{b}$ majorizes $\underline{a}$ iff:

$$\sum_{i=1}^k a_i \leq \sum_{i=1}^k b_i, \tag{22}$$

for any $k = 1 : n$.

If this holds to be true then we denote this fact as $\underline{a} \preceq \underline{b}$.

We say that $\underline{b}$ majorizes multiplicatively $\underline{a}$ iff for any k, and $k = 1 : n$:

$$\prod_{i=1}^k (a_i + 1) \leq \prod_{i=1}^k (b_i + 1). \tag{23}$$

If this holds to be true then we denote this fact as $\underline{a} \preceq_m \underline{b}$.

Let F be any function (continuous, but not necessarily) on the interval $[0, 1]$. Let us recall the well known result, see i.e. [2].

Lemma 2. *Let as assume that f is a continuous, increasing and convex function on $\mathbb{R}$. If $\underline{a} \preceq \underline{b}$ then $f(\underline{a}) \preceq f(\underline{b})$.*

It is clear from the very definition that $\underline{a} \preceq_m \underline{b}$ iff $\log(\underline{a} + 1) \preceq \log(1 + \underline{b})$.

Proposition 6. *Let $\underline{a}, \underline{b} \in C^{\geq}$ and let us assume $\underline{a} \preceq_m \underline{b}$. Let f be continuous, increasing function such that the composition $f \circ \exp(x)$ is convex on a suitable domain. Then $f(\underline{a}) \preceq f(\underline{b})$.*

Proof. Fixing $k, k = 1 : n$, the following holds to be true:

$$\prod_{i=1}^k (a_i + 1) \leq \prod_{i=1}^k (b_i + 1). \tag{24}$$

Taking log of both side we obtain

$$\sum_{i=1}^k \log(1 + a_i) \leq \sum_{i=1}^k \log(1 + b_i). \tag{25}$$

Applying Lemma 2, we get

$$\sum_{i=1}^k f(e \cdot a_i) \leq \sum_{i=1}^k f(e \cdot b_i), \tag{26}$$

and the symbol e represents the Euler constant. $\square$

In particular, taking constant function $f(x) = x$, we conclude:

Corollary 1. *Let $\underline{a}, \underline{b} \in C^{\geq}$ and let us assume that $\underline{a} \preceq_m \underline{b}$, then $\underline{a} \preceq \underline{b}$.*

The last result says that each linear chain of the semi-order relation $\preceq_m$ in $E(H)$ is contained in some linear chain of the semi-order. It means that the semi-order $\preceq_m$ is finer than those induced by $\prec$.

Theorem 1. *Let $\mathcal{H}$ be a separable Hilbert space and let Q_1 and Q_2 be states on $\mathcal{H}$. Then any $\preceq$-maximal element in $E(\mathcal{H})$ is also $\preceq_m$-maximal.*

Proof. If $\sigma(Q_1) \preceq_m \sigma(Q_2)$ then $\sigma(Q_1) \prec \sigma(Q_2)$. Let Q^* be a $\prec$-maximal in $E(H)$, and let as assume that there exists $Q\#$ such that $Q^* \preceq_m Q\#$ and the contradiction is present.   □

3   Implementation and Example in Python Environment

In this part of the article, we present a basic information concerning the EntDetector package. The most important assumptions, about the implementation and realised functionalities, are described in Sect. 3.1. The examples of our package's use are shown in Sect. 3.2.

In Sect. 3.3, we join the spectral and Schmidt decomposition for pure and mixed states. Our package carries out the decomposition which, as it was shown in Sect. 2.1, allows characterising a bipartite quantum state by the level of entanglement e.g. for isotropic states, as shown in Sect. 3.4.

3.1   Implementation Assumptions

One of the fundamental assumptions about the EntDetector implementation was the choice of Python programming language because of its accessibility and simplicity. Another crucial issue is the prospect of utilizing other known libraries for quantum computation, like qiskit [1] and QuTiP [16], which are available in Python. We also assume that the realisation of numerical calculations are performed by the packages NumPy [13] and SciKit [24].

The mentioned simplicity of the EntDetector's usage may be shown on the example of an access to basic functions generating quantum states. We have prepared representations of pure states (e.g. Bell, GHZ, W states) and density matrices (e.g. Horodecki (2×4) and (3×3) bound entangled states, isotropic states) which are invoked in the following way:

```
- q = create_qubit_bell_state(), q = create_ghz_state(d, n),
- q = create_wstate(n), qden = create_bes_horodecki_24_state(),
- qden = create_bes_horodecki_33_state(),
- qden = create_isotropic_state(p, d, n),
```

Fig. 1. Execution of an example calculating left, right, and full Gram matrices for the product system generated by the states $|0\rangle$ and $|+\rangle$. The example was run in Spyder 4.x editor, system distribution: Linux Ubuntu 20.04.01 LTS, environment Anaconda 4.x, Python version 3.8 64bit

where d stands for the dimension of a quantum unit, n represents the number of qudits, and p is a real value between 0 and 1. The variable q contains a vector state, and qden encloses a density matrix. The functions names are quite complex, but they describe the purpose of usage clearly (Fig. 1).

The package offers basic functions for estimating entanglement's level, e.g. computing entropy, values of Concurrence and Negativity measures, calculating monotones for the systems of two (there also exist functions for three, and four) qubits:

- v = entropy(qden), v = concurrency(qden), v = negativity(qden),
- v = monotone_for_two_qubit_system(qden),

where qden represents a density matrix, and the result is assigned to the variable v. Types of arguments for presented functions are objects defined in the NumPy library, so it is easy to perform calculation joining functions from different packages (written in Python) dedicated to quantum computing, e.g. QuTIP.

The EntDetector includes a new tools like Gram matrices generation and a joined spectral-Schmidt decomposition:

- tblsas = create_spectral_and_schmidt_data(qden, schmidt_shape),
- right, left, full = gram_matrices_of_vector_state(v, d1, d2).

Naturally, we are not able to signal all implemented functions here. A current version of the package, directly related to this paper, is available at [22]. However, a future versions of the package and documentation are going to be published at main repository of our project in [21].

Remark 4. It should be emphasised once more that the main goal of the EntDetector package is to provide a user-friendly set of functions which allow analysing entanglement level with the help of known methods, e.g. the Negativity measure or Schmidt decomposition. It is also important to introduce some new methods, like density matrices expressed as Gramians what gives us a prospect of entanglement examination with the use of Gram volume and SaSD method, proposed in this work. The EntDetector should supplement other existing packages, like qiskit and QuTIP, with additional functions dedicated to entanglement analysis (e.g. SaSD tables, Gramian matrices).

3.2   Simple Examples

The first example that we would like to present, is the Schmidt decomposition of an entangled state – more precisely: one of so-called EPR pairs: $|\psi\rangle = \frac{1}{\sqrt{2}}(|00\rangle + |11\rangle)$. To do it, we need to generate the mentioned state, perform the decomposition, and check if two Schmidt coefficients could be obtained. The code realising this task has the following form:

```
from entdetector import *

q = create_qubit_bell_state()
schmidt_shp=(2, 2)
s,e,f = schmidt_decomposition_for_vector_pure_state(q, schmidt_shp)
```

The function executing Schmidt decomposition requires introducing dimensions of the decomposition. The size of EPR vector is 4, so the decomposition's dimensions are 2×2 (given as a variable: schmidt_shp=(2, 2)). A reconstruction of quantum states is realised by a function:

```
qrebuild = reconstruct_state_after_schmidt_decomposition(s, e, f)
```

The Schmidt coefficients values are assigned to the variable s. The basis vectors are stored in e and f. Naturally, in the example with the EPR pair, we obtain two Schmidt coefficients equal to numeric value 0.70710678.

3.3   Spectral and Schmidt Decomposition

We described the properties of spectral decomposition and the Schmidt decomposition in Sect. 2.1. Let us generate the pure state $q = q0 \otimes qplus$:

```
q0 = create_qubit_zero_state()
qplus = create_qubit_plus_state()
q = np.kron(q0, qplus)
```

where q0 $= |0\rangle$ and qplus $= |+\rangle$. Now, we can calculate both decompositions directly:

```
schmidt_shape=(2, 2)
qden = vector_state_to_density_matrix( q )
sas_tbl = create_sas_table_data(qden, schmidt_shape)
```

As the result, we obtain the following SaSD given in Eq. (9):

```
>>> print_sas_table( sas_tbl )
1.0   0.0 | 0.9999
1.0   0.0 | 0.0
1.0   0.0 | 0.0
1.0   0.0 | 0.0
```

A SaSD table for the Bell state $q = \frac{1}{\sqrt{2}}(|00\rangle + |11\rangle)$ is generated by:

```
q = create_qubit_bell_state()
qden = vector_state_to_density_matrix( q )
sas_tbl = create_sas_table_data(qden, schmidt_shape)
```

has the form:

```
>>> print_sas_table( sas_tbl )
0.70710678 0.70710678 | 0.9999
1.0        0.0         | 0.0
1.0        0.0         | 0.0
0.70710678 0.70710678 | 0.0
```

3.4   Decomposition of Two-Qubit Isotropic State

SaSD tables allow analysing the isotropic state given as:

$$\rho = (p \cdot |\psi_+\rangle\langle\psi_+|) + (1-p)\frac{1}{d^2} \cdot (\mathbb{I}_d \otimes \mathbb{I}_d), \tag{27}$$

where in our case we assume that p is real and $-\frac{1}{d^2-1} \leq p \leq 1$. The parameter d is the dimension of a quantum unit utilized to construct the maximally entangled state $|\psi_+\rangle$ (and its density matrix $|\psi_+\rangle\langle\psi_+|$). The state ρ in general is entangled when $\frac{1}{d+1} < p \leq 1$, and in a contrary, separable if $-\frac{1}{d^2-1} \leq p \leq \frac{1}{d+1}$.

The SaSD table for the isotropic state with $d = 2$ and $p = 0$ is calculated by:

```
p=0.0
q = create_qubit_bell_state()
qdentmp = np.outer(q, q)
qden = (p * qdentmp) + ((1-p) * 0.25 * np.eye(4))
schmidt_shape=(2, 2)
sas_tbl = create_sas_table_data(qden, schmidt_shape)
```

and the result as SaSD table is:

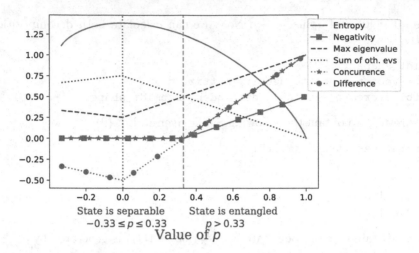

Fig. 2. The value of the maximal eigenvalue, the sum of others eigenvalues (denoted as sum of oth. evs), and also their difference for two-qubit isotropic state. The figure also shows the values of other quantum entanglement measures like Negativity and Concurrence. It is possible to point out that the Concurrence measure and the difference between the max eigenvalue and the sum of others eigenvalues are equal for $p \geq 0.33$

```
>>> print_sas_table( sas_tbl )
1.0 0.0 | 0.25
1.0 0.0 | 0.25
1.0 0.0 | 0.25
1.0 0.0 | 0.25
```

It should be noted that for parameter $p = 0$ the examined isotropic state takes a form of the maximally mixed state.

A SaSD table analysis enables entanglement detection in a given isotropic state for $d = 2$. Let us generate the exemplary tables for $p = 0.25$ and $p = 0.75$:

```
p=0.25                              p=0.75
0.7071 0.7071 | 0.5125              0.7071 0.7071 | 0.8125
0.9265 0.3760 | 0.1625              1.0    0.0    | 0.0625
0.9070 0.4209 | 0.1625              1.0    0.0    | 0.0625
0.8243 0.5661 | 0.1625              0.7071 0.7071 | 0.0625
```

The numeric analysis shows that we can point out the greatest eigenvalue correlated with the row having two non-zero coefficients what allows delineating the border between entangled and separable states. This phenomenon is depict on the plot in Fig. 2.

4   Conclusions

In the article, we presented a package supporting work with qubit and qudit systems in terms of detecting entanglement in bipartite systems. The most important assumption was to provide convenient tools in the form of computational functions that perform the necessary decompositions, as well as entanglement detection functions.

In addition to providing support for known tools in the field of entanglement research, such as the Negativity and Concurrence measures, our package also offers new tools useful in the field of entanglement, such as the left and right Gramian of a given quantum state. EntDetector also offers an additional tool called SaSD. The information obtained with SaSD allows us, for example, to calculate the entropy for the studied quantum state.

The package has also the ability to examine the presence of entanglement in the indicated parts of a quantum register. The current procedure enables, for example, detecting entanglement, e.g. in graph states, and to verify whether a given examined state contains correctly entangled qubits.

Acknowledgments. We would like to thank for useful discussions with the *Q-INFO* group at the Institute of Control and Computation Engineering (ISSI) of the University of Zielona Góra, Poland. We would like also to thank to anonymous referees for useful comments on the preliminary version of the paper. The numerical results were done using the hardware and software available at the "GPU μ-Lab" located at the Institute of Control and Computation Engineering of the University of Zielona Góra, Poland.

References

1. Abraham, H., et al.: QISKIT: an open-source framework for quantum computing (2019). https://doi.org/10.5281/zenodo.2562110
2. Bengtsson, I., Życzkowski, K.: Geometry of Quantum States: An Introduction to Quantum Entanglement, 2nd edn. Cambridge University Press, Cambridge (2017). https://doi.org/10.1017/9781139207010
3. Bennett, C.H., Bessette, F., Brassard, G., Salvail, L., Smolin, J.: Experimental quantum cryptography. J. Cryptol. **5**(1), 3–28 (1992). https://doi.org/10.1007/BF00191318
4. Brassard, G., Braunstein, S.L., Cleve, R.: Teleportation as a quantum computation. Physica D **120**(1), 43–47 (1998). https://doi.org/10.1016/S0167-2789(98)00043-8
5. Einstein, A., Podolsky, B., Rosen, N.: Can quantum-mechanical description of physical reality be considered complete? Phys. Rev. **47**, 777–780 (1935). https://doi.org/10.1103/PhysRev.47.777
6. Gharibian, S.: Strong NP-hardness of the quantum separability problem. Quant. Inf. Comput. **10**(3 & 4), 343–360 (2010)
7. Gühne, O., Tóth, G.: Entanglement detection. Phys. Rep. **474**(1), 1–75 (2009). https://doi.org/10.1016/j.physrep.2009.02.004
8. Gielerak, R., Sawerwain, M.: A Gramian approach to entanglement in bipartite finite dimensional systems: the case of pure states. Quant. Inf. Comput. **20**(13 and 14), 1081–1108 (2020). https://doi.org/10.26421/QIC20.13-1
9. Gielerak, R., Sawerwain, M.: In preparations (2021)
10. Gisin, N., Thew, R.: Quantum communication. Nat. Photonics **1**, 165–171 (2007). https://doi.org/10.1038/nphoton.2007.22
11. Golub, G.H., van Loan, C.F.: Matrix Computations, 4th edn. Johns Hopkins University Press, Baltimore (2013)

12. Gurvits, L.: Classical deterministic complexity of Edmonds' problem and quantum entanglement. In: Proceedings of the Thirty-Fifth Annual ACM Symposium on Theory of Computing, STOC 2003, pp. 10–19. Association for Computing Machinery, New York (2003). https://doi.org/10.1145/780542.780545

13. Harris, C.R., Millman, K.J., et al.: Array programming with NumPy. Nature **585**(7825), 357–362 (2020). https://doi.org/10.1038/s41586-020-2649-2

14. Horodecki, M., Horodecki, P., Horodecki, R.: Separability of mixed states: necessary and sufficient conditions. Phys. Lett. A **223**(1), 1–8 (1996). https://doi.org/10.1016/S0375-9601(96)00706-2

15. Horodecki, R., Horodecki, P., Horodecki, M., Horodecki, K.: Quantum entanglement. Rev. Mod. Phys. **81**, 865–942 (2009). https://doi.org/10.1103/RevModPhys.81.865

16. Johansson, J., Nation, P., Nori, F.: QuTiP: an open-source python framework for the dynamics of open quantum systems. Comput. Phys. Commun. **183**(8), 1760–1772 (2012). https://doi.org/10.1016/j.cpc.2012.02.021

17. Johnston, N.: QETLAB: A MATLAB toolbox for quantum entanglement, version 0.9 (2016). https://doi.org/10.5281/zenodo.44637

18. Kolaczek, D., Spisak, B.J., Woloszyn, M.: The phase-space approach to time evolution of quantum states in confined systems: the spectral split-operator method. Int. J. Appl. Math. Comput. Sci. **29**(3), 439–451 (2019). https://doi.org/10.2478/amcs-2019-0032

19. MATLAB: 9.7.0.1190202 (R2019b). The MathWorks Inc., Natick, Massachusetts (2018)

20. Nielsen, M.A., Chuang, I.L.: Quantum Computation and Quantum Information: 10th Anniversary Edition, 10th edn. Cambridge University Press, Cambridge (2011)

21. Sawerwain, M., Wiśniewska, J., Wróblewski, M., Gielerak, R.: GitHub repository for EntDectector package (2021). https://github.com/qMSUZ/EntDetector

22. Sawerwain, M., Wiśniewska, J., Wróblewski, M., Gielerak, R.: Source code of EntDetector: entanglement detecting toolbox for bipartite quantum states (2021). https://doi.org/10.5281/zenodo.4643878

23. Tóth, G.: QUBIT4MATLAB V3.0: a program package for quantum information science and quantum optics for MATLAB. Comput. Phys. Commun. **179**(6), 430–437 (2008). https://doi.org/10.1016/j.cpc.2008.03.007

24. Virtanen, P., Gommers, R., Oliphant, T., et al.: SciPy 1.0: fundamental algorithms for scientific computing in python. Nat. Methods **17**, 261–272 (2020). https://doi.org/10.1038/s41592-019-0686-2

On Decision Support for Quantum Application Developers: Categorization, Comparison, and Analysis of Existing Technologies

Daniel Vietz[✉][iD], Johanna Barzen[iD], Frank Leymann[iD], and Karoline Wild[iD]

Institute of Architecture of Application Systems, University of Stuttgart,
Universitätsstraße 38, 70569 Stuttgart, Germany
{Daniel.Vietz,Johanna.Barzen,Frank.Leymann,
Karoline.Wild}@iaas.uni-stuttgart.de

Abstract. Quantum computers have been significantly advanced in recent years. Offered as cloud services, quantum computers have become accessible to a broad range of users. Along with the physical advances, the landscape of technologies supporting quantum application development has also grown rapidly in recent years. However, there is a variety of tools, services, and techniques available for the development of quantum applications, and which ones are best suited for a particular use case depends, among other things, on the quantum algorithm and quantum hardware. Thus, their selection is a manual and cumbersome process. To tackle this challenge, we introduce a categorization and a taxonomy of available tools, services, and techniques for quantum application development to enable their analysis and comparison. Based on that we further present a comparison framework to support quantum application developers in their decision for certain technologies.

Keywords: Quantum software development · Quantum computing technologies · Quantum cloud services · Decision support

1 Introduction

Quantum computing promises to solve many problems more efficiently or precisely than it is possible with classical computers. In recent years, the number of quantum hardware vendors, such as IBM, Rigetti, or D-Wave has steadily increased. Via the cloud, quantum computing capacities are made publicly available. Not only hardware vendors offer various quantum cloud services, e.g., IBM via IBM Quantum Experience (IBMQ) [25], but also established cloud providers such as Amazon Web Services (AWS) have added quantum cloud services that facilitate executing quantum algorithms on quantum hardware to their portfolio.

Typically, a quantum application comprises not only the implementation of a quantum algorithm, but also pre- and post-processing components [32]. The

© Springer Nature Switzerland AG 2021
M. Paszynski et al. (Eds.): ICCS 2021, LNCS 12747, pp. 127–141, 2021.
https://doi.org/10.1007/978-3-030-77980-1_10

development of quantum applications, however, differs significantly from classical application development [52] and currently depends heavily on the used hardware. Due to the growing number of hardware vendors, service providers, and constantly improving quantum hardware, the software landscape for the development and execution of quantum applications is also growing steadily: Almost each quantum cloud provider offers a Software Development Kit (SDK) in order to compile and run quantum applications on their corresponding hardware, such as Qiskit [1] for IBMQ [25] and Ocean [13] for D-Wave Leap [14]. But there are also vendor-agnostic SDKs, such as XACC [33] and ProjectQ [51] that are able to connect to quantum cloud services of different providers. In addition, there are libraries, such as Pennylane [8], that not only enable the connection to a variety of providers but also offer specific algorithms for certain problem classes, e.g., machine learning. For the implementation of a quantum algorithm, different programming languages can also be used, for which specific compilers and transpilers are required. Finally, for integrating pre- and post-processing components with quantum algorithms, orchestration tools, such as Orquestra [57], or extensions for existing workflow languages, such as QuantMe [53], are proposed.

Thus, a variety of tools, services, and techniques is available that can be used for developing quantum applications. However, which of these fit best for a certain use case and how they can be combined depends on (i) the quantum cloud service provider, (ii) the quantum hardware used for the execution, and (iii) the implemented quantum algorithm itself. Furthermore, developer preferences and capabilities, such as the programming language and available tutorials, also play an important role in the decision which tools to use. Due to the variety of possibilities and a missing overview and characterization, it is difficult to compare individual tools and services. The decision for certain tools and services is complex since it requires a lot of knowledge and understanding of their implemented concepts. Therefore, identifying suitable software tools for realizing a certain use case is a manual and cumbersome process.

To tackle these issues, this paper introduces (i) a categorization of currently available technologies and provides (ii) a taxonomy for quantum application development. Based on the proposed categorization and taxonomy, we further introduce (iii) a comparison framework that enables to identify and compare different tools, services, and techniques and, thus, provides decision support to a certain degree. For this, we analyzed various technologies and literature and experimented with several tools and services. The categorization gives an overview of different kinds of technologies and identifies the different building blocks of current quantum application development. The taxonomy further provides a broad view on the different aspects that need to be considered in quantum application development. It enables to understand, analyze, and compare different tools, services, and techniques. We also describe the dependencies and relationships of the different categories in the comparison framework to identify interoperabilities of different tools and services.

After having covered the basic principles and related work in Sect. 2, Sect. 3 provides a detailed problem statement. Section 4 introduces the categorization,

Sect. 5 the taxonomy, and Sect. 6 the prototypical comparison framework developed in the context of this work. Finally, Sect. 7 gives a conclusion and an outlook on future work.

2   Fundamentals and Related Work

The development of quantum applications differs significantly from the development of classical applications [52]. A quantum application typically contains the implementation of a quantum algorithm, pre- and post-processing components, and additional glue code for the execution of the quantum algorithm on a quantum computer. The development of quantum applications is supported by a wide range of different technologies. However, which tools, services, and techniques shall be used for a particular quantum application depends on several factors.

First, it depends on the used quantum hardware. On the one hand, SDKs that enable the implementation and execution of quantum applications are often tailored to the quantum computers of certain vendors and thereby limit the execution on the respective hardware. E.g., Qiskit [1] (IBM), Strawberry Fields [29] (Xanadu), and Ocean [13] (D-Wave) are each designed for their own hardware and by default do not allow quantum applications to run on other hardware. On the other hand, the physical limitations of current quantum hardware play a central role when implementing quantum algorithms. Today's quantum computers are "noisy", i.e., the computational results are not completely accurate, and their size is of "intermediate scale". Thus, they are called Noisy Intermediate Scale Quantum (NISQ) computers [43]. Selecting the best quantum computer for a specific use case is an important task in the current NISQ-era [47] and some approaches, such as TriQ [39] and t|ket⟩ [48], offer compilers with hardware-specific optimization in order to use available hardware in the best possible way.

Available libraries are also important when selecting specific SDKs since they come with pre-implemented algorithms that can be adapted to custom use cases, such as Pennylane [8] provides different libraries in the area of machine learning. On top of that, technologies for the integration with classical applications are becoming increasingly important since hybrid applications are emerging as most promising. A hybrid quantum-classical application comprises quantum components as well as classical components. For the integration of these components, orchestration approaches, such as Orquestra [57] and QuantMe [53] can be used.

In different papers certain aspects of the technology landscape of quantum application development have already been analyzed. Hassija et al. [22], for example, describe the overall landscape of quantum computing, identify the key players, and compare their technologies. LaRose [30] compare different SDKs in terms of their requirements, syntax, library support, and simulation abilities. Quantum programming languages have also been studied in terms of paradigms and features (e.g., [18] and [23]). Fingerhuth et al. [16] identify available open-source quantum software projects and their accompanying website[1] lists many available

---

[1] https://qosf.org.

technologies. There are also other websites that list various tools and cloud services [45]. However, none of them provides an insight on the dependencies and interrelationships, nor does any of them provide decision support. Gill et al. [20] provide a taxonomy in the area of quantum computing, however, it focuses on the algorithmic characteristics of tools and libraries.

The existing papers and websites are a first step to compare available tools, services, and techniques for the development of quantum applications. However, so far only certain aspects have been considered and the decision support for quantum application developers is not focused by any work known to us. Nevertheless, the importance of comparison and decision support frameworks has already proven in several other domains, such as for service provider selection [5,15], and deployment automation technologies [56].

3   Problem Statement

As shown in the previous section, a variety of different tools, services, and techniques exist for the development of quantum applications. The decision on which SDK and which libraries to use must be made early in the development phase—both for the implementation of classical and quantum applications. For the development of quantum applications, however, this decision restricts very early on which quantum hardware and which additional libraries can be used. Hence the portability of quantum applications is currently very limited. Therefore, the first research question (RQ) is as follows:

> **RQ 1:** *What types of tools, services, and techniques enable the development and execution of quantum applications and how can they be categorized in order to understand, analyze, and compare them?*

Although a detailed categorization and characterization of available tools, services, and techniques is a good basis for the analysis and comparison of technologies, quantum application developers must be supported in their decision for specific tools, services, and techniques. Therefore, the second RQ is as follows:

> **RQ 2:** *How can quantum application developers be supported in the decision for certain tools, services, and techniques?*

To address the introduced RQs, we investigated several technologies and derived a categorization and taxonomy for quantum application development as well as a comparison framework, all introduced in the following sections.

4   Overview and Categorization of Existing Technologies for Quantum Application Development

Based on a systematic literature study on concepts, technologies, and best practices for integrating quantum applications with classical applications (including

ACM Digital Library, arXiv, IEEE XPlore, Science Direct, Springer Link, and Wiley Online Library) and a review of related websites [16,44,45], we have identified various technologies for the development of quantum applications.

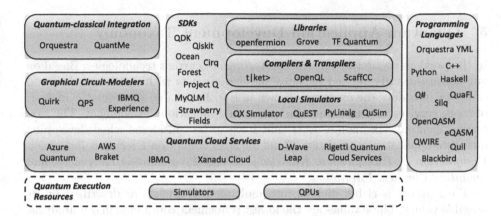

Fig. 1. Categories of quantum application development technologies with examples.

Figure 1 shows the categorization of existing quantum application development technologies derived from the analysis that we describe in the following. At the bottom of Fig. 1 the category *Quantum Execution Resources* is shown that includes *Quantum Processing Units (QPUs)* as well as *Simulators*. Access to these *Quantum Execution Resources* is typically provided via the cloud by *Quantum Cloud Services* which are not limited to hardware access, e.g., *Graphical Circuit-Modelers* are also often provided as services.

Graphical Circuit-Modelers (left in Fig. 1) enable to model a sequence of operations (called gates) to be applied to the specified qubits. *SDKs*, which are either provided by quantum cloud service providers or by third-party providers, offer advanced developer tools. They can include *Libraries* that provide implementations of algorithms from different areas, such as chemistry [34], cryptography [41], and machine learning [10]. Furthermore, *SDKs* contain *Compilers and Transpilers*, such as the quilc compiler [50] as part of the Forest SDK [46]. Finally, *SDKs* often include a *Local Simulator* to simulate the execution locally, e.g., MyQLM [7] includes the pylinalg [6] simulator. However, Libraries, Compilers, Transpilers, and Simulators are not necessarily part of an SDK but can also be available as standalone technologies, such as t|ket⟩ [48] and ScaffCC [27].

In order to integrate quantum applications with classical applications, there are further tools, such as Orquestra [57], allowing to model the control and data flow between the different components required for pre- and post-processing as well as the execution of the quantum algorithm. This forms the workflow that orchestrates classical and quantum application components.

Finally, the programming languages used for implementing quantum applications are considered as a separate category (on the right of Fig. 1). Since the different tools and services, such as the SDKs, Compilers, Transpilers, and Quantum Cloud Services, support different languages, the language plays an important role when considering the compatibilities between different tools and services. In the next section the different characteristics are discussed in detail.

5 Quantum Application Development Taxonomy

In the previous section we identified categories of current technologies. Based on our literature study, related websites and experiments with various services and tools, we introduce the taxonomy shown in Fig. 2 to enable a systematic analysis of tools, services, and techniques in quantum application development. The taxonomy identifies six main aspects that have to be considered when developing quantum applications: *Quantum Cloud Services*, *Quantum Execution Resources*, *Compilation & Transpilation*, *Knowledge Reuse*, *Programming Languages*, and *Quantum-Classical Integration*.

Each aspect is either divided into multiple sub-aspects or described by its possible values. These values are the lowest refinement considered in our analysis. We abstract from further refinements since we aim to provide a broad overview of current quantum application development. The following subsections describe each aspect in detail.

5.1 Quantum Cloud Services

The *Quantum Cloud Services* aspect identifies at which layer services are available and how these services can be accessed, therefore, the two sub-aspects *Service Model* and *Access Methods* are considered.

In general, three different kinds of *Service Models* can be distinguished, namely *Infrastructure as a Service (IaaS)*, *Platform as a Service (PaaS)*, and *Software as a Service (SaaS)* [36]. *IaaS* offerings provide access to processing, storage, networks, and other fundamental computation resources. Quantum cloud services, such as AWS Braket [3], IBMQ [25], and D-Wave Leap [14], offer quantum computation capabilities as a service and can be assigned to IaaS. *PaaS* offerings provide an application hosting environment to host applications developed using certain programming languages, libraries, services, and tools. A popular platform technology in quantum computing often hosted as a service is Jupyter Notebook, e.g., offered as a service by IBM [25]. Jupyter Notebooks combine source code, console instructions, and documentation in a single document-styled format and, thus, are often used for tutorials. *SaaS* offerings provide users a fully managed software. For example, graphical circuit modelers, such as Quirk [19], belong to this category.

In order to access a cloud service, providers offer different *Access Methods*: *SDKs*, are a typical way by which quantum cloud service providers offer access to their services. For example, Qiskit [1] and Forest [46] offer the ability to execute written source code on the respective quantum cloud services. Furthermore, access can be provided via *Web Services*. For example, IBMQ [25] provides

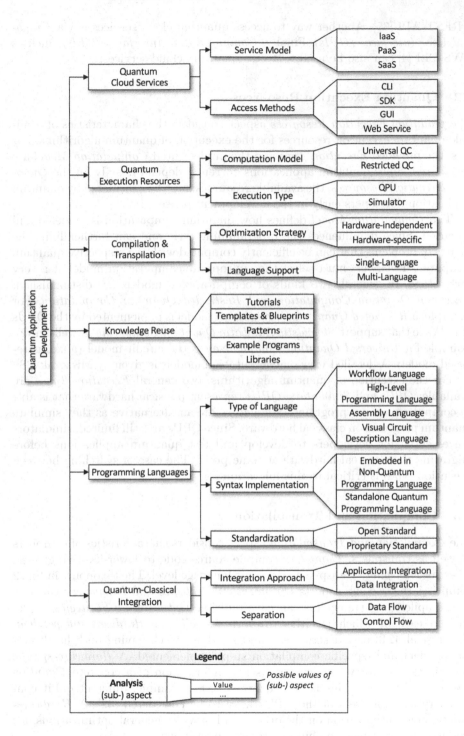

Fig. 2. Taxonomy of quantum application development. Notation is based on [54].

a REST-API [35]. Another way to access quantum cloud services is via *Graphical User Interfaces (GUIs)*. Finally, *Command-Line Interfaces (CLIs)*, such as AWS CLI [4] can also be used to access quantum cloud services.

5.2 Quantum Execution Resources

The *Quantum Execution Resources* aspect considers the characteristics of available quantum execution resources for the execution of quantum algorithms. For this, the provided *Execution Type* and the implemented *Computation Model* are considered. Since quantum applications currently depend heavily on the *Quantum Execution Resources*, the available resources must be considered by quantum application developers early in the development phase.

The *Computation Model* defines how quantum computation is modeled and executed. Since it influences the programming style and can further limit the classes of problems that can be efficiently computed with the respective quantum computer, developers must select the appropriate computation model at a very early stage. In general, two kinds of computational models are distinguished: *Universal Quantum Computation* and *Restricted Quantum Computation*. For example, a *Restricted Quantum Computation* model is implemented by the QPUs of D-Wave that support *Stoquastic Adiabatic Quantum Computing*. A well-known example for *Universal Quantum Computation* is the circuit-model (a.k.a. gate-based model). A detailed view on computation models is given by Miszczak [38].

For the execution of quantum algorithms, two general *Execution Types* are available: *QPUs* and *Simulators*. *QPUs* represent physical hardware that is able to compute quantum programs. *Simulators* are an alternative as they simulate quantum programs on classical hardware. Since QPUs are still limited, simulators are essential for developers to develop and test quantum applications before migrating them to real hardware at some point. The ease of switching between execution types is critical to this migration.

5.3 Compilation and Transpilation

The *Compilation and Transpilation* aspect considers characteristics of compilers as well as transpilers. *Compilers* compile source code to lower-level languages, whereas *Transpilers* transpile on the same language level. The taxonomy in Fig. 2 comprises the two sub-aspects *Optimization Strategies* and *Language Support*.

Compilers and transpilers can have multiple *Optimization Strategies* implemented each of which is either *Hardware-specific* or *Hardware-independent*. Häner et al. [24], e.g., describe an end-to-end approach having both hardware-independent and -specific compilation steps implemented. A *Hardware-specific* optimization strategy uses properties and information of a concrete *Quantum Execution Resource*. This is needed because, for example, different qubit connectivity and gate sets of the QPUs have to be considered [32]. A *Hardware-independent* optimization on the other hand provides general optimizations, for example, to rearrange, combine, or remove operations.

Although, theoretically, language functionalities could be abstracted from, the dependency on the language is still high and *Language Support* is an important aspect. Some compilers and transpilers are only built for exactly one input and one output language, such as quilc [50] and ScaffCC [27], and others can consume various different languages or produce various output languages, such as t|ket⟩ [48] and XACC [33].

5.4   Knowledge Reuse

Since the implementation of quantum algorithms from scratch requires a lot of knowledge, there are different ways for *Knowledge Reuse*. A rather abstract way is given by *Tutorials* that can be offered via websites, documents, or packed within a software. *Patterns*, such as proposed by Leymann [31] and Weigold et al. [55], provide an abstract view on common problems and their solutions [2]. Thus, they offer technology-independent knowledge to specific problems. Furthermore, *Templates and Blueprints* are scaffold implementations that provide a basic structure for further source code. *Example Programs*, which are, e.g., provided as Jupyter notebooks, allow insights into the implementation of other applications and can be adapted to own custom use cases. Finally, concrete implementations of algorithms can also be provided as *Libraries* which can be used as subroutines in a quantum application [10,34,41].

5.5   Programming Languages

The *Programming Language* aspect considers the characteristics of available programming languages for quantum applications. For this, the *Type of Language*, the *Syntax Implementation*, and *Standardization* are considered.

Currently, four types of programming languages can be distinguished: *Workflow Languages*, *High-level Programming Languages*, *Assembly Languages*, and *Graphical Circuit Description Languages*. *Assembly Languages*, such as OpenQASM [12], eQASM [17], QWIRE [42], and Quil [49], are low level and provide a textual representation of every operation the quantum computer is to perform. *High-level Programming Languages*, such as Quipper [21] and Q# [37], are machine independent and provide high-level language features, such as loops and recursion. *Workflow Languages*, such as offered in Orquestra [57], allow to model the control flow of (hybrid quantum-classical) applications. *Graphical Circuit Description Languages* provide graphical representations of quantum circuits.

For the *Syntax Implementation* we distinguish between two cases. On the one hand, there are programming languages that have their own independent syntax, such as Silq [9] and Scaffold [26] Thus, they are *Standalone Quantum Programming Languages*. On the other hand, a programming language can also be embedded into another programming language, for example, Quipper [21] is embedded in Haskell, and Qiskit [1], PyQuil [28], and Cirq [11] offer programming languages embedded in Python.

Standardization is important for the interoperability of different tools and services and is considered in the last aspect. *Open Standards* (a.k.a. *de-jure standards*) are developed or adopted by standards organizations. For example, BPMN [40] is an Open Standard of the Object Management Group. *Proprietary Standards* (a.k.a. *de-facto standards*) evolve from vendor-specific solutions and are widely accepted by a broad base of users. Since OpenQASM [12] is supported by many tools and services, including Qiskit [1], t|ket⟩ [48], XACC [33], Project Q [51], and Cirq [11], it has a wide acceptance and can be considered as de-facto standard for describing circuits. However, most of current programming languages for quantum application development do not implement any standard.

5.6 Quantum-Classical Integration

Hybrid applications that integrate quantum and classical components are increasingly appearing as the most promising solutions at present and in the near future. Thus, the last aspect of Fig. 2 considers *Quantum-Classical Integration*.

In general, two *Integration Approaches* can be distinguished. On the one hand, *Data Integration* consolidates data from different sources to provide applications with a uniform access to this data. On the other hand, *Application Integration* integrates different applications on a functional level.

Furthermore, we characterize integration approaches by their ability to separate business functions from *Data Flow* and *Control Flow*. Technologies, such as Orquestra [57] and QuantMe [53], enable business functions to be defined separately from data- and control flow. With SDKs, such as Qiskit [1], integration must be implemented in the source code and, therefore, does not enable separation of business functions and control flow. In general, *Quantum-Classical Integration* is still largely neglected, although it is essential for the realization of future applications.

6   Comparison Framework

In the previous sections we have introduced a categorization (Sect. 4) and taxonomy (Sect. 5) to characterize existing tools, services, and techniques for quantum application development. Based on these results, in this section we introduce a comparison framework that supports developers in selecting suitable technologies. The comparison framework enables the comparison of characteristics and dependencies between tools and services of different categories.

Figure 3 shows an excerpt of the comparison framework[2] considering exemplary Software Development Kits (SDKs) and Quantum Cloud Services (QCS). An SDK, identified by its name, is available for certain programming languages. Furthermore, since an SDK contains compilers and transpilers that can support various input and output languages, these are separately listed. Besides the availability of a local simulator, which is provided by all examples in Fig. 3, it is finally listed which QCS are directly supported. The category of QCS is another

---

[2] The framework can be found at http://www.github.com/UST-QuAntiL/Qverview.

category exemplary sown in Fig. 3. It contains the name, access method, input language, service model, and available quantum execution resources[3]. All further columns of both categories are hidden in Fig. 3 but can be found in the comparison framework.

The comparison framework implements multi-criteria filtering for all attributes of each category. For example, if Python is chosen as programming language and the availability of a local simulator is required, only Qiskit [1], Strawberry Fields [29], and Ocean [13] will be displayed in the example in Fig. 3. Furthermore, the comparison framework enables to identify interoperabilities between different categories via cross-category filtering. When selecting IBMQ [25] as quantum cloud service, the comparison framework automatically exposes all interoperable SDKs. In addition to Qiskit [1], which is the SDK provided by IBMQ [25], XACC [33], for example, can also be used to execute quantum applications there, as shown in Fig. 3. However, when combined with the programming language filter, only Qiskit would be shown.

The comparison framework supports a multi-criteria cross-category analysis of current technologies for quantum application development. Thus, it (i) provides an overview of technologies of different categories, (ii) enables filtering, comparison and analysis of technologies within each category, and (iii) enables identification of cross-category interoperabilities.

	Name	Programming Language	Compile & Transpile Input	Output	Local Sim.	QCS
SDK	Qiskit	Python, Javascript	OpenQASM	OpenQASM	true	IBMQ, AQT
	Strawberry Fields	Python	-	Blackbird	true	Xanadu
	XACC	C++	Quil, OpenQASM	XASM, Quil, etc.	true	IBMQ, Rigetti, DW Leap
	Ocean	Python	BQM	BQM	true	D-Wave LEAP

	Name	Access Methods	Input	Service Model	Quantum Execution Resources
QCS	IBMQ	SDK:Qiskit, GUI, REST	OpenQASM	IaaS	IBM, Sim.
	Xanadu	SDK:Strawberry Fields	Blackbird	IaaS	Xanadu, Sim.
	DW Leap	SDK:Ocean, GUI, REST	BQM	IaaS	D-Wave, Sim.
	AWS Braket	SDK, GUI, CLI	braket.Circuit	IaaS	IonQ, Rigetti, D-Wave, Sim.

Fig. 3. Excerpt of the comparison framework with exemplary SDKs and QCS.

7 Conclusion and Future Work

Currently, there are strong dependencies between the tools and services used to develop quantum applications and the hardware on which they will run. Due to the limited portability, it is important that quantum application developers identify the appropriate tools and services at an early stage. To make a first step towards decision support for quantum application developers, we have introduced in this paper (i) a categorization, (ii) a taxonomy, and (iii) a comparison

---

[3] QPUs are grouped by their respective vendor.

framework. This is based on an investigation of a variety of technologies and publications to provide guidance for quantum application developers.

With the taxonomy, we have introduced several aspects that need to be considered when selecting specific tools, services, and techniques. While the comparison framework provides guidance in the complex landscape of quantum technologies, it does not yet provide full-automated decision support. In the next step, we therefore plan to further extend our comparison framework to also comprise library support for certain algorithms. This is crucial in order to make a decision based on the problem at hand. The comparison framework so far allows filtering based on the presented taxonomy. In addition, we plan to incorporate weight factors, which will allow categories to be weighted differently.

Acknowledgments. This work was partially funded by the BMWi project *PlanQK (01MK20005N)* as well as the WM BW project *SEQUOIA*.

References

1. Abraham, H., et al.: Qiskit: An Open-source Framework for Quantum Computing (2019). https://doi.org/10.5281/zenodo.2562110
2. Alexander, C., Ishikawa, S., Silverstein, M.: A Pattern Language: Towns, Buildings, Construction. Oxford University Press, Oxford (1977)
3. Amazon Web Services Inc.: AWS Braket (2021). https://aws.amazon.com/braket
4. Amazon.com Inc.: aws-cli (2020). https://github.com/aws/aws-cli
5. Andrikopoulos, V., Gómez Sáez, S., Leymann, F., Wettinger, J.: Optimal distribution of applications in the cloud. In: Jarke, M., et al. (eds.) CAiSE 2014. LNCS, vol. 8484, pp. 75–90. Springer, Cham (2014). https://doi.org/10.1007/978-3-319-07881-6_6
6. Atos SE: qat.pylinalg: Python Linear-algebra simulator (2020). https://myqlm.github.io/myqlm_specific/qat-pylinalg.html
7. Atos SE: MyQLM (2021). https://atos.net/en/lp/myqlm
8. Bergholm, V., et al.: PennyLane: Automatic differentiation of hybrid quantum-classical computations (2020). arXiv preprint arXiv:1811.04968
9. Bichsel, B., Baader, M., Gehr, T., Vechev, M.: Silq: a high-level quantum language with safe uncomputation and intuitive semantics. In: Proceedings of the 41st ACM SIGPLAN Conference on Programming Language Design and Implementation, pp. 286–300. PLDI 2020. Association for Computing Machinery (2020). https://doi.org/10.1145/3385412.3386007
10. Broughton, M., et al.: TensorFlow Quantum: A Software Framework for Quantum Machine Learning (2020). arXiv preprint arXiv:2003.02989
11. Developers, Cirq: Cirq (2021). https://doi.org/10.5281/zenodo.4062499
12. Cross, A.W., Bishop, L.S., Smolin, J.A., Gambetta, J.M.: Open Quantum Assembly Language (2017). arXiv preprint arXiv:1707.03429
13. D-Wave Systems Inc.: dwave-ocean-sdk (2021). https://github.com/dwavesystems/dwave-ocean-sdk
14. D-Wave Systems Inc.: Leap (2021). https://dwavesys.com/take-leap
15. Farshidi, S., Jansen, S., de Jong, R., Brinkkemper, S.: A decision support system for cloud service provider selection problem in software producing organizations. In: 2018 IEEE 20th Conference on Business Informatics (CBI), vol. 01, pp. 139–148 (2018). https://doi.org/10.1109/CBI.2018.00024

16. Fingerhuth, M., Babej, T., Wittek, P.: Open source software in quantum computing. PLoS One **13**(12), e0208561 (2018). https://doi.org/10.1371/journal.pone.0208561

17. Fu, X., et al.: eQASM: an executable quantum instruction set architecture. In: 2019 IEEE International Symposium on High Performance Computer Architecture (HPCA), pp. 224–237 (2019). https://doi.org/10.1109/HPCA.2019.00040

18. Garhwal, S., Ghorani, M., Ahmad, A.: Quantum programming language: a systematic review of research topic and top cited languages. Arch. Comput. Methods Eng. **28**(2), 289–310 (2019). https://doi.org/10.1007/s11831-019-09372-6

19. Gidney, C., Marwaha, K., Haugeland, J., ebraminio, Kalra, N.: Quirk: Quantum Circuit Simulator (2021). https://algassert.com/quirk

20. Gill, S.S., et al.: Quantum Computing: A Taxonomy, Systematic Review and Future Directions (2020). arXiv preprint arXiv:2010.15559

21. Green, A.S., Lumsdaine, P.L., Ross, N.J., Selinger, P., Valiron, B.: Quipper: a scalable quantum programming language. In: Proceedings of the 34th ACM SIGPLAN Conference on Programming Language Design and Implementation, pp. 333–342. PLDI 2013. Association for Computing Machinery (2013). https://doi.org/10.1145/2491956.2462177

22. Hassija, V., et al.: Present landscape of quantum computing. IET Quant. Commun. **1**(2), 42–48 (2020). https://doi.org/10.1049/iet-qtc.2020.0027

23. Heim, B., et al.: Quantum programming languages. Nat. Rev. Phys. **2**(12), 709–722 (2020). https://doi.org/10.1038/s42254-020-00245-7

24. Häner, T., Steiger, D.S., Svore, K., Troyer, M.: A software methodology for compiling quantum programs. Quant. Sci. Technol. **3**(2) (2018). https://doi.org/10.1088/2058-9565/aaa5cc

25. IBM: IBM Quantum Experience (2021). https://quantum-computing.ibm.com

26. Javadi-Abhari, A., et al.: Scaffold: Quantum Programming Language. Princeton University, NJ, Department of Computer Science, Technical report (2012)

27. Javadi-Abhari, A., et al.: ScaffCC: a framework for compilation and analysis of quantum computing programs. In: Proceedings of the 11th ACM Conference on Computing Frontiers, CF 2014. Association for Computing Machinery (2014). https://doi.org/10.1145/2597917.2597939

28. Karalekas, P.J., et al.: PyQuil: Quantum programming in Python (2020). https://doi.org/10.5281/zenodo.3631770

29. Killoran, N., et al.: Strawberry fields: a software platform for photonic quantum computing. Quantum **3**, 129 (2019). https://doi.org/10.22331/q-2019-03-11-129

30. LaRose, R.: Overview and comparison of gate level quantum software platforms. Quantum **3**, 130 (2019). https://doi.org/10.22331/q-2019-03-25-130

31. Leymann, F.: Towards a pattern language for quantum algorithms. In: Feld, S., Linnhoff-Popien, C. (eds.) QTOP 2019. LNCS, vol. 11413, pp. 218–230. Springer, Cham (2019). https://doi.org/10.1007/978-3-030-14082-3_19

32. Leymann, F., Barzen, J.: The bitter truth about gate-based quantum algorithms in the NISQ era. Quant. Sci. Technol. 1–28 (2020). https://doi.org/10.1088/2058-9565/abae7d

33. McCaskey, A.J., Lyakh, D.I., Dumitrescu, E.F., Powers, S.S., Humble, T.S.: XACC: a system-level software infrastructure for heterogeneous quantum–classical computing. Quant. Sci. Technolo. **5**(2) (2020). https://doi.org/10.1088/2058-9565/ab6bf6

34. McClean, J.R., et al.: OpenFermion: The Electronic Structure Package for Quantum Computers (2019). arXiv preprint arXiv:1710.07629

35. McKay, D.C., et al.: Qiskit Backend Specifications for OpenQASM and OpenPulse Experiments (2018). arXiv preprint arXiv:1809.03452
36. Mell, P., Grance, T.: The NIST definition of cloud computing. Technical report. NIST SP 800–145, National Institute of Standards and Technology (2011). https://doi.org/10.6028/NIST.SP.800-145
37. Microsoft: Q# Language (2021). https://github.com/microsoft/qsharp-language
38. Miszczak, J.A.: Models of quantum computation and quantum programming languages. Bull. Pol. Acad. Sci. Tech. Sci. **59**(3), 305–324 (2011). https://doi.org/10.2478/v10175-011-0039-5
39. Murali, P., Baker, J.M., Javadi-Abhari, A., Chong, F.T., Martonosi, M.: Noise-adaptive compiler mappings for noisy intermediate-scale quantum computers. In: Proceedings of the 24th International Conference on Architectural Support for Programming Languages and Operating Systems, pp. 1015–1029, ASPLOS 2019. Association for Computing Machinery (2019). https://doi.org/10.1145/3297858.3304075
40. OMG: Business Process Model and Notation (BPMN) Version 2.0. Object Management Group (OMG) (2011)
41. Open Quantum Safe Project: liboqs (2021). https://openquantumsafe.org/liboqs/
42. Paykin, J., Rand, R., Zdancewic, S.: QWIRE: a core language for quantum circuits. In: Proceedings of the 44th ACM SIGPLAN Symposium on Principles of Programming Languages, pp. 846–858, POPL 2017. Association for Computing Machinery (2017). https://doi.org/10.1145/3009837.3009894
43. Preskill, J.: Quantum computing in the NISQ era and beyond. Quantum **2**, 79 (2018). https://doi.org/10.22331/q-2018-08-06-79
44. Quantiki: QC simulators (2021). https://quantiki.org/wiki/list-qc-simulators
45. Quantum Computing Report: Tools (2021). https://quantumcomputingreport.com/tools/
46. Rigetti Computing: Forest SDK (2019). https://pyquil-docs.rigetti.com/
47. Salm, M., Barzen, J., Breitenbücher, U., Leymann, F., Weder, B., Wild, K.: The NISQ analyzer: automating the selection of quantum computers for quantum algorithms. In: Dustdar, S. (ed.) SummerSOC 2020. CCIS, vol. 1310, pp. 66–85. Springer, Cham (2020). https://doi.org/10.1007/978-3-030-64846-6_5
48. Sivarajah, S., Dilkes, S., Cowtan, A., Simmons, W., Edgington, A., Duncan, R.: t|ket⟩: a retargetable compiler for NISQ devices. Quant. Sci. Technol. **6**(1) (2020). https://doi.org/10.1088/2058-9565/ab8e92
49. Smith, R.S., Curtis, M.J., Zeng, W.J.: A Practical Quantum Instruction Set Architecture (2017). arXiv preprint arXiv:1608.03355
50. Smith, R.S., Peterson, E.C., Davis, E.J., Skilbeck, M.G.: quilc: An Optimizing Quil Compiler (2020). https://doi.org/10.5281/zenodo.3677537
51. Steiger, D.S., Häner, T., Troyer, M.: ProjectQ: an open source software framework for quantum computing. Quantum **2**, 49 (2018). https://doi.org/10.22331/q-2018-01-31-49
52. Weder, B., Barzen, J., Leymann, F., Salm, M., Vietz, D.: The quantum software lifecycle. In: Proceedings of the 1st ACM SIGSOFT International Workshop on Architectures and Paradigms for Engineering Quantum Software (APEQS 2020), pp. 2–9. ACM (2020). https://doi.org/10.1145/3412451.3428497
53. Weder, B., Breitenbücher, U., Leymann, F., Wild, K.: Integrating quantum computing into workflow modeling and execution. In: Proceedings of the 13th IEEE/ACM International Conference on Utility and Cloud Computing (UCC 2020), pp. 279–291. IEEE Computer Society (2020). https://doi.org/10.1109/UCC48980.2020.00046

54. Weerasiri, D., Barukh, M.C., Benatallah, B., Sheng, Q.Z., Ranjan, R.: A taxonomy and survey of cloud resource orchestration techniques. ACM Comput. Surv. **50**(2) (2017). https://doi.org/10.1145/3054177

55. Weigold, M., Barzen, J., Salm, M., Leymann, F.: Data encoding patterns for quantum computing. In: Proceedings of the 27th Conference on Pattern Languages of Programs. The Hillside Group (2021, accepted for publication)

56. Wurster, M., et al.: The essential deployment metamodel: a systematic review of deployment automation technologies. SICS Softw. Intensive Cyber-Phys. Syst. **35**, 63–75 (2019). https://doi.org/10.1007/s00450-019-00412-x

57. Zapata Computing: Orquestra (2021). https://zapatacomputing.com/orquestra/

Quantum Asymmetric Encryption Based on Quantum Point Obfuscation

Chuyue Pan, Tao Shang$^{(\boxtimes)}$, and Jianwei Liu

School of Cyber Science and Technology, Beihang University, Beijing 100083, China
shangtao@buaa.edu.cn

Abstract. Quantum obfuscation means encrypting the functionality of circuits or functions by quantum mechanics. It works as a form of quantum computation to improve the security and confidentiality of quantum programs. Although some quantum encryption schemes have been discussed, any quantum asymmetric scheme based on quantum obfuscation is not still proposed. In this paper, we construct an asymmetric encryption scheme based on quantum point function, which applies the advantages of quantum obfuscation to quantum public-key encryption. As a start of the study on applications of quantum obfuscation to asymmetric encryption, our work will be helpful in the future quantum obfuscation theory and will therefore promote the development of quantum computation.

Keywords: Quantum computation · Quantum asymmetric encryption · Quantum obfuscation

1 Introduction

Quantum computation combines ideas of classical information theory, computer science, and quantum physics [1]. It introduces some quantum concepts including quantum information, quantum algorithms and quantum error correction, etc. Among the existing quantum algorithms, quantum obfuscation, which means encrypting the functionality of circuits or functions by quantum mechanics, is an emergent branch to improve the security and confidentiality of quantum information. It is developed from the concept of classical obfuscation.

Classical obfuscation drives from code obfuscation in software engineering. It means reorganizing and processing the released program so that the processed code has the same function as previous one. In 2001, Barak et al. [2] first introduced the concept of obfuscation into cryptography and formally defined three properties of obfuscation. In 2004, Lynn et al. [3] put forward the first positive result of obfuscation and gave several provable schemes of point obfuscation based on complex access control under the random oracle model. In 2005, Goldwasser et al. [4] proved that obfuscation with auxiliary input cannot be realized no matter whether the auxiliary input is independent of obfuscation programs. In 2014, Alagic et al. [5] proposed a quantum obfuscator based on quantum

© Springer Nature Switzerland AG 2021
M. Paszynski et al. (Eds.): ICCS 2021, LNCS 12747, pp. 142–148, 2021.
https://doi.org/10.1007/978-3-030-77980-1_11

topological computation. In 2016, Alagic et al. [6] formally put forward the definition of quantum obfuscation which is a form of quantum computation to protect quantum circuits. In 2019, Shang et al. [7,8] initiated the obfuscatibility of quantum point function and proposed the indistinguishability(IND)-secure quantum symmetric encryption scheme based on point obfuscation.

In this paper, we construct an asymmetric encryption scheme based on quantum point obfuscation. We combine the advantages of quantum obfuscation with asymmetric encryption to achieve indistinguishability security. Here, quantum point function is just an instantiation of quantum obfuscation. Asymmetric encryption schemes of other quantum functions still remain widely open.

2   Related Works

Definition 1. *If there exists (O, δ) in a QPT algorithm of indistinguishability obfuscation, then the following three conditions hold:*

1. Functional equivalence: *the obfuscation result $O(C)$ is interpreted as $\delta_{O(C)}$, which holds the same functionality as the input circuit C:*

$$\|\delta_{O(C)} - C\| \leq negl(n). \tag{1}$$

2. Polynomial slowdown: *the length of the obfuscator $O(C)$ must be limited to polynomial qubits, which refers to*

$$\|O(C)\| = poly(n). \tag{2}$$

3. Indistinguishability: *for any $\rho_n \in R_n$, $\sigma_n \in S_n$, there are three types of indistinguishability:*
   - *Perfect indistinguishability: $\rho_n = \sigma_n$.*
   - *Statistical indistinguishability: $\|\rho_n - \sigma_n\| \leq negl(n)$.*
   - *Computational indistinguishability: for any QPT interpreter δ, $\|\delta_{\rho_n} - \delta_{\sigma_n}\| \leq negl(n)$.*

Among the formula above, the interpreter $\delta(C)$ refers to a compiler interpreting the functionality of the circuit C from $O(C)$.

Definition 2. *A quantum point function $U_{\alpha,\beta}$ with a general output is*

$$U_{\alpha,\beta} : |x, 0^n\rangle \mapsto |x, P_{\alpha,\beta}(x)\rangle. \tag{3}$$

where $\alpha \in \{0,1\}^n$, $\beta \in \{0,1\}^n \backslash 0^n$, and $P_{\alpha,\beta}$ is a classical point function with a multi-bit output working as

$$P_{\alpha,\beta}(x) = \begin{cases} \beta & if \ x = \alpha \\ 0^n & otherwise \end{cases} \tag{4}$$

By means of constructive proof, Shang et al. [7,8] demonstrated the obfuscatability of the quantum point function with a general output.

3 Quantum Asymmetric Encryption Scheme

3.1 Basic Idea

Quantum asymmetric encryption scheme based on quantum obfuscation can be constructed as follows. Firstly, we implement operation of qubit rotation and measurement before we obtain a quantum state. Then we interact the quantum state with the obfuscated quantum point function to get the qubit $|0\rangle$ or $|1\rangle$. We consequently encrypt the coefficients of the quantum state of point obfuscation by the asymmetric encryption algorithm.

Definition 3. *Single-qubit rotation is a one-way function with a quantum trapdoor. The qubit is located in the x-z plane of three-dimensional Bloch sphere. The eigenstate $|0\rangle$ is located in the positive half axis of z-axis and the eigenstate $|1\rangle$ in the negative half. Single-qubit rotation is actually a rotation transformation around y-axis. That is*

$$\hat{\gamma} = i(|1\rangle\langle 0| - |0\rangle\langle 1|). \tag{5}$$

The eigenstates are transformed into quantum superposition states on x-z plane, and the sum probabilities of eigenstate $|0\rangle$ and $|1\rangle$ is 1 which can be written as trigonometric function:

$$|\varphi\rangle = \cos\frac{\alpha}{2}|0\rangle + \sin\frac{\alpha}{2}|1\rangle. \tag{6}$$

Supposing $\hat{\gamma} = i(|1\rangle\langle 0| - |0\rangle\langle 1|)$, we have

$$|\varphi_b(\alpha_k)\rangle = \cos\frac{b\alpha_k}{2}|0\rangle + \sin\frac{b\alpha_k}{2}|1\rangle = e^{\frac{ib\alpha_k\hat{\gamma}}{2}}|0\rangle = R(b\alpha_k)|0\rangle. \tag{7}$$

The first result $|\varphi_{b_i,m_i}(\alpha_k)\rangle_i$ is transformed by the second rotation on the basis of superposition state. When the output state $|m_i\rangle$ of quantum point obfuscation is $|0\rangle$, rotation about angle 0 is performed in Hilbert space. When the $|m_i\rangle$ is $|1\rangle$, rotation about angle π is performed. We can show the process with the formula

$$|\varphi_{b_i,m_i}(\alpha_k)\rangle_i = R(m_i\pi)|\varphi_{b_i}(\alpha_k)\rangle_i. \tag{8}$$

The previous result is inversely processed by qubit rotation shown in Fig. 1 and is subsequently measured after decryption. That is, if the information to be encrypted is $|0\rangle$, it will approach the positive half axis of z-axis with high probability after rotation. If it is $|1\rangle$, it will approach the negative half with high probability. After measurement, we can get specific plaintext m_i.

3.2 Scheme

Key Generation. Firstly, we randomly select an integer string $b = (b_1, b_2, \cdots, b_K)$ of length K, a rotation angle $\alpha_k = \frac{2\pi}{2^k} = \frac{\pi}{2^{k-1}}$ where k is a positive integer, and the eigenstate $|0\rangle^{\otimes d}$ of d qubits to generate the corresponding public and private keys. Here we introduce a randomly generated $2(d+1)$ bits string $s = \{0,1\}^{2d+2}$

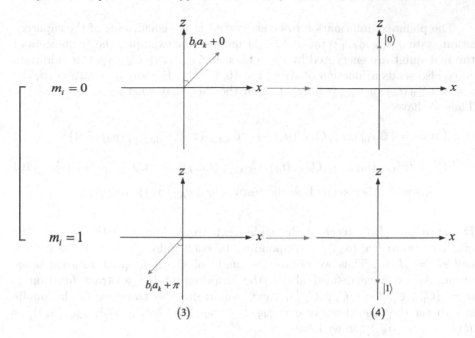

Fig. 1. Two dimensional planes of second qubit rotation.

and another random string $t = \{0,1\}^{2d+2}$ of the same length which is only known by sender (the public key holder). We define $C_{a,b}$ as a classical one-way trapdoor function and its trapdoor is $u = (C_{0,0}^{-1}, C_{0,1}^{-1}, \cdots, C_{K,0}^{-1}, C_{K,1}^{-1})$, where $C_{a,b}^{-1}$ represents the inverse function of $C_{a,b}$. The algorithm based on function $C_{a,b}$ is easy to generate but difficult to inverse unless trapdoor u is used. Supposing $a = 0, 1, \cdots, K$, the sender chooses $C_{a,0}$ or $C_{a,1}$ to encrypt the coefficients x or y of the superposition quantum state output from quantum obfuscation.

The private key of the encryption scheme is divided into two parts: $\{k, b\}$ and $u = (C_{0,0}^{-1}, C_{0,1}^{-1}, \cdots, C_{K,0}^{-1}, C_{K,1}^{-1})$. The former is used to decrypt the quantum obfuscated state so as to obtain the specific angle of the inverse rotation operation while the latter is used to decrypt the quantum state after the second rotation.

Encryption. Alice wants to send Bob a multi-qubit string $|m\rangle = |(m_1, m_2, \cdots, m_d)\rangle$. Firstly, we need to compare d and K, if $d > K$, we need to extend the length of K in the public key. For the ith qubit $|m_i\rangle$, we interact it with quantum point function and obtain $|0\rangle$ or $|1\rangle$ after obfuscation. Then we implement qubit rotation operation to get $|\varphi_{b_i}(\alpha_k)\rangle_i = R(b_i\alpha_k)|0\rangle_i$ and consequently implement qubit rotation operation for the second time to get $q_i = |\varphi_{b_i,m_i}(\alpha_k)\rangle_i = R(m_i\pi)|\varphi_{b_i}(\alpha_k)\rangle_i$. Finally, we encrypt the quantum superposition state result $q_i = x|0\rangle + y|1\rangle$. The specific encryption steps are described as follows.

The plaintext information to be encrypted is the coefficients of the superposition state $q = q_d q_{d-1} \cdots q_0$ of $(d+1)$ qubits. For example, the coefficients of the first qubit are encrypted by the sender as $C_{0,t_0}(x_0), C_{0,t_1}(y_0)$. In addition, Alice also sends a function of string $t = \{0, 1\}^{2d+2}$. For any $a = 0, 1, \cdots, d$, the use of $C_{a,0}(t_a)$ or $C_{a,1}(t_a)$ depends on the fact that whether s_a is $|0\rangle$ or $|1\rangle$. Thus we have

$$En(q) = \{C_{0,t_0}(x_0), C_{0,t_1}(y_0), \cdots, C_{d,t_{2d}}(x_d), C_{d,t_{2d+1}}(y_d), F(t)\}, \quad (9)$$

$$F(t) = \{C_{0,s_0}(t_0), \cdots, C_{d,s_d}(t_d), C_{0,s_{d+1}}(t_{d+1}), \cdots, C_{d,s_{2d+1}}(t_{2d+1})\}. \quad (10)$$

In this way, Alice sends Bob the result $\{|\varphi_{b(PK)}(\alpha_k)\rangle, En(q)\}$.

Decryption. Bob receives the ciphertext from Alice. Firstly, we get the quantum state $|\varphi_{b_i}(\alpha_k)\rangle_i$ corresponding to each qubit $|m_i\rangle$ with the private key $sk = \{k, b\}$. Thus we obtain the angle of the first qubit rotation operation. As we have defined above, the trapdoor of the abstract function is $u = (C_{0,0}^{-1}, C_{0,1}^{-1}, \cdots, C_{k,0}^{-1}, C_{k,0}^{-1})$. Next, we utilize the trapdoor in the public key to get the coefficients of $q = q_d q_{d-1} \cdots q_0$. We know $q_i = |\varphi_{b_i,m_i}(\alpha_k)\rangle_i = R(m_i\pi) |\varphi_{b_i}(\alpha_k)\rangle_i$, so we have

$$F^{-1}(t) = \{C_{0,s_0}^{-1}(t_0), \cdots, C_{d,s_d}^{-1}(t_d), C_{1,s+2}^{-1}(t_{d+2}), \cdots, C_{d,s_{2d+1}}^{-1}(t_{2d+1})\}, \quad (11)$$

$$Dn(q) = \{C_{0,t_0}^{-1}(x_0), C_{0,t_1}^{-1}(y_1), \cdots, C_{d,t_{2d}}^{-1}(x_d), C_{d,t_{2d+1}}^{-1}(y_d), F^{-1}(t)\}. \quad (12)$$

At this time, we measure $R(b_i\alpha_k)_i^{-1} |\varphi_{b_i,m_i}(\alpha_k)\rangle_i$ on the x-z plane of the Bloch sphere. If the qubit is on the positive half axis of the Z axis, the output of quantum point function is $|0\rangle$. If it is on the negative half, the output is $|1\rangle$. At this time, Alice replaces s with t and publishes the new private key sk.

4   Security Analysis

4.1   Key Updating

This scheme updates s in the public key with the classical bit string $t = \{0, 1\}^{2d+2}$ generated randomly each time. Alice generates t randomly, and Bob needs to use the private key $(C_{0,0}^{-1}, C_{0,1}^{-1}, \cdots, C_{K,0}^{-1}, C_{K,1}^{-1})$ to calculate t, and then obtain the coefficients of the quantum state $q_i = |\varphi_{b_i,m_i}(\alpha_k)\rangle_i = R(m_i\pi) |\varphi_{b_i}(\alpha_k)\rangle_i$.

Because t is not published to the public, Bob can verify the identity information of the encrypting party while using the private key. The encryption and decryption scheme can be carried out in two directions and circularly.

4.2   Indistinguishablility Security

Theorem 1. *If there exist quantum black-box obfuscation and secure quantum one-way trapdoor function, there also exists a quantum asymmetric encryption scheme which satisfies indistinguishability chosen-plaintext attack (IND-CPA) security.*

Proof. *A quantum polynomial time interpreter δ with only black-box access to En_{sk} can be used to simulate the access of any QPT adversary. Because of the access to the quantum encryption circuit, the interpreter can select a random number to be used for encryption.*

For any QPT adversary, $A = (R, R^{-1})$, $u = \left|\varphi_{b(PK)}(\alpha_k)\right\rangle \otimes R(m_j)$, $v = \left|\varphi_{b(PK)}(\alpha_k)\right\rangle \left(\left|0^d\right\rangle \left\langle 0^d\right| \otimes m_i\right)$, we have

$$
\begin{aligned}
&\left| \Pr\left\{ R^{-1}\left[PK \otimes R(m_i)\right] = 1\right\} - \Pr\left\{ R^{-1}\left[PK\left(\left|0^d\right\rangle \left\langle 0^d\right| \otimes m_i\right)\right] = 1\right\}\right| \\
&= \left|\Pr\left\{R^{-1}\left[u, O\left(U_{b,k}\right)\right] = 1\right\} - \Pr\left\{R^{-1}\left[v, O\left(U_{b,k}\right)\right] = 1\right\}\right| \\
&\leq \sum_k \left|\Pr\left\{R^{-1}[u, \beta(b)] = 1\right\} - \Pr\left\{R^{-1}[v, \beta(b)] = 1\right\}\right| \cdot \Pr\left\{R_1^{-1}\left[u, O\left(U_{b,k}\right) = \beta(b)\right]\right\}.
\end{aligned}
\tag{13}
$$

Here R_1^{-1} is the subline of R^{-1}. Owing to the virtual black-box property, there is

$$
\left|\Pr\left[R^{-1}\left(O\left(U_{b,k}\right)\right) = 1\right] - \Pr\left[S^{U_{b,k}}\left(\left|0^d\right\rangle\right) = 1\right]\right| \leq \text{negl}(n)
\tag{14}
$$

And we have

$$
\begin{aligned}
&\sum_k \left|\Pr\left\{R^{-1}[u, \beta(b)] = 1\right\} - \Pr\left\{R^{-1}[v, \beta(b)] = 1\right\}\right| \cdot \Pr\left\{R_1^{-1}\left[u, O\left(U_{b,k}\right) = \beta(b)\right]\right\} \\
&\leq \sum_k \left|\Pr\left\{R^{-1}[u, \beta(b)] = 1\right\} - \Pr\left\{R^{-1}[v, \beta(b)] = 1\right\}\right| \cdot \left|\Pr\left\{S^{U_{b,k}}\left(\left|0^d\right\rangle\right) = b\right\} + \text{negl}(n)\right|
\end{aligned}
\tag{15}
$$

When S under the quantum-accessible random oracle accesses $U_{b,k}$ successfully, $\beta(b) = b_k$, otherwise $\beta(b) = 0$. So

$$
\begin{aligned}
&\sum_k \left|\Pr\left\{R^{-1}[u, \beta(b)] = 1\right\} - \Pr\left\{R^{-1}[v, \beta(b)] = 1\right\}\right| \cdot \left|\Pr\left\{S^{U_{b,k}}\left(\left|0^d\right\rangle\right) = b\right\} + \text{negl}(n)\right| \\
&= \left|\Pr\left\{R^{-1}[u, b_k] = 1\right\} - \Pr\left\{R^{-1}[v, b_k] = 1\right\}\right| \cdot \left|\Pr\left\{S^{U_{b,k}}\left(\left|0^d\right\rangle\right) = b\right\} + \text{negl}(n)\right| \\
&\quad + \left|\Pr\left\{R^{-1}[u, 0] = 1\right\} - \Pr\left\{R^{-1}[v, 0] = 1\right\}\right| \cdot \left|\Pr\left\{S^{U_{b,k}}\left(\left|0^d\right\rangle\right) = 0\right\} + \text{negl}(n)\right|
\end{aligned}
\tag{16}
$$

Owing to $\Pr\left\{S^{U_{bk}}\left(\left|0^K\right\rangle\right) = b_k\right\} = \text{poly}(n)/2^n \leq \text{negl}(n)$ and IND-security of one-time pad, we have

$$
\begin{aligned}
&\left|\Pr\left\{R^{-1}[u, 0] = 1\right\} - \Pr\left\{R^{-1}[v, 0] = 1\right\}\right| \\
&= \left| \Pr\left\{R^{-1}\left[\left|\varphi_{b(PK)}(\alpha_k)\right\rangle \otimes R(m_i)\right] = 1\right\} \right. \\
&\quad \left. - \Pr\left\{R^{-1}\left[\left|\varphi_{b(PK)}(\alpha_k)\right\rangle \left(\left|0^d\right\rangle \left\langle 0^d\right| \otimes m_i\right)\right] = 1\right\} \right| \\
&= \text{negl}(n)
\end{aligned}
\tag{17}
$$

Thus we have

$$\left| \Pr\left\{ R^{-1}\left[PK \otimes R\left(m_i\right)\right] = 1 \right\} - \Pr\left\{ R^{-1}\left[PK\left(\left|0^d\right\rangle\left\langle 0^d\right| \otimes m_i\right)\right] = 1 \right\} \right|$$
$$\leq \left| \Pr\left\{ R^{-1}\left[u, b_k\right] = 1 \right\} - \Pr\left\{ R^{-1}\left[v, b_k\right] = 1 \right\} \right| \cdot \left| \operatorname{negl}(n) + \operatorname{negl}(n) \right| \qquad (18)$$
$$+ \operatorname{negl}(n) \cdot \left| \Pr\left\{ S^{U_{b,k}}\left(\left|0^d\right\rangle\right) = 0 \right\} + \operatorname{negl}(n) \right| = \operatorname{negl}(n)$$

In conclusion, the asymmetric encryption scheme of quantum obfuscation satisfies indistinguishability security.

5   Conclusion

In this paper, we presented an asymmetric encryption scheme based on quantum point obfuscation. It not only achieves indistinguishability security without classical cryptography, but also solves the problem of key management in symmetric encryption. This work promotes the research on quantum computation and provides quantum obfuscation as a more secure method to achieve confidentiality of cryptography.

Acknowledgments. This project was supported by the National Natural Science Foundation of China (No. 61971021, 61571024) and the National Key Research and Development Program of China (No. 2016YFC1000307) for valuable helps.

References

1. Steane, A.: Quantum computing. Rep. Prog. Phys. **61**(2), 117 (1997)
2. Lynn, B., Prabhakaran, M., Sahai, A.: Positive results and techniques for obfuscation. In: Cachin, C., Camenisch, J.L. (eds.) EUROCRYPT 2004. LNCS, vol. 3027, pp. 20–39. Springer, Heidelberg (2004). https://doi.org/10.1007/978-3-540-24676-3_2
3. Barak, B., et al.: On the (im)possibility of obfuscating programs. In: Kilian, J. (ed.) CRYPTO 2001. LNCS, vol. 2139, pp. 1–18. Springer, Heidelberg (2001). https://doi.org/10.1007/3-540-44647-8_1
4. Goldwasser, S., Kalai, Y.T.: On the impossibility of obfuscation with auxiliary input. In: Foundations of Computer Science. FOCS 2005. IEEE Symposium on, pp. 553–562. IEEE (2005)
5. Alagic, G., Jefery, S., Jordan, S.: Circuit obfuscation using braids. In: Proceedings of 9th Conference on Theory of Quantum Computation, Communication and Cryptography (TQC), May 21–23, 2014, Singapore, pp. 141–160 (2014). https://doi.org/10.4230/LIPIcs.TQC.2014.141
6. Alagic, G., Fefferman, B.: On Quantum Obfuscation (2016). 2016 ArXiv: Quantum Physics. arXiv Preprint ArXiv: 1602.01771
7. Shang, T., Chen, R., Liu, J.: On the obfuscatability of quantum point functions. Quant. Inf. Process. **18**, 55 (2019). https://doi.org/10.1007/s11128-019-2172-2
8. Chen, R., Shang, T., Liu, J.: IND-secure quantum symmetric encryption based on point obfuscation. Quant. Inf. Process. **18**(6), 161 (2019). https://doi.org/10.1007/s11128-019-2280-z

Index Calculus Method for Solving Elliptic Curve Discrete Logarithm Problem Using Quantum Annealing

Michał Wroński[✉] [iD]

Military University of Technology, Kaliskiego Street 2, Warsaw, Poland
michal.wronski@wat.edu.pl

Abstract. This paper presents an index calculus method for elliptic curves over prime fields using quantum annealing. The relation searching step is transformed into the QUBO (Quadratic Unconstrained Boolean Optimization) problem, which may be efficiently solved using quantum annealing, for example, on a D-Wave computer. Unfortunately, it is hard to estimate the complexity of solving the given QUBO problem using quantum annealing. Using Leap hybrid sampler on the D-Wave Leap cloud, we could break ECDLP for the 8-bit prime field. The most powerful general-purpose quantum computers nowadays would break ECDLP at most for a 6-bit prime using Shor's algorithm. In presented approach, the Semaev method of construction of decomposition base is used, where the decomposition base has a form $\mathcal{B} = \left\{ x : 0 \leq x \leq p^{\frac{1}{m}} \right\}$, with m being a fixed integer.

Keywords: Cryptanalysis · Index calculus on elliptic curves · Quantum annealing · QUBO · D-Wave

1 Introduction

Quantum computing is a significant branch of modern cryptology. There are known several quantum algorithms supporting cryptanalysis of public-key schemes. In the last few years, notable progress in this field has been established. The two approaches of quantum computing for cryptography are the most popular nowadays. The first one is an approach using quantum annealing, which is used in D-Wave computers. The second one is the general-purpose quantum computing approach. The first approach has limited applications, where mainly QUBO and Ising problems may be solved using such quantum computers. On the other hand, several cryptographical problems may be translated to the QUBO problem, for example, integer factorization. The quantum factorization record belonged to the D-Wave computer for some time, where Dridi and Alghassi [3] factorized integer $200,099$. This result was later beaten by Jiang et al. [6], where $376,289$ was factorized, and by Wang et al. [11], where they factorized 20-bit integer $1,028,171$. Furthermore, general-purpose quantum computers have limited resources. The most potent Intel, IBM, and Google quantum computers

© Springer Nature Switzerland AG 2021
M. Paszynski et al. (Eds.): ICCS 2021, LNCS 12747, pp. 149–155, 2021.
https://doi.org/10.1007/978-3-030-77980-1_12

have 49, 53, and 72 qubits, respectively [4,5,10]. Resources of general quantum computers are nowadays too small to solve real-world cryptography problems.

This paper shows how to transform the relation searching step for the index calculus method on elliptic curves into the optimization problem. The relation searching problem is transformed into the QUBO (Quadratic Unconstrained Boolean Optimization) problem, where constraints exchanged by penalties added to the objective function.

Even though Shor's quantum algorithm for factorization, discrete logarithm problem, and elliptic curves discrete logarithm problem is known to be efficient (it works in polynomial time), present quantum computers (even the most powerful) can solve ECDLP defined on at most 6-bit prime field $\mathbb{F}_p$. Using the index calculus method and transformation of relations searching step to the QUBO problem, we solved the biggest ECDLP problem using the quantum method nowadays. Using Leap hybrid sampler on the D-Wave Leap cloud, we broke ECDLP for elliptic curve defined over 8-bit prime field $\mathbb{F}_p$, where $p = 251$.

2   Index Calculus Method on Elliptic Curves Using Quantum Annealing

2.1   Index Calculus and Summation Polynomials

One of the most interesting index calculus method on elliptic curve was firstly described by Semaev in [9], where he introduced summation polynomials. The summation polynomials have been defined there for elliptic curve in short Weierstrass form $E/\mathbb{F}_p : y^2 = x^3 + Ax + B$. Semaev summation polynomials have roots, when curve points sum to $\mathcal{O}$. The 2-nd Semaev polynomial is given by

$$f_3(x_1, x_2) = x_1 - x_2. \tag{1}$$

Using elementary methods it is also possible to find 3-rd Seamev polynomial as

$$f_3(x_1, x_2, x_3) = x_2^2 x_3^2 - 2x_1 x_2 x_3^2 + x_1^2 x_3^2 - 2x_1 x_2^2 x_3 - 2x_1^2 x_2 x_3$$
$$-2Ax_2 x_3 - 2Ax_1 x_3 - 4Bx_3 + x_1^2 x_2^2 - 2Ax_1 x_2 - 4Bx_2 - 4Bx_1 + A^2. \tag{2}$$

For any $m \geq 4$ and $m - 3 \geq k \geq 1$, one can find $f_m(x_1, \ldots, x_m)$ as

$$f_m(x_1, \ldots, x_m) = Res(f_{m-k}(x_1, \ldots, x_{m-k-1}, x), f_{k+2}(x_{m-k}, \ldots, x_m, x)). \tag{3}$$

$m + 1$-th Summation polynomials $f_{m+1}(x_1, \ldots, x_m, x_R)$ is equal to zero iff there exist $y_1, \ldots, y_m$, for which every point of the form (x_i, y_i), where $i = \overline{1, m}$, lies on an elliptic curve and their sum $(x_1, y_1) + \cdots + (x_m, y_m)$ is equal to $(x_R, y_R) \in E(\overline{\mathbb{K}})$. Point (x_R, y_R) is computed as $[\alpha]P + [\beta]Q$ for some randomly chosen α and β. The roots of polynomial f_{m+1} should be found with a high probability if x_i is bounded by $p^{\frac{1}{m}}$.

.

2.2   Transformation of Relation Searching Problem into the QUBO Problem

Let us define the Semaev approach of relation searching

Problem 1

$$
\begin{cases}
f_{m+1}(x_1, \ldots, x_m, x_R) = 0, \\
0 \le x_1 \le p^{\frac{1}{m}}, \\
\ldots \\
0 \le x_m \le p^{\frac{1}{m}},
\end{cases}
\tag{4}
$$

where $f_{m+1}(x_1, \ldots, x_m, x_R)$ is $m+1$-th Semaev polynomial.

We consider an approach of Semaev of relations searching given by Problem 1. To formulate the QUBO problem, we have to have some function to minimize, and we should not have any constraints. Problem 1 consists only of constraints. So to transform, one needs to take the following steps.

1. At first, all variables need to be changed to binary form. If variable z is from interval $U = \overline{a, b}$, the easiest way is to find surjection $g : 2^l \to U$, where $z = g(z_1, \ldots z_l) = a + \sum_{i=0}^{l-1} (2^i z_i) + ((b-a)+1-2^l)z_l$, and $l = \lfloor \log_2 (b-a) \rfloor$. This idea may be found in [2].
2. After transforming each of the variables, one has to make substitutions in polynomial f_{m+1}.
3. In the next step, we linearize the equation $f_{m+1}(x_1, \ldots, x_m, x_R) = 0$, so all monomials of the degree of bigger than 1 have to be substituted using new variables $x_i x_j = u_k$. Additionally, these constraints also need to be changed to the penalty to add it to the QUBO problem, but penalties will be added later. Each penalty monomial of the form $x_i x_j x_l$ will be constructed, according to [6], in the following way $x_i x_j x_l \to u_k x_l = x_l u_k + 2(x_i x_j - 2u_k(x_i + x_j) + 3u_k)$.
4. In the next step, it is necessary to convert equation $f_{m+1}(x_1, \ldots, x_m, x_R) = 0$, which is a modular equation, to equation over $\mathbb{Z}$. To make such transformation, it is necessary to write $f_{m+1}(x_1, \ldots, x_m, x_R) - vp = 0$, where v is some positive integer. One can bound v if he previously computes the maximal value of f_{m+1} over $\mathbb{Z}$. Let us assume that maximal value of f_{m+1} over $\mathbb{Z}$ is equal to f_{max}. Then bit length of v is less or equal to $\log_2 \left\lfloor \frac{f_{max}}{p} \right\rfloor + 1$. Of course, we should also transform v into binary form, as presented in Step 1.
5. Finally, to use constraint that $f_{m+1}(x_1, \ldots, x_m, x_R) - vp = 0$ in a minimization problem, one has to use this constraint as a penalty, where the right values of variables result in that such penalty is equal to 0, and the penalty is bigger than 0 otherwise. So one needs to construct this penalty as $(f_{m+1}(x_1, \ldots, x_m, x_R) - vp)^2$. Because $f_{m+1}(x_1, \ldots, x_m, x_R)$ was previously linearized, in $(f_{m+1}(x_1, \ldots, x_m, x_R) - vp)^2$ only monomials of degree 2 may exist. Moreover, now one can add penalties obtained during linearization in the previous step. Each penalty is multiplied by a square of maximal coefficient appearing in $f_{m+1}(x_1, \ldots, x_m, x_R)$, to get a high probability that minimal energy will be obtained only for a proper solution.

After making the steps above, one obtains the QUBO problem, which has to be minimized.

Therefore, the number of variables in the QUBO problem is equal to $2\left(\left\lfloor\log_2 p^{\frac{1}{m}}\right\rfloor + 1\right) + \lfloor\log_2 v\rfloor + 1 + c$, where c is the number of auxiliary variables obtained during linearization.

Now we will estimate how many variables one needs to solve Problem 1 for $m = 2$. Let us note that (after transformation to binary form) f_3 is polynomial of degree 4 of $2l$ boolean variables, where $l = \left\lfloor\log_2 p^{\frac{1}{m}}\right\rfloor + 1$. It means that $(f_{m+1}(x_1, \ldots, x_m, x_R) - vp)^2$ is polynomial of degree 8 of $2\left(\left\lfloor\log_2 p^{\frac{1}{m}}\right\rfloor + 1\right) + \lfloor\log_2 v\rfloor + 1$ variables. Let us try to estimate the maximal value of v. Let us note that after transformation to boolean variables, in $f_{m+1}(x_1, \ldots, x_m, x_R)$ appear all possible combinations of monomials, where is one monomial of degree 0, $2l$ monomials of degree 1, $3l^2 - 2l$ monomials of degree 2, $2l(l^2 - l)$ monomials of degree 3, and $(l^2 - l)^2$ monomials of degree 4. Summing up, we have $(l^2 - l)^2 + 2l(l^2 - l) + (3l^2 - 2l) + 2l + 1 = l^4 + 2l^2 + 1 = (l^2 + 1)^2$ monomials at most. It means that $v_{max} = \left\lfloor\frac{(l^2+1)^2(p-1)}{p}\right\rfloor < (l^2 + 1)^2$.

For $m = 2$, if one wants to linearize polynomial $(f_{m+1}(x_1, \ldots, x_m, x_R) - vp)^2$ (penalties should be stored and added later, but there will not be necessary to obtain more auxiliary variables), then f_{m+1} will consist of at most $(l^2 + 1)^2$ variables. It means that $(f_{m+1}(x_1, \ldots, x_m, x_R) - vp)^2$ will require at most $(l^2 + 1)^2 + \lfloor\log_2 (l^2 + 1)^2\rfloor + 1$ variables, because v is equal to at most $v_{max} = (l^2 + 1)^2$ and the bit length of v is then equal to $\lfloor\log_2 (l^2 + 1)^2\rfloor + 1$.

In the case of $m \geq 3$, we can estimate the total number of variables in the following way. At first, let us note that f_{m+1} is polynomial of $m + 1$ variables, but in this case, the last variable (x_R) is fixed, so it means that we have m variables. For each variable in $m + 1$-th Semaev polynomial, this polynomial is of degree 2^{m-1}. As previously, let us denote the bit length of each variable by l. Then, after transformation to binary form, f_{m+1} will consist of at most $\left(\sum_{i=0}^{2^{m-1}} l^i\right)^m$ monomials. If one wants to linearize this polynomial (penalties should be stored and added later, but there will not be necessary to obtain more auxiliary variables), then f_{m+1} will consist of at most $s = \left(\sum_{i=0}^{2^{m-1}} l^i\right)^m$ variables. It means that $(f_{m+1}(x_1, \ldots, x_m, x_R) - vp)^2$ will require at most $s + \lfloor\log_2 s\rfloor + 1$ variables, because v is equal to at most s and the bit length of v is equal to $\lfloor\log_2 s\rfloor + 1$.

Lagrange coefficient in penalties is equal to $2\left(Coeff_{max}\right)^2$, where $Coeff_{max}$ is the maximal coefficient of polynomial $f_{m+1}(x_1, \ldots, x_m, x_R)$. It is possible to obtain a QUBO problem in such a form that a minimal solution is always equal to 0, including offset. In such case, Lagrange coefficient of polynomial $f_{m+1}(x_1, \ldots, x_m, x_R)$ should be equal to $v_{max}p < \left(\sum_{i=0}^{2^{m-1}} l^i\right)^m p$. The disadvantage of this method is that one obtains larger coefficients in the QUBO problem. Therefore, it is harder to find the optimal solution because the minimal energy gap will be proportionally smaller.

Moreover, the QUBO problem may be (efficiently) solved on computers using quantum annealing like, for example, a D-Wave computer. An example of applying the method presented above for an elliptic curve $E/\mathbb{F}_{13} : y^2 = x^3 + 2x + 4$, where $\#E(\mathbb{F}_p) = 17$ is presented in https://github.com/Michal-Wronski/ECDLP-index-calculus-using-QUBO/blob/main/QUBO_Example.pdf.

3   Experiments, Results, and Discussion

We analyzed an approach when the relation searching problem is transformed into the QUBO problem. We did several experiments for different sizes of p. In each experiment, we looked for a relationship for a given random point R. We used Magma Computational Algebra System (http://magma.maths.usyd.edu.au/magma/) for precomputations up to obtaining problems in the QUBO form. We used the D-Wave Leap cloud (cloud.dwavesys.com/leap/), which allows us to access the D-Wave computer remotely. Using this environment and D-Wave hybrid Leap sampler, we found a discrete logarithm on the elliptic curve $E/\mathbb{F}_p : y^2 = x^3 + Ax + B$, over 8-bit prime $p = 251$, where $A = 1, B = 4$. The curve E order is equal to $\#E(\mathbb{F}_p) = 271$, which is prime. The generator is equal to $P = (128, 44)$, and the resulting point is equal to $Q = [k]P = (95, 73)$. We used the previous section index calculus method applying the QUBO problem, with $m = 2$, to find discrete logarithm $\log_P Q = k = 157$. It is, of course, a tiny example. Still, one should note that according to [7] and [8], breaking ECDLP for the elliptic curve over an 8-bit prime would require 75 and 88 logical qubits, respectively, which is more than the biggest quantum computers nowadays have.

Using these estimations and number of variables, for $m = 2$, being not greater than $(l^2 + 1)^2 + \lfloor \log_2 (l^2 + 1)^2 \rfloor + 1$, we obtained that, using D-Wave, it would be (optimistically) possible to break ECDLP for 23-bit prime at most, interpreting our QUBO problem as dense (what seems to be more accurate in this case) and for 64-bit prime at most, interpreting our QUBO problem as general. Let us note that according to D-Wave Advantage documentation (https://www.dwavesys.com/d-wave-two-system), the maximal number of variables for dense problems is equal to $20,000$ and for general problems is equal to $1,000,0000$. One should also note that these are only theoretical expectations because the problem is minimal energy gap, which is proportionally smaller while p is growing. It results in that for larger p and larger m, the probability of obtaining minimal solution instead of suboptimal decreases.

Figure 1 presents how many variables obtained the QUBO problem require, depending on the bit length of p and given m. Let us note that for $m = 3$ and $m = 4$, the number of variables grows very fast. Thus, solving ECDLP using the index calculus method for these values of m (or larger) and D-Wave is impractical nowadays.

Furthermore, the size of the QUBO problem for relation searching, however, is polynomial to $\log_2 p$, but it grows very fast when m grows, and therefore, this approach seems to be impractical for solving real problems.

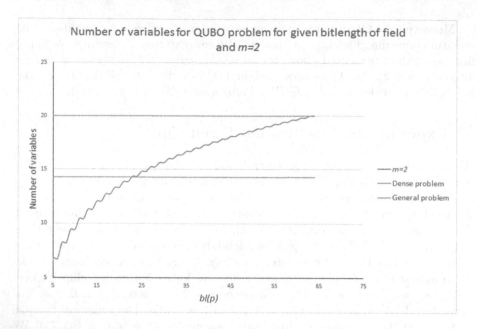

Fig. 1. Maximal total number of variables of QUBO problem for relation searching for $m = 2$ using D-Wave.

4   Conclusion and Further Works

Searching for a single relation depends on the computational complexity of solving a given optimization problem. It is possible to transform the relation searching problem into the QUBO problem. Solving the QUBO problem is exponential using classical algorithms [1], but the complexity of solving the QUBO problem using quantum annealing is unknown. Using the D-Wave Leap cloud, we broke ECDLP for the 8-bit prime field. Moreover, it would be possible to use D-Wave to break ECDLP for a maximally 23-bit prime field in a very optimistic case. However, it is still a small size problem compared to classical computers. It seems that it is far beyond the abilities of general-purpose quantum computers available nowadays.

The efficiency of the approach using QUBO may be increased by applying the following improvements:

- applying the different algorithm of quadratization of the polynomial $(f_{m+1}(x_1, \ldots, x_m, x_R) - vp)^2$, which would result in less number of total variables or smaller connectivity between variables,
- modifying a method of translation of relation searching step to the QUBO problem to obtain the larger value of minimal energy gap and thus decreasing the probability of obtaining suboptimal solution instead of the optimal solution,

– manual embedding of the given QUBO problem to the D-Wave Advantage computer. Using D-Wave hybrid Leap sampler, one cannot control how the problem is decomposed and if the solution is obtained classically or quantumly. Moreover, automatic embedding may give improper solutions.

It seems that if improvements above would be possible to apply, the index calculus method using QUBO would be much more efficient and could be solved on D-Wave for larger fields. Summing up, this approach seems to have potential and more research in this area should be done.

Acknowledgments. I would like to thank Christophe Petit for his apt comments and advice during the preparation of this paper.

References

1. Borle, A., Lomonaco, S.J.: Analyzing the quantum annealing approach for solving linear least squares problems. In: Das, G.K., Mandal, P.S., Mukhopadhyaya, K., Nakano, S. (eds.) WALCOM 2019. LNCS, vol. 11355, pp. 289–301. Springer, Cham (2019). https://doi.org/10.1007/978-3-030-10564-8_23
2. Chen, Y.A., Gao, X.S., Yuan, C.M.: Quantum algorithm for optimization and polynomial system solving over finite field and application to cryptanalysis. arXiv preprint arXiv:1802.03856 (2018)
3. Dridi, R., Alghassi, H.: Prime factorization using quantum annealing and computational algebraic geometry. Sci. Rep. **7**, 43048 (2017)
4. Greene, T.: Google reclaims quantum computer crown with 72 qubit processor (2018). https://thenextweb.com/artificial-intelligence/2018/03/06/google-reclaims-quantum-computer-crown-with-72-qubit-processor
5. Intel: The future of quantum computing is counted in qubits (2018). https://newsroom.intel.com/news/future-quantum-computing-counted-qubits/#gs.iiybkc
6. Jiang, S., Britt, K.A., McCaskey, A.J., Humble, T.S., Kais, S.: Quantum annealing for prime factorization. Sci. Rep. **8**(1), 1–9 (2018)
7. Proos, J., Zalka, C.: Shor's discrete logarithm quantum algorithm for elliptic curves. arXiv preprint quant-ph/0301141 (2003)
8. Roetteler, M., Naehrig, M., Svore, K.M., Lauter, K.: Quantum resource estimates for computing elliptic curve discrete logarithms. In: Takagi, T., Peyrin, T. (eds.) ASIACRYPT 2017, Part II. LNCS, vol. 10625, pp. 241–270. Springer, Cham (2017). https://doi.org/10.1007/978-3-319-70697-9_9
9. Semaev, I.A.: Summation polynomials and the discrete logarithm problem on elliptic curves. IACR Cryptol. ePrint Arch. **2004**, 31 (2004)
10. Shankland, S.: IBM's new 53-qubit quantum computer is its biggest yet (2019). https://www.cnet.com/news/ibm-new-53-qubit-quantum-computer-is-its-biggest-yet/
11. Wang, B., Hu, F., Yao, H., Wang, C.: Prime factorization algorithm based on parameter optimization of Ising model. Sci. Rep. **10**(1), 1–10 (2020)

Simulations of Flow and Transport:
Modeling, Algorithms and Computation

Multi-phase Compressible Compositional Simulations with Phase Equilibrium Computation in the *VTN* Specification

Tomáš Smejkal$^{(\boxtimes)}$ and Jiří Mikyška

Czech Technical University in Prague, Faculty of Nuclear Sciences and Physical Engineering, Department of Mathematics, Trojanova 13, 120 00 Prague 2, Czech Republic
`tomas.smejkal@fjfi.cvut.cz`

Abstract. In this paper, we present a numerical solution of a multi-phase compressible Darcy's flow of a multi-component mixture in a porous medium. The mathematical model consists of mass conservation equation of each component, extended Darcy's law for each phase, and an appropriate set of the initial and boundary conditions. The phase split is computed using the phase equilibrium computation in the *VTN*-specification (known as VTN-flash). The transport equations are solved numerically using the mixed-hybrid finite element method and a novel iterative IMPEC scheme [1]. We provide two examples showing the performance of the numerical scheme.

Keywords: Compositional simulations · Multi-phase flow · Phase equilibrium computation · Mixed-hybrid finite element method · VTN-flash · VTN-stability · Iterative IMPEC · Darcy's flow

1 Introduction

The mathematical modeling of compositional flow in a porous medium is an important topic in chemical engineering and has many applications in the industry, e.g., CO_2 sequestration or enhanced oil recovery. The mathematical model has to include a transport equation for each component in the mixture and a thermodynamical model describing the local equilibrium behavior.

The work was supported by the project Development and application of advanced methods for mathematical modeling of natural and industrial processes using high-performance computing, project no. SGS20/184/OHK4/3T/14 of the Student Grant Agency of the Czech Technical University in Prague, 2020–2022, the project Multiphase flow, transport, and structural changes related to water freezing and thawing in the subsurface of the Czech Science Foundation, project no. 21-09093S, and the project Center for advanced applied science, project no. CZ.02.1.01/0.0/0.0/16-019/0000778, supported by the Operational Programme Research, Development and Education, co-financed by the European Structural and Investment Funds and the state budget of the Czech Republic, 2018–2023.

© Springer Nature Switzerland AG 2021
M. Paszynski et al. (Eds.): ICCS 2021, LNCS 12747, pp. 159–172, 2021.
https://doi.org/10.1007/978-3-030-77980-1_13

In literature, two main approaches for solving the transport equations are common. The first approach, known as IMPEC [2], solves the equations in two steps. First, the pressure equation is solved implicitly to get the pressure field. Then, the concentrations of the first $n - 1$ components are updated explicitly using the pressure from the previous step. The concentration of the last component is updated using the previous ones, the total concentration, and the equation of state. The conservation of mass holds for the $n - 1$ components. However, for the last n-th component, the conservation of mass does not hold [1]. Chen et al. [1] presented a novel iterative IMPEC scheme where the conservation of mass of all components is guaranteed. An alternative to the IMPEC approach is the one of Young and Stephenson [3], where a method based on the Newton-Raphson iterations is used.

Concerning the thermodynamical model, traditionally, the PTN approach (constant pressure, temperature, and moles) [4,5] is used to determine the composition of equilibrium phases. No matter how wide-spread the PTN-specification is, the approach has some limitations [6,7], e.g., the equilibrium state of the system is not always determined uniquely. Alternatively, the VTN approach (constant volume, temperature, and moles) [6,8,9] can be used to determine the equilibrium state. Since most equations of state are given explicitly in pressure, i.e., $p = p(T, V, N_1, \ldots, N_n)$, the VTN-approach has some benefits, e.g., the inversion of the equation of state does not have be performed, and the equilibrium states are uniquely determined.

In this work, we are interested in modeling of the compositional flow with the use of the phase equilibrium computation. One approach, is using the IMPEC method with PTN approach, e.g., [10–12]. Alternatively, the VTN approach can be used. In our previous work [7], we use a fully implicit scheme, which gives the pressure field directly. However, this approach is computationally intensive. Therefore, we are proposing an alternative method based on the IMPEC strategy using the VTN-specification.

In this paper, we present a new numerical solution of the multi-phase compositional model. The solution is based on a novel iterative IMPEC scheme [1] that was originally developed for the single-phase compositional flow. In this paper, we extend this method to multi-phase problems. The stabilization of the numerical scheme is ensured by an upwind technique. The chemical equilibrium is computed locally on each finite element using the VTN-phase stability testing and VTN-phase equilibrium computation. In this approach, the Helmholtz free energy density of the system is minimized to obtain the equilibrium state. We are using the Newton-Raphson method with line-search and the modified Cholesky decomposition to find the minimum [8,13].

The structure of this paper is as follows. In Sect. 2, the physical and mathematical model describing compressible multi-phase multi-component compositional flow will be presented. In Sect. 3, the numerical solution will be given. In Sect. 4, examples showing the performance will be presented. In Sect. 5, the results are discussed, and some conclusions are drawn.

2   Physical and Mathematical Model

2.1   Physical Model

In this paper, the studied system is a fixed porous medium filled with a multi-component fluid. The porous medium in our interests are the hydrocarbon reservoirs. Based on the injection and the boundary condition, we study the flow of this multi-component multi-phase fluid through the fixed porous medium.

2.2   Transport Equations

Consider a mixture of n components with a constant temperature T. The mass balance equation for component $i \in \widehat{n}$ (the symbol $\widehat{n}$ represents a set of positive integers not exceeding n) is

$$\frac{\partial(\phi c_i)}{\partial t} + \boldsymbol{\nabla} \cdot \mathbf{q}_i = f_i, \tag{1}$$

where ϕ [-] is the porosity, c_i [mol m^{-3}] is the molar concentration (density) of the i-th component, $\mathbf{q}_i$ [mol m^{-2} s^{-1}] is the flux of the i-th component, and f_i [mol m^{-3} s^{-1}] is the source/sink of the i-th component. For a multi-phase system without diffusion, the flux $\mathbf{q}_i$ can be expressed as

$$\mathbf{q}_i = \left(\sum_{\alpha=1}^{\varPi} c_{\alpha,i} \mathbf{u}_\alpha \right), \tag{2}$$

where $c_{\alpha,i}$ [mol m^{-3}] is the concentration of the i-th component in phase α, $\varPi$ is the number of phases presented in the phase split, and $\mathbf{u}_\alpha$ [m s^{-1}] is the velocity of phase α. The relation between concentrations c_i and $c_{\alpha,i}$ is presented in Sect. 2.4. The velocity of each phase is modelled using Darcy's law

$$\mathbf{u}_\alpha = -\lambda_\alpha \mathbf{K} \left(\boldsymbol{\nabla} p - \rho_\alpha \mathbf{g} \right), \tag{3}$$

where λ_α [kg^{-1} m s] is the mobility of phase α, $\mathbf{K}$ [m^2] is the intrinsic permeability tensor, p [Pa] is the pressure, ρ_α [kg m^{-3}] is the mass density of phase α, and $\mathbf{g}$ [m s^{-2}] is the gravity acceleration. The mobility and the density are calculated using

$$\lambda_\alpha = \frac{k_{r\alpha}(S_\alpha)}{\eta_\alpha \left(T, c_{\alpha,1}, \ldots, c_{\alpha,n} \right)}, \quad \rho_\alpha = \sum_{i=1}^{n} c_{\alpha,i} M_i, \tag{4}$$

where $k_{r\alpha}$ [-] is the relative permeability, S_α [-] is the saturation, η_α [kg m^{-1} s^{-1}] is the dynamic viscosity, and M_i [kg mol^{-1}] is the molar weight of the i-th component. In this work, we are using a linear model to compute the relative permeability:

$$k_{r\alpha}(S_\alpha) = S_\alpha. \tag{5}$$

The dynamic viscosity η_α is calculated using the Lohrenz, Bray and Clark model [14]. The mathematical model has to be supplemented with an equation which connects the concentrations and the pressure:

$$p = p^{(eq)}(c_1, \ldots, c_n). \tag{6}$$

Details are in Sect. 2.4. Using Eq. (6), the mass conservation (1), and the chain rule

$$\frac{\partial p}{\partial t} = \sum_{i=1}^{n} \frac{\partial p^{(eq)}}{\partial c_i} \frac{\partial c_i}{\partial t}, \tag{7}$$

the equation known-as pressure equation can be derived. In the *VTN*-formulation the pressure equation reads as

$$\phi \frac{\partial p}{\partial t} + \sum_{i=1}^{n} \Theta_i (\boldsymbol{\nabla} \cdot \mathbf{q}_i - f_i) = 0, \tag{8}$$

where $\Theta_i = \frac{\partial p^{(eq)}}{\partial c_i}$.

2.3 Fluxes Definition

In this section, we define fluxes needed for the description of the numerical scheme. Let

$$\mathbf{q}_{\alpha,i} = c_{\alpha,i} \mathbf{u}_\alpha \tag{9}$$

be the flux of the i-component of phase α. Then, the flux of phase α is

$$\mathbf{q}_\alpha = \sum_{i=1}^{n} \mathbf{q}_{\alpha,i} = c_\alpha \mathbf{u}_\alpha, \tag{10}$$

where $c_\alpha = \sum_{i=1}^{n} c_{\alpha,i}$ is the total concentration of phase α. Lastly, the total flux $\mathbf{q}$ is defined as

$$\mathbf{q} = \sum_{\alpha=1}^{\Pi} \mathbf{q}_\alpha = \sum_{\alpha=1}^{\Pi} c_\alpha \mathbf{u}_\alpha, \tag{11}$$

and the total velocity $\mathbf{u}$ as

$$\mathbf{u} = \sum_{\alpha=1}^{\Pi} \mathbf{u}_\alpha. \tag{12}$$

Inserting Eq. (3) into previous equation results in

$$\mathbf{u} = -\lambda \mathbf{K} \left(\boldsymbol{\nabla} p - \rho^{(avg)} \mathbf{g} \right), \tag{13}$$

where the total mobility λ and the average density $\rho^{(avg)}$ are defined as

$$\lambda = \sum_{\alpha=1}^{\Pi} \lambda_\alpha, \quad \rho^{(avg)} = \frac{\sum_{\alpha=1}^{\Pi} \lambda_\alpha \rho_\alpha}{\lambda}. \tag{14}$$

As the tensor $\mathbf{K}$ is positive definite, its inversion exists, and the gradient ∇p can be expressed from Eq. (13) as

$$\nabla p = -\lambda^{-1} \mathbf{K}^{-1} \mathbf{u} + \rho^{(avg)} \mathbf{g}. \tag{15}$$

Inserting previous equation into Darcy's law (3) results in

$$\mathbf{u}_\alpha = \lambda^{-1} \lambda_\alpha \left(\mathbf{u} - \sum_{\beta=1}^{\Pi} \lambda_\beta (\rho_\beta - \rho_\alpha) \mathbf{K} \mathbf{g} \right). \tag{16}$$

Therefore, the flux of the i-th component in phase α is

$$\mathbf{q}_{\alpha,i} = c_{\alpha,i} \lambda^{-1} \lambda_\alpha \left(\mathbf{u} - \sum_{\beta=1}^{\Pi} \lambda_\beta (\rho_\beta - \rho_\alpha) \mathbf{K} \mathbf{g} \right). \tag{17}$$

2.4 Phase Stability Testing and Phase Equilibrium Calculation

Depending on the mixture's temperature and concentrations, the state can be in one or more phases. The phase stability testing [6] and phase equilibrium computation [8] is used to determine the number of phases and the composition of each phase described by $c_1, \ldots, c_n$ and T. In the VTN-phase stability testing the goal is to predict whether a given state is stable or if this state is unstable and splitting will occur. The VTN-phase equilibrium computation is used to determine the composition of the equilibrium state. The problem can be defined as an optimization task minimizing the objective function

$$a^{(\Pi)} \left(\mathbf{c}^{(1)}, \ldots, \mathbf{c}^{(\Pi)}, \mathbf{S} \right) = \sum_{\alpha=1}^{\Pi} S_\alpha a \left(c_{\alpha,1}, \ldots, c_{\alpha,n} \right) \tag{18}$$

subject to

$$\sum_{\alpha=1}^{\Pi} S_\alpha = 1, \quad \sum_{\alpha=1}^{\Pi} S_\alpha c_{\alpha,i} = c_i^*, \quad i \in \widehat{n}, \tag{19}$$

where a is the Helmholtz free energy density (for details see, e.g., [15,16]), $\mathbf{S} = (S_1, \ldots, S_\Pi)^T$ and $\mathbf{c}^{(\alpha)} = (c_{\alpha,1}, \ldots, c_{\alpha,n})^T$ for $\alpha \in \widehat{\Pi}$. The necessary equilibrium conditions are [8]

$$p(c_{\alpha,1}, \ldots, c_{\alpha,n}) = p(c_{\beta,1}, \ldots, c_{\beta,n}), \quad \forall \alpha, \beta \in \widehat{\Pi}, \alpha \neq \beta, \tag{20}$$

$$\mu_i(c_{\alpha,1}, \ldots, c_{\alpha,n}) = \mu_i(c_{\beta,1}, \ldots, c_{\beta,n}), \quad \forall \alpha, \beta \in \widehat{\Pi}, \alpha \neq \beta, \forall i \in \widehat{n}, \tag{21}$$

where μ_i is the chemical potential of the i-th component. If the state is in one phase ($\Pi = 1$), the equilibrium pressure $p^{(eq)}$ (6) is given by the equation of state

$$p^{(eq)}(c_1, \ldots, c_n) = p^{(EOS)}(c_1, \ldots, c_n). \tag{22}$$

In this work, we use the Peng-Robinson equation of state [17] in the following form

$$p^{(EOS)}(c_1, \ldots, c_n) = \frac{\sum_{i=1}^n c_i RT}{1 - \sum_{i=1}^n b_i c_i} - \frac{\sum_{i,j=1}^n a_{ij} c_i c_j}{1 + 2\sum_{i=1}^n b_i c_i - \left(\sum_{i=1}^n b_i c_i\right)^2}, \tag{23}$$

where a_{ij}, b_i are parameters. See [15,17] for details. On the other hand, if the equilibrium state is in $\Pi > 1$ phases, the equilibrium pressure $p^{(eq)}$ is given by

$$p^{(eq)}(c_1, \ldots, c_n) = p^{(EOS)}(c_{\alpha,1}, \ldots, c_{\alpha,n}), \tag{24}$$

for an arbitrary $\alpha \in \widehat{\Pi}$ since the pressures of each phase in the phase equilibrium are equal (see Eq. (20)).

2.5   Initial and Boundary Conditions

Now, let us summarize the equations and define the initial and boundary conditions. Let $\Omega \subset \mathbb{R}^d$ be a bounded domain and J is a time interval. In $J \times \Omega$, we solve Eqs. (1) and (8) for $p = p(t, \mathbf{x})$ and $c_i = c_i(t, \mathbf{x})$, $i \in \widehat{n}$. The fluxes $\mathbf{q}_i$ are given by Eq. (2), and the velocities $\mathbf{u}_\alpha$ are computed using Darcy's law (3). The composition of the multi-phase state is determined by solving the optimization problem given by (18) and (19). The mathematical model has to be equipped with initial conditions and an appropriate set of boundary conditions. The initial conditions read as

$$c_i(0, \mathbf{x}) = c_i^{(ini)}, \quad \forall \mathbf{x} \in \Omega, \forall i \in \widehat{n}, \tag{25}$$

$$p(0, \mathbf{x}) = p^{(eq)}\left(c_1^{(ini)}, \ldots, c_n^{(ini)}\right), \quad \forall \mathbf{x} \in \Omega. \tag{26}$$

Moreover, we impose the following boundary conditions

$$p(t, \mathbf{x}) = p^{(D)}(t, \mathbf{x}), \mathbf{x} \in \Gamma_p, t \in J, \tag{27}$$

$$\mathbf{q}_i(t, \mathbf{x}) \cdot \mathbf{n}(\mathbf{x}) = 0, \mathbf{x} \in \Gamma_q, t \in J, \tag{28}$$

where $\mathbf{n}$ is the unit outward normal vector to the boundary $\partial\Omega$, $\Gamma_p \cup \Gamma_q = \partial\Omega$, and $\Gamma_p \cap \Gamma_q = \emptyset$.

3   Numerical Solution

In this work, we assume that the computation domain Ω is a 2D rectangular domain. We use a triangulation $\tau_\Omega = \left\{ K_i; i \in \widehat{N}_{el} \right\}$, where N_{El} is the number of elements. Moreover, we denote N_{Si} the number of sides.

3.1 Disretization of Darcy's Law

On each element $K \in \tau_\Omega$, we shall approximate $\mathbf{u}$ in the lowest order Raviar-Thomas space $\mathbf{RT}_0(K)$ [18,19]

$$\mathbf{u}(t,\mathbf{x}) = \sum_{E \in \partial K} u_{K,E}(t)\mathbf{w}_{K,E}(\mathbf{x}), \tag{29}$$

where $\mathbf{w}_{K,E}$ are the basis functions and $u_{K,E}$ is the velocity across the side E in the outward direction with respect to K. Multiplying Eq. (15) with function $\mathbf{w}_{K,E'}$, integrating over element $K \in \tau_\Omega$, and using the Gauss-Ostrogradski theorem results in the weak formulation of Darcy's law

$$\widehat{p}_{K,E'} - p_K = -\lambda_K^{-1} \sum_{E \in \partial K} u_{K,E}(t) \int_K (\mathbf{K}^{-1}\mathbf{w}_{K,E}) \cdot \mathbf{w}_{K,E'} d\mathbf{x} \tag{30}$$

$$+ \rho_K^{(avg)} \int_K \mathbf{g} \cdot \mathbf{w}_{K,E'} d\mathbf{x},$$

where we have denoted the average pressures on element K by p_K, average traces of the pressures on side E by $\widehat{p}_{K,E}$, and the average density on element K by $\rho_K^{(avg)}$. Denoting

$$B_{E,E'}^K - \int_K (\mathbf{K}^{-1}\mathbf{w}_{K,E}) \ \mathbf{w}_{K,E'} d\mathbf{x}, \quad C_{E'}^K - \int_K \mathbf{g} \cdot \mathbf{w}_{K,E'} d\mathbf{x}, \tag{31}$$

Eq. (30) reads as

$$\sum_{E \in \partial K} u_{K,E}(t) B_{E,E'}^K = \lambda_K \left(p_K - \widehat{p}_{K,E'} + \rho_K^{(avg)} C_{E'}^K \right). \tag{32}$$

This equation can be inverted and the velocities $u_{K,E}$ are expressed

$$u_{K,E}(t) = \lambda_K \left(D_E^K p_K - \sum_{E' \partial K} (B^K)_{E,E'}^{-1} \widehat{p}_{K,E'} + F_E^K \rho_K^{(avg)} \right), \tag{33}$$

where

$$D_E^K = \sum_{E' \in \partial K} (B^K)_{E,E'}^{-1}, \quad F_E^K = \sum_{E' \in \partial K} (B^K)_{E,E'}^{-1} C_{E'}^K. \tag{34}$$

Now, we will use continuity assumptions. If side E is not on the boundary, then,

$$u_{K',E} + u_{K,E} = 0, \tag{35}$$

$$\widehat{p}_{K,E} = \widehat{p}_{K',E} =: \widehat{p}_E, \tag{36}$$

where $K' \cap K = E$. If side E is on the boundary, then

$$\widehat{p}_{K,E} = p_E^{(D)}, \text{ for } E \in \Gamma_p, \tag{37}$$

$$u_{K,E} = 0, \text{ for } E \in \Gamma_q. \tag{38}$$

Therefore, in Eq. (33), the velocities $u_{K,E}(t)$ can be eliminated and the only unknowns are $p_K, \widehat{p}_E$. If $E \notin \partial\Omega$, Eq. (35) implies

$$0 = \sum_{K \supset E} \lambda_K \left(D_E^K p_K - \sum_{E' \in \partial K} \left(B^K\right)_{E,E'}^{-1} \widehat{p}_{E'} + F_E^K \rho_K^{(avg)} \right). \tag{39}$$

On the other hand, if $E \in \partial\Omega$, then

$$\widehat{p}_E = p_E^{(D)}, \text{ if } E \in \Gamma_p, \tag{40}$$

$$-\lambda_K D_E^K p_K + \sum_{E' \in \partial K} \lambda_K \left(B^K\right)_{E,E'}^{-1} \widehat{p}_{E'} = \lambda_K F_E^K \rho_K^{(avg)}, \text{ if } E \in \Gamma_q. \tag{41}$$

Previous Eqs. (39)–(41) form a system of linear equation for the unknowns $\widehat{p}_E, p_K$:

$$\mathbf{R}_1 \mathbf{p} + \mathbf{R}_2 \widehat{\mathbf{p}} = \mathbf{L}_1, \tag{42}$$

where $\mathbf{R}_1 \in \mathbb{R}^{N_{Si}, N_{El}}$, $\mathbf{R}_2 \in \mathbb{R}^{N_{Si}, N_{Si}}$, $\mathbf{L}_1 \in \mathbb{R}^{N_{Si}}$.

3.2   Discretization of Pressure Equation

Integrating the pressure Eq. (8) over an element $K \in \tau_\Omega$, and using the divergence theorem results in

$$0 = \phi_K |K| \frac{\mathrm{d}p_K}{\mathrm{d}t} + \sum_{i=1}^{n} \Theta_i \sum_{E \in \partial K} q_{i,K,E} - \sum_{i=1}^{n} \Theta_i \int_K f_i \mathrm{d}\mathbf{x} \tag{43}$$

Using $q_{i,K,E} = \sum_{\alpha=1}^{\Pi} q_{\alpha,i,K,E}$, Eq. (17), and the backwards Euler scheme, the previous equation can be approximated by

$$0 = \phi_K |K| \frac{p_K^{m+1} - p_K^m}{\Delta t} - \sum_{i=1}^{n} \Theta_i^{m+1} \int_K f_i^{m+1} \mathrm{d}\mathbf{x} + p_K^{m+1} X_K^{m+1}$$
$$+ \sum_{E' \in \partial K} \widehat{p}_{E'}^{m+1} Y_{E'}^{m+1} + Z_K^{m+1}, \tag{44}$$

where

$$X_K = \sum_{i=1}^{n} \sum_{E \in \partial K} \sum_{\alpha=1}^{\Pi(K)} \Theta_i c_{\alpha,i,K} \lambda_{\alpha,K} D_E^K \tag{45}$$

$$Y_{E'} = \sum_{i=1}^{n} \sum_{E \in \partial K} \sum_{\alpha=1}^{\Pi(K)} -\Theta_i c_{\alpha,i,K} \lambda_{\alpha,K} \left(B_K\right)_{E,E'}^{-1}, \tag{46}$$

$$Z_K = \sum_{i=1}^{n} \sum_{E \in \partial K} \sum_{\alpha=1}^{\Pi(K)} \lambda_K^{-1} \Theta_i c_{\alpha,i,K} \lambda_{\alpha,K} \left(\lambda_K F_E^K \rho_K^{avg} \right.$$
$$\left. - \sum_{\beta=1}^{\Pi(K)} \lambda_{\beta,K} (\rho_{\beta,K} - \rho_{\alpha,K}) F_E^K \right), \tag{47}$$

where $c_{\alpha,i,K}$ is the average concentration of the i-th component in phase α on element K. Equation (44) forms a system of linear equation for the unknown p_K^{m+1} and $\widehat{p}_{E'}^{m+1}$

$$\mathbf{R}_3\mathbf{p} + \mathbf{R}_4\widehat{\mathbf{p}} = \mathbf{L}_2, \tag{48}$$

where $\mathbf{R}_3 \in \mathbb{R}^{N_{El},N_{El}}$, $\mathbf{R}_4 \in \mathbb{R}^{N_{El},N_{Si}}$, and $\mathbf{L}_2 \in \mathbb{R}^{N_{El}}$. To conclude, combining Eqs. (42) and (48) gives the final system for the pressure field

$$\begin{pmatrix} \mathbf{R}_3 & \mathbf{R}_4 \\ \mathbf{R}_1 & \mathbf{R}_2 \end{pmatrix} \begin{pmatrix} \mathbf{p} \\ \widehat{\mathbf{p}} \end{pmatrix} = \begin{pmatrix} \mathbf{L}_2 \\ \mathbf{L}_1 \end{pmatrix}. \tag{49}$$

The matrix $\mathbf{R}_3$ is diagonal, therefore, its inversion $\mathbf{R}_3^{-1}$ is readily available. Multiplying Eq. (48) with $\mathbf{R}_3^{-1}$ gives

$$\mathbf{p} = \mathbf{R}_3^{-1}\mathbf{L}_2 - \mathbf{R}_3^{-1}\mathbf{R}_4\widehat{\mathbf{p}} \tag{50}$$

Therefore, the unknowns $\mathbf{p}$ can be eliminated from the system (49), and only the pressure traces $\widehat{\mathbf{p}}$ are computed using

$$(\mathbf{R}_2 - \mathbf{R}_1\mathbf{R}_3^{-1}\mathbf{R}_4)\widehat{\mathbf{p}} = \mathbf{L}_1 - \mathbf{R}_1\mathbf{R}_3^{-1}\mathbf{L}_2. \tag{51}$$

In this work, we use the C++ numerical library Armadillo [20,21] to solve system (51). Having the pressure traces $\widehat{\mathbf{p}}$, the pressures $\mathbf{p}$ and consequently, the discrete velocities $u_{K,E}$ are computed using Eqs. (50) and (33), respectively.

3.3   Solution of Transport Equations

Having the pressure field, the concentrations are updated using the explicit finite-volume method. Integrating Eq. (1) over $K \in \tau_\Omega$, using the divergence theorem, and the Euler scheme results with an approximation of Eq. (1):

$$c_{i,K}^{m+1} = c_{i,K}^m + \frac{\Delta t}{\phi|K|}\left(|K|f_{i,K}^m - \sum_{E\in\partial K} q_{i,K,E}^m\right), \tag{52}$$

where $c_{i,K}$ and $f_{i,K}$ are the average concentration and source/sink of the i-th component on element K, respectively. The fluxes $q_{i,K,E}$ are calculated using the upwind scheme

$$q_{i,K,E} = \begin{cases} \sum_{\alpha\in\widehat{\Pi^+}(K,E)} q_{\alpha,i,K,E} - \sum_{\beta\in\widehat{\Pi^+}(K,E)} q_{\beta,i,K',E}, & \forall E \notin \partial\Omega, \\ \sum_{\alpha\in\widehat{\Pi^+}(K,E)} q_{\alpha,i,K,E}, & \forall E \in \Gamma_p, \\ 0, & \forall E \in \Gamma_q. \end{cases} \tag{53}$$

where $\widehat{\Pi^+}(K,E) = \left\{\alpha \in \widehat{\Pi}(K); q_{\alpha,i,K,E} > 0\right\}$ for $E \in \partial K$ and

$$q_{\alpha,i,K,E} = c_{\alpha,i,K}\lambda_K^{-1}\lambda_{\alpha,K}\left(u_{K,E} - \sum_{\beta=1}^{\Pi(K)} \lambda_{\beta,K}\left(\rho_\beta - \rho_\alpha\right)F_{K,E}\right), \tag{54}$$

where $u_{K,E}$ is given by Eq. (33).

3.4 Algorithm for One Time Step Δt

Now, we present the full numerical algorithm. This iterative IMPEC algorithm is based on numerical scheme presented in [1]. Having solution on time-level t_m, the solution on time level t_{m+1} is computed using the following algorithm.

1. Set $l = 0$ and $p_K^{m+1,0} = p_K^m$, $c_{i,K}^{m+1,0} = c_{i,K}^m$, $\Theta_{i,K}^{m+1,0} = \frac{\partial p^{(eq)}}{\partial c_i}\left(c_{1,K}^m, \dots, c_{n,K}^m\right)$ for $K \in \tau_\Omega, i \in \widehat{n}$.

2. Set $l = l + 1$.

3. On each element $K \in \tau_\Omega$, compute $c_{\alpha,i,K}^{m+1,l-1}$ and $S_{\alpha,K}^{m+1,l-1}$ by solving the phase equilibrium computation given by Eqs. (18)–(19) with initial concentration $c_{1,K}^{m+1,l-1}, \dots, c_{n,K}^{m+1,l-1}$. In this work, we are using numerical solution presented in [13].

4. On each element $K \in \tau_\Omega$, update $\lambda_K^{m+1,l-1}$ and $\rho_K^{(avg),m+1,l-1}$ using Eqs. (4) and (14) with values $c_{\alpha,i,K}^{m+1,l-1}$ and $S_{\alpha,K}^{n+1,l-1}$ computed in the previous step.

5. Find $p_K^{m+1,l}$ and $u_{K,E}^{m+1,l}$ by solving system (49) with the concentrations $c_{\alpha,i,K}^{m+1,l-1}$, coefficients $\Theta_{i,K}^{m+1,l-1}$, total mobility $\lambda_K^{m+1,l-1}$, and average density $\rho_K^{(avg),m+1,l-1}$.

6. On each element $K \in \tau_\Omega$, for all $i \in \widehat{n}$ update $c_{i,K}^{m+1,l}$ explicitly by

$$c_{i,K}^{m+1,l} = c_{i,K}^m + \frac{\Delta t}{\phi|K|}\left(|K|f_{i,K}^m - \sum_{E \in \partial K} q_{i,K,E}^{m+1,l-1}\right), \tag{55}$$

where the flux $q_{i,K,E}^{m+1,l-1}$ is evaluated using the velocity $u_{K,E}^{m+1,l}$ and concentrations $c_{\alpha,i,K}^{m+1,l-1}$.

7. On each element $K \in \tau_\Omega$, for all $i \in \widehat{n}$ update $\Theta_{i,K}^{m+1,l}$ by

$$\Theta_{i,K}^{m+1,l} = \frac{p^{(eq)}(\mathbf{c}^{(1)}) - p^{(eq)}(\mathbf{c}^{(2)})}{c_{i,K}^{m+1,l} - c_{i,K}^m}, \tag{56}$$

where

$$\mathbf{c}^{(1)} = \left(c_{1,K}^{m+1,l}, \dots, c_{i,K}^{m+1,l}, c_{i+1,K}^m, \dots, c_{n,K}^m\right)^T, \tag{57}$$

$$\mathbf{c}^{(2)} = \left(c_{1,K}^{m+1,l}, \dots, c_{i-1,K}^{m+1,l}, c_{i,K}^m, \dots, c_{n,K}^m\right)^T. \tag{58}$$

To compute the pressures, the phase equilibrium computation is used to determine the number of phases and the equilibrium pressure.

8. Check convergence. If the convergence criteria are met, set

$$p_K^{m+1} = p_K^{m+1,l}, \quad c_{i,K}^{m+1} = c_{i,K}^{m+1,l}, \quad \forall K \in \tau_\Omega, \forall i \in \widehat{n}, \tag{59}$$

and terminate the Algorithm. Otherwise, go to step 2. In this work, we terminate the algorithm if the maximum number of iterations $l_{\max}$ is reached or the criterion

$$\max\left\{\frac{\left\|p^{m+1,l}-p^{m+1,l-1}\right\|}{\left\|p^{m+1,l}\right\|}, \sum_{i=1}^{n}\frac{\left\|c_i^{m+1,l}-c_i^{m+1,l-1}\right\|}{\left\|c_i^{m+1,l}\right\|}, \right.$$

$$\left. \sum_{i=1}^{n}\frac{\left\|\Theta_i^{m+1,l}-\Theta_i^{m+1,l-1}\right\|}{\left\|\Theta_i^{m+1,l}\right\|}\right\} < \varepsilon, \quad (60)$$

is fullfilled. In previous equation $\|\cdot\|$ is the $L^2(\Omega)$ norm and ε is a given tolerance.

4 Numerical Results

In this section, we provide two numerical examples. In both examples, the computation domain Ω is a square domain of size 50×50 m with porosity $\phi = 0.2$ and isotropic permeability $\mathbf{K} = k = 9.87 \times 10^{-15}$ m^2, i.e., 10 mD. Moreover, we use a triangular mesh with $2 \times 20 \times 20$ elements, i.e., total 400 elements are used. The final time in both examples is $t_{\text{final}} = 150$ days. The ε tolerance is set to $\varepsilon = 10^{-8}$, and the maximum number of inner iterations is set to $l_{\max} = 30$. For the computation, a computer with Intel(R) Core(TM) i7-8700 (3.20 GHz) processor was used.

4.1 Example 1: C_1 Injection

In the first example, we simulate the injection of methane (C_1) into a horizontal (i.e., no gravity) reservoir. The reservoir is initially filled with a mixture of 95% propane and 5% methane at a constant pressure $p = 6.9$ MPa and temperature $T = 311$ K. The mixture with 95% of the methane is injected at the right bottom corner. The rate of the injection is 125.33 m^2 per day at atmospheric pressure and temperature 293 K. In Table 1, the parameters for the Peng-Robinson equation of state are presented. The binary interaction coefficient is $\delta_{C_1-C_3} = 0.0365$. The boundary of the domain is impermeable except for the outflow corner where pressure $p = 6.9$ MPa is maintained. The time step was set to $\Delta t = 3000$ s. In Fig. 1, the iso-lines of methane mole fraction at different times are depicted. The values are from 0.05 to 0.95 with a step size 0.1. Moreover, in Fig. 1, the two-phase region is depicted in the black color. The total computation time was 2.5 h.

4.2 Example 2: CO_2 Injection

In the second example, we simulate the injection of carbon dioxide (CO_2) into a vertical (i.e., with gravity) reservoir. The reservoir is initially filled with pure propane at a constant pressure $p = 5$ MPa and temperature $T = 311$ K. The CO_2 is injected at the right bottom corner. The rate of the injection is 125.33 m^2 per day at atmospheric pressure and temperature 293 K. In Table 1, the parameters for the Peng-Robinson equation of state are presented. The binary interaction coefficient is $\delta_{C_1-C_3} = 0.15$. The boundary of the domain is impermeable except for the outflow corner where pressure $p = 5$ MPa is maintained. To reach convergence, the time step had to be decreased to $\Delta t = 500$ seconds and the maximum number of iteration increased to $l_{max} = 50$. In Fig. 2, the iso-lines of carbon dioxide mole fraction at different times are depicted. The values are from 0.05 to 0.95 with step size 0.1. Moreover, in Fig. 2, the two-phase region is depicted in the black color. The total computation time was approximately 24 h.

Table 1. Component properties.

Component	T_{crit} [K]	P_{crit} [MPa]	ω [-]	M [g mol^{-1}]	V_{crit} [m^3 kg^{-1}]
C_1	190.56	4.599	0.011	16.0	6.10639×10^{-3}
C_3	369.83	4.248	0.153	44.096	4.53554×10^{-3}
CO_2	304.14	7.375	0.239	44.0	2.13589×10^{-3}

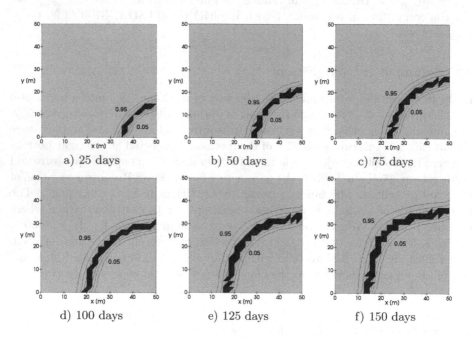

Fig. 1. The iso-lines of methane mole fraction in different times. The values are from 0.05 to 0.95 with step size 0.1. The two-phase area is depicted in the black color. Example 1: C_1 injection.

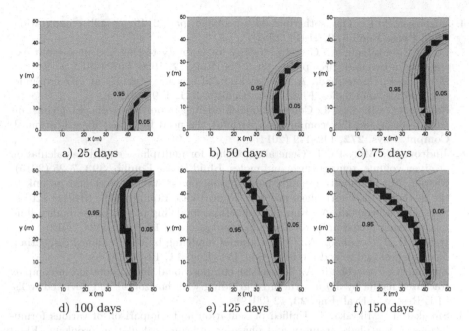

Fig. 2. The iso-lines of carbon dioxide mole fraction in different times. The values are from 0.05 to 0.95 with step size 0.1. The two-phase area is depicted in the black color. Example 2: CO_2 injection.

5   Conclusion

In this paper, we presented a new numerical solution of multi-phase compositional flow in a porous medium. The numerical solution is based on mixed-hybrid finite element method and a novel iterative IMPEC scheme. Unlike in tradition solvers, the local thermodynamical behaviour is determined by the phase equilibrium computation in the VTN-specification. Using this specification, unpleasant properties such as non-uniqueness of the equilibrium states are avoided. We provided two examples showing the performance of the numerical scheme. In the second example, the time step has to be significantly decreased to reach convergence. Investigation of this phenomenon is our current research.

References

1. Chen, H., Fan, X., Sun, S.: A fully mass-conservative iterative IMPEC method for multicomponent compressible flow in porous media. J. Comput. Appl. Math. **362**, 1–21 (2019)
2. Acs, G., Doleschall, S., Farkas, E.: General purpose compositional model. Soc. Pet. Eng. **25**, 543–553 (1985)
3. Young, L.C., Stephenson, R.E.: A generalized compositional approach for reservoir simulation. Soc. Petrol. Eng. J. **23**, 727–742 (1983)

4. Michelsen, M.L.: The isothermal flash problem, part 2. Phase-split computation. Fluid Phase Equilib. **9**, 21–40 (1982)
5. Li, Z., Firoozabadi, A.: General strategy for stability testing and phase-split calculation in two and three phases. Soc. Pet. Eng. **17**, 1096–1107 (2012)
6. Mikyška, J., Firoozabadi, A.: Investigation of mixture stability at given volume, temperature, and moles. Fluid Phase Equilib. **321**, 1–9 (2012)
7. Polívka, O., Mikyška, J.: Compositional modeling in porous media using constant volume flash and flux computation without the need for phase identification. J. Comput. Phys. **272**, 149–179 (2014)
8. Jindrová, T., Mikyška, J.: General algorithm for multiphase equilibria calculation at given volume, temperature, and moles. Fluid Phase Equilib. **393**, 7–25 (2015)
9. Castier, M.: Helmholtz function-based global phase stability test and its link to the isothermal-isochoric flash problem. Fluid Phase Equilib. **379**, 104–111 (2014)
10. Sun, S., Firoozabadi, A., Kou, J.: Numerical modeling of two-phase binary fluid mixing using mixed finite elements. Comput. Geosci. **16**, 1101–1124 (2012)
11. Hoteit, H., Firoozabadi, A.: Compositional modeling by the combined discontinuous Galerkin and mixed methods. Soc. Pet. Eng. **11**, 19–34 (2006)
12. Zidane, A., Firoozabadi, A.: Two-phase compositional flow simulation in complex fractured media by 3D unstructured gridding with horizontal and deviated wells. SPE Reservoir Eval. Eng. **23**, 12 (2019)
13. Smejkal, T., Mikyška, J.: Unified presentation and comparison of various formulations of the phase stability and phase equilibrium calculation problems. Fluid Phase Equilib. **476**, 61–88 (2018)
14. Lohrenz, J., Bray, B.G., Clark, C.R.: Calculating viscosities of reservoir fluids from their compositions. J. Petrol. Technol. **16**(10), 1171–1176 (1964)
15. Firoozabadi, A.: Thermodynamics and Applications of Hydrocarbons Energy Production. McGrew-Hill, New York (2015)
16. Smejkal, T., Mikyška, J.: Efficient solution of linear systems arising in the linearization of the VTN-phase stability problem using the Sherman-Morrison iterations. Fluid Phase Equilibria **527**, 112832 (2021). https://doi.org/10.1016/j.fluid.2020.112832
17. Peng, D.-Y., Robinson, D.B.: A new two-constant equation of state. Ind. Eng. Chem. Fundam. **15**, 59–64 (1976)
18. Brezzi, F., Fortin, M.: Mixed and Hybrid Finite Element Methods, vol. 15. Springer Science & Business Media (2012). https://doi.org/10.1007/978-1-4612-3172-1
19. Raviart, P.-A., Thomas, J.: Mathematical Aspects of Finite Element Methods. Springer (1977). https://doi.org/10.1007/BFb0064451
20. Eddelbuettel, D., Sanderson, C.: RcppArmadillo: accelerating R with high-performance C++ linear algebra. Comput. Stat. Data Anal. **71**, 1054–1063 (2014)
21. Sanderson, C., Curtin, R.: Armadillo: a template-based C++ library for linear algebra. J. Open Source Softw. **1**(2), 26 (2016)

A Three-Level Linearized Time Integration Scheme for Tumor Simulations with Cahn-Hilliard Equations

Maciej Smołka[ID], Maciej Woźniak[✉][ID], and Robert Schaefer[ID]

Institute of Computer Science, AGH University of Science and Technology,
Kraków, Poland
{smolka,macwozni,schaefer}@agh.edu.pl
https://informatyka.agh.edu.pl/

Abstract. The paper contains an analysis of a three-level linearized time integration scheme for Cahn-Hilliard equations. We start with a rigorous mixed strong/variational formulation of the appropriate initial boundary value problem taking into account the existence and uniqueness of its solution. Next we pass to the definition of two time integration schemes: the Crank-Nicolson and a three-level linearized ones. Both schemes are applied to the discrete version of Cahn-Hilliard equation obtained through the Galerkin approximation in space. We prove that the sequence of solutions of the mixed three level finite difference scheme combined with the Galerkin approximation converges when the time step length and the space approximation error decrease. We also recall the verification of the second order of this scheme and its unconditional stability with respect to the time variable. A comparative scalability analysis of parallel implementations of the schemes is also presented.

Keywords: Isogeometric analysis · Time-integration schemes · Tumor simulations · Cahn-Hilliard equations

1   Introduction

Cahn-Hilliard equations are widely used to describe the temporal evolution of two phases of a system engaged in the phase transition, like a solidifying liquid. It is a system of equations that can be reduced to one equation that is first order in time and fourth order in space. When using the finite element method this high spatial order requires the use of smooth spatial basis functions, like the ones coming from the isogeometric analysis (IGA). And, in fact, IGA has been successfully applied to the solution of Cahn-Hilliard equations, cf. [1,2]. On the other hand, Cahn-Hilliard equations has been applied to model the tumor growth as well, cf. [3,4]. The complexity of such problems is significant because they involve dynamic chemical and biological processes occurring in living tissues with interactions between cellular and vascular levels. The Cahn-Hilliard equations are applied to model interfaces between blood vessels and host tissue. There

© Springer Nature Switzerland AG 2021
M. Paszynski et al. (Eds.): ICCS 2021, LNCS 12747, pp. 173–185, 2021.
https://doi.org/10.1007/978-3-030-77980-1_14

are already several models of the tumor evolution [5,6] utilizing the isogeometric analysis concept. In this paper we follow on the approach utilizing Cahn-Hilliard equations described in paper [6]. To solve numerically the equations we propose an adaptation of the linearized three-level time integration scheme presented in [7], that on one hand is implicit and on the other hand makes it possible to use a direct solver at every time step. The approach presented in this paper is an alternative to the one described in [2].

2 Strong and Weak Formulations of Cahn-Hilliard Equations

This section presents the strong and weak formulations for the Cahn-Hilliard equations, based on [8]. As a strong one we consider the following Cauchy problem: Find $u \in C^1(0, T; C^4(\Omega))$ such that

$$u_t = \nabla \cdot (B(u)\nabla(-\gamma\Delta u + \Psi'(u))) \text{ on } \Omega_T = [0, T] \times \Omega \text{ and}$$
$$u(0, x) = u_0(x) \text{ on } \Omega, \tag{1}$$

where Ω is an open subset of $\mathbb{R}^n$, $n = 2, 3$ with smooth boundary, $\gamma > 0$ is a positive constant. The scalar field u is the difference of the two fluid phase concentrations. It belongs to $u \in [-1, 1]$. The non-negative $B(u) \geq 0$ is the diffusional mobility, and $\Psi(u)$ is the homogeneous free energy. Following [8] we introduce the Ginzburg-Landau free energy

$$\mathcal{E}(u) = \int_\Omega \left(\frac{\gamma}{2}|\nabla u|^2 + \Psi(u)\right) dx, \tag{2}$$

that allows us to monitor the stability of the numerical simulation. Namely, it is supposed to constantly decrease. Let us consider the following boundary conditions

$$u = 0 \text{ in } (0, T) \times \Gamma_D, \quad \frac{\partial u}{\partial \mathbf{n}} = 0 \text{ on } \Omega_T,$$
$$\mathbf{n} \cdot (B(u)\nabla(-\gamma\Delta u + \Psi'(u))) = 0 \text{ in } (0, T) \times (\partial\Omega \setminus \Gamma_D), \tag{3}$$

for $\Gamma_D \subset \partial\Omega$ with $\sigma(\Gamma_D) > 0$. Next, following [8], we define

$$B(u) = 1 - u^2,$$
$$\Psi(u) = \frac{\theta}{2}((1 + u)\log(1 + u) + (1 - u)\log(1 - u)) + 1 - u^2, \tag{4}$$

where $\theta = 1.5$.

In order to pass to the weak formulation, let us define the following Hilbert space

$$V = \left\{v \in H^2(\Omega) : \mathbf{tr}(v) = 0 \text{ on } \Gamma_D \text{ and } \frac{\partial u}{\partial \mathbf{n}} = 0 \text{ on } \partial\Omega \setminus \Gamma_D\right\} \tag{5}$$

with the inner product inherited from $H^2(\Omega)$, where **tr** is the Γ_D-related trace operator on $H^2(\Omega)$. Next, following [1,8] we introduce the space differential operator $A : H^2(\Omega) \rightarrow (H^2(\Omega))'$, such that

$$\langle A(u), w \rangle = \int_\Omega (\gamma \Delta u \nabla \cdot (B(u)\nabla w) + (B\Psi'')(u)\nabla u \cdot \nabla w) \, dx,$$
$$\forall w \in H^2(\Omega). \tag{6}$$

We also introduce the simple dualizing operator $\tau : H^1(\Omega) \rightarrow (H^1(\Omega))'$ such that

$$\langle u, w \rangle = \int_\Omega u \cdot w \, dx, \quad \forall w \in H^1(\Omega). \tag{7}$$

and the time-derivative operator $\cdot_t : C^1(0, T; H^1(\Omega)) \rightarrow C(0, T; (H^1(\Omega))')$

$$\langle u_t(t), w \rangle = \left\langle \frac{\partial u}{\partial t}(t), w \right\rangle, \quad \forall w \in H^1(\Omega), \forall t \in [0, T]. \tag{8}$$

Then, we are able to introduce the second variational equation preserving the classical, Frechet derivative with respect to the time variable: We seek for $u \in C^1(0, T; V)$ such that

$$\langle u_t(t), w \rangle + \langle A(u(t)), w \rangle = 0 \;\; \forall w \in V, \;\; \forall t \in [0, T] \text{ and}$$
$$u(0, x) = u_0(x) \text{ a.e. on } \Omega. \tag{9}$$

The above weak formulation of Cahn-Hilliard equation with boundary conditions (3) can be rewritten in a brief, dual form

$$u_t(t) + A(u(t)) = 0, \; u(0) = u_0. \tag{10}$$

3   Semi-discrete Galerkin Formulation

Let us introduce the sequence of approximation finite dimensional spaces $\{X_n\}$, such, that $X_{n_1} \subset X_{n_2} \subset V, \forall n_2 > n_1$, moreover $\overline{\bigcup_n X_n} = V$ in the strong topology induced from $H^2(\Omega)$. The sequence $\{X_n\}$ can be obtained in particular by using the Finite Element Method for creating the base for the first subspace X_{n_0}, and the proper adaptive policy for obtaining the consecutive spaces $X_m, m > n_0$ (see e.g. [9]).

Now, we are able to introduce the sequence of Galerkin problems with a continuous time leading to find $u_n \in C^1(0, T; X_n)$ so, that

$$(u_n)_t(t) + A(u_n(t)) = 0, \; u_n(0) = u_0. \tag{11}$$

For the sake of simplicity we assume that $u_0 \in \bigcap_n X_n$. As far as we use the same notation for the time derivative and A operators as in (10), they are now the restrictions of operators used there to the space $C^1(0, T; X_n)$.

4   Finite Difference Schemes

In order to solve approximately the semi-discrete Galerkin Eq. (11) in the particular space $C^1(0, T; X_n)$ we can apply a finite-difference scheme along the time variable $t \in [0, T]$. Because explicit schemes, like Euler, are unstable in the case of Cahn-Hilliard equation, in the sequel we will consider only implicit schemes.

We introduce a mesh in the time domain $S_\tau = \{i\tau; i = 0, \ldots, K\} \subset [0, T]$, where $K\tau = T$ and τ stands for the length of the time step. Let $g \in C(0, T; X_n)$ be an arbitrary function. We will denote by $g_\tau = g|S_\tau$ the restriction of g to the mesh S_τ, so that $g_\tau^i = g(t\tau) \in X_n, i = 0, \ldots, K$ and $g_\tau = \{g_\tau^i\} \in (X_n)^{K+1}$.

Let us consider the following Crank-Nicolson integration scheme for the semi-continuous variational formulations of the Cahn-Hilliard equation (11):

We are looking for $u_{n\tau} : S_\tau \to X_n$ such that

$$\left\langle \frac{u_{n\tau}^{i+1} - u_{n\tau}^i}{\tau}, w \right\rangle + \tfrac{1}{2} \langle A(u_{n\tau}^{i+1}) + A(u_{n\tau}^i), w \rangle = 0,$$

$$u_{n\tau}^0 = u_0, \quad \forall w \in X_n, \quad i = 0, \ldots, K. \tag{12}$$

This scheme is unconditionally stable with respect to the time step τ. Unfortunately, the price for this property is very high, because it requires solving nonlinear variational equation at each time step. It is also worth noticing that the Crank-Nicolson scheme can be extended to the so-called generalized α-scheme presented in [10], where the time integration step can be adapted.

Let us assume now that we additionally know a solution $u_{(-1)} \in X_n$ to (11) at the time instance $-\tau$. In other words, we have double initial conditions $u_{n\tau}^{-1} = u_{(-1)}, u_{n\tau}^0 = u_0 \in X_n$. We may then define the following three-level linearized integration scheme:

We are looking for $u_{n\tau} : \{-\tau\} \cup S_\tau \to X_n$ such that

$$\left\langle \frac{u_{n\tau}^{i+2} - u_{n\tau}^i}{2\tau}, w \right\rangle + \left\langle \tfrac{1}{2} \left. DA \right|_{u_{n\tau}^{i+1}} (u_{n\tau}^{i+2} + u_{n\tau}^i - 2u_{n\tau}^{i+1}) + A(u_{n\tau}^{i+1}), w \right\rangle = 0,$$

$$u_{n\tau}^{-1} = u_{(-1)}, \; u_{n\tau}^0 = u_0, \; \forall w \in X_n, \; i = -1, 0, 1, \ldots, K. \tag{13}$$

In contrast to the Crank-Nicolson scheme, we can compute the next-step solution $u_{n\tau}^{i+2}$ by solving linear equation, instead of the nonlinear one, which is usually much more cheaper. The double initial conditions does not cause any problem because if the single initial condition $u_{(-1)} \in X_n$ is given, then $u_0 \in X_n$ can be approximated using, e.g., a single step of the Crank-Nicolson scheme (12).

Now, let us denote by $\{\eta_i\}, i = 1, \ldots, n, n < +\infty$ an arbitrary basis in X_n convenient for solving (11). The solution $u_{n\tau}$ of both mixed schemes can be represented as a sequence of real vectors $\{\alpha^j\}, j = 1, \ldots, K$, so that $\alpha^j \in \mathbb{R}^n, u_{n\tau}^j = \sum_{i=1}^n \alpha_i^j \eta_i$. Moreover, the basis vectors $\eta_i, i = 1, \ldots, n$ will be used as test functions $w \in X_n$. The details of solving equations resulting from both schemes can be found in many books and papers (see e.g. [9]).

The scheme (13) becomes a sequence of linear systems with the $n \times n$ matrices $(\mathbf{1} + \tau \mathbf{M}^j), \; j = 1, \ldots, K$, where $\mathbf{1}$ is the Gram matrix and $\mathbf{M}^j$ denotes the matrix associated with the differential $DA|u_{n\tau}^{j-1}$, both computed with respect to the selected basis in X_n.

Finally, the simulation of cancer growth by Cahn-Hilliard equation using the proposed three-level linearized scheme will follow the simple algorithm (see Listing 1).

```
1   BEGIN
2   Choose the space  Xn ⊂ V  and  its  basis
        {ηi}, i = 1, . . . , n ;
3   Choose the  initial  time  step  τ ;
4   Compute  initial  condition  u(−1) ∈ Xn  by  projecting
        u(0)  on  Xn ;
5   Compute  u0 ∈ Xn  from  equation  (12) ;
6   FOR  j = 1, . . . , K
7       Compute  unτ^j ∈ Xn  from  equation  (13) ;
8   ENDFOR
9   IF  the  solution  is  unsatisfactory
10      SWITCH  //perform 1,  2,  or   both
11          (1) Decrease  the  time  step  τ ;
12          (2) Improve  the  space  Xn  according  some
                adaptation  rule  (see  [9]) ;
13      ENDSWITCH
14      GOTO  4;
15  ENDIF
16  END
```

Listing 1. The algorithm implementing three-level finite-difference scheme (13) for Cahn-Hilliard equation.

The linearized three-level scheme can be also reformulated in a way similar to the generalized α-scheme [10], so we can adjust the time integration step adaptively. We can utilize either direct or iterative linear solvers for the computations in every time step.

5   Mathematical Properties of the Linearized Three Level Scheme

Two important asymptotic features of the mixed Galerkin/three-level linearized scheme (13) were studied.

Observation 1. *If both B and Ψ'' are positive constants functions ($B \equiv b > 0$ and $\Psi'' \equiv c > 0$), then Problem (11) has the unique solution in $C^1(0, T; X_n)$ for any u_0 and n and the sequence of Galerkin solutions to (11) converges to the solution of (10), i.e. $\|u_n - u\|_{C(0,T;L^2(\Omega))}$ for $n \to +\infty$.*

The above observation follows immediately from the Observation 4 in Appendix.

Observation 2. *Let us assume, that the time network is regular, i.e. $S = \{t_{-1} = -\tau, t_1 = 0, t_2 = \tau, \ldots, t_{K\tau} = K\tau = T\}$ and we know $u(-\tau)$ and $u(0) = u_0$ being the values of the exact solution to (10) for all τ; $T_l > \tau > 0$, for some positive constant $T_l > 0$. If moreover the assumptions of Observation 1 hold, then:*

1. *Three-level linearized scheme (13) applied for the semi-discrete Galerkin formulation of the Cahn-Hilliard Eq. (11) has the unique solution $u_{n\tau}^i$ for each time step $i = -1, 0, 1, \ldots, K$.*
2. *The sequence of solutions $\{u_{n\tau}\}$ to the mixed Galerkin/three-level linearized scheme of solving Cahn-Hilliard equation converges to the solution u of the exact variational formulation (10), i.e.*

$$\lim_{n \to +\infty, \tau \to 0} \|u_{n\tau} - u\|_\tau = 0,$$

where $\|u\|_\tau = \max\{\|u(i\tau)\|_{L^2(\Omega)}, i = 1, 2, \ldots, K\}$.

The above observation is the simple issue of the Observation 5 in Appendix.

It is worth to notice, that the three-level linearized scheme (13) is unconditionally stable with respect to the time step τ, because its convergence was proven without any assumed dependencies between the time and space (Galerkin) approximations.

Moreover, it can be proven, that the three-level linearized scheme (13) applied for the approximate, semi-discrete (11) variational formulations of Cahn-Hilliard equations is of the second order with respect to the time step τ (see [11]).

6 Scalability Analysis

Most approaches to the Cahn-Hilliard equation, including Crank-Nicolson integration schemes (12), result in a sequence of nonlinear algebraic systems. The presented linearized three-level scheme results in a sequence of linear systems, thus its dominating part of computational cost at each time step (namely multifrontal solver applied for solving the linear system) is comparable to the cost of solving a linear variational elliptic problem with a linear operator. Paper [5] describes such a particular case using a L^2-projection scheme. Both cases (L^2 projection equation for elliptic problem and one step linearized three level scheme for Cahn-Hilliard) possess exactly the same matrix size and structure, thus the same computational cost. All computations for both cases presented in this paper were performed with IGA-FEM [12], which utilizes B-Spline basis functions for the approximation in space domain.

The first goal we try to obtain by numerical experiments is the analysis of software scalability obtained by implementing both schemes (12) and (13). To analyze scalability of the entire scheme it is enough to analyze scalability in a single step, because in each step the most expensive part (i.e., the multifrontal solver execution) is repeated in the same manner and possess almost identical computational cost. In case of scaling Crank-Nicolson scheme computation, only

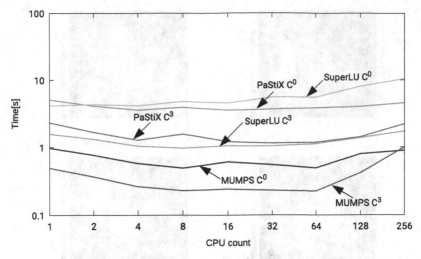

Fig. 1. Comparison of scalability for quartics C^3 used for the first formulation with quadratics C^0 used for the second formulation, over the mesh with 128×128 elements.

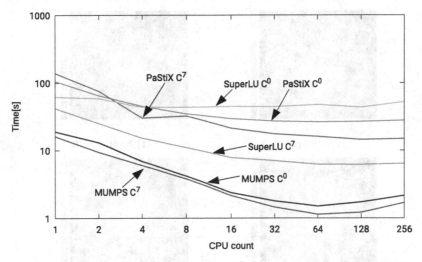

Fig. 2. Comparison of scalability for octics C^7 used for the first formulation with quartics C^0 used for the second formulation, over the mesh with 256×256 elements.

one Newton iteration of solving nonlinear algebraic system was included. The experiment observes the course of computational time regression with respect to increasing processor count.

In order to make computations more convenient in PetIGA interface, each of variational Eqs. (12) and (13) was reformulated to equivalent form: from the single fourth order PDEs (or H^2 weak formulation) down to the system of two second order PDEs (or a system of two H^1 weak problems).

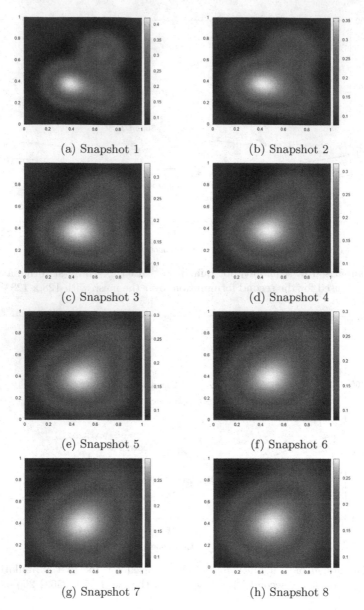

(a) Snapshot 1   (b) Snapshot 2

(c) Snapshot 3   (d) Snapshot 4

(e) Snapshot 5   (f) Snapshot 6

(g) Snapshot 7   (h) Snapshot 8

Fig. 3. Snapshots from the tumor growth simulations with the Cahn-Hilliard equation.

The computations were performed on a distributed-memory Linux cluster. One processor per node is utilized, up to 256 STAMPEDE Linux cluster nodes. The PETSC interface [13–15] delivers multiple solvers including MUMPS [16–18], SuperLU [19,20] and PaSTiX [21]. All of benchmarks presented in this paper were performed on a L^2 projection problem coded in PetIGA [22] interface (an

IGA-FEM overlay to PETSC). The interface and solvers control execution in concurrent environment. The aim was to examine and compare scalability and computational cost of different combinations of schemes and solvers (implementations).

Tests were executed for the number of processors increasing from 1 to 256. Various types of B-Spline basis functions [12] resulted in various global regularity utilized. Quartics C^3 used for the first formulation are compared with quadratics C^0 used for the second formulation, over the mesh with 128×128 elements, see Fig. 1. Octics C^7 used for the first formulation are compared with quartics C^0 used for the second formulation, over the mesh with 256×256 elements, see Fig. 2. It can be noted that for both cases, the MUMPS solver for the second formulation (for C^0 basis) outperforms all the other solvers with increased number of processors.

Preliminary results are presented in Fig. 3. The software implementing three-level linearized scheme scales in the same manner as Crank-Nicolson for all tested multi-frontal solvers. Moreover it can be noted that higher continuity B-Spline basis functions tend to compute faster for the same mesh sizes.

The second goal of numerical experiments is to compare both schemes by simulating benchmarks of single tumor growth. Both simulations performed for 8000 time steps present similar final images of tumor cell concentration. It can be concluded that utilizing three-level linearized time integration scheme doesn't degrade the accuracy in comparison with state-of-the-art Crank-Nicolson time integration scheme.

7   Conclusions

The theoretical results presented in this paper complete the formal analysis of the mixed three-level linearized finite-difference-Galerkin numerical scheme applied for solving the Cahn-Hiliard equation. We have obtained theorems guaranteeing the existence and the uniqueness of solutions to the exact continuous/variational problem (8), (10) and its semi-discrete Galerkin version (11). We have also shown the convergence of the solutions of (11) to the exact solution when the Galerkin approximation error decays. Next, we have proved the convergence of the solutions of three-level linearized scheme (13) when both space and time approximations are improved ($n \to \infty$, $\tau \to 0$) and that the scheme is unconditionally stable with respect to the time variable. Additionally, we refer to the proof of the second order of (13) with respect to the time variable.

The scheme is flexible to incorporate different linear solvers and well dedicated to particular B-Spline basis functions. The software implementing three-level linearized scheme scales in the same manner as Crank-Nicolson for all tested multi-frontal solvers. Moreover it can be noted that higher continuity B-Spline basis functions tend to compute faster for the same mesh sizes. It can be also observed, that for both cases, the MUMPS solver for the second formulation (for C^0 basis) outperforms all the other solvers with increased number of processors.

Both, theoretical and experimental results presented in this paper show, that the proposed three level linearized time integration scheme is an advantageous tool for solving initial boundary value problems for Cahn-Hiliard equations. Theorems and observations proved that the scheme is as well conditioned as Crank-Nicolson one, concerning convergence, stability and order. Numerical results show that utilizing three-level linearized time integration scheme doesn't degrade the accuracy in comparison with state-of-the-art time integration scheme - Crank-Nicolson one.

The main advantage of the three-level integration scheme over Crank-Nicolson is lower computational cost. It always requires only one linear system to be solved within each time step. In case of non-linear Cahn-Hiliard equation Crank-Nicolson scheme (12) may require solving multiple linear system within each time step.

The effective numerical model of the Cahn-Hiliard equation is crucial for simulating tumor growth abd then is helpful by the medical diagnosis and therapy of this group of heavy diseases.

The future work may involve incorporating the Cahn-Hilliard based models with supermodeling approach [23–25].

Acknowledgement. The Authors are thankful for support from the funds assigned to AGH University of Science and Technology by the Polish Ministry of Science and Higher Education.

Appendix: Convergence of the Mixed 3-level Linearized Scheme

In this section we prove the convergence of the 3-level scheme (13) . Crucial properties of operator A are its continuity and coercivity, which are used to prove the convergence of the numerical schema. They are formulated in the following way: there exist positive constants m and M and function ζ satisfying $\zeta(s) \to +\infty (s \to +\infty)$ such that for every $u, v \in V$ we have

$$\|A(u) - A(v)\|_{V'} \leq M \|u - v\|_V \tag{14a}$$

$$\langle A(u) - A(v), u - v \rangle_{V' \times V} \geq m \|u - v\|_V^2 \tag{14b}$$

$$\langle A(u), u \rangle_{V' \times V} \geq \zeta(\|u\|_V) \|u\|_V \tag{14c}$$

A sample case when the above conditions hold is shown in the following observation.

Observation 3. *Assume that both B and Ψ'' are positive constants ($B \equiv b > 0$ and $\Psi'' \equiv c > 0$). Then, conditions (14) hold.*

Proof. This is in fact a linear case, i.e.

$$\langle A(u), w \rangle = \int_\Omega (\gamma b \Delta u \Delta w + bc \nabla u \nabla w) \, dx.$$

Therefore,

$$|\langle A(u), w\rangle| \leq \gamma b\|\Delta u\|_{L^2(\Omega)}\|\Delta w\|_{L^2(\Omega)} + bc\|\nabla u\|_{L^2(\Omega;\mathbb{R}^n)}\|\nabla w\|_{L^2(\Omega;\mathbb{R}^n)},$$

which yields (14a) with, e.g., $M = b(\gamma + c)$. Moreover,

$$\langle A(u), u\rangle = \int_\Omega \left(\gamma b(\Delta u)^2 + bc|\nabla u|^2\right) dx$$
$$= \gamma b\|\Delta u\|_{L^2(\Omega)}^2 + bc\|\nabla u\|_{L^2(\Omega;\mathbb{R}^n)}^2,$$

which, together with an appropriate version of Poincaré inequality, gives us (14c). Finally, in this case, it is easy to see that (14b) is a consequence of (14c).

Let assume now, that we know the solution to (10) in some interval $[-T_l, 0]$ for $T_l > 0$. We can introduce the time grids

$$S_\tau = \{i\tau;\ i = 1, 2, \ldots, K;\ \tau < t_0,\ K\tau = T\}. \tag{15}$$

The arbitrary function $g : [-T_l, T] \to X_n((X_n)')$ can be restricted to S_τ, then we obtain the grid function $g_\tau = \{g_\tau^{-1} = g(-\tau), g_\tau^0 = g(0), g_\tau^1 = g(\tau), g_\tau^2 = g(2\tau), \ldots, g_\tau^K = g(K\tau)\}$.

Let us denote by V_n and V_n' the vector spaces being the the restrictions of $C(-T_l, T; X_n)$ and $C(-T_l, T; X_n')$ to the network S_τ equipped with the norms:

$$\|g_\tau\|_{K\tau} = \max\{\|g_\tau^i\|_{H^2(\Omega)}, i = -1, 0, 1, 2, \ldots, K\},$$
$$\|g_\tau\|'_{K\tau} = \max\{\|g_\tau^i\|_{H^{-2}(\Omega)}, i = -1, 0, 1, 2, \ldots, K\}, \tag{16}$$

respectively.

We are ready now to define the time grid operator $R_\tau : V_n \to V_n'$ as the collection of coordinate operators

$$(R_\tau(g_\tau))^i = \frac{g_\tau^{i+1} - g_\tau^i}{2\tau} + A(g_\tau^i) + \frac{1}{2}DA\big|_{g_\tau^{i+1}}(g_\tau^{i+2} - 2g_\tau^i + g_\tau^{i-1}) \tag{17}$$

associated with the three-level linearized scheme (13). The mixed Galerkin three-level linearized scheme discrete problem can be formulated as follows:

Let us assume, that the exact solution u of (10) is well-known and continuous with respect to the time variable on the interval $[-T_l, 0]$ and moreover $\forall t \in [-T_l, 0]\ u(t) \in \bigcap_n X_n$. We are looking for $u_{n\tau} \in V_n$ that satisfies

$$R_\tau(u_{n\tau}) = 0 \quad \text{and} \quad u_{n\tau}^{-1} = u(-\tau),\ u_{n\tau}^0 = u_0. \tag{18}$$

Notice, that for the sake of simplicity the notation of the operator R_τ is polymorphic in the same way as the notation of A here, i.e. denote the families of operators for all n.

Solving nonlinear parabolic variational equations of type (10) by using mixed Galerkin 3-leveled linearized schema was intensively studied in papers [26, 27]. Observation 3 states the fact that operator A satisfies the assumptions of Theorems 1 and 2 from paper [27]. In particular, Theorem 1 in [27] implies that:

Observation 4. *Under the assumptions of Observation 3 the following statements hold:*

1. *Problem (10) has the unique solution in $L^2(0, T; V) \cap C(0, T; L^2(\Omega))$ for any u_0.*

2. *Problem (11) has the unique solution in $C^1(0, T; X_n)$ for any u_0 and n.*

3. $\|u_n - u\|_{C(0,T;L^2(\Omega))}$ *for $n \to +\infty$.*

Moreover, taking into account Theorem 2 in [27] we have:

Observation 5. *The problem (18) has the unique solution for any $u(-\tau)$ and u_0, moreover*

$$\lim_{n \to +\infty, \tau \to 0} \|u_{n\tau} - u\|_\tau = 0$$

where $\|u\|_\tau = \max\{\|u(i\tau)\|_{L^2(\Omega)}, i = 1, 2, \ldots, K\}$.

References

1. Gómez, H., Calo, V.M., Bazileves, Y., Hughes, T.J.R.: Isogeometric analysis of the Cahn-Hilliard phase-field model. Comput. Methods Appl. Mech. Eng. **197**, 4333–4352 (2008)
2. Gómez, H., Hughes, T.J.R.: Provably unconditionally stable, second-order time-accurate, mixed variational methods for phase-field models. J. Comput. Phys. **230**, 5310–5327 (2011)
3. Hawkins-Daarud, A., Prudhomme, S., van der Zee, K.G., Oden, J.T.: Bayesian calibration, validation, and uncertainty quantification of diffuse interface models of tumor growth. J. Math. Biol. **67**, 1457–1485 (2012). https://doi.org/10.1007/s00285-012-0595-9
4. Wu, X., van Zwieten, G.J., van der Zee, K.: Stabilized second-order convex splitting schemes for Cahn-Hilliard models with application to diffuse-interface tumor-growth models. Numer. Methods Biomech. Eng. **30**(3), 180–203 (2014)
5. Łoś, M., Kłusek, A., Hassaan, M.A., Pingali, K., Dzwinel, W., Paszyński, M.: Parallel fast isogeometric L2 projection solver with GALOIS system for 3D tumor growth simulations. Comput. Methods Appl. Mech. Eng. **343**, 1–22 (2019)
6. Puzyrev, V., Łoś, M., Gurgul, G., Calo, V.M., Dzwinel, W., Paszyński, M.: Parallel splitting solvers for the isogeometric analysis of the Cahn-Hilliard equation. Comput. Methods Biomech. Biomed. Eng. **22**(16), 1269–1281 (2019)
7. Woźniak, M., Smołka, M., Cortes, A., Paszyński, M., Schaefer, R.: Scalability of direct solver for non-stationary Cahn-Hilliard simulations with linearized time integration scheme. Procedia Comput. Sci. **80**, 834–844 (2016)
8. Elliott, C.M., Garcke, H.: On the Cahn-Hilliard equation with degenerate mobility. SIAM J. Math. Anal. **27**, 404–423 (1996)
9. Demkowicz, L., Kurtz, J., Pardo, D., Paszyński, M., Rachowicz, W., Zdunek, A.: Computing with hp Finite Elements. II. Frontiers: Three-Dimensional Elliptic and Maxwell Problems with Applications. Chapman&Chall/CRC, Taylor&Francis Group Boca Raton, London, New York (2007). ISBN-13: 978-1584886723, ISBN-10: 1584886722

10. Jansen, K.E., Whiting, C.H., Hulbert, G.M.: A generalized-α method for integrating the filtered Navier-Stokes equations with a stabilized finite element method. Comput. Methods Appl. Mech. Eng. **190**, 305–319 (2000)
11. Woźniak, M., Smołka, M., Cortes, A., Paszyński, M., Schaefer, R.: Scalability of direct solver for non-stationary Cahn-Hilliard simulations with linearized time integration scheme. Procedia Comput. Sci. **80**, 834–844 (2016)
12. Austin Cottrell, J., Hughes, T.J.R., Bazilevs, Y.: Isogeometric Analysis: Toward Integration of CAD and FEA. Wiley, Hoboken (2009)
13. Balay, S. et al.: PETSc (2014). http://www.mcs.anl.gov/petsc
14. Balay, S., et al.: PETSc User Manual, Argonne National Laboratory ANL-95/11 - Revision 3.4 (2013)
15. Balay, S., Gropp, W.D., Curfman McInnes, L., Smith, B.F.: Efficient management of parallelism in Object Oriented Numerical Software Libraries. In: Arge, E., Bruaset, A.M., Langtangen, H.P. (eds.) Modern Software Tools in Scientific Computing. Birkhäuser, Boston (1997). https://doi.org/10.1007/978-1-4612-1986-6_8
16. Amestoy, P.R., Duff, I.S.: Multifrontal parallel distributed symmetric and unsymmetric solvers. Comput. Methods Appl. Mech. Eng. **184**, 501–520 (2000)
17. Amestoy, P.R., Duff, I.S., Koster, J., L'Excellent, J.Y.: A fully asynchronous multifrontal solver using distributed dynamic scheduling. SIAM J. Matrix Anal. Appl. **1**(23), 15–41 (2001)
18. Amestoy, P.R., Guermouche, A., L'Excellent, J.-Y., Pralet, S.: Hybrid scheduling for the parallel solution of linear systems. Comput. Methods Appl. Mech. Eng. **2**(32), 136–156 (2001)
19. Li, X.S.: An overview of SuperLU: algorithms, implementation, and user interface. TOMS Trans. Math. Softw. **31**(3), 302–325 (2005)
20. Li, X.S., Demmel, J.W., Gilbert, J.R., Grigori, L., Shao, M., Yamazaki, I.: SuperLU Users' Guide, Lawrence Berkeley National Laboratory, LBNL-44289 (1999). http://crd.lbl.gov/xiaoye/SuperLU/
21. Hénon, P., Ramet, P., Roman, J.: PaStiX: a high-performance parallel direct solver for sparse symmetric definite systems. Parallel Comput. **28**(2), 301–321 (2002)
22. Collier, N., Dalcin, L., Calo, V.M.: PetIGA: high-performance isogeometric analysis. arxiv:1305.4452 (2013)
23. Dzwinel, W., Kłusek, A., Paszyński, M.: A concept of a prognostic system for personalized anti-tumor therapy based on supermodeling. Procedia Comput. Sci. **108C**, 1832–1841 (2017)
24. Dzwinel, W., Kłusek, A., Vasilyev, O.V.: Supermodeling in simulation of melanoma progression. Procedia Comput. Sci. **80**, 999–1010 (2016)
25. Siwik, L., Łoś, M., Kłusek, A., Dzwinel, W., Paszyński, M., Pingali, K.: Supermodeling of tumor dynamics with parallel isogeometric analysis solver. arXiv:1912.12836
26. Schaefer, R., Sędziwy, S.: Filtration in cohesive soils. Part II - Numerical approach, Computer Assisted Mechanics and Engineering Sciences (CAMES), vol. 6, pp. 15–26 (1999)
27. Schaefer, R., Sędziwy, S.: Filtration in cohesive soils. Part I - The mathematical model, Computer Assisted Mechanics and Engineering Sciences (CAMES), vol. 6, pp. 1–13 (1999)

Poroelasticity Modules in DarcyLite

Jiangguo Liu[1] and Zhuoran Wang[2(✉)]

[1] Department of Mathematics, Colorado State University,
Fort Collins, CO 80523, USA
liu@math.colostate.edu

[2] School of Mathematics (Zhuhai), Sun Yat-sen University,
Zhuhai, Guangdong 519082, China
wangzhr25@mail.sysu.edu.cn

Abstract. This paper elaborates on design and implementation of code modules for finite element solvers for poroelasticity in our Matlab package DarcyLite [15]. The Biot's model is adopted. Both linear and nonlinear cases are discussed. Numerical experiments are presented to demonstrate the accuracy and efficiency of these solvers.

Keywords: Biot's model · Enriched Lagrangian finite elements · Poroelasticity · Quadrilateral meshes · Weak Galerkin finite element methods

1 Introduction

Poroelasticity problems exist widely in the real world, e.g., drug delivery, food processing, petroleum reservoirs, and tissue engineering. These problems involve fluid flow in porous media that are elastic and deform due to fluid pressure. The Biot's model for linear and nonlinear poroelasticity has been well accepted [4,7,12,20,30]. It couples solid displacement $\mathbf{u}$ and fluid pressure p through the following partial differential equations (PDEs) on a bounded domain Ω for a time period $[0, T]$:

$$\begin{cases} -\nabla \cdot (2\mu\varepsilon(\mathbf{u}) + \lambda(\nabla \cdot \mathbf{u})\mathbf{I}) + \alpha\nabla p = \mathbf{f}, \\ \partial_t (\alpha\nabla \cdot \mathbf{u} + c_0 p) + \nabla \cdot (-\mathbf{K}(\mathbf{u})\nabla p) = s, \end{cases} \quad (1)$$

where $\varepsilon(\mathbf{u}) = \frac{1}{2} \left(\nabla \mathbf{u} + (\nabla \mathbf{u})^T \right)$ is the strain tensor, $\sigma(\mathbf{u}) = 2\mu\,\varepsilon(\mathbf{u}) + \lambda(\nabla \cdot \mathbf{u})\mathbf{I}$ the stress tensor, $\lambda > 0, \mu > 0$ the Lamé constants, $\mathbf{f}$ a given body force, $\mathbf{K}$ a conductivity/permeability tensor, s the fluid source, α (≈ 1) the Biot-Williams constant, and $c_0 \geq 0$ the constrained storage capacity. Furthermore, the total stress is defined as

$$\widetilde{\sigma}(\mathbf{u}, p) = \sigma - \alpha\,p\,\mathbf{I}. \quad (2)$$

Liu was partially supported by US National Science Foundation grant DMS-1819252.
Wang was partially supported by grant 74120-18841215 from Sun Yat-sen University.

Dirichlet and Neumann boundary conditions for solid are posed as

$$\mathbf{u}|_{\Gamma_D^{\mathcal{E}}} = \mathbf{u}_D, \qquad (\tilde{\sigma}\mathbf{n})|_{\Gamma_N^{\mathcal{E}}} = \mathbf{t}_N, \tag{3}$$

whereas Dirichlet and Neumann boundary conditions for fluid are posed as

$$p|_{\Gamma_D^p} = p_D, \qquad (-\mathbf{K}\nabla p) \cdot \mathbf{n}|_{\Gamma_N^p} = u_N, \tag{4}$$

where $\mathbf{n}$ is the outward unit normal vector to $\partial\Omega$, which has a non-overlapping decomposition $\partial\Omega = \Gamma_D^{\mathcal{E}} \cup \Gamma_N^{\mathcal{E}}$ for solid and another non-overlapping decomposition $\partial\Omega = \Gamma_D^p \cup \Gamma_N^p$ for fluid. As for initial conditions, we have

$$p(\mathbf{x}, 0) = p_0, \qquad \mathbf{u}(\mathbf{x}, 0) = \mathbf{u}_0. \tag{5}$$

Usually, $\mathbf{u}_0 = \mathbf{0}$, i.e., there is no deformation at the beginning of the simulation.

Finite element methods (FEMs) are common tools for solving the Biot's model. Depending on the unknown quantities to be solved, poroelasticity solvers are usually grouped into 3 types:

- *2-field*: Solid displacement and fluid pressure are to be solved;
- *3-field*: Solid displacement, fluid pressure and velocity are to be solved;
- *4-field*: Solid stress & displacement, fluid pressure & velocity are to be solved.

A major issue in numerical solvers for poroelasticity is the poroelasticity locking, which usually appears as nonphysical pressure oscillations or deteriorating convergence rates in displacement errors. This happens when the porous media are low-permeable or nearly incompressible ($\lambda \to \infty$) [7, 21, 30].

Early on, the continuous Galerkin (CG) FEMs were applied respectively to solve for displacement and pressure. But it was soon recognized that such solvers were subject to poroelasticity locking and the 2-field approach was nearly abandoned. The mixed finite element methods can be used to solve for pressure and velocity simultaneously and meanwhile coupled with a FEM for linear elasticity that is free of Poisson-locking. Therefore, the 3-field approach has been the main stream [4, 18–20, 27, 28]. The 4-field approach is certainly worth of investigation, but it may involve too many unknowns (degrees of freedom) [29].

The weak Galerkin (WG) finite element methods [23] have emerged as a new class of numerical methods with nice features that can be applied to a wide variety of problems including Darcy flow and linear elasticity [10, 13]. Certainly, WG solvers can be developed for linear poroelasticity [12], they are free of poroelasticity locking but may involve more degrees of freedom. There have also been efforts on developing HDG methods for the Biot's model [6].

This paper elaborates on code modules for poroelasticity recently added to our `DarcyLite` package [15], which has gained some popularity. We shall discuss five solvers with a variety of discretization schemes for linear elasticity and Darcy flow. A poroelasticity problem may be solved as a monolithic systems (MS) or through operator splitting (OS). This paper explains the mathematical ideas behind these solvers, and their implementation with consideration of modularity and code re-usability. Numerical experiments are presented to demonstrate their use, accuracy, and efficiency.

We focus on 2-dim problems with a quasi-uniform triangular mesh $\mathcal{T}_h$ or convex quadrilateral mesh $\mathcal{E}_h$. For ease of presentation, we consider a uniform temporal partition of $[0, T]$ with $\Delta t = T/N$ and $t_n = n\Delta t$ for $0 \leq n \leq N$.

2   Solver I: A 2-Field Penalty-Free Weak Galerkin Finite Element Solver for Quadrilateral Meshes

Now we consider linear poroelasticity (1) in which $\mathbf{K}$ actually does not depend on $\mathbf{u}$. Then the variational form reads as

$$\begin{cases} 2\mu(\varepsilon(\mathbf{u}), \varepsilon(\mathbf{v})) + \lambda(\nabla \cdot \mathbf{u}, \nabla \cdot \mathbf{v}) - \alpha(p, \nabla \cdot \mathbf{v}) = (\mathbf{f}, \mathbf{v}) + \langle \mathbf{t}_N, \mathbf{v} \rangle_{\Gamma_N^{\mathcal{E}}}, \\ \alpha(\partial_t \nabla \cdot \mathbf{u}, q) + c_0(\partial_t p, q) + (\mathbf{K}\nabla p, \nabla q) = (s, q) - \langle u_N, q \rangle_{\Gamma_N^{\mathcal{D}}} \end{cases} \quad (6)$$

with incorporation of boundary and initial conditions.

We consider WG finite element discretization for both linear elasticity and Darcy flow. $WG(P_0, P_0; AC_0)$ for Darcy flow has been investigated in [17]. Here we briefly discuss $WG(P_0^2, P_0^2; AC_0^2)$ finite elements for linear elasticity.

The classical Raviart-Thomas spaces $RT_{[k]}(k \geq 0)$ for rectangles have some limitations. The recently developed Arbogast-Correa mixed finite elements are designed for more general convex quadrilaterals [3]. We shall use the lowest-order AC_0 space, which has a local basis [17] as shown below

$$\begin{bmatrix} 1 \\ 0 \end{bmatrix}, \quad \begin{bmatrix} 0 \\ 1 \end{bmatrix}, \quad \begin{bmatrix} X \\ Y \end{bmatrix}, \quad \mathcal{P}_E \begin{bmatrix} \hat{x} \\ -\hat{y} \end{bmatrix},$$

where $X = x - x_c, Y = y - y_c$, (x_c, y_c) is the element center, $(\hat{x}, \hat{y})$ are the coordinates in the reference element $[0, 1]^2$, and $\mathcal{P}_E$ is the Piola transformation.

Let E be a convex quadrilateral. and $AC_0^2(E)$ be the space of order-2 matrices whose row vectors are in $AC_0(E)$. We consider a typical discrete weak function $\mathbf{v} = \{\mathbf{v}^\circ, \mathbf{v}^\partial\} \in WG(P_0^2, P_0^2)$. Its **discrete weak gradient** $\nabla_w \mathbf{v}$ is reconstructed in $AC_0^2(E)$ via integration by parts

$$(\nabla_w \mathbf{v}, \tau) = \langle \mathbf{v}^\partial, \tau \mathbf{n} \rangle_{E^\partial} - (\mathbf{v}^\circ, \nabla \cdot \tau)_{E^\circ}, \qquad \forall \tau \in AC_0^2(E). \quad (7)$$

Its **discrete weak divergence** $\nabla_w \cdot \mathbf{v}$ is reconstructed in $P_0(E)$ as

$$(\nabla_w \cdot \mathbf{v}, \phi) = \langle \mathbf{v}^\partial, \phi \mathbf{n} \rangle_{E^\partial} - (\mathbf{v}^\circ, \nabla \phi)_{E^\circ}, \qquad \forall \phi \in P_0(E). \quad (8)$$

Solver I as a time-marching 2-field finite element scheme for (6) reads as

$$\begin{cases} \mathcal{A}_h^{\mathcal{E}}(\mathbf{u}_h^{(n)}, \mathbf{v}) - \mathcal{B}_h(p_h^{(n)}, \mathbf{v}) = \mathcal{F}_h^{\mathcal{E}}(\mathbf{v}), \\ \mathcal{B}_h(\mathbf{u}_h^{(n)}, q) + \mathcal{A}_h^{\mathcal{D}}(p_h^{(n)}, q) = \mathcal{F}_h^{\mathcal{D}}(q), \end{cases} \quad (9)$$

where the bilinear forms on the left-hand sides are defined as

$$\begin{cases} \mathcal{A}_h^{\mathcal{E}}(\mathbf{u}_h^{(n)}, \mathbf{v}) = \sum_{E \in \mathcal{E}_h} 2\mu(\varepsilon_w(\mathbf{u}_h^{(n)}), \varepsilon_w(\mathbf{v}))_E + \lambda(\nabla_w \cdot \mathbf{u}_h^{(n)}, \nabla_w \cdot \mathbf{v})_E, \\ \mathcal{A}_h^{\mathcal{D}}(p_h^{(n)}, q) = \sum_{E \in \mathcal{E}_h} c_0(p_h^{(n),\circ}, q^\circ)_{E^\circ} + \Delta t(\mathbf{K}\nabla_w p_h^{(n)}, \nabla_w q)_E, \\ \mathcal{B}_h(\mathbf{u}_h^{(n)}, q) = \sum_{E \in \mathcal{E}_h} \alpha(\nabla_w \cdot \mathbf{u}_h^{(n)}, q^\circ)_{E^\circ}. \end{cases} \qquad (10)$$

The linear forms on the right-hand sides are defined as

$$\begin{cases} \mathcal{F}_h^{\mathcal{E}}(\mathbf{v}) = \sum_{E \in \mathcal{E}_h} (\mathbf{f}^{(n)}, \mathbf{v}^\circ)_{E^\circ} + \sum_{e \in \Gamma_N^{\mathcal{E}}} \langle \mathbf{t}_N, \mathbf{v}^\partial \rangle_e, \\ \mathcal{F}_h^{\mathcal{D}}(q) = \sum_{E \in \mathcal{E}_h} \Delta t(s^{(n)}, q^\circ) + c_0(p_h^{(n-1),\circ}, q^\circ)_{E^\circ} + \alpha(\nabla_w \cdot \mathbf{u}_h^{(n-1)}, q^\circ)_{E^\circ} \\ \qquad - \sum_{e \in \Gamma_N^{\mathcal{D}}} \Delta t \langle u_N, q^\partial \rangle_e. \end{cases} \qquad (11)$$

Unlike the methods in [12], this WG solver does not need stabilization for either elasticity or Darcy flow. The degrees of freedom (DOFs) at each time step are

$$3\,\#\text{Elements} + 3\,\#\text{Edges}.$$

3   Solver II: A 3-Field CG+MFEM Solver for Triangular Meshes

In the 3-field approach, Darcy velocity $\mathbf{q}$ is used. The PDEs take the form

$$\begin{cases} -\nabla \cdot (2\mu\varepsilon(\mathbf{u}) + \lambda(\nabla \cdot \mathbf{u})\mathbf{I}) + \alpha\nabla p = \mathbf{f}, \\ \mathbf{K}^{-1}\mathbf{q} + \nabla p = 0, \\ \partial_t(\alpha\nabla \cdot \mathbf{u} + c_0 p) + \nabla \cdot \mathbf{q} = s. \end{cases} \qquad (12)$$

We shall need spaces $\mathbf{V} = \mathbf{H}^1(\Omega)$, $\mathbf{V}^0 = \mathbf{H}_0^1(\Omega)$, $\mathbf{W} = H(\text{div}, \Omega)$, $\mathbf{W}^0 = H_0(\text{div}, \Omega)$, $S = L_0^2(\Omega)$. The variational problem seeks solutions $\mathbf{u} \in \mathbf{V}$, $\mathbf{q} \in \mathbf{W}$, $p \in S$ such that for any $\mathbf{v} \in \mathbf{V}^0$, $\mathbf{w} \in \mathbf{W}^0$ and $q \in S$, there holds

$$\begin{cases} 2\mu(\varepsilon(\mathbf{u}), \varepsilon(\mathbf{v})) + \lambda(\nabla \cdot \mathbf{u}, \nabla \cdot \mathbf{v}) - \alpha(p, \nabla \cdot \mathbf{v}) = (\mathbf{f}, \mathbf{v}) + \langle \mathbf{t}_N, \mathbf{v} \rangle, \\ (\mathbf{K}^{-1}\mathbf{q}, \mathbf{w}) - (p, \nabla \cdot \mathbf{w}) = -\langle p_D, \mathbf{w} \cdot \mathbf{n} \rangle, \\ \alpha(\nabla \cdot (\partial_t \mathbf{u}), q) + (\nabla \cdot \mathbf{q}, q) + c_0(\partial_t p, q) = (s, q). \end{cases} \qquad (13)$$

Again initial conditions are omitted for ease of presentation.

As presented in [30], one considers a triangular mesh $\mathcal{T}_h$ for spatial discretization and the implicit Euler for temporal discretization. One utilizes the 1st order Bernardi-Raugel element space $\mathbf{V}_h$ for displacement discretization. The mixed FE pair (RT_0, P_0) is used for discretization of Darcy flow. The velocity/flux FE spaces are denoted as $\mathbf{W}_h$ and $\mathbf{W}_h^0$, for which the edge-based basis functions are used [2,14]. This is especially convenient for handling the Neumann boundary conditions.

Let $\mathbf{u}_h^{(n)}, \mathbf{u}_h^{(n-1)} \in \mathbf{V}_h$ be the approximations to solid displacement at time moments t_n and t_{n-1}, respectively. Similarly, Let $\mathbf{q}_h^{(n)}, \mathbf{q}_h^{(n-1)} \in \mathbf{W}_h$ be the approximations to Darcy velocity. Let $p_h^{(n)}, p_h^{(n-1)} \in S_h$ be the approximations to fluid pressure at time moments t_n and t_{n-1}.

Combined with the implicit Euler discretization, one establishes the following time-marching scheme, for any $\mathbf{v} \in \mathbf{V}_h^0, \mathbf{w} \in \mathbf{W}_h^0, q \in S_h^0$,

$$
\begin{cases}
\mathcal{A}_h^{\mathcal{E}}(\mathbf{u}_h^{(n)}, \mathbf{v}) & - \mathcal{B}_h(p_h^{(n)}, \mathbf{v}) = \mathcal{F}_h^{\mathcal{E}}(\mathbf{v}), \\
\quad \mathcal{A}_h^{\mathcal{D}}(\mathbf{q}_h^{(n)}, \mathbf{w}) - \mathcal{B}_h^{\mathcal{D}}(p_h^{(n)}, \mathbf{w}) = \mathcal{F}_h^{\mathcal{D},1}(\mathbf{w}), \\
\mathcal{B}_h(\mathbf{u}_h^{(n)}, q) + \mathcal{B}_h^{\mathcal{D}}(\mathbf{q}_h^{(n)}, q) + \mathcal{C}_h^{\mathcal{D}}(p_h^{(n)}, q) = \mathcal{F}_h^{\mathcal{D},2}(q),
\end{cases}
\tag{14}
$$

where

$$
\begin{cases}
\mathcal{A}_h^{\mathcal{E}}(\mathbf{u}_h^{(n)}, \mathbf{v}) = 2\mu(\varepsilon(\mathbf{u}_h^{(n)}), \varepsilon(\mathbf{v})) + \lambda(\overline{\nabla \cdot \mathbf{u}_h^{(n)}}, \overline{\nabla \cdot \mathbf{v}}), \\
\mathcal{B}_h(p_h^{(n)}, \mathbf{v}) = \alpha(p_h^{(n)}, \overline{\nabla \cdot \mathbf{v}}),
\end{cases}
\tag{15}
$$

and

$$
\begin{cases}
\mathcal{A}_h^{\mathcal{D}}(\mathbf{q}_h^{(n)}, \mathbf{w}) = \Delta t\,(\mathbf{K}^{-1}\mathbf{q}_h^{(n)}, \mathbf{w}), \\
\mathcal{B}_h^{\mathcal{D}}(\mathbf{q}_h^{(n)}, q) = \Delta t\,(\nabla \cdot \mathbf{q}_h^{(n)}, q), \\
\mathcal{C}_h^{\mathcal{D}}(p_h^{(n)}, q) = c_0(p_h^{(n)}, q).
\end{cases}
\tag{16}
$$

Additionally,

$$
\begin{cases}
\mathcal{F}_h^{\mathcal{E}}(\mathbf{v}) = \sum_{T \in \mathcal{T}_h} (\mathbf{f}^{(n)}, \mathbf{v})_T + \sum_{e \in \Gamma_N^{\mathcal{E}}} \langle \mathbf{t}_N, \mathbf{v} \rangle_e, \\
\mathcal{F}_h^{\mathcal{D},1}(\mathbf{w}) = - \sum_{e \in \Gamma_D^{\mathcal{D}}} \Delta t \langle p_D, \mathbf{w} \cdot \mathbf{n} \rangle_e, \\
\mathcal{F}_h^{\mathcal{D},2}(q) = \sum_{T \in \mathcal{T}_h} \Delta t(s^{(n)}, q)_T + c_0(p_h^{(n-1)}, q)_T + \alpha(\overline{\nabla \cdot \mathbf{u}_h^{(n-1)}}, q)_T.
\end{cases}
\tag{17}
$$

Note $\overline{\nabla \cdot \mathbf{v}}$ is the elementwise average that represents the reduced integration technique. The above two equations are further augmented with appropriate boundary and initial conditions. This results in a large monolithic system at each time step. The DOFs at each time step are

$$2 \,\#\text{Nodes} + \#\text{Elements} + 2 \,\#\text{Edges}.$$

4   Solver III and IV: 2-Field CG+WG Solvers for Triangular and Quadrilateral Meshes

The MFEM(RT_0, P_0) discretization for Darcy flow in Solver II can be replaced by WG($P_0, P_0; RT_0$) discretization. This results in a new 2-field solver, which is easier in implementation, based on our experience. This is labeled as Solver III in this series. Here we provide a brief description of the scheme.

Let $\mathcal{T}_h$ be a quasi-uniform triangular mesh. We use BR_1 for displacement discretization (in linear elasticity), as done in [30]. However, we use $WG(P_0, P_0; RT_0)$ for pressure discretization (in Darcy flow) [13,16]. Let $\mathbf{V}_h$ be the space of BR_1 shape functions on $\mathcal{T}_h$ and $\mathbf{V}_h^0$ be its subspace with vanishing values on solid Dirichlet boundary. Similarly, S_h denotes the space of $WG(P_0, P_0)$ shape functions on $\mathcal{T}_h$ and S_h^0 be its subspace with vanishing values on fluid Dirichlet boundary. Treatment of initial and boundary conditions involve appropriate interpolation and/or projection operators into respective finite element spaces [26]:

$$\mathbf{u}_h^{(0)} = \mathbf{P}_h \mathbf{u}_0, \qquad p_h^{(0)} = Q_h p_0,$$

and

$$\mathbf{u}_h^{(n)}|_{\Gamma_D^{\mathcal{E}}} = \mathbf{P}_h \mathbf{u}_D, \qquad p_h^{(n,\partial)}|_{\Gamma_D^{\mathcal{D}}} = Q_h^{\partial}(p_D).$$

Solver III as a time-marching finite element scheme is formulated as

$$\begin{cases} \mathcal{A}_h^{\mathcal{E}}(\mathbf{u}_h^{(n)}, \mathbf{v}) - \mathcal{B}_h(p_h^{(n)}, \mathbf{v}) = \mathcal{F}_h^{\mathcal{E}}(\mathbf{v}), \\ \mathcal{B}_h(\mathbf{u}_h^{(n)}, q) + \mathcal{A}_h^{\mathcal{D}}(p_h^{(n)}, q) = \mathcal{F}_h^{\mathcal{D}}(q), \end{cases} \tag{18}$$

for any $\mathbf{v} \in \mathbf{V}_h^0$ and any $q \in S_h^0$. The bilinear forms are defined as

$$\begin{cases} \mathcal{A}_h^{\mathcal{E}}(\mathbf{u}_h^{(n)}, \mathbf{v}) = \sum_{T \in \mathcal{T}_h} 2\mu(\varepsilon(\mathbf{u}_h^{(n)}), \varepsilon(\mathbf{v}))_T + \lambda(\overline{\nabla \cdot \mathbf{u}_h^{(n)}}, \overline{\nabla \cdot \mathbf{v}})_T, \\ \mathcal{A}_h^{\mathcal{D}}(p_h^{(n)}, q) = \sum_{T \in \mathcal{T}_h} \Delta t \left(\mathbf{K}\nabla_w p_h^{(n)}, \nabla_w q\right)_T + c_0(p_h^{(n),\circ}, q^\circ)_{T^\circ}, \\ \mathcal{B}_h(\mathbf{u}_h^{(n)}, q) = \sum_{T \in \mathcal{T}_h} \alpha(\nabla \cdot \mathbf{u}_h^{(n)}, q^\circ)_{T^\circ}. \end{cases} \tag{19}$$

The linear forms are defined as

$$\begin{cases} \mathcal{F}_h^{\mathcal{E}}(\mathbf{v}) = \sum_{T \in \mathcal{T}_h} (\mathbf{f}^{(n)}, \mathbf{v})_T + \sum_{e \in \Gamma_N^{\mathcal{E}}} \langle \mathbf{t}_N, \mathbf{v} \rangle_e, \\ \mathcal{F}_h^{\mathcal{D}}(q) = \sum_{T \in \mathcal{T}_h} \Delta t(s^{(n)}, q^\circ)_{T^\circ} + c_0(p_h^{(n-1),\circ}, q^\circ)_{T^\circ} + \alpha(\nabla \cdot \mathbf{u}_h^{(n-1)}, q^\circ)_{T^\circ} \\ \quad - \sum_{e \in \Gamma_N^{\mathcal{D}}} \Delta t \langle u_N, q^\partial \rangle_e. \end{cases} \tag{20}$$

It is interesting to see that for each time moment t_n, the discrete linear system (18) has the same size as the discrete linear system (14).

Solver IV for quadrilateral meshes is similar to Solver III (for triangular meshes). But quadrilateral meshes are equally versatile as triangular meshes in accommodation of complicated domain geometry but may need less degrees of freedom for discretization. In certain cases, quadrilateral meshes may hold advantages in alignment with geometric and physical features in the problems to be solved [9]. With these considerations, Solver IV in two versions for poroelasticity has already been developed in [8,26].

The later version of Solver IV in [26] applies to general convex quadrilateral meshes. It uses the newly developed AC_0 space in [3] for Darcy flow discretization, which includes the rectangular RT_0 as a special case. For discretization of linear elasticity on quadrilaterals, the BR_1 or enriched Lagrangian elements EQ_1 are used. Therefore, for Solver IV, one just needs slight modification in Eq. (18–20):

- Replace the triangular mesh $\mathcal{T}_h$ by a quadrilateral mesh $\mathcal{E}_h$;
- Replace the triangular BR_1 elements by quadrilateral BR_1 elements (the enriched Lagrangian elements EQ_1) [11];
- Replace triangular $WG(P_0, P_0; RT_0)$ by $WG(P_0, P_0; AC_0)$ for quadrilaterals.

The DOFs for Solver IV is also

$$2\#Nodes + \#Elements + 2\#Edges,$$

but there are less elements and edges in a quadrilateral mesh.

5   Solver V Based on Operator-Splitting for Problems with Dilation-Dependent Permeability

Solver V is developed on top of Solver IV but aims at nonlinear poroelasticity in which permeability may depend on dilation. We adopt the approach of operator splitting (OS), namely, linear elasticity and Darcy problems are solved separately within Gauss-Seidel iterations.

For ease of presentation, we consider a convex quadrilateral mesh $\mathcal{E}_h$. We use EQ_1 or BR_1 finite elements for elasticity discretization [11] and $WG(P_0, P_0; AC_0)$ for discretization of Darcy flow [17], as in Solver IV. Treatment of initial and boundary conditions is similar to that in Solver III or IV.

Solver V as a time-marching finite element scheme is formulated as

$$\begin{cases} \mathcal{A}_h^{\mathcal{E}}(\mathbf{u}_h^{(n)}, \mathbf{v}) - \mathcal{B}_h(p_h^{(n)}, \mathbf{v}) & = \mathcal{F}_h^{\mathcal{E}}(\mathbf{v}), \\ \mathcal{B}_h(\mathbf{u}_h^{(n)}, q) + \mathcal{A}_h^{\mathcal{D}}(p_h^{(n)}, q; \mathbf{u}_h^{(n)}) & = \mathcal{F}_h^{\mathcal{D}}(q), \end{cases} \tag{21}$$

for any $\mathbf{v} \in \mathbf{V}_h^0$ and any $q \in S_h^0$, where the FE spaces have definitions similar to those for Solver IV. The bilinear forms $\mathcal{A}_h^{\mathcal{E}}(\mathbf{u}_h^{(n)}, \mathbf{v})$ and $\mathcal{B}_h(\mathbf{u}_h^{(n)}, q)$ have similar definitions as in Eq. (19). But the bilinear form $\mathcal{A}_h^{\mathcal{D}}(p_h^{(n)}, q; \mathbf{u}_h^{(n)})$ depends on the numerical displacement as shown below

$$\mathcal{A}_h^{\mathcal{D}}(p_h^{(n)}, q; \mathbf{u}_h^{(n)}) = \sum_{E \in \mathcal{E}_h} \Delta t \left(\mathbf{K}(\mathbf{u}_h^{(n)}) \nabla_w p_h^{(n)}, \nabla_w q \right)_E + c_0(p_h^{(n),\circ}, q^\circ)_{E^\circ}, \tag{22}$$

The linear forms $\mathcal{F}_h^{\mathcal{E}}(\mathbf{v})$ and $\mathcal{F}_h^{\mathcal{D}}(q)$ have definitions similar to those in (19).

However, (21) is a nonlinear discrete system about $\mathbf{u}_h^{(n)}, p_h^{(n)}$. This will be solved via operator-splitting or a Gauss-Seidel type iterative procedure as shown below

$$
\begin{cases}
\mathcal{A}_h^{\mathcal{E}}(\mathbf{u}_h^{(n,k)}, \mathbf{v}) = \mathcal{F}_h^{\mathcal{E}}(\mathbf{v}) + \mathcal{B}_h(p_h^{(n,k-1)}, \mathbf{v}), \\
\mathcal{A}_h^{\mathcal{D}}(p_h^{(n,k)}, q; \mathbf{u}_h^{(n,k)}) = \mathcal{F}_h^{\mathcal{D}}(q) - \mathcal{B}_h(\mathbf{u}_h^{(n,k)}, q).
\end{cases}
\tag{23}
$$

As shown later in Sect. 6, a typical nonlinear case is a dilation-dependent permeability, e.g.,

$$
\mathbf{K}(\mathbf{u}) = (1 + a\nabla \cdot \mathbf{u})\mathbf{K}_0,
\tag{24}
$$

where a is a small constant and $\mathbf{K}_0$ is a reference permeability. This requires calculation of elementwise averages of dilation (divergence of displacement). It is clear that for Solver IV and hence Solver V, such quantities are readily available.

6   Matlab Implementation of Poroelasticity Solvers

For Matlab implementation of the poroelasticity solvers discussed in this paper and other similar solvers, we emphasize code modularity. This has implication in two aspects.

(i) Each module has its entirety and a well-designed interface to the calling module. Each module fulfills a well-defined scientific computing task that is clearly separated from other tasks.

(ii) Code modules for similar solvers share uniformity and common features. Some code segments are conveniently portable or can be re-used after simple modification.

Besides mesh preparation and presentation of results (physical quantities of interest and errors when exact solutions are known, etc.), a typical finite element solver usually involves

– Element-wise or edge-wise integration and even node-wise evaluation;
– Assembly of element-wise stiffness matrices and source-type vectors;
– Incorporation and enforcement of boundary conditions;
– Modification of (non-)linear systems due to boundary conditions;
– Solvers for linear or nonlinear systems.

For instance,

– Solver I, IV, V share the same modules for mesh preparation and presentation of quantities of interest;
– Solver I, IV, V share many common modules of $WG(P_0, P_0; AC_0)$ for pressure discretization and Darcy velocity computation;
– Solver II, III share common modules for triangle Bernardi-Raugel elements;
– The modules for triangular and quadrilateral BR_1 elements share the same structure and many common features.

Many modules previously developed in our DarcyLite package for weak Galerkin FEMs for Darcy flow and linear elasticity are also re-used.

7 Numerical Experiments

This section presents numerical examples to demonstrate the accuracy and robustness of the finite element solvers for poroelasticity studied in this paper. We shall focus on Solver I, III, and V. Solver II is essentially equivalent to Solver III but the latter seems to have some convenience in implementation. Some numerical experiments on Solver IV can be found in [26].

Example 1 (Locking-Free). Here $\Omega = (0,1)^2$. Analytical solutions for displacement and pressure are given as

$$\mathbf{u} = \sin\left(\frac{\pi}{2}t\right)\left(\begin{bmatrix} \sin^2(\pi x)\sin(2\pi y) \\ -\sin^2(\pi y)\sin(2\pi x) \end{bmatrix} + \frac{1}{1+\lambda}\begin{bmatrix} \sin(\pi x)\sin(\pi y) \\ \sin(\pi x)\sin(\pi y) \end{bmatrix}\right),$$

$$p = \sin\left(\frac{\pi}{2}t\right)\frac{\pi}{1+\lambda}\sin(\pi(x+y)).$$

Table 1. Ex.1 ($\lambda = 1$): Errors and convergence rates of numerical solutions obtained from Solver I (WG+WG) on rectangular meshes

$1/h$	$1/\Delta t$	$\|\mathbf{u} - \mathbf{u}_h\|_{L_2(L_2)}$	Rate	$\|p - p_h^\circ\|_{L_2(L_2)}$	Rate	$\|\mathbf{q} - \mathbf{q}_h\|_{L_2(L_2)}$	Rate
8	8	1.2757E−1	−	1.3289E−1	−	4.2093E−1	−
16	16	6.1993E−2	1.04	6.4829E−2	1.03	2.0427E−1	1.04
32	32	3.0529E−2	1.02	3.1964E−2	1.02	1.0056E−1	1.02
64	64	1.5147E−2	1.01	1.5863E−2	1.01	4.9881E−2	1.01

Table 2. Ex.1 ($\lambda = 10^6$): Errors and convergence rates of numerical solutions obtained from Solver I (WG+WG) on rectangular meshes

$1/h$	$1/\Delta t$	$\|\mathbf{u} - \mathbf{u}_h\|_{L_2(L_2)}$	Rate	$\|p - p_h^\circ\|_{L_2(L_2)}$	Rate	$\|\mathbf{q} - \mathbf{q}_h\|_{L_2(L_2)}$	Rate
8	8	1.2042e−01	−	2.6577e−07	−	8.4154E−7	−
16	16	5.8469e−02	1.04	1.2965e−07	1.03	4.0848E−7	1.04
32	32	2.8786e−02	1.02	6.3926e−08	1.02	2.0110E−7	1.02
64	64	1.4281e−02	1.01	3.1727e−08	1.01	9.9761E−8	1.01

Table 3. Ex.1 ($\lambda = 1$): Numerical results of Solver III (CG+WG) on triangular meshes

$h = \Delta t$	$\|\mathbf{u} - \mathbf{u}_h\|_{L_2(L_2)}$	$\|\mathbf{u} - \mathbf{u}_h\|_{L_\infty(H^1)}$	$\|\sigma - \sigma_h\|_{L_2(L_2)}$	$\|p - p_h^\circ\|_{L_2(L_2)}$	$\|\mathbf{q} - \mathbf{q}_h\|_{L_2(L_2)}$
1/4	6.533E−2	1.401E+0	1.775E+0	1.613E−1	8.760E−1
1/8	1.485E−2	6.648E−1	8.230E−1	7.696E−2	4.186E−1
1/16	3.551E−3	3.265E−1	3.964E−1	3.745E−2	2.038E−1
1/32	8.727E−4	1.624E−1	1.948E−1	1.845E−2	1.004E−1
1/64	2.177E−4	8.113E−2	9.662E−2	9.160E−3	4.986E−2
Conv.rate	2.05	1.02	1.04	1.03	1.03

Table 4. Ex.1 ($\lambda = 10^6$): Numer. results of Solver III (CG+WG) on triangular meshes

$h = \Delta t$	$\|\mathbf{u} - \mathbf{u}_h\|_{L_2(L_2)}$	$\|\mathbf{u} - \mathbf{u}_h\|_{L_\infty(H^1)}$	$\|\sigma - \sigma_h\|_{L_2(L_2)}$	$\|p - p_h^\circ\|_{L_2(L_2)}$	$\|\mathbf{q} - \mathbf{q}_h\|_{L_2(L_2)}$
1/4	6.502E$-$2	1.395E$+$0	1.814E$+$0	3.226E$-$7	1.752E$-$6
1/8	1.485E$-$2	6.631E$-$1	8.537E$-$1	1.539E$-$7	8.372E$-$7
1/16	3.557E$-$3	3.259E$-$1	4.132E$-$1	7.491E$-$8	4.076E$-$7
1/32	8.727E$-$4	1.622E$-$1	2.033E$-$1	3.691E$-$8	2.009E$-$7
1/64	2.163E$-$4	8.102E$-$2	1.008E$-$1	1.832E$-$8	9.972E$-$8
Conv.rate	2.05	1.02	1.03	1.03	1.03

It is interesting to see that $\nabla \cdot \mathbf{u} = p$ and $\nabla \cdot \mathbf{u} \to 0$ as $\lambda \to \infty$. Dirichlet boundary conditions for both displacement and pressure are specified on the whole boundary using the exact solutions. Furthermore, $\mathbf{K} = \kappa \mathbf{I}$. Direct calculations show that

$$
\begin{aligned}
\mathbf{f} = -\nabla \cdot \tilde{\sigma} = &- \sin\left(\frac{\pi}{2}t\right)\left(2\mu\pi^2 \begin{bmatrix} (1 - 4\sin^2(\pi x))\sin(2\pi y) \\ -(1 - 4\sin^2(\pi y))\sin(2\pi x) \end{bmatrix} \right. \\
&- \frac{2\mu}{1+\lambda}\pi^2 \begin{bmatrix} \sin(\pi x)\sin(\pi y) \\ \sin(\pi x)\sin(\pi y) \end{bmatrix} + \left. \frac{\lambda+\mu-\alpha}{1+\lambda}\pi^2 \begin{bmatrix} \cos(\pi(x+y)) \\ \cos(\pi(x+y)) \end{bmatrix} \right)
\end{aligned}
\tag{25}
$$

and

$$
s = \left((\alpha + c_0)\cos\left(\frac{\pi}{2}t\right)\frac{\pi}{2} + \sin\left(\frac{\pi}{2}t\right)\kappa(2\pi^2)\right)\frac{\pi}{1+\lambda}\sin(\pi(x+y)).
\tag{26}
$$

For numerical simulations, we set $\kappa = 1$, $\mu = 1$, $\alpha = 1$, $c_0 = 0$, and $T = 1$. To examine the locking-free property of these solvers, we consider $\lambda = 1$ and $\lambda = 10^6$, respectively.

We examine errors in displacement ($\mathbf{u} - \mathbf{u}_h$), stress ($\sigma - \sigma_h$), pressure ($p - p_h^\circ$), and Darcy velocity ($\mathbf{q} - \mathbf{q}_h$). For Solver I, see results in Tables 1 and 2. For Solver III, see results in Tables 3 and 4. Clearly, the convergence rates do not deteriorate as λ is increased from 1 to 10^6. In other words, these new 2-field solvers based on the weak Galerkin methodology are locking-free.

Example 2 (Model's Problem). This is a frequently tested benchmark that has known analytical solutions. See [1,5,12,18,22]. The problem involves a poroelastic rectangular slab with extent $2a$ in the x-direction and extent $2b$ in the y-direction being sandwiched by two rigid plates at the top and the bottom. Two forces of magnitude $2F$, pointing to the slab, are applied at the top and bottom plates, respectively. Due to the rigidity of the plates, the slab remains in contact with the two plates. Thus the vertical displacement at the top and bottom are uniform. The initial condition for displacement is $\mathbf{u}(x, y, 0) = \mathbf{0}$. Based on symmetry in the problem, we choose the center of the slab as the origin and consider the upper-right quadrant. The Mandel's problem is thus posed for the domain $\Omega = (0, a) \times (0, b)$ for a time period $[0, T]$.

The boundary conditions for the solid and fluid are, see Fig. 1(a),

(i) Symmetry or partial Dirichlet: $u_1 = 0$ for $x = 0$; $u_2 = 0$ for $y = 0$;
(ii) Neumann or traction-free: $\tilde{\sigma}\mathbf{n} = \mathbf{0}$ for $x = a$;

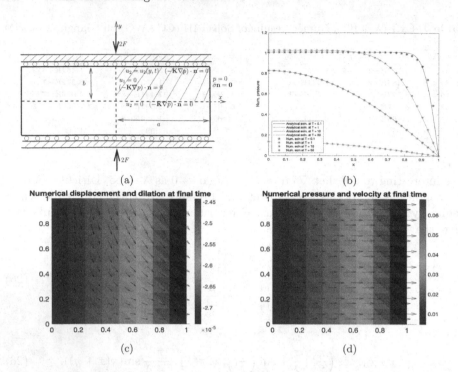

Fig. 1. Mandel's problem ($\nu = 0.4$) solved by Solver III on uniform triangular meshes. **(a)** Problem illustration; **(b)** Numerical pressure on the center line ($y = 0$) at final time $T = 50$ with $h = 1/32$; **(c)** Numerical displacement and dilation with $h = 1/8$; **(d)** Numerical pressure and velocity with $h = 1/8$.

(iii) Specially, for $y = b$ (the top side), it is subject to the traction condition $\tilde{\sigma}\mathbf{n} = [0, -2F]$ along with the "rigid plate" constraint, which requires u_2 stays the same for the whole top side;

(iv) Dirichlet: $p = 0$ for $x = a$ (drained);

 (v) Neumann or no-flow: $(-\mathbf{K}\nabla p) \cdot \mathbf{n} = 0$ for $x = 0, y = 0, y = b$.

An easier but equivalent treatment for (iii) is to impose a partial Dirichlet boundary condition for u_2 using the known exact solution for displacement [5, 22, 26].

Example 3 (A Nonlinear Problem). Here we consider an example in which the permeability depends on the dilation. In particular,

$$\mathbf{K}(\mathbf{u}) = (1 + a\nabla \cdot \mathbf{u})\,\kappa\,\mathbf{I}, \tag{27}$$

where a is a small constant and κ is a reference permeability. Furthermore, we consider a case with known analytical solutions for displacement and pressure:

$$\mathbf{u} = \sin\left(\frac{\pi}{2}t\right)\begin{bmatrix} \sin(\pi x)\sin(\pi y) \\ \sin(\pi x)\sin(\pi y) \end{bmatrix}, \quad p = \sin\left(\frac{\pi}{2}t\right)(1 + \cos(\pi y)). \tag{28}$$

This allows examination of accuracy and efficiency of Solver V. Furthermore, $\Omega = (0,1)^2$, $T = 1$, $\lambda = \mu = 1$, $\kappa = 1$, $a = 0.1$, $c_0 = 0$, and $\alpha = 1$. Dirichlet boundary conditions are posed for both displacement and pressure.

Table 5. Ex.3 (Solver V): Convergence rates of errors of numerical solutions obtained from combining EQ_1 and $WG(P_0, P_0; AC_0)$ on rectangular meshes with $\Delta t = h$

$1/h$	$\|p - p_h^\circ\|$	Rate	$\|q - q_h\|$	Rate	$|u - u_h|$	Rate	$\|\sigma - \sigma_h\|$	Rate	Runtime
4	5.119E−1	–	1.366E+0	–	1.260E−1	–	1.503E−1	–	0.59 s
8	2.528E−1	1.01	6.527E−1	1.06	6.023E−2	1.06	6.398E−2	1.23	1.26 s
16	1.260E−1	1.00	3.177E−1	1.03	2.931E−2	1.03	2.944E−2	1.11	4.12 s
32	6.297E−2	1.00	1.565E−1	1.02	1.444E−2	1.02	1.413E−2	1.05	26.57 s
64	3.148E−2	1.00	7.770E−2	1.01	7.169E−3	1.01	6.931E−3	1.02	333.46 s

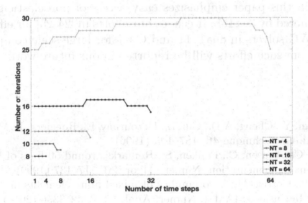

Fig. 2. Example 3: numbers of Gauss-Seidel iterations during time-marching

We test Solver V on rectangular meshes with $\Delta t = h$. Following the common practices, we examine the discrepancy (difference) of two successive approximate solutions within the Gauss-Seidel iteration. We set discrepancy threshold as $\delta = 10^{-12}$ and check whether the following conditions are satisfied:

$$\|u_h^{(n,k)} - u_h^{(n,k-1)}\|_{L_2} < \delta,$$
$$\|p_h^{(n,k)} - p_h^{(n,k-1)}\|_{L_2} < \delta. \tag{29}$$

Table 5 demonstrates good performance of Solver V on this nonlinear poroelasticity problem. For these particular choices of parameters, the numbers of Gauss-Seidel iterations during the time-marching are reported in Fig. 2.

8   Concluding Remarks

In this paper, we have discussed five different finite element solvers for linear and nonlinear poroelasticity problems, along with `Matlab` implementation of these solvers. It is demonstrated that weak Galerkin finite element methods can be well integrated with other types of finite element methods. This is also reflected in the modularity of our code development. Under the guidelines discussed in this paper, more modules for finite element solvers for poroelasticity can be integrated into our code package `DarcyLite` [15].

These poroelasticity modules can also provide computed physical quantities that are needed in finite element solvers for other physical processes. It is particularly interesting to see integration of these poroelasticity solvers with (mass, positivity) property-preserving transport solvers in development of numerical simulators for transport in poroelastic media. This is currently under our investigation and will be reported in our future work.

The work in this paper emphasizes easy access of poroelasticity solvers on the platform offered by `Matlab`. It echoes our efforts in [24,25] for efficient implementation of WG solvers in `deal.II` and C++ for large-scale computing tasks. More results from such efforts will be reported in our future work.

References

1. Abousleiman, Y., Cheng, A.D., Cui, L., Detournay, E., Roegiers, J.: Mandel's problem revisted. Geotechnique **46**, 187–195 (1996)
2. Alberty, J., Carstensen, C., Funken, S.: Remarks around 50 lines of Matlab: short finite element implementation. Numer. Algor. **20**, 117–137 (1999)
3. Arbogast, T., Correa, M.: Two families of mixed finite elements on quadrilaterals of minimal dimension. SIAM J. Numer. Anal. **54**, 3332–3356 (2016)
4. Berger, L., Bordas, R., Kay, D., Tavener, S.: Stabilized lowest-order finite element approximation for linear three-field poroelasticity. SIAM J. Sci. Comput. **37**, A2222–A2245 (2015)
5. Correa, M.R., Murad, M.A.: A new sequential method for three-phase immiscible flow in poroelastic media. J. Comput. Phys. **373**, 493–532 (2018)
6. Fu, G.: A high-order HDG method for the Biot's consolidation model. Comput. Math. Appl. **77**(1), 237–252 (2019)
7. Haga, J., Osnes, H., Langtangen, H.: On the causes of pressure oscillations in low permeable and low compressible porous media. Int. J. Numer. Anal. Meth. Geomech. **36**, 1507–1522 (2012)
8. Harper, G., Liu, J., Tavener, S., Wang, Z.: A two-field finite element solver for poroelasticity on quadrilateral meshes. Lecturer Notes in Computer Science, vol. 10862, pp. 76–88 (2018)
9. Harper, G., Liu, J., Tavener, S., Wildey, T.: Coupling arbogast-correa and bernardi-raugel elements to resolve coupled Stokes-Darcy flow problems. Comput. Meth. Appl. Mech. Engrg. **373**, 113469 (2021)
10. Harper, G., Liu, J., Tavener, S., Zheng, B.: Lowest-order weak Galerkin finite element methods for linear elasticity on rectangular and brick meshes. J. Sci. Comput. **78**, 1917–1941 (2019)

11. Harper, G., Wang, R., Liu, J., Tavener, S., Zhang, R.: A locking-free solver for linear elasticity on quadrilateral and hexahedral meshes based on enrichment of Lagrangian elements. Comput. Math. Appl. **80**, 1578–1595 (2020)

12. Hu, X., Mu, L., Ye, X.: Weak Galerkin method for the Biot's consolidation model. Comput. Math. Appl. **75**, 2017–2030 (2018)

13. Lin, G., Liu, J., Mu, L., Ye, X.: Weak Galerkin finite element methdos for Darcy flow: anistropy and heterogeneity. J. Comput. Phys. **276**, 422–437 (2014)

14. Lin, G., Liu, J., Sadre-Marandi, F.: A comparative study on the weak Galerkin, discontinuous Galerkin, and mixed finite element methods. J. Comput. Appl. Math. **273**, 346–362 (2015)

15. Liu, J., Sadre-Marandi, F., Wang, Z.: DarcyLite: a Matlab toolbox for Darcy flow computation. Proc. Comput. Sci. **80**, 1301–1312 (2016)

16. Liu, J., Tavener, S., Wang, Z.: The lowest-order weak Galerkin finite element method for the Darcy equation on quadrilateral and hybrid meshes. J. Comput. Phys. **359**, 312–330 (2018)

17. Liu, J., Tavener, S., Wang, Z.: Penalty-free any-order weak Galerkin Fems for elliptic problems on quadrilateral meshes. J. Sci. Comput. **83**, 47 (2020)

18. Phillips, P., Wheeler, M.: A coupling of mixed with continuous Galerkin finite element methods for poroelasticity I: the continuous in time case. Comput. Geosci. **11**, 131–144 (2007)

19. Phillips, P., Wheeler, M.: A coupling of mixed with continuous Galerkin finite element methods for poroelasticity II: the-discrete-in-time case. Comput. Geosci. **11**, 145–158 (2007)

20. Phillips, P., Wheeler, M.: A coupling of mixed with discontinuous Galerkin finite element methods for poroelasticity. Comput. Geosci. **12**, 417–435 (2008)

21. Phillips, P.J., Wheeler, M.F.: Overcoming the problem of locking in linear elasticity and poroelasticity: an heuristic approach. Comput. Geosci. **13**, 5–12 (2009)

22. Rodrigo, C., Gaspar, F., Hu, X., Zikatanov, L.: Stability and monotonicity for some discretizations of the Biot's consolidation model. Comput. Meth. Appl. Mech. Engrg. **298**, 183–204 (2016)

23. Wang, J., Ye, X.: A weak Galerkin finite element method for second order elliptic problems. J. Comput. Appl. Math. **241**, 103–115 (2013)

24. Wang, Z., Harper, G., O'Leary, P., Liu, J., Tavener, S.: Deal.II implementation of a weak Galerkin finite element solver for Darcy flow. Lec. Notes Comput. Sci. **11539**, 495–509 (2019)

25. Wang, Z., Liu, J.: Deal.II implementation of a two-field finite element solver for poroelasticity. Lecturer Notes Computer Science, vol. 12143, pp. 88–101 (2020)

26. Wang, Z., Tavener, S., Liu, J.: Analysis of a 2-field finite element solver for poroelasticity on quadrilateral meshes. J. Comput. Appl. Math. **393**, 113539 (2021)

27. Wheeler, M., Xue, G., Yotov, I.: Coupling multipoint flux mixed finite element methods with continuous Galerkin methods for poroelasticity. Comput. Geosci. **18**, 57–75 (2014)

28. Yi, S.Y.: A coupling of nonconforming and mixed finite element methods for Biot's consolidation model. Numer. Meth. PDEs **29**, 1749–1777 (2013)

29. Yi, S.Y.: Convergence analysis of a new mixed finite element method for Biot's consolidation model. Numer. Meth. PDEs **30**, 1189–1210 (2014)

30. Yi, S.Y.: A study of two modes of locking in poroelasticity. SIAM J. Numer. Anal. **55**, 1915–1936 (2017)

Mathematical Modeling
of the Single-Phase Multicomponent
Flow in Porous Media

Petr Gális[(✉)] [iD] and Jiří Mikyška[iD]

Department of Mathematics, Faculty of Nuclear Sciences and Physical Engineering,
Czech Technical University in Prague, Prague, Czech Republic
`galispet@fjfi.cvut.cz`

Abstract. A numerical scheme of higher-order approximation in space
for the single-phase multicomponent flow in porous media is presented.
The mathematical model consists of Darcy velocity, transport equations
for components of a mixture, pressure equation and associated relations
for physical quantities such as viscosity or density. The discrete problem
is obtained via discontinuous Galerkin method for the discretization of
transport equations with the combination of mixed-hybrid finite element
method for the discretization of Darcy velocity and pressure equation
both using higher-order approximation. Subsequent problem is solved
with the fully mass-conservative iterative IMPEC method. Numerical
experiments of 2D flow are carried out.

Keywords: Compositional flow · Mixed-hybrid finite element
method · Discontinuous Galerkin method

1 Introduction

The compositional modeling has many applications in various disciplines ranging from petroleum engineering (oil recovery, CO2 sequestration) to geochem-
ical engineering (groundwater contamination, radioactive waste storage in the
subsurface). Modeling of such phenoma is therefore of a great importance. In
this work we consider the single-phase flow of miscible and compressible multicomponent fluids in porous media. We follow an approach of the previous
works of [6,7,12,13] based on the combination of the mixed-hybrid finite element (MHFEM) for the approximation of the pressure and velocity fields and
discontinuous Galerkin method (DG) for the approximation of transport equations. Hoteit and Firoozabadi [7] and Moortgat, Sun and Firoozabadi [13] used a

The work was supported by the Czech Science Foundation project no. 21-09093S Multiphase flow, transport, and structural changes related to water freezing and thawing
in the subsurface, by the Ministry of Education, Youth and Sports of the Czech Republic under the OP RDE grant number CZ.02.1.01/0.0/0.0/16_019/0000778 Centre for
Advanced Applied Sciences, and by the Student Grant Agency of the Czech Technical
University in Prague, grant no. SGS20/184/OHK4/3T/14.

M. Paszynski et al. (Eds.): ICCS 2021, LNCS 12747, pp. 200–214, 2021.
https://doi.org/10.1007/978-3-030-77980-1_16

combination of MHFEM for the pressure equation and higher-order DG method for the transport equations. Although they used piecewise linear basis functions for the concentrations, they used only piecewise constant functions for the pressures and $\mathbb{RT}_0$ (i.e. first-order) approximation for the velocity field. They also used a different form of the pressure equation which seems to be more complicated than the one used in [8]. In contrast to classical IMPEC schemes, used in the previous works and which are known to have a mass-conservation problem, Chen, Fan and Sun [8] have rewritten the pressure equation with only one additional parameter to be determined and proposed a new fully mass-conservative IMPEC scheme where conservation of mass for all components holds true. So far only the first-order approximation for pressure and velocity field has been utilized in the models. In this work we show how to extend these ideas for the higher-order framework. In contrast to previous works we apply higher order scheme not only for the transport of the species but also for the pressure and the velocity fields.

2   Mathematical Model

Consider single-phase compressible flow of fluid of n_c components at constant temperature T [K] in a bounded domain $\Omega \subset \mathbb{R}^2$ with porosity ϕ [-]. In this work we assume that the porosity does not depent on time, i.e. we have $\phi = \phi(\mathbf{x})$. Neglecting diffusion, the transport of the components is described by the following equations

$$\frac{\partial (\phi c_i)}{\partial t} + \nabla \cdot (c_i \mathbf{v}) = f_i, \quad i = 1, ..., n_c, \tag{1}$$

where c_i [mol$\cdot$m^{-3}] are molar concentrations of components, f_i [mol$\cdot$m$^{-3}\cdot$s^{-1}] are source/sink terms and $\mathbf{v}$ [m$\cdot$s^{-1}] is the velocity field described by the Darcy's law

$$\mathbf{v} = -\mu^{-1}\mathbf{K}\left(\nabla p - \rho \mathbf{g}\right), \tag{2}$$

where p [Pa], is the pressure field, μ [kg$\cdot$m$^{-1}\cdot$s^{-1}] is the dynamic viscosity, ρ [kg$\cdot$m^{-3}] is the density of fluid, $\mathbf{K}$ [m^2] is the medium permeability tensor and $\mathbf{g}$ [m$\cdot$s^{-2}] is the gravity acceleration vector. Equations (1) and (2) are coupled with generally nonlinear dependencies

$$p = p(c_1, ..., c_{n_c}, T), \quad \mu = \mu(c_1, ..., c_{n_c}, T), \quad \rho = \rho(c_1, ..., c_{n_c}). \tag{3}$$

to be found in [11] or [15]. Using the chain rule $\frac{\partial p}{\partial t} = \sum_{i=1}^{n_c} \frac{\partial p}{\partial c_i} \frac{\partial c_i}{\partial t}$ and transport Eqs. (1), we derive an equation for the pressure field

$$\phi \frac{\partial p}{\partial t} + \sum_{i=1}^{n_c} \theta_i \left[\nabla \cdot (c_i \mathbf{v}) - f_i\right] = 0, \tag{4}$$

where parameters $\theta_i = \theta_i(c_1, ..., c_{n_c})$ are defined as $\theta_i = \left(\frac{\partial p}{\partial c_i}\right)_{c_j \neq c_i}$ as in [8]. Let $I \subset \mathbb{R}$ be a time interval. The initial and boundary conditions are given by

$$
\begin{aligned}
c_i(0, \mathbf{x}) &= c_i^0(\mathbf{x}), & \mathbf{x} \in \Omega, \ i = 1, ..., n_c, \\
c_i(t, \mathbf{x}) &= c_i^D(\mathbf{x}), & \mathbf{x} \in \Gamma_c, \ t \in I, \ i = 1, ..., n_c, \\
p(t, \mathbf{x}) &= p^D(t, \mathbf{x}), & \mathbf{x} \in \Gamma_p, \ t \in I, \\
\mathbf{v}(t, \mathbf{x}) \cdot \mathbf{n}(\mathbf{x}) &= v^N(t, \mathbf{x}), & \mathbf{x} \in \Gamma_v, \ t \in I,
\end{aligned}
\tag{5}
$$

where $\mathbf{n}$ is the outward unit normal vector to the boundary $\partial\Omega$, $\Gamma_p \cup \Gamma_v = \partial\Omega$ and $\Gamma_p \cap \Gamma_v = \emptyset$. Note that the initial pressure field is obtained from the equation of state (3) by substituting c_i^0, $i = 1, ..., n_c$. Further, we define the inflow part of boundary $\Gamma_c(t) = \{\mathbf{x} \in \partial\Omega \mid \mathbf{v}(t, \mathbf{x}) \cdot \mathbf{n}(\mathbf{x}) < 0\}$ on which Dirichlet-type conditions c_i^D must be prescribed. On $\Gamma_c \cap \Gamma_p$ the following constraint must be satisfied $p^D = p(c_1^D, ..., c_{n_c}^D, T)$. The source terms f_i in (1) and (4) are usually expressed via injection rate r [m$^3 \cdot$ s^{-1}] as $f_i = c_i^{inj} r / V^{inj}$ where c_i^{inj} is the amount of i-th component injected in some part of the domain with the volume V^{inj}. Similarly, Neumann boundary condition v^N in (5) is often expressed as $v^N = r / A^{inj}$ where A^{inj} is the area through which the mixture is injected.

3   Numerical Model

A discrete form of the system of Eqs. (1)–(3) and (4), (5) is obtained using the mixed-hybrid finite element method for the Darcy's law and the pressure equation and discontinuous Galerkin method for the transport equations. We consider a polygonal domain $\Omega \subset \mathbb{R}^2$ covered with a conforming triangulation $\mathcal{T}_h$. Let us denote by $\mathcal{E}_h$ the set of all edges in the triangulation $\mathcal{T}_h$. For the $K \in \mathcal{T}_h$ and $E \in \mathcal{E}_h$ we denoted $|K|$ and $|E|$ the measures of the element K and edge E, respectively. The triangulation consists of n_k triangle elements and n_e edges.

3.1   Discretization of Darcy's Law

The velocity field is approximated in the Raviart-Thomas space $\mathbb{RT}_1(K)$ locally on the element $K \in \mathcal{T}_h$ as

$$
\mathbf{v}_K(t, \mathbf{x}) = \sum_{j=1}^{8} v_{K,j}(t) \, \mathbf{w}_{K,j}(\mathbf{x}),
\tag{6}
$$

where $\mathbb{RT}_1(K) = \text{span}\{\mathbf{w}_{K,j}\}_{j=1}^{8}$ and $v_{K,j}$ are associated degrees o freedom. The definition of the $\mathbb{RT}_1(K)$ space is taken over from [3]. The basis $\{\hat{\mathbf{w}}_j\}_{j=1}^{8}$ on the reference element $\hat{K}$ (see Fig. 1) is obtained via the following moments

$$
N_s^\alpha(\hat{\mathbf{w}}_j) = \int_{e_\alpha} (\hat{\mathbf{w}}_j \cdot \mathbf{n}_\alpha) \, p_s, \quad \forall p_s \in \mathbb{P}_1(e_\alpha), \ s = 1, 2, \ \alpha = 1, 2, 3,
$$

$$
M_r(\hat{\mathbf{w}}_j) = \int_{\hat{K}} (\hat{\mathbf{w}}_j \cdot \mathbf{q}_r), \quad \forall \mathbf{q}_r \in [\mathbb{P}_0(\hat{K})]^2, \ r = 1, 2,
$$

where $\mathbb{P}_1(e_\alpha) = \text{span}\{p_s\}_{s=1}^2$ is the space of linear polynomials defined on the edge e_α of the reference element and $[\mathbb{P}_0(\hat{K})]^2 = \text{span}\{\mathbf{q}_r\}_{r=1}^2$ is the space of vector-valued constant polynomials. The discrete form of Darcy's law is obtained by multiplying (2) by a basis function $\mathbf{w}_{K,m}$, integrating over the element K and using Green's theorem

$$v_{K,m} = \mu_K^{-1}\left(\sum_{j=1}^{3}\sum_{l=1}^{8} \left[\alpha_{m,l}^K \beta_{l,j}^K\right] p_{K,j} - \sum_{E\in\partial K}\sum_{s=1}^{2}\sum_{l=1}^{8} \left[\alpha_{m,l}^K \chi_{l,s}^{K,E}\right] \hat{p}_{E,s} \right.$$
$$\left. +\rho_K \sum_{l=1}^{8} \alpha_{m,l}^K \gamma_l^K \right), \quad m = 1,...,8, \tag{7}$$

where μ_K, ρ_K denote the mean values of viscosity and density over the element K, respectively. In the derivation of (7) we used the following approximation of the pressure and pressure trace on the element K and edge E

$$p(t,\mathbf{x})|_K = \sum_{j=1}^{3} p_{K,j}(t)\, \Phi_{K,j}(\mathbf{x}), \quad p(t,\mathbf{x})|_E = \sum_{s=1}^{2} \hat{p}_{E,s}(t)\, \varphi_s^E(\mathbf{x}), \tag{8}$$

where $\mathbb{P}_1(K) = \text{span}\{\Phi_{K,j}\}_{j=1}^3$ and $\mathbb{P}_1(E) = \text{span}\{\varphi_j^E\}_{j=1}^2$ are spaces of linear polynomials defined on the element K and on the edge E, respectively. The coefficients in (7) are given by

$$\tilde{\alpha}_{m,j}^K = \int_K \mathbf{w}_{K,m}\cdot\mathbf{K}^{-1}\mathbf{w}_{K,j}, \quad \beta_{m,j}^K = \int_K \Phi_{K,j}\,(\nabla\cdot\mathbf{w}_{K,m}),$$
$$\chi_{m,s}^{K,E} = \int_E \varphi_s^E\,(\mathbf{w}_{K,m}\cdot\mathbf{n}_E^K), \quad \gamma_m^K = \int_K \mathbf{g}\cdot\mathbf{w}_{K,m}, \tag{9}$$

where the coefficients $\alpha_{i,j}^K$ are elements of the inverse matrix $\left(\tilde{\alpha}^K\right)^{-1}$. In order to compute integrals in (9) on the reference element we use the following affine transformation of variables $F_K : \hat{K} \to K$ and the transformation of the vector-valued function which preserves normal components

$$F_K(s,t) = \begin{pmatrix} x_1 \\ y_1 \end{pmatrix} + J_{F_K}\begin{pmatrix} s \\ t \end{pmatrix}, \quad J_{F_K} = \begin{pmatrix} x_2 - x_1, x_3 - x_1 \\ y_2 - y_1, y_3 - y_1 \end{pmatrix},$$
$$\mathbf{w}_{K,m}(x,y) = \frac{1}{|J_{F_K}|} J_{F_K}\cdot\hat{\mathbf{w}}_m(s,t), \quad (s,t) = F_K^{-1}(x,y) \in \hat{K}, \tag{10}$$

where (x_i, y_i), $i = 1, 2, 3$ are the coordinates of vertices of the physical element K and $|J_{F_K}|$ is the determinant of the matrix J_{F_K}. Continuity of the normal component of the velocity field along the edges shared by neighboring elements is enforced through the conditions [2]

$$\sum_{K\in\mathcal{T}_h} \langle\mathbf{v}\cdot\mathbf{n}^K, \varphi_1^E\rangle_{\partial K} = 0, \quad \sum_{K\in\mathcal{T}_h} \langle\mathbf{v}\cdot\mathbf{n}^K, \varphi_2^E\rangle_{\partial K} = 0 \quad \forall E \in \mathcal{E}_h, \tag{11}$$

where $\langle f,g\rangle_{\partial K} = \int_{\partial K} fg$. Denoting by $\sum_{K\in E}$ the sum over adjacent elements to the given edge E, continuity conditions (11) can be further simplified to

$$\sum_{K\in E} v_{K,LI(K,E,s)}\, \chi^{K,E}_{LI(K,E,s),s} = 0, \quad s = 1,2, \quad \forall E \in \mathcal{E}_h, \tag{12}$$

where $LI(K,E,s) \in \{1,2,3,4,5,6\}$ is the local index of the velocity degree of freedom for $s \in \{1,2\}$ on the edge E with respect to the element K, see Fig. 1. As the degrees of freedom $v_{K,7}$ and $v_{K,8}$ are associated with basis functions $\mathbf{w}_{K,7}$ and $\mathbf{w}_{K,8}$, which have zero normal component along the boundary ∂K of the element K, these do not play any role when enforcing the continuity constraint. The discrete form of the boundary and initial conditions (5) reads as

$$\begin{aligned}
\hat{p}_{E,s} &= p_s^D|_E, \quad s = 1,2, \quad \forall E \subset \Gamma_p, \\
v_{K,LI(K,E,s)} &= v_s^N|_E, \quad s = 1,2, \quad \forall E \subset \Gamma_v,\ E \in \partial K.
\end{aligned} \tag{13}$$

The velocity can be eliminated substituting (7) into (12) and (13) resulting in the system of linear algebraic equations for pressures p and pressure traces $\hat{p}$

$$\begin{aligned}
\sum_{K\in E}\sum_{m=1}^{3}\sum_{l=1}^{8} \Big[\mu_K^{-1} \chi^{K,E}_{r,j} \alpha^K_{r,l}\, \beta_{l,m}\Big] p_{K,m} &- \sum_{s=1}^{2}\sum_{K\in E}\sum_{F\in\partial K}\sum_{l=1}^{8}\Big[\mu_K^{-1}\chi^{K,E}_{r,j}\alpha^K_{r,l}\chi^{K,F}_{l,s}\Big]\hat{p}_{F,s} \\
&= v_j^N|_{(E\cap\Gamma_v)} - \sum_{K\in E}\mu_K^{-1}\rho_K\,\chi^{K,E}_{r,j}\sum_{l=1}^{8}\alpha^K_{r,l}\,\gamma^K_l, \quad j = 1,2, \quad \forall E \not\subset \Gamma_p, \\
\hat{p}_{E,j} &= p_j^D|_E, \quad j = 1,2, \quad \forall E \subset \Gamma_p,
\end{aligned} \tag{14}$$

where $r = LI(K,E,j)$.

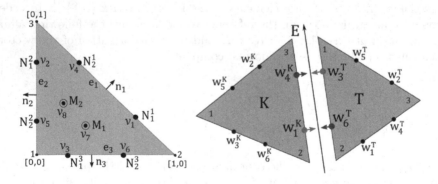

Fig. 1. Reference element and degrees of freedom (left). Local indexing of the neighboring triangles sharing an edge (right). For this setup values of indexing function are $LI(K,E,1) = 1$, $LI(K,E,2) = 4$, $LI(T,E,1) = 3$, $LI(T,E,2) = 6$.

3.2   Discretization of Pressure and Transport Equations

In the discontinuous Galerkin discretization we use linear approximation of the concentrations in the space $\mathbb{P}_1(K)$

$$c_i(t,\mathbf{x})|_K = \sum_{j=1}^{3} c_{K,i,j}(t)\, \Phi_{K,j}(\mathbf{x}). \tag{15}$$

We recall that we assume $\phi = \phi(\mathbf{x})$. Multiplying (1) by the basis function $\Phi_{K,m} \in \mathbb{P}_1(K)$, integrating over the element K and using Green's theorem, we derive a discrete form of (1) for $i = 1, ..., n_c$ and $m = 1, 2, 3$

$$\frac{\mathrm{d}c_{K,i,m}}{\mathrm{d}t} = \frac{1}{\phi_K} \sum_{q=1}^{3} \eta^K_{m,q} \left(F^K_{i,q} - \sum_{j=1}^{8} v_{K,j} \left[\sum_{E \in \partial K} \delta^{K,E}_{i,q,j} - \sum_{l=1}^{3} \tau^K_{q,j,l}\, c_{K,i,l} \right] \right), \tag{16}$$

where the following definitions were used

$$\tilde{\eta}^K_{m,j} = \int_K \Phi_{K,m}\Phi_{K,j}, \qquad \delta^{K,E}_{i,m,j} = \int_E \hat{c}_{K,i,E}\left(\mathbf{w}_{K,j} \cdot \mathbf{n}^K_E\right)\Phi_{K,m},$$
$$F^K_{i,m} = \int_K f_i\, \Phi_{K,m}, \qquad \tau^K_{m,j,l} = \int_K \Phi_{K,l}\left(\mathbf{w}_{K,j} \cdot \nabla\Phi_{K,m}\right). \tag{17}$$

The coefficients $\eta^K_{i,j}$ are elements of the inverse matrix $\left(\tilde{\eta}^K\right)^{-1}$. Note that coefficients δ and F are time dependent. In (17) the quantity $\hat{c}_{K,i,E}$ is the upwinded value of the concentration c_i on the edge E with respect to the element K and velocity $\mathbf{v}$.

A discrete version of the pressure Eq. (4) is derived in a similar manner as the discrete transport equation with additional substitution of (7) into $v_{K,j}$

$$\frac{\mathrm{d}p_{K,m}}{\mathrm{d}t} = \sum_{f=1}^{3} \sigma^K_{m,f}\, p_{K,f} + \sum_{s=1}^{2} \sum_{E \in \partial K} \lambda^{K,E}_{m,s}\, \hat{p}_{E,s} + \Gamma^K_m + \Sigma^K_m, \quad m = 1,2,3, \tag{18}$$

where we used

$$\sigma^K_{m,f} = \frac{-1}{\phi_K \mu_K} \sum_{q=1}^{3} \eta^K_{m,q} \sum_{j=1}^{8} \sum_{l=1}^{8} \alpha^K_{l,j}\, \beta^K_{j,f} \sum_{i=1}^{n_c} \theta_{K,i}\, \omega_{K,i,q,j},$$

$$\lambda^{K,E}_{m,s} = \frac{1}{\phi_K \mu_K} \sum_{q=1}^{3} \eta^K_{m,q} \sum_{j=1}^{8} \sum_{l=1}^{8} \alpha^K_{l,j}\, \chi^{K,E}_{l,s} \sum_{i=1}^{n_c} \theta_{K,i}\, \omega_{K,i,q,j},$$

$$\Gamma^K_m = \frac{-\rho_K}{\phi_K \mu_K} \sum_{q=1}^{3} \eta^K_{m,q} \sum_{j=1}^{8} \sum_{l=1}^{8} \alpha^K_{l,j}\, \gamma^K_l \sum_{i=1}^{n_c} \theta_{K,i}\, \omega_{K,i,q,j},$$

$$\Sigma^K_m = \frac{1}{\phi_K} \sum_{q=1}^{3} \eta^K_{m,q} \sum_{i=1}^{n_c} \theta_{K,i}\, F^K_{i,q}, \qquad \omega_{K,i,q,j} = \sum_{F \in \partial K} \delta^{K,F}_{i,q,j} - \sum_{r=1}^{3} \tau^K_{q,j,r}\, c_{K,i,r} \tag{19}$$

and by $\theta_{K,i}$ we denoted the mean value of the coefficient θ_i over the element K.

3.3 Iterative IMPEC

For the solution of the given non-linear problem we use the fully mass-conservative iterative IMPEC method proposed in [8]. We made the following change in the algorithm. In this work the pressure is initialized using the EOS from the given concentrations $p^n = p(c_{i=1,...,n_c}^n)$ as opposed to [8], where the pressure p^n is taken as the solution pressure from the previous time step. This way, the error from the discretization will not cumulate as the simulation goes on. Given a solution $(c_1^n, ..., c_{n_c}^n)$ at the $n-th$ time level, to obtain pressure field and concentrations at the $(n+1)-th$ time level we proceed iteratively as follows.

1. Set $l = 0$ and $p^{n+1,0} = p^n = p(c_1^n, ..., c_{n_c}^n)$, $c_i^{n+1,0} = c_i^n$, for $i = 1, ..., n_c$. The boundary values are evaluated at the $(n+1)-th$ time level. Values of θ_i^{n+1} are initially estimated using the EOS as $\theta_i^{n+1,0} = \frac{\partial p}{\partial c_i}(c_1^n, ..., c_{n_c}^n)$.
2. Repeat
   (a) Set $l = l + 1$.
   (b) The following problem is solved with the MHFEM for pressures $p^{n+1,l}$, pressure traces $\hat{p}^{n+1,l}$ and velocity $\mathbf{v}^{n+1,l}$ (for details see below)

$$
\phi \frac{p^{n+1,l} - p^n}{\Delta t} + \sum_{i=1}^{n_c} \theta_i^{n+1,l-1} \left[\nabla \cdot \left(c_i^{n+1,l-1} \mathbf{v}^{n+1,l} \right) - f_i^{n+1} \right] = 0,
$$

$$
\mathbf{v}^{n+1,l} = -\frac{1}{\mu^{n+1,l-1}} \mathbf{K} \left[\nabla p^{n+1,l} - \rho^{n+1,l-1} \mathbf{g} \right],
$$
(20)

   (c) Values of $c_i^{n+1,l}$ for $i = 1, ..., n_c$ are explicitly updated as (for details see below)

$$
\phi \frac{c_i^{n+1,l} - c_i^n}{\Delta t} + \nabla \cdot \left(c_i^{n+1,l-1} \mathbf{v}^{n+1,l} \right) = f_i^{n+1}.
$$
(21)

   (d) A slope limiter for $c_i^{n+1,l}$, $i = 1, ..., n_c$, is applied.
   (e) Parameters $\theta_i^{n+1,l}$ are updated using the difference formula

$$
\theta_i^{n+1,l} = \frac{p(\xi_i) - p(\eta_i)}{c_i^{n+1,l} - c_i^n},
$$
(22)

   where
$$
\xi_i = (c_1^{n+1,l}, ..., c_{i-1}^{n+1,l}, c_i^{n+1,l}, c_{i+1}^n, ..., c_{n_c}^n),
$$
$$
\eta_i = (c_1^{n+1,l}, ..., c_{i-1}^{n+1,l}, c_i^n, c_{i+1}^n, ..., c_{n_c}^n).
$$

   Computation of $p(\cdot)$ in the nominator (22) is based on the EOS.

(f) Iteration stops when $\max\{E_p,\ E_c,\ E_\theta\} < \delta$ where

$$
E_p = \frac{\|p^{n+1,l} - p^{n+1,l-1}\|_{L_2(\Omega)}^2}{\|p^{n+1,l}\|_{L_2(\Omega)}^2},
$$

$$
E_c = \sum_{i=1}^{n_c} \frac{\|c_i^{n+1,l} - c_i^{n+1,l-1}\|_{L_2(\Omega)}^2}{\|c_i^{n+1,l}\|_{L_2(\Omega)}^2}, \tag{23}
$$

$$
E_\theta = \sum_{i=1}^{n_c} \frac{\|\theta_i^{n+1,l} - \theta_i^{n+1,l-1}\|_{L_2(\Omega)}^2}{\|\theta_i^{n+1,l}\|_{L_2(\Omega)}^2}.
$$

If the criterion is met we then set $p^{n+1} = p^{n+1,l}$, $\mathbf{v}^{n+1} = \mathbf{v}^{n+1,l}$ and $c_i^{n+1} = c_i^{n+1,l}$ for $i = 1, ..., n_c$. Otherwise we go back to step (a).

We now describe the step 2 of the algorithm in more details. In (b), as the pressures $p^{n+1,l}$ can be eliminated, the system is solved for pressure traces $\hat{p}^{n+1,l}$ only by means of the Eqs. (14) and (18) with the time discretization as in (20). The underlying system of linear equations for pressure traces $\hat{p}^{n+1,l}$ have the followig form

$$
p^{n+1,l} + \mathbf{D}^{-1}\mathbf{H}_1\ \hat{p}_1^{n+1,l} + \mathbf{D}^{-1}\mathbf{H}_2\ \hat{p}_2^{n+1,l} = \mathbf{D}^{-1}\mathbf{G},
$$

$$
\mathbf{R}_1\ p^{n+1,l} - \mathbf{M}_{1,1}\ \hat{p}_1^{n+1,l} - \mathbf{M}_{1,2}\ \hat{p}_2^{n+1,l} = \mathbf{V}_1, \tag{24}
$$

$$
\mathbf{R}_2\ p^{n+1,l} - \mathbf{M}_{2,1}\ \hat{p}_1^{n+1,l} - \mathbf{M}_{2,2}\ \hat{p}_2^{n+1,l} = \mathbf{V}_2,
$$

from which the elimination of the pressures is apparent. The matrices in (24) can be deduced from (14) and (18), see e.g. [6]. The velocity field $\mathbf{v}^{n+1,l}$ is then computed according to (6) by evaluating (7). The mean values of $\mu^{n+1,l-1}$ and $\rho^{n+1,l-1}$ are computed at $(c_1^{n+1,l-1}, ..., c_{n_c}^{n+1,l-1})$. In (c), the semi-discrete equation (16) is used with the time discretization as in (21). Such time discretization leads to an explicit scheme for $c_{K,i}^{n+1,l}$, because only $c_{K,i}^{n+1,l-1}$ are present. Upwind values $\hat{c}_{K,i,E}^{n+1,l-1}$ in (20) and (21) are evaluated as follows

$$
\hat{c}_{K,i,E}^{n+1,l-1}(\mathbf{x}) = \begin{cases} c_{K,i}^{n+1,l-1}(\mathbf{x}), & \mathbf{v}_K^{n+1,l-1} \cdot \mathbf{n}_E^K(\mathbf{x}) \geq 0, & \mathbf{x} \in E \\ c_{T,i}^{n+1,l-1}(\mathbf{x}), & \mathbf{v}_K^{n+1,l-1} \cdot \mathbf{n}_E^K(\mathbf{x}) < 0, & \mathbf{x} \in E \notin \partial\Omega, \\ c_i^{D,n+1}(\mathbf{x}), & \mathbf{v}_K^{n+1,l-1} \cdot \mathbf{n}_E^K(\mathbf{x}) < 0, & \mathbf{x} \in E \subset \Gamma_c. \end{cases} \tag{25}
$$

Note that concentrations $c_{K,i}^{n+1,l-1}$ or $c_{T,i}^{n+1,l-1}$ are computed using (15) on the elements K or T, respectively. Evaluation of the difference formula in (e) is discussed in the Sect. 4.4. For slope limiting procedure in (d), we refer the reader to [9]. We note that the choice of the slope limiter greatly affects the convergence behavior of the iterative IMPEC algorithm.

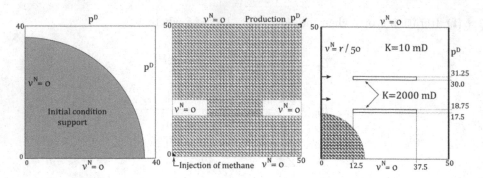

Fig. 2. Structure of the computational meshes for Example 1 (left), Example 2 (middle) and Example 3 (right).

4   Numerical Examples

4.1   Example 1

Example 1 serves to verify higher-order approximation of the numerical scheme with the use of experimental order of convergence (EOC) analysis. Let us consider the following problem

$$\frac{\partial c}{\partial t} + \nabla \cdot (c\mathbf{v}) = 0, \quad \mathbf{v} = -2\nabla p, \tag{26}$$

with the equation of state of the form $p(c) = c$ and initial and boundary condition

$$\begin{aligned} c(t_0, \mathbf{x}) &= B_2(t_0, \mathbf{x}), && \mathbf{x} \in \Omega, \\ p(t, \mathbf{x}) &= B_2(t, \mathbf{x}), && \mathbf{x} \in \Gamma_p,\ t \in (t_0, t_1), \\ \mathbf{v}(t, \mathbf{x}) \cdot \mathbf{n}(\mathbf{x}) &= 0, && \mathbf{x} \in \Gamma_v,\ t \in (t_0, t_1), \end{aligned} \tag{27}$$

where $\Omega = [0, 40] \times [0, 40]\ m^2$, $t_0 = 7500s$, $t_1 = 45000s$. Function $B_m = B_m(t, \mathbf{x})$ is the well-known Barenblatt solution given by

$$B_m(t, \mathbf{x}) = \max \left\{ 0, t^{-\alpha} \left(\Lambda - \frac{\alpha(m-1)}{2dm} \frac{|\mathbf{x}|^2}{t^{2\alpha/d}} \right)^{\frac{1}{m-1}} \right\}. \tag{28}$$

In (28) we choose $d = 2$, $m = 2$, $\alpha = (m - 1 + 2/d)^{-1} = 1/2$ and $\Lambda = 1$. We further define the Dirichlet and Neumann boundaries as

$$\begin{aligned} \Gamma_p &= \{ (x, 40) \cup (40, y) \mid x \in (0, 40),\ y \in (0, 40) \}, \\ \Gamma_v &= \{ (0, y) \cup (x, 0) \mid y \in (0, 40),\ x \in (0, 40) \}. \end{aligned} \tag{29}$$

The EOC and errors are included in the Table 1. The computational mesh is parametrized with a parameter $h \in \mathbb{N}$ as $n_k = 2 \times 4h \times 4h$ (structured triangular mesh). We choose $\Delta t \sim h^{-2}$ so that the error from the time discretization does not interfere with the space discretization. The error is computed by interpolating (28) into the basis of $\mathbb{P}_1(K)$ for each $K \in \mathcal{T}_h$. Tolerance for the stopping criterion (23) is chosen as $\delta = 1.49 \times 10^{-12}$.

Table 1. Experimental order of convergence and errors at time t_1 for Example 1.

h	Δt	$\|E_h\|_{L_1}$	EOC_1	$\|E_h\|_{L_2}$	EOC_2	$\|E_h\|_{L_\infty}$	EOC_∞
4	18.75	5.2228×10^{-3}		1.3792×10^{-4}		3.6610×10^{-6}	
8	4.6875	1.1837×10^{-3}	2.1415	3.1263×10^{-5}	2.1413	8.6644×10^{-7}	2.0791
16	1.1719	2.9109×10^{-4}	2.0238	7.6769×10^{-6}	2.0258	2.1072×10^{-7}	2.0398
32	2.9297×10^{-1}	7.3970×10^{-5}	1.9764	1.9481×10^{-6}	1.9785	5.1956×10^{-8}	2.0199
64	7.3242×10^{-2}	1.7480×10^{-5}	2.0812	4.6140×10^{-7}	2.0780	1.2901×10^{-8}	2.0099
128	1.8311×10^{-2}	5.2721×10^{-6}	1.7293	1.3844×10^{-7}	1.7368	3.2146×10^{-9}	2.0047

4.2 Example 2

In Example 2, we will try to reproduce numerical results in [15]. Let us consider a reservoir $\Omega = [0,50] \times [0,50]\ m^2$ with porosity $\phi = 0.2$ and permeability $\mathbf{K} = 10^{-14}\mathbf{I}\ m^2$ at initial pressure $p = 5 \times 10^6$ Pa and temperature $T = 397$ K in a horizontal position with $\mathbf{g} = (0,0)$ m/s^2 or vertical position with $\mathbf{g} = (0,-9.81)$ m/s^2 initially filled with propane. In the corner $\{(x,y)|\ 0 \leq x \leq 1.25,\ 0 \leq y \leq 1.25 - x\}$ pure methane with concentration (molar density) $c^{inj} = 42.2896$ mol/m^3 is injected with the injection rate $r = 3.90625 \times 10^{-4}$ m^3/s. Mixture of propane and methane is produced on the boundary $\{(x,50)|\ 48.75 \leq x \leq 50\} \cup \{(50,y)|\ 48.75 \leq y \leq 50\}$ where pressure $p = 5 \times 10^6$ Pa is maintained. The rest of the boundary is impermeable, i.e. zero Neumann condition $v^N = 0$ is imposed. Relevant data for the Peng-Robinson EOS are taken over from [15] and listed in Table 2. The binary interaction coefficient of the methane-propane is $k_{12} = 0.0365$. The mesh consists of $2 \times 40 \times 40$ triangular elements. The time step is chosen constant $\Delta t = 6000s$ for both horizontal and vertical case. Tolerance for the stopping criterion (23) is chosen as $\delta = 1.49 \times 10^{-7}$.

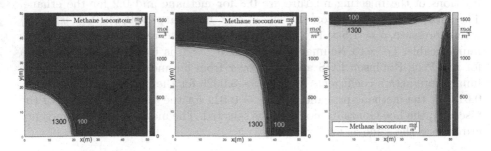

Fig. 3. Contours of methane from Example 2 with $\mathbf{g} = (0,0)$ at the time $t = 6 \times 10^6 s$ (left), $t = 24 \times 10^6 s$ (middle) and $t = 48 \times 10^6 s$ (right).

4.3 Example 3

In Example 3, we will try to reproduce numerical results of the Example 6.5 in [8]. Let us consider a reservoir $\Omega = [0,50] \times [0,50]\ m^2$ with porosity $\phi = 0.2$ and

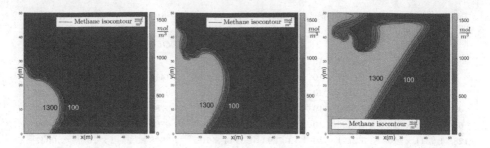

Fig. 4. Contours of methane from Example 2 with $\mathbf{g} = (0, -9.81)$ at the time $t = 6 \times 10^6 s$ (left), $t = 12 \times 10^6 s$ (middle) and $t = 21.6 \times 10^6 s$ (right).

Table 2. Peng-Robinson EOS parameters for the Example 2.

Component	p_{ci} [Pa]	T_{ci} [K]	V_{ci} [m³mol⁻¹]	M_i [kg mol⁻¹]	ω_i [-]
1 (methane)	$4.58373 \cdot 10^6$	$1.89743 \cdot 10^2$	$9.897054 \cdot 10^{-5}$	$1.62077 \cdot 10^{-2}$	$1.14272 \cdot 10^{-2}$
2 (propane)	$4.248 \cdot 10^6$	$3.6983 \cdot 10^2$	$2.000001 \cdot 10^{-4}$	$4.40962 \cdot 10^{-2}$	$1.53 \cdot 10^{-1}$

permeability distribution

$$\mathbf{K} = \begin{cases} 2000 \text{ mD} \times \mathbf{I}, & \{(x,y)|x \in (12.5, 37.5), \ y \in (17.5, 18.75) \cup (30, 31.25)\} \\ 10 \text{ mD} \times \mathbf{I}, & \text{elsewhere,} \end{cases}$$

at initial pressure $p = 1.5 \times 10^5$ Pa and temperature $T = 554.8$ K initially filled with propane. On the whole west boundary $\{(0,y)|\ 0 \leq y \leq 50\}$ the mixture of methane and ethane with total concentration (molar density) $c^{inj} = 32.527$ mol/m³ is injected with the injection rate $r = 3.17 \times 10^{-6}$ m³/s. Molar fractions of the injecting mixture are 0.8 for methane and 0.2 for the ethane. The mixture is produced on the whole east boundary $\{(50,y)|\ 0 \leq y \leq 50\}$ where pressure $p = 1.5 \times 10^5$ Pa is maintained. The rest of the boundary is impermeable, i.e. zero Neumann condition $v^N = 0$ is imposed. Relevant data for the Peng-Robinson EOS are taken over from [5] and listed in Table 3. The binary interaction coefficients are $k_{12} = -0.0026$ for the methane-ethane, $k_{13} = 0.014$ for the methane-propane and $k_{23} = 0.011$ for the ethane-propane. For the viscosity, the Lee-Gonzalez model [10] is selected. The mesh and the convergence criterion is the same as for Example 2.

4.4 Update of θ Parameter

When computing parameters θ_i, we have to deal with special case $c_i^{n+1,l} \to c_i^n$ for which the denominator in (22) approaches zero. In [8] this is treated in the following perturbation manner. If $|c_i^{n+1,l}| < \varepsilon$ and $|c_i^{n+1,l} - c_i^n| < \varepsilon$, that is $c_i^{n+1,l}$ and c_i^n are both close to zero, we then set $c_i^{n+1,l} := c_i^n + \varepsilon$. Else if $|c_i^{n+1,l} - c_i^n| < \varepsilon |c_i^{n+1,l}|$, that is $c_i^{n+1,l}$ and c_i^n are close to each other, we set

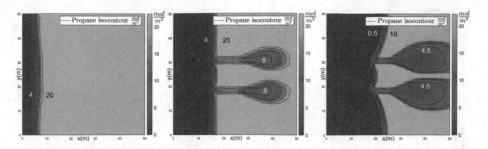

Fig. 5. Contours of propane from Example 3 at the time $t = 17.4 \times 10^6 s$ (left), $t = 52.2 \times 10^6 s$ (middle) and $t = 86.4 \times 10^6 s$ (right).

Table 3. Peng-Robinson EOS parameters for the Example 3.

Component	p_{ci} [Pa]	T_{ci} [K]	V_{ci} [m³kg⁻¹]	M_i [kg mol⁻¹]	ω_i [-]
1 (methane)	$4.604 \cdot 10^6$	$1.9058 \cdot 10^2$	$6.17284 \cdot 10^{-3}$	$1.62077 \cdot 10^{-2}$	$0.04348 \cdot 10^{-1}$
2 (ethane)	$4.880 \cdot 10^6$	$3.0542 \cdot 10^2$	$4.92611 \cdot 10^{-3}$	$3.070 \cdot 10^{-2}$	$1.0109 \cdot 10^{-1}$
3 (propane)	$4.250 \cdot 10^6$	$3.6982 \cdot 10^2$	$4.608295 \cdot 10^{-3}$	$4.40962 \cdot 10^{-2}$	$1.5788 \cdot 10^{-1}$

$c_i^{n+1,l} := c_i^n + \varepsilon c_i^{n+1,l}$. These perturbations are used for the parameter update step only. When all the updates are done, the values of $c_i^{n+1,l}$ are reset to their original values. The ε value is chosen as the square root of machine precision. In this paper, we try to pursuit this problem with the use of

$$\lim_{c_i^n \to c_i^{n+1,l}} \theta_i^{n+1,l} = \left(\frac{\partial p}{\partial c_i} \right) (c_1^{n+1,l}, ..., c_{i-1}^{n+1,l}, c_i^{n+1,l}, c_{i+1}^n, ..., c_{n_c}^n). \qquad (30)$$

In Fig. 6 we compare both approaches by plotting the numbers of iterations which are needed to converge at each time level in the step 2 of the iterative IMPEC algorithm in Example 2. The tolerance for the stopping criterion (23) is chosen as $\delta = 1.49 \times 10^{-7}$. From the given comparison we concluded, that the formula (30) can be utilized as well as the perturbation treatment given in [8]. If we choose stronger tolerance for the stopping criterion (e.g. $\delta = 1.49 \times 10^{-8}$), then the iterative IMPEC algorithm would have trouble satisfying stopping criterion for E_θ as can be seen in Fig. 7. The problem is that even though concentrations converge with respect to the stopping criterion, the error E_θ does not further decrease.

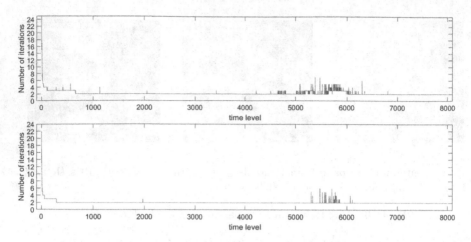

Fig. 6. Number of iteration needed to converge at each time level using perturbation (top) and limit (30) (bottom) approaches. Tolerance $\delta = 1.49 \times 10^{-7}$ is chosen.

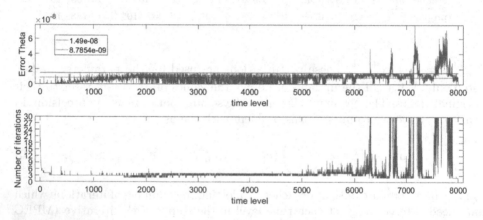

Fig. 7. Error E_θ (top) and number of iteration needed to converge (bottom) at each time level for the tolerance $\delta = 1.49 \times 10^{-8}$. In the later stage of the simulation the convergence criterion is not satisfied for many time steps as the E_θ does not decrease.

5   Conclusions

In this work we have shown how to extend modeling of the multicomponent compressible single-phase Darcy flow in porous media based on the combination of MHFEM and DG methods to a higher-order approximation scheme. The main extension is the use of the higher-order Raviart-Thomas space for the approximation of the velocity field. The system of nonlinear algebraic equations for concentration, pressure and velocity fields obtained by combining the MHFEM and DG is solved with the fully mass-conservative iterative IMPEC scheme in which the given equation of state is incorporated through the pressure equation with an additional nonlinear parameter. Three numerical experiments were performed. The first example gave a proof of the expected second order approximation in space through the computation of EOC. The second and the third examples were taken over from the literature to verify the correctness of the code. In the second example the outcome was positive and we obtained the same results. In the third example we obtained different results from the original authors. We believe that such outcome is only a consequence of different interpretation of the source terms. In the experiments, the proposed derivative approach was used in the update of the pressure equation parameters which slightly improved the convergence of the method. In the future work, we would like to improve the current model with the inclusion of a diffusive term in the transport equations and with the higher-order time discretization.

References

1. Ács, G., Doleschall, S., Farkas, É.: General purpose compositional model. Soc. Petrol. Eng. J. **25**, 543–553 (1985)
2. Brezzi, F., Douglas, J., Fortin, M., Marini, L.D.: Efficient rectangular mixed finite elements in two and three space variables. Modélisation mathématique et Analyse numérique **21**, 581–604 (1987)
3. Brezzi, F., Fortin, M.: Mixed and Hybrid Finite Element Methods. Springer Series in Computational Mathematics, vol. 15 (1991). https://doi.org/10.1007/978-1-4612-3172-1
4. Boffi D., Brezzi, F., Fortin, M.: Mixed Finite Element Methods and Applications. Springer Series in Computational Mathematics, vol. 44. Springer Science & Business Media, (2013). https://doi.org/10.1007/978-3-642-36519-5
5. Firoozabadi A.: Thermodynamics of Hydrocarbon Reservoirs. McGraw-Hill Education, New York (1999)
6. Hoteit, H., Firoozabadi, A.: Multicomponent fluid flow by discontinuous Galerkin and mixed methods in unfractured and fractured media. Water Resour. Res. (2005). https://doi.org/10.1029/2005WR004339
7. Hoteit H., Firoozabadi A.: Compositional Modeling By the Combined Discontinuous Galerkin and Mixed Methods, Society of Petroleum Engineers (2006)
8. Chen, H., Fan, X., Sun, S.: A fully mass-conservative iterative IMPEC method for multicomponent compressible flow in porous media. J. Comput. Appl. Math. **362**, 1–21 (2019)
9. Barth T., Jespersen D.: The design and application of upwind schemes on unstructured meshes, 27th Aerospace Sciences Meeting (1989)
10. Lee, A.L., Gonzalez, M.H., Eakin, B.E.: The viscosity of natural gases. J. Petrol. Technol. **18**, 997–1000 (1966)
11. Lohrenz, J., Bray, B.G., Clark, C.R.: Calculating viscosities of reservoir fluids from their compositions. J. Petrol. Technol. **16**, 1171–1176 (1964)
12. Moortgat, J., Firoozabadi, A.: Mixed-hybrid and vertex-discontinuous-Galerkin finite element modeling of multiphase compositional flow on 3D unstructured grids. J. Comput. Phys. **315**, 476–500 (2016)
13. Moortgat, J., Sun, S., Firoozabadi, A.: compositional modeling of three-phase flow with gravity using higher-order finite element methods. Water Resour. Res. **47** (2011). https://doi.org/10.1029/2010WR009801

14. Polívka O., Mikyška J.: Combined mixed-hybrid finite element-finite volume scheme for computation of multicomponent compressible flow in porous media. Numerical Mathematics and Advanced Applications. Springer-Verlag, pp. 559–567 (2011). https://doi.org/10.1007/978-3-642-33134-3_59
15. Polívka O., Mikyška J.: Numerical simulation of multicomponent compressible flow in porous medium. J. Math Ind. **3**, 53–60 (2013)

An Enhanced Finite Element Algorithm for Thermal Darcy Flows with Variable Viscosity

Loubna Salhi[1]($\boxtimes$), Mofdi El-Amrani[2], and Mohammed Seaid[3,4]

[1] Laboratory Modeling Simulation and Data Analysis, University Mohammed VI Polytechnic, Benguerir, Morocco
Loubna.salhi@um6p.ma
[2] Mathematics and Applications Laboratory, FST, Abdelmalek Essaadi University, Tangier, Morocco
[3] Department of Engineering, University of Durham, South Road, Durham DH1 3LE, UK
[4] International Water Research Institute, University Mohammed VI Polytechnic, Benguerir, Morocco

Abstract. This paper deals with the development of a stable and efficient unified finite element method for the numerical solution of thermal Darcy flows with variable viscosity. The governing equations consist of coupling the Darcy equations for the pressure and velocity fields to a convection-diffusion equation for the heat transfer. The viscosity in the Darcy flows is assumed to be nonlinear depending on the temperature of the medium. The proposed method is based on combining a semi-Lagrangian scheme with a Galerkin finite element discretization of the governing equations along with an robust iterative solver for the associate linear systems. The main features of the enhanced finite element algorithm are that the same finite element space is used for all solutions to the problem including the pressure, velocity and temperature. In addition, the convection terms are accurately dealt with using the semi-Lagrangian scheme and the standard Courant-Friedrichs-Lewy condition is relaxed and the time truncation errors are reduced in the diffusion terms. Numerical results are presented for two examples to demonstrate the performance of the proposed finite element algorithm.

Keywords: Thermal Darcy flows · Unified finite elements · Semi-Lagrangian method · Moving fronts

1 Introduction

In the present study, given a bounded two-dimensional domain $\Omega \subset \mathbb{R}^2$ with Lipschitz continuous boundary Γ and a time interval $[0, T]$, we focus on solving a time-dependent heat equation coupled with the Darcy equations. For all $(\mathbf{x}, t)$ in the domain $\Omega \times [0, T]$, the governing equations read

© Springer Nature Switzerland AG 2021
M. Paszynski et al. (Eds.): ICCS 2021, LNCS 12747, pp. 215–229, 2021.
https://doi.org/10.1007/978-3-030-77980-1_17

$$\nu(\Theta)\boldsymbol{u} + \nabla p = \boldsymbol{f}(\Theta), \qquad \text{in} \quad \Omega,$$
$$\nabla \cdot \boldsymbol{u} = 0, \qquad \text{in} \quad \Omega, \tag{1a}$$
$$\frac{D\Theta}{Dt} - \kappa \nabla^2 \Theta = g(\boldsymbol{x}, t), \qquad \text{in} \quad \Omega, \tag{1b}$$

where $\boldsymbol{u}$ is the velocity field, p the pressure and Θ the fluid temperature. In the Darcy equations (1a), ν is the variable viscosity and $\boldsymbol{f}$ the force density which both are dependent on the temperature. In the heat equation (1b), κ is the thermal diffusivity coefficient supposed to be a positive constant, g an external source term and

$$\frac{D\Theta}{Dt} = \frac{\partial \Theta}{\partial t} + \boldsymbol{u} \cdot \nabla \Theta, \tag{1c}$$

is the total derivative which measures the temperature rate of change along the trajectories of the fluid particles known by characteristic curves. The system of Eqs. (1a)–(1b) is equipped with an initial condition

$$\Theta(\boldsymbol{x}, 0) = \Theta_0(\boldsymbol{x}), \qquad \text{in} \quad \Omega, \tag{1d}$$

as well as given boundary conditions. In view of simplification, homogeneous Dirichlet boundary conditions are considered for the pressure and temperature, while a no-slip boundary condition is prescribed on the velocity:

$$p = 0 \quad \text{and} \quad \boldsymbol{u} \cdot \boldsymbol{n} = 0, \qquad \text{on} \quad \Gamma, \tag{1e}$$
$$\Theta = 0, \qquad \text{on} \quad \Gamma, \tag{1f}$$

where $\boldsymbol{n}$ is the unit outward normal vector on the boundary Γ. Notice that the present study easily extends to different types of boundary conditions without major conceptual changes in the formulation. It should be stressed that the system of Eqs. (1) has been widely used in the literature to model several applications of fluid mechanics such as transport of contaminants in saturated zones and aquifers, heat explosion problems in the chemical industry, nuclear waste and carbon dioxide geological storage, see [2,10,11,20,26,27] among others. In bio-medical fields, the model has also been used to study the so-called bio-heat transfer, the arterial and venous blood flows by considering the human body as a deformable porous medium, see for instance [21,22,30]. While the mathematical theory and existence of a weak solution for the problem (1) is well developed in the literature [9,17], the numerical resolution is still challenging for several reasons. Especially, when the diffusion term is negligible in comparison with the convective term. In such a case, the numerical solution give rise to serious computational difficulties by generating either non-physical oscillations or numerical dissipation in the presence of steep fronts and shocks, see for instance [12,13,24,28].

Numerical techniques used to deal with these difficulties include Eulerian finite element methods which are usually easy to implement. However, most Eulerian methods use fixed grids and incorporate some upstream weighting in their formulations to stabilize spatial discretization. In addition, time truncation errors dominate their solutions and are subject to the Courant-Friedrichs-Lewy (CFL) stability conditions, which impose a sever restriction on the size of the time steps taken in numerical simulations. In the current work, we use a semi-Lagrangian finite element method to deal with the transport equation in (1). This class of methods has been used for solving many convection-dominated flow problems, see for example [12,13,23,24,26,27]. The main advantage in these methods is that they allow to convert the temperature equation in (1) from an Eulerian description to a semi-Lagrangian one in terms of the particle trajectories, also known as characteristics, and to treat the transport part (1c) separately in the finite element discretization. Thus, the time derivative and the convection terms are combined as a directional derivative along the particles trajectories, leading to a characteristic time-stepping procedure. This results in a substantial reduction in the computational cost and in the time truncation errors. Moreover, the semi-Lagrangian scheme offers the possibility of using time steps that exceed those allowed by the stability CFL condition for the conventional Eulerian methods, see [14–16,26] for further details. Next, to improve the accuracy of the enhanced method, we use a stabilization technique proposed by [5] for spatial discretization of Darcy equations. The main advantage of this technique is that it allows the use of equal-order finite element approximations for all solutions in the problem and thus, it does not require the use of mixed formulations such as those widely employed in the literature, see for instance [1,25]. In [5], it was shown that the stabilization method is unconditionally stable and it allows to achieve optimal accuracy with respect to solution regularity.

This paper is organized as follows. Formulation of the proposed semi-Lagrangian finite element method is presented in Sect. 2. Section 3 is devoted to numerical results for two test examples for coupled Darcy-heat problems. Concluding remarks are presented in Sect. 4.

2   Semi-Lagrangian Finite Element Method

Let $\Omega_h \subset \bar{\Omega} = \Omega \cup \Gamma$ denotes a quasi-uniform partition of Ω into triangular finite elements $\mathcal{K}_j$ with the partition step h. We define the conforming finite element space for the temperature and pressure as

$$V_h = \left\{ v_h \in C^0(\Omega) : \quad v_h|_{\mathcal{K}_j} \in P_k(\mathcal{K}_j), \quad \forall \, \mathcal{K}_j \in \Omega_h \right\},$$

where $P_k(\mathcal{K}_j)$ is the space of complete polynomials of degree k, $k \geq 2$, on each element $\mathcal{K}_j$. We also define the conforming finite element space $\mathbf{V}_h = (V_h)^2$ for the velocity field. For time discretization, we divide the time interval $[0, t_N]$ into N equal subintervals $[t_n, t_{n+1}]$ with length $\Delta t = t_{n+1} - t_n$ for $n = 0, 1, \ldots, N$. Then, we formulate the finite element solutions to $\mathbf{u}^n(\mathbf{x})$, $p^n(\mathbf{x})$ and $\Theta^n(\mathbf{x})$ as

$$\mathbf{u}_h^n(\mathbf{x}) = \sum_{j=1}^M U_j^n \circ \varphi_j(\mathbf{x}), \quad p_h^n(\mathbf{x}) = \sum_{j=1}^M P_j^n \phi_j(\mathbf{x}), \quad \Theta_h^n(\mathbf{x}) = \sum_{j=1}^M T_j^n \phi_j(\mathbf{x}),$$
$$(2)$$

were the symbol $\circ$ denotes the hadamard product that produces vectors through element-by-element multiplication of the original two vectors. In (2), U_j^n, P_j^n and T_j^n are the corresponding nodal values of $\mathbf{u}_h^n(\mathbf{x})$, $p_h^n(\mathbf{x})$ and $\Theta_h^n(\mathbf{x})$ respectively defined by $U_j^n = u_h^n(x_j)$, $P_j^n = p_h^n(x_j)$ and $T_j^n = \Theta_h^n(x_j)$, with $\{x_j\}_{j=1}^M$ being the set of mesh points in the partition Ω_h, $\{\varphi_j\}_{j=1}^M = \{(\phi_j, \phi_j)\}_{j=1}^M$ and $\{\phi_j\}_{j=1}^M$ are the basis vectors and functions of $\mathbf{V}_h$ and V_h respectively given by the Kronecker delta symbol.

Next, we define the functional spaces that are useful for the existence and uniqueness of the solution of problem (1). We introduce the Hilbert space

$$H_0^1(\Omega) = \left\{ v \in H^1(\Omega) : \quad v|_\Gamma = 0 \right\}.$$

We also define the space $L_0^2(\Omega)$ of all square integrable functions with vanishing mean as

$$L_0^2(\Omega) = \left\{ w : \Omega \longrightarrow \mathbb{R} : \quad \int_\Omega w d\Omega = 0 \right\}.$$

We define the following inner product and norm in $L^2(\Omega)$:

$$(w, v) = \int_\Omega wv \, d\Omega \quad \text{and} \quad \|v\|_{L^2(\Omega)} = (v, v)^{\frac{1}{2}}, \quad \forall \, w, v \in L^2(\Omega).$$

We recall the standard space:

$$\mathbf{H}_0 \left(div, \Omega \right) = \left\{ \mathbf{v} \in \mathbf{H}(div, \Omega) : \quad (\mathbf{v} \cdot \mathbf{n})\Big|_\Gamma = 0 \right\}.$$

To approximate the velocity and pressure solutions of the Darcy equations (1a), we shall need the equal-order finite element pair $(\mathbf{S}_h, Q_h)$ defined as

$$\mathbf{S}_h = \mathbf{V}_h \cap \mathbf{H}_0 \left(div, \Omega \right) \quad \text{and} \quad Q_h = V_h \cap L_0^2(\Omega). \tag{3}$$

We also introduce the necessary finite element space R_h to approximate the temperature solution of the heat equation (1b) as

$$R_h = V_h \cap H_0^1(\Omega). \tag{4}$$

Notice that the spaces introduced above are necessary to prove the existence and uniqueness of the solution of problem (1), see for instance [4,5,26].

2.1   Solution of the Darcy-Heat Problem

As in most finite element methods, we start with the weak formulation that reads as: Find (u, p, Θ) in $\mathbf{H}_0 \left(div, \Omega \right) \times L_0^2(\Omega) \times H_0^1(\Omega)$ such that

$$\int_\Omega \nu(\Theta) u \cdot s \, d\Omega - \int_\Omega p \nabla \cdot s \, d\Omega = \int_\Omega f(\Theta) \cdot s \, d\Omega, \quad \forall \, s \in \mathbf{H}_0(div, \Omega),$$

$$\tag{5a}$$

$$\int_\Omega q \, \nabla \cdot u \, d\Omega = 0, \quad \forall \, q \in L_0^2(\Omega),$$

$$\int_\Omega \frac{D\Theta}{Dt} r \, d\Omega + \kappa \int_\Omega \nabla \Theta \cdot \nabla r \, d\Omega = \int_\Omega gr \, d\Omega, \quad \forall \, r \in H_0^1(\Omega). \tag{5b}$$

Note that it is evident to prove that any triplet $(\boldsymbol{u}, p, \Theta)$ in $\mathbf{H}_0(div, \Omega) \times L_0^2(\Omega) \times H_0^1(\Omega)$ solving the problem (1) in the sense of distributions in Ω is a solution of the weak problem (5), see [8,26] for more details. To approximate the velocity, pressure and temperature solutions of system (5a), we use the equal-order finite element spaces $\mathbf{S}_h$, Q_h and R_h defined in (3) and (4), respectively. Note that equations (5a) can also be rewritten as

$$A\left(\boldsymbol{u}_h, \mathbf{s}_h\right) - B\left(p_h, \mathbf{s}_h\right) = \mathcal{L}_f\left(\mathbf{s}_h\right), \qquad \forall \, \mathbf{s}_h \in \mathbf{S}_h,$$

$$B\left(q_h, \boldsymbol{u}_h\right) = 0, \qquad \forall \, q_h \in Q_h, \tag{6}$$

where $\mathcal{A}$, $\mathcal{B}$ are the bilinear forms and $\mathcal{L}_f$ is the linear form defined as

$$\mathcal{A}(\boldsymbol{u}_h, \mathbf{s}_h) = \int_\Omega \nu(\Theta) \boldsymbol{u}_h \cdot \mathbf{s}_h \, d\Omega, \quad B(p_h, \mathbf{s}_h) = \int_\Omega p_h \nabla \cdot \mathbf{s}_h \, d\Omega,$$

$$\mathcal{L}_f(\mathbf{s}_h) = \int_\Omega \boldsymbol{f}(\Theta_h) \cdot \mathbf{s}_h \, d\Omega.$$

It should also be noted that stable and accurate solutions of the discrete problem (6) are obtained for discrete spaces $\mathbf{S}_h$ and Q_h satisfying the known discrete inf-sup condition [7]. However, the velocity-pressure space $(\mathbf{S}_h, Q_h)$ does not verify the inf-sup condition associated with the mixed form (6), see [5,6,19] for further details. Then, the discrete weak problem is not stable, which makes it necessary to opt for a stabilization technique. To deal with it, we use a polynomial pressure-projection stabilization method which makes that the pair $(\mathbf{S}_h, Q_h)$ verifies a stabilized form of the inf-sup condition [4,5]. Thus, the stabilized weak form of equations (5a) reads as : Find $(\boldsymbol{u}_h, p_h) \in \mathbf{S}_h \times Q_h$ such that

$$A\left(\boldsymbol{u}_h, \mathbf{s}_h\right) - B\left(p_h, \mathbf{s}_h\right) = \mathcal{L}_f\left(\mathbf{s}_h\right), \qquad \forall \, \mathbf{s}_h \in \mathbf{S}_h,$$

$$B\left(q_h, \boldsymbol{u}_h\right) = \mathcal{D}\left(p_h, q_h\right), \qquad \forall \, q_h \in Q_h, \tag{7}$$

where $\mathcal{D}$ is the bilinear form defined as

$$\mathcal{D}\left(p_h, q_h\right) = \int_\Omega \left(p_h - \Pi_{k-1} p_h\right) \left(q_h - \Pi_{k-1} q_h\right) \, d\Omega.$$

Here, Π_{k-1} is the projection operator $\Pi_{k-1} : L^2(\Omega) \longrightarrow [P]_{k-1}$ defined as

$$\Pi_{k-1}(p) = \arg\min \frac{1}{2} \int_\Omega \left(\Pi_{k-1} q - p\right)^2 \, d\Omega, \quad \forall q \in [P]_{k-1}, \tag{8}$$

with $[P]_{k-1}$ being the discontinuous polynomial space:

$$[P]_{k-1} = \left\{ q \in L^2(\Omega) : \left. q \right|_{K_j} \in P_{k-1}(K_j), \ \forall K_j \in \Omega_h \right\}.$$

Next, we use the modified method of characteristics to solve the heat equation (5b). The main idea is to treat the transport term (1c) of equation (5b) in Lagrangian, and separately in the finite element discretization. Then, the new temperature solution is approximated at each time subinterval $[t_n, t_{n+1}]$ using the characteristic curves, also known as the departure points, associated with the material derivative (1c). These characteristic curves are the solutions of the ordinary differential equations

$$\frac{d\boldsymbol{X}(\boldsymbol{x}, t_{n+1}; t)}{dt} = \boldsymbol{u}\left(\boldsymbol{X}(\boldsymbol{x}, t_{n+1}; t), t\right), \qquad \forall\, (t, \boldsymbol{x}) \in [t_n, t_{n+1}] \times \bar{\Omega},$$

$$\boldsymbol{X}(\boldsymbol{x}, t_{n+1}; t_{n+1}) = \boldsymbol{x}. \tag{9}$$

The existence and uniqueness of the solution of (9) for all times t are established, see for instance [18]. To obtain the departure points $\{\boldsymbol{X}_{hj}^n\}$ for each mesh point $\boldsymbol{x}_j$, $j = 1, \ldots, M$, we use the algorithm proposed in [29] which accurately solves (9) with a second-order accuracy. We write the solution of (9) in the form of

$$\boldsymbol{X}_{hj}^n = \boldsymbol{x}_j - \boldsymbol{\alpha}_{hj}, \qquad j = 1, \ldots, M, \tag{10}$$

where the displacement $\boldsymbol{\alpha}_{hj}$ is calculated by the iterative procedure

$$\boldsymbol{\alpha}_{hj}^{(0)} = \frac{\Delta t}{2}\left(3\boldsymbol{u}_h^n(\boldsymbol{x}_j) - \boldsymbol{u}_h^{n-1}(\boldsymbol{x}_j)\right),$$

$$\boldsymbol{\alpha}_{hj}^{(k+1)} = \frac{\Delta t}{2}\left(3\boldsymbol{u}_h^n\left(\boldsymbol{x}_j - \frac{1}{2}d_{hj}^{(k)}\right) - \boldsymbol{u}_h^{n-1}\left(\boldsymbol{x}_j - \frac{1}{2}d_{hj}^{(k)}\right)\right), \qquad k = 0, 1, \ldots. \tag{11}$$

To evaluate values of the approximate velocities in (11), we first identify the mesh element $\widehat{\mathcal{K}}_j$ where $\boldsymbol{x}_j - \frac{1}{2}\boldsymbol{\alpha}_{hj}^{(k)}$ resides. Then, a finite element interpolation on $\widehat{\mathcal{K}}_j$ is performed according to (2). Thus, assuming that the pairs $(\boldsymbol{X}_{hj}^n, \widehat{\mathcal{K}}_j)$ along with the mesh point values $\{T_j^n\}$ are known for all $j = 1, \ldots, M$, we can approximate the values $\{\widehat{T}_j^n\}$ by

$$\widehat{T}_j^n := \Theta_h^n(\boldsymbol{X}_{hj}^n) = \sum_{k=1}^{M} T_k \phi(\boldsymbol{X}_{hj}^n). \tag{12}$$

The solution $\{\widehat{\Theta}_h^n\}$ of the heat equation (1b) is then obtained by

$$\widehat{\Theta}_h^n(\boldsymbol{x}) = \sum_{j=1}^{M} \widehat{T}_j^n \phi_j(\boldsymbol{x}). \tag{13}$$

Notice that (12) and (13) are respectively, the local and global approximations of the solutions Θ_h^n at the departure points $\boldsymbol{X}_{hj}^n$.

2.2   Time Integration Procedure

For time integration, we use a second-order semi-implicit Crank-Nicolson scheme. Then, we obtain the discretization of the Darcy-heat problem (5) as: Find $(u_h^{n+1}, p_h^{n+1}, \Theta_h^{n+1})$ in $S_h \times Q_h \times R_h$ such that

$$\int_\Omega \nu(\widehat{\Theta}_h^n) u_h^{n+1} \cdot s_h \, d\Omega - \int_\Omega p_h^{n+1} \nabla \cdot s_h \, d\Omega = \int_\Omega f(\widehat{\Theta}_h^n) \cdot s_h \, d\Omega, \quad \forall \, s_h \in S_h,$$

$$\int_\Omega q_h \, \nabla \cdot u_h^{n+1} \, d\Omega = \int_\Omega \left(p_h^{n+1} - \Pi_{k-1} p_h^{n+1} \right) \left(q_h - \Pi_{k-1} q_h \right) \, d\Omega, \quad \forall \, q_h \in Q_h,$$

$$\tag{14a}$$

$$\int_\Omega \frac{\Theta_h^{n+1} - \widehat{\Theta}_h^n}{\Delta t} r_h \, d\Omega + \frac{\kappa}{2} \int_\Omega \nabla \Theta_h^{n+1} \cdot \nabla r_h \, d\Omega = \int_\Omega g_h^n r_h \, d\Omega$$

$$+ \frac{\kappa}{2} \int_\Omega \nabla \widehat{\Theta}_h^n \cdot \nabla r_h \, d\Omega, \quad \forall \, r \in R_h,$$

$$\tag{14b}$$

where $\widehat{\Theta}_h^n$ are the characteristics curves obtained by (13) and g_h^n the function given by

$$g_h^n = \frac{1}{\Delta t} \int_{t_n}^{t_{n+1}} g_h(s) \, ds.$$

For the existence and uniqueness of the solution of (14), we consider the following assumptions:

ASSUMPTION 1. *The functions ν, f and g are assumed to verify:*

1. *ν is Lipschitz continuous and there exist two strictly positive constants ν_1 and ν_2 such that*

$$\nu_1 \leq \nu(\xi) \leq \nu_2, \quad \forall \, \xi \in \mathbb{R}.$$

2. *f is Lipschitz with respect to its variable Θ, i.e.,*

$$\|f(\Theta)\|_{L^\infty(0,T;L^2(\Omega)^d)} \leq c_f \, \|\Theta\|_{L^\infty(0,T;L^2(\Omega)^d)},$$

where c_f is a positive constant.
3. *$g \in L^2 \left(0, T; L^2(\Omega) \right)$.*

ASSUMPTION 2. *For all $p \in L^2(\Omega)$, the operator Π_{k-1} defined in (8) is assumed to satisfy:*

1. *$\Pi_{k-1} : L^2(\Omega) \longrightarrow L^2(\Omega)$ is continuous and*

$$\|\Pi_{k-1} p\|_{L^2(\Omega)} \leq c \, \|p\|_{L^2(\Omega)}, \tag{15}$$

where c is a positive constant independent of h.
2. *The properties of Π_{k-1} must be augmented by the approximation*

$$\|p - \Pi_{k-1} p\|_{L^2(\Omega)} \leq c' h \, |p|_{H^1(\Omega)}, \tag{16}$$

where c' is a positive constant independent of h.

Thus, for the a priori bounds for the numerical solutions, we have the following Theorem:

THEOREM 1. *At each time step t_n and for given $\widehat{\Theta}_h^n \in R_h$, problem (14) has a unique solution $(u_h^{n+1}, p_h^{n+1}, \Theta_h^{n+1})$ in $S_h \times Q_h \times R_h$ that verifies the following bounds:*

$$\left\| u_h^{n+1} \right\|_{L^2(\Omega)^d}^2 \leq \left(\frac{c_f}{\nu_1} \right)^2 \left\| \widehat{\Theta}_h^n \right\|_{L^2(\Omega)^d}^2 + ch^2 \left| p_h^{n+1} \right|_{H^1(\Omega)}^2, \tag{17}$$

$$\left\| \Theta_h^{n+1} \right\|_{L^2(\Omega)} - \left\| \widehat{\Theta}_h^n \right\|_{L^2(\Omega)} \leq \Delta t \left\| g_h^n \right\|_{L^2(\Omega)}, \tag{18}$$

where c is a positive constant independent of h.□

PROOF. It is clear that the Darcy equations (7) have a unique solution since they satisfy the stabilized inf-sup condition [4,5]. Let a and b be two real numbers. For any positive real number ϵ, we have the well-known Young's inequality

$$ab \leq \frac{1}{2\epsilon} a^2 + \frac{1}{2} \epsilon b^2. \tag{19}$$

By testing equations (14a) with $s_h = u_h$ and $q_h = p_h$ and using the Cauchy-Schwarz inequality and (19) with $\epsilon = \frac{\nu_1}{c_f}$, along with Assumptions 1 and 2, we immediately derive (17).

Next, knowing $u_h^n \in S_h$ and thus $\widehat{\Theta}_h^n$, the heat equation (14b) admits also a unique solution $\Theta_h^{n+1} \in R_h$. Thus, if we take $r_h = \Theta_h^{n+1} + \widehat{\Theta}_h^n$ in (14b), we obtain and use the Cauchy-Schwarz and triangle inequalities, we easily get the inequality (18). ∎

It is clear that the heat equation (14b) gives rise to the linear system form

$$\left([M] + \frac{\Delta t}{2} [S] \right) T^{n+1} = [M] \left(\widehat{T}^n + \Delta t\, G^n \right) - \frac{\Delta t}{2} [S] \widehat{T}^n, \tag{20}$$

with $T^{n+1} = \left(T_1^{n+1}, \ldots, T_M^{n+1} \right)^T$, $\widehat{T}^n = \left(\widehat{T}_1^n, \ldots, \widehat{T}_M^n \right)^T$ and $G^n = (g_1^n, \ldots, g_M^n)^T$ being the source term vector. Here, [M] and [S] are the mass and stiffness matrices whose elements are given respectively by

$$m_{ij} = \int_\Omega \phi_j \phi_i \, d\Omega, \qquad s_{ij} = \int_\Omega \nabla \phi_j \nabla \phi_i \, d\Omega, \qquad i, j = 1, \ldots, M.$$

Thus, the solution of the coupled Darcy-heat problem (1) can be reformulated in matrix form as

$$\begin{pmatrix} [A] & [B] & [O] \\ [B]^T & [D] & [O] \\ [O] & [O] & [\widehat{M}] \end{pmatrix} \begin{pmatrix} \mathbb{U} \\ \mathbb{P} \\ \mathbb{T} \end{pmatrix} = \begin{pmatrix} \mathbb{F} \\ \mathbb{O} \\ \widehat{\mathbb{R}} \end{pmatrix}, \tag{21}$$

Algorithm 1

1: Assemble and store the matrices associated to the linear system (21).
2: **while** $t_{n+1} \leq T$ **do**
3:      **for** each mesh element $\mathcal{K}_j$ **do**
4:            Calculate the departure point $\boldsymbol{X}_j^n$ using the algorithm (11).
5:            Search-locate the mesh element $\widehat{\mathcal{K}}_j$ where $\boldsymbol{X}_j^n$ belongs.
6:            Evaluate the solutions $\widehat{\Theta}_j^n = \Theta^n(\boldsymbol{X}_j^n)$ using (12).
7:      **end for**
8:      Assemble and store the right-hand sides associated to the linear system (21).
9:      Compute the solution of (1) by solving the linear system (21).
10: **end while**

---

where $\mathbb{U}$, $\mathbb{P}$ and $\mathbb{T}$ are M-valued vectors with unknowns entries $\mathcal{U}_j^n$, $\mathcal{P}_j^n$ and $\mathcal{T}_j^n$ $(j = 1, \ldots, M)$, respectively, as defined in (2). In (21), $[\mathbf{O}]$ is the M square zero matrix, $[\mathbf{A}]$, $[\mathbf{B}]$ and $[\mathbf{D}]$ are square $M \times M$-valued matrices whose elements entries, according to (7), are respectively given by

$$a_{ij} = \int_\Omega \varphi_j \cdot \varphi_i \, d\Omega, \qquad b_{ij} = -\int_\Omega \phi_j (\nabla \cdot \varphi_i) \, d\Omega,$$

$$d_{ij} = \int_\Omega (\phi_j - \Pi_{k-1}\phi_j)(\phi_i - \Pi_{k-1}\phi_i) \, d\Omega,$$

where $i, j = 1, \ldots, M$, and $\left[\widehat{\mathbf{M}}\right]$ is the matrix given by $\left[\widehat{\mathbf{M}}\right] = [\mathbf{M}] + \frac{\Delta t}{2}[\mathbf{S}]$. In the right-hand side of (21), $\mathbb{F}$ is the M-valued vector with entries

$$F_j = \int_\Omega \boldsymbol{f}(\widehat{\mathcal{T}}_j^n) \cdot \varphi_i \, d\Omega, \qquad i, j = 1, \ldots, M,$$

$\mathbb{O}$ is the zero vector in $\mathbb{R}^M$ and $\widehat{\mathbb{R}}$ is the right-hand side of (20).

To summarize, the implementation of the proposed semi-Lagrangian unified element method for solving the coupled Darcy-heat equations (1) is carried out following the steps in Algorithm 1. Note that, since only the right-hand side of the linear system (21) changes at subsequent time steps, it is convenient to use a Cholesky factorization at the first time step thus the solution is reduced into subsequent forward and backward substitutions. This can significantly increase the efficiency when a large number of time steps is required, compared to updating the matrix and fully solving the system at every time step.

3   Numerical Experiments

To validate the accuracy and performance of the proposed semi-Lagrangian finite element method, we present numerical results for two coupled Darcy-heat problems. All the computations are performed using unstructured triangular meshes with different element densities and by using the quadratic P_2 elements for all the solutions in the model (1). The obtained linear system of algebraic equations (21) is solved using the conjugate gradient solver with incomplete Cholesky decomposition. Moreover, all stopping criteria for iterative solvers are set to 10^{-7} which is small enough to guarantee that truncation errors in the algorithm dominate the total numerical errors.

3.1 Accuracy Coupled Darcy-Heat Problem

In this test example, We consider the following time-dependent coupled Darcy-heat equations

$$
\begin{aligned}
\nu(\Theta)\boldsymbol{u} + \nabla p &= \big(\Theta + f(x,y,t)\big)\boldsymbol{j}, & (x,y,t) &\in \Omega \times [0,T], \\
\nabla \cdot \boldsymbol{u} &= 0, & (x,y,t) &\in \Omega \times [0,T], \quad (22) \\
\frac{\partial \Theta}{\partial t} + \boldsymbol{u} \cdot \nabla\Theta - \kappa\nabla^2\Theta &= g(x,y,t), & (x,y,t) &\in \Omega \times [0,T],
\end{aligned}
$$

supplemented with the following initial condition

$$
\Theta(x,y,0) = 0 \qquad (x,y) \in \Omega. \tag{23}
$$

In (22), $\boldsymbol{j} = (0,1)^\top$ is the unit vector in the upward direction, $\Omega = [0,3] \times [0,3]$, and $\nu(\Theta) = \Theta + 1$. The functions $f(x,y,t)$ and $g(x,y,t)$ are calculated such that the exact solution of (22) is given by

$$
\boldsymbol{u}(x,y,t) = e^{-t/4}\,\mathrm{curl}\psi, \qquad p(x,y,t) = (t+1)\cos(\frac{\pi}{3}x)\cos(\frac{\pi}{3}y),
$$

$$
\Theta(x,y,t) = \sin(t)x^2(x-3)^2y^2(y-3)^2,
$$

where the function ψ is defined as

$$
\psi(x,y,t) = e^{-3\big((x-1)^2+(y-1)^2\big)}.
$$

In our numerical simulations, the diffusion coefficient value is $\kappa = 5 \times 10^{-4}$ and the time step $\Delta t = 0.05$. Table 1 shows the averaged number of iterations in the linear solver, the relative L^2-errors and convergence rates at time $t = 1$ for the pressure p, the velocity $\boldsymbol{u} = (u,v)^\top$ and the temperature Θ using different structured meshes with uniform step h. It is obvious that, increasing the mesh density in the numerical simulations results in a decrease in the number of iterations needed for the linear solver and in the relative L^2-error for all variables and thus, a good approximation for the pressure, velocity and temperature solutions at the time in question. As expected, the proposed semi-Lagrangian finite element method converges at about the same rate for all meshes and for all solutions confirming a second-order accuracy.

Table 1. Relative L^2-error and convergence rates obtained for the pressure, velocity and temperature solutions in the accuracy test example of the coupled Darcy-heat problem at time $t = 1$.

h	Iter	Pressure p		Velocity u		Velocity v		Temperature Θ	
		L^2-error	Rate	L^2-error	Rate	L^2-error	Rate	L^2-error	Rate
$\frac{1}{16}$	14.2	1.897561E-03	—	5.88932E-01	—	6.85811E-01	—	1.59171E-02	—
$\frac{1}{32}$	13.6	5.26369E-04	1.85	1.56710E-01	1.91	1.18123E-01	1.92	4.17711E-03	1.93
$\frac{1}{64}$	12.1	1.44001E-04	1.87	4.14113E-02	1.92	3.03099E-02	1.93	1.05886E-03	1.98
$\frac{1}{128}$	11.1	3.75290E-05	1.94	1.06439E-02	1.96	7.73669E-03	1.97	2.66556E-04	1.99
$\frac{1}{256}$	10.7	9.18917E-06	2.03	2.64259E-03	2.01	1.90754E-03	2.02	6.57216E-05	2.02

3.2 Moving Thermal Front Past an Array of Cylinders

In this second example, we consider a problem of moving thermal front in a channel past an array of circular cylinders. We present the numerical solution of the system of equations (1) in a channel of length $L = 4$ and height $H = 1$. The channel consists of 16 circular cylinders with equal diameter and uniformly distributed in the second quarter of the channel at the domain $[1, 2] \times [0, 1]$. In [3], a similar computational domain was considered but with square-shaped obstacles, and in order to study the incompressible Navier-Stokes equations. Here, no-slip boundary condition is imposed at all cylinder walls. The left and right vertical walls are respectively, at dimensionless temperatures $\Theta = 0.5$ and $\Theta = -0.5$ whereas, the top and bottom walls are insulated. Initially, the flow is at cold rest $i.e.$, $\boldsymbol{u} = \boldsymbol{0}$ and $\Theta = -0.5$. In our simulations, a non-linear viscosity defined by $\nu(\Theta) = \sin(\Theta) + 2$ is used in (1), the source terms $\boldsymbol{f} = \boldsymbol{0}$ and $g = 0$. In order to ensure accuracy and efficiency in the numerical method, we use the unstructured triangular mesh depicted in Fig. 1 with 12224 elements and 25295 nodes in our simulations. To approximate the temperature, velocity and pressure solutions, we use the quadratic P_2 finite elements. In Fig. 2 and Fig. 3, we display the results obtained for the temperature and velocity fields at four different instants namely, $t = 1$, $t = 3$, $t = 5$ and $t = 7$. To examine effects of diffusion in the moving thermal front past the cylinders, we present numerical results for three different values taken for the diffusion coefficient, $\kappa = 10^{-2}$, 10^{-3} and 5×10^{-4}. For a better insight, Fig. 4 illustrates the vertical cross-sections at $x = 2.05$ of the temperature and the u-velocity at time $t = 5$ using the considered diffusion coefficient. It is clear that the cross-sections in Fig. 4 show good symmetry in the numerical simulations. Similar features, not presented here, have been obtained at other locations in the channel using different values of κ. The results clearly illustrate the influence of the diffusion variation on both the temperature patterns and velocity fields.

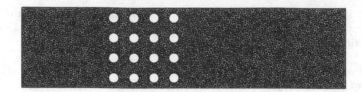

Fig. 1. Mesh used for the moving thermal front past an array of 16 cylinders of equal radius $R = 0.125$ uniformly distributed in a channel of length $L = 4$ and height $H = 1$. Here, the mesh contains 12224 triangular elements and 25295 nodes.

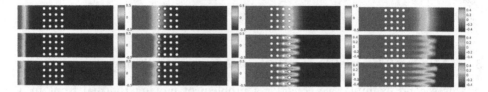

Fig. 2. Results for temperature for $\nu(\Theta) = \sin(\Theta) + 2$. From left to right: $t = 1$, $t = 3$, $t = 5$ and $t = 7$. From top to bottom: $\kappa = 0.01$, $\kappa = 0.001$, and $\kappa = 0.0005$.

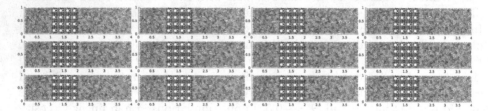

Fig. 3. Results for velocities for $\nu(\Theta) = \sin(\Theta) + 2$. From left to right: $t = 1$, $t = 3$, $t = 5$ and $t = 7$. From top to bottom: $\kappa = 0.01$, $\kappa = 0.001$, and $\kappa = 0.0005$.

Indeed, when the diffusion coefficient κ decreases, the transport speed increases with the size of the thermal front exhibiting steep gradients with different magnitudes, thin boundary layer, and separating shear layers. It is also worth noting that when the diffusion value is set to $\kappa = 5 \times 10^{-4}$, the flow becomes convection-dominated and steep fronts along with shock solutions appear in the temperature solution. For the considered Darcy-heat problem, these results obviously show that the small complex structures of the temperature being well captured by the proposed semi-Lagrangian finite element method. In fact, the computed solutions remain stable and highly accurate even when a relatively coarse mesh is used, and the numerical resolution does not require the use of small time steps and mixed finite element discretizations.

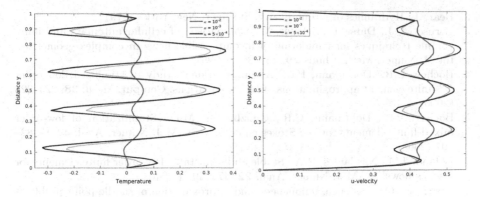

Fig. 4. Vertical cross-sections obtained in the channel at $x = 2.05$ for the temperature (left plot) and the u-velocity (right plot) for the Darcy-heat problem at time $t = 5$ using different diffusion coefficient values $\kappa = 10^{-2}, 10^{-3}$ and 5×10^{-4}.

4 Conclusions

In this study, we have presented a semi-Lagrangian finite element method for the numerical solution of time-dependent Darcy-heat problems. The governing equations consist of a nonlinear Darcy problem, with a variable viscosity, for the flow field and pressure coupled with a time-dependent convection-diffusion equation for the heat transfer. The enhanced method consists of coupling the semi-Lagrangian approach with a Galerkin finite element discretization on unstructured grids. To stabilize the solutions, we use a polynomial pressure-projection stabilization approach enabling the use of equal-order finite element approximations for all solutions in the coupled problem. The proposed method avoids mixed finite element formulations which generally require a higher computational cost for mesh generation and element matrix assembly. In our numerical simulations, we use the P_2-type polynomials to formulate the finite element solutions. However, the method can also be extended to the use of higher order polynomials based on a similar formulation. Numerical results have been presented for a test example with known exact solutions. The method has also been applied for solving a moving thermal front problem past an array of cylinders using different diffusion values. The presented results support our expectations for an accurate and stable behaviour for all transport regimes considered. Future work will concentrate on the extension of this method to Darcy-heat problems in three-dimensional domains using high-order finite element discretizations on unstructured meshes.

References

1. Amanbek, Y., Singh, G., Pencheva, G., Wheeler, M.F.: Error indicators for incompressible Darcy flow problems using enhanced velocity mixed finite element method. Comput. Methods Appl. Mech. Eng. **363**, 112884 (2020)

2. Bear, J.: Hydraulics of Groundwater. Springer, New York (1979)
3. Bernsdorf, J., Durst, F., Schäfer, M.: Comparison of cellular automata and finite volume techniques for simulation of incompressible flows in complex geometries. Int. J. Numer. Meth. Fluids **29**, 251–264 (1999)
4. Bochev, C.R., Dohrmann, P.B.: A computational study of stabilized, low-order C^0 finite element approximations of Darcy equations. Comput. Mech. **38**, 223–323 (2006)
5. Bochev, P.B., Dohrmann, C.R., Gunzburger, M.D.: Stabilization of low-order mixed finite elements for the Stokes equations. SIAM J. Numer. Anal. **44**(1), 82–101 (2006)
6. Boland, J.M., Nicolaides, R.A.: Stable and semistable low order finite elements for viscous flows. SIAM J. Numer. Anal. **22**, 474–492 (1985)
7. Brezzi, F.: On existence, uniqueness, and approximation of saddle-point problems arising from Lagrangian multipliers. RAIRO Model. Math. Anal. Numer. **21**, 129–151 (1974)
8. Chalhoub, N., Omnes, P., Sayah, T., El Zahlaniyeh, R.: Full discretization of time dependent convection–diffusion–reaction equation coupled with the Darcy system. Calcolo **57**(1), 1–28 (2019). https://doi.org/10.1007/s10092-019-0352-1
9. Chen, Z., Ewing, R.: Mathematical analysis for reservoir models. SIAM J. Math. Anal. **30**, 431–453 (1999)
10. De Marsily, G.: Quantitative Hydrogeology: Groundwater Hydrology for Engineers. Academic Press, New York (1986)
11. Dejam, M., Hassanzadeh, H.: Diffusive leakage of brine from aquifers during CO_2 geological storage. Adv. Water Res. **111**, 36–57 (2018)
12. Douglas, T.F., Russell, J.: Numerical methods for convection dominated diffusion problems based on combining the method of characteristics with finite elements or finite differences. SIAM J. Numer. Anal. **19**, 871–885 (1982)
13. El-Amrani, M., Seaid, M.: Numerical simulation of natural and mixed convection flows by Galerkin-characteristic method. Int. J. Numer. Meth. Fluids **53**(12), 1819–1845 (2007)
14. El-Amrani, M., Seaid, M.: An L^2-projection for the Galerkin-characteristic solution of incompressible flows. SIAM J. Sci. Comput. **33**(6), 3110–3131 (2011)
15. El-Amrani, M., Seaid, M.: A Galerkin-characteristic method for large-eddy simulation of turbulent flow and heat transfer. SIAM J. Sci. Comput. **30**(6), 2734–2754 (2008)
16. El-Amrani, M., Seaïd, M.: A finite element semi-Lagrangian method with l2 interpolation. Int. J. Numer. Methods Eng. **90**(12), 1485–1507 (2012)
17. Feng, X.: On existence and uniqueness results for a coupled system modeling miscible displacement in porous media. J. Math. Anal. Appl. **194**, 883–910 (1995)
18. Foicas, C., Guillopé, C., Temam, R.R.: Lagrangian representation of the flow. J. Diff. Eqn. **57**, 440–449 (1985)
19. Gunzburger, M.: Finite Element Methods for Viscous Incompressible Flows. Academic Press, Boston (1989)
20. Halassi, A., Joundy, J., Salhi, L., Taik, A.: A meshfree method for heat explosion problems with natural convection in inclined porous media. MATEC Web Conf. **241**, 01019 (2018)
21. Khaled, A.-R.A., Vafai, K.: The role of porous media in modeling flow and heat transfer in biological tissues. Int. J. Heat Mass Transf. **46**(26), 4989–5003 (2003)
22. Nield, D.A., Bejan, A.: Convection in Porous Media, 2nd edn. Springer, New York (1999). https://doi.org/10.1007/978-1-4757-3033-3

23. Notsu, H., Rui, H., Tabata, M.: Development and L2-analysis of a single-step characteristics finite difference scheme of second order in time for convection-diffusion problems. J. Algorithms Comput. Technol. **7**, 343–380 (2013)
24. Pironneau, O.: On the transport-diffusion algorithm and its applications to the Navier-Stokes equations. Numer. Math. **38**, 309–332 (1982)
25. Rui, H., Zhang, J.: A stabilized mixed finite element method for coupled Stokes and Darcy flows with transport. Comput. Methods Appl. Mech. Eng. **315**, 169–189 (2017)
26. Salhi, L., El-Amrani, M., Seaid, M.: A Galerkin-characteristic unified finite element method for moving thermal fronts in porous media. J. Comput. Appl. Math. p. 113159 (2020)
27. Salhi, L., El-Amrani, M., Seaid, M.: A stabilized semi-Lagrangian finite element method for natural convection in Darcy flows. Comput. Math. Methods p. e1140 (2021)
28. Seaid, M.: Semi-Lagrangian integration schemes for viscous flows. Comp. Methods Appl. Math. **4**, 392–409 (2002)
29. Temperton, A., Staniforth, C.: An efficient two-time-level semi-Lagrangian semi-implicit integration scheme. Q. J. Roy. Meteor. Soc. **113**, 1025–1039 (1987)
30. Xuan, Y.M., Roetzel, W.: Bioheat equation of the human thermal system. Chem. Eng. Technol. **20**(4), 268–276 (1997)

Multilevel Adaptive Lagrange-Galerkin Methods for Unsteady Incompressible Viscous Flows

Abdelouahed Ouardghi[1,2]([✉]), Mofdi El-Amrani[2], and Mohammed Seaid[1,3]

[1] International Water Research Institute, University Mohammed VI Polytechnic, Benguerir, Morocco
abdelouahed.ouardghi@um6p.ma
[2] Mathematics and Applications Laboratory, FST, Abdelmalek Essaadi University, Tangier, Morocco
[3] Department of Engineering, University of Durham, South Road, Durham DH1 3LE, United Kingdom
m.seaid@durham.ac.uk

Abstract. A highly efficient multilevel adaptive Lagrange-Galerkin finite element method for unsteady incompressible viscous flows is proposed in this work. The novel approach has several advantages including (i) the convective part is handled by the modified method of characteristics, (ii) the complex and irregular geometries are discretized using the quadratic finite elements, and (iii) for more accuracy and efficiency a multilevel adaptive L^2-projection using quadrature rules is employed. An error indicator based on the gradient of the velocity field is used in the current study for the multilevel adaptation. Contrary to the h-adaptive, p-adaptive and hp-adaptive finite element methods for incompressible flows, the resulted linear system in our Lagrange-Galerkin finite element method keeps the same fixed structure and size at each refinement in the adaptation procedure. To evaluate the performance of the proposed approach, we solve a coupled Burgers problem with known analytical solution for errors quantification then, we solve an incompressible flow past two circular cylinders to illustrate the performance of the multilevel adaptive algorithm.

Keywords: Incompressible Navier-Stokes equations · Finite element method · Lagrange-Galerkin method · L^2-projection · Adaptive algorithm

1 Introduction

Unsteady incompressible viscous flows use in their modelling the Navier-Stokes equations with the property that the convective terms are distinctly more dominant than the diffusive terms especially when the Reynolds numbers reach high values. At high Reynolds numbers, the convective term is known to be a source of

© Springer Nature Switzerland AG 2021
M. Paszynski et al. (Eds.): ICCS 2021, LNCS 12747, pp. 230–243, 2021.
https://doi.org/10.1007/978-3-030-77980-1_18

computational difficulties and nonphysical oscillations. In addition, sharp fronts, shocks, vortex shedding and boundary layers are among other difficulties that most Eulerian finite element methods fail to resolve accurately. In general, Eulerian finite element methods employ fixed meshes along with some upstream weightings in their formulations to stabilize the discretization. Examples of Eulerian finite element methods include the Petrov-Galerkin, streamline diffusion, discontinuous Galerkin methods and also many other methods of the high-order reconstructions from computational fluid dynamics such as isogeometric analysis, see for example [1,2,5,13,14,21,23]. However, the main drawback of these methods for solving the convection-dominated problems is the stability conditions which impose a severe restriction on the size of the time steps taken in the numerical simulations. Lagrange-Galerkin finite element methods have the potential to efficiently solve convection-dominated flow problems, see for example [3,6,8,22]. The main idea in these methods lies on reformulating the governing equations in terms of the Lagrangian coordinates as defined by the particle trajectories (or characteristics) associated with the problem under study. In this case, the time derivative and the advection operator are combined in a total directional derivative along the characteristics which can be integrated using a semi-Lagrangian time stepping. In [17], an L^2-projection on the finite element space is used for the evaluation of solutions at the departure points. The performance of the L^2-projection Lagrange-Galerkin finite element method has been assessed in [9,10] for several convection-dominated problems and the incompressible Navier-Stokes equations at high Reynolds numbers. Comparisons between the conventional and the L^2-projection Lagrange-Galerkin finite element methods have also reported in these references and the L^2-projection has demonstrated higher accuracy and stronger stability than the conventional method. However, for practical applications in the incompressible viscous flows, these methods may become computationally very demanding due to the dense quadratures required for the L^2 projection.

The objective of the present work is to develop a class of multilevel adaptive Lagrange-Galerkin finite element methods for the numerical solution of incompressible Navier-Stokes equations. The advantage of this approach is the use of multiple quadratures in the L^2-projection in the numerical solution. This yields considerable efficiency gains to be made since the matrix in the linear system is fixed and reused throughout the time stepping procedure. One other objective is also to implement a multilevel adaptive algorithm for enrichments using the gradient of the velocity field as an error indicator. In contrast to the gradient-based h-adaptive finite element methods as those investigated in [1,2,2,13,16,18], the linear systems in the proposed Lagrange-Galerkin finite element method keep the same structure and size at each adaptation step. Indeed, an initial coarse mesh is needed for the gradient-based h-adaptive methods to compute a primary solution for estimation of the gradient. This allows for error accumulations due to the coarse mesh used in the approximation and the computational cost becomes prohibitive due to multiple interpolations between adaptive meshes. The performance of the proposed method is assessed using several test problems for incompressible viscous flows. For various parameters like the Reynolds numbers, multilevel adaptation and mesh refinements, results of the adaptive Lagrange-Galerkin finite element method are compared with those computed using the fixed approach.

This paper is organized as follows. In Sect. 2 we present the formulation of the Lagrange-Galerkin finite element method. This section includes the implementation of the L^2-projection procedure for the convection stage. Section 3 is devoted to the development of a multilevel adaptive Lagrange-Galerkin finite element method. In this section we also discuss the criteria used for adaptation. In Sect. 4, we examine the numerical performance of the proposed method using several examples of incompressible Navier-Stokes flows. Our new approach is demonstrated to enjoy the expected efficiency as well as the accuracy. Concluding remarks are summarized in Sect. 5.

2   Enriched Lagrange-Galerkin Finite Element Method

In the present study, we are concerned with solving the incompressible Navier-Stokes equations reformulated in a dimensionless form as

$$\nabla \cdot \boldsymbol{u} = 0,$$
$$\frac{D\boldsymbol{u}}{Dt} + \nabla p - \frac{1}{Re}\Delta u = \mathbf{f}, \tag{1}$$

where $\boldsymbol{u} = (u, v)^{\top}$ is the velocity field, p the pressure, $\mathbf{f}$ a source term, Re the Reynolds number and $\frac{D\cdot}{Dt}$ the total derivative defined as

$$\frac{D\boldsymbol{u}}{Dt} := \frac{\partial \boldsymbol{u}}{\partial t} + \boldsymbol{u} \cdot \nabla \boldsymbol{u} = \mathbf{0}, \tag{2}$$

In order to solve the incompressible Navier-Stokes Eqs. (1)–(2), the Lagrange-Galerkin finite element method solves separately at each time step, the convective part (2) then the Stokes Eqs. (1). A quasi-uniform partition $\Omega_h \subset \overline{\Omega}$ composed of triangular elements $\mathcal{K}_k$ is considered for the finite element discretization. The generated triangles are configured in such a manner, that there are no empty spaces between two elements and that they do not overlap. It is well known that, for such a problem the mixed Taylor-Hood finite elements P_2-P_1 is used for the conforming finite element spaces. Furthermore, it has been shown that for this mixed formulation the discrete velocity and pressure solutions satisfy the inf-sup condition, see for instance [6]. For the time discretization, the time interval $[0, T]$ is partitioned into a set of sub-intervals $[t_n, t_{n+1}]$ with fixed length $\Delta t = t_{n+1} - t_n$ for $n \geq 0$. In the rest of this paper, the notation $w_h^n := w(\boldsymbol{x}_h, t_n)$ is used to denote the value of a given function w at time t_n and the mesh point $\boldsymbol{x}_h$. Using this notation, the solutions $\boldsymbol{u}^n(\boldsymbol{x})$ and $p^n(\boldsymbol{x})$ are formulated in their associated finite element spaces as

$$\boldsymbol{u}_h^n(\boldsymbol{x}) = \sum_{j=1}^{M_v} \boldsymbol{U}_j^n \phi_j(\boldsymbol{x}), \qquad p_h^n(\boldsymbol{x}) = \sum_{l=1}^{M_p} P_l^n \psi_l(\boldsymbol{x}), \tag{3}$$

where $\{\phi_j\}_{j=1}^M$ and $\{\psi_l\}_{l=1}^{M_p}$ are the set of the well known global nodal basis functions of the velocity and the pressure, respectively. In (3), M_v and M_p represent the number of the grid points of the velocity and pressure, respectively. Hence, the semi-Lagrangian solution of the convection problem (2) is formulated for all mesh points x_j, $j = 1, \cdots, M_v$ as

$$U_j^{n+1} = u^{n+1}(x_j) = u^n(\boldsymbol{X}_j^n) =: \widehat{U}_j^n, \tag{4}$$

where $\boldsymbol{X}_j^n = \boldsymbol{X}_j(t_n)$ is the departure point defined at time t_n of a physical particle that will attain the grid point x_j at time t_{n+1}. Here, $\boldsymbol{X}_j(t)$ is the characteristic curve associated to the Eq. (2) for the mesh point x_j which is the solution of the following backward ordinary differential equations

$$\frac{d\boldsymbol{X}_j(t)}{dt} = u(\boldsymbol{X}_j(t), t), \qquad \boldsymbol{X}_j(t_{n+1}) = x_j, \qquad j = 1, \ldots, M_v. \tag{5}$$

To evaluate the solution of the Eq. (5) we use a second-order extrapolation method based on the mid-point rule, details on these procedures can be found in [8,10,20] among others. It is worth mentioning that, the evaluated departure point $\boldsymbol{X}_j^n$ does not generally match with any of the mesh points. Consequently, a search-locate algorithm is needed to allocate the mesh element $\widehat{\mathcal{K}}_j$ where the departure point $\boldsymbol{X}_j^n$ belongs, see for example [7,10]. Thus, the finite element solution $\widehat{U}^n$ can be evaluated at the departure point $\boldsymbol{X}_j^n$ as

$$\widehat{U}_j^n := u^n\left(\boldsymbol{X}_j^n\right) = \sum_{i=1}^N u^n(\widehat{x}_i)\varphi_i\left(\boldsymbol{X}_{hj}^n\right), \tag{6}$$

where $\{\varphi_i\}_{i=1}^N$ are the local shape functions defined on the host element $\widehat{\mathcal{K}}_j$, N is the number of nodes which define the velocity mesh points, and $\{\widehat{x}_i\}_{i=1}^N$ are the vertices of the element $\widehat{\mathcal{K}}_j$. Using the Eqs. (4) and (6), the global solution obtained suing the conventional Lagrange-Galerkin finite element method can be expressed as

$$u^{n+1}(\mathbf{x}) = \sum_{j=1}^{M_v} \widehat{U}_j^n \phi_j(\mathbf{x}). \tag{7}$$

Note that as in most numerical methods, the accuracy of the conventional semi-Lagrangian finite element method depends on the computational mesh used in the simulations. Moreover, it has been proved in [10] that if the computational mesh is not sufficiently fine, the conventional semi-Lagrangian finite element method fails to accurately resolve the sharp gradients generated by the convective terms. In the present work, we aim to introduce local enrichments using the L^2−projection to improve the accuracy of considered the semi-Lagrangian finite element method without refining the meshes.

2.1 L^2-projection for Local Enrichments

In the present section, we formulate the L^2-projection presented in [9,10] as a local enrichment technique for the convection Eq. (2). Thus, the weak formulation can be achieved by multiplying Eq. (4) by the finite element basis functions ϕ_i and integrating over the domain Ω as

$$\int_\Omega \boldsymbol{u}^{n+1}(\boldsymbol{x})\phi_i(\boldsymbol{x})d\boldsymbol{x} = \int_\Omega \boldsymbol{u}^n\left(\boldsymbol{\mathcal{X}}^n\right)\phi_i(\boldsymbol{x})\,d\boldsymbol{x}, \qquad i = 1,\dots,M. \tag{8}$$

The weak form (8) can be expressed in a matrix-vector form as

$$[\mathbf{M}]\left\{\boldsymbol{U}^{n+1}\right\} = \left\{\mathbf{r}^n\right\}, \tag{9}$$

where $[\mathbf{M}]$ is the Lagrange-Galerkin finite element mass matrix with entries $m_{ij} = \int_\Omega \phi_j(\boldsymbol{x})\phi_i(\boldsymbol{x})d\boldsymbol{x}$, $\boldsymbol{U}^{n+1}$ the vector of the unknown nodal values of the solution with entries U_j^{n+1}, and $\mathbf{r}^n$ the known right-hand side with entries r_i^n defined as

$$r_i^n = \int_\Omega \boldsymbol{u}^n\left(\boldsymbol{\mathcal{X}}^n\right)\phi_i(\boldsymbol{x})\,d\boldsymbol{x} = \sum_{k=1}^{N_e} \int_{\mathcal{K}_k} \boldsymbol{u}^n\left(\boldsymbol{\mathcal{X}}_h^n\right)\phi_i(\boldsymbol{x})\,d\boldsymbol{x}, \tag{10}$$

where N_e is the total number of the mesh elements. The quadrature rule is used to evaluate the integrals $m_{i,j}$ and r_i as

$$m_{ij} \approx \sum_{k=1}^{N_e}\sum_{q=1}^{N_{k,Q}} \omega_{q,k}\phi_j(\boldsymbol{x}_{q,k})\phi_i(\boldsymbol{x}_{q,k}), \qquad r_i \approx \sum_{k=1}^{N_e}\sum_{q=1}^{N_{k,Q}} \omega_{q,k}\widehat{\boldsymbol{U}}_{q,k}^n\phi_i(\boldsymbol{x}_{q,k}) \tag{11}$$

with $\boldsymbol{x}_{q,k}$ are the quadrature points of the element $\mathcal{K}_k$ and $\omega_{q,k}$ their associated weights. Here, $\boldsymbol{\mathcal{X}}_{q,k}^n$ are the departure points associated with $\boldsymbol{x}_{q,k}$ computed using (5), and $N_{k,Q}$ is the total number of quadrature points in the element $\mathcal{K}_k$. Hence, $\widehat{\boldsymbol{U}}_{q,k}^n = \boldsymbol{u}_h^n(\boldsymbol{\mathcal{X}}_{q,k}^n)$ is the solution evaluated at the departure point $\boldsymbol{\mathcal{X}}_{q,k}^n$ which can be evaluated according to (6) as

$$\widehat{\boldsymbol{U}}_{q,k}^n = \sum_{i=1}^N \boldsymbol{u}_h^n(\widehat{\boldsymbol{x}}_i)\varphi_i(\boldsymbol{\mathcal{X}}_{q,k}^n). \tag{12}$$

The approximations in (11) can be enriched by adjusting the number of quadrature points $N_{k,Q}$ either globally in the entire mesh or locally at each element in the computational domain. In the current work, the Dunavant quadrature rules studied in [4] are used. A distribution of Dunavant quadrature points is illustrated in Fig. 1 for $N_{k,Q} = 6, 12, 25, 52$ and 70.

3   Multilevel Adaptive Enrichments

In many incompressible viscous flows, the solution involves sharp gradients, localized eddies and shear layers specially when the Reynolds number attends high

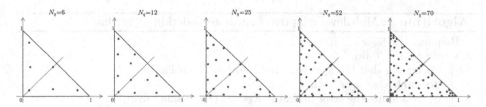

Fig. 1. A schematic distribution of quadrature points used in the L^2-projection method for global and local enrichments.

values. To accurately resolve these flow features, the enriched Lagrange-Galerkin finite element method introduced in Sect. 2 may require very fine meshes and high number of quadrature points particularly in regions where the solution gradients become very high. In the present work, to avoid uniform enrichment in the entire computational mesh, we propose an adaptive local enrichment to speed up the algorithm. The main idea of this adaptive technique is to refine the number of quadrature points $N_{k,Q}$ in mesh elements where the solution gradient attends high values and unrefine otherwise according to a given criterion. In practice, to perform this adaptation one needs an error estimator or error indicator along with a given tolerance to adapt the quadrature accordingly. Gradient-based estimators have been widely used in the literature in h-adaptive finite element methods for incompressible Navier-Stokes equations, see for example [1,2,2,13,16,18,19,21]. However, most of gradient-based h-adaptive algorithms employ an initial coarse mesh to compute a primary solution for estimating the gradient. As a consequence, error accumulation occurs due to the coarse mesh used in the approximation and computational cost becomes prohibitive due to multiple interpolations between adaptive meshes. In [2], the gradient of the velocity field is used as indicator for mesh adaptation to study vortex shedding in incompressible flows. The results presented demonstrate that this adaptation procedure for dynamic refinement and unrefinement is fully operational. Here, we use similar techniques and the normalized gradient of the velocity is employed as an adaptation criterion for the local enrichment of each element in the computational domain as

$$Err^{n+1}(\mathcal{K}_k) = \frac{\left\|\nabla u_{\mathcal{K}_k}^{n+1}\right\|}{\displaystyle\max_{j=1}^{N_e}\left\|\nabla u_{\mathcal{K}_j}^{n+1}\right\|}, \tag{13}$$

where $u_{\mathcal{K}_k}^{n+1}$ is the solution on element $\mathcal{K}_k$ at time t_{n+1} and $\left\|\nabla u_{\mathcal{K}_k}^{n+1}\right\|$ is the L^2-norm of the gradient of $u_{\mathcal{K}_k}^{n+1}$ defined as

$$\left\|\nabla u_{\mathcal{K}_k}^{n+1}\right\| = \left(\int_{\mathcal{K}_k} \nabla u^{n+1} \cdot \nabla u^{n+1}\, dx + \int_{\mathcal{K}_k} \nabla v^{n+1} \cdot \nabla v^{n+1}\, dx\right)^{\frac{1}{2}}. \tag{14}$$

Algorithm 1: Multilevel adaptive Lagrange-Galerkin algorithm

---

1 Require: $\{\varepsilon_m\}_{m=0,1,\ldots,4}$;
2 **while** $t_{n+1} \leq T$ **do**
3   Assuming that the approximated solution U^n is known;
4   **foreach** *element* $\mathcal{K}_k$ **do**
5   Compute the error indicator $Err^{n+1}(\mathcal{K}_k)$ using (13);
6   **foreach** $m = 0, 1, 2, 3$ **do**
7   **if** $\varepsilon_m \leq Err^{n+1}(\mathcal{K}_k) \leq \varepsilon_{m+1}$ **then**
8                  $N_{k,Q} = N_{k,q_m}$;
9   **end**
10   **end**
11   **end**
12   Generate the quadrature pair $(\boldsymbol{x}_{q,k}, \omega_{q,k})$, $q = 1, \ldots, N_{k,Q}$;
13   Evaluate the L^2-projection mass matrix $[\mathbf{M}]$ using left part of (11);
14   **foreach** *element* $\mathcal{K}_k$ **do**
15   **foreach** *quadrature point* $\boldsymbol{x}_{q,k}$, $q = 1, \ldots, N_{k,Q}$ **do**
16   Calculate the departure point $\boldsymbol{X}_{k,q}^n$;
17   Search for the element $\widehat{\mathcal{K}}_{q,k}$ where $\boldsymbol{X}_{q,k}^n$ resides;
18   Compute the value of $\widehat{U}_{q,k}^n$ using the equation (12);
19   **end**
20   **end**
21   Compute the element right-hand side r_i^n using equation (11);
22   Assemble the vector $\mathbf{r}^n$;
23   Solve the resulted linear system (9);
24   Update the solution $\boldsymbol{u}_h^{n+1}$ at time t_{n+1} using equation (7);
25 **end**

Using the finite element discretization on the element $\mathcal{K}_k$ the velocity gradient (14) can be reformulated as

$$\|\nabla \boldsymbol{u}_{\mathcal{K}_k}^{n+1}\| = \left(\left(\boldsymbol{U}_{\mathcal{K}_k}^{n+1}\right)^\top \mathbf{S}_{\mathcal{K}_k} \boldsymbol{U}_{\mathcal{K}_k}^{n+1} + \left(\boldsymbol{V}_{\mathcal{K}_k}^{n+1}\right)^\top \mathbf{S}_{\mathcal{K}_k} \boldsymbol{V}_{\mathcal{K}_k}^{n+1}\right)^{\frac{1}{2}}. \quad (15)$$

where $\boldsymbol{U}_{\mathcal{K}_k}^{n+1} = (U_1^{n+1}, \ldots, U_N^{n+1})^\top$, $\boldsymbol{V}_{\mathcal{K}_k}^{n+1} = (V_1^{n+1}, \ldots, V_N^{n+1})^\top$, and $\mathbf{S}_{\mathcal{K}_k}$ is the elementary stiffness matrix evaluated at the element $\mathcal{K}_k$. It should be noted that the adaptation criterion (13) is a gradient-based error indicator which is evaluated at time t_{n+1} from the known solutions at time t_n due to the backward property of the modified method of characteristics. Normalization of the error indicator is used to keep its values bounded in the interval $[0, 1]$.

Hence, the multilevel adaptation procedure we propose in this study is performed as follows: given a sequence of three real numbers $\{\varepsilon_m\}$ such that $0 = \varepsilon_0 < \varepsilon_1 < \varepsilon_2 < \varepsilon_3 < \varepsilon_4 = 1$. If an element $\mathcal{K}_k$ satisfies the condition

$$\varepsilon_m \leq Err^{n+1}(\mathcal{K}_k) \leq \varepsilon_{m+1}, \qquad m = 0, 1, 2, 3,$$

then $\mathcal{K}_k$ is enriched with the quadrature rule $(\boldsymbol{x}_{k,q}, w_{k,q})$ with $q = 1, 2 \ldots, N_{k,q_m}$. Here, the values of $\{\varepsilon_1, \varepsilon_2, \varepsilon_3\}$ can be interpreted as tolerances to be set by the user resulting into a three-level refining. Note that the number of levels m and the values of tolerances $\{\varepsilon_m\}$ in the above adaptive enriched Lagrange-Galerkin finite element method are problem dependent and their discussions is postponed for Sect. 4 where numerical examples are discussed. The steps used in the proposed adaptive enriched Lagrange-Galerkin finite element method for solving the convection stage are summarized in Algorithm 1. Note that other adaptive criteria as those used in h-, p- and hp-adaptivity in [1,13,16,18,19] can also be implemented in our algorithm without major conceptual modifications. A posteriori error estimations as those developed in [5,12] can also be adopted for our enriched Lagrange-Galerkin finite element methods.

4   Numerical Results

In this section we examine the accuracy of the new enriched Lagrange-Galerkin finite element method introduced in the above sections using two examples of incompressible flow problems. For the first example the analytical solution is known, so that we can evaluate the relative L^1-error and L^2-error at time t_n as

$$
L^1\text{-error} = \frac{\int_\Omega \left| u_h^n - u_{\text{exact}}^n \right| d\boldsymbol{x}}{\int_\Omega \left| u_{\text{exact}}^n \right| d\boldsymbol{x}}, \qquad L^2\text{-error} = \frac{\left(\int_\Omega \left| u_h^n - u_{\text{exact}}^n \right|^2 d\boldsymbol{x} \right)^{\frac{1}{2}}}{\left(\int_\Omega \left| u_{\text{exact}}^n \right|^2 d\boldsymbol{x} \right)^{\frac{1}{2}}},
$$

where u_{exact}^n and u_h^n are respectively, the exact and numerical solutions at the gridpoint $\boldsymbol{x}_h$ and time t_n. In all the computations reported in this section, the resulting linear systems of algebraic equations are solved using the conjugate gradient solver with incomplete Cholesky decomposition. In addition, all stopping criteria for iterative solvers were set to 10^{-6}, which is small enough to guarantee that the algorithm truncation errors dominate the total numerical errors. All the computations were performed on an Intel® Core(TM) i7-7500U @ 2.70 GHz with 16 GB of RAM.

4.1   Viscous Burgers Flow Problem

To evaluate the accuracy of the proposed Lagrange-Galerkin finite element approach, the coupled viscous Burgers flow problem is considered. It should be noted that, the coupled viscous Burgers system is a suitable form of the incompressible Navier-Stokes equations. Thus, we solve the following system

$$
\frac{D\boldsymbol{u}}{Dt} - \frac{1}{Re}\Delta\boldsymbol{u} = \boldsymbol{0}, \tag{16}
$$

in the squared domain $\Omega = [0,1] \times [0,1]$. Initial and boundary conditions for this example are obtained from the analytical solution studied in [11]

$$
u(x,y,t) = \frac{3}{4} - \frac{1}{g(x,y,t)}, \qquad v(x,y,t) = \frac{3}{4} + \frac{1}{g(x,y,t)}, \tag{17}
$$

where

$$g(x,y,t) = 4\left(1 + \exp\left(-\frac{(4x - 4y + t)Re}{32}\right)\right)$$ (18)

It is wroth mentioning that only results of the component u are presented in this section, and the results of the component v are similar to those of u.

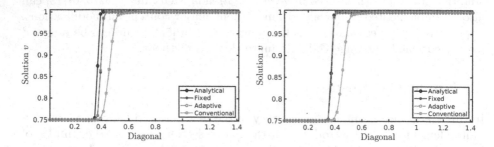

Fig. 2. Cross-sections at the main diagonal $y = 1 - x$ of the solution u obtained for the viscous Burgers problem at time $t = 2$ and $Re = 10^3$ on a mesh with $h = \frac{1}{64}$ using $N_{k,Q} = 12$ (left) and $N_{k,Q} = 52$ (right).

Table 1. Results for viscous Burgers problem obtained by the fixed and adaptive Lagrange-Galerkin finite element methods on a mesh with $h = \frac{1}{128}$ using different quadratures at time $t = 2$. The CPU times are given in seconds.

Re	$N_{k,Q}$	Fixed			Adaptive		
		L^1-error	L^2-error	CPU	L^1-error	L^2-error	CPU
10^2	12	3.907E-04	4.534E-04	94.05	3.951E-04	4.480E-04	45.12
	25	1.902E-04	2.108E-04	129.20	1.923E-04	2.182E-04	50.00
	52	8.229E-05	9.317E-05	220.00	7.249E-05	0.978E-05	67.26
10^3	12	7.683E-04	8.651E-03	113.32	7.629E-04	8.628E-03	62.35
	25	6.094E-04	5.089E-03	150.81	6.207E-04	5.145E-03	69.00
	52	3.051E-04	2.254E-03	270.43	3.012E-04	2.234E-03	85.23
10^4	12	7.063E-03	4.120E-03	120.32	6.500E-03	3.939E-03	65.42
	25	3.524E-03	2.834E-02	131.74	3.924E-03	2.998E-02	73.27
	52	1.921E-03	1.939E-02	223.00	1.807E-03	1.865E-02	88.54

Cross-sections of the obtained results at the diagonal of equation $y = 1 - x$ are presented in Fig. 2. These results are computed on a mesh with $h = \frac{1}{64}$ for the Reynold number $Re = 10^3$ using two different numbers of quadratures namely, $N_{k,Q} = 6$ and $N_{k,Q} = 52$. It can be shown from Fig. 2 that by increasing the number of quadrature points $N_{k,Q}$ globally or locally in the considered mesh yields to an improve in the accuracy of the results calculated using the proposed

Lagrange-Galerkin finite element method with fixed or multilevel adaptive projection. From the same figure it can be clearly shown that, the conventional semi-Lagrangian finite element method suffers from an excessive numerical diffusion while the proposed Lagrange-Galerkin finite element method resolves successfully the shock.

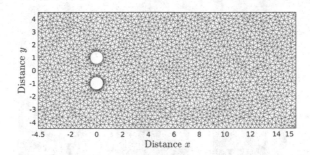

Fig. 3. Computational mesh used for the flow past two circular cylinders.

For the error quantification, a comparison between the proposed fixed and adaptive Lagrange-Galerkin finite element approaches is also performed for this example. The obtained L^1 error, L^2 error and CPU times are presented in Table 1 using different numbers of quadrature points for $Re = 100$, $Re = 1000$ and $Re = 10000$ at time $t = 2$. With reference to error norms and for all selected numbers of quadrature points, both the fixed and adaptive approaches generate similar results for all considered Reynolds numbers Re. Moreover, increasing the number of quadrature points leads to a significant increase in the accuracy of the studied approaches. In terms of CPU times, it is clear from Table 1 that the CPU times of the adaptive method are lower than those of the fixed method. For example, the CPU time of the adaptive approach is about 64%, 63% and 62% less than the CPU time of the fixed approach for $Re = 100$, 1000 and 10^4, respectively. Note that this reduction in the CPU times becomes large using fine meshes. As expected, the adaptive enriched method is more efficient than its fixed counterpart.

4.2   Flow Past Two Circular Cylinders

To illustrate the performance of the proposed multilevel adaptive Lagrange-Galerkin finite element method we solve the problem of a viscous flow past two circular cylinders in a channel. A similar configuration is used in [15]. In our computations, two circular cylinders with diameter $D = 1$ are immersed vertically in a viscous incompressible flow entering the channel with a uniform velocity $u_\infty = 1$. The Reynolds number for this test case is defined as $Re = Du_\infty/\nu$, with ν is the kinematic viscosity. We perform computations with the mixed formulation P_2-P_1 using the unstructured mesh composed of 4372 elements, 9063 velocity nodes and 2345 pressure nodes, see Fig. 3. The main purpose of this problem is

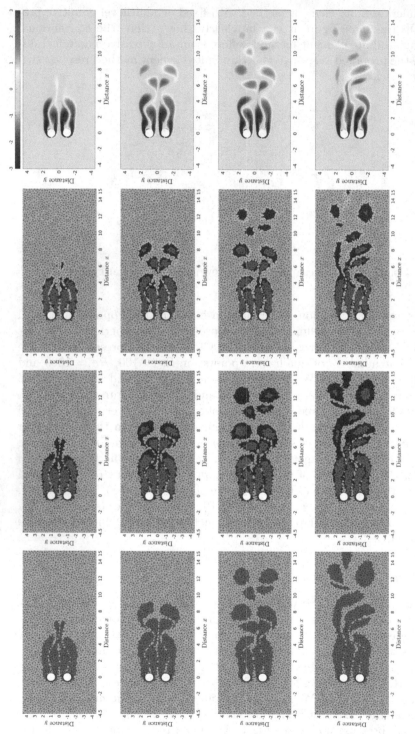

Fig. 4. Adaptive quadrature points using single-level enrichments (first column), two-level enrichments (second column), three-level enrichments (third column) and vorticity snapshots (fourth column) for the flow past two circular cylinders at Reynolds number $Re = 100$ and time $t = 7.4$ (first row), $t = 12.8$ (second row), $t = 17.4$ (third row) and $t = 35.7$ (fourth column). For a better visualization, red, blue and green colors are used for elements with single-level, two-level and three-level enrichments, respectively. (Color figure online)

to show the capability of the multilevel adaptive approach to accurately capture these steep gradients and vortex shedding exhibit by the numerical solutions at low computational costs. In our simulation, we consider single-level, two-level and three-level adaptive procedure using $\varepsilon_1 = 0.065$, $\varepsilon_2 = 0.17$ and $\varepsilon_3 = 0.3$. Initially, the number of quadrature points is $N_{k,Q} = 6$ in each element and we use $N_{k,Q} = 70$ for the single-level adaptive approach, $(N_{k,Q} = 52, N_{k,Q} = 70)$ for the two-level adaptive approach and $(N_{k,Q} = 12, N_{k,Q} = 52, N_{k,Q} = 70)$ for the three-level adaptive approach. At time $t = 35.7$, the total number of quadrature points used for fixed approach is 306040 whereas, the total number of quadrature points used for the single-level, the two-level and the three-level approaches are 85066, 79245 and 65751, respectively.

The distribution of quadrature points using the single-level, two-level and three-level adaptive Lagrange-Galerkin finite element methods at four different times $t = 7, 4$, 12.8, 17.4 and 35.7 and the vorticity snapshots obtained using the three-level method are presented in Fig. 4. In the distribution of quadrature points, three different colors are used to identify the mesh element with quadrature points for each level of adaptivity. Notice that the single-level and two-level methods produce results similar to those of the three-level method with different CPU time. For this reason, the vorticity results obtained using the single-level and two-level approaches are not displayed in this figure. It can be shown from Fig. 4 that the flow past two cylinders exhibits areas with large vorticity and vortex shedding. Consequentially, elements with high level of adaptivity are generated in the mesh. Moreover, the proposed Lagrange-Galerkin approach successfully captures the small complex structures of the flow and the eddies over the cylinders. This is because the proposed approach adapts the quadrature points where it is needed according to the used error indicator.

5 Concluding Remarks

A multilevel adaptive Lagrange-Galerkin finite element method is developed in this paper for efficiently solving the incompressible viscous flow problems on unstructured meshes. The proposed method combines the modified method of characteristics, the finite element method and an adaptive procedure based on L^2-projection using quadrature rules. Therefore, it benefits from the advantages of all combined procedures to ensure the efficiency and the accuracy of the proposed adaptive algorithm for incompressible viscous flows. Moreover, the considered multilevel adaptive algorithm increases the number of quadrature points where it is needed according to an error indicator without refining the computational mesh during the time integration procedure. As a result and contrary to other adaptive finite element methods, the resulted linear systems in the proposed Lagrange-Galerkin finite element method preserve the same fixed structure and size during the adaptation process. The gradient of the velocity is used as feature-based error indicator for the proposed adaptation technique. We demonstrate using several numerical test examples that the proposed algorithm can capture the flow features on coarse meshes and with a significant reduction in the computational requirement. Future work will concentrate on solving coupled flow-transport and natural convection problems to simulate the transport and dispersion of pollutant in seas. The extension of the proposed multilevel adaptive Lagrange-Galerkin finite element method for the incompressible viscous flow problems in three space dimensions is also under consideration and results will published in near future.

References

1. Ahrabi, B.R., Anderson, W.K., Newman III, J.C.: An adjoint-based hp-adaptive stabilized finite-element method with shock capturing for turbulent flows. Comput. Methods Appl. Mech. Eng. **318**, 1030–1065 (2017)
2. De Sampaio, P., Lyra, P.R.M., Morgan, K., Weatherill, N.P.: Petrov-Galerkin solutions of the incompressible Navier-Stokes equations in primitive variables with adaptive remeshing. Comput. Methods Appl. Mech. Eng. **106**(1–2), 143–178 (1993)
3. Douglas Jr., J., Russell, T.F.: Numerical methods for convection-dominated diffusion problems based on combining the method of characteristics with finite element or finite difference procedures. SIAM J. Numer. Anal. **19**(5), 871–885 (1982)
4. Dunavant, D.: High degree efficient symmetrical gaussian quadrature rules for the triangle. Int. J. Numer. Meth. Eng. **21**(6), 1129–1148 (1985)
5. Durango, F., Novo, J.: A posteriori error estimations for mixed finite element approximations to the Navier-Stokes equations based on newton-type linearization. J. Comput. Appl. Math. **367**, 112429 (2020)
6. El-Amrani, M., Seaid, M.: Convergence and stability of finite element modified method of characteristics for the incompressible Navier-Stokes equations. J. Numer. Math. **15**(2), 101–135 (2007)
7. El-Amrani, M., Seaid, M.: Numerical simulation of natural and mixed convection flows by Galerkin-characteristic method. Int. J. Numer. Meth. Fluids. **53**(12), 1819–1845 (2007)
8. El-Amrani, M., Seaid, M.: An essentially non-oscillatory semi-Lagrangian method for tidal flow simulations. Int. J. Numer. Meth. Eng. **81**(7), 805–834 (2010)
9. El-Amrani, M., Seaid, M.: An L^2-projection for the Galerkin-characteristic solution of incompressible flows. SIAM J. Sci. Comput. **33**(6), 3110–3131 (2011)
10. El-Amrani, M., Seaid, M.: A finite element semi-Lagrangian method with L^2 interpolation. Int. J. Numer. Meth. Eng. **90**(12), 1485–1507 (2012)
11. Fletcher, C.A.: Generating exact solutions of the two-dimensional Burgers equations. Int. J. Numer. Meth. Fluids. **3**(3), 213–216 (1983)
12. Khan, A., Kanschat, G.: A robust a posteriori error estimator for divergence-conforming discontinuous Galerkin methods for the oseen equation. SIAM J. Numer. Anal. **58**(1), 492–518 (2020)
13. Min, C., Gibou, F.: A second order accurate projection method for the incompressible Navier-Stokes equations on non-graded adaptive grids. J. Comput. Phys. **219**(2), 912–929 (2006)
14. Nielsen, P.N., Gersborg, A.R., Gravesen, J., Pedersen, N.L.: Discretizations in isogeometric analysis of Navier-Stokes flow. Comput. Methods Appl. Mech. Eng. **200**(45–46), 3242–3253 (2011)
15. Patel, C.G., Sarkar, S., Saha, K.S.: Mixed convective vertically upward flow past side-by-side square cylinders at incidence. Int. J. Heat Mass Transf. **127**, 927–947 (2018)

16. Patro, S.K., Selvam, R.P., Bosch, H.: Adaptive h-finite element modeling of wind flow around bridges. Eng. Struct. **48**, 569–577 (2013)
17. Pironneau, O.: On the transport-diffusion algorithm and its applications to the Navier-Stokes equations. Numer. Math. **38**(3), 309–332 (1982)
18. Popinet, S.: Gerris: a tree-based adaptive solver for the incompressible Euler equations in complex geometries. J. Comp. Phys. **190**(2), 572–600 (2003)
19. Selim, K., Logg, A., Larson, M.G.: An adaptive finite element splitting method for the incompressible Navier-Stokes equations. Comput. Methods Appl. Mech. Eng. **209**, 54–65 (2012)
20. Temperton, C., Staniforth, A.: An efficient two-time-level Galerkin-characteristics semi-implicit integration scheme. Quart. J. Roy. Meteor. Soc. **113**, 1025–1039 (1987)
21. Wang, F., Ling, M., Han, W., Jing, F.: Adaptive discontinuous Galerkin methods for solving an incompressible Stokes flow problem with slip boundary condition of frictional type. J. Comput. Appl. Math. **371**, 112700 (2020)
22. Xiu, D., Karniadakis, G.: A semi-Lagrangian high-order method for Navier-Stokes equations. J. Comput. Phys. **172**(2), 658–684 (2001)
23. Zhao, L., Zhu, H., Mao, J., Zhang, H., Peng, D., Li, T.: A generalized simple implicit interpolation scheme in CFD for non-conforming meshes. Comput. Fluids **198**, 104390 (2020)

Numerical Investigation of Transport Processes in Porous Media Under Laminar, Transitional and Turbulent Flow Conditions with the Lattice-Boltzmann Method

Tobit Flatscher[ID], René Prieler[✉][ID], and Christoph Hochenauer[ID]

Institute of Thermal Engineering, Graz University of Technology, Inffeldgasse 25/B, 8010 Graz, Austria
rene.prieler@tugraz.at

Abstract. In the present paper the mass transfer in porous media under laminar, transitional and turbulent flow conditions was investigated using the *lattice-Boltzmann method (LBM)*. While previous studies have applied the LBM to species transport in complex geometries under laminar conditions, the main objective of this study was to demonstrate its applicability to *turbulent internal flows* including the *transport of a scalar quantity*. Thus, besides the resolved scalar transport, an additional turbulent diffusion coefficient was introduced to account for the subgrid-scale turbulent transport. A *packed-bed of spheres* and an *adsorber geometry based on μCT scans* were considered. While a two-relaxation time (TRT) model was applied to the laminar and transitional cases, the Bhatnagar-Gross-Krook (BGK) collision operator in conjunction with the Smagorinsky turbulence model was used for the turbulent flow regime. To validate the LBM results, simulations under the same conditions were carried out with ANSYS Fluent v19.2. It was found that the *pressure drop* over the height of the packed-bed were in close accordance to empirical correlations. Furthermore, the comparison of the calculated *species concentrations* for all flow regimes showed good agreement between the LBM and the results obtained with Ansys Fluent. Subsequently, the proposed extension of the Smagorinsky turbulence model seems to be able to predict the scalar transport under turbulent conditions.

Keywords: Turbulent scalar transport · Lattice-Boltzmann method · Porous media · Laminar flow · Turbulent flow · Turbulent relaxation time

We sincerely thank *Markus Pieber* for the PBG simulation and his assistance with the Fluent finite-volume simulations as well as *Dr. Bernd Oberdorfer* from Österreichisches Gießerei-Institut (ÖGI), Leoben (Austria), for the generation of the raw μCT scans. This work was financially supported by the *Austrian Research Promotion Agency (FFG)*, 'Simulation der Wärmetransportvorgänge in Hochtemperaturprozessen und porösen Medien mittels lattice-Boltzmann Methode' (project 872619, eCall 22609361).

© Springer Nature Switzerland AG 2021
M. Paszynski et al. (Eds.): ICCS 2021, LNCS 12747, pp. 244–257, 2021.
https://doi.org/10.1007/978-3-030-77980-1_19

1   Introduction

In the chemical and process industry, as well as energy applications, porous media are widely used to ensure that chemical reactions, species transport and heat transfer meet certain specifications. Tubular fixed-bed reactors are a common reactor type, where cylindrical tubes filled with catalyst pellets are continuously flowed through by a (reactive) fluid. In the process the fluid takes part in endothermic (exothermic) surface reactions that require an effective heat transfer into (out of) the system. The designer wishes to understand, quantify and control the chemical and physical phenomena inside the porous media in order to maintain a stable process. Over the last decade numerical methods have established themselves as effective tools to obtain estimates and partial insight into isolated phenomena and simplified processes that can't be directly accessed by experiments. In particular simulations of fluid flows through fully resolved geometries using Computational Fluid Dynamics (CFD) may contribute to a more fundamental understanding of the topic by taking local flow effects and their impact on the reaction process into account. [17] Such methods may be used to determine the effect of particles and their arrangement on macroscopic phenomena such as heat transfer, pressure drop, axial dispersion, surface reactions etc.

Traditionally a fluid is modelled as a continuum - a dense viscous bulk - that is dominated by the conservation of mass, momentum and energy on a macroscopic level - and may be described by non-linear partial differential equations in the macroscopic unknowns, the Navier-Stokes equations. For an *incompressible Newtonian fluid* ($\rho = const.$), which is assumed for the rest of this paper, the energy equation decouples to a simple advection-diffusion equation and the mass and momentum conservation equations take a very favourable form:

$$\sum_{j \in \mathcal{D}} \frac{\partial u_j}{\partial x_j} = 0 \quad (1a) \qquad \frac{\partial u_i}{\partial t} + \sum_{j \in \mathcal{D}} u_j \frac{\partial u_i}{\partial x_j} = -\frac{1}{\rho} \frac{\partial p}{\partial x_i} + \sum_{j \in \mathcal{D}} \frac{\partial}{\partial x_j} \left(\nu \frac{\partial u_i}{\partial x_j} \right) \quad (1b)$$

For a numerical simulation the governing equations the continuous equations have to be transformed into a system of algebraic equation by discretising in time and space. Although *conventional CFD methods*, such as the finite volume method (FVM) on unstructured grids, have become a standard, they have several *limitations* in particular in modelling microscopic phenomena, grid generation for complex geometries as well as regarding computational speed and parallel scalability.

Over the years multiple particle-based methods have emerged, where the fluid is represented by discrete particles in the form of single atoms, molecules or artificial clusters of molecules instead of a continuum that interact on a short range with each other. One of the most prominent, the *incompressible lattice Boltzmann method (LBM)* was *initially investigated* in this context by several authors as an alternative to directly continuum-based methods due to its flexibility and computational efficiency. In particular Zeiser et al. initially investigated flows through porous beds of spherical particles at low Reynolds numbers [8,26] with the basic Bhatnagar-Gross-Krook (BGK) collision operator, [2] that suffers from a viscosity dependent wall position and instabilities for higher Reynolds (Re) numbers, but it was soon discarded due to the lack of a consistent inclusion of heat transfer as well as the *inherently transient nature* and therefore, an unjustifiable computational burden for steady-state creeping flows. Since then only Caulkin et al. have investigated the LBM for flows with non-spherical particles [4] and thus, *high-Reynolds number flows through porous beds with the LBM have never been considered.* With the advent of LBM large-eddy turbulent models [16] and novel collision operators, such as the entropic [18] and cumulant [9] collision operators, the simulation of turbulent external fluid flows using the LBM has made significant advances in recent years. Some of the resulting studies also included the *turbulent* transport of a scalar quantity (e.g. [25]) but - to the best of the authors' knowledge - in all previous studies this was implemented through a coupled Finite Difference Method (FDM) simulation.

1.1 Objectives of the Present Study

Contrary to the aforementioned publications the present paper tried to demonstrate the abilities of LBM to predict the transitional and turbulent fluid flow and species transport in realistic porous geometries. Laminar, transitional and turbulent conditions were considered in order to point out that *modern LBM is not limited to low Reynolds numbers.* Instead in particular its application to *turbulent flows in porous media is computationally very attractive* and can be carried out on consumer-grade hardware. In the process we went on to illustrate that the *additional turbulent transport of a scalar quantity can be considered directly in LBM* by coupling a second population for advection-diffusion to the hydrodynamic population with a turbulent Schmidt number. This way it is not necessary to implement a FDM solver but instead most of the LBM code can be re-used.

The present paper is therefore structured as follows: The brief introduction to the numerical modelling with LBM including the used LES model (Sect. 2) is followed by a short paragraph about the particle Reynolds number and the considered packed-beds (Sect. 3). Then the numerical simulation of the fluid flow and species transport under laminar, transitional and turbulent conditions in the packed-bed of spheres is discussed and results obtained by LBM are compared to ones from commercial FVM code (Sect. 4.1). Finally in Sect. 4.2 a simple trick is presented in order to increase the simulation domain of a realistic adsorber bed and as a proof of concept a realistic porous adsorber bed with a tube-to-particle ratio of $r_{tp} \approx 13.39$ is investigated (Sect. 4.2).

2 Numerical Approach - Incompressible Lattice-Boltzmann Method

2.1 LBM for Computational Fluid Dynamics

Ludwig Boltzmann introduced an evolution equation for dilute gases based on an extended concept of density, the single-particle probability distribution

$f(\boldsymbol{x}, \boldsymbol{\xi}, t) := \frac{dN_p}{d\boldsymbol{x}\,d\boldsymbol{\xi}}$ and an equation describing the propagation due to the free motion and collisions of such a distribution, the Boltzmann equation, in 1872. [3] Macroscopic variables like density and velocity emerge as expected values (moments) from the distribution. Since continuum-based flows are assumed to be close to equilibrium, the macroscopic behaviour for dense fluids ($Kn := l_{mfp}/L \to 0$), given by the Navier-Stokes equations, can be recovered by applying an asymptotic *perturbation analysis* ($\epsilon \propto Kn$), $f = \sum_{n=0}^{\infty} \epsilon^n f^{(n)}$ referred to as *Chapman-Enskog expansion*, [5] around the equilibrium $f^{(0)} := f^{(eq)}$ and recombining the conserved quantities.

After the success of the closely-related *cellular automata*, the framework of the Boltzmann equation was used to construct a physically consistent *fictional gas* with convenient numerical properties: [20] In the second-order accurate *lattice-Boltzmann* method (LBM) the Boltzmann equation is discretised in space $\mathcal{D} := \{x, y, z\}$ into regular velocity subsets, *lattices* $\mathcal{L} := \{c_\alpha\}$, that only interact with their *nearest neighbourhood*, defined by $\Delta \boldsymbol{x}_\alpha := \{\Delta_t\, c_\alpha\}$. The time Δ_t and spatial step Δ_x are chosen to unity ("lattice units") respectively and the populations now travel with a *single speed* $c := \Delta_x/\Delta_t$ on a discrete *equidistant grid* and collide with populations from neighbouring nodes. All continuous variables have to be mapped to discrete space while retaining exactness for the main hydrodynamic quantities. They may be constructed using expected values, moments of the discrete distribution functions,

$$\rho = \Delta^3 \sum_{\alpha \in \mathcal{L}} f_\alpha \qquad (2a) \qquad\qquad \rho_0 u_i = \Delta^3 \sum_{\alpha \in \mathcal{L}} c_{i\alpha} f_\alpha \qquad (2b)$$

where Δ is introduced as a velocity element for the consistency of physical units only: The sum reflects a quadrature of the continuous integral.

In this study the three-dimensional D3Q19-lattice [22] is used

$$\boldsymbol{c}_\beta = \begin{bmatrix} c_{x\beta} \\ c_{y\beta} \\ c_{z\beta} \end{bmatrix} = c \begin{bmatrix} 0 & 1 & 0 & 0 & 1 & 1 & 1 & 1 & 0 & 0 \\ 0 & 0 & 1 & 0 & 1 & -1 & 0 & 0 & 1 & 1 \\ 0 & 0 & 0 & 1 & 0 & 0 & 1 & -1 & 1 & -1 \end{bmatrix} \qquad (3)$$

where $\beta \in [0, 9]$ and their opposite directions $c_{\bar{\beta}} = -c_\beta$ for $\bar{\beta} \in [1, 9]$ form the 19 lattice velocities $\alpha := \{\beta, \bar{\beta}\}$. The length of different vectors c_α is accounted for by weighting each direction with the corresponding weights

$$w_\beta = \begin{bmatrix} \frac{1}{3} & \frac{1}{18} & \frac{1}{18} & \frac{1}{18} & \frac{1}{36} & \frac{1}{36} & \frac{1}{36} & \frac{1}{36} & \frac{1}{36} & \frac{1}{36} \end{bmatrix}^T \qquad (4)$$

which can be interpreted as corrected particle masses. The resulting discrete evolution equation can be separated into two steps: The collision of all particle distributions at a node and the streaming to its neighbouring nodes. The collision step is modelled through a linear relaxation towards a *discrete Maxwell-Boltzmann equilibrium* obtained by a *low Mach number expansion* where - for consistency - the terms of higher order stemming from pressure fluctuations $\Delta p = (\rho - \rho_0)/c_s^2$ are neglected as well (incompressible equilibrium) [14]

$$f_\alpha^{(eq)}(\boldsymbol{x}, t) = \frac{w_\alpha}{\Delta^3} \left[\rho + \rho_0 \Big(\frac{(\boldsymbol{c}_\alpha \cdot \boldsymbol{u})}{c_s^2} + \frac{(\boldsymbol{c}_\alpha \cdot \boldsymbol{u})^2}{2c_s^4} - \frac{\boldsymbol{u}^2}{2c_s^2} \Big) \right] + \mathcal{O}(Ma^2). \qquad (5)$$

$c_s := \sqrt{\left(\frac{\partial p}{\partial \rho}\right)_T} = \sqrt{R_m T}$ corresponds to the iso-thermal speed of sound of an ideal gas that in case of the most common discretisations can be found to correspond to $c_s = c/\sqrt{3}$. For the turbulent large-eddy simulations the BGK collision operator with a single relaxation time τ^+ [2]

$$f_\alpha(\boldsymbol{x} + \Delta_t\, \boldsymbol{c}_\alpha, t + \Delta_t) = f_\alpha(\boldsymbol{x}, t) + \frac{\Delta_t}{\tau^+}(f_\alpha^{(eq)} - f_\alpha) + \mathcal{O}(\Delta_x^2) + \mathcal{O}(\Delta_t^2) \quad (6)$$

was used while for the simulations considering laminar flow conditions the two-relaxation time (TRT) model was chosen [11], which applies an eigen-decomposition and relaxes even (+) and odd (−) hydrodynamic moments $\sum_\alpha c_{i\alpha}^n f_\alpha$ at individual rates τ^+ and τ^-.

Through a Chapman-Enskog expansion [14] the conservation equations for incompressible flow containing an error term $\mathcal{O}(Ma^3)$ can be found to govern this discrete system where the equation of state is given by $p = \rho c_s^2$ and the kinematic viscosity can be redefined absorbing the emerging discretisation error as

$$\nu = \frac{\mu}{\rho_0} = \left(\frac{\tau^+}{\Delta_t} - \frac{1}{2}\right) c_s^2 \Delta_t. \quad (7)$$

Now this numerical scheme can be used to *approximate the solution of incompressible fluid flows by allowing controlled compressibility*. Neglecting body forces the only characteristic number of relevance for single component fluid flow is the *Reynolds number* that is used to set the relaxation time by enforcing the viscosity $\nu_0 = (U\, L)\, /Re$. The corresponding temporal resolution is set by choosing the characteristic simulation velocity U within the stability limit $Ma = U/c_s << 1$. The simulation units can be converted to physical units according to the *law of similarity*.

Smagorinsky Large-Eddy Turbulence Model. The Smagorinsky *subgrid-scale model* is an eddy-viscosity based model, which applies low-pass filtering of the governing equations while smaller unresolved scales are modelled using an *additional isotropic dissipation*, the turbulent viscosity ν_T. As the explicit LBM algorithm on a regular grid naturally uses a high temporal and spatial resolution and furthermore the strain-rate can be calculated locally without interpolation (see Eq. 8)

$$S_{ij} := \frac{1}{2}\left(\frac{\partial u_i}{\partial x_j} + \frac{\partial u_j}{\partial x_i}\right) = -\frac{1}{2\rho_0 c_s^2 \tau} \Pi_{ij}^{(1)} + \mathcal{O}(Ma^3) \quad (8)$$

this model can be included fairly easily into an existing LBM simulation by including an additional local turbulent relaxation time τ_T (see Eqs. 9 and 10, where C is the Smagorinsky parameter) [16].

$$\nu = \left(\frac{\tau}{\Delta_t} - \frac{1}{2}\right) c_s^2 \Delta_t = \underbrace{\left(\frac{\tau^+}{\Delta_t} - \frac{1}{2}\right) c_s^2 \Delta_t}_{\nu_0} + \underbrace{\tau_T\, c_s^2}_{\nu_T} \quad (9)$$

$$\tau_T \approx \frac{1}{2} \left(\sqrt{\tau^{+\,2} + \frac{2\sqrt{2}(C\Delta_{\bar{x}})^2}{\rho_0 c_s^4} \sqrt{\sum_{i,j \in \mathcal{D}} \Pi_{ij}^{(1)} \, \Pi_{ij}^{(1)}}} - \tau^+ \right) \tag{10}$$

The first-order contribution $\Pi_{ij}^{(1)}$ to the incompressible momentum flux tensor $\Pi_{ij} := \rho u_i u_j - p\delta_{ij} + 2\mu S_{ij}$ is given by Eq. 11, where the first order contribution $f_\alpha^{(1)}$ is approximated by the entire non-equilibrium part $f_\alpha^{(neq)} := f_\alpha - f_\alpha^{(eq)}$ from the previous time step.

$$\Pi_{ij}^{(1)} = \Delta^3 \sum_{\alpha \in \mathcal{L}} c_{i\alpha} c_{j\alpha} f_\alpha^{(1)} \tag{11}$$

2.2 LBM for Mass Transfer

The momentum equation can be seen as an advection-diffusion equation for the momentum with a corresponding "diffusion coefficient", the viscosity ν. Similarly an artificial algorithm exhibiting the desired macroscopic behaviour on a continuum level can be constructed for other quantities that could be seen as advected by a flow field and diffused. Similar to the hydrodynamic flow this can be done for the mass-fraction $\Upsilon := m_{comp}/m_{tot}$ using a second population g_α while enforcing the basic property $\Upsilon := \Delta^3 \sum_{\alpha \in \mathcal{L}} g_\alpha$. With a relatively simple Chapman-Enskog expansion [7] the following correlation of the diffusion coefficient and the anti-symmetric relaxation time λ^- may be found

$$D = \left(\frac{\lambda^-}{\Delta_t} - \frac{1}{2} \right) c_s^2 \Delta_t. \tag{12}$$

Such a scalar quantity can be useful to determine *dispersion properties of porous media*, e.g. by calculating the cumulative residence time distribution $F(t)$ by means of *virtual tracer step-experiments*.

Turbulent Scalar Transport. Analogous to turbulent fluid flows, also turbulent scalar transport leads to a closure problem. The simplest way of modelling the unknown term $\partial \overline{u_j' \Upsilon'}/\partial x_j$ [24] is by assuming an additional constant turbulent diffusion coefficient D_T that is modelled according to the *Reynolds analogy* with a *turbulent Schmidt number* $Sc_T := 0.2 - 2.5$ [12], given in Eq. 13.

$$D_T := \frac{\nu_T}{Sc_T} \tag{13}$$

Therefore we propose an extension to the advection-diffusion LBM in an analogous way to fluid flow with an additional turbulent relaxation time λ_T for mass-transfer (see Eq. 14).

$$D = \left(\frac{\lambda}{\Delta_t} - \frac{1}{2} \right) c_s^2 \Delta_t = \underbrace{\left(\frac{\lambda^-}{\Delta_t} - \frac{1}{2} \right) c_s^2 \Delta_t}_{D_0} + \underbrace{\lambda_T c_s^2}_{D_T} \tag{14}$$

Based on these equations the *additional turbulent relaxation time for scalar transport* λ_T equals to a rescaled turbulent relaxation time of the fluid flow τ_T. Assuming an equal speed of sound of the hydrodynamic and diffusion lattice this results in Eq. 15.

$$\lambda_T = \frac{\tau_T}{Sc_T} \tag{15}$$

For the simulations the product of Smagorinsky constant and filter width is specified as $C\,\Delta_{\overline{x}} = 0.15$ while the laminar and turbulent Schmidt numbers are set to $Sc = 1$ and $Sc_T = 0.7$, respectively.

3   Porous Media - Simulation Domain

Confined flows in porous geometries are inherently different from free flows. Many characteristics such as velocity distributions depend on the precise bed morphology. For particle flows a specific Reynolds number, the particle Reynolds number, can be defined as

$$Re_p := \frac{U\,D_s}{\nu_0} \tag{16}$$

where the characteristic velocity U is the unperturbed velocity at some distance from the particle and the characteristic length is the diameter of a sphere of equivalent volume D_s. To define the different flow regimes Dybbs and Edwards [6] introduced the interstitial Reynolds number $Re_\phi := |\boldsymbol{u}|\,D_s/\nu_0$, which for isotropic media degenerates to $Re_\phi \approx Re_p/\phi$. The different regimes are then given by: *Viscous flow* ($Re_\phi \lesssim 1$), *steady laminar* inertial regime ($10 \lesssim Re_\phi \lesssim 150$), *unsteady laminar* inertial regime with oscillating behaviour ($150 \lesssim Re_\phi \lesssim 300$) and unsteady chaotic *turbulent* flow ($Re_\phi \gtrsim 300$).

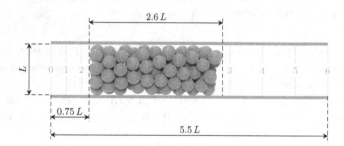

Fig. 1. Geometry made of spheres with its inlet (left) and six planes where the species mass fraction was tracked

Packed-Bed of Spheres. In order to obtain an artificial porous medium, a tube ($D_t = 15\,\mathrm{mm}$) was filled virtually with 105 spheres with a diameter of $D_s = 4\,\mathrm{mm}$ leading to a tube-to-particle ratio $r_{tp} := D_t/D_p$ of 3.75. For this purpose, the Packed Bed Generator PBG V.2 for Blender by B. Partopour and

A.G. Dixon [21] was used. The generated geometry, resulting in a void fraction of $\phi = 0.45$, is displayed in Fig. 1. For the simulation with Fluent v19.2 the numerical grid consists of 8.5 million cells while for the LBM simulation meshes with a characteristic length of 146 and 196 lattice units were used. Tracer step-experiments were simulated at four different Reynolds numbers in the case of laminar viscous ($Re_p = 1$), steady laminar ($Re_p = 10$), unsteady transitional ($Re_p = 100$) and unsteady turbulent ($Re_p = 1000$) conditions. At the inlet a *block velocity profile* (in the LBM simulation with a characteristic velocity of $U_{lb} = 0.005$) was imposed in the form of a *Guo's non-equilibrium extrapolation boundary condition* [13] and a *Dirichlet boundary condition* (given concentration value) for the species with an *anti-bounce-back* boundary condition (ABB) [10]. At the outlet Guo's method was chosen to enforce a *constant pressure* and the *copying of all the populations* of the neighbouring fluid node was used to impose a second-order accurate *zero-gradient outlet* for the mass transfer population. Solid walls were modelled for both, the fluid flow and the species, with *half-way bounce-back* (HW-BB) [15].

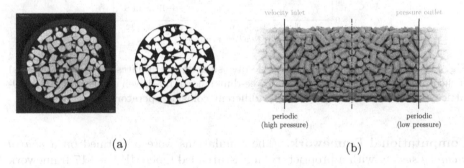

(a)   (b)

Fig. 2. Subfigure a: Cross-section of the adsorber using μCT scanning technique (left) and post-processed image (right), Subfigure b: Resulting computational domain obtained by mirroring and boundary conditions for the adsorber with about 11 million nodes

Realistic Adsorber Geometry. The second geometry considered with the LBM was a realistic adsorber bed made of hopcalite pellets. Cross-sections of the adsorber were obtained using μCT scans as given in Fig. 2a. The pellets are irregular cylinders with a diameter of approx. 1.12 mm and heights that vary between 0.83 mm and 3.88 mm (average 2.12 mm, standard deviation 0.74 mm), conforming to a tube-to particle ratio r_{tp} of around 13.39. The average equivalent spherical diameter (diameter of a sphere with the same volume as the particle) corresponds to $D_s = 1.56$ mm. Grid generation for a simulation based on FVM (Fluent v19.2) was no longer possible, however the regular grid applied in the LBM was created by voxelising the scans of the cross-sections. Ideally one would use periodic boundary conditions to determine the transport properties of such porous media. However, this is generally not possible as start and end do not fit neatly together. We therefore mirrored the domain as can be seen in Fig. 2b resulting in an undisturbed flow field with a constant void fraction. In order to obtain a realistic periodic simulation we started the simulation with a *constant velocity inlet and pressure outlet* at a given Reynolds number of $Re_p \approx 5$, calculated the pressure drop due to the porous medium and then imposed *periodic pressure drop boundaries* according to Kim and Pitsch [19]. This led to a realistic velocity profile as well as an accurate pressure drop corresponding to the chosen particle Reynolds number.

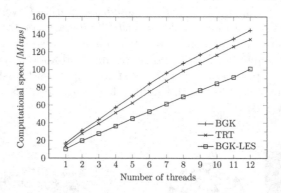

Fig. 3. Parallel scaling in million double-precision lattice updates per second (Mlups) of the proposed implementation for a three-dimensional lid-driven cavity and a $D3Q19$-lattice on a 12-core desktop system for different collision operators

Computational Framework. The simulations were performed on a *shared memory system* with a proprietary multi-threaded OpenMP C++17 framework that makes use of several optimisations: *Grid merging* for multiple interacting populations is combined with a *row-major linear memory layout* and additional optional array *padding* to match the cache line size of 64 Bytes and minimise false sharing. The padding can also be used to store a *logical mask for sparse domains* such as the considered porous media. For more predictable branch prediction behaviour every cell (if not excluded by the logical mask) performs collision and streaming while boundary conditions are imposed locally afterwards. In order to reduce the memory bandwidth for the memory-bound LBM algorithms an *A-A access pattern*, [1] where even time steps perform local collisions with a reverse read and odd steps a combined streaming-collision-streaming step with a reverse write, and 3-way spatial *loop blocking* were introduced. Furthermore to maximise the performance all these features were implemented by means of macros, templates and inline functions which are already known to the compiler at compilation time. With this implementation on a *twelve-core* Intel i9-7920X *processor* for all collision operators a *parallel scalability* of *around 90%* was obtained (see Fig. 3). A reduced framework for generic lattices that uses these optimisations and additionally manual AVX2/AVX512 vector intrinsics implementations can be found on the Github repository https://github.com/2b-t/LB-t.

4   Results and Discussion

In this section the results obtained with the in-house LBM code and FVM-based Fluent are compared with special consideration of the species transport within the packed-bed domain (see Sect. 4.1). Simulations in the packed-bed were carried out under laminar, transitional and turbulent flow regime. Since many previous studies focused on scalar transport under laminar conditions, the results presented here are focused on the turbulent fluid flow and species transport, where the applicability of modelling the turbulent scalar transport using an additional turbulent relaxation time for the mass transfer were validated. In Sect. 4.2 the LBM is also applied for the simulation within a adsorber, where a simulation with the FVM was no longer possible, again highlighting the advantage of the LBM for complex geometries.

4.1   Packed-Bed of Spheres

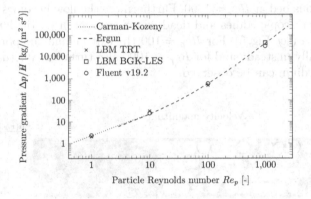

Fig. 4. Pressure drop in the packed-bed of spheres

Within this section the simulation results will be analysed considering the (i) pressure drop in the domain, (ii) the velocity profile and (iii) observation of the species concentrations at the 6 planes depicted in Fig. 1.

Pressure Drop. The pressure drop caused by the porous domain was tracked (see Fig. 4) and compared to the *empirical correlations* proposed by *Carman-Kozeny* and *Ergun* (approaches are summarized in [23]). For all considered Reynolds numbers the pressure drops determined with LBM correspond well with the empirical correlations and the Fluent simulations. For the LBM simulations up to the transitional state ($Re_p = 100$) the TRT model was used while for $Re_p = 1000$ the LES model was necessary to keep the simulations stable and describe the unresolved scales. For the FVM simulations at the highest

Reynolds number the commercial software Fluent with a $k - \omega$ URANS turbulence model was used. In all simulations the computation time of the two
approached was similar but the LBM simulations naturally required a time-step
that is 60 times smaller and therefore the much more accurate large-eddy turbulence model can be used at virtually no additional cost. A FVM large-eddy
simulation on the other hand is not computationally feasible in particular on
consumer-grade hardware like the one used in terms of this study.

Velocity Profile. In order to analyse the flow properties inside the porous
medium the time-averaged instantaneous velocity distributions for each particle
Reynolds number were calculated. For all Reynolds numbers we obtained *peak
axial velocities* that are around *twelve times higher* than the inlet velocity and
radial velocities that are roughly six times higher than the inlet velocity (see
also Fig. 5). Furthermore, while for low Reynolds number flow some parts of the
fluid might move slowly while others move significantly faster, only around 1.5%
of the fluid inside the porous medium move in opposite direction to the main
flow. With rising Reynolds number the fraction of backflow rises to up to 14% of
the entire porous bed at $Re_p = 1000$. Furthermore the flow becomes increasingly
unsteady: macroscopic eddies and dead water emerge, in particular behind the
porous medium (see Fig. 5). For $Re_p = 100$ the exit behind the porous medium
becomes visually unsteady and for $Re_p = 1000$ also turbulent fluctuations inside
the porous medium can be observed.

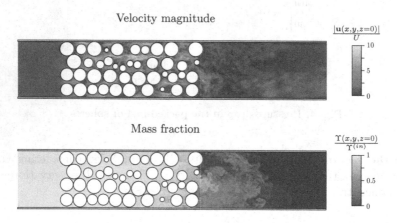

Fig. 5. Normalised instantaneous velocity field (top) and mass fraction (bottom) in a
cross-section of the large-eddy LBM simulation for $Re_p = 1000$

Species Transport in the Packed-Bed Domain. Axial Péclet numbers for
gaseous flows are known to be of the magnitude of the molecular diffusion.
Commonly in the literature values around $Pe_{ax} \approx 2$ are found. This leads to
curves of the mass fractions that in dimensional coordinates are comparably

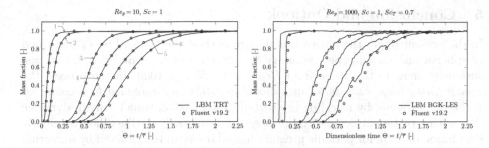

Fig. 6. Mass-flow averaged cumulative residence time distribution $F(t)$ over dimensionless time Θ (where $\bar{\tau}$ is the mean residence time) at different cross-sections obtained by LBM (solid lines) and Ansys Fluent (markers) for two different particle Reynolds numbers Re_p

similar in shape almost independent of the Reynolds number (see Fig. 6). The turbulent fluctuations and macroscopic eddies lead to fluctuations in the mass fractions but the trend of the mass fraction closely resembles the ones generated by the Fluent simulations for the two considered Reynolds numbers $Re_p = 10$ and $Re_p = 1000$.

4.2   Realistic Adsorber Geometry

Since the μCT scan of this particular porous medium is characterised by a higher void fraction in one corner of the domain, the LBM simulation showed an increased species transport in this region (see red box in Fig. 7). Due to mass continuity the higher mass flow in the short-circuit region leads to a stagnant flow in the center of the bed. This effect represents a significant deviation from the ideal plug flow, assumed in many simplified models to predict the species transport in porous media. To reveal such effects, advanced simulation methods such as the proposed LBM model are necessary. The backflow across the entire porous medium was found to only account for 1.29% of the fluid cells for the considered Reynolds number of $Re_p \approx 5$.

Fig. 7. Instantaneous iso-mass fraction $\Upsilon = 0.6$ in realistic adsorber geometry obtained my mirroring for $Re_p \approx 5$ - A short-circuit (red box) leading to a bi-modal distribution is clearly visible on top. (Color figure online)

5   Conclusion and Outlook

In the present paper the lattice-Boltzmann method was applied to a packed-bed of spheres and an adsorber geometry for the case of *laminar, transitional and turbulent flows* in the range from $Re_p = 1$ to $Re_p = 1000$. In the process the *Smagorinsky large-eddy turbulence model in LBM was extended to account for turbulent diffusion* by rescaling the turbulent relaxation time for hydrodynamic flow with a turbulent Schmidt number and it was briefly outlined how a suitable simulation domain for periodic pressure boundaries can be obtained by mirroring the μCT scan. The obtained results agreed well with the empirical correlations for the *pressure drop* and the species transport predicted by the commercial software Fluent. The application of LBM clearly reaches far beyond the laminar simulations in the creeping and low-Reynolds number regime and is a powerful tool for transitional and turbulent simulations inside porous media.

References

1. Bailey, P., Myre, J., Walsh, S., Lilja, D., Saar, M.: Accelerating lattice-boltzmann fluid flow simulations using graphics processors, pp. 550–557 (September 2009). https://doi.org/10.1109/ICPP.2009.38
2. Bhatnagar, P.L., Gross, E.P., Krook, M.: A model for collision processes in gases. i. small amplitude processes in charged and neutral one-component systems. Phys. Rev **94**(3), 511–525 (1954). https://doi.org/10.1103/PhysRev.94.511
3. Boltzmann, L.: Weitere Studien über das Wärmegleichgewicht unter Gas-molekülen. In: Brush, S.G. (ed.) Kinetische Theorie II: Irreversible Prozesse Einführung und Originaltexte, pp. 115–225. WTB Wissenschaftliche Taschenbücher, Vieweg+Teubner Verlag, Wiesbaden (1970). https://doi.org/10.1007/978-3-322-84986-1_3
4. Caulkin, R., Jia, X., Fairweather, M., Williams, R.: Predictions of porosity and fluid distribution through non spherical-packed columns. AIChE J. **58**(5), 1503–1512 (2012). https://doi.org/10.1002/aic.12691
5. Chapman, S., Cowling, T.G., Burnett, D.: The Mathematical Theory of Non-uniform Gases: An Account of the Kinetic Theory of Viscosity, Thermal Conduction and Diffusion in Gases. Cambridge University Press, Cambridge (1990)
6. Dybbs, A., Edwards, R.V.: A new look at porous media fluid mechanics – Darcy to turbulent. In: Bear, J., Corapcioglu, M.Y. (eds.) Fundamentals of Transport Phenomena in Porous Media, pp. 199–256. NATO ASI Series, Springer, Dordrecht (1984). https://doi.org/10.1007/978-94-009-6175-3_4
7. Flekkøy, E.G.: Lattice Bhatnagar-Gross-Krook models for miscible fluids. Phys. Rev. E **47**(6), 4247–4257 (1993). https://doi.org/10.1103/PhysRevE.47.4247
8. Freund, H., Bauer, J., Zeiser, T., Emig, G.: Detailed simulation of transport processes in fixed-beds. Ind. Eng. Chem. Res. **44**(16), 6423–6434 (2005). https://doi.org/10.1021/ie0489453
9. Geier, M., Sch, M.: The cumulant Lattice-Boltzmann equation in three dimensions: theory and validation. Comput. Math. Appl. **70**(4), 507–547 (2015). https://doi.org/10.1016/j.camwa.2015.05.001
10. Ginzburg, I.: Generic boundary conditions for Lattice-Boltzmann models and their application to advection and anisotropic dispersion equations. Adv. Water Resour. **28**(11), 1196–1216 (2005). https://doi.org/10.1016/j.advwatres.2005.03.009

11. Ginzburg, I.: Two-relaxation-time Lattice-Boltzmann scheme: about parametrization, velocity, pressure and mixed boundary conditions. Commun. Comput. Phys. **3**, 427–478 (2008)
12. Gualtieri, C., Angeloudis, A., Bombardelli, F., Jha, S., Stoesser, T.: On the values for the turbulent Schmidt number in environmental flows. Fluids **2**(2), 17 (2017). https://doi.org/10.3390/fluids2020017
13. Guo, Z.L., Zheng, C.G., Shi, B.C.: Non-equilibrium extrapolation method for velocity and pressure boundary conditions in the Lattice-Boltzmann method. Chin. Phys. **11**(4), 366–374 (2002). https://doi.org/10.1088/1009-1963/11/4/310
14. He, X., Luo, L.S.: Lattice-Boltzmann model for the incompressible Navier-Stokes equation. J. Stat. Phys. **88**(3), 927–944 (1997). https://doi.org/10.1023/B:JOSS.0000015179.12689.e4
15. He, X., Zou, Q., Luo, L.S., Dembo, M.: Analytic solutions of simple flows and analysis of non-slip boundary conditions for the Lattice-Boltzmann BGK model. J. Stat. Phys. **87**(1), 115–136 (1997). https://doi.org/10.1007/BF02181482
16. Hou, S., Sterling, J., Chen, S., Doolen, D.G.: A Lattice-Boltzmann Subgrid model for high reynolds number flows. Fields Inst. Commun. **6**, 151–166 (1994)
17. Jurtz, N., Kraume, M., Wehinger, G.D.: Advances in fixed-bed reactor modeling using particle-resolved computational fluid dynamics (CFD). Rev. Chem. Eng. **35**(2), 139–190 (2019). https://doi.org/10.1515/revce-2017-0059
18. Karlin, I.V., Gorban, A.N., Succi, S., Boffi, V.: Maximum entropy principle for lattice kinetic equations. Phys. Rev. Lett. **81**(1), 6–9 (1998). https://doi.org/10.1103/PhysRevLett.81.6
19. Kim, S., Pitsch, H.: A generalized periodic boundary condition for Lattice-Boltzmann method simulation of a pressure-driven flow in a periodic geometry. Phys. Fluids **19**, 108101 (2007). https://doi.org/10.1063/1.2780194
20. McNamara, G.R., Zanetti, G.: Use of the Boltzmann equation to simulate lattice-gas automata. Phys. Rev. Lett. **61**(20), 2332–2335 (1988). https://doi.org/10.1103/PhysRevLett.61.2332
21. Partopour, B., Dixon, A.G.: An integrated workflow for numerical generation and meshing of packed-beds of non-spherical particles: applications in chemical reaction engineering. iN: 2017 AIChE Annual Meeting (November 2017)
22. Qian, Y.H., D'Humières, D., Lallemand, P.: Lattice BGK models for Navier-Stokes equation. Europhys. Lett. (EPL) **17**(6), 479–484 (1992). https://doi.org/10.1209/0295-5075/17/6/001
23. Rhodes, M.: Introduction to Particle Technology, 2nd edn. Wiley, Chichester (April 2008)
24. Roberts, P.J.W., Webster, D.R.: Turbulent Diffusion. Environmental Fluid Mechanics: Theories and Applications, pp. 7–45 (2002). https://cedb.asce.org/CEDBsearch/record.jsp?dockey=0277542
25. Uphoff, S.: Development and Validation of Turbulence Models for Lattice-Boltzmann Schemes. Ph.D. Thesis, Technische Universität Braunschweig (January 2013). https://publikationsserver.tu-braunschweig.de/receive/dbbs_mods_00055260
26. Zeiser, T., et al.: Analysis of the flow field and pressure drop in fixed-bed reactors with the help of Lattice-Boltzmann simulations. Philos. Trans. Ser. A Math. Phys. Eng. Sci. **360**, 507–520 (2002). https://doi.org/10.1098/rsta.2001.0945

A Study on a Marine Reservoir and a Fluvial Reservoir History Matching Based on Ensemble Kalman Filter

Zelong Wang[1,2,3,4(✉)], Xiangui Liu[1,2,3], Haifa Tang[3], Zhikai Lv[3], and Qunming Liu[3]

[1] University of the Chinese Academy of Sciences, Beijing 100049, China
wangzelong18@mails.ucas.ac.cn
[2] Institute of Porous Flow and Fluid Mechanics, Chinese Academy of Sciences, Langfang 065007, China
[3] PetroChina Research Institute of Petroleum Exploration and Development, Beijing 100083, China
[4] CNPC International Ltd., Beijing 100034, China

Abstract. In reservoir management, utilizing all the observed data to update the reservoir models is the key to make accurate forecast on the parameters changing and future production. Ensemble Kalman Filter (EnKF) provides a practical way to continuously update the petroleum reservoir models, but its application reliability in different reservoirs types and the proper design of the ensemble size are still remain unknown. In this paper, we mathematically demonstrated Ensemble Kalman Filter method; discussed its advantages over standard Kalman Filter and Extended Kalman Filter (EKF) in reservoir history matching, and the limitations of EnKF. We also carried out two numerical experiments on a marine reservoir and a fluvial reservoir by EnKF history matching method to update the static geological models by fitting bottom-hole pressure and well water cut, and found the optimal way of designing the ensemble size. A comparison of those the two numerical experiments is also presented. Lastly, we suggested some adjustments of the EnKF for its application in fluvial reservoirs.

Keywords: History matching · EnKF · Marine reservoir · Fluvial reservoir

1 Introduction

History matching is the act of adjusting a model of a reservoir until it closely reproduces the past behavior of a petroleum reservoir, which is a crucial component in reservoir management. Once a geological model has been history matched, it can be used to simulate future reservoir behavior with a higher degree of confidence, particularly if the adjustments are constrained by known geological properties in the reservoir. With the advancement of the computer science, automatic history matching uses computer algorithms to solve the optimization problem based on the reasonable objective function. By applying proper history matching approaches, geological settings of the reservoir can be preserved.

© Springer Nature Switzerland AG 2021
M. Paszynski et al. (Eds.): ICCS 2021, LNCS 12747, pp. 258–267, 2021.
https://doi.org/10.1007/978-3-030-77980-1_20

Ensemble Kalman Filter (EnKF) works with an ensemble of the reservoir models, with its inherent forecasting and updating process (correct the forecasted data with the new measurements), the EnKF gives a suite of realizations of reservoir models which are consistent with the prior geological settings and the dynamic data. Recently, the EnKF has gain increasing attention for real time reservoir management and history matching, using data from the permanent down whole sensors. However, there are still great challenges of using the EnKF. For example, with limited computational power, a small ensemble size is viable in the practical history matching, but a small ensemble size also leads to an erroneous result. EnKF also may fail to detect facies boundaries.

2   Theoretical Formulations

2.1   Kalman Filter

We first review the formulations of Kalman filter, which is a set of mathematical equations that provides an efficient recursive estimation of the states of a process, in a way that with minimized mean and standard variances.

For a linear stochastic model

$$x_k = Ax_{k-1} + Bu_{k-1} + w_{k-1} \tag{1}$$

with a measurement

$$z_k = Hx_k + v_k \tag{2}$$

where variable x_k is the state variable at time k, matrix A is the state transition matrix, B is the control operator, z_k is the measurement vector at time k and H is the measurement operator. The random variables w_{k-1} and v_k represent the process and measurement noise respectively, and they are assumed to be independent, white, with a standard Gaussian distribution with the noise covariance Q and R respectively.

Define $\hat{x}_k^- \in R^n$ (note the "super minus") to be our a prioristate estimate at step k given knowledge of the process prior to step k, and to be our a posteriori state estimate at step k given measurement. We can then define a priori and posteriori estimate errors as

$$e_k^- = x_k - \hat{x}_k^- \tag{3}$$

$$e_k = x_k - \hat{x}_k \tag{4}$$

The priori estimate error covariance and the a posteriori estimate error covariance are

$$P_k^- = E[e_k^- e_k^{-T}] \tag{5}$$

$$P_k = E[e_k e_k^T] \tag{6}$$

The equation that computes an a posteriori state estimate $\hat{x}_k^-$ as a linear combination of an a priori estimate $\hat{x}_k^-$ and a weighted difference between an actual measurement z_k and a measurement prediction $H\hat{x}_k^-$ as shown below:

$$\hat{x}_k = \hat{x}_k^- + K_k\left(z_k - H\hat{x}_k^-\right) \tag{7}$$

where the Kalman gain K is

$$K_k = P_k^- H^T \left(HP_k^- H^T + R\right)^{-1} \tag{8}$$

Kalman filter is very suitable for the linear problems, and it is a quite effective in several aspects: it supports estimations of past, present, and even future states, and it can do so even when the precise nature of the modeled system is unknown. However, Kalman filter only works for linear models and it has two main drawbacks: large dimensionality and error covariance propagation.

2.2 Ensemble Kalman Filter

Some adjustments to Kalman filter have been raised in order to apply Kalman filter in non-linear problems, and eliminate the pertaining drawbacks of the standard Kalman filter. A Kalman filter that linearizes about the current mean and covariance is referred to as an extended Kalman filter (EKF), but EKF only works to weakly non-linear problems and the computational power is even more than standard Kalman filter. For the complex non-linear problems with a large number of variables, like history matching in reservoir management, ensemble Kalman filter, which is first introduced by Evensen, provides a plausible solution.

The EnKF is a Monte Carlo method based on Markov chain approach. First, it samples many realizations from the prior probability density function. Second, for each realization, it use the model forecast function (in history matching, the forecast function is the reservoir simulator) to estimate the dynamic data at the next time step. Third, it uses those predicted realizations to calculate the approximation of the predicted covariance.

Similar to Kalman filter, the EnKF also consists of two sets of equations: the state forecast equation, and the update equation. For nonlinear models, the time update equations are no longer linear.

$$\hat{x}_k^- = f(\hat{x}_{k-1}) + w_{k-1} \tag{14}$$

Where $f(x)$ is the forecast function, the other notations keep the same as they are explained in Kalman filter. The measurement is assumed to be a linear relationship by adding a Gaussian white noise.

$$z_k = Hx_k + v_k \tag{15}$$

Then the update equation can be expressed as

$$\hat{x}_k = \hat{x}_k^- + K(z_k - K\hat{x}_k^-) \tag{16}$$

Here, the Kalman gain is defined the same as it is in Kalman filter. But to calculate the Kalman gain, in Monte Carlo approach, it is different from its linear case. For N realizations, the unbiased covariance is

$$P_k^- = \frac{1}{N-1}(\hat{x}_k^- - \overline{\hat{x}_k^-})(\hat{x}_k^- - \overline{\hat{x}_k^-})^T \tag{17}$$

Where "—" over denotes the statistical mean. The covariance is in a large dimension for a typical highly underdetermined history matching problem, and it takes much computational power and storage to get it. However, it is not necessary to explicitly having it. Because we substitute the estimated covariance into Eq. (16), and therefore K can be expressed as

$$K = \frac{1}{N-1}(\hat{x}_k^- - \overline{\hat{x}_k^-})(\hat{x}_k^- - \overline{\hat{x}_k^-})^T H^T \bullet [H^T \frac{1}{N-1}(\hat{x}_k^- - \overline{\hat{x}_k^-})[(\hat{x}_k^- - \overline{\hat{x}_k^-})^T H^T] + R]^{-1} \tag{18}$$

For a highly underdetermined problem, to reduce the dimension, when calculation K, combine the term $(\hat{x}_k^- - \overline{\hat{x}_k^-})^T$ and H^T together as $[(\hat{x}_k^- - \overline{\hat{x}_k^-})^T H^T]$, so K can also be expressed as

$$K = \frac{1}{N-1}(\hat{x}_k^- - \overline{\hat{x}_k^-})[(\hat{x}_k^- - \overline{\hat{x}_k^-})^T H^T] \bullet [H^T \frac{1}{N-1}(\hat{x}_k^- - \overline{\hat{x}_k^-})[(\hat{x}_k^- - \overline{\hat{x}_k^-})^T H^T] + R]^{-1} \tag{19}$$

In this way, without calculating the estimated covariance P_k^- explicitly, the computational power is reduced significantly. There is an inversion in calculating the Kalman gain, and the term $(\hat{x}_k^- - \overline{\hat{x}_k^-})^T H^T$ might be singular, therefore, by applying singular value decomposition, the pseudo-inversion might be required. The sampling error is inevitable for large problems, because in order to avoid prohibitive forward simulations, ensemble size N is usually much less than state or sometimes even observation size.

3 Numerical Experiments Setup

To study the EnKF history matching method on its application limitations and to find the optimal design of the ensemble size. We used the EnKF to update two reservoir types: one is a marine reservoir, which has a smooth distribution of permeability; the other is a fluvial reservoir, which has distinct facies boundaries. In experiments, we used the grid block-oriented parameterization and the reservoirs are represented by two 50 by 50 Cartesian grid blocks models. Figure 1 shows the permeability model of the reference marine reservoir and the reference fluvial reservoir respectively.

3.1 Build Prior Models for the Reservoirs

Both prior realizations are generated from the hard data by proper geostatistical methods. We used Sequential Gaussian simulation, and we used a training image and by sequential indicator simulation and to establish the priors of the fluvial reservoir. Figure 2 represents the one of the prior ensemble of the two reservoirs respectively.

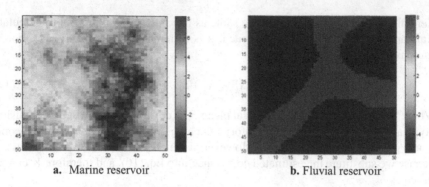

a. Marine reservoir   **b.** Fluvial reservoir

Fig. 1. The reference permeability of the marine reservoir and the fluvial reservoir.

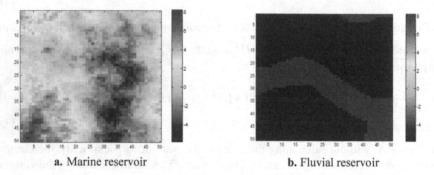

a. Marine reservoir   **b.** Fluvial reservoir

Fig. 2. The reference permeability of the marine reservoir and the fluvial reservoir.

3.2 Build Prior Dynamic Data

Typically, the dynamic data of a petroleum reservoir includes oil production, bottom-hole pressure, well water cut, oil saturation, water saturation. In this study, we used the bottom hole pressure (BHP) and water cut (WWCT) as the dynamic data. The well configuration is an inverted 9-point pattern, which consists an injector in the center and 8 producers at the edge of the reservoir. For both reservoirs, load the true permeability data for the simulator. Add the proper white noise to the outputs form the simulator. The standard deviation is 0.48 and 0.07 for BHP and WWCT noise. In this way, the observation data is generated for this project. See Fig. 3 and Fig. 4 accordingly.

4 Numerical Experiments Results

4.1 Result of EnKF History Matching for the Marine Reservoir

After using EnKF for the history matching of a marine reservoir, with the ensemble size of 10 the results are shown in Fig. 5 the computational time is 65 s; with the ensemble size of 30, the results are shown in Fig. 6, and the computational time is 106 s.

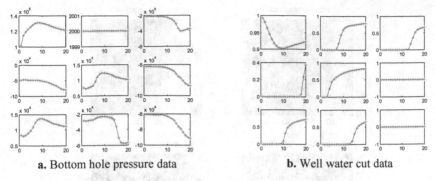

a. Bottom hole pressure data   **b.** Well water cut data

Fig. 3. The observation data of the marine reservoir.

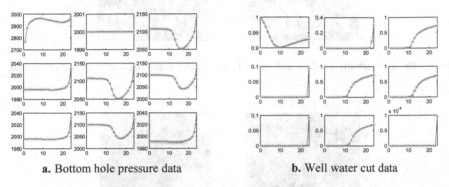

a. Bottom hole pressure data   **b.** Well water cut data

Fig. 4. The observation data of the fluvial reservoir.

With a larger ensemble size, the results are very likely to be improved. However, it also took longer computational time. We designed different ensemble sizes, and plotted the result in Fig. 7 shows the relationship between ensemble size and computational time. For a practical problem, to balance the time and accuracy, an ensemble size of 100 is a popular choice, however, it is still not large enough to eliminate the errors. Figure 8 shows the permeability model updates with the ensemble size of 100. Generally, the EnKF is more efficient than the traditional gradient based method.

Figure 9 shows the history matched BHP plot and WWCT plot respectively. The red line is the read observation data, and the green lines are the simulated dynamic data with the prior ensemble.

We can see the reliability of EnKF is highly dependent on the size of the ensemble. If it is so small that the ensemble is not statistically representative the system is said to be under sampled. Under sampling causes three major problems: inbreeding, filter divergence and spurious correlation.

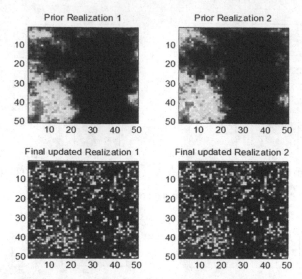

Fig. 5. The permeability model result of the EnKF history matching with the ensemble of 10.

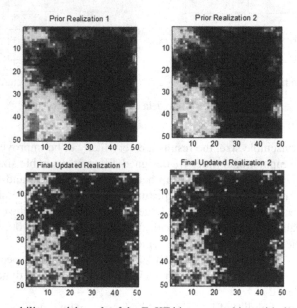

Fig. 6. The permeability model result of the EnKF history matching with the ensemble of 30.

Inbreeding refers to a problem that the analysis error covariance are inherently under-estimated after each of the observation assimilations. The Kalman gain uses a ratio of the error covariance of the forecast state and the error covariance of the observations to calculate how much emphasis or weight should be placed on the background state and how much weighting should be given to the observations. Therefore without a proper weighting for the two terms, then the adjustment of the forecast state will be incorrect.

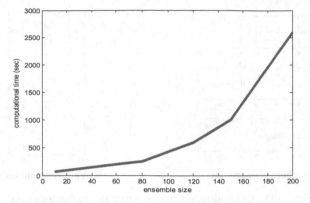

Fig. 7. The plot of ensemble size of which computational time of history matching.

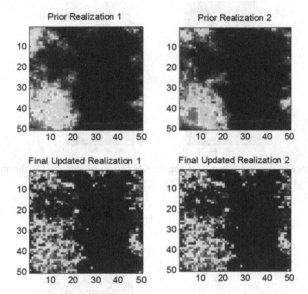

Fig. 8. The permeability model result of the EnKF history matching with the ensemble of 100.

4.2   Result of EnKF History Matching for the Fluvial Reservoir

Similarly, applying EnKF into history matching for a fluvial reservoir, with a binary permeability system, the results are shown in Fig. 10.

It is clearly indicated that even the prior has a binary permeability, only applying EnKF in history, the results will a system be the continuous permeability distribution. If we put an arbitrary threshold for the shale face and forced it into a binary system, the results still fallacious, since we lose the conductivity feature and lose continuity, as the Fig. 11. shows.

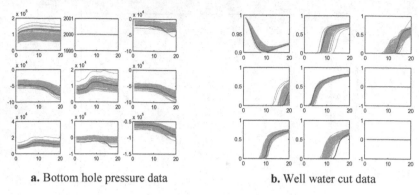

a. Bottom hole pressure data   **b.** Well water cut data

Fig. 9. The EnKF history matching result for the dynamic data

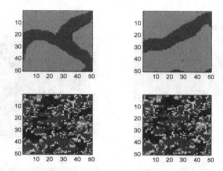

Fig. 10. The permeability model result of the EnKF history matching for the fluvial reservoir with the ensemble of 100.

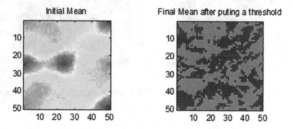

Fig. 11. The mean of the initial and updated fluvial reservoir realizations.

5   Conclusions and Recommendations

The EnKF provides a framework for real-time updating and prediction in reservoir simulation models. Every time new observations are available and are assimilated there is an improvement of the model parameters and of the associated dynamic data.

The EnKF works very well for marine reservoir which has a continuous and smooth distribution of permeability, when the prior has an efficient ensemble size, and the prior models can capture the conductivity features of the real reservoir.

To overcome the problem of under sampling, it is recommended to apply the following methods:

1) Implement covariance inflation, which can correct an underestimation in the forecast error covariance matrix. The aim is to increase the forecast error covariance by inflating, for each ensemble member, the deviation of the background error from the ensemble mean by factor. In the experiments of Whitaker and Hamill the optimal values of the inflation factors were 7% for the EnKF.
2) Covariance localization is a process of cutting off longer range correlations in the error covariance at a specified distance. It is a method of improving the estimate of the forecast error covariance. It is ordinarily achieved by applying a Schur product to the forecast error covariance matrix.

Standard EnKF can be applied in history matching for reservoirs with a continuous and smooth permeability distribution. However, for fluvial reservoirs, despite for its large computational requirement, Gradient based methods, with the first norm in its objective function, works well, and easy to implement.

Acknowledgments. This work is financially supported by Chinese National Science and Technology Major Project (2016ZX05015-005). The first author also gratefully acknowledged the University of Southern California for facilitating the research, as well as Jafarpour's professional guidance.

References

1. Kalman, R.E.: A new approach to linear filtering and prediction problems. Trans. ASME-J. Basic Eng. **82**, 35–45 (1960)
2. Evensen, G.: Sequential data assimilation with a non-linear quasi-geostrophic model using Monte Carlo methods to forecast error statistics. J. Geophys. Res. **99**, 10143–10162 (1994)
3. Evensen, G.: The ensemble Kalman filter: theoretical formulation and practical implementation. Ocean Dyn. **53**, 343–367 (2003)
4. Oliver, D.S., Chen, Y.: Recent progress on reservoir history matching: a review. J. Comput. Geosci. **15**, 185–221 (2011)
5. Deutsch, C.V., Journel, A.G.: GSLIB Geostatistical Software Library and User's Guide, pp. 119–147. Oxford University Press, New York (1998)
6. Yong, H., He, W., Guo, B.: Combining sedimentary forward modeling with sequential Gauss simulation for fine prediction of tight sandstone reservoir. Mar. Pet. Geol. **112**, 1–15 (2020)
7. Chen, Y., Oliver, D.S., Zhang, D.: Efficient ensemble-based closed-loop production optimization. SPE J. **14**(4), 634–645 (2009)
8. Aanonsen, S., Nævdal, G., Oliver, D.S., Reynolds, A.A., Vallès, B.: The ensemble Kalman filter in reservoir engineering-a review. SPE J. **14**, 393–412 (2009)
9. Jafarpour, B., Mclaughlin, D.B.: History matching with an ensemble Kalman filter and discrete cosine parameterization. Comput. Geosci. **12**(2), 227–244 (2008)
10. Wang, Z., Liu, X., Tang, H., et al.: Geophysical and production data history matching based on ensemble smoother with multiple data assimilation. Comput. Model. Eng. Sci. **123**(2), 873–893 (2020)

Numerical Simulation of Free Surface Affected by Submarine with a Rotating Screw Moving Underwater

Masashi Yamakawa[1(✉)], Kohei Yoshioka[1], Shinichi Asao[2], Seiichi Takeuchi[2], Atsuhide Kitagawa[1], and Kyohei Tajiri[1]

[1] Kyoto Institute of Technology, Matsugasaki, Sakyo-ku, Kyoto 606-8585, Japan
yamakawa@kit.ac.jp
[2] College of Industrial Technology, Nishikoya, Amagasaki, Hyogo 661-0047, Japan

Abstract. We conducted a numerical simulation of the free surface affected by the diving movement of an object such as a submarine. We have already proposed a computation method that combines the moving grid finite volume method and a surface height function method. In this case, the dive movement was expressed only as a traveling motion, not as a deformation. To express the deformation of a body underwater, the unstructured moving grid finite volume method and sliding mesh approach are combined. The calculation method is expected to be suitable for a computation with high versatility. After the scheme was validated, it was put to practical use. The free surface affected by a submarine with a rotating screw moving underwater was computed using the proposed method. Owing to the computation being for a relatively shallow depth, a remarkable deformation of the free surface occurred. In addition, the movement of the submarine body had a more dominant effect than a screw rotation on changing the shape of the free water surface.

Keywords: Free surface · Moving grid · Submarine

1 Introduction

Studying how the shape of the free water surface is affected by the movement of submerged bodies is very useful and interesting from not only an engineering perspective but also a computational science perspective. For example, such studies could be carried out in the preliminary design of the shape and placement of wave activated power generators [1]. In addition, it would be useful for designing underwater exploration submersibles [2]. Furthermore, a lot of basic research on interactions between the free surface and the motion of submerged objects have been reported [3]. Benusiglio et al. [4] investigated the drag and shape of waves caused by a moving small sphere under the water. But, as there are a lot of complicated flows with the free surface, from the perspectives on experimental equipment cost and flow reproducibility, the applied flow fields have had to be simple. In a previous study using numerical simulations, Kwag et al. [5] investigated the shape of the free surface when varying the distance between a three

© Springer Nature Switzerland AG 2021
M. Paszynski et al. (Eds.): ICCS 2021, LNCS 12747, pp. 268–281, 2021.
https://doi.org/10.1007/978-3-030-77980-1_21

dimensional airfoil and the surface and when varying the attack angle of the airfoil to the surface. In addition, Moonesun et al. [6] compared computational results and experimental results of the interaction between the free water surface and a submersible in rectilinear motion. However, the reproducibility of the motion itself and of the resulting shapes upon modeling were limited.

In general, computational methods for the interface can be classified into two broad categories: interface tracking methods [7] and interface capturing methods. In the interface tracking method as represented by the Arbitrary Lagrangian-Eulerian method, an interface is expressed directly by moving and deforming a mesh according to the motion of the interface. On the other hand, in the interface capturing method represented by the Volume of Fluid method or the level set method, an interface is expressed indirectly using a function indicating the interface on a fixed mesh. The interface capturing method is suitable for expressing a separation or large deformation of the interface. Thus, it is often used to solve such interface problems. However, it is difficult to express the interface with a moving body, because it is not easy for the method to maintain the calculation accuracy for a flow around the moving body. For this reason, this study used the interface tracking method, in which it is relatively easy to maintain the calculation accuracy around an interface. The method is used together with the unstructured moving grid finite volume method [8, 9] and the moving computational domain method to avoid calculation failures. Furthermore, the combination of these methods permits removal of body movement restrictions and generation of flexible meshes. However, the method has not been applied to a flow around a body with a complicated motion such as an oscillatory heaving motion and rotational motion yet. Thus, in this paper, it was applied to a free surface affected by a submarine with a rotating screw moving underwater.

2 Numerical Approach

2.1 Governing Equations

The governing equations are composed of the following three-dimensional (3D) continuity equation and Navier-Stokes equation for incompressible flows written in conservation law form.

$$\frac{\partial u}{\partial x} + \frac{\partial v}{\partial y} + \frac{\partial w}{\partial z} = 0 \tag{1}$$

$$\frac{\partial \mathbf{q}}{\partial t} + \frac{\partial \mathbf{E}}{\partial x} + \frac{\partial \mathbf{F}}{\partial y} + \frac{\partial \mathbf{G}}{\partial z} - \frac{1}{Re}\left(\frac{\partial \mathbf{E_v}}{\partial x} + \frac{\partial \mathbf{F_v}}{\partial y} + \frac{\partial \mathbf{G_v}}{\partial z}\right) = \mathbf{H_g} \tag{2}$$

where,

$$\mathbf{q} = \begin{pmatrix} u \\ v \\ w \end{pmatrix}, \mathbf{E} = \begin{pmatrix} u^2 + p \\ uv \\ uw \end{pmatrix}, \mathbf{F} = \begin{pmatrix} uv \\ v^2 + p \\ vw \end{pmatrix}, \mathbf{G} = \begin{pmatrix} uw \\ vw \\ w^2 + p \end{pmatrix}, \tag{3}$$

$$\mathbf{E_v} = \begin{pmatrix} u_x \\ v_x \\ w_x \end{pmatrix}, \mathbf{F_v} = \begin{pmatrix} u_y \\ v_y \\ w_y \end{pmatrix}, \mathbf{G_v} = \begin{pmatrix} u_z \\ v_z \\ w_z \end{pmatrix}, \mathbf{H_g} = \begin{pmatrix} 0 \\ 0 \\ -\frac{1}{Fr^2} \end{pmatrix}.$$

Here, $\mathbf{q}$ is a conservative quantity, $\mathbf{E}$, $\mathbf{F}$ and $\mathbf{G}$ are inviscid flux vectors in the x, y, z directions, $\mathbf{E_v}$, $\mathbf{F_v}$ and $\mathbf{G_v}$ are viscous flux vectors, $\mathbf{H_g}$ is a gravity flux vector, and u, v, w are velocity components in the x, y, z directions respectively. p is pressure, Re is Reynolds number, Fr is Froude number, while the x, y and z subscripts represent the differential in each direction.

2.2 Numerical Scheme

The free surface and rotating object are expressed as a moving mesh using the moving grid finite volume method. The method estimates a control volume in the unified space-time domain. So, to express 3D movement, the method uses a four-dimensional (4D) domain to satisfy a geometric conservation law as well as a physical conservation law. As the discretization of the method, Eq. (2) is separated into a velocity vector term and a pressure vector term as shown in Eq. (4).

$$\hat{\mathbf{E}} = \mathbf{E} - \mathbf{P}_1, \hat{\mathbf{F}} = \mathbf{F} - \mathbf{P}_2, \hat{\mathbf{G}} = \mathbf{G} - \mathbf{P}_3 \tag{4}$$

where,

$$\hat{\mathbf{E}} = \begin{pmatrix} u^2 \\ uv \\ uw \end{pmatrix}, \hat{\mathbf{F}} = \begin{pmatrix} uv \\ v^2 \\ vw \end{pmatrix}, \hat{\mathbf{G}} = \begin{pmatrix} uw \\ vw \\ w^2 \end{pmatrix}, \mathbf{P}_1 = \begin{pmatrix} p \\ 0 \\ 0 \end{pmatrix}, \mathbf{P}_2 = \begin{pmatrix} 0 \\ p \\ 0 \end{pmatrix}, \mathbf{P}_3 = \begin{pmatrix} 0 \\ 0 \\ p \end{pmatrix} \tag{5}$$

Equation (2) can be rewritten as follows by using the above equations.

$$\frac{\partial \mathbf{q}}{\partial t} + \frac{\partial \hat{\mathbf{E}}}{\partial x} + \frac{\partial \hat{\mathbf{F}}}{\partial y} + \frac{\partial \hat{\mathbf{G}}}{\partial z} - \frac{1}{Re}\left(\frac{\partial \mathbf{E}_v}{\partial x} + \frac{\partial \mathbf{F}_v}{\partial y} + \frac{\partial \mathbf{G}_v}{\partial z}\right) + \frac{\partial \mathbf{P}_1}{\partial x} + \frac{\partial \mathbf{P}_2}{\partial y} + \frac{\partial \mathbf{P}_3}{\partial z} = \mathbf{H}_g \tag{6}$$

The equation is separated into Eq. (7) and (8) in order to perform the fractional step method.

$$\frac{\partial \mathbf{q}}{\partial t} + \frac{\partial \hat{\mathbf{E}}}{\partial x} + \frac{\partial \hat{\mathbf{F}}}{\partial y} + \frac{\partial \hat{\mathbf{G}}}{\partial z} - \frac{1}{Re}\left(\frac{\partial \mathbf{E}_v}{\partial x} + \frac{\partial \mathbf{F}_v}{\partial y} + \frac{\partial \mathbf{G}_v}{\partial z}\right) = \mathbf{H}_g \tag{7}$$

$$\frac{\partial \mathbf{q}}{\partial t} + \frac{\partial \mathbf{P}_1}{\partial x} + \frac{\partial \mathbf{P}_2}{\partial y} + \frac{\partial \mathbf{P}_3}{\partial z} = 0 \left(\because 0 = [0\,0\,0]^{\mathrm{T}}\right) \tag{8}$$

Equation (7) is integrated over the control volume in the unified space-time domain and can be written as

$$\int_{\Omega}\left[\frac{\partial \mathbf{q}}{\partial t} + \frac{\partial \hat{\mathbf{E}}}{\partial x} + \frac{\partial \hat{\mathbf{F}}}{\partial y} + \frac{\partial \hat{\mathbf{G}}}{\partial z} - \frac{1}{Re}\left(\frac{\partial \mathbf{E}_v}{\partial x} + \frac{\partial \mathbf{F}_v}{\partial y} + \frac{\partial \mathbf{G}_v}{\partial z}\right)\right]d\Omega = \int_{\Omega} \mathbf{H}_g d\Omega \tag{9}$$

The equation is rewritten in terms of a divergence integral over a volume V_{Ω} in the 4D domain.

$$\int_{\Omega}\left[\left(\frac{\partial}{\partial x}, \frac{\partial}{\partial y}, \frac{\partial}{\partial z}, \frac{\partial}{\partial t}\right)\left\{\left(\hat{\mathbf{E}} - \frac{1}{Re}\mathbf{E}_v\right), \left(\hat{\mathbf{F}} - \frac{1}{Re}\mathbf{F}_v\right), \left(\hat{\mathbf{G}} - \frac{1}{Re}\mathbf{G}_v\right), \mathbf{q}\right\}\right]d\Omega = V_{\Omega}\mathbf{H}_g \tag{10}$$

Using Gauss' theorem, the equation can be written as

$$\oint_{\Omega} \left[\left\{ \left(\hat{\mathbf{E}} - \frac{1}{Re}\mathbf{E}_v \right), \left(\hat{\mathbf{F}} - \frac{1}{Re}\mathbf{F}_v \right), \left(\hat{\mathbf{G}} - \frac{1}{Re}\mathbf{G}_v \right), \mathbf{q} \right\} (n_x, n_y, n_z, n_t) \right] ds = V_{\Omega} \mathbf{H}_g \tag{11}$$

$$\sum_{l=1}^{6} \left[\mathbf{q} n_t + \left(\hat{\mathbf{E}} - \frac{1}{Re}\mathbf{E}_v \right) n_x + \left(\hat{\mathbf{F}} - \frac{1}{Re}\mathbf{F}_v \right) n_y + \left(\hat{\mathbf{G}} - \frac{1}{Re}\mathbf{G}_v \right) n_z \right]_l = V_{\Omega} \mathbf{H}_g \tag{12}$$

Finally, the discretization for the governing equation is as follows.

$$\mathbf{q}^{n+1}(n_t)_6 + \mathbf{q}^n(n_t)_5 + \sum_{l=1}^{4} \left[\mathbf{q}^{n+1/2} n_t + \mathbf{\Phi}^{n+1/2} - \mathbf{\Psi}^{n+1/2} \right]_l = V_{\Omega} \mathbf{H}_g \tag{13}$$

where, n_x, n_y, n_z, and n_t are normal unit vectors in the x, y, z and t directions, respectively, and $\mathbf{\Phi}$ and $\mathbf{\Psi}$ are as follows.

$$\mathbf{\Phi} = \hat{\mathbf{E}} n_x + \hat{\mathbf{F}} n_y + \hat{\mathbf{G}} n_z \tag{14}$$

$$\mathbf{\Psi} = \frac{1}{Re} \left(\mathbf{E}_v n_x + \mathbf{F}_v n_y + \mathbf{G}_v n_z \right) \tag{15}$$

The pressure Eq. (8) is discretized as

$$\left(\mathbf{q}^{n+1} - \mathbf{q}^* \right)(n_t)_6 + \sum_{l=1}^{4} \left(\tilde{\mathbf{P}}_1 n_x + \tilde{\mathbf{P}}_2 n_y + \tilde{\mathbf{P}}_3 n_z \right)_l = 0 \tag{16}$$

Equations (13) and (16) are iteratively solved using the lower-upper symmetric-Gauss-Seidel (LU-SGS) method [10] and using the bi-conjugate gradient stabilized (Bi-CGSTAB) method [11], respectively. Here, the convective flux vectors are evaluated with the second-order upwind difference scheme. The viscous-flux and pressure vectors are evaluated with the central difference scheme.

3 Application to Submarine with Rotating Screw

3.1 Sliding Mesh Approach

A sliding mesh approach [12] was used to express the rotating screw in the simulation around a submarine. Here, the embedded sub computational domain rotates in the main domain. So far, we have used this approach only for compressible flows. Here, we devise an efficient approach for an incompressible flow.

The sliding approach, which divides up the computational domain and slides its boundary, is a moving grid method. The physical values between domains are interpolated through the boundary surface. The sliding surface is dealt with as a moving boundary. To satisfy the geometric conservation law on the moving boundary, a conservative quantity $\mathbf{q}_{\text{ghost}i}$ is obtained as a boundary condition. A schematic diagram of the sliding surface is shown in Fig. 1, and Eq. (17) defines $\mathbf{q}_{\text{ghost}i}$ using the overlapping area S_{ij}.

$$\mathbf{q}_{\text{ghost}i} = \frac{\sum_{j \in i} \mathbf{q}_j S_{ij}}{\sum_{j \in i} S_{ij}} \tag{17}$$

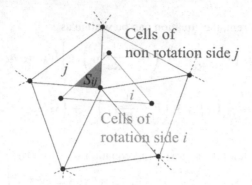

Fig. 1. Schematic diagram of sliding surface.

3.2 Verification of Sliding Mesh Approach

A uniform flow on a sliding mesh was computed to verify the sliding mesh approach for an incompressible environment. Figure 2 shows the rotating embedded domain (Domain 2) set in the main domain (Domain 1), while Fig. 3 shows the computational mesh, including the embedded rotating mesh in the main mesh and a horizontal cross section.

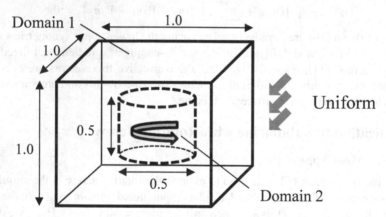

Fig. 2. Computational domain.

The mesh was generated using MEGG3D [13]. The total number of elements was 153,301. A uniform flow ($u = v = w = 1.0$) was set as the initial condition. The velocities on all boundaries were fixed to be a uniform flow ($u = v = w = 0$). Regarding the other computational conditions, the angular velocity of the embedded domain was set to 1.0, and the Reynolds number was set to 100.

Figure 4 shows the history of the velocity error in the rotating embedded mesh system. The error is defined in Eq. (18). Here, *imax* is the number of cell in the computational domain. The order of the error is 10^{-15}, which indicates machine zero. Thus, the geometric conservation law is satisfied between the rotating embedded mesh and fixed main mesh.

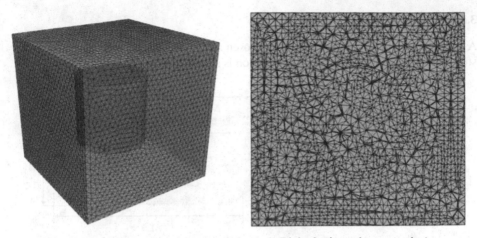

Fig. 3. Computational mesh (Left: whole mesh, Right: horizontal cross section).

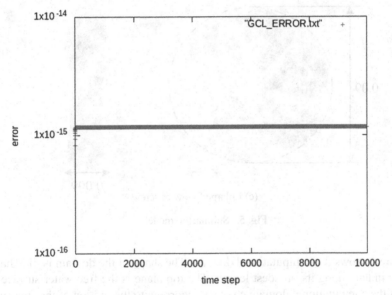

Fig. 4. History of velocity error.

$$Error = \left\{ \left(\sum_{i=1}^{imax} (u_i - 1.0)^2 + \sum_{i=1}^{imax} (v_i - 1.0)^2 + \sum_{i=1}^{imax} (w_i - 1.0)^2 \right) / (3 \times imax) \right\}^{\frac{1}{2}}$$

(18)

3.3 Submarine Model

A submarine with a rotating screw was chosen as a complicated shape and motion for study. A simplified model of the computation is shown in Fig. 5.

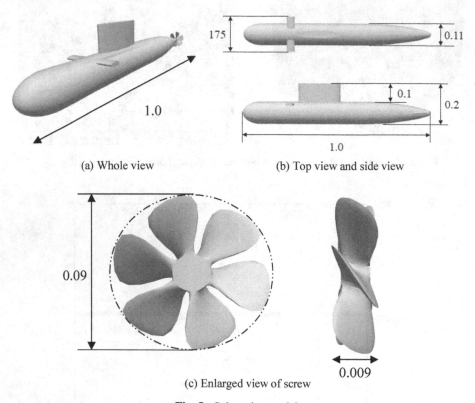

(a) Whole view   (b) Top view and side view

(c) Enlarged view of screw

Fig. 5. Submarine model.

Figure 6 shows the computational domain. The shape of the domain is like that of a bean cut in half along its broadest length. The top plane is the free water surface. The shape of the computational domain changes according to the motion of the free surface caused by the movement of the submarine. The shapes of the boundaries other than the top plane are fixed. The submarine is placed at a depth of 1.0, as shown in Fig. 6.

Cross-sections of mesh around the submarine and the whole computational domain are shown in Fig. 7. The figure illustrates the fine mesh around the submarine and the screw.

The meshes around the screw are shown in Fig. 8. A cylindrical mesh around the screw is embedded in the whole mesh, and it is placed at the rear of the submarine. The cylindrical mesh can be rotated using the sliding mesh approach in accordance with the rotation of the screw.

(a) Shape of computational domain

(b) Side view and front view of computational domain

Fig. 6. Computational domain.

(a) Cross section of mesh around submarine

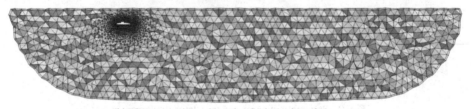

(b) Cross-section of mesh of whole domain

Fig. 7. Cross-section of mesh of submarine and whole domain.

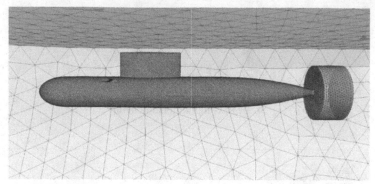

(a) Submarine surface mesh and embedded mesh around the screw

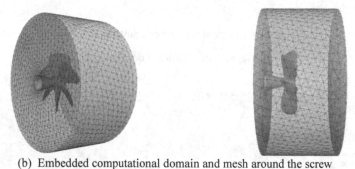

(b) Embedded computational domain and mesh around the screw

Fig. 8. Embedded mesh around the screw and placement for the submarine.

3.4 Computation for Translator Movement

The flow around the submarine for translator movement with a rotating screw was computed. The motion of the submarine was in a straight line keeping an initial depth of 1.0, as shown in Fig. 9. The computational conditions are listed in Table 1.

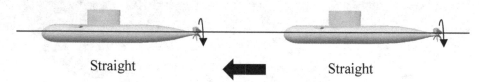

Straight   Straight

Fig. 9. Schematic diagram of translator movement.

Figures 10 and 11 show velocity contours in the x-dirction and z-direction in side-view cross section. The ups and downs of the free water surface increase as time proceeds. In Fig. 11, a negative velocity appears behind the submarine as an effect of the rotating screw. The heights of the free surfaces are shown in Fig. 12. The height of the surface behind the submarine recovers after a drop.

Table 1. Computational conditions for translator movement.

Total number of elements		1,158,625
Reynolds number Re		100
Froude number Fr		2.0
Time step Δt		0.001
Acceleration a		0.2 $(t \leq 5.0)$ 0 $(t > 5.0)$
Initial conditions		Velocity: $u = v = w = 0.0$ Pressure: Determined from height
Boundary condition	Free surface	Velocity: Extrapolation Pressure:$p = 0.0$
	Forward	Velocity: $u = v = w = 0.0$ Pressure: Determined from height
	Submarine (with screw)	Velocity: Non-slip Pressure: Neumann
	Sliding surface	Sliding boundary condition
	Others	Velocity: Extrapolation Pressure: Determined from height

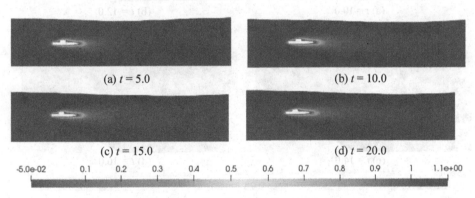

(a) $t = 5.0$ (b) $t = 10.0$

(c) $t = 15.0$ (d) $t = 20.0$

-5.0e-02   0.1   0.2   0.3   0.4   0.5   0.6   0.7   0.8   0.9   1   1.1e+00

Fig. 10. x-direction velocity contours.

3.5 Computation for Translator Movement with Rising Motion

In addition to the translator movement, the rising motion of the submarine was simulated. A schematic diagram of the motion is shown in Fig. 13. The calculation conditions are listed in Table 2.

Figures 14 and 15 respectively show the velocity contours in the x-direction and z-direction in the case of a rising submarine. The velocity contours are similar to the previous case. However, in the rising process at $t = 7.0$, the velocity in the z-direction increases around the submarine. Figure 16 shows top views of the free surface behind the

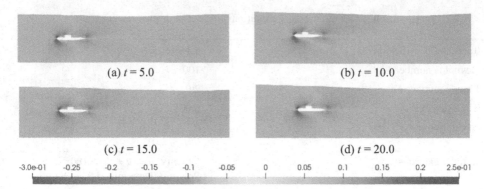

(a) $t = 5.0$ (b) $t = 10.0$

(c) $t = 15.0$ (d) $t = 20.0$

-3.0e-01   -0.25   -0.2   -0.15   -0.1   -0.05   0   0.05   0.1   0.15   0.2   2.5e-01

Fig. 11. z-direction velocity contours.

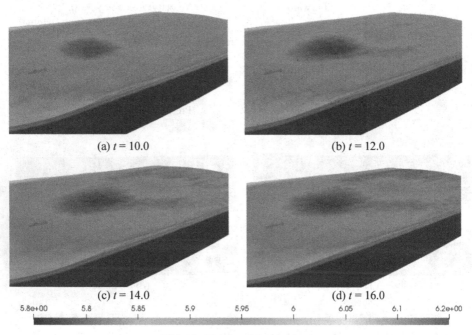

(a) $t = 10.0$ (b) $t = 12.0$

(c) $t = 14.0$ (d) $t = 16.0$

5.8e+00   5.8   5.85   5.9   5.95   6   6.05   6.1   6.2e+00

Fig. 12. Height of free surface.

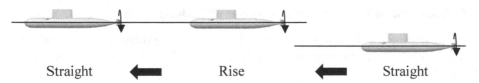

Straight   ⬅   Rise   ⬅   Straight

Fig. 13. Schematic diagram of translator movement with rising motion.

Table 2. Computational conditions for translator movement with rising motion.

Total number of elements		1,158,625
Reynolds number Re		100
Froude number Fr		2.0
Time step Δt		0.001
Acceleration a		0.2 ($t \le 5.0$) 0 ($t > 5.0$)
Initial conditions		Velocity: $u = v = w = 0.0$ Pressure: Determined from height
Submarine rise		$z = A\sin\left(\omega t - \frac{\pi}{2}\right)$ (A $= 0.2$, $\omega = \frac{\pi}{3.75}$) ($5.0 < t \le 9.0$)
Boundary condition	Free surface	Velocity: Extrapolation Pressure: $p = 0.0$
	Forward	Velocity: $u = v = w = 0.0$ Pressure: Determined from height
	Submarine (with screw)	Velocity: Non-slip Pressure: Neumann
	Sliding surface	Sliding boundary condition
	Others	Velocity: Extrapolation Pressure: Determined from height

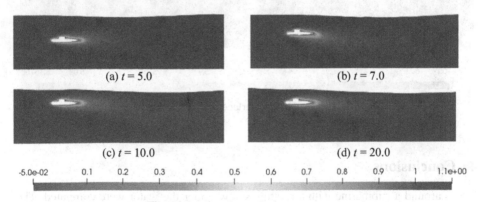

(a) $t = 5.0$ (b) $t = 7.0$

(c) $t = 10.0$ (d) $t = 20.0$

-5.0e-02   0.1   0.2   0.3   0.4   0.5   0.6   0.7   0.8   0.9   1   1.1e+00

Fig. 14. x-direction velocity contours.

submarine comparing the translator movement and rising motion. In the case of the rising motion, a deep drop in the free surface behind the submarine is clearly seen. Although the effect of the rotating screw itself might not great, it is obvious that the motion of the object under the water affects the movement of the free surface. Furthermore, the simulation demonstrates the possibility of conducting useful complicated computations with the free surface.

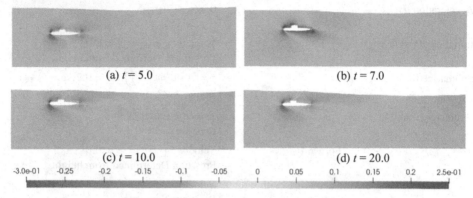

(a) $t = 5.0$ (b) $t = 7.0$

(c) $t = 10.0$ (d) $t = 20.0$

-3.0e-01   -0.25   -0.2   -0.15   -0.1   -0.05   0   0.05   0.1   0.15   0.2   2.5e-01

Fig. 15. z-direction velocity contours.

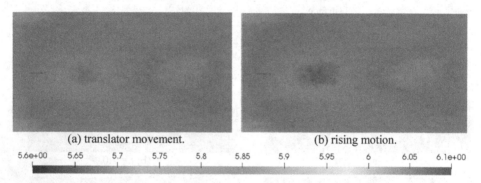

(a) translator movement.   (b) rising motion.

5.6e+00   5.65   5.7   5.75   5.8   5.85   5.9   5.95   6   6.05   6.1e+00

Fig. 16. Top view of free surface behind the submarine.

4   Conclusions

Flows around a submarine with a rotating screw under the water were computed. The flows included the free water surface. To solve such a complicated combination of flows, the unstructured moving grid finite volume method and the surface height function method were used. This combined computational method could capture the uniform flow on the rotating cylindrical sliding mesh. In this test, the geometric conservation law was satisfied. The method was then applied to free surface flows around a submarine with a rotating screw. The results demonstrated the potential of a valid computation for a complicated flow field.

Acknowledgments. This publication was subsidized by the Takahashi Industrial and Economic Research Foundation.

References

1. Drew, B., et al.: A review of wave energy converter technology. Proc. Inst. Mech. Engineers Part A J. Power Energy **223**(8), 887–902 (2009)
2. Mansoorzadeh, S., Javanmard, E.: An investigation of free surface effects on drag and lift coefficients of an autonomous underwater vehicle (AUV) using computational and experimental fluid dynamics methods. J. Fluids Struct. **51**, 161–171 (2014)
3. Dean, R.G., Dalrymple, R.A.: Water wave mechanics for engineers and scientists. Adv. Ser. Ocean Eng. **2**, 212–260 (1991)
4. Benusiglio, A., et al.: Wave drag on a submerged sphere. Phys. Fluids **27**(072101), 1–11 (2015)
5. Kwag, S., et al.: Numerical computation of 3-D free surface flows by N-S solver and detection of sub-breaking. J. Soc. Naval Arch. Jpn. **1989**(166), 9–16 (1989)
6. Moonesun, M., et al.: Technical notes on the near surface experiments of submerged submarine. Int. J. Maritime Technol. **5**, 41–54 (2016)
7. Hirt, C.W., Nichols, B.D.: Volume of fluid (VOF) method for the dynamics of free boundaries. J. Comput. Phys. **39**, 201–225 (1981)
8. Yamakawa, M., et al.: Numerical simulation for a flow around body ejection using an axisymmetric unstructured moving grid method. Comput. Therm. Sci. **4**(3), 217–223 (2012)
9. Yamakawa, M., et al.: Optimization of knee joint maximum angle on dolphin kick. Phys. Fluids **32**, 067105 (2020)
10. Yoon, S., et al.: Lower-upper symmetric-Gauss-Seidel method for the Euler and Navier-Stokes equation. AIAA J. **26–9**, 1025–1026 (1998)
11. Vorst, H.: BI-CGSTAB: a fast and smoothly converging variant of BI-CG for the solution of nonsymmertic linear system. SIAM J. Sci. Stat. Comput. **13–2**, 631–644 (1992)
12. Takii, A., et al.: Six degrees of freedom flight simulation of tilt-rotor aircraft with nacelle conversion. J. Comput. Sci. **44**, 101164 (2020)
13. Ito, Y.: Challenges in unstructured mesh generation for practical and efficient computational fluid dynamics simulations. Comput. Fluids **85**, 47–52 (2013)

Modeling and Simulation of Atmospheric Water Generation Unit Using Anhydrous Salts

Shereen K. Sibie[1], Mohamed F. El-Amin[2(✉)], and Shuyu Sun[3]

[1] Renewable Energy Engineering, College of Engineering, Effat University, Jeddah 21478, Kingdom of Saudi Arabia
shsibie@effatuniversity.edu
[2] Energy Research Lab., College of Engineering, Effat University, Jeddah 21478, Kingdom of Saudi Arabia
momousa@effatuniversity.edu
[3] Division of Physical Sciences and Engineering (PSE), King Abdullah University of Science and Technology (KAUST), Thuwal, Jeddah 23955-6900, Kingdom of Saudi Arabia

Abstract. The atmosphere contains 3400 trillion gallons of water vapor, which would be enough to cover the entire earth in 1 inch of water. Air humidity is available everywhere, and it acts as a great alternative as a renewable reservoir of water known as atmospheric water. Atmospheric water harvesting system efficiency depends on the sorption capacity of water based on the adsorption phenomenon. Using anhydrous salts is an efficient process for capturing and delivering water from ambient air, especially at a low relative humidity as low as 15%. A lot of water-scarce countries like Saudi Arabia have much annual solar radiation and relatively high humidity. This study is focusing on modeling and simulating the water absorption and release of the anhydrous salt copper chloride ($CuCl_2$) under different relative humidity to produce atmospheric drinking water in scarce regions.

Keywords: Atmospheric water · Anhydrous salts · Absorption · Relative humidity · Modeling · Simulation

1 Introduction

About four billion people, two-thirds of the world population, suffer from water scarcity [1]. And about 13 sextillions (10^{21}) liters of water vapor exist in the atmosphere [2]. The atmospheric water, which is considered a substantial renewable reservoir of water and enough to meet every person's needs on the planet, is unfortunately ignored [3]. Atmospheric water usually exists in three types: fog, water vapor in the air, and clouds. Cloud and fog are all made up of tiny drops of water, typically with a diameter from 1 to 40 mm, compared with the size of rain droplets varying from 0.5 to 5 mm. Still, the concentration of water droplets in fog is usually larger. Water vapor is a recyclable natural resource with the potential to water the world's arid regions [4]. Real challenges are facing Saudi Arabia due to the depletion of the rapidly used nonrenewable groundwater.

© Springer Nature Switzerland AG 2021
M. Paszynski et al. (Eds.): ICCS 2021, LNCS 12747, pp. 282–288, 2021.
https://doi.org/10.1007/978-3-030-77980-1_22

In Saudi Arabia, the water is extremely scarce due to the arid climate conditions. The high water demand in the agriculture sector is exacerbating the water scarcity situation in the Kingdom. Urban water and sanitation services incur a high cost to the government [5]. Besides, there is always plenty of water available in the atmosphere, even in very dry desert regions [6].

Methods to harvest water from humid air are known [3–10], and currently, 25 countries worldwide are capturing atmospheric water droplets from the fog for the remote villages[8–12]. However, the fog harvesting technique requires a frequent presence of high relative humidity (RH), typically 100% in the air, making it viable in limited locations. Witch restrains this technology to be applied in any other site unless it has a very high RH [13, 14]. Further, There have been efforts to meet limited success to harvest water vapor from a low RH air to produce water with a self-sustained energy source [3, 15]. In 2017, Yaghi and Wang et al. demonstrated water harvesting by vapor adsorption using a porous metal-organic framework that works in low RH of 20% and delivers water using low-grade heat natural sunlight assisted by photothermal material [16]. More recently, Li and Shi et al. have fabricated an all-in-one bilayered composite disk device to integrate water vapor collection, and photothermal assisted water release using anhydrous salts Copper chloride ($CuCl_2$), Copper sulfate ($CuSO_4$), and Magnesium sulfate ($MgSO_4$) with low relative humidity as low as 15% [17].

This paper will model and simulate atmospheric water generation unit using anhydrous salts; Copper chloride ($CuCl_2$) salt to produce atmospheric water from thin air. This study explores and investigates the potential of harvesting atmospheric water using a cost-effective material like anhydrous salt. Also, identify the potential of modeling and simulation of the absorption and release of water from ambient air using anhydrous salt to produce atmospheric water.

2 Mathematical Modeling

In this work, we consider the model developed by Cesek et al. [18], which was originally used to describe the vapor absorption in porous cellulosic fiber web. Fick's first law can describe the one-dimensional diffusion of molecules in the perpendicular direction of the porous material into thin layer (Fig. 1) as:

$$-D\frac{dC}{dt} = C_0\varphi\left(\frac{dx}{dt}\right)^2 \tag{1}$$

where C is the actual vapor concentration and C_0 is the saturated vapor concentration. D is the diffusion coefficient, and φ is the relative humidity, and x is the thickness of the absorbent material. Now, let us assume that:

$$Y = Y_e\frac{x}{x_{ra}} \tag{2}$$

Here Y_e is the moisture content of the condensed component at time $t \rightarrow \infty$, while Y is the moisture content of the condensed component at any time t and x_{ra} is half of the thickness.

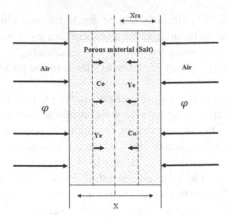

Fig. 1. Schematic diagram of the system under consideration.

The variation of diffusion and absorption in the porous material is proportional to the relative concentration variation of the condenser water inside the absorbent, which is described as:

$$D\frac{dY}{dt} = \frac{C_0\varphi M\varepsilon}{\rho_p}\left(\frac{dx}{dt}\right)^2 \tag{3}$$

Where M is the molecular weight of the water, ρ_p Is the apparent density, and ε is the porosity of the salt. Assuming that the concentration gradient on the boundary of the porous material is constant, the velocity of the penetration on the boundary becomes:

$$\frac{dx}{dt} = k^{0.5}\left(\frac{x}{t^{\frac{d+1}{2}}}\right) \tag{4}$$

Where k is the proportionality coefficient that will be taken $(k = 1)$ in our calculations. The parameter d characterizes a uniformity of the stratified structure of porous material. When $d = 1$, the porous material is uniformly stratified. However, if $d > 1$ or $d < 1$, the porous material is non-uniformly stratified [18]. From the above equations, a simplified differential equation can be obtained as:

$$D\frac{dY}{dt} = \frac{kC_{0\varphi}M\varepsilon Y^2}{\rho p}\left(\frac{x_{ra}}{Y_e}\right)^2\frac{1}{t^{d+1}} \tag{5}$$

Which is a separable equation that can be solved easily to give [18]:

$$Y = \frac{Y_e t^d}{\left(\frac{\varphi x_{ra}^2}{D_{ps}dY_e}\right) + t^d} \tag{6}$$

such that $D_{ps} = D\rho_p/kC_0M\varepsilon$.

3 Model Validation

Equation (6) is used to validate our computed model, and this section presents a comparison between the current computed water content against the experimental data obtained from the literature [17].

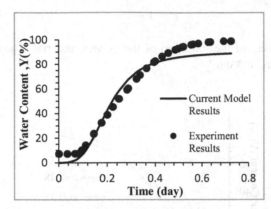

Fig. 2. Comparison between the current computed water content against the experimental data [17].

In Fig. 2, the current computed model presents the current moisture content in the Copper chloride ($CuCl_2$), under RH 15%, it reaches the maximum saturation rate of 88.748 (Kg L/ Kg S) after 17.4 h of hydration. Compared with the experimental results [17] with an 8.3% maximum absolute error. Table 1 presents the parameter used for the current model simulation.

Table 1. The initial data for the current model are listed in Table 1

Parameters	Value	Unit
Ye	0.9	(Kg L/ Kg S)
t	1	(Day)
d	3.4	(Dimensionless)
φ	15	Relative humidity in %
x_{ra}	0.0015	(m)
D_{ps}	0.002720402	$m^2 kg\, S/kg\, Lday^d$

4   Results

4.1   Sensitivity of the Porosity (ε)

The ε resembles the total porosity of the pore sample, as it can be calculated on knowing the apparent density of the porous material ρ_P [18].

$$\varepsilon = 1 - \frac{\rho_P}{\rho_s} \tag{7}$$

where ρ_s is the density of the solid part of the porous material. The Copper chloride ($CuCl_2$) has a density of 3386 $\frac{Kg}{m3}$.

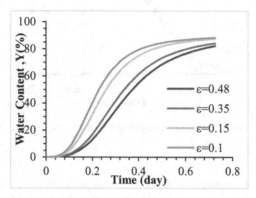

Fig. 3. Water content curves of $CuCl_2$ under different porosity ε.

All parameters for the current model were constant (Table1) except ε was tested with different values shown in (Fig. 3). As it shows, the more uniform the salt, the more water content the salt holds. The smaller the pore volume (smaller porosity), the faster the salt diffusion rate due to the decrease in the diffusion path, which increases the water content in the salt.

4.2   Sensitivity of the Thickness (x_{ra})

The x_{ra} resembles half-thickness of the sample, as this parameter has an essential effect on the absorption rate behavior on the Copper chloride ($CuCl_2$). The initial values are 1.5 mm, as other values were used in the (Fig. 4) to observe the water content's behavior in the salt.

The thickness of the salt plays an important role in the time and water content in the salt. As the thickness decreases, it allows the salt to absorb the water vapor faster as the water content increases.

4.3   Sensitivity of the Uniformity of the Stratified Structure (d)

The d resembles the uniformity of stratified structure of porous web material, for the Copper chloride ($CuCl_2$) the initial value was d = 3.2 (dimensionless). The results in

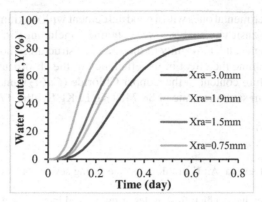

Fig. 4. Water vapor absorption curves of $CuCl_2$ under the different thickness.

(Fig. 5) showed more weight of the water is gained with less time in the salt when the salt is more uniform. The salt takes more time and gains less weight of the water content when it is less uniform. As the uniformity of the stratified structure of the CuCl2 plays a role in the kinetics of the process.

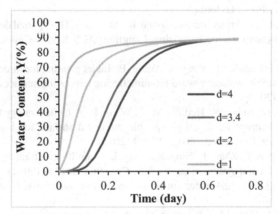

Fig. 5. Water vapor absorption curves of $CuCl_2$ under different uniformity of the stratified structure (d).

5 Conclusion

In conclusion, atmospheric water is considered a huge renewable reservoir, as recently researchers have been developing different techniques and materials to absorb water from the atmosphere. Extracting water vapor from thin air using anhydrous salt is one of the efficient and cost-effective materials up-to-date. It is economical and practical. This paper was devoted to model and simulate water vapor absorption from the air using anhydrous salt copper chloride ($CuCl_2$). An analytical model was developed. The results are compared to experimental ones with a good agreement with maxiumum absolute error of 88.3%; different sensitivity analyses were obtained, including different porosity, the thickness of the sample, and uniformity of the stratified structure. Those key parameters were selected to examine the effect of the efficiency of the absorbent materials. It was found that the moisture content in the Copper Chloride ($CuCl_2$), under RH of 15%, it reaches the maximum saturation rate of 88.748 (Kg L/ Kg S) after 17.4 h of hydration.

References

1. Mekonnen, M., Hoekstra, A.: Four billion people facing severe water scarcity. Sci. Adv. **2**, e1500323 (2016)
2. Gleick, P.H.: Water and conflict: fresh water resources and international security. Int. Secur. **18**(1), 79–112 (1993)
3. Wahlgren, R.V.: Atmospheric water vapour processor designs for potable water production: a review. Water Res. **35**(1), 1–22 (2001)
4. "Practical water production from desert air|Science Advances Homepage. https://advances. sciencemag.org/content/4/6/eaat3198
5. National-Water-Strategy-Homepage.   https://www.mewa.gov.sa/en/Ministry/Agencies/The WaterAgency/Topics/Pages/Strategy.aspx
6. Schneider, S., Root, T., Mastrandrea, M.: Encyclopedia of Climate and Weather, 2nd edn. Oxford University Press, Oxford (2011)
7. Park, K.-C., Chhatre, S., Srinivasan, S., Cohen, R., Mckinley, G.: Optimal design of permeable fiber network structures for fog harvesting. Langmuir ACS J. Surf. Colloids **29**, 132–13277 (2013)
8. Zhang, L., Wu, J., Hedhili, M., Yang, X., Wang, P.: Inkjet printing for direct micropatterning of a superhydrophobic surface: toward biomimetic fog harvesting surfaces. J. Mater. Chem. A **3**, 2844–2852 (2014)
9. Wang, Y., Zhang, L., Jinbo, W., Hedhili, M.N., Wang, P.: A facile strategy for the fabrication of a bioinspired hydrophilic-superhydrophobic patterned surface for highly efficient fog-harvesting. J. Mater. Chem. A **3**, 18963–18969 (2015)
10. Ju, J., Bai, H., Zheng, Y., Zhao, T., Fang, R., Jiang, L.: A multi-structural and multi-functional integrated fog collection system in cactus. Nat. Commun. **3**, 1247 (2012)
11. Zheng, Y., et al.: Directional water collection on wetted spider silk. Nature **463**, 640–643 (2010)
12. Kim, G., Gim, S.J., Cho, S.M., Koratkar, N., Oh, I.K.: Graphene films: wetting-transparent graphene films for hydrophobic water-harvesting surfaces. Adv. Mater. **26**, 5070 (2014)
13. Olivier, J., Rautenbach, C.J.: The implementation of fog water collection systems in South Africa. Atmos. Res. **64**, 227–238 (2002)
14. Estrela, M., Valiente, J., Corell, D., Millán, M.: Fog collection in the Western Mediterranean Basin (Valencia region, Spain). Atmos. Res. - ATMOS RES **87**, 324–337 (2008)
15. Ji, J.G., Wang, R.Z., Li, L.X.: New composite adsorbent for solar-driven freshwater production from the atmosphere. Desalination **212**(1), 176–182 (2007)
16. Kim, H., et al.: Water harvesting from air with metal-organic frameworks powered by natural sunlight. Science **356**(6336), 430–434 (2017)
17. Li, R., Shi, Y., Shi, L., Alsaedi, M., Wang, P.: Harvesting water from air: using anhydrous salt with sunlight. Environ. Sci. Technol. **52**, 5398–5406 (2018)
18. Češek, B., Milichovský, M., Potucek, F.: Kinetics of vapour diffusion and condensation in natural porous cellulosic fibre web. Int. Sch. Res. Notic. **2011**, Article ID 794306 (2011)

Smart Systems: Bringing Together Computer Vision, Sensor Networks and Machine Learning

Improving UWB Indoor Localization Accuracy Using Sparse Fingerprinting and Transfer Learning

Krzysztof Adamkiewicz, Piotr Koch, Barbara Morawska, Piotr Lipiński ⓘ, Krzysztof Lichy$^{(\boxtimes)}$ ⓘ, and Marcin Leplawy

Lodz University of Technology, 116 Żeromskiego Street, 90-924 Lodz, Poland
krzysztof.lichy@p.lodz.pl

Abstract. Indoor localization systems become more and more popular. Several technologies are intensively studied with application to high precision object localization in such environments. Ultra-wideband (UWB) is one of the most promising, as it combines relatively low cost and high localization accuracy, especially compared to Beacon or WiFi. Nevertheless, we noticed that leading UWB systems' accuracy is far below values declared in the documentation. To improve it, we proposed a transfer learning approach, which combines high localization accuracy with low fingerprinting complexity. We perform very precise fingerprinting in a controlled environment to learn the neural network. When the system is deployed in a new localization, full fingerprinting is not necessary. We demonstrate that thanks to the transfer learning, high localization accuracy can be maintained when only 7% of fingerprinting samples from a new localization are used to update the neural network, which is very important in practical applications. It is also worth noticing that our approach can be easily extended to other localization technologies.

Keywords: Indoor localization · Transfer learning · UWB

1 Introduction

Nowadays, it is possible to localize objects around the world using GPS. It is a low-cost and easy to use solution available for everybody. However, it is still challenging to localize objects in GPS-denied environments located in buildings or underground. It is necessary to use other technologies to make it possible, such as dead reckoning, LiDAR-based, magnetic-field-based, or radio-frequency-based technologies [1]. Here we focus on radio-based ultra-wideband (UWB) technology, which is especially promising for indoor localization. It combines relatively low cost and good localization accuracy, especially when compared to WIFI or Bluetooth. Such systems can be used in small-scale areas, where full coverage can easily be assured by a small number of anchors. Nevertheless, the concepts presented here are general and can be applied to other localization technologies. Despite its relatively good performance, UWB localization accuracy can be improved using fingerprinting [2] as localization errors in buildings and underground

© Springer Nature Switzerland AG 2021
M. Paszynski et al. (Eds.): ICCS 2021, LNCS 12747, pp. 291–302, 2021.
https://doi.org/10.1007/978-3-030-77980-1_23

areas are often caused by systematic errors. Systematic errors are difficult to reduce, as they are usually caused by the surrounding environment, especially in radio-based localization systems. They cannot be reduced by using well-known filtering techniques such as Kalman filters or particle filters.

Here we introduce the algorithm which takes advantage of fingerprinting and Transfer learning to reduce systematic errors and Kalman filtering to reduce random errors. We demonstrate that it can improve the localization accuracy by 31% compared to the state of the art UWB localization system – Pozyx [5].

2 Related Studies

There are several UWB-based localization system, to name only: Pozyx [5, 10], Zebra UWB Technology [6], DecaWave [11], BeSpoon [7], Ubisense [8], NXP's automotive UWB [9, 12]. According to [13], especially after taking over DecaWave, it performs best in terms of localization accuracy of all above. The localization accuracy of the UWB system can be improved in many ways, but reflections and multi-path propagation make high accuracy localization very challenging. In all possible concepts, two approaches are mainly used for this purpose: deterministic and probabilistic. Both of them follow the fingerprinting as a source of reference data [14], and the localization improvement is achieved thanks to matching the current measurement with this in a previously prepared database [14]. In the deterministic approach, the measurement' mathematical model is known [15]. As a result, the first possible way to achieve it is to find the closest database fingerprint location using a relevant similarity metric for comparison. For this purpose, it is possible to use large numbers of distance metrics (e.g. Euclidean distance, taxicab geometry, etc.) [16]. The second one uses deterministic methods represented by machine learning algorithms such as the support vector machine (SVM) [16], decision trees [18], and k-nearest neighbors (KNN) [19]. Another approach includes probabilistic algorithms, which use measurement values represented as a probability distribution. In such a case, the output localization is calculated using the signal' statistical properties based on the current online measurements and fingerprinting results from the database. The probabilistic approach is more expensive than deterministic. It can also provide higher accuracy even with the increasing number of lost samples or incompatible data [20].

Both deterministic and probabilistic localization approaches require fingerprinting data [16]. Creating such a database is a big challenge since acquiring it can take a lot of time and effort, especially for large-scale environments and for the reference points' parameters determined manually [16]. It raises costs with the expected localization accuracy. Several methods were proposed to reduce it. One of them reduces the number of fingerprinting points and then use interpolation and extrapolation methods to recover missing data [22]. We essentially introduce a similar approach, but we apply it to UWB technology and use Kalman Filter (KF) and Transfer Learning (TL) instead of the hidden Markov model. Other techniques use unsupervised learning methods for this purpose, see [23]. The authors of this article apply artificial intelligence to improve WiFi-based localization. As a result, the accuracy of the localization system is relatively low, despite using fingerprinting.

It is possible to use other solutions that can improve the results [2]. The first one is a median filter, which is an efficient nonlinear tool for removing outliers. Its efficiency depends on the window size. Unfortunately, it introduces the delay, which increases the localization error [24]. We apply a short median filter as an outline detector in our algorithm. The second one is the Autoregressive-moving-average (ARMA) filter. It specifies that the output depends linearly on a set of previous input samples. Generally, it is an infinite impulse response filter without delay, making it very useful during the localization improvement process [24]. We demonstrate that it performs worse, despite its properties, than the algorithm introduced here, in terms of localization accuracy. The third one is the k-nearest neighbors' algorithm (k-NN), which is often applied in localization algorithms [25]. In our experiment, it also performed worse than our algorithm. The fourth one is the edge detection algorithm, which detects the leading edge of the UWB pulse signal to determine the time of arrival (TOA) with higher accuracy even under low signal-to-noise ratio (SNR) conditions [27]. As this approach requires direct modifications in UWB hardware, it is beyond the scope of this article. The fifth one is the Kalman filter, a real-time solution widely used in indoor and outdoor positioning systems for signal filtering and data fusion. It effectively reduces Gaussian errors, but it has also proved to perform well in other conditions [3]. We use it as one of the reference methods in our research. Other algorithms use machine learning (ML) methods. They are used in data analyzing, data processing, autonomous driving, and much more. Of course, it has found its adaptation in indoor localization [31]. We essentially adapt and extend these concepts to improve UWB localization accuracy. Another solution is the use of deep learning methods. Unfortunately, they require a large amount of training data [32] or external sensors [33]. Therefore, they are not applicable in our case, as we want to reduce the effort necessary to improve the accuracy and use UWB and inertial measurement unit (IMU) sensors solely. Interesting examples on this topic can also be found in [35, 36].

3 The Algorithm

Our localization algorithm combines the Kalman Filter (KF) filter to reduce stochastic, Gaussian localization errors followed by a neural network (NN) to minimize the deterministic localization errors. In order to reduce the effort necessary to perform fingerprinting, we use transfer learning to adapt the network to the new operating environment of the UWB system. Unlike [21], we do not focus on distinguishing LOS, NLOS, MP signals but on determining deterministic and stochastic components of the signal, which we believe is a more general approach in terms of measurement theory.

In UWB-based localization, the leading research in localization accuracy improvement focuses on the analysis of line-of-sight (LOS), non-line-of-sight (NLOS), and multi-path analysis (MP) [28]. Here we introduce a more general approach. According to the convention of the 1998 ANSI/ASME guidelines [29], the errors that arise in the measurement process can be categorized into systematic and random errors. Therefore, instead of identifying LOS, NLOS, and MP signals, we apply separate tools for reducing systematic and random errors separately despite their origin. It leads to a distance measurement filtration algorithm, which is shown in

It takes raw localization data (x_i, y_i) from the UWB system. The first step removes outliers by removing the samples that exceed the theoretically maximal velocity of the object. Second, we use the Kalman Filter (KF) to remove random errors. We consider the KF to be the best choice here, as it is the mathematically optimal filter for removing random error with Gaussian distribution for robot movement at fixed speed along the track [29], which is considered in this paper. KF cannot remove systematic errors, which means that after applying KF filtration, the signal still contains systematic errors. Therefore, after KF filtration, we use a neural network to remove systematic errors. As a result, we obtain corrected localization (x'_i, y'_i), which better reflects the object' actual localization (Fig. 1).

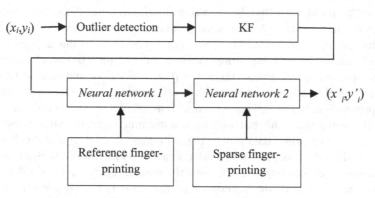

Fig. 1. Localization algorithm.

The system' critical element is the neural network. We train it using fingerprinting results. Obtaining the best results requires performing fingerprinting in each localization where the system is deployed to acquire neural network training data. We use transfer learning to avoid it, which reduces the fingerprinting density in new localizations where the system operates. We train the *neural network 1* using dense reference fingerprinting (see). Next, we apply transfer learning by updating *neural network 1* using the more simple *neural network 2*, which is trained using just a few fingerprinting points in a new localization. It facilitates system deployment in real-life conditions without losing system accuracy. Our experiments demonstrate that our approach can significantly improve the localization accuracy and reduce the effort necessary to deploy the UWB system in different localizations, thanks to transfer learning. A similar results, but for static measurements, has been introduced in [28].

4   Experiment

Our research focused on a dynamic localization algorithm that localizes the object in movement using the UWB Pozyx localization system. Pozyx is a localization system that uses a DW1000 chip and STM32F401 ARM Cortex M4. We used four anchors and a tag of Pozyx development kit. The tag is shield compatible with the Arduino board, which was used to capture the measurement data. The Arduino board communicates with

Pozyx over Inter-Integrated Circuit. The tag can determine its position and motion data from an accelerometer, a gyroscope, a compass, and a pressure sensor. The anchor is not compatible with Arduino and communicates over the serial port. We used a dedicated *pypozyx* Python library to configure the system parameters: channel, bitrate, and function. We divided the experiment into four parts:

1. static fingerprinting in reference, indoor localization - *testbed 1*,
2. static fingerprinting in another localization, in which the system operates - *testbed 2*,
3. dynamic localization of the robot in motion on *testbed 1*,
4. dynamic localization of the robot in motion on *testbed 2*.

The fingerprinting was used to capture the data, which was then used for training the *neural network 1*. The localization of the fingerprinting points together with the points' ID's are marked with green dots in Fig. 2. The localization of anchors A0, A1, A2, A3 is marked with red triangles. Please note that the fingerprinting area goes beyond the rectangle bounded by the anchors. Even though this is incompatible with the Pozyx documentation, the system's localization accuracy in the rectangle area and beyond is at the same level. For each fingerprinting point from Fig. 2 on both tracks, we collected 200 measurements. The measurements from testbed 1 were then used to train the *neural network 1* from

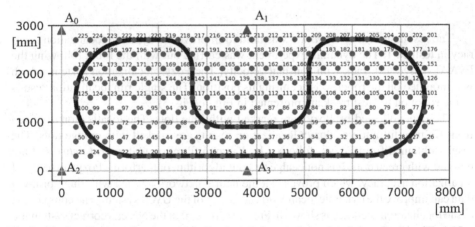

Fig. 2. The localization of anchors (red triangles), fingerprinting points (green dots), and EvAAL-based track (black lines) in both testbeds. (Color figure online)

Our dynamic experiments used the EvAAL-based test track from [31], which is marked as a black curve in Fig. 2. We shrank the original EvAAL track to fit the size of our laboratories. Finding the reference localization of the object in dynamic localization tests is challenging as reference localization accuracy must be more precise than the localization of the UWB system, which requires the reference's centimeter accuracy. We achieved this by using MakeBlock Robot mBot V1.1, which followed the line on the floor using an optical sensor at constant linear speed. It leads to the reference measurement variance below 1cm, which is sufficient when considering that UWB dynamic localization accuracy is about 50 cm, according to our measurement results. We captured the measurements data for six laps. Half of the lap passes were captured for the robot moving clockwise, the other half for the robot moving counterclockwise. The localization sampling frequency was around 16 Hz.

We carried out the experiment using the same Pozyx development kit, test track, and relative anchors localization in two different testbeds localized in two different laboratories. The fingerprinting results from the first testbed were used to teach the neural network. We used the TensorFlow library to optimize the neural network architecture and learn it. The resultant NN parameters are summarized in Table 1.

Table 1. The architecture of *neural network 1*.

Layer	Number of neurons	Activation function
1	2	tanh
2	82	selu
3	152	relu
4	2	relu
5	2	relu

The resultant *neural network 1* was used to improve the dynamic localization accuracy in testbed 1. The dynamic experiment was carried out with the robot following the EvAAL-based test track with constant velocity. The Pozyx localization system, which was mounted on the robot, was acquiring its position. The single run localization results of the robot on the test track are presented in Fig. 4 (Fig. 3).

The data collected by Pozyx are marked with blue dots. The localization obtained from Kalman filtration of Pozyx localization results are marked with orange dots. The results obtained from Pozyx, followed by Kalman filtration and *neural network 1* are marked with green dots. It is noticeable that our algorithm outperforms Pozyx in terms of localization accuracy, which proves that the neural network learned with fingerprinting data can improve the dynamic localization accuracy of the UWB system. The comparison of the localization accuracy is shown in Fig. 4. It proves that the NN can correct systematic errors, which cannot be removed using KF.

Next, we repeated the fingerprinting on testbed 2, which was situated in the laboratory with different size and similar shape as in testbed 1.. In this scenario, the distance from walls to the track was around twice more significant than in testbed 1. In this case,we didn't use all the fingerprinting points from testbed 2 to learn the *neural network 2*. We increased the density of fingerprinting gradually. The *neural network 2* was updated in consecutive steps using fingerprinting results of higher density. In each step, we performed neural network architecture optimization and learning using TensorFlow. The resultant NN2 responsible for transfer learning consisted of 4 layers, consisting of 22, 62, 12, 2 neurons, respectively. The list of chosen points in each step and the localization

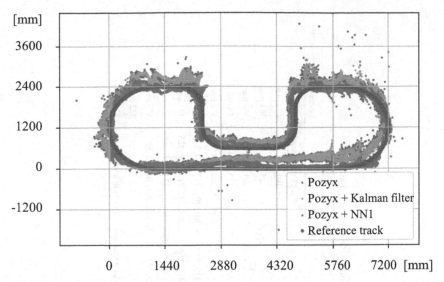

Fig. 3. The robot's localization results for EvAAL-based track, laboratory number one, run one; gray dots – Pozyx localization results, red dots - our algorithm. (Color figure online)

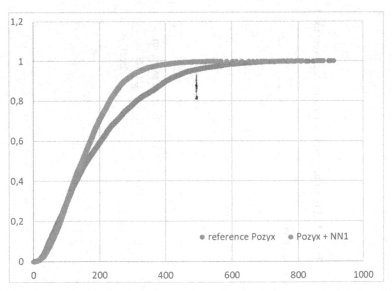

Fig. 4. Cumulative Distribution Error for the dynamic experiment on testbed 1. Comparison of Poxyx localization and Pozyx + NN1.

Table 2. List of fingerprinting points that were used to train the neural network in testbed two.

Step	List of localization points/algorithm	Localization error at 95%
0	Pozyx (reference)	472 mm
1	13, 101, 125, 213	470 mm
2	13, 19, 25,113, 125, 213, 225	365 mm
3	1, 13, 25, 101, 125, 201, 213, 225	360 mm
4	13, 19, 25, 113, 119, 125, 213, 219, 225	330 mm
5	1, 13, 25, 101, 113, 125, 201, 213, 225	355 mm
6	1, 7, 13, 19, 25, 101, 107, 113, 119, 125, 201, 207, 213, 219, 225	323 mm
7	1, 7, 13, 19, 25, 54, 60, 66, 72, 101, 107, 113, 119, 125, 154, 160, 166, 172, 201, 207, 213, 219, 225	323 mm
8	1, 5, 9,13, 17, 21, 25, 51, 55, 59, 63, 67, 71, 75, 101, 105, 109, 113, 117, 121, 125, 151, 155, 159, 163, 167, 171, 175, 201, 206, 209, 213, 217, 221, 225	323 mm
9	1, 3, …, 25, 51, 53, …, 75, 101, 103, …, 125, 151, 153, …, 175, 201, 203, … 225,	323 mm
10	1, 3, 5, 7, …, 225,	322 mm
11	All fingerprinting points in testbed two	323 mm

error of the 95% test samples is shown in Table 2. Corresponding CDF functions for each step are presented in Fig. 5. Please notice that the localization accuracy is much below the accuracy declared in Pozyx documentation. Localization accuracy does not change significantly in steps 6 – 11, which means that localization accuracy achieved thanks to *neural network 1* can be maintained for lower fingerprinting density thanks to transfer learning.

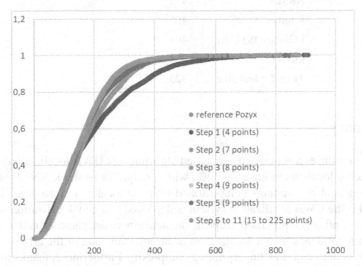

Fig. 5. Cumulative Distribution Error for the dynamic experiment on testbed 2. Localization results for steps from Table 2 and *neural network 2*.

The results presented in both Table 2 and demonstrate that the UWB Pozyx system's localization accuracy can be improved using KF filtering and fingerprinting-based neural network. Furthermore, if the neural network is trained in a well-controlled environment, it can be easily updated by making very sparse fingerprinting, using just 7% of the samples necessary to train the neural network. The main purpose of this process was to reduce the number of points required in fin-gerprinting. As demonstrated in Fig. 5, it is possible to achieve the same location ac-curacy for the 15 points as for the 225 points. As a result, the cost of fingerprinting can be significantly reduced while maintaining location accuracy. In our experiments, we used a uniform, sparse fingerprinting grid. However, the localization accuracy might be improved when using higher grid density in the vicinity of obstacles. The reduction of the fingerprinting points is very desirable in practical applications, as it reduces the costs of system deployment. The comparison of our algorithm with other algorithms that can be applied to improve the Poxyx localization accuracy is presented in Table 3. The localization error of our transfer learning-based algorithm is 32% lower when compared to the Pozyx system. It is also 31% lower than Kalman filtering, 30% lower than median filtering, 20% lower than ARMA filtering, 17% lower than the LOS/NLOS algorithm introduced in [21], and 7% lower than the k-NN algorithm introduced in [2].

Table 3. Comparison of the transfer learning-based algorithm with median filtering, ARMA filtering, Kalman filter, LOS/NLOS algorithm from [21], the k-NN algorithm from [2].

Localization algorithm	Localization error [mm]
Pozyx orig	472
Median	462
ARMA	415
Kalman	465
LOS/NLOS [21]	401
k-NN [2]	354
Transfer learning	**323**

5 Summary

This paper introduces a neural network-based algorithm for UWB localization improvement, reducing localization error by 32%, which outperforms the k-NN algorithm by 7%. The practical experiments demonstrated that the localization accuracy could be maintained if the UWB localization system is deployed in a new localization with only 7% of fingerprinting samples used to update the neural network parameters. It means that our approach combines higher localization accuracy than state of the art UWB Pozyx localization system with low fingerprinting complexity. Furthermore, thanks to transfer learning, it also reduces the learning time required to train the neural network, which is very important in practical applications. Our research has also proved that the combination of Kalman filter, to reduce random localization errors, with a neural network, to minimize systematic error, is an efficient approach to improve indoor localization accuracy.

References

1. Mautz, R.: Indoor positioning technologies, Habilitation Thesis submitted to ETH Zurich Application for Venia Legend in Positioning and Engineering Geodesy, Institute of Geodesy and Photogrammetry, Department of Civil, Environmental and Geomatic Engineering, ETH Zurich (2012)
2. Subedi, S., Pyun, Y.: Practical fingerprinting localization for indoor positioning system by using beacons. J. Sens. **2017**, Article ID 9742170 (2017)
3. Zarchan, P., Musoff, H.: Fundamentals of Kalman Filtering: A Practical Approach. American Institute of Aeronautics and Astronautics, Incorporated (2000). ISBN 978-1-56347-455-2
4. Zand, G., Taherkhani, M., Safabakhsh, R.: Exponential Natural Particle Filter (2015). arXiv: 1511.06603
5. Pozyx Homepage. https://www.pozyx.io, Accessed 11 Jan 2021
6. Zebra UWB Homepage. https://www.zebra.com/pl/pl.html, Accessed 11 Jan 2021
7. BeSpoon Homepage. https://ubisense.com/dimension4/, Accessed 11 Jan 2021
8. Ubisense HomePage. https://www.decawave.com, Accessed 11 Jan 2021

9. NXP's automotive UWB Homepage. http://bespoon.com/shop/en/3-products, Accessed 11 Jan 2021
10. Mimoune, K.M., Ahriz, I., Guillory, J.: Evaluation and improvement of localization algorithms based on UWB pozyx system. In: 2019 International Conference on Software, Telecommunications and Computer Networks (SoftCOM) (2019)
11. DecaWave Homepage, Product Information EVB1000. Overview of EVB1000 Evaluation Board (2013). https://www.decawave.com/product/evk1000-evaluation-kit/, Accessed 11 Jan 2021
12. Ruiz, A.R.J., Granja, F.S.: Comparing Ubisense, BeSpoon, and DecaWave UWB location systems. In: Indoor Performance Analysis IEEE Transactions on Instrumentation and Measurement. IEEE (2017).
13. Dabove, P., Di Pietra, V., Piras, M., Jabbar, A.A., Kazim, S.A.:Indoor positioning using Ultra-wide band (UWB) technologies: positioning accuracies and sensors' performances. In: IEEE/ION Position, Location and Navigation Symposium (PLANS), Monterey, CA (2018)
14. Suining He, S., Gary Chan H.: WiFi fingerprint-based indoor positioning: recent advances and comparisons. In: IEEE Communications Surveys & Tutorials Year. IEEE (2016)
15. Dunn, P.F., Davis, M.P.: Measurement and Data Analysis for Engineering and Science. CRC Press, Boca Raton (2017). ISBN 9781138050860
16. Djosic, S., Stojanovic, I., Jovanoic, M., Nikolic, T., Djrdjevic, G.: Fingerprinting-assisted UWB-based localization technique for complex indoor environments. University of Nis, Faculty of Electronic Engineering (2020). https://doi.org/10.1016/j.eswa.2020.114188
17. Cai, Y., Rai, S.K., Yu, H.: Indoor positioning by distributed machine-learning based data analytics on smart gateway network. In: 2015 International Conference on Indoor Positioning and Indoor Navigation (IPIN) (2020)
18. Banitaan, S., Azzeh, M., Nassif, S.K.: User movement prediction: the contribution of machine learning techniques. In: 15th IEEE International Conference on Machine Learning and Applications (ICMLA) (2016)
19. Bahl, P., Padmanabhan, V.N.: RADAR: an in-building RF-based user location and tracking system. In: Proceedings IEEE INFOCOM 2000, Conference on Computer Communications, Nineteenth Annual Joint Conference of the IEEE Computer and Communications Societies (2020)
20. Bisio, I., Lavagetto, F., Marchese, M., Sciarrone, A.: Performance comparison of a probabilistic fingerprint-based indoor positioning system over different smartphones, In: 2013 International Symposium on Performance Evaluation of Computer and Telecommunication Systems (SPECTS), (2013).
21. Sang, C.L., Steinhagen, B., Homburg, J.D., Adams, M., Hesse, M., Rückert, U.: Identification of NLOS and multi-path conditions in UWB localization using machine learning methods. In: Applied Science 2020 (2020). https://doi.org/10.3390/app10113980
22. Chai, X., Yang, Q.: Reducing the calibration effort for probabilistic indoor location estimation. IEEE Trans. Mob. Comput. **6**(6), 649–662 (2007)
23. Chintalapudi, K.I., Venkata, A.P.: Indoor localization without the pain. In: Proceedings of the Annual International Conference on Mobile Computing and Networking, (MOBICOM) (2010)
24. IEE Colloquium on Kalman Filters: Introduction, Applications and Future Developments (Digest No. 27). In: IEE Colloquium on Kalman Filters: Introduction, Applications and Future Developments (1989)
25. Brockwell, P.J., Davis, R.A.: Time Series: Theory and Methods, 2nd ed. Springer, New York (2009).
26. Taneja, S., Gupta, C., Goyal, K., Gureja, D.: An enhanced k-nearest neighbor algorithm using information gain and clustering. In: 2014 Fourth International Conference on Advanced Computing & Communication Technologies, Rohtak (2014)

27. Sreenivasulu, P., Sarada, J., Dhanesh G.K.: An accurate UWB based localization system using modified leading edge detection algorithm. In: Ad Hoc Networks, vol. 97 (2020)
28. Mimoune, K., Ahriz, I., Guillory, J.: Evaluation and improvement of localization algorithms based on UWB pozyx system. In: 2019 International Conference on Software, Telecommunications and Computer Networks (SoftCOM), Split, Croatia (2019)
29. Test Uncertainty Performance Test Codes, The American Society of Mechanical Engineers, An American Nationa Standard ASME PTC 19.1–2013 (2013)
30. Kalman, R.E.: A new approach to linear filtering and prediction problems. J. Basic Eng. **82**(1), 35–45 (1960)
31. Porti F., Sangjoon, P., Ruiz, A.R., Barsocchi, P.: Comparing the performance of indoor localization systems through the EvAAL framework. In: Sensors 2017, vol. 17 (2017)
32. D'Aloia, M., et al.: IoT indoor localization with AI technique. In: 2020 IEEE Interna-Workshop on Metrology for Industry 4.0 & IoT, Roma, Italy (2020)
33. Zhang, W., Sengupta, R., Fodero, J., Li, X.: DeepPositioning: intelligent fusion of pervasive magnetic field and WiFi fingerprinting for smartphone indoor localization via deep learning. In: 2017 16th IEEE International Conference on Machine Learning and Applications (ICMLA), Cancun (2017)
34. Bai, X., Huang, M., Prasad, N.R., Mihovska, A.D.: A survey of image-based indoor localization using deep learning. In: 22nd International Symposium on Wireless Personal Multimedia Communications (WPMC), Lisbon, Portugal (2019)
35. Glonek, G., Wojciechowski A.: Kinect and IMU sensors imprecisions compensation method for human limbs tracking. In: International Conference on Computer Vision and Graphics, ICCVG 2016, Poland (2016)
36. Daszuta, M., Szajerman, D., Napieralski, P.: New emotional model environment for navigation in a virtual reality. Open Phys. **18**(1) (2020)

Effective Car Collision Detection
with Mobile Phone Only

Mateusz Paciorek[1]([✉]), Adrian Kłusek[1], Piotr Wawryka[1], Michał Kosowski[1],
Andrzej Piechowicz[1], Julia Plewa[1], Marek Powroźnik[1], Wojciech Wach[2],
Bartosz Rakoczy[1], Aleksander Byrski[1], Marcin Kurdziel[1], and Wojciech Turek[1]

[1] AGH University of Science and Technology, Krakow, Poland
mpaciorek@agh.edu.pl
[2] Institute of Forensic Research, Kraków, Poland

Abstract. Despite fast progress in the automotive industry, the number
of deaths in car accidents is constantly growing. One of the most important challenges in this area, besides crash prevention, is immediate and
precise notification of rescue services. Automatic crash detection systems
go a long way towards improving these notifications, and new cars currently sold in developed countries often come with such systems factory
installed. However, the majority of life threatening accidents occur in
low-income countries, where these novel and expensive solutions will not
become common anytime soon. This paper presents a method for detecting car collisions, which requires a mobile phone only, and therefore can
be used in any type of car. The method was developed and evaluated
using data from real crash tests. It integrates data series from various
sensors using an optimized decision tree. The evaluation results show
that it can successfully detect even minor collisions while keeping the
number of false positives at an acceptable level.

Keywords: Vehicle safety · Collision detection · Sensor data
processing · Decision tree

1 Motivation

According to the World Health Organization [9] 1.35 million people died in
2016 in road traffic. What is even more disturbing, this number has been constantly growing over the last 20 years. Despite the fast technological development
observed in the automotive industry, providing efficient safety features is still a
great challenge. Development of widespread, affordable means of improving road
safety and post-crash care should receive more attention.

Severe road accidents, which threaten human lives, require immediate help of
medical rescuers. The need for minimizing the time required to receive help after
an accident is obvious, therefore a lot of effort has been put into organizing efficient rescue services. An immediate notification with detailed information about
the location and expected consequences of the accident is a crucial element—one

© Springer Nature Switzerland AG 2021
M. Paszynski et al. (Eds.): ICCS 2021, LNCS 12747, pp. 303–317, 2021.
https://doi.org/10.1007/978-3-030-77980-1_24

which can be improved by using automated collision-detection and classification systems. Such a solution would be especially important when victims are not able to call for help by themselves.

Solutions supporting emergency calls and automated crash detection are not a novel concept. In fact, the need for and the advantages of such systems are already recognized by many legislator worldwide—an interesting overview of acts, both planned and in force, can be found in a report created by the ITS mobility cluster in Northern Germany [16]. The most advanced and strict regulations have been introduced in the Russian Federation where all cars registered after 2017 (new and imported) have to be equipped with the Accident Emergency Response System, also called ERA-GLONASS [19]. A similar system is being introduced in the European Union where new cars have to be fitted with an automatic emergency call system [11] called eCall. Simpler solutions, which do not interfere with a car's internal systems, are introduced by some insurance companies.

These are definitely valuable steps towards the reduction of deaths on roads. However, before these legal acts have any significant effects, many people will die. Older cars are still present on European roads, not mentioning less-developed countries, where the majority of fatal accidents occur.

The aim of the presented research is to develop a method for detecting car collisions which could be used in both new and older cars and could become a widespread, common solution. There are three main requirements that have to be met:

- independence of cars embedded systems which allows applicability in any car,
- reliability, which provides high accuracy in detecting collisions together with a low number of false alarms,
- availability and very low price which would motivate people to use the solution on a daily basis.

In the paper we are presenting the whole process of developing such a solution. After analyzing the existing approaches, the first step was to obtain data from real collisions. Data from 22 various crashes, recorded with different sensors, has been carefully analyzed in order to verify the solutions described in literature. The conclusions let us identify major issues and challenges, and create a novel method for detecting collisions. The method has been developed and carefully evaluated, providing very promising results.

2   Existing Approaches

Detection of car accidents is a problem that may be solved using either car-based or environmental sensors. Environmental sensors are usually based on cameras and the image is processed in order to signal an accident to some monitoring person (see, e.g. [6]). Of course environmental sensors are not mobile, unless mounted on a helicopter or a drone, so their usability is limited to the vicinity of well-known dangerous spots on the road. A real flexibility and usability can be delivered by telematics systems mounted inside the car participating in traffic.

Focusing on car-based crash detection methods, one can easily distinguish the following types according to the sensing system used:

- car-internal,
- dedicated device (beacon-based),
- universal device (smartphone-based).

A very good example of a **car-internal system** is eCall. This system must be mounted in all new cars sold in the EU starting from March 31st, 2018. The system can be activated either manually or automatically (monitoring e.g. the airbag status) and the infrastructure of mobile phone networks has undergone appropriate adaptation in order to support this system EU-wide (cf. the EC Regulation [11], making the vehicle fit for eCall, and EC Directive [10], making the public infrastructure fit for eCall).

In Russia, a fully interoperable system called ERA-GLONASS is being deployed with the aim to require an eCall terminal and a GPS/GLONASS receiver in new vehicles by 2015–2017. As ERA-GLONASS was the first system that started to operate, eCall is based on its technology [3]. At the same time, in North America, a similar service is provided by GM via their OnStar service [8].

Considering **beacon-based systems**, those are usually used by insurance companies and the installation of such a system in the car usually affects the price of the insurance. Selecting a third-party sensor makes it possible to utilize crash-detection facilities in older cars where no internal systems are present. A good example is the Octo system, which puts together a smartphone and a dedicated sensor, attached to the windshield. The system provides services like fleet monitoring, crash detection, driving-style modelling and even gamification in order to properly motivate the drivers [7]. A similar solution is the IoT DriveWell tag offered by Cambridge Mobile Telematics [13] or the PZU-GO offered by PZU [5]. Bosch offers a Telematic eCall Plug that can be configured for monitoring of driving style or for signalling a crash using the eCall connectivity [4].

The presented solutions, based on embedded car sensors or dedicated sensing devices, are gaining popularity in well-developed countries among experienced drivers. At the same time the majority of severe car accidents involve different drivers – often less experienced and using older cars. These people simply cannot afford the existing safety features. A reliable **Smartphone-based** accident detection system can become a solution to this situation.

Some research in this area also exists. Thompson et al. [14] describe an accident detection and reporting system that uses smartphone accelerometers to detect collisions. To decrease the number of false-positive detections, they trigger an alarm only when a GPS sensor indicates that the smartphone is moving faster than 15 mi per hour and when the recorded acceleration exceeds 4G. When a collision is detected, the smartphone app sends a notification to the central server. The notification includes location data and collision characteristics, such as the recorded acceleration and the vehicle's speed. In a follow-up work [17] this system was extended to include acoustic data in the collision detection procedure. The follow-up work also reports an evaluation against a publicly available dataset with acceleration readings from real accidents. Zaldivar et al. [18] describe an Android application that detects car collisions by monitoring smartphone accelerometer readings and the airbag status reported by a Bluetooth-connected OBD2 interface. An accident is suspected when the accelerometer readings exceed 5G or the airbags are triggered. The accident alarm can be canceled by the smartphone owner within 1 min of the triggering event. Afterwards an accident notification is send via a text or an e-mail message. Finally, Amin et al. [1] describe and evaluate a collision detection procedure that employs accelerometer readings and GPS location data. However, unlike the works mentioned previously, they use a dedicated accelerometer unit and a dedicated GPS receiver rather than smartphone sensors.

One of the real-world smartphone-only crash detection and notification applications is the SOSmart automatic crash detection app [12] designed by a startup based in Santiago, Chile. The authors claim that their crash detection algorithms are based on data gathered by the National Highway Traffic Safety Administration [15]. They need GPS data in order to detect a collision and have implemented detection of smartphone dropping, claiming that the acceleration readings in the case of an accident reaches hundreds of G while a fall leads to a reading of several G. There is no publicly available data (nor are there research articles) regarding the actual efficacy of the proposed application. Its last version was released to the Apple iTunes store on January 30th, 2017 (as the app is already unavailable through Google Play), and the user reviews are unfortunately very unfavourable, therefore while we fully acknowledge the effort of this startup company, we are unable to treat this application as a reliable reference point for presenting our research results.

The experiments presented in this paper show that it is not possible to create a credible crash-detection method by using only phone accelerometer readings threshold. The limitations and variety of sensors installed in contemporary smartphones require far more sophisticated algorithms, which will be presented in the following sections.

3   Crash Tests and Data Acquisition

To acquire the necessary data, we prepared dedicated mobile applications. The applications were intended to run as a background service and collect all possible information available on the mobile device. The following list enumerates and describes data gathered by our software:

- GPS - latitude, longitude, bearing and speed,
- accelerometer - acceleration for three coordinate axes,
- gyroscope - rate of rotation for three coordinate axes,
- magnetometer - magnetic field for three coordinate axes,
- gravity - force of gravity for three coordinate axes,
- rotation vector - orientation of the device.

It is important to note that many smartphone models are not equipped with all the sensors. To have reliable representation of a typical driver, we gathered phone usage statistics from the Internet. Based on that research, we selected 20 leading phone models from all popular manufacturers. The selected models varied from the cheapest ones up to flagship models. Thanks to this variety of phones, we have the access to sensor chips which vary up to the ranges and thresholds of measurements.

The applications were used in our day-to-day life for a few months. During that time many different types of drives were recorded: traffic jams while driving to work in a big city, casual rides through towns and country roads, and long journeys on highways. Selected parts of the collected data were then used as a counter example for recorded crashes.

In order to collect reliable set of data from accidents we conducted series of crash tests using real cars. In order to verify the smartphone measurements, we installed a professional device, the PicDAQ5 [2] (Data Aquisition Platform) designed by the Dr. Steffan Datentechnik for vehicle dynamics and crash test research. The basic technical data of the device:

- two 3-axial accelerometers (± 1.5 g and $\pm$ 200 g) for vehicle dynamics and impact testing respectively,
- one 3-axial angular velocity gyro-sensor ($\pm$ 300 deg/s for roll, pitch and yaw),
- 15 analog input channels (12 bit resolution),
- four digital inputs,
- user selectable sampling rate up to 1 kHz per channel (in our measurements: 500 Hz),
- 15 analog input channels are alternative usable for wheel revolutions, steering wheel angle and other measurements, 5 Hz GPS receiver,
- PC-based analysis software – PocketDAQ Analyzer.

The first set of experiments was conducted without drivers inside cars. Three vehicles were used to perform tests that were intended to imitate casual accidents in a parking lot, light clashes with one stationary car, and finally a head-on collision (see Fig. 1).

Fig. 1. Head-on collision at 40.3 kph.

Because of the fact that there was no stuntman, we had limited control over the vehicles. Moreover, those tests could not be performed at very high speeds. During the crashes, the smartphones were placed in storage compartments near the gear shift and on the doors. There was always one device mounted using a mount holder on the windshield.

The second event was set up by our partner, PZU (the largest insurance company in Poland and at the same time one of the EU-wide players in the insurance market), at a race course in Poznań, the largest track in Poland. All experiments were performed by professional stuntmen. During those tests we had an opportunity to collect data from different scenarios. The first one was driving onto a curb at a high speed. That scenario was followed by a rear collision and a head-on collision with a tree. (Fig. 2). The final test was a side collision at the highest speed recorded. The phones were mounted similarly as in the first crash event.

Fig. 2. Results of the collision with a tree (27.7 kph).

To sum up, in both experiments we have collected data from 22 unique crash events. We have prepared 53 crash examples collected by different phones, including 36 crash examples gathered by phones with gyroscope.

4   Proposed Solution

4.1   Data Analysis

The collected raw data covered a time frame that included the vehicles approaching one another before the collision and, in most cases, moving away from each other afterwards as well. To be able to analyze only those readings representing the behavior of the vehicles during a crash, we removed the unnecessary portions of the data using recorded footage of the events. In most cases the time span of useful data covers 1–3 s, of which 200–400 ms represent the time of physical contact between the vehicles.

Accelerometer is the most commonly used sensor in this application, as it is designed to measure the forces acting on the object to which it is attached. A sample of accelerometer data collected from one of the collisions can be seen in Fig. 3. In this specific event, one vehicle was stationary with the handbrake engaged and its front oriented towards the second, moving vehicle. The collision occurred at a speed of 20 kph, the cars hit each other with the left corners of their hoods. Subfigures of Fig. 3 represent the magnitudes of acceleration vectors recorded during the same event by different devices:

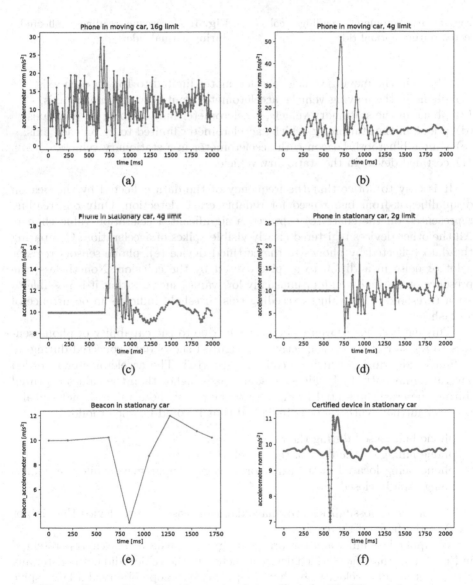

Fig. 3. Comparison of accelerometer data collected with different devices

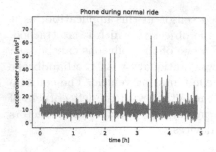

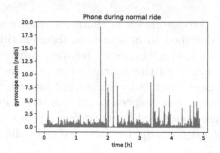

Fig. 4. Accelerometer readings collected during normal ride

Fig. 5. Gyroscope readings collected during normal ride

(a) phone in the moving vehicle, accelerometer limited to 16g on each axis,
(b) phone in the moving vehicle, accelerometer limited to 4g on each axis,
(c) phone in the stationary vehicle, accelerometer limited to 4g on each axis,
(d) phone in the stationary vehicle, accelerometer limited to 2g on each axis,
(e) simple bluetooth beacon with accelerometer in a stationary vehicle,
(f) certified device in the stationary vehicle.

It is easy to notice that the frequency of the data collected by the beacon disqualifies it from being used for reliable crash detection. Only one reading represented on the chart (e) captured a significant acceleration value change. All the other devices registered clearly visible spikes of acceleration. Comparing the data collected by phones to the certified device (e), phone sensors register a lot of noise in addition to a spike caused by the collision. Nonetheless, the presence of unusually high or unusually low values suggests that it is possible to set a threshold – any values exceeding this threshold indicate an occurrence of a crash.

This method has proven to be incorrect due to the capability of phone sensors to register other vehicle activities. Figure 4 shows data collected during an ordinary ride, during which no collision occurred. The accelerometer recorded several significantly high values, in some cases higher than the values captured during the crash presented in Fig. 3. The source of those values is not certain, however further synthetic tests indicated that it could be one of following:

– phone being used during the ride,
– phone falling on the floor of the vehicle,
– phone being located in the car door storage compartment while the door is being rapidly closed.

All of these possibilities are unavoidable in case of any device that is not fixed to the body of the car.

We applied a similar analysis process to gyroscope data. The charts presented in Fig. 6 show that the certified device attached to the vehicle did not register any significant angular velocity (b), but the phone gyroscope observed a large spike (a). These result were promising, showing that an unattached device is capable

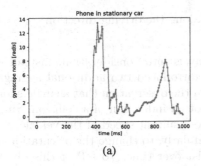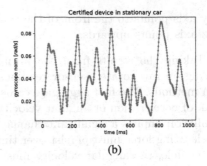

(a)   (b)

Fig. 6. Comparison of gyroscope data collected on different devices

of registering much more significant angular motion compared to a more massive vehicle. Unfortunately, an analysis of the readings from an ordinary ride shown in Fig. 5 suggests that the threshold method would yield numerous false positives in this case as well.

4.2 Data Filtering and Aggregation

Before inputting collected data to any detection algorithm, we had to change the coordinate system of the collected data. With the device positioned flat on a surface with the screen facing upwards, the axes of the initial system are oriented as follows:

- x axis points to the right of the screen,
- y axis points to the top of the screen,
- z axis points upwards.

Since all the sensor axes are fixed in reference to the phone, a movement of the device causes a movement of the axes in relation to the vehicle. To counteract these changes, we used the virtual "rotation vector" sensor built into most modern devices. This sensor represents a rotation quaternion that, if applied to any other 3-axis sensor, reorients its axes to the following:

- x axis points to the geographic east,
- y axis points to the geographic north,
- z axis points upwards (away from the center of Earth).

To further clarify, those two systems are identical when the phone is positioned perfectly flat with its screen facing upwards, its display's top edge facing north and its display's right side facing east.

However, the x and y axes of the resulting system are still independent from the axes of the vehicle. As the final step, we rotated the coordinate system around the z axis to obtain the following axes in relation to the vehicle:

- x axis points to the right side of vehicle (perpendicular to the driving direction),

– y axis points to the front of the vehicle (along the driving direction),
– z axis points upwards.

Taking the results from the initial analysis into consideration, instead of relying on raw rotated sensor readings, we introduced two additional aggregation methods. The first aggregate comes from the observation that a single short peak of acceleration or angular velocity is not an indication of a collision. Since a collision implies a significant change in the resultant speed of the vehicle, the accelerating force must persist over time. Similarly, to change the orientation of the vehicle, the angular velocity must persist over time. To reflect this observation, we aggregated accelerometer and gyroscope readings into a sequence of integrals calculated from each axis in a sliding time window.

Given $[v_1, v_2, ..., v_n]$ - values along a single axis, $[t_1, t_2, ..., t_n]$ - times of collecting those values, we calculate trapezoidal integrals as in (1) and sum integrals within a time window of size k as in (2).

$$tr_i = \frac{(v_{i+1} + v_i) \cdot (t_{i+1} - t_i)}{2} \tag{1}$$

$$trw_i^k = \sum_{j=i}^{i+k-1} tr_j \tag{2}$$

The second aggregate is inspired by the observation that in multiple cases both the peak and the integral are reduced by an unrestricted motion of the phone. The same resultant loss of velocity is distributed over a longer time. As a result, sensors generated an oscillating pattern that did not reach values as large as the single peak in other cases, but the change of amplitude between any two consecutive readings was noticeable. As a result, we the introduced second method of aggregation which captured abrupt changes in data.

Using the same notation: $[v_1, v_2, ..., v_n]$—values along a single axis, $[t_1, t_2, ..., t_n]$—times of collecting those values, we calculate average absolute derivatives over time as in (3) and sum these values within a time window of size k as in (4).

$$ds_i = \frac{|v_{i+1} - v_i|}{t_{i+1} - t_i} \tag{3}$$

$$dsw_i^k = \sum_{j=i}^{i+k-1} ds_j \tag{4}$$

4.3   Collision Detection Method

Following the introduction of aggregation methods we attempted the threshold approach once again, but to no avail. Neither one of the aggregates carried enough information to precisely determine whether a crash has occurred or not.

As a next attempt, we took another observation into consideration. Phone sensors are limited in the maximal value that can be registered along any single axis. As a result, devices with lower limits may experience "clipping" of readings if the actual value would exceed the limit. To address this issue we grouped devices by the limits of their sensors and calculated thresholds for each group separately. Again, we were able to identify all the collisions, but at the cost of a significant number of false positive indications.

Since the method of calculating thresholds for aggregates did not yield satisfying results, we decided to use a decision tree. This method accepts a series of labeled examples and, based on the observed relations, creates a ruleset for assigning labels to new, unlabeled examples. The single example consists of a number of values, each representing one feature of the observed phenomenon, while a label represents the class of the observation.

Although this method itself also operates on thresholds on the low level, it is able to produce complex rules involving several input values. Moreover, the decision tree is able to identify the most important features - something that we utilized later.

Due to the fact that a significant fraction of modern phones is not equipped with a gyroscope, we created two sets of features, which implied two different decision trees. The set of features for an accelerometer-only tree included the following 17 features:

- raw accelerometer—x axis, y axis, z axis, magnitude,
- accelerometer integrals—x axis, y axis, z axis, magnitude,
- accelerometer average derivatives—x axis, y axis, z axis, magnitude,
- absolute accelerometer values—x axis, y axis, z axis,
- accelerometer threshold and frequency.

The magnitude of the absolute values is identical to the magnitude of raw data, therefore it is not included to avoid duplication. The set of features for a gyroscope and accelerometer tree included additional 17 features calculated using gyroscope values in the same manner. Our set of labels included two classes: crash and non-crash.

As a first iteration, we trained a decision tree with a depth limit of 8 levels. This tree was created in order to identify the most important features. In this process we learned which features are crucial in determining the class of the observation. Then, we repeated the learning process, but in this iteration we limited the depth to 3 levels and reduced the feature set to the 8 most significant ones obtained in the previous iteration. This process was repeated for the both tree types: including and excluding gyroscope. As a final step, we tested the 3-level trees, and we present our results in the next section.

5   Results

In order to verify the effectiveness of the decision tree and compare it to the threshold-based algorithm, we collected necessary measurements from the crash tests described in Sect. 3. For negative samples (data without crashes) we used measurements obtained in two long rides between cities in Poland. Half of this data was used as negative examples for training and the rest was used for evaluation. In these road experiments we utilized several mobile devices with different accelerator and gyroscope limits.

First, we present the results for the threshold-based detection. In Fig. 7 we plot the maximum value of acceleration integral registered in each crash (per each involved device). In Fig. 8 we present the histogram of acceleration integrals during normal rides. Finally, in Fig. 9 we plot an average number of false positives per hour of normal ride versus crash detection sensitivity. Note that in this and subsequent plots we consider any sequence of detections shorter than 5 s as one positive detection. This reflects the fact that during accidents, or false positives due to e.g. poor road conditions, violent phone movements may span an interval lasting several seconds. Results in Fig. 9 demonstrate that threshold based detection is unreliable due to excessive number of false alarms.

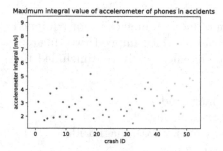

Fig. 7. Maximum value of accelerometer integral registered by devices in crashes

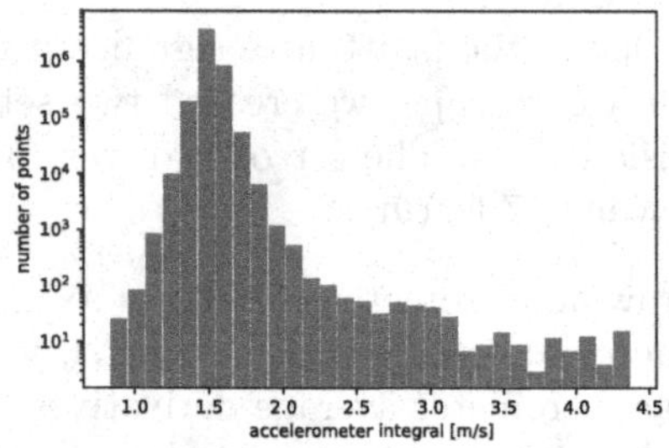

Fig. 8. Histogram of accelerometer integral values during normal rides

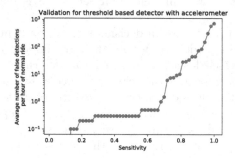

Fig. 9. Sensitivity vs. number of false positives for threshold based crash detector (using accelerometer data only)

Because the number of collected crash data instances is limited, we opted to evaluate the performance of decision tree detector with leave-one-out cross validation. Therefore, each detector was trained on all but one crash data instance and then used to classify that held-out positive example. In each case we used the same set of negative examples. Also, all trained decision trees were run on test data from normal rides, where we counted the number of false positive detections. Figure 10 report an average number of false positive detections per hour of normal ride versus crash detection sensitivity for detectors that use only the accelerometer data. Results for accelerometer and gyroscope sensors are reported in Fig. 11.

As we can see, decision tree detector gives vastly better results than simple threshold based detection. For a sensitivity of 80%, the threshold-based detector gives around 10 false positive detections per hour of driving. For a decision tree detector this number is around 0.6 when using accelerometer data and around 0.5 for devices with accelerometer and gyroscope.

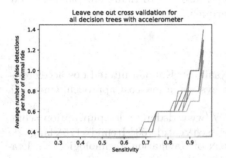 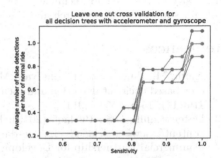

Fig. 10. Sensitivity vs. number of false positives for decision tree detectors that uses only the accelerometer sensor.

Fig. 11. Sensitivity vs. number of false positives for decision tree detectors that uses both accelerometer and gyroscope.

It is important to mention here that the positive data collected in our tests came from recorded accidents in which speeds varied from 12 kph to 43.5 kph, while the negative data came from drives where speeds reached up to 140 kph. The increased number of false positives at high sensitivity, shown in Figs. 9, 10 and 11 may result directly from the very small margin between the interface of negative and positive examples in which readings from the accelerometer and gyroscope at low speed accidents may resemble some readings from fast car driving.

6   Conclusions

Although we have dealt with certain problems of data quality and spotted significant differences between the data readings in different smartphones, we are sure that these problems can be overcome by processing methods for the reduction of the inevitable false-positives detection.

It is important to note that in the real world application of the presented solution, false positives can be accepted and mitigated by introducing a dedicated loopback approach: the emergency notification service should first try to contact the driver. It is quite obvious that false-negatives might do much more damage and therefore a reasonable attitude must be maintained towards the classification efficacy of the proposed system.

In the future we are planning to further extend the application, perform more real-world tests based on individual drivers and fleets and extend the features of the application, focusing on driving-style modelling and displaying suggestions relevant to the driver when necessary based on their current driving style and the features of the route.

Acknowledgment. This research was partially supported by the Polish National Center for Research and Development under the project no. TANGO2/340869/NCBR/2017 and by the funds of Polish Ministry of Science and Higher Education assigned to AGH University of Science and Technology. We would also like to thank our partners from PZU for supporting the research and the experiments.

References

1. Amin, M., Ibne Reaz, M., Bhuiyan, M., Nasir, S.: Kalman filtered gps accelerometer based accident detection and location system: a low-cost approach. Curr. Sci. **106**(11), 1548–1554 (2014)
2. Datentechnik, D.S.: Picdaq5 (2019). http://www.dsd.at/index.php?option=com_content&view=article&id=327:pic-daq-deutsch&catid=53&Itemid=175&lang=de
3. Commercial Partnership for Development N of Navigation Technologies, U.: Era-glonass (2015). http://en.glonassunion.ru/era-glonass/
4. Devices, B.C., GmbH, S.: Telematics ecall plug (2019). https://www.bosch-connectivity.com/products/telematics-ecall-plug/
5. Group, P.: Pzu go (2019). https://www.pzu.pl/pzugo
6. Jiansheng, F.: Vision-based real-time traffic accident detection. In: Proceeding of the 11th World Congress on Intelligent Control and Automation, pp. 1035–1038 (2014). https://doi.org/10.1109/WCICA.2014.7052859
7. Octo: Insurance telematics solutions for personal auto (2019). https://www.octousa.com/personal-insurance-telematics-solutions/
8. OnStar: Welcome to onstar (2019). https://www.onstar.com/us/en/home/
9. World Health Organization: Global status report on road safety 2018: Summary (2018). https://www.who.int/violence_injury_prevention/road_safety_status/2018/en/
10. European Parliament: European commission decision 585/2014/eu on the deployment of the interoperable eu-wide ecall services. https://eur-lex.europa.eu/legal-content/EN/TXT/?uri=CELEX:32014D0585 (2015)
11. European Parliament: European commission regulation 2015/758 on ecall type-approval ecall amending directive 2007/46/ec. https://eur-lex.europa.eu/legal-content/EN/TXT/?uri=CELEX:32015R0758 (2015)
12. SpA, S.: Automatic car crash detection app (2015). http://www.sosmartapp.com/
13. Telematics, C.M.: The drivewell tag (2019). https://www.cmtelematics.com/drivewell-tag/
14. Thompson, C., White, J., Dougherty, B., Albright, A., Schmidt, D.C.: Using smartphones to detect car accidents and provide situational awareness to emergency responders. In: Mobile Wireless Middleware, Operating Systems, and Applications, pp. 29–42 (2010)

15. U.S. Department of Transportation: National highway traffic safety administration (2019). https://www.nhtsa.gov/
16. Visser, M.: Overview on emergency call worldwide - status 2016 (2016). https://www.its-mobility.de/download/eCallDays2016/presentations/Visser_Gemalto_eCallDays_2016.pdf
17. White, J., Thompson, C., Turner, H., Dougherty, B., Schmidt, D.C.: Wreckwatch: Automatic traffic accident detection and notification with smartphones. Mob. Netw. Appl. **16**(3), 285 (2011)
18. Zaldivar, J., Calafate, C.T., Cano, J.C., Manzoni, P.: Providing accident detection in vehicular networks through obd-ii devices and android-based smartphones. In: 2011 IEEE 36th Conference on Local Computer Networks, pp. 813–819 (2011). https://doi.org/10.1109/LCN.2011.6115556
19. Öörni, R., Meilikhov, E., Korhonen, T.O.: Interoperability of ecall and era-glonass in-vehicle emergency call systems.IET Intell. Transp. Systems **9**(6), 582–590 (2015)

Corrosion Detection on Aircraft Fuselage with Multi-teacher Knowledge Distillation

K. Zuchniak[1,3](), W. Dzwinel[1], E. Majerz[1], A. Pasternak[1], and K. Dragan[2]

[1] Department of Computer Science, AGH University of Science and Technology,
Cracow, Poland
zuchniak@agh.edu.pl
[2] Department of Airworthiness, Air Force Institute of Technology, Warsaw, Poland
[3] Neuralbit Technologies sp. z o. o, Cracow, Poland

Abstract. The procedures of non-destructive inspection (NDI) are employed by the aerospace industry to reduce operational costs and the risk of catastrophe. The success of deep learning (DL) in numerous engineering applications encouraged us to check the usefulness of autonomous DL models also in this field. Particularly, in the inspection of the fuselage surface and search for corrosion defects. Herein, we present the tests of employing convolutional neural network (CNN) architectures in detecting small spots of corrosion on the fuselage surface and rivets. We use a unique and difficult dataset consisting of 1.3×10^4 images (640×480) of various fuselage parts from several aircraft types, brands, and service life. The images come from the non-invasive DAIS (D-Sight Aircraft Inspection System) inspection system, which can be treated as an analog image enhancement device. We demonstrate that our novel DL ensembling scheme, i.e., multi-teacher/single-student knowledge distillation architecture, allows for 100% detection of the images representing the "moderate corrosion" class on the test set. Simultaneously, we show that the proposed ensemble classifier, when used for the whole dataset with images representing various stages of corrosion, yields significant improvement in the classification accuracy in comparison to the baseline single ResNet50 neural network. Our work is the contribution to a relatively new discussion of deep learning applications in the fast inspection of the full surface of an aircraft fuselage but not only its fragments.

Keywords: Aircraft maintenance · Deep learning · Ensemble learning · Knowledge distillation · Fuselage corrosion detection · DAIS system

1 Introduction

Corrosion, fatigue, and corrosion-fatigue cracking are the most common types of structural problems experienced in the aerospace industry. To ensure flight safety of aircraft structures, it is necessary to have regular maintenance by using visual methods of non-destructive inspection (NDI) [16]. Traditionally, visual

© Springer Nature Switzerland AG 2021
M. Paszynski et al. (Eds.): ICCS 2021, LNCS 12747, pp. 318–332, 2021.
https://doi.org/10.1007/978-3-030-77980-1_25

inspections are conducted by human operators that scan the aircraft fuselage looking for corrosion, cracks, and incidental damage. However, this is a costly and time-consuming procedure apt to be subjected to human mistakes caused by mental fatigue and boredom.

In the last decade, various image processing algorithms have been applied in the field of aircraft inspection [18]. However, these algorithms work well only in controlled environments. They often fail in more complex real-world scenarios due to noisy and complex backgrounds. Therefore, used together with the classical machine learning models, fine-tuned image processing techniques are strongly biased by the type of datasets considered.

The success of deep learning in many domains of science and engineering, particularly, the efficacy of various convolutional neural network (CNN) architectures in producing amazingly accurate data models for images (in terms of classification, object recognition, semantic segmentation and others) encouraged us to test this technology as an autonomous support for the inspection system of wide-area surface of the aircraft fuselage. The data for our research come from the imaging acquisition system DAIS (D-Sight Aircraft Inspection System) [11] widely used by the Polish air force and collected by the Air Force Institute of Technology (AFIT). DAIS images are able to enhance the hidden corrosion spots invisible to the naked eye in similar lighting conditions.

To decrease the costs simultaneously increasing the reliability of this time-consuming procedure, herein we propose to support it by the autonomous system based on advanced neural network architectures. From application point of view, the main target of this research is to improve and partially automate aircraft fuselage inspections. Additionally, the research aspect of this work, not directly related to the domain of aircraft inspection, is the use of knowledge distillation as an ensemble learning aggregation mechanism. We can summarize our contributions as follows:

1. We have tested several CNN architectures on DAIS images and estimate their various degree of usefulness in recognition of corroded fuselage rivets.
2. To solve the problems with high data inhomogeneity - data coming from many types of airships of various ages, very typical ones in many inspection/fault detection systems - we propose using the ensemble learning concept to increase generality and to deal with overfitting. Moreover, we modified the knowledge distillation framework [10], to allow its aggregation from the whole *multi − teacher* ensemble into only one *student* model.
3. We have developed a novel method for mimicking ensemble output by the knowledge distillation employing a multi-teacher/single-student network. This type of knowledge distillation allows training a single CNN of a similar accuracy as the CNN ensemble but requiring more modest resources (i.e. storage size and shorter response time). The experiments show on the test data that it yields superior accuracy among other tested CNNs architectures.

Summing up, we demonstrate that proposed CNN architectures have sufficient classification power to be considered as a valuable support in the wide-area inspection of the aircraft fuselage.

In the following section, we shortly present the main idea of the DAIS image acquisition system and the dataset that is the subject of our study. Then we describe the methodology proposed, i.e., (1) the ensemble learning scheme and (2) a new knowledge distillation variant based on the multi-teacher/single-student approach. Next, we present a detailed description of the experiments and discuss their results. Our approach and its results can be confronted with the state-of-the-art DL applications in the aircraft inspection in the Related Work section. Finally, we summarize the conclusions and suggest future research goals.

2   Methodology

To deal with overfitting problem, resulting from the relatively small number of examples and the high complexity of our data set, we use ensemble learning. We trained several models whose aggregate predictions were more accurate compared to a single model. On the other hand, the use of ensemble increases computational complexity of our solution, which is also treated as a big disadvantage. To solve this problem, we use the knowledge distillation, transferring the objective knowledge of the entire ensemble to the weights space of only one model. Ensembling models of various types (e.g., formal mathematical models and data models), usually lead to better results, i.e., better approximations, predictions or classification accuracy [20]. However, this is not for free but at the expense of the increase of model storage&time complexity. Therefore, the high demand for computational resources required by big data models (such as ensembled DNNs) is still a challenging problem.

The proposed methodology combines ensembling and knowledge distillation approaches into a new multi-teacher/single-student approach. In the following subsections, we present shortly its principles on the background of ensembling and knowledge distillation techniques.

2.1   Ensemble Learning

The main purpose of ensembling the models is to increase the accuracy of predictions [20]. Especially, in the cases of very fine image details and high uncertainty caused by inhomogeneity of data. This is just the case to be encountered in the detection of corrosion on small fuselage rivets from very inhomogeneous data coming from many types of airships of various ages. Thus, the data set consists of many "fuzzy" small subsets having a specific structure (fine graining) while we are looking for common attributes of corrosion (coarse-graining) neglecting fine structural details of data. Therefore, the ensembling of neural nets, each trained on a different homogeneous part of the baseline data set, is an encouraging idea to be applied in the diagnostic of aeronautical structures. The benefits of this DNNs architecture in the context of the aircraft fuselage inspection is presented in the seminal publication [18].

On the other hand, ensemble learning can increase the computational complexity of the predictive model. In the classic implementation of ensembling of neural networks (e.g. bagging), each single sub-model is generated in an independent training process. Consequently, N independent models increases the computational complexity of the classifier N times both in the training and inference phase. In general, however, it is not exactly the truth. The sub-nets can be pre-trained in considerably shorter time than a single baseline model [3,6,22] or the architecture of the sub-models can be much simpler than the baseline model.

2.2 Knowledge Distillation

The main idea behind the knowledge distillation is that the simpler *student* model mimics the complex *teacher* model resulting with its better interpretation or obtaining a simpler and competitive black box with similar or even superior performance. In this way, the knowledge inscribed in the *teacher* model weights is compressed and transferred into the parameter space of the *student's* model. This technique was popularized by Hinton et al. in [10]. In the standard training process of a classifier, the loss function is closely related to the data labels. In the case of knowledge distillation, the loss function has a second component related to the distance between *teacher* and *student* output logits. More details and variants of knowledge distillation are presented in the survey paper by Gou et al. [7]. Versatility resulting from the concept of knowledge distillation comes primarily from the lack of requirements for types of the *teacher* and *student* models. This technique is most often used to compress machine learning models based on neural networks whose architectures are very similar, differing only in the number of layers and neurons and weights in each layer. On the other hand, there are no formal requirements as to the type of ML model used. It is possible to distill knowledge between machine learning models of completely different structure, type and principle of operation. It is sufficient to ensure that both models have the same output and input structure.

2.3 Multi-teacher/Single-Student Network

In our approach, we assumed the lack of formal restrictions of knowledge distillation (comparing hidden layers activations requires consistency between *teacher* and *student* architectures). We treat the entire ensemble (composed of sub-nets trained on unique, randomly generated subsets of training dataset) as the *teacher* model. As result, we can use knowledge distillation as an ensemble *decision fusion* scheme. The *student* model learns to mimic predictions of the whole ensemble of the *teacher* sub-models.

There have been several studies that utilize knowledge distillation as ensemble aggregation [2,25]. However, there are a few important differences between those and our approaches. The major modification assumes to transfer the decision about the aggregation of individual sub-models from the stage before knowledge distillation to the *student* model itself. The main task of the *student* model is not to imitate some aggregation of *teachers'* outputs, e.g., averaging them, but all individual *teachers* "cooperate" during training synchronously developing a more sophisticated common response. We also decided not to include in the loss function the factor representing the similarity between *teacher* and *student* in internal NN layers. Our loss function forces only mimicking the *teacher* output predictions by the *student* model. This approach gives more flexibility in terms of knowledge distillation between models of a different type (similar architecture is not required), which we plan to use in the future. The temperature parameter T in *softmax* probabilities

$$P_i = \frac{e^{\frac{y_i}{T}}}{\sum_{k=1}^{n} e^{\frac{y_k}{T}}} \tag{1}$$

is set to 1 what means the loss function utilizes unchanged *softmax* output.

We analyzed three variants of the multi-teacher/single-student architecture, with different sub-models prediction-aggregation schemes.

1. **Prediction averaging** - Currently used approach [24] consists in averaging the ensemble predictions before the *teacher* output is included in the *student* loss function. The *student* model learns to mimic the average response of the $multi - techer$ ensemble (Fig. 1 upper).
2. **Mimic of prediction geometric center** - In the training process, the output of the *student* model is compared with the predictions of all N *teachers* individually. The *student* model learns to mimic predictions of several *teachers* simultaneously. However, since bringing prediction too closely to a single *teacher* output increases part of the loss function responsible for mimicking other *teachers*, *student* model output settles in the geometric center of all the *teachers'* predictions (Fig. 1 center).
3. **Independent mimicking of all the N teachers** - In contrast to the model presented above the *student* model does not produce a single output, but N outputs - where N is equal to the number of *teachers*, each is characterized by an independent set of trainable weights, see Fig. 1 lower. The last layer or the last few layers may be separated. Each of these independent outputs in the training process is compared with its assigned *teacher* output. It should be noted that the convolution part responsible for the feature extraction is common. However, the weights of the last layer (or last few layers) responsible for classifications are specific to each *teacher*. This way, the model does not learn to mimic the aggregation of all *teacher's* outputs but actually generates N independent predictions linked to each *teacher*.

As shown in [24] there are many loss function definitions, different distance matrices, distillation strategies *et cetera*. We decided to use hard *ground true* labels and hard ensemble outputs, instead of light labels (in which the labels do not have the entire probability assigned to one class, but it is partially "fuzzified" to other classes). We also used the Kullback-Leibler divergence (KLD) to determine the distance between *teacher* and the *student* models. Below we present respective equations determining loss function for $teacher - student$ models variants described above.

$$Loss_{avg} = \alpha \sum_{i=1}^{D} \bar{y}_i \cdot log(\frac{\bar{y}_i}{\tilde{y}_i}) - (1 - \alpha) \cdot \sum_{i=1}^{D} y_i log(\tilde{y}_i) \tag{2}$$

$$Loss_{geo} = \alpha \frac{1}{N} \sum_{i=1}^{D} \sum_{j=1}^{N} y_{ij} \cdot log(\frac{y_{ij}}{\tilde{y}_i}) - (1-\alpha) \cdot \sum_{i=1}^{D} y_i log(\tilde{y}_i) \qquad (3)$$

$$Loss_{ind} = \frac{1}{N} \sum_{j=1}^{N} \cdot (\alpha \sum_{i=1}^{D} y_{ij} \cdot log(\frac{y_{ij}}{\tilde{y}_{ij}}) - (1-\alpha) \cdot \sum_{i=1}^{D} y_i log(\tilde{y}_{ij})), \qquad (4)$$

where D is the *student* output size (number of classes), N is number of *teachers*, $\tilde{y}_i$ is the i-th scalar value in the *student* model output, y_i is the corresponding target value, $\bar{y}_i$ is the corresponding average of *teachers* model output, y_{ij} is the corresponding j-th *teacher* output, and $\tilde{y}_{ij}$ is the i-th scalar value in j-th output of *student* model output (independent mimicking variant). The α weight setting proportion between the expression associated with knowledge distillation (first sum in equations) and the standard loss function connected with data *ground truth* (second sum), is the process controlling parameter, increasing this parameter, increases student imitation loss in the total loss function. Figure 1 demonstrates the block diagrams of all multi-teacher/single-student networks described above in the order presented in the text.

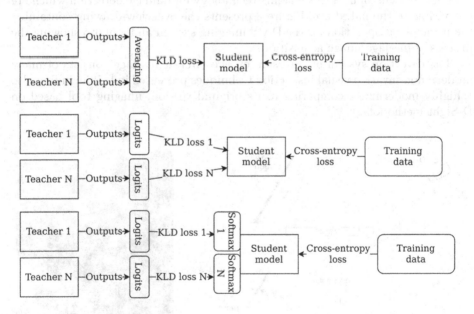

Fig. 1. The schemes of the multi-teacher/single-student models employed in this paper: *prediction averaging* model (upper), the model *mimic of prediction geometric center* (middle), *independent mimicking of all the N teachers* (bottom).

3   DAIS Data

3.1   DAIS System

D-Sight [9,13] is an optical double-pass retroreflection surface inspection technique created by Diffracto Ltd from Canada. It is a patented method of visualizing very small surface distortions outside the plane, such as dents and corrosion. The D-Sight optical system consists of a retroreflective screen, camera, a light source, and a tested fuselage fragment (Fig. 2). The light from a standard divergent source is reflected off the sample. The surface of the sample must be reflective. The reflected light is then shone onto a reflective screen, which consists of many semi-silvered glass spheres (typical diameter 60 μm). This screen tries to redirect all incident light rays at the same angle to the starting point of reflection on the sample surface. However, the screen is not perfectly reflective and actually returns a divergent cone of light rather than a single beam at the same angle. It is this imperfection of the reflective screen that creates the D-Sight effect. The light is reflected again by the sample and collected by a camera slightly away from the light source.

DAIS system [5] uses this imaging technology for damage detection which are not visible to the naked eye. Figure 2 presents the overview drawing containing the principle of operation for the DAIS imaging system and a photo showing the process of fuselage image acquisition.

The detection system of corrosion on the aircraft fuselage consists of many modern non-invasive visual inspection techniques presented in [17] including also a highly modernized, comparing to its original version, imaging tool based on D-Sight methodology.

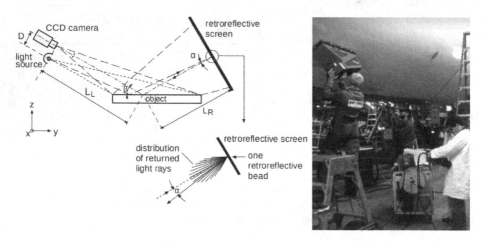

Fig. 2. Left: a scheme of DAIS imaging system operation, from [9] Right: the image demonstrating the process of captioning aircraft images.

3.2 Fuselage Corrosion Dataset

Thanks to the Air Force Institute of Technology (Warsaw, Poland), we got access to data representing the images acquired by using D-Sight technology. We received about 1.3×10^4 labeled images (640×480 pixels). The labels include the testing year, the anonymous *id* of the aircraft, and the label representing the extent of corrosion damage.

Figure 3 shows the data details, i.e., the frequency distribution of samples according to the year of technical examination and an aircraft *id*. Sample images from the DAIS system are also shown. We aim to classify the images according to the strength of the identified damage. Due to the imbalanced data set, we decided to consider this problem as a binary classification: "no damage" and "damage detected". The original images come in 640×480 resolution, however, following the guidelines of the authors of the models used, we have reduced the resolution to 320×240 - for training and inference speed up. The tests, carried out while training the models at full resolution, showed a minimal decrease in the classification accuracy [15].

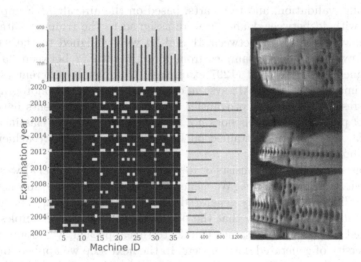

Fig. 3. Left: Distribution of the image examples in DAIS dataset with machine id and inspection year, Right: DAIS samples. Top: no corrosion, Center: light corrosion, Bottom: moderate corrosion. Total number of examples in the data set by class: no corrosion and no damage:6431, light corrosion:6040, moderate corrosion:578, strong corrosion:0, minor damage:26. Histograms presents marginal distribution according to machine ID and examination year dimension.

4 Results and Discussion

4.1 Hardware and Software Setup

The computations were performed on the Prometheus supercomputer (288th on top500 list (June 2020); HP Apollo 8000, Xeon E5-2680v3 12C 2.5 GHz, Infiniband FDR, HPE Cyfronet Poland). We used just one node (Intel Xeon E5-2680 v3, 2.5 GHz) and two Nvidia V100 GPU accelerators on the cluster dedicated to deep learning. In the computations We used TensorFlow framework [1].

4.2 Experiment Description

Using previous analyzes, we have selected ResNet50 [8] as the baseline CNN architecture. We show in the supplementary materials [15] that this architecture produces the best and more stable results comparing to the others.

ResNet50 training setup are as follows: ADAM optimizer[12], learning rate $= 0.001$, batch size $= 128$, number of epochs $= 150$. The dataset was split into training, validation, and test parts, based on the aircraft id. Samples from machines with id between 1 and 30 were assigned to the training data set (10 534 examples), from $id =$ between 31 and 34 were assigned to the validation set (1463 examples) while samples from aircraft with id between 35 and 37 were assigned to the test set (1297 examples). Completely different physics of acquiring images from the DAIS system compared to standard photography was reason of resigning from use of transfer learning. Our experiments have shown that use of pre-trained models does not improve the quality of classification on DAIS data, we test models trained on ImageNet100 [4], and we achieve lower classification accuracy.

The ensemble classifier consists of ResNet50 sub-nets (*teachers*) was trained on different training subsets. We generated many ResNet50 sub-nets, each trained on different, randomly generated subset of training data. The examples were generated in such a way that the percentage of common examples for any two selected subsets was the same. Thanks to this approach, we obtain the maximum diversity of generated data subsets. In the next step we applied knowledge distillation to aggregated (trained) ensemble of *teachers* into a single *student* model. We chose a number of sub-models $N = 5$ and *coverage* factor equal to 0.7 (defined as the size of the training subset relative to the entire training set).

4.3 Corrosion Detection

We tested and compared three main approaches (based on ResNet50 architecture) described in previous section in order to develop corrosion classifier. Depending on the threshold level, We can modify the trade-off between the number of false negatives and false positives. Figure 4a shows precision and recall metrics depending on the threshold value. We define the threshold as the minimum value of the probability of assigning a sample to the "corrosion detected" class. Images labeled as "corrosion detected" came from several more specific classes representing various degrees of material failure. We conducted the analysis for these specific corrosion classes, comparing how the models dealt with samples labeled as "light corrosion" and "moderate corrosion". The results appeared to be very promising. On the test set our models were able to recognize 100% "moderate corrosion" samples (with the appropriate threshold level). Unfortunately, the test set of examples with "moderate corrosion" is limited to only 79 examples. On the other hand, from the application point of view, the detection of stronger examples of corrosion is the most important and can be the positive test for the usefulness and reliability of our detection algorithm. For safety reasons, in the operation of the autonomous corrosion detection system, the detection of stronger corrosion samples is crucial. It should also be remembered that the whole data set was manually labeled by experts and this may be the reason for the existence of some bias (incorrect markings for pairs "no corrosion" - "light corrosion"). Figure 4b demonstrates different recall curves for specific corrosion levels.

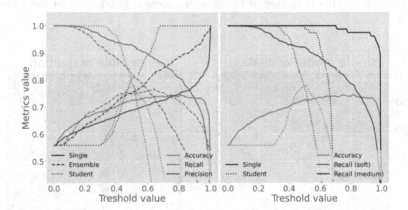

Fig. 4. Left: Precision-recall characteristics for the models considered. The intersection of recall and precision lines is at the highest point for the *student* model. Right: Precision-recall characteristic with separation for "light" and "moderate" corrosion levels. The "moderate corrosion" samples are much better recognized by the models. We achieved 97.5%–100% detection of corrosion on this level.

To determine the appropriate threshold values for a fair comparison of the various methods, we assessed them independently for each model. The maximum classification accuracy achieved on the validation set was the selection criterion for the thresholds. Then we calculated the remaining metrics on the test set. For the thresholds selected in this way, the geometric *student* model achieves detection of 100% "moderate corrosion" class while the ensemble and single models get 97.5%. It gives also superior results on the other metrics. The results are collected in Table 1.

Table 1. Accuracy, recall, precision and F1 score matrices obtained by tested classifiers. Complexity* is expressed as a relative value, where 1 means complexity level of a single ResNet50 base model

Used model	Threshold	Accuracy	Recall	Precision	F1 Score	Com*
Single	0.89	73.6%	73.96%	77.75%	75.81%	1
Ensemble	0.62	76.25%	74.81%	81.24%	77.89%	5
Averaging student	0.54	74.14%	69.63%	81.43%	75.07%	1
Geometric student distillation	0.47	**76.63%**	**84.26%**	76.39%	**80.13%**	1
Multi output student	0.51	74.3%	69.78%	**81.6%**	75.23%	1

To visualize, which areas of analyzed images influence the decision on corrosion classification we use the Grad-CAM [21] method. The algorithm employs the cumulative gradients calculated in back-propagation which are treated as "weights" to explain network decisions. It can be seen that the greatest activations are generated on the riveting line (see Fig. 5). As hidden corrosion occurs on the rivets, so this behavior of the model shows a good level of data understanding.

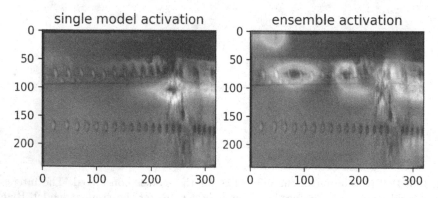

Fig. 5. Grad-CAM activations for a single baseline model (left) and the ensemble (right) model. The activation map for ensemble is much wider. From the explainability point of machine learning models, we can determine that ensemble takes more factors into account when generating predictions.

Additionally, we used t-SNE [14] data embedding method to visualize the localization of samples from the test set in 2D space. The *Single* model and the geometric *student* model were compared. The feature vectors are collected from the output of the global max-pooling layer, which follows the last convolutional layer. The resulting feature vector had 2048 dimensions. Figure 6 shows this feature vector embedding into a 2D space for visualization purposes. It is easy to observe a strong separation between the "moderate corrosion" and "no corrosion" classes. The "light corrosion" class lies in the middle area and partly overlaps "no corrosion" class. This result coincides with the classification metrics achieved by the model for individual classes. Data points were normalized to better cover plot canvas. We calculated Silhouette coefficient [19] (*Single*: .0015, *student*: .0359) to quantitatively show that *student* produce better clustering (higher coefficient score means better clusters class separation).

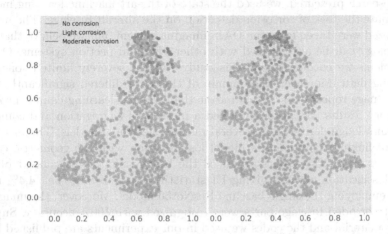

Fig. 6. DAIS samples embedding by using t-SNE. The *Single* model and the geometric *student* model were compared, respectively

5   Related Works

Very few works on aircraft fuselage inspection using modern DNN architectures can be found in the literature. In [23] a deep learning-based framework is proposed for automatic damage detection in aircraft engine borescope inspection. It utilizes the state-of-the-art NN model called Fully Convolutional Networks (FCN) to identify and locate damages from borescope images. This framework can successfully identify two major types of damages, namely cracks and burn, and extract ROI regions on these images with high accuracy. In [16] the authors present the system for crack detection on the aircraft fuselage based on high-resolution drone images.

The similar, in spirit, work to that presented here is described in [18]. The authors had a modest dataset consisting of images from a borescope inspection of aircraft propeller blade bores. Due to the limited size of the data, they used the transfer learning by pre-training a convolutional neural network on the large ImageNet, assuming that the low-order features will be the same for both datasets. It is shown that the ensemble method improves inspection accuracy over conventional single CNN. However, the borescope is designed to assist visual inspection of narrow, difficult-to-reach cavities but not to cover big areas of aircraft fuselages. Thus, the data used in this system are completely different from those considered in this paper. However, the success of the application of CNN's ensemble was the inspiration of our paper.

6   Conclusions and Future Work

In the research presented, we used the state-of-the-art machine learning models to automate the task of corrosion detection on the aircraft fuselage. The images we analyzed were taken from the DAIS imaging system, but we believe that this methodology can be also applied with other visual inspection systems. One of the problems we encountered in this study was the severely limited collection of training data. Moreover, the domain of this data differed significantly from popular image repositories, which meant that transfer learning cannot produce satisfactory results. We have developed a method of aggregation and compression of knowledge derived from several machine learning models. The proposed variant of linking the *student's* model loss function with the geometric center of *teachers* ensemble outputs produced the best results in the context of corrosion classification efficacy, giving F1 statistics equal to 80%, i.e., 4,4% more than by employing the single baseline ResNet50 model. Moreover, the images of the most corroded fuselage parts were recognized with 100% accuracy. Supplementary material and the codes we used in our experiments are published here: (https://github.com/ZuchniakK/DAISCorrosionDetection).

In the nearest future, we intend to expand our dataset, which will allow us to generate more accurate models and perform better and more certain validation. We also plan further work on the NN model compression and quantization to enable its implementation directly in DAIS hardware. In the future we intend to test our proposed multi-teacher ensembling framework on the other difficult data sets such as medical images.

The *Geometric student distillation* method we proposed generated the best performing models, but the differences are small. We believe that student generation methods can still be improved and will be the subject of our further research.

Acknowledgements. The research was supported by the funds assigned to AGH University of Science and Technology by the Polish Ministry of Science and Higher Education and in part by PL-Grid Infrastructure. We thank dr Jerzy Komorowski (former General Manager for National Research Council Canada) and professor dr Stan Matwin (Director of Institute for Big Data Analytics, Dalhousie University) for their contribution to this research.

References

1. Abadi, M., et al.: TensorFlow: large-scale machine learning on heterogeneous systems (2015). http://tensorflow.org/, software available from tensorflow.org
2. Asif, U., Tang, J., Harrer, S.: Ensemble knowledge distillation for learning improved and efficient networks. arXiv preprint arXiv:1909.08097 (2019)

3. Bukowski, L., Dzwinel, W.: Supernet-an efficient method of neural networks ensembling. arXiv preprint arXiv:2003.13021 (2020)
4. Deng, J., Dong, W., Socher, R., Li, L.J., Li, K., Fei-Fei, L.: Imagenet: a large-scale hierarchical image database. In: 2009 IEEE Conference on Computer Vision and Pattern Recognition, pp. 248–255. IEEE (2009)
5. Forsyth, D., Komorowski, J., Gould, R., Marincak, A.: Automation of enhanced visual NDT techniques. In: Proceedings 1st Pan-American Conference for NDT, Toronto, Canada, pp. 107–117 (1998)
6. Garipov, T., Izmailov, P., Podoprikhin, D., Vetrov, D.P., Wilson, A.G.: Loss surfaces, mode connectivity, and fast ensembling of DNNs. In: Advances in Neural Information Processing Systems, pp. 8789–8798 (2018)
7. Gou, J., Yu, B., Maybank, S.J., Tao, D.: Knowledge distillation: a survey. arXiv preprint arXiv:2006.05525 (2020)
8. He, K., Zhang, X., Ren, S., Sun, J.: Deep residual learning for image recognition. In: Proceedings of the IEEE Conference on Computer Vision and Pattern Recognition, pp. 770–778 (2016)
9. Heida, J., Bruinsma, A.: D-sight technique for rapid impact damage detection on composite aircraft structures (1997)
10. Hinton, G., Vinyals, O., Dean, J.: Distilling the knowledge in a neural network. arXiv preprint arXiv:1503.02531 (2015)
11. Jerzy, P., Komorowski, R.W., Gould, D.L.S.: Application of diffracto sight to the nondestructive inspection of aircraft structures. Rev. Prog. Quant. Nondestr. Eval. **12**, 449 (2012)
12. Kingma, D.P., Ba, J.: Adam: a method for stochastic optimization. arXiv preprint arXiv:1412.6980 (2014)
13. Komorowski, J.P., Gould, R.W., Simpson, D.L.: Synergy between advanced composites and new NDI methods. Adv. Perf. Mater. **5**(1–2), 137–151 (1998)
14. Van der Maaten, L., Hinton, G.: Visualizing data using t-sne. J. Mach. Learn. Res. **9**(11), 2579–2605 (2008)
15. Majerz, E., Pasternak, A.: System for analyzing damage to the surface of aircraft stryctures using convolutional neural networks. Bachelor's thesis, AGH University of Science and Technology (2021). https://github.com/majerzemilia/engineering-thesis
16. Malekzadeh, T., Abdollahzadeh, M., Nejati, H., Cheung, N.M.: Aircraft fuselage defect detection using deep neural networks. arXiv preprint arXiv:1712.09213 (2017)
17. Niepokólczycki, A., Leski, A., Dragan, K.: Review of aeronautical fatigue investigations in poland (2013–2014). Fatigue Aircraft Struct. **2016**(8), 5–48 (2016)
18. Ren, I.: An Ensemble Machine Vision System for Automated Detection of Surface Defects in Aircraft Propeller Blades. Ph.D. thesis, Georgia Institute of Technology (2020)
19. Rousseeuw, P.J.: Silhouettes: a graphical aid to the interpretation and validation of cluster analysis. J. Comput. Appl. Math. **20**, 53–65 (1987)
20. Sagi, O., Rokach, L.: Ensemble learning: a survey. Wiley Interdisc. Rev. Data Mining Knowl. Disc. **8**(4), e1249 (2018)
21. Selvaraju, R.R., Cogswell, M., Das, A., Vedantam, R., Parikh, D., Batra, D.: Gradcam: Visual explanations from deep networks via gradient-based localization. In: Proceedings of the IEEE International Conference on Computer Vision, pp. 618–626 (2017)

22. Sendera, M., Duane, G.S., Dzwinel, W.: Supermodeling: the next level of abstraction in the use of data assimilation. In: Krzhizhanovskaya, V.V., et al. (eds.) ICCS 2020. LNCS, vol. 12142, pp. 133–147. Springer, Cham (2020). https://doi.org/10.1007/978-3-030-50433-5_11
23. Shen, Z., Wan, X., Ye, F., Guan, X., Liu, S.: Deep learning based framework for automatic damage detection in aircraft engine borescope inspection. In: 2019 International Conference on Computing, Networking and Communications (ICNC), pp. 1005–1010. IEEE (2019)
24. Wang, L., Yoon, K.J.: Knowledge distillation and student-teacher learning for visual intelligence: a review and new outlooks. arXiv preprint arXiv:2004.05937 (2020)
25. Zhu, X., Gong, S., et al.: Knowledge distillation by on-the-fly native ensemble. In: Advances in Neural Information Processing Systems, pp. 7517–7527 (2018)

Warm-Start Meta-Ensembles
for Forecasting Energy Consumption
in Service Buildings

Pedro J.S. Cardoso[1,2]([✉]) [ID], Pedro M.M. Guerreiro[2] [ID], Jânio Monteiro[2,3] [ID],
André S. Pedro[2] [ID], and João M.F. Rodrigues[1,2] [ID]

[1] LARSyS, Universidade do Algarve, Faro, Portugal
{pcardoso,jrodrig}@ualg.pt
[2] ISE, Universidade do Algarve, Faro, Portugal
{pmguerre,jmmontei,aspedro}@ualg.pt
[3] INESC-ID, Lisbon, Portugal

Abstract. Energy Management Systems are equipments that normally perform the individual supervision of power controllable loads. With the objective of reducing energy costs, those management decisions result from algorithms that select how the different working periods of equipment should be combined, taking into account the usage of the locally generated renewable energy, electricity tariffs etc., while complying with the restrictions imposed by users and electric circuits. Forecasting energy usage, as described in this paper, allows to optimize the management being a major asset.

This paper proposes and compares three new meta-methods for forecasts associated to real-valued time series, applied to the buildings energy consumption case, namely: a meta-method which uses a single regressor (called Sliding Regressor – SR), an ensemble of regressors with no memory of previous fittings (called Bagging Sliding Regressor – BSR), and a warm-start bagging meta-method (called Warm-start Bagging Sliding Regressor – WsBSR). The novelty of this framework is combination of the meta-methods, warm-start ensembles and time series in a forecast framework for energy consumption in buildings. Experimental tests done over data from an hotel show that, the best accuracy is obtained using the second method, though the last one has comparable results with less computational requirements.

Keywords: Warm-start ensembles · Meta ensembles · Decision tree regressors · Energy consumption forecasting · Time series

This work was supported by the Portuguese Foundation for Science and Technology (FCT), project LARSyS - FCT Project UIDB/50009/2020 and by the European Union, under the FEDER (Fundo Europeu de Desenvolvimento Regional) and INTERREG programs, in the scope of the T2UES (0517_TTUES_6_E) project.

M. Paszynski et al. (Eds.): ICCS 2021, LNCS 12747, pp. 333–346, 2021.
https://doi.org/10.1007/978-3-030-77980-1_26

1   Introduction

In 2016, the European Union presented a package of measures with the aim of providing a stable legislative framework to facilitate the transition process to renewable energy. In this context, Regulation (EU) 2018/1999 [12], required that all member states should prepare and submit to the European Commission, a National Energy and Climate Plan (NECP), with a medium-term perspective (Horizon 2021–2030). E.g., Portugal's main goal in the NECP is to become neutral in greenhouse gas emissions by 2050, which requires to comply with trajectories that lead to a reduction in greenhouse gas emissions between 85 and 90% by 2050. To achieve this goal, the greatest reduction in emissions will have to be achieved in the current decade, with decreases ranging between 45 and 55%. One of the several goals defined by Portugal in its NECP [14], was that until 2030, 80% of the electric energy that is consumed, should come from renewable energy sources. Such a high percentage of self-sufficiency cannot be achieved by solely generating more energy. It also requires consuming less energy and adapting the consumption pattern with the generation levels, based on Demand Response (DR) measures. A reduction in consumption can either be accomplished using deterministic or data-driven methods [10]. Deterministic methods are mainly used in the project phase of new buildings, allowing the proper design of the building structure and materials to be used. Data-driven methods are mostly used in functional buildings, identifying the normal consumption levels in existing infrastructures, allowing the identification of its consumption targets, or the definition of a set of levels for the evaluation of the consumption profile.

In terms of DR, Energy Management Systems (EMSs) [18,40] are the equipment that normally performs the individual supervision of shiftable/power controllable loads. Optimized management decisions result from algorithms that select how the different working periods of equipment should be combined, observing the generated energy, energy tariffs etc., while complying with the restrictions imposed by users and electric circuits [24]. E.g., decisions can be made using mathematical optimization, model predictive control or heuristic control, with several methods requiring a look into the future, i.e., forecasting energy generation and the building's consumption, before deciding how loads should be scheduled to work[6,23]. Similarly, the detection of anomalies can be exposed if the real consumption deviates from the one that was forecasted, and many situations exists in which minor changes in consumption can indicate a serious problem. In any case, it might be important to have a human to judge alarms, as unpredicted values can be normal due to naturally extraordinary events or anomalous due to machine failure, malfunctioning of a sensor etc. So, either when using DR or performing an assessment of the efficiency of buildings, a prediction of the consumption is a tool to help decision makers in theirs tasks.

In this context, large service buildings, such as hotels, shopping centers, hospitals, schools, offices, public buildings etc., have great variability in their consumption of resources, such as energy or water. Their energy consumption values depend on many variables (e.g., occupation, outside temperature, solar radiation, appliances settings or building occupation) making it very difficult to predict those values with precision. Having a human, for example a specialized engineer, to analyze each of these situations is obviously unaffordable from a technical and economic point of view. The solution is to develop an artificial intelligence system, that can learn over time what is considered normal and detect what is abnormal for each building. Machine learning algorithms will make it possible to combine different variables such as time of day, day of week, temperature, radiation, occupation, and electricity consumption to predict the normal consumption of the building and the respective expected deviation. Distinct methods can be used to obtain this [37].

This paper proposes and compares three new meta-methods for the forecast associated to real-valued time series, namely, energy consumption in buildings. With a short and a full memory variant, the meta-methods will be supported on well known regression methods to implement a sliding window solution which will forecast the energy consumption of an hotel at certain instants. The first method, called Sliding Regressor – SR, is somehow a standard method which uses a single regressor given as a parameter, being fitted with the latest data just before forecasting is required. Similar in the fitting moment, the second method, called Bagging Sliding Regressor – BSR, uses an ensemble of regressors with no memory of previous fittings. The third method, called Warm-start Bagging Sliding Regressor – WsBSR, also uses an ensemble of regressors, however, it is distinct from the BSR by maintaining the previous regressors in memory. Preserving the ensemble idea, the latter allows to fit less regressors, a step with high computational cost, while using a broad number of regressors to make the forecasts. The use of the WsBSR is therefore a possibility as it achieves slightly worst results but with a fraction of the computational requirements. The paper's main contribution to the state of the art is the combination of the proposed meta-methods, warm-start ensembles, and time series in a forecast framework for energy consumption in buildings.

The remaining paper is structured as follow. Section 2 describes the problem and presents a brief summary of the state of the art. The third section explains the proposed methods. Experimental results are given in Sect. 4 and the last section presents a conclusion and future work.

2   Preliminary Considerations

This section describes the problem and presents a brief state of the art in the resolution of forecasting real-valued time series problems.

2.1   Problem Description

Time series regression are methods for forecasting a future numeric value based on historical responses. Time series regression can help to understand and predict the behavior of dynamic systems from observed data, being commonly used for modeling and forecasting economic, financial and biological systems [26].

This paper proposes a set of meta-methods to predict the energy consumption of buildings. Independently of the number and type of parameters that might differ significantly depending on the available data (e.g., the number and type of sensors that equip the building), it is assumed that we are given a set of N_F dependent variables or features, $(x_1, x_2, \ldots, x_{N_F})$, and want to forecast an independent variable or target, y. Furthermore, for experimental purposes but generalizable, it will be assumed that observations are indexed with a timestamp and the target will be the energy consumption of a building. For training and fitting purpose, it is also assumed that observations will be made inside a time window $\mathcal{W} = [t_s, t_f]$, between the initial, t_s, and final, t_f, instants. Since the observations are made in discrete instants, N_O observations are assumed in the interval $\mathcal{W}$, at moments $\mathcal{T} = \{t_s = t_1, t_2, \ldots, t_f = t_{N_O}\} \subset \mathcal{W}$, being $\Omega = \{(X_t, y_t) : t \in \mathcal{T}\}$ the set of observations, where $X_t = (x_{t,1}, x_{t,2}, \ldots, x_{t,N_f})$ are the features values and y_t the corresponding target value. To simplify the description, it is also assumed that observation are taken at regular intervals of time, δ, i.e., $\mathcal{T} = \{t_1, t_1 + \delta, t_1 + 2\delta, \ldots, t_1 + (N_O - 1)\delta\} \subset \mathcal{W}$.

In the training phase, methods will forecast future consumption from a specific moment in time, t_p, and the following setup and goals are considered: (a) Methods will forecast n consumptions with a granularity δ in the period $\Delta_{t_p} = [t_p + \delta, t_p + n\delta]$, i.e., the methods will do forecasts for instants $\mathcal{F}_{t_p} = \{t_{p+1}, t_{p+2}, \ldots, t_{p+n}\} = \{t_p + \delta, t_p + 2\delta, \ldots, t_p + n\delta\}$. (b) During the tuning phase, developers can adjust $\mathcal{F}_{t_p}$ to known (future) values of the dependent (e.g., temperature, occupation) and independent variables, included in the observation set Ω. This allows to use metrics to identify the best conjunction of parameters (see Sect. 2.3). In the production phase, methods will be fed with forecasted values for the independent variables and predict the dependent one. (c) In the fitting and training phases, methods will have available a time window of historical data, with timestamps in $[t_p - \delta_W, t_p]$. Depending on the size of the time window, i.e., the interval of data used to fit the regressors, two types of methods can be considered, namely: the ones that use all historical data, called full memory (FM) methods, and the ones which only use "recent" data, called short memory (SM) methods. δ_W is a parameter which allows to define how long should we go into the past. Obviously, the FM methods can be considered as a sub case of the SM methods as δ_W can be as big as desired.

2.2   Forecasting Time Series Methods

In many energy efficient situations, the data obtained from water consumption, electricity, outdoor temperature, occupation or solar radiation, refer to the same time and place are characterized as multimodal data in the field of Machine Learning. In many of these cases, it is difficult to determine how these different types of data relate or how one modality relates to another. Distinct methods have been defined to perform the prediction of consumption. Globally these methods can be classified in time series versus non-time series techniques.

Time series techniques analyse the data, recorded over equal intervals of time, extracting statistical information and other characteristics from it. Time series solutions can be classified in univariate versus multivariate methods. It is considered an univariate time series if there is a single sequence of values in an observation, while if exists multiple sequences of values in an observation, we have a multivariate time series. Several methods specified for univariate time series analysis were used in energy prediction, for instance the Autoregressive Integrated Moving Average (ARIMA) [37], Case-Based Reasoning [22], Support Vector Machines (SVM) [8,25], Artificial Neural Networks (ANN) [13,20], Grey prediction models [9], Moving Average [16,32], Exponential Smoothing [5] or Fuzzy Time Series [19,29].

Energy consumption normally vary according to other variables like date-time, outside/inside temperatures, humidity, solar radiation and building occupancy. When several time series variables are evaluated together, multivariate time series should be used. Some of the models used in multivariate analysis include the Vector Auto-Regressive [7] method, the vector ARIMA [35], Vector Autoregressive Moving Average [17] and the Bayesian Vector Autoregression [15]. While all these models have been used in predictions, some have not been used in energy consumption forecast.

Besides time series methods, other methods have also been used in energy prediction. These include Regression Analysis [33], Decision Trees [34], and k-Nearest Neighbours (KNN) algorithms [36]. For instance, in [34] a comparison is made between Regression Analysis, Decision Trees and Neural Networks when predicting electricity energy consumption. In [36] the KNN method was applied to the prediction of energy consumption in residential buildings.

Many studies have emphasized the superior performance of Ensemble and Hybrid models [28], as for instance [1,38], which led to the development of different solutions in energy [4,31]. For instance, in [31] an evolutionary multi-objective ensemble learning solution was tested for the prediction of Electricity Consumption, in [4] an ensemble learning framework was created for anomaly detection in energy consumption of buildings, and in [11] a stacking ensemble learning was proposed for short-term prediction of energy consumption. In [21] the authors compared ANN and SVM with an Hybrid Method that combines both, concluding the later achieves the best accuracy. A survey of time series prediction applications using SVM is presented in [30]. A tree-based ensemble method with warm-start gradient for short-term load forecasting is proposed in [39].

Nevertheless, to the best of the authors' knowledge, no meta-method framework which combines bagging, warm-start, and forecasting of energy time series was ever proposed.

2.3 Scoring Methods

To compare the different models, some metrics must be calculated. These metrics measure the distance between the prediction of the model and the real observations [2]. In regression this might be relatively straightforward as we are comparing real numbers, but several metrics are available, each one with different strengths. Considering a set of observation $\mathcal{O} = \{(X_t, y_t) : t \in \{1, 2, \ldots, m\}\} \subset \Omega$ and a function $pred_M$ associated to model M, that predicts y_t given X_t, $\hat{y}_t = pred_M(X_t)$, it is possible to define several performance metrics. For instance, $MAE = \frac{1}{m} \sum_{t=1}^{m} |y_t - \hat{y}_t|$ defines the Mean Absolute Error, $MSE = \frac{1}{m} \sum_{t=1}^{m} (y_t - \hat{y}_t)^2$ the Mean Square Error, $MAPE = \frac{1}{m} \sum_{t=1}^{m} |y_t - \hat{y}_t|/|y_t|$ the

---

Algorithm 1. Forecasting with the Sliding Regressor method – SR

Require: Forecasting instant (t_p); Unfitted regressor (R); Set of observations (Ω);
   Size of the training window (δ_W); Set of features values $(\mathcal{X} = \{X_t : t \in \mathcal{F}_{t_p}\})$.
1: $\mathcal{D} \leftarrow$ select data from Ω with timestamp in $[t_p - \delta_W, t_p]$.
2: Fit the regressor R using $\mathcal{D}$.
3: **return** $\{(t, pred_R(X_t)) : X_t \in \mathcal{X}\}$ ▷ *Predict target using regressor R*

Mean Absolute Percentage Error, and $R^2 = 1 - \sum_{t=1}^{m} |y_t - \hat{y}_t| / \sum_{t=1}^{m} |y_t - \bar{y}_t|$, where $\bar{y}_t = \frac{1}{m} \sum_{t=1}^{m} y_t$, defines the Coefficient of Determination score.

3 Proposed Methods

This section presents the three proposed meta-methods, namely: Sliding Regressor, Bagging Sliding Regressor, and Warm-start Bagging Sliding Regressor. The objective was to build a method independent framework, combining bagging and warm-start, for the forecasting of energy consumption time series.

3.1 Sliding Regressor Method – SR

The Sliding Regressor (SR) meta-method, in its broad sense, is a traditional regression method that fits a model using given data and a mandatory parameterized regressor (e.g., Decision Tree regressor or Lasso regressor) [2]. The name was chosen thinking that the method will use a sliding time window when applied to the forecast of time series. Two variants can be considered: the full memory method (SR-FM) will use all available data to train the model, while the short memory model (SR-SM) will use a interval of data (the size is a parameter of the method) previous to the instance in time when predictions/forecasts are to be made.

In operation phase, given an initial instant t_p from which predictions are to be made, the SR method will forget any previous fits and refit the regressor using data in the interval $[t_p - \delta_W, t_p]$ (as previously stated, δ_W is a parameter which allows to define how long should we look in the past for data). Then, the fitted method will forecast the consumption in a set of future instants, $\mathcal{F}_{t_p} = \{t'_1, t'_2, \ldots, t'_n\}$. To make the forecast, the fitted model needs the values of the features in those instants $\mathcal{X} = \{X_t : t \in \mathcal{F}_{t_p}\}$. In this case, features might be known in advance or be forecasted themselves. For instance, it will be necessary to forecast the temperature or occupation of the building for the next predicting period (if those are features to be considered). Algorithm 1 sketches the procedure. This method was implemented in order to have a base line for the methods proposed in the next sections (Sects. 3.2 and 3.3).

3.2 Bagging Sliding Regressor Method – BSR

The Bagging Sliding Regressor (BSR) method extends the SR method by using a bag/ensemble of estimators. On other words, instead of using a single regressor

---

Algorithm 2. Forecasting with the Bagging Sliding Regressor method – BSR

Require: Forecasting instant (t_p); Unfitted regressor (R); Set of observations (Ω); Size of the training window (δ_W); Set of features values $(\mathcal{X} = \{X_t : t \in \mathcal{F}_{t_p}\})$; Size of the bag of regressors (N_p); Percentage of data use to fit regressors (p).

1: $\mathcal{R} \leftarrow \emptyset$ ▷ *Bag of regressors*
2: **for** $i \in \{1, 2, \ldots, N_p\}$ **do**
3:     $\mathcal{D} \leftarrow$ randomly select $p\%$ of data from Ω with timestamp in $[t_p - \delta_W, t_p]$.
4:     $R_i \leftarrow$ clone of regressor R
5:     Fit regressor R_i using $\mathcal{D}$
6:     $\mathcal{R} \leftarrow \mathcal{R} \cup \{R_i\}$
7: **end for**
8: **return** $\left\{ \left(t, \frac{1}{N_p} \sum_{R \in \mathcal{R}} pred_R(X_t) \right) : X_t \in \mathcal{X} \right\}$

---

at every prediction instant, t_p, the BSR fits N_p regressors and for each forecast returns the mean value of the predictions forecasted by each of those regressors. Like the SR, the BSR method has full memory (BSR-FM) and short memory (BSR-SM) variants, which are implemented using parameter δ_W. Furthermore, the BSR method is parameterized by the percentage/fraction of data, p, used to fit each of the regressors. If $p < 1$ the regressors will be fitted with distinct sets of data, as data is selected before each of the regressors is fitted. Algorithm 2 sketches the procedure. Although the method can be used in continuous operation, it can also be applied with full potential in any fixed moment in time (becoming a traditional ensemble method), since it is independent of previous fits, which differs from the following method.

3.3 Warm-Start Bagging Sliding Regressor Method – WsBSR

The Warm-Start Bagging Sliding Regressor (WsBSR) considers that the system is fitted in a continuous operation. On other words, a bag of regressors $\mathcal{R}$ is maintained and amplified in regular periods, which can coincide with the forecast moments. So, when forecast are to be made, N_r new regressors are fitted, either in full or short memory (WsBSR–FM or WsBSR–SM) variants, and those regressors are added to an existing bag of regressors. Then, to make a forecast, N_p regressors are selected from the bag using their ages as weights (younger models have higher probability of being chosen) obtaining $\mathcal{R}' \subset \mathcal{R}$. Finally, the forecast value is the mean value of the individual predictions, $\frac{1}{N_p} \sum_{R \in \mathcal{R}'} pred_R(X_t)$, for some set of feature values X_t. Algorithm 3 sketches the procedure.

This section proposed three meta-methods that computational complexity. Next section will experimentally try to conclude on the utility of using one over the others.

4 Data and Experiments

For the experimental setup it was used data collected in the Alto da Colina hotel. The Alto da Colina hotel [3] is a 4-star aparthotel located in Albufeira,

---

Algorithm 3. Forecasting with the Warm-Start Bagging Sliding Regressor method – WsBSR

Require: Forecasting instant (t_p); Unfitted regressor (R); Set of observations (Ω);
   Size of the training window (δ_W); Set of features values $(\mathcal{X} = \{X_t : t \in \mathcal{F}_{t_p}\})$;
   List of already fitted regressors ordered by age $(\mathcal{R} = [R_1, R_2, \ldots, R_k])$; Number of
   regressors to fit (N_r); Number of regressors to use in the predictions $(N_p, N_p \geq N_r)$;
   Percentage of data use to fit regressors (p).
1: **for** $i \in \{1, 2, \ldots, N_r\}$ **do**
2:     $\mathcal{D} \leftarrow$ randomly select $p\%$ of data from Ω with timestamp in $[t_p - \delta_W, t_p]$.
3:     $R_i \leftarrow$ clone of regressor R.
4:     Fit regressor R_i using $\mathcal{D}$ and append the new model to $\mathcal{R}$.
5: **end for**
6: $\mathcal{R}' \leftarrow$ Select N_p models from $\mathcal{R}$ using their ages as weights (younger models have
   higher probability of being chosen).
7: **return** $\left\{ \left(t, \frac{1}{N_p} \sum_{R \in \mathcal{R}'} pred_R(X_t) \right) : X_t \in \mathcal{X} \right\}$

---

in the south of Portugal. It is comprised of 174 apartments and contains several facilities, including four outdoor and one indoor swimming pools, a football field, tennis court, gymnasium, a small water park for children and several bars and restaurants, most of these, relying in electric supply. The only exception is the water heating system, that relies in a combination between solar panels (108 Solar Thermal panels) and gas, through two propane boilers.

Data was collected every 15 min between January 1st and October 30th, 2017, with the exception of some periods where the system failed to collect some of the features, being those observations discarded. After cleaning up the data, 23.854 observations were considered with the following features: hour, wind velocity, exterior temperature, number of registered guests, weekday and energy consumption. Figure 1 shows the daily mean consumption (left axis) and number of guests (right axis) for the period in analysis. Although we don't have data after October (the hotel closes in the Autumn), it is observable that the hotel suffers from a estival effect with its higher consumption in the summer months.

4.1   Experimental Setup

For experimental purposes, it was decided to have a daily based forecast with the following setup and goals. To analyse the methods it was decided to run them for the full period of known data (January 1st to October 30th, 2017). More precisely, at midnight the method will predict the consumption for the following day or days, defined by a δ_T parameter. In the first phase, it was decided to set $\delta_T = 1$ which, given the gaps in the data acquired, corresponds to 237 predicted days. Since each day has 96 readings (made each 15 min), the forecast was made for those known values, allowing us to score our predictions (see Sect. 2.3). Regressor will have available a time window of data defined by δ_W (depending on its size, it is used the full or the short memory methods). Finally, the chosen metrics are applied to the forecasts. Algorithm 4 summarizes the procedure.

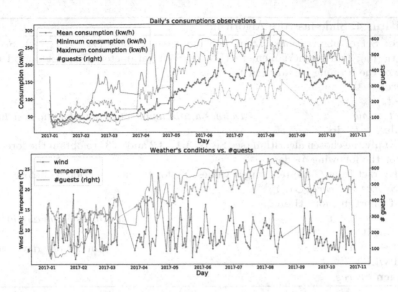

Fig. 1. Variation of the daily consumption and number of guests in the analyzed period (top); Variation in the weather conditions (mean exterior temperature, in °C, and mean wind velocity, in km/h) related with the number of guests (bottom).

The proposed methods are meta-regressors since they require a regressor to make forecasts. In this experimental setup, it was decided to use Decision Tree (DT) Regressors [2,34] for its precision and simplicity. This was a thought decision although it is known that these methods are not adequate for extrapolation, which might be the case whenever higher or lower values of consumption are reached. So, it was used the DT implementation in Scikit-learn [27] with two parameterizations: (a) maximum depth – with nodes expanded until all leaves are pure and (b) limited depth – with a maximum depth of 5 and minimum number of samples required to split an internal node equal to 10. Furthermore, before applying the methods it was also decided to analyse two transformation to the data, namely: polynomial features with degree, $p_d \in \{1, 2, 3\}$, and the scaling of the features to specific ranges, $\mathcal{I} = \{None, [--1, 1], [0, 1], [0, 100]\}$, where $None$ means no transformation. Other parameters are summarized in Table 1. Finally, experiments were run on a Intel(R) Core(TM) i7-4770 CPU @ 3.40GHz with 16Mb or RAM and the Kubuntu 20.10 operating system.

4.2   Computational Results

Considering the use of DT with maximum depth, Table 2 presents for each of the three methods the 10 best R^2 means (μ_{R^2}) and corresponding standard deviations (σ_{R^2}) values, computed considering the 237 forecasted days (with $96 = 24 \times 4$ forecasts *per* day). Best results are achieved by BSR method with the 10 higher mean R^2 values standing between 0.884 and 0.900. The second best method is WsBSR, followed by SR in third. Regarding parameters, it is

---

Algorithm 4. Slide-fit-score Algorithm

Require: Set of observations (Ω) in the full time window ($\mathcal{W} = [t_s, t_f]$); Scoring
metric (ϕ); Full or short memory (FM or SM) parameter; Forecast period (δ_T);
Size of the training window (δ_W).
1: $Scores \leftarrow \emptyset$
2: $s \leftarrow t_s$
3: $t \leftarrow t_s + \delta_W$ ▷ *Defines a minimum set of data for the first fitting*
4: **while** $t \leq t_f$ **do**
5:     Apply the chosen algorithm (see Sects. 3.1, 3.2 and 3.3) to obtain the forecasts,
       $\hat{Y}$, for the following δ_T days.
6:     Extract from Ω the real target values, Y, for the same period as before.
7:     $Scores \leftarrow Scores \cup \phi(Y, \hat{Y})$
8:     **if** Short memory **then**
9:         $s \leftarrow s + 1$ ▷ *Moves s one day forward*
10:    **end if**
11:    $t \leftarrow t + 1$ ▷ *Moves t one day forward*
12: **end while**
13: **return** $Scores$

Table 1. Parameters used in the experimental phase ([†]– used in SR, [‡]– used in BSR,
and [*] - used in WsBSR).

Parameter	Values	Observation
δ_W	7	Number of days to fit in the short memory case[†,‡,*]
δ_T	1	Number of days to score[†,‡,*]
p	[0.1, 0.5, 1]	Percentage of data used to fit regressor[‡,*]
p_d	[1, 2, 3]	Polynomial feature transformation degree[†,‡,*]
$\mathcal{I}$	[$None$, [−1, 1], [0, 1], [0, 100]]	Scaling ranges of the features values[†,‡,*]
N_p	[5, 10, 50]	Number of regressors to make prediction[‡,*]
N_r	[1, 5, 10]	Number of new regressors to fit each time[*]

observable that using polynomial features with degree 2 seems to provide the
best results. The scaling to some interval does not show to have big influence
as results are mixed. In terms of memory, the short memory variant looks to
provide the best results, as 20 of the 30 results presented were achieved using it.
Furthermore, for the tested values, it seems also advisable to use all (i.e., 100%)
available data to fit the regressor in the short memory case and 50% of it in the
full memory case. Referring to the number of regressors in the bagging methods,
in the BSR it looks like "the more the merrier" while in the WsBSR using a
large number of regressors in the predictions seems to be worse as, probably, this
obliges to use more older regressors. As a side note, to understand the magnitude
of the errors, the second best run of BSR (an unscaled case, with $\mu_{R^2} = 0.899$)
had $\mu_{MAE} = 4.82$ ($\sigma_{MAE} = 3.1$) and $\mu_{MAPE} = 0.04$ ($\sigma_{MAPE} = 0.02$).

Table 3 presents the same metric values but now considering the use of DT
with limited depth of 5 and minimum number of samples required to split an
internal node equal to 10. As somehow expected, the value of R^2 got worse but

Table 2. Top 10 R^2 results for the case of the DT with full depth.

Table 3. Top 10 R^2 results for the case of the DT with maximum depth of 5.

Method	μ_{R^2}	σ_{R^2}	p_d	$\mathcal{I}$	Memory	p	N_p	N_r
SR	0,771	0,288	2	[0, 100]	SM			
	0,766	0,308	2	None	SM			
	0,751	0,274	3	[−1, 1]	SM			
	0,745	0,338	2	[−1, 1]	SM			
	0,736	0,491	2	[0, 1]	SM			
	0,714	0,516	2	[−1, 1]	FM			
	0,690	0,460	3	[0, 100]	SM			
	0,679	0,422	3	[0, 1]	SM			
	0,676	0,678	2	[0, 100]	FM			
	0,675	0,782	2	[0, 1]	FM			
BSR	0,900	0,141	2	[0, 100]	SM	1	50	
	0,899	0,142	2	None	SM	1	50	
	0,898	0,142	2	[0, 1]	SM	1	50	
	0,897	0,143	2	[−1, 1]	SM	1	50	
	0,893	0,163	2	[0, 100]	FM	0.5	50	
	0,890	0,173	2	[0, 1]	FM	0.5	50	
	0,890	0,178	2	[−1, 1]	FM	0.5	50	
	0,887	0,184	2	None	FM	0.5	50	
	0,885	0,155	2	[0, 100]	SM	1	10	
	0,884	0,160	2	None	SM	1	10	
WsBSR	0,838	0,185	2	[0, 1]	SM	1	10	5
	0,833	0,174	2	None	SM	1	10	5
	0,833	0,174	2	[0, 100]	SM	1	10	5
	0,826	0,205	2	[−1, 1]	SM	1	10	5
	0,809	0,184	2	[0, 100]	SM	0.5	10	5
	0,804	0,221	2	[0, 1]	SM	0.5	10	5
	0,802	0,214	2	None	SM	0.5	10	5
	0,796	0,228	2	[0, 1]	FM	0.5	10	5
	0,795	0,277	2	None	FM	0.5	10	5
	0,793	0,263	2	[−1, 1]	FM	0.5	10	5

Method	μ_{R^2}	σ_{R^2}	p_d	$\mathcal{I}$	Memory	p	N_p	N_r
SR	0,607	0,409	2	[]	SM			
	0,577	0,477	2	[0, 1]	SM			
	0,567	0,526	2	[−1, 1]	SM			
	0,536	0,468	2	[0, 100]	SM			
	0,519	0,461	3	[0, 1]	SM			
	0,502	0,551	3	[0, 100]	SM			
	0,492	0,478	3	None	SM			
	0,422	1,593	3	[−1, 1]	SM			
	0,351	0,751	2	[0, 1]	FM			
	0,349	0,775	2	None	FM			
BSR	0,796	0,220	2	[]	SM	1	50	
	0,795	0,223	2	[0, 100]	SM	1	50	
	0,794	0,226	2	[−1, 1]	SM	1	50	
	0,794	0,228	2	[0, 1]	SM	1	50	
	0,767	0,264	2	[−1, 1]	SM	1	10	
	0,767	0,266	2	[0, 100]	SM	1	10	
	0,766	0,277	2	None	SM	1	10	
	0,765	0,278	2	[0, 1]	SM	1	10	
	0,763	0,295	2	[0, 1]	SM	0.5	50	
	0,761	0,301	2	None	SM	0.5	50	
WsBSR	0,710	0,249	2	[]	SM	1	10	5
	0,709	0,272	2	[0, 100]	SM	1	10	5
	0,699	0,262	2	[0, 1]	SM	1	10	5
	0,691	0,286	2	[]	SM	0.5	10	5
	0,685	0,323	2	[0, 100]	SM	0.5	10	5
	0,682	0,353	2	[−1, 1]	SM	1	10	5
	0,680	0,275	2	[0, 1]	SM	0.5	10	5
	0,666	0,306	2	[−1, 1]	SM	0.5	10	5
	0,660	0,289	2	[0, 1]	SM	1	50	10
	0,652	0,307	2	[0, 100]	FM	0.1	10	5

the methods maintained the relative ranking between them. Again the usage of polynomial features of degree two seems to be the more appropriate and no definitive conclusion can be taken about the scaling of the features. The short memory solution shows to be a better option, when compared with the full memory one. Regarding the number of regressors to use in the ensemble cases, results were not conclusive for BSR but, using 50 allowed to achieve the best results. On the other hand, using only 10 regressors can be a good enough forecast for a decision maker, saving considerable computational resources. For the WsBSR method, a similar conclusion as previously can be taken, i.e., using a larger number of regressors in the predictions seems to worsen the results (probably for the obligation of using older regressors).

5 Conclusion and Future Work

Energy is one of the largest parcels in the operation of service buildings. The possibility to forecast the energy consumption of unmovable loads, added up with the fixed ones, gives decision makers the chance to plan ahead the positioning of the movable charges. These can be later optimized taking into consideration many factors as energy production from renewable sources or energy prices.

In this paper, three meta-methods are analyzed to perform the referred forecast. Established the regressor to be used with the meta-method (e.g., Decision Tree regressor), the first (SR) can be seen as a traditional forecast method, while the second (BSR) and third (WsBSR) use ensembles of regressors to make the forecasts. The difference between them comes from the fact that the last uses a warm-start procedure, adding new regressors when required, which are complemented by the ones already fitted in the past. Over the elected parameters, the BSR method has shown a better accuracy but with higher computational cost in the fitting phase. The use of the WsBSR is therefore a possibility as it has shown slightly worse results but with a fraction of the computational cost (in a typical run, for similar set of parameters, WsBSR took approximately 25% of the time required by BSR to run the full simulation).

In terms of future work, other methods besides the decision trees are to be used. Furthermore, the usage of heterogeneous methods in the ensemble will be tried, i.e., the usage of the same method but with different parameters and also distinct methods. In this case, it is also intended that the selection of the forecasting methods (which ones to fit and which ones to use in the forecast) will be tuned on run time. Finally, integration of data assimilation techniques also seem a very promising field of research.

References

1. Ahmad, A., Hassan, M., Abdullah, M., Rahman, H., Hussin, F., Abdullah, H., Saidur, R.: A review on applications of ANN and SVM for building electrical energy consumption forecasting. Renew. Sustain. Energy Rev. **33**, 102–109 (2014). https://doi.org/10.1016/j.rser.2014.01.069
2. Alpaydin, E.: Introduction to Machine Learning. MIT press, Cambridge (2020)
3. Alto da Colina Hotel: https://www.alfagar.com/alfagar-alto-da-colina.html, Accessed 22 Jan 2021
4. Araya, D.B., Grolinger, K., ElYamany, H.F., Capretz, M.A., Bitsuamlak, G.: An ensemble learning framework for anomaly detection in building energy consumption. Energy Build. **144**, 191–206 (2017). https://doi.org/10.1016/j.enbuild.2017.02.058
5. Bindiu, R., Chindris, M., Pop, G.: Day-ahead load forecasting using exponential smoothing. Sci. Bull. "Petru Maior" Univ. Targu Mures **6**, 89 (2009)
6. Cabrita, C.L., Monteiro, J.M., Cardoso, P.J.S.: Improving energy efficiency in smart-houses by optimizing electrical loads management. In: 2019 1st International Conference on Energy Transition in the Mediterranean Area (SyNERGY MED). IEEE (2019). https://doi.org/10.1109/synergy-med.2019.8764140

7. Chandramowli, S., Lahr, M.L.: Forecasting New Jersey's electricity demand using auto-regressive models. SSRN Electron. J. (2012). https://doi.org/10.2139/ssrn.2258552
8. Chen, T.T., Lee, S.J.: A weighted LS-SVM based learning system for time series forecasting. Inf. Sci. **299**, 99–116 (2015). https://doi.org/10.1016/j.ins.2014.12.031
9. Chiang, J., Wu, P., Chiang, S., Chang, T., Chang, S., Wen, K.: Introduction to Grey System Theory. Gao-Li Publication, Taiwan (1998)
10. Deb, C., Zhang, F., Yang, J., Lee, S.E., Shah, K.W.: A review on time series forecasting techniques for building energy consumption. Renew. Sustain. Energy Rev. **74**, 902–924 (2017). https://doi.org/10.1016/j.rser.2017.02.085
11. Divina, F., Gilson, A., Goméz-Vela, F., Torres, M.G., Torres, J.: Stacking ensemble learning for short-term electricity consumption forecasting. Energies **11**(4), 949 (2018). https://doi.org/10.3390/en11040949
12. Eurepean Union: Regulation (Eu) 2018/1999 of the European Parliament and of the Council (2018). https://eur-lex.europa.eu/legal-content/EN/TXT/?uri=uriserv%3AOJ.L_.2018.328.01.0001.01.ENG
13. Gómez, J., Molina-Solana, M.: Towards self-adaptive building energy control in smart grids. In: NeurIPS 2019 Workshop Tackling Climate Change with Machine Learning. Vancouver, Canada (2019). https://www.climatechange.ai/NeurIPS2019_workshop.html
14. Governo Português: Plano nacional de energia e clima 2021–2030 (PNEC 2030) (2020). https://apambiente.pt/_zdata/Alteracoes_Climaticas/Mitigacao/PNEC/PNECPT_TemplateFinal201930122019.pdf
15. Joutz, F.L., Maddala, G.S., Trost, R.P.: An integrated bayesian vector auto regression and error correction model for forecasting electricity consumption and prices. J. Forecast. **14**(3), 287–310 (1995). https://doi.org/10.1002/for.3980140310
16. Karim, S.A.A., Alwi, S.A.: Electricity load forecasting in UTP using moving averages and exponential smoothing techniques. Appl. Math. Sci. **7**, 4003–4014 (2013). https://doi.org/10.12988/ams.2013.33149
17. Kascha, C.: A comparison of estimation methods for vector autoregressive moving-average models. Econ. Rev. **31**(3), 297–324 (2012). https://doi.org/10.1080/07474938.2011.607343
18. Lee, D., Cheng, C.C.: Energy savings by energy management systems: a review. Renew. Sustain. Energy Rev. **56**, 760–777 (2016). https://doi.org/10.1016/j.rser.2015.11.067
19. Lee, W.J., Hong, J.: A hybrid dynamic and fuzzy time series model for mid-term power load forecasting. Int. J. Electric. Power Energy Syst. **64**, 1057–1062 (2015). https://doi.org/10.1016/j.ijepes.2014.08.006
20. Li, K., Hu, C., Liu, G., Xue, W.: Building's electricity consumption prediction using optimized artificial neural networks and principal component analysis. Energy Build. **108**, 106–113 (2015). https://doi.org/10.1016/j.enbuild.2015.09.002
21. Liu, Z., et al.: Accuracy analyses and model comparison of machine learning adopted in building energy consumption prediction. Energy Explor. Exploit. **37**(4), 1426–1451 (2019). https://doi.org/10.1177/0144598718822400
22. Monfet, D., Corsi, M., Choinière, D., Arkhipova, E.: Development of an energy prediction tool for commercial buildings using case-based reasoning. Energy Build. **81**, 152–160 (2014). https://doi.org/10.1016/j.enbuild.2014.06.017
23. Monteiro, J., Cardoso, P.J.S., Serra, R., Fernandes, L.: Evaluation of the human factor in the scheduling of smart appliances in smart grids. In: Stephanidis, C., Antona, M. (eds.) UAHCI 2014. LNCS, vol. 8515, pp. 537–548. Springer, Cham (2014). https://doi.org/10.1007/978-3-319-07446-7_52

24. Monteiro, J., Eduardo, J., Cardoso, P.J.S., Ao, J.S.: A distributed load scheduling mechanism for micro grids. In: 2014 IEEE International Conference on Smart Grid Communications (SmartGridComm). IEEE (2014). https://doi.org/10.1109/smartgridcomm.2014.7007659
25. Nie, H., Liu, G., Liu, X., Wang, Y.: Hybrid of ARIMA and SVMs for short-term load forecasting. Energy Procedia **16**, 1455–1460 (2012). https://doi.org/10.1016/j.egypro.2012.01.229
26. Paolella, M.S.: Linear Models and Time-Series Analysis: Regression. ARMA and GARCH. John Wiley & Sons, ANOVA (2018)
27. Pedregosa, F., et al.: Scikit-learn: machine learning in Python. J. Mach. Learn. Res. **12**, 2825–2830 (2011)
28. Rokach, L.: Ensemble Learning. WSPC (2019). https://www.ebook.de/de/product/35671842/lior_rokach_ensemble_learning.html
29. Sadaei, H.J., Enayatifar, R., Abdullah, A.H., Gani, A.: Short-term load forecasting using a hybrid model with a refined exponentially weighted fuzzy time series and an improved harmony search. Int. J. Electric. Power Energy Syst. **62**, 118–129 (2014). https://doi.org/10.1016/j.ijepes.2014.04.026
30. Sapankevych, N.I., Sankar, R.: Time series prediction using support vector machines: a survey. IEEE Comput. Intell. Mag. **4**(2), 24–38 (2009). https://doi.org/10.1109/MCI.2009.932254
31. Song, H., Qin, A.K., Salim, F.D.: Evolutionary multi-objective ensemble learning for multivariate electricity consumption prediction. In: 2018 International Joint Conference on Neural Networks (IJCNN). IEEE (2018). https://doi.org/10.1109/ijcnn.2018.8489261
32. Taylor, J.W.: Triple seasonal methods for short-term electricity demand forecasting. Eur. J. Oper. Res. **204**(1), 139–152 (2010). https://doi.org/10.1016/j.ejor.2009.10.003
33. Tso, G.: A study of domestic energy usage patterns in Hong Kong. Energy **28**(15), 1671–1682 (2003). https://doi.org/10.1016/s0360-5442(03)00153-1
34. Tso, G.K., Yau, K.K.: Predicting electricity energy consumption: a comparison of regression analysis, decision tree and neural networks. Energy **32**(9), 1761–1768 (2007). https://doi.org/10.1016/j.energy.2006.11.010
35. Vu, K.M.: The ARIMA and VARIMA time series: their modelings. AuLac Technologies Inc., Analyses and Applications (2007)
36. Wahid, F., Kim, D.: A prediction approach for demand analysis of energy consumption using k-nearest neighbor in residential buildings. Int. J. Smart Home **10**(2), 97–108 (2016). https://doi.org/10.14257/ijsh.2016.10.2.10
37. Wang, X., Meng, M.: A hybrid neural network and ARIMA model for energy consumption forecasting. J. Comput. **7**(5) (2012). https://doi.org/10.4304/jcp.7.5.1184-1190
38. Wang, Z., Srinivasan, R.S.: A review of artificial intelligence based building energy use prediction: contrasting the capabilities of single and ensemble prediction models. Renew. Sustain. Energy Rev. **75**, 796–808 (2017). https://doi.org/10.1016/j.rser.2016.10.079
39. Zhang, Y., Wang, J.: Short-term load forecasting based on hybrid strategy using warm-start gradient tree boosting. J. Renew. Sustain. Energy **12**(6) (2020). https://doi.org/10.1063/5.0015220
40. Zia, M.F., Elbouchikhi, E., Benbouzid, M.: Microgrids energy management systems: a critical review on methods, solutions, and prospects. Appl. Energy **222**, 1033–1055 (2018). https://doi.org/10.1016/j.apenergy.2018.04.103

Supporting the Process of Sewer Pipes Inspection Using Machine Learning on Embedded Devices

Mieszko Klusek and Tomasz Szydlo[✉]

Institute of Computer Science, AGH University of Science and Technology,
Krakow, Poland
tomasz.szydlo@agh.edu.pl

Abstract. We are currently seeing an increasing interest in using machine learning and image recognition methods to support routine human-made processes in various application domains. In the paper, the results of the conducted research on supporting the sewage network inspection process with the use of machine learning on embedded devices are presented. We analyze several image recognition algorithms on real-world data, and then we discuss the possibility of running these methods on embedded hardware accelerators.

Keywords: IoT · Machine learning · Embedded devices

1 Introduction

Supporting processes and decision-making using machine vision and artificial intelligence is becoming more popular in many industries and everyday lives. Cars that support the driver or cameras that suggest what settings to choose for the current scenery become everyday life. One of the processes that can be improved using the above techniques is a visual inspection of sewer networks. Every day, the average operator inspects several hundred meters of sewage networks using robots with installed cameras, where he constantly observes the acquired image and provides information on the condition of pipes, damage present in the network, and structural elements [3]. Long working hours and their monotony may have a negative impact on the quality of the inspections carried out. According to J. Dirksen et al. [2] on average, the operator ignores 25% of defects during the inspection, so additional support by informing the operator in real-time about automatically detected objects or status changes can improve the quality of the work performed.

The inspection process can be supported by machine learning. Unfortunately, the use of cloud computing services is usually not possible due to limited network connectivity at the inspection site. The development of hardware accelerators for launching artificial intelligence models has recently made it possible to use it on embedded devices with limited resources that operators work with on a daily

© Springer Nature Switzerland AG 2021
M. Paszynski et al. (Eds.): ICCS 2021, LNCS 12747, pp. 347–360, 2021.
https://doi.org/10.1007/978-3-030-77980-1_27

basis. We have conducted research to analyse the possibility of using machine vision methods to support the process of visual inspection of sewage networks on embedded devices.

This paper presents a comprehensive approach to the ML-based automatic detection of defects and structural elements of sewage networks based on video material analysis from the network inspection. The possibility of using the methods of classification, detection, and image segmentation was analyzed. The considered data set includes photos from different cameras; thus their quality and resolution are varied. The photos are single frames showing defects or structural elements, wherein the case of other works described in the literature was mostly a large number of consecutive video frames. Finally, two image analysis methods were used - segmentation and classification for which performance studies were carried out for many network architectures. The research was conducted to analyze these algorithms' execution time based on deep convolutional neural networks on embedded devices using hardware accelerators.

The organization of the paper is as follows. The Sect. 2 describes the related work and Sect. 3 discusses the research methodology. Sections 4 and 5 analyses the machine learning methods for video processing. The Sect. 6 describes the evaluation, while Sect. 7 concludes the paper.

2   Related Work

The section presents issues related to the context of the research work. The purpose of the sewage network inspection process is justified and the equipment used for this purpose is discussed. The concept of observation is introduced, i.e. a description of a damage, structural element or conditions inside a section of the sewage network, which is an elementary component of the inspection description.

2.1   Pipe Inspection Process

The main method of inspection of sewer networks is video inspection using CCTV cameras. It is performed periodically by companies providing such services to assess the network's quality and plan possible repairs. The horizontal part of the network inspected is called a section and usually connects two wells - vertical elements of the sewer pipes infrastructure.

The detail of the process and its steps differ depending on the standard in force in a given country or region. Nevertheless, the main idea of the process is common to all standards. For example, the standard described by MSCC4 [11] defines the process carried out using a device equipped with a camera, selected depending on the size of the pipe and the expected water filling inside. Small cameras are used for the smallest pipes, pushed by a cable transmitting video material (the so-called push camera). For larger pipes, controlled travelling robots are used, optionally with a raised structure that allows the camera's centring in pipes of larger diameter. For pipes with an expected high level of water inside, floating robots are used (e.g. *Proteus Float Raft*), and for the largest ones, prototypes of flying drones are being developed.

The nomenclature of observations may differ depending on the applicable standard. Particular standards[1] also differently define the degree of detail of the description of a given observation, while the defects and structural elements that are described by these observations are the same for all standards. Based on this, a high-level part of the description of these observations can be distinguished, which is common to all standards. Among the observations, we can distinguish those corresponding to defects, structural elements and the conditions prevailing inside.

2.2 Machine Learning Methods Used in the Inspection Process

J.B. Haurum and T.B. Moeslund [5] analyzed the results of publications on the automation of visual inspection of sewage networks from the last 25 years. Initial work is based on image analysis using methods such as morphological operations [16], oriented gradient histogram (*HOG*) [4], and support vector machines (*SVM*) [8]. Some of them deal with the subject of segmentation of some classes (e.g. cracks) using classical methods of image segmentation.

Along with the development of convolutional networks, subsequent works indicate their use for image analysis, initially using classification, object detection, and segmentation. The authors point out the problem of comparing works by differentiating the data set, the number of classes and considered metrics. For this reason, it is difficult to determine the best solution at a given moment, therefore, in the following part, selected works using deep neural networks describing the latest solutions from 2017–2020 will be analyzed.

M. Wang and J.C.P Cheng [17] proposed a solution based on the detection of observations in photos using the *Faster R-CNN* model. Their dataset included photos containing 4 classes of observations. A year later, the same authors [18] propose usage of the proprietary network architecture called *DilaSeg* for semantic segmentation of three classes of observations. The dataset contains the extracted video frames and segmentation masks for each of the photos. This publication shows an increase in efficiency and a decrease in inference time compared to the *FCN-8s* architecture. The authors point out that the cost of computation is rarely taken into account in the work so far, and this is a key factor that should be taken into account when implementing a solution for embedded devices.

D. Meijer et al. [9] propose a solution that uses image classification and detects 12 different classes of observations. The dataset contains photos taken with the same camera from 30 various inspections, 0.8% of these photos show defects. The publication uses the proprietary convolutional network architecture, the authors have shown that it achieves higher efficiency than the previous solutions. The emphasis was placed on the validation of the solution. It was proposed to introduce new metrics so that it was possible to assess the measure of possible performance improvement in real scenarios in addition to the classification effectiveness.

---

[1] For example, the Polish standard PNEN13508 or the American NASSCO PACP-6.

Q. Xie et al. [19] proposed using a two-level hierarchical deep convolutional network for the classification of sewage network defects. The first binary level distinguishes between images representing all the considered defects from those without defects, and the second distinguishes between the considered defects. Both models share the network structure in addition to the last output layer, and the network architecture itself is a proprietary solution containing 3 convolutional layers, 3 of *max-pooling* type and 3 fully connected. The dataset consists of photos with 16 types of defects, while only the 6 most common ones were selected for training the model. The presented network results were better than the knowledge transfer using popular pre-trained models such as *VGG-16*, *Inception-V3*, and *Resnet*.

Kunzel et al. [6] describe the solution of semantic segmentation of a full scan of a network segment made with the 360° camera. The image is considered a high-quality image made of multiple 360° photos showing the entire tube unfolded onto a flat area. Successive clippings of this photo are transferred to the model. The network structure is based on *FRRN* [12] architecture. The dataset is a scan of 111 pipes with a length of 4.6 km, 6 classes of observation were considered.

The solution presented by Yin et al. [20] uses the *YOLOv3* model to detect objects. The dataset contains photos with 6 classes of observation. The authors indicate that this solution is able to analyze video in real-time and surpasses previous solutions both in terms of detection efficiency and execution time. The emphasis was also placed on validating the entire video material's solution, not just on individual frames.

Table 1. Comparison of image analysis methods in the problem of observation detection.

Paper	Classic methods			Methods using neural networks		
	Morphological operations	HOG	SVM	Classification	Object detection	Semantic segmentation
[4]	X	X	X			
[8]	X		X			
[16]	X					
[17]					X	
[18]						X
[9]				X		
[19]				X		
[6]						X
[20]					X	

The methods used in selected works have been collected in the Table 1. Currently, no works have been found that would deal with the topic of implementing such a solution for embedded devices. The hardware platform is omitted in the works, it is mentioned that a PC, supercomputer or cloud services were used. The emphasis is on algorithms, not the possibility of their implementation on the target platform.

3   Methodology

The research work aims to verify the feasibility of implementing a solution for the automatic detection of observations in sewage networks based on video material for embedded devices. The methodology of the research is shown in Fig. 1.

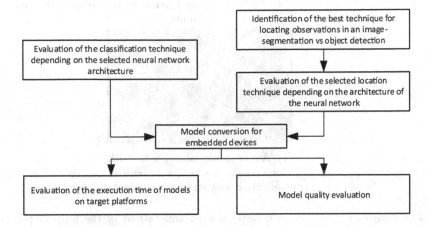

Fig. 1. The course of subsequent stages of research

The dataset used in the research had over 20,000 photos. It contained two types of annotations - polygon coordinates defining the position for 22 classes (examples in Fig. 2) and classification labels for 5 classes (some examples are presented in Fig. 3).

Classes containing location information in the form of polygon coordinates have been grouped into nine more general classes because of their visual similarity. The assignment and the number of classes obtained in this way are presented in the Table 2. Due to the conditions inside the pipes, the picture quality is mostly poor. Many of them are fuzzy and out of focus, yet still contain human-readable information.

4   Selection of Neural Network Architecture for the Classification Problem

The neural network architectures examined in this work were selected based on their effectiveness on the *ImageNet* dataset. Additionally, the number of model parameters was also taken into account, which translates into its complexity.

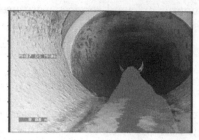

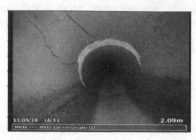

(a) Photo example with annotations: water and defective cross joint

(b) Photo example with annotations: defective transverse joint and axial cracks without discontinuity

(c) An example of a photo with annotations: water, joint, built-in connection and attached sediments

Fig. 2. Examples of data set visualization with annotations in the form of polygon coordinates defining objects' location. Light blue is the color of water, light green is the defective transverse joint, the dark blue is the added joint, the red is the axial fracture without breaking the continuity, the pink is the joint, and the green is the sticky sediment. (Color figure online)

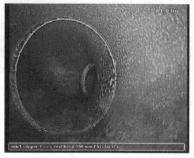

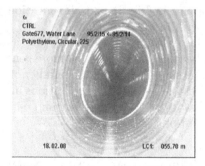

(a) Example of a photo with a right deviation label (same as for other deviations)

(b) Example of a photo with a deformation label

Fig. 3. Examples of photos from the dataset with classification labels

Table 2. The form and number of grouped classes with annotations in the form of polygon coordinates defining the location of objects

Grouped class	The class in the original dataset	Count
Fracture	Axial fracture without discontinuity	5523
	Spiral fracture without breaking the continuity	
	Round fracture without breaking the continuity	
	Axial fracture with discontinuity	
	Spiral fracture with discontinuity	
	Round crack with discontinuity	
Break	Break	717
Missing wall fragments	Missing wall fragments	776
Roots	Independent, fine roots	2224
	Pile roots	
	Complex mass of the roots	
Accumulation of material	Attached settlements	6562
	Postponed settlements	
	Other obstacles	
Faulty joint	Defective longitudinal joint	2946
	Defective cross joint	
	Defective angle joint	
Joint	Joint	10062
connection	Original connection	2175
	Built-in connection	
	Incoming connection	
Water	Water	13656

Therefore, the size and time of inference execution are important for implementation on embedded devices with limited resources. The most diverse models were selected from the available in the *Keras* module of the *Tensorflow* library - *InceptionResNetV2* [10], *Xception* [1] and *MobileNetV2* [15].

The averaged values of the metrics obtained for the trained networks are presented in the Table 3. The value of the $F1$ metric for the analyzed models is similar, but the *MobileNetV2* has the smallest network size, thus it will be used in further experiments as the architecture of choice for the image classification problem.

Table 3. Average metric values of the trained classification models.

Model	F1	Recall	Precision
InceptionResNetV2	0.80	0.78	0.81
Xception	0.79	0.74	0.85
MobileNetV2	0.81	0.78	0.84

5   Selection of the Technique for Locating the Observations

This section describes the course and results of the research carried out in order to choose the technique of locating the observations on the photos. Two classes of solutions were considered - methods of object detection and semantic segmentation.

In the case of object detection, the possibility of running the learned model on embedded devices was taken into account. The lightweight architecture *YOLOv2* (You Only Look Once) [13] was selected for the study for the object detection problem. In the case of semantic segmentation, the architectures and frameworks were selected from the *segmentation-models* package, which implements selected networks using the *Tensorflow* library. The following network architectures were examined - *U-Net* [14] and *FPN* [7]. They were used with encoders - *ResNet101*, *EfficientNetB3* and *MobileNet*-V2. Among the architectures compared, better results were obtained using the *FPN* network, so in further experiments it will be used as the chosen architecture for the problem of semantic segmentation.

Table 4. Quality scores after converting the results from the models: *FPN* and *YOLOv2*.

Score	Semantic segmentation			Object detection
	FPN + MobileNetV2	FPN + EfficientNetB3	FPN + ResNet101	YOLOv2
Precision	0.59	0.63	0.69	0.89
Recall	0.89	0.89	0.89	0.22
F1	0.71	0.74	0.71	0.35

In order to select the localization technique, the results obtained for semantic segmentation and object detection were compared. The results are presented in the Table 4. In the comparison, networks that perform semantic segmentation fared much better. Therefore, in further experiments, only semantic segmentation will be considered as a localization technique.

6   Implementation of Trained Models on Embedded Devices

This section presents the results of experiments with three hardware accelerators: Intel NCS2[2] (Fig. 4a), Google Coral[3] (Fig. 4b), Nvidia Jetson Nano[4] (Fig. 4c). The neural network architectures selected in the previous sections were used to train models that need to be converted to run them with hardware accelerators.

---

[2] https://movidius.github.io/ncsdk/.
[3] https://coral.ai/products/.
[4] https://developer.nvidia.com/embedded/jetson-nano-developer-kit.

Both the metrics of the models after the conversion operation and their execution time were examined. The tools used and the target numerical representation of the models' quantized weights for individual accelerators have been collected in the Table 5. The versions of the tools used are presented in the Table 6.

(a) Intel NCS2 accelerator connected via USB

(b) Google Coral development kit

(c) Nvidia Jetson development kit

Fig. 4. Hardware accelerators used in research

Table 5. The tools used for model conversion and the target numerical representation of the quantized model weights for individual hardware accelerators.

Accelerator	Tools used	Representation of quantized weights
Google Coral	Tensorflow lite converter + Edge TPU compiler	INT8
Nvidia Jetson Nano	Nvidia TensorRT	FP16
Intel NCS2	Intel OpenVINO	FP16

Table 6. Versions of the tools used to convert the models.

Tool	Version
Tensorflow lite converter	2.2
Edge TPU compiler	14.1.317412892
Nvidia TensorRT	7.0
Intel OpenVINO	2020.3.194

The values of the studied models' metrics before and after conversion are presented in the Table 7 for segmentation and in Table 8 for classification. No metric values for *FPN* are given with *EfficientNetB3* scaffold after converting to *Google Coral* because the compilation of the model to an executable form with *Edge TPU* could not be performed due to an internal compiler error. In the case of semantic segmentation, the values of the Intersection-over-Union (IOU)

Table 7. Metrics of semantic segmentation models before and after conversion to executable form on individual hardware accelerators.

Model	Accelerator	IOU	F1	Recall	Precision
FPN + ResNet101	–	0.55	0.77	0.88	0.69
	Coral	0.18	0.61	0.53	0.73
	Jetson Nano	0.55	0.78	0.88	0.70
	NCS2	0.55	0.78	0.88	0.70
FPN + EfficientNetB3	–	0.55	0.74	0.89	0.63
	Coral	–	–	–	–
	Jetson Nano	0.55	0.74	0.89	0.63
	NCS2	0.52	0.71	0.90	0.59
FPN + MobileNetV2	–	0.48	0.71	0.89	0.59
	Coral	0.48	0.72	0.84	0.63
	Jetson Nano	0.48	0.71	0.89	0.59
	NCS2	0.48	0.70	0.90	0.57

Table 8. Classification model metrics before and after conversion to executable form on individual hardware accelerators.

Model	Accelerator	F1	Recall	Precision
MobileNetV2	–	0.81	0.78	0.84
	Coral	0.80	0.76	0.85
	Jetson Nano	0.81	0.78	0.84
	NCS2	0.81	0.78	0.83

metric was also calculated as the area of overlap divided by the area of union between the predicted segmentation and the original data.

Finally, the models were tested on the representative video showing a recording of a complete inspection of a section of the sewage network. For each model, the average analysis time of a single video frame from the test recording is calculated using the formula:

$$t = \frac{\sum_{i=1}^{N} t_i}{N}, \tag{1}$$

where N is the number of frames in the video and t_i is the time to analyze ith frame. The procedure was repeated for each tested hardware accelerator.

In the case of *Google Coral* and *Nvidia Jetson*, the trained models after conversion were run on dedicated development kits, while *Intel NCS2* was connected to the *Raspberry Pi 3B* minicomputer. On each platform, the video was played from a file using *OpenCV*. A comparison of resources available on all platforms is presented in the Table 9.

Table 10 shows the average time of analysis of one frame using the analyzed segmentation models, the Table 11 shows the classification model results. Missing results for the *FPN* with *EfficientNetB3* and platform *Google Coral* model are due to the same build error when examining segmentation metrics. Also noteworthy is the much longer execution time for segmentation models on *Google Coral*, due to the fact that some of the operations performed within the model cannot be compiled for execution on *Edge TPU* and are instead performed on the CPU.

Table 9. Comparison of the available resources of the platforms on which the experiments are performed.

Hardware platform	CPU	RAM
Google Coral Dev Board	ARM Cortex-A53 4 × 1,8 GHz	1GB LPDDR4
Nvidia Jetson Nano	ARM Cortex-A57 4 × 1,43 GHz	4GB LPDDR4
Raspberry Pi 3B	Broadcom BCM2837 4 × 1,2 GHz	1GB LPDDR2

Table 10. Average time to analyze a single frame using trained segmentation models on each platform.

Model	Hardware platform	Time [ms]
FPN + ResNet101	Google Coral Dev Board	4036
	Nvidia Jetson Nano	234
	Raspberry Pi 3B + Intel NCS2	415
FPN + EfficientNetB3	Google Coral Dev Board	DNR
	Nvidia Jetson Nano	264
	Raspberry Pi 3B + Intel NCS2	480
FPN + MobileNetV2	Google Coral Dev Board	3954
	Nvidia Jetson Nano	121
	Raspberry Pi 3B + Intel NCS2	324

Table 11. Average time to analyze a single frame using a trained classification model on each platform.

Model	Hardware platform	Time [ms]
MobileNetV2	Google Coral Dev Board	7.21
	Nvidia Jetson Nano	95.42
	Raspberry Pi 3B + Intel NCS2	44.65

For comparison, additional runtime tests were carried out on one of the platforms without the use of acceleration, using only the CPU. For this purpose,

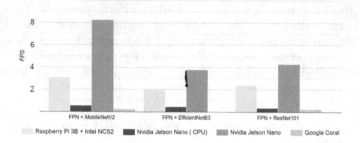

Fig. 5. The number of frames per second that can be processed for the segmentation models

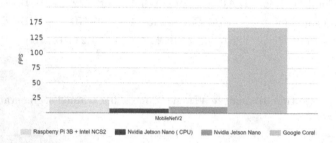

Fig. 6. The number of frames per second that can be processed for the classification model

Nvidia Jetson was used due to the largest amount of RAM available. The execution time was converted into the number of frames that could be analyzed per second - *FPS* (frames per second), and the results are presented in Fig. 5 and Fig. 6 respectively for segmentation and classification models.

7   Summary and Future Work

The values of the model metrics after conversion to the executable form with the use of hardware accelerators and the obtained execution times indicate that it is possible to implement the discussed solution on embedded devices, moreover, the implementation would not be possible without the use of hardware acceleration, as shown in the graphs in Fig. 5 and Fig. 6. Among the tested devices, only *Google Coral* is not suitable for use for too long of the segmentation models. All other configurations of models and accelerators have a total execution time of less than 1s, which is the limit value, since the [20] assumption is made that the observations are visible on the video material for at least one second.

Among the examined methods of locating objects in photos - object detection and semantic segmentation, the use of segmentation gives better results. From the investigated *FPN* and *U-Net* architectures, better results were obtained for *FPN*. Despite the smallest size, the best of the studied architectures for the classification problem turned out to be *MobileNetV2*.

Directions for further development are possible both in the context of the detection mechanism and validation of the usability of the solution. This paper does not deal with the aspect of calculating the water level, although water is detected as one of the segmentation classes. A mask that is detected with fairly high accuracy (metric $IoU = 0.79$ for $FPN/ResNet101$) can be used to calculate the water level using conventional image analysis methods. In addition to the water level detection, based on segmentation masks, it is possible to calculate other observations' parameters, such as the connection diameter.

Another aspect is the more detailed classification of the classes of observations detected. The classes present in the original dataset have been grouped into more general ones, so having the results of the grouped class segmentation, one can cut out a part of the original photo limited by the segmentation mask for a given class and forward it to smaller, more specialized classifiers that will refine the detection results.

An important element that can be further developed is the extension of validation of the solution and conducting experiments with camera operators. Dividing them into two groups - with and without software support-and the subsequent analysis of their work results will help answer the question of how the generated prompts affect the quality and time of inspections.

Acknowledgment. The research presented in this paper was partially supported by the funds assigned to AGH University of Science and Technology by the Polish Ministry of Science and Higher Education.

References

1. Chollet, F.: Xception: deep learning with depthwise separable convolutions. In: 2017 IEEE Conference on Computer Vision and Pattern Recognition (CVPR), pp. 1800–1807 (2017)
2. Dirksen, J., et al.: The consistency of visual sewer inspection data. Struct. Infrastruct. Eng. **9**(3), 214–228 (2013)
3. Gay, L.F., Bayat, A.: Productivity improvement of sewer CCTV inspection through time study and route optimization. J. Construct. Eng. Manage. **141**(6), 04015009 (2015)
4. Halfawy, M., Hengmeechai, J.: Automated defect detection in sewer closed circuit television images using histograms of oriented gradients and support vector machine. Autom. Construct. **38**, 1–13 (2014)
5. Haurum, J., Moeslund, T.: A survey on image-based automation of CCTV and SSET sewer inspections. Autom. Construct. **111**, 103061 (2020)
6. Kunzel, J., Werner, T., Eisert, P., Waschnewski, J.: Automatic analysis of sewer pipes based on unrolled monocular fisheye images. In: 2018 IEEE Winter Conference on Applications of Computer Vision (WACV), pp. 2019–2027 (2018)
7. Lin, T., Dollár, P., Girshick, R., He, K., Hariharan, B., Belongie, S.: Feature pyramid networks for object detection. In: 2017 IEEE Conference on Computer Vision and Pattern Recognition (CVPR), pp. 936–944 (2017)
8. Mashford, J., Rahilly, M., Davis, P., Stewart, B.: A morphological approach to pipe image interpretation based on segmentation by support vector machine. Autom. Construct. **19**(7), 875–883 (2010)

9. Meijer, D., Scholten, L., Clemens, F., Knobbe, A.: A defect classification methodology for sewer image sets with convolutional neural networks. Autom. Construct. **104**, 281–298 (2019)

10. Nguyen, L., Lin, D., Lin, Z., Cao, J.: Deep CNNs for microscopic image classification by exploiting transfer learning and feature concatenation, pp. 1–5 (05 2018)

11. Orman, N., Lambert, J.E.: Manual of Sewer Condition Classification, 4th edn. WRc, Runcorn (2004)

12. Pohlen, T., Hermans, A., Mathias, M., Leibe, B.: Full-resolution residual networks for semantic segmentation in street scenes. In: Computer Vision and Pattern Recognition (CVPR), 2017 IEEE Conference on (2017)

13. Redmon, J., Divvala, S., Girshick, R., Farhadi, A.: You only look once: unified, real-time object detection. In: 2016 IEEE Conference on Computer Vision and Pattern Recognition (CVPR), pp. 779–788 (2016)

14. Ronneberger, O., Fischer, P., Brox, T.: U-Net: convolutional networks for biomedical image segmentation. In: Navab, N., Hornegger, J., Wells, W.M., Frangi, A.F. (eds.) MICCAI 2015, Part III. LNCS, vol. 9351, pp. 234–241. Springer, Cham (2015). https://doi.org/10.1007/978-3-319-24574-4_28

15. Sandler, M., Howard, A., Zhu, M., Zhmoginov, A., Chen, L.: Mobilenetv2: inverted residuals and linear bottlenecks. In: 2018 IEEE/CVF Conference on Computer Vision and Pattern Recognition, pp. 4510–4520 (2018)

16. Swarnalatha, P., Kota, M., Resu, N.R., Srivasanth, G.: Automated assessment tool for the depth of pipe deterioration. In: 2009 IEEE International Advance Computing Conference, pp. 721–724 (2009)

17. Wang, M., Cheng, J.: Development and improvement of deep learning based automated defect detection for sewer pipe inspection using faster R-CNN. Adv. Comput. Strat. Eng. **10864**, 171–192 (2018)

18. Wang, M., Cheng, J.: Semantic segmentation of sewer pipe defects using deep dilated convolutional neural network. In: 36th International Symposium on Automation and Robotics in Construction (2019)

19. Xie, Q., Li, D., Xu, J., Yu, Z., Wang, J.: Automatic detection and classification of sewer defects via hierarchical deep learning. IEEE Trans. Autom. Sci. Eng. **16**(4), 1836–1847 (2019)

20. Yin, X., Chen, Y., Bouferguebe, A., Zaman, H., Al-Hussein, M., Kurach, L.: A deep learning-based framework for an automated defect detection system for sewer pipes. Autom. Construct. 109, 102967 (2020)

Explanation-Driven Model Stacking

Szymon Bobek[1]([⊠]) [ID], Maciej Mozolewski[2], and Grzegorz J. Nalepa[1] [ID]

[1] Jagiellonian Human-Centered Artificial Intelligence Laboratory (JAHCAI)
and Institute of Applied Computer Science, Jagiellonian University,
31-007 Kraków, Poland
{szymon.bobek,grzegorz.j.nalepa}@uj.edu.pl
[2] Edrone Ltd., Kraków, Poland
m.mozolewski@edrone.me

Abstract. With advances of artificial intelligence (AI), there is a growing need for provisioning of transparency and accountability to AI systems. These properties can be achieved with eXplainable AI (XAI) methods, extensively developed over the last few years with relation for machine learning (ML) models. However, the practical usage of XAI is limited nowadays in most of the cases to the feature engineering phase of the data mining (DM) process. We argue that explainability as a property of a system should be used along with other quality metrics such as accuracy, precision, recall in order to deliver better AI models. In this paper we present a method that allows for weighted ML model stacking and demonstrates its practical use in an illustrative example.

Keywords: Explainability · Machine learning · Optimization

1 Introduction

Recent advancements in black-box machine learning models such as deep neural networks and their applications to sensitive areas such as medical and law applications, or industry 4.0 provoked a discussion on the accountability and transparency of AI systems [3]. Although the concept of explanation of decisions of AI systems has long tradition that dates back to times of knowledge-based AI systems [16], it has been extensively developed over the last decade to facilitate new algorithms.

A large portfolio of XAI algorithms is now available for data scientists and system engineers, which includes model-agnostic methods such as Lime [12], Shap [10], Anchor [13], and model specific solutions like GradCAM or DeepLift for neural networks [17]. However, the methodology of incorporating them into the classic data mining and machine learning pipeline is not clearly stated. In this paper we argue that the explainability (or intelligibility) is a property of a system as a whole and should be considered as an important factor in designing and evaluating such a system. This requires the explanation to be quantified with respect to some criterion.

In this paper we show how the quality of explanation and the quality of machine learning model can be fused in order to produce a model that combines

© Springer Nature Switzerland AG 2021
M. Paszynski et al. (Eds.): ICCS 2021, LNCS 12747, pp. 361–371, 2021.
https://doi.org/10.1007/978-3-030-77980-1_28

property of both, resulting in a model of good quality with explanations of good quality. In order to measure the quality of explanations we used *InXAI framework*[1] we developed, that provides objective metrics such as consistency, stability and perturbational accuracy loss [18]. We show how to combine these measures along with standard metrics for ML model performance (i.e. accuracy, F1, precision, recall, etc.) within a Bayesian optimization framework based on the SMAC toolkit [7] that stacks multiple ML models into one meta-model. We demonstrate the feasibility of our solution in an illustrative example.

The rest of the paper is organized as follows. In Sect. 2 we discuss the role of XAI methods in standard ML/DM pipeline and its potential usage as a criteria for model selection. Formal definition of our approach for explanation-driven model stacking is given in Sect. 3. In Sect. 4 we demonstrate the usage of our approach on illustrative, reproducible example. Finally in Sect. 5 we summarize the original contribution and indicate future works.

2   Role of XAI in the Machine Learning Pipeline

Explainable AI aims at bringing transparency to the decision making process of automated systems. Along with the development of deep neural networks and other black-box machine learning methods, it has been extensively developed over the last decade. Both the recent GDPR EU regulation [6] and the DARPA-BAA-16-53 program on XAI [4] catalysed the progress in this field.

Although the general concept of explainable decision making is clear, the underlying methods and specific goals differ depending on who is the addressee of the explanation. Similarly, the location of the explanation mechanism in the pipeline of developing AI systems will be different depending on the end-user. In GDPR and DARPA documents the role of the end-user is emphasised, as the final recipient of the explanation. In such a case the *explainability* will be considered more of the property of an AI system as a whole and can be defined as a capability of the system to be understood. In the history of AI systems such a property was most often called intelligibility [9]. It was provided by building systems with frameworks that supported that feature inherently [1,5,15]. Nowadays it is addressed also by dedicated methods such as conversational recommender systems [8]. However, such approaches are crafted for the purpose of the specific problem and do not generalize well to other cases.

Most recent advancements in XAI focus mostly on generating explanation in a way that is mostly used by data scientists and domain experts to validate the correctness of the decision model (e.g. bias analysis), or to enable the adoption of the decision support systems in sensitive areas by building trust via explanations (e.g. medical diagnosis decision support systems). In both of the cases evaluation is done manually either via user-experience studies or by observational studies. This is why it is difficult to incorporate the XAI methods within machine learning and data mining pipelines, which is a highly automated process.

---

[1] See: https://github.com/sbobek/inxai.

To address these challenges, several approaches for automated evaluation of XAI methods were proposed. There were attempts to provide methodological approach for evaluation and verification of explanation results [11,18]. Among many qualitative approaches there are also ones that allow for quantitative evaluation. In [14] measures such as fidelity, consistency and stability were coined, that can be used for a numerical comparison of methods. In [19] the aforementioned measures were used to improve overall explanations. In [2] a measure that allows us to capture stability or robustness of explanations was introduced. However, all of the solution provide only human-based evaluation procedure that does not produce objectively comparative results among different explanaibility methods. This limits their usage to use cases where expert-based analysis is the only one desired, discarding the possibility of including them in an automated pipeline and using their results as optimization parameters.

Our *InXAI framework* implements several metrics from aforementioned works, and allows include them in classic machine learning pipeline that is consistent with scikit-learn API[2]. This opens a possibility to use XAI metrics as any other machine learning model performance indicators and use them as model-selection attributes. In the next section we demonstrate how to use it along with Bayesian optimization framework to stack several machine learning models that finally yields high accuracy and good explainability properties.

3   Optimization of the Explanation-Driven Meta Model

Model selection is an important stage in building any AI system. Usually it is governed by the mechanisms that are based on comparison of standard metrics for machine learning models such as accuracy for classification or R2 score for regression. In this section we demonstrate how additional metrics associated with explainability can be combined with standard measures in order to facilitate model selection, but also with the model stacking mechanism.

Let us consider an example of a simple binary classifier. One can train several classifiers. These models will vary in terms of performance metrics, such as Accuracy, Logarithmic Loss, F1 Score to name a few. Depending on a specific data mining problem, it may also be important to take into account explainability.

Instead of choosing only one of the models, several "component/unit" models can be combined altogether to obtain a meta-model, so that specific performance metric remains at a decent level. At the same time, one may want to optimize XAI metrics, such as Stability, Perturbational Accuracy Loss or Consistency of the meta-model, emphasizing a specific aspect of explainability. The simplest way to obtain a meta-model for binary classifiers will be a weighted sum of k component models. Training of unit models is done independently. Suppose that the model predicts class 0 or 1. Each of those models predicts the probability of an instance $x^{(i)}$ belonging to a given class Q denoted as $P_k(Q|x^{(i)})$. Prediction P_{mm} of meta-model can therefore be defined as a weighted sum of predictions probability of its components and is given by Eq. (1).

---

[2] See https://scikit-learn.org.

$$\mathbb{P}_{mm}(Q|x^{(i)}) = \frac{\sum_k \mathbb{P}_k(Q|x^{(i)})w_k}{\sum_k w_k} \tag{1}$$

$$\sum_k w_k > 0; \; w_k \geq 0$$

On such a meta-model, result of the classification for observation $x^{(i)}$ is straightforward and can be defined as $\underset{Q}{\mathrm{argmax}} \; \mathbb{P}_{mm}(Q|x^{(i)})$.

Where w_k is a weight associated with the model k and reflects the importance of that model in calculating global prediction. Such weight can be calculated as shares in the quality metric of a particular model (e.g. accuracy). In the following sections we show how w_k can be determined with the usage of XAI quality metrics and SMAC optimization framework.

3.1   Metrics of Explainability

Meta model defined in Eq. (1) can be considered as a black-box mechanism, and easily used along with any ML quality metrics and XAI quality metrics. In the following section we briefly discuss selected XAI metrics that are used in our solution to measure the overall model performance in terms of quality of explanations.

Stability. It expresses to what extent are the explanations for similar observations similar to each other. This metric is specific for a single model. It is based on Local Lipschitz Continuity metric [2].

AUC Perturbational Accuracy Loss. This metric describes how accuracy metric changes along with increasing disruptions in predictor values. Like stability, this metric is defined for a single model. Perturbations are expressed as a percentage of random changes made to the data set. The smaller the weight of a variable given by a local explainer model, the larger perturbations are applied. The significance of the feature can be determined by permutation importance method (e.g. *Permutation Importance* from ELI5 package[3]).

Meta-model Inner Consistency. Consistency answers the question of how similar are the explanations of two or more different ML models that were trained on the same data set. It is a measure obtained on set o explanations $\{\Phi^{e_{m_1}}, \ldots, \Phi^{e_{m_k}}\}$ generated for k models and is defined by Eq. (2).

$$C(\Phi^{m_1}, \Phi^{m_2}, \ldots, \Phi^{m_k}) = \frac{1}{\underset{a,b \in 1,2,\ldots,k}{\max} ||\Phi^{m_a} - \Phi^{m_b}||_2 + 1} \tag{2}$$

---

[3] See: https://eli5.readthedocs.io/en/latest/blackbox.

The measure is applicable when one compares two or more (with the InXAI framework extension) different models. However, for the sake of clarity we limit the discussion to a single meta-model. Thus, we propose the *Inner meta-model consistency* measure given by Eq. (3). Note that Φ^{m_k} is an explanation generated with any explainer (e.g. SHAP, LIME) for model k and is a matrix of i rows and n columns reflecting number of observations and number of features in a dataset respectively. Therefore the consistency of a meta-model C_{mm} is a vector of i elements.

$$C_{mm} = C\left(\frac{w_1}{\sum_k w_k}\,\Phi^{m_1},\, \frac{w_2}{\sum_k w_k}\,\Phi^{m_2},\, \dots,\, \frac{w_1}{\sum_k w_k}\,\Phi^{m_k}\right) \qquad (3)$$

For a meta-model constructed as a weighted sum of unit models, performance metrics will have lower bound equal to the metric of the weakest component model in the regard of the given metric. Therefore, by optimizing the weights of the meta-model components in terms of the selected XAI metric, one can be sure that the performance metrics for the resulting model will not fall below the expected level. The level is determined on the basis of the initial selection of the component models. This can be demonstrated on the example of the area under the ROC curve metric. Consider the first model m_1, which predicts the class on the basis of a random throw of an unbiased die ($P(0) = 0$ or $P(0) = 1$ with equal probability). For balanced classes, the area under the ROC curve for m_1 will be equal to 0.5. Let's assume that the second model m_2 will have an area under the ROC curve greater than 0.5. Let m_1 be included in the meta-model with weight w_1, and m_2 has a corresponding weight w_2. Then the area under the ROC curve of the meta-model will be greater than 0.5, as long as $w_2 > 0$.

3.2   Selection of Weights of Unit Models in Meta-model

To combine XAI metrics, the formula for Loss function L_{mm} for meta-model, allowing to put more emphasis on a given metric, depending on the course of the experiment, was developed. To control the extent to which a given XAI metric is important for optimisation, importance meta-parameters were introduced. For AUC Perturbational Accuracy Loss of meta-model ($AUCx_{mm}$), the γ_{auc} parameter was used. For stability (S_{mm}) γ_s and for consistency (C_{mm}) γ_c were used, respectively. The idea behind those parameters is to put more emphasis on the given metric by taking a given metric to the power of the parameter >1. Stability and consistency are vectors, thus mean value across all observations were used. For details see Eq. (4).

$$L_{mm} = \frac{AUCx_{mm}{}^{\gamma_{auc}}}{\overline{S_{mm}}{}^{\gamma_s} \cdot \overline{C_{mm}}{}^{\gamma_c}} \qquad (4)$$

Where $\overline{S_{mm}} = \frac{\sum_i^N S_{mm}{}^i}{N}$ and $\overline{C_{mm}} = \frac{\sum_i^N C_{mm}{}^i}{N}$ are defined as average stability and consistency on the dataset for selected models.

The next section provides an evaluation scenario of the framework using an illustrative example.

4 Evaluation on a Case Study

In this section we demonstrate how the formal representation of framework given in Sect. 3 can be operationalized and enclosed into a working module. For the sake of clarity we demonstrate the solution on a synthetic, reproducible example.[4]

4.1 Synthetic Dataset and ML Models

For the purpose of a demonstration, a synthetic example was used. Dataset with two interleaving half circles was generated with the *sklearn* library. It contains 200 observations and is visualised on Fig. 1. There are two predictor variables. The test set consisted of 33% of the observations. Models trained on the dataset are summarized in Table 1.

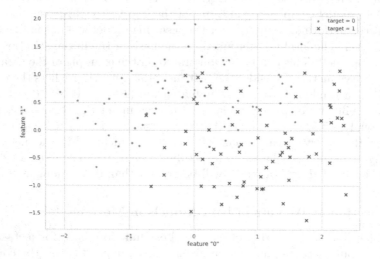

Fig. 1. Dataset with 2 classes and 2 predictors.

4.2 XAI Metrics on Unit Models

Metrics were obtained according to methodology presented in Sect. 3.1 with the use of the *InXAI* framework. SHAP values were used as a local explainer. Stability per unit model is presented on Fig. 2. The highest stability characterizes SVM Classifier models, followed by RandomForest, with XGBoost in the middle and CatBoost as least performant. Consistency between models was calculated pairwise and is depicted in Fig. 3. High level of Consistency was shown only between RandomForest and SVM models. In terms of AUC (Area Under Curve)

---

[4] For source code see: https://github.com/mozo64/inxai/blob/main/examples/xai_on_synth_data/XAI-boost-on-syntetic-data-v4.ipynb.

Table 1. Summary of trained models.

Model	Model abbreviation	Accuracy score	F1-score	
			class "0"	class "1"
SVMClassifier with RBF kernel	svc_radial	0.76	0.78	0.73
SVMClassifier with linear kernel	svc_lin	0.82	0.83	0.80
XGBClassifier	xgbc	0.74	0.76	0.72
RandomForestClassifier	rfc	0.74	0.77	0.70
CatBoostClassifier	ctbc	0.65	0.66	0.65

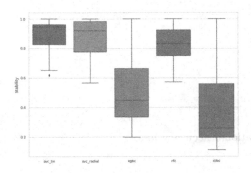

Fig. 2. Stability per unit model.

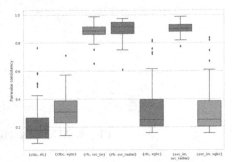

Fig. 3. Pairwise consistency.

for Perturbational Accuracy Loss, most models performed equally, with the one exception of CatBoost. This model was also the weakest in terms of initial – not perturbed – Accuracy. The summary is presented on Fig. 4.

4.3 Optimization-Driven Weight Selection

It is worth noting that computation of XAI metrics is very computational intensive operation that does not scale well to large datasets. This is why optimization of parameters that are based on such metrics has to be performed wisely. In order to assure the feasibility of the approach the *SMAC* framework[5] was used for optimization of the Loss function L (finding the minimum) that is based on the model-based Bayesian optimization algorithm.

The first optimization run began with user-defined weights for the component models. Initial weights for component models were the same in all experiments. They were computed as follows:

$$w_k = \frac{1}{\gamma_{auc}^k}$$

and afterwards re-scaled to sum up to 100%. In the course of the experiment, the optimizer created a meta-model linking individual unit models according to

---

[5] See: https://automl.github.io/.

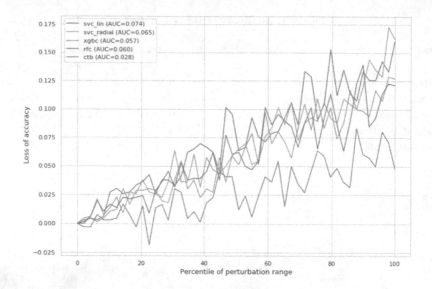

Fig. 4. AUC perturbational accuracy loss.

Eq. (1). SMAC optimised the weights for all 5 unit models with weights in the uniform range from 0.0 to 1.0 and scale them so that they sum up to 100%. The Loss function L given by the formula Eq. (4) was minimized with different values of meta-parameters. 30 iterations per experiment were performed.

To speed up computations, two of the metrics – Stability and AUC acc. loss – were approximated with Eq. (5) and Eq. (6), respectively. However, this approximation was used only during the SMAC optimization process. The final result of the experiment was calculated with obtained weights for component models, using the exact equations from the Sect. 3.1.

$$S_{approx} = \frac{\sum_k S_k \, w_k}{\sum_k w_k} \tag{5}$$

$$AUCx_{approx} = \frac{\sum_k AUCx_k \, w_k}{\sum_k w_k} \tag{6}$$

4.4   Results

The summary of experiment runs is depicted in Table 2. In the run #3a the impact on Stability was comparable to run #4a (where we optimised for Inner meta-model consistency). Thus importance of Stability was further increased in run #3b. Likewise, in the run #4a the result for the Inner meta-model consistency was the same as in run #3a. So we increased the importance of the Inner meta-model consistency even further. Overall one can see impact on meta-model XAI metrics in terms of AUC acc. loss, Stability and Consistency, what was the aim of the study. Meta-model Accuracy stayed on a decent level, as expected.

Table 2. Results of experiment runs.

#	Meta-parameter			Weights for models after optimization					Metrics			
	AUC acc. loss	Stability	Consistency	xgbc	rfc	ctbc	svc_lin	svc_radial	Model acc.	AUC acc. loss	Stability	Consistency
1	1.0	1.0	1.0	.000087	.363524	.000031	.272740	.363619	0.76	0.060	0.872	0.895
2	**3.0**	1.0	1.0	.000042	.499893	.000025	**.000004**	.500035	0.73	**0.048**	0.858	0.862
3a	1.0	**3.0**	1.0	.000007	.315697	.000021	.312844	.371430	0.77	0.062	**0.874**	0.899
3b	1.0	**5.0**	1.0	.000021	**.000013**	.000020	.499952	.499993	0.77	0.059	**0.887**	0.871
4a	1.0	1.0	**3.0**	.000062	.318573	.000074	.310580	.370711	0.77	0.064	0.874	**0.899**
4b	1.0	1.0	**5.0**	.000026	.293124	.000037	**.350562**	**.356252**	0.77	0.067	0.876	**0.902**

XGBoost and CatBoost Classifier models in all runs got low, negligible weights in the meta-model. It is related to low Stability of those models, in comparison to other models (see Fig. 2). *Loss function* penalised low Stability in all experiment runs. Optimisation for AUC acc. loss (#2) resulted in the lowest weight for SVM Classifier with Linear kernel. This classifier has the worst performance in terms of AUC acc. loss – see: Fig. 4. Best Stability (#3b) was when the lowest weight was applied to RandomForest Classifier. This model has slightly worse Stability than both SVM Classifiers (see: Fig. 2). The best result when optimising for Inner meta-model consistency (#4b) was for SVM Classifier models (Linear kernel, RBF kernel) having similar weights. This is in line with the fact that those 2 models have highest pairwise consistency (see: Fig. 3).

5 Summary and Future Works

In this paper we presented a method that allows us to combine XAI quality metrics along with standard ML evaluation metrics in order to provide an optimization framework that maximizes both ML performance and XAI quality within a single meta-model. We demonstrated that in our approach there is no need to compromise on performance metrics, such as accuracy, as the meta-model preserves the quality of its components.

Also the *naive* approximation of Stability and AUC Perturbational Accuracy Loss for meta-model was given. This approximation is more effective in terms of CPU needed for computations. We also introduced the concept of the *inner meta-model consistency*, which shows its usefulness, as it promotes higher weights for models which were pairwise consistent.

The idea of an ensemble meta-model is worth further research. Firstly, especially valuable will be generalization on a multi-classifier problem and regression problem. In the future works, we can also check whether other XAI metrics than AUC Perturbational Accuracy Loss, Stability and Inner meta-model consistency could be optimized in this framework. Secondly, it is also necessary to validate the method on real life examples. Finally, the idea of combining different explanations for one ML model into one meta-explanation is worth exploring. So instead of weighted sum of unit models, one could put together different local explainers (e.g. SHAP, LIME), to create one meta-explainer, optimized for specific XAI metrics.

One of the limitations of the presented framework is that it only provides model weighting using comparative evaluation metrics among several models/explainers. It does not assure that the explanations generated by the explainer are correct nor feasible to the end user. It only takes into account their performance on the measurable, objective aspects. The fit to the expectations is another research topic that not yet has been operationalized in our framework.

Acknowledgements. The paper is funded from the PACMEL project funded by the National Science Centre, Poland under CHIST-ERA programme (NCN 2018/27/Z/ST6/03392). The authors are grateful to ACK Cyfronet, Krakow for granting access to the computing infrastructure built in the projects No. POIG.02.03.00-00-028/08 "PLATON - Science Services Platform" and No. POIG.02.03.00-00-110/13 "Deploying high-availability, critical services in Metropolitan Area Networks (MAN-HA)".

References

1. Almeida, A., Lopez-de Ipina, D.: Assessing ambiguity of context data in intelligent environments: towards a more reliable context managing systems. Sensors **12**(4), 4934–4951 (2012). http://www.mdpi.com/1424-8220/12/4/4934
2. Alvarez-Melis, D., Jaakkola, T.S.: On the robustness of interpretability methods (2018)
3. Barredo Arrieta, A., et al.: Explainable artificial intelligence (XAI): concepts, taxonomies, opportunities and challenges toward responsible AI. Inf. Fusion **58**, 82–115 (2020)
4. DARPA: Broad agency announcement - explainable artificial intelligence (XAI). DARPA-BAA-16-53 (Aug 2016)
5. Dey, A.K.: Modeling and intelligibility in ambient environments. J. Ambient Intell. Smart Environ. **1**(1), 57–62 (2009)
6. Goodman, B., Flaxman, S.: European union regulations on algorithmic decision-making and a "right to explanation". arXiv preprint arXiv:1606.08813 (2016)
7. Hutter, F., Hoos, H.H., Leyton-Brown, K.: Sequential model-based optimization for general algorithm configuration (extended version). Technical report. TR-2010-10, University of British Columbia, Department of Computer Science (2010). http://www.cs.ubc.ca/~hutter/papers/10-TR-SMAC.pdf
8. Jannach, D., Manzoor, A., Cai, W., Chen, L.: A survey on conversational recommender systems (2020)
9. Lim, B.Y., Dey, A.K., Avrahami, D.: Why and why not explanations improve the intelligibility of context-aware intelligent systems. In: Proceedings of the SIGCHI Conference on Human Factors in Computing Systems, pp. 2119–2128, CHI 2009. ACM, New York (2009). https://doi.org/10.1145/1518701.1519023
10. Lundberg, S.M., Lee, S.I.: A unified approach to interpreting model predictions. In: Proceedings of the 31st International Conference on Neural Information Processing Systems, pp. 4768–4777. NIPS2017, Curran Associates Inc. (2017)
11. Mohseni, S., Zarei, N., Ragan, E.D.: A multidisciplinary survey and framework for design and evaluation of explainable AI systems (2020)
12. Ribeiro, M.T., Singh, S., Guestrin, C.: "Why should i trust you?": explaining the predictions of any classifier. In: Proceedings of the 22nd ACM SIGKDD International Conference on Knowledge Discovery and Data Mining, pp. 1135–1144, KDD 2016. Association for Computing Machinery, New York (2016). https://doi.org/10.1145/2939672.2939778

13. Ribeiro, M.T., Singh, S., Guestrin, C.: Anchors: high-precision model-agnostic explanations. In: AAAI Publications, Thirty-Second AAAI Conference on Artificial Intelligence (2018)
14. Robnik-Šikonja, M., Bohanec, M.: Perturbation-based explanations of prediction models. In: Zhou, J., Chen, F. (eds.) Human and Machine Learning. HIS, pp. 159–175. Springer, Cham (2018). https://doi.org/10.1007/978-3-319-90403-0_9
15. Roy, N., Das, S.K., Julien, C.: Resource-optimized quality-assured ambiguous context mediation framework in pervasive environments. IEEE Trans. Mob. Comput. **11**(2), 218–229 (2012). http://dblp.uni-trier.de/db/journals/tmc/tmc11.html#RoyDJ12
16. Schank, R.C.: Explanation: A first pass. In: Kolodner, J.L., Riesbeck, C.K. (eds.) Experience, Memory, and Reasoning, pp. 139–165. Lawrence Erlbaum Associates, Hillsdale (1986)
17. Selvaraju, R.R., Das, A., Vedantam, R., Cogswell, M., Parikh, D., Batra, D.: Grad-cam: Why did you say that? visual explanations from deep networks via gradient-based localization. CoRR abs/1610.02391 (2016). http://arxiv.org/abs/1610.02391
18. Sokol, K., Flach, P.A.: Explainability fact sheets: A framework for systematic assessment of explainable approaches. CoRR abs/1912.05100 (2019)
19. Yeh, C.K., Hsieh, C.Y., Suggala, A.S., Inouye, D.I., Ravikumar, P.: On the (in)fidelity and sensitivity for explanations (2019)

Software Engineering
for Computational Science

I/O Associations in Scientific Software: A Study of SWMM

Zedong Peng[1], Xuanyi Lin[1], Nan Niu[1(✉)], and Omar I. Abdul-Aziz[2]

[1] University of Cincinnati, Cincinnati, OH 45221, USA
{pengzd,linx7}@mail.uc.edu, nan.niu@uc.edu
[2] West Virginia University, Morgantown, WV 26506, USA
oiabdulaziz@mail.wvu.edu

Abstract. Understanding which input and output variables are related to each other is important for metamorphic testing, a simple and effective approach for testing scientific software. We report in this paper a quantitative analysis of input/output (I/O) associations based on co-occurrence statistics of the user manual, as well as association rule mining of a user forum, of the Storm Water Management Model (SWMM). The results show a positive correlation of the identified I/O pairs, and further reveal the complementary aspects of the user manual and user forum in supporting scientific software engineering tasks.

Keywords: Scientific software · User manual · User forum · Association rule mining · Storm Water Management Model (SWMM)

1 Introduction

The behavior of scientific software, e.g., a seismic wave propagation [11], is typically a function of a large input space with hundreds of variables. Similarly, the output space is often large with many variables to be computed. Rather than requiring stimuli from the users in an interactive mode, scientific software executes once the input values are entered as a batch [32].

The large input/output (I/O) spaces are common for the scientific understanding of complex phenomena like climate change. However, the size and complexity have been recognized as challenges for software testing [15], especially for selecting test cases from a large input space and for determining the corresponding outputs to examine.

Relating I/O is fundamental to metamorphic testing, which is considered to be a simple and effective approach for testing scientific software [8]. The prototypical example is the trigonometric function: $sine(x)$ [13]. The exact value of $sine(x)$ may be unavailable due to floating-point computations. Metamorphic testing uses properties like $sine(x)=sine(\pi-x)$ to test any implementation without having to know the concrete values of either $sine(x)$ or $sine(\pi-x)$.

While the I/O relations are clear in the above example, namely, changing the input of an angle relates to the output of the angle's sine value, determining the

© Springer Nature Switzerland AG 2021
M. Paszynski et al. (Eds.): ICCS 2021, LNCS 12747, pp. 375–389, 2021.
https://doi.org/10.1007/978-3-030-77980-1_29

I/O associations at the system level, rather than at the unit level, is difficult due to the size, complexity, and batch execution mode. The scientific software of our study, for example, has over 800 input variables and over 150 output variables. Tracking the I/O dependencies in the source code (e.g., via program slicing or define-use data relationships) can face scalability issues.

In this paper, we investigate I/O associations in the user manual and user forum of a scientific software system: the Storm Water Management Model (SWMM) [30] developed and maintained by the U.S. Environmental Protection Agency (EPA) for five decades. We manually identify the I/O variables from the SWMM user manual [26], and analyze their degrees of association based on the co-occurrence statistics. We further mine the I/O variables' association rules from one of the largest SWMM user forums with approximately 2,000 contributors and 17,000 posts [21]. Comparing the I/O associations reveals the complementary aspects of the user manual and the user forum, suggesting concrete ways to exploit metamorphic testing for scientific software's quality assurance.

The contributions of our work lie in the quantification of I/O associations from how the scientific software is introduced by the development team to the users, and how the end users discuss the actual software usages among themselves. In what follows, we provide background information and introduce SWMM in Sect. 2. Section 3 presents our quantification and comparison of the SWMM I/O associations, Sect. 4 discusses the implications of our results, and finally, Sect. 5 draws some concluding remarks and outlines future work.

2   Background

2.1   Metamorphic Relations and I/O Associations

Metamorphic testing requires properties like $sine(x)=sine(\pi-x)$ to guide the testing process. These properties represent necessary conditions for the software to behave correctly, and are referred to as *metamorphic relations* (MRs). Each MR consists of two parts: (1) an input transformation that can be used to generate new test cases from existing test data, and (2) an output relation that compares the outputs produced by a pair of test cases. As shown in Fig. 1, establishing an MR is about connecting a particular input with a corresponding output, and then asserting how such an I/O pair co-changes.

Constructing MRs is an essential task in metamorphic testing. The early work by Chen *et al.* [4], for example, relied on researchers' domain knowledge to manually create one MR and further illustrated the MR's effectiveness via testing a program that solves an elliptic partial differential equation with Dirichlet boundary conditions. Murphy *et al.* [17] made one of the first attempts to enumerate six MR classes applicable to numerical and collection-like inputs.

Although numerical MRs may be suitable for computational units like the trigonometric functions, system testing in which the scientific software is tested as a whole likely requires different MRs. Our work on integrating two different scientific software systems [7,12,14], for instance, shows the importance of understanding the entire software's inputs, outputs, and their relationships. Next is

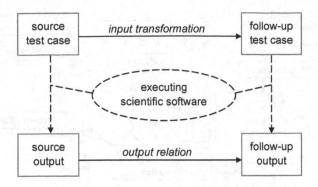

Fig. 1. A metamorphic relation (MR) consists of an *input* transformation (e.g., from x to $\pi - x$) and the associated *output* relation (e.g., equivalence relation).

an introduction of the scientific software whose I/O associations are the focal points of our study.

2.2 Storm Water Management Model (SWMM)

The Storm Water Management Model (SWMM) [30], created by the U.S. Environmental Protection Agency (EPA), is a dynamic rainfall-runoff simulation model that computes runoff quantity and quality from primarily urban areas. The development of SWMM began in 1971 and since then the software has undergone several major upgrades.

The most current implementation of the model is version 5.1.015 which was released in July 2020. Figure 2 shows a screenshot of SWMM running as a Windows application. The computational engine, which implements hydraulic modeling, pollutant load estimation, etc. is written in C/C++ with about 46,300 lines of code. This size is considered to be medium (between 1,000 and 100,000 lines of code) according to Sanders and Kelly's study of scientific software [27].

The users of SWMM include hydrologists, engineers, and water resources management specialists who are interested in the planning, analysis, and design related to storm water runoff, combined and sanitary sewers, and other drainage systems in urban areas. Thousands of studies worldwide have been carried out by using SWMM, such as land use [1, 6] and stormwater modeling [3].

3 I/O Associations in SWMM

The wide adoption of SWMM in supporting critical tasks of urban planning and environment protection makes it important for the development team at EPA to introduce the software to its users via a user manual [26]. In fact, producing the user manual is not only a common practice among scientific software developers [18], but also a requirement mandated by agencies like EPA [29] and the U.S. Geological Survey (USGS) [31]. For software evolved over many years, the

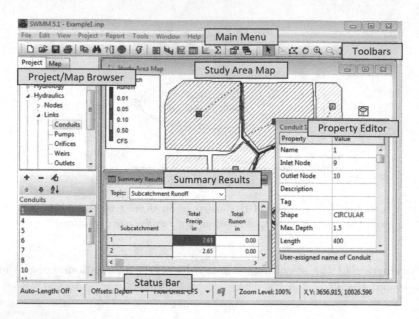

Fig. 2. SWMM running as a Windows application, annotated with functional areas in the graphical user interface.

documentation generated by end users themselves, such as user forums, builds a massive resource which has gradually become informative and comprehensive [22]. This section thus reports our analysis of SWMM's user manual in Sect. 3.1 and a user forum in Sect. 3.2. We then compare the I/O associations from these sources in Sect. 3.3, and discuss the threats to validity in Sect. 3.4.

3.1 Co-Occurrence Statistics in User Manual

The SWMM user manual (version 5.1) is a 353-page PDF document written by a core developer and environmental scientist at EPA [26]. It contains 12 chapters and 5 appendices, covering software installation and configuration steps, SWMM's conceptual model, working with map and objects (e.g., conduits of Fig. 2), running a simulation, viewing results (e.g., subcatchment runoff summary of Fig. 2), and detailed information about units of measurement, properties of visual objects, and error and warning messages. The user manual is such a comprehensive document that it remains relevant for the different sub-versions of SWMM 5.1 (5.1.010–5.1.015) since 2015.

Building on the recent work [24], we manually identified the I/O variables from SWMM's user manual. Two researchers independently performed the variable identification in a randomly chosen chapter, and Cohen's kappa between their results was 0.87 indicating an almost perfect agreement [5]. We attribute this high inter-rater agreement to the clarity of SWMM's user manual. The two researchers then individually identified the variables for the rest of the user manual. In total, 807 input and 164 output variables were identified and the manual

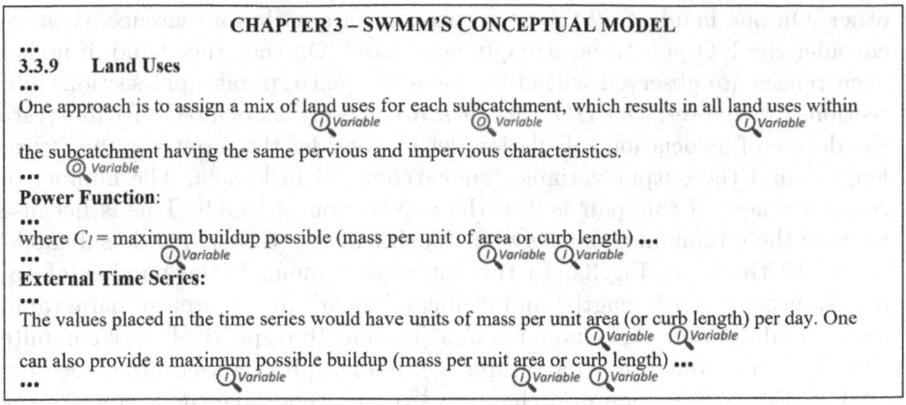

CHAPTER 3 – SWMM'S CONCEPTUAL MODEL

...

3.3.9   Land Uses

...

One approach is to assign a mix of land uses for each subcatchment, which results in all land uses within
(*I*) *Variable*   (*O*) *Variable*   (*I*) *Variable*

the subcatchment having the same pervious and impervious characteristics.
...   (*O*) *Variable*

Power Function:

...

where C_l = maximum buildup possible (mass per unit of area or curb length) ...
...   (*I*) *Variable*   (*O*) *Variable* (*I*) *Variable*

External Time Series:

...

The values placed in the time series would have units of mass per unit area (or curb length) per day. One
(*I*) *Variable* (*I*) *Variable*

can also provide a maximum possible buildup (mass per unit area or curb length) ...
...   (*I*) *Variable*   (*O*) *Variable* (*I*) *Variable*

...

(a) I/O variables in natural language text

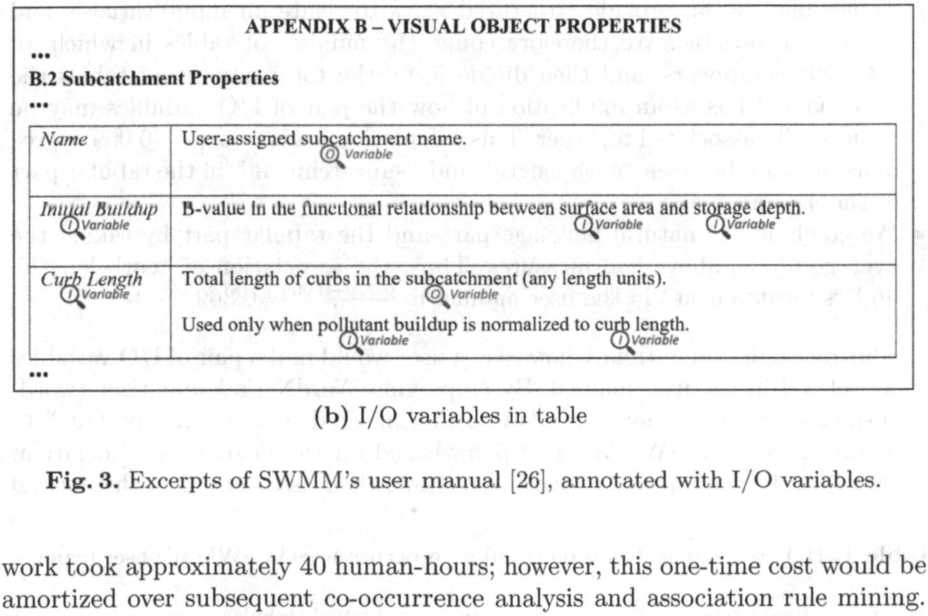

APPENDIX B – VISUAL OBJECT PROPERTIES

...

B.2 Subcatchment Properties

...

Name (*O*) *Variable*	User-assigned subcatchment name.
Initial Buildup (*I*) *Variable*	B-value in the functional relationship between surface area and storage depth. (*I*) *Variable* (*I*) *Variable*
Curb Length (*I*) *Variable*	Total length of curbs in the subcatchment (any length units). (*O*) *Variable* Used only when pollutant buildup is normalized to curb length. (*I*) *Variable* (*I*) *Variable*

...

(b) I/O variables in table

Fig. 3. Excerpts of SWMM's user manual [26], annotated with I/O variables.

---

work took approximately 40 human-hours; however, this one-time cost would be amortized over subsequent co-occurrence analysis and association rule mining. We also share the data of our work, including the I/O variables, in the institutional digital preservation site Scholar@UC [19,23] to facilitate replication.

Figure 3 shows the excerpts of SWMM's user manual, annotated with the input ('*I*') and output ('*O*') variables. To explore the I/O associations, we distinguish their appearances in the natural language texts (cf. Fig. 3a) and in the structured tables (cf. Fig. 3b). We measure the extent to which an input variable is discoverable together with an output variable as follows.

- Natural language text is hierarchical: a chapter has one or more sections or sub-sections, a section or sub-section has one or more paragraphs, and a paragraph has one or more sentences. We therefore use the hierarchical information to calculate how closely related a pair of I/O variables are to each other. On one hand, if *all* the co-occurrences are within a sentence, then we consider the I/O pair to be strongly associated. On the other hand, if no co-occurrences are observed within the same sentence, paragraph, section/sub-section, or chapter, the I/O variables are loosely associated. To illustrate the degree of association calculation, let us consider the input variable "curb length" and the output variable "subcatchment" in Fig. 3a. The number of co-occurrences of this pair is 2 in the sub-section of §3.3.9. This is because we take the minimum count between "curb length" (3 times) and "subcatchment" (2 times) in Fig. 3a. In the entire user manual, the number of co-occurrences of "curb length" and "subcatchment" in a sentence, paragraph, section/sub-section, and chapter is 3, 3, 16, and 16 respectively. We compute the ratios of sentence over paragraph ($\frac{3}{3}$), paragraph over section/sub-section ($\frac{3}{16}$), section/sub-section over chapter ($\frac{16}{16}$), and then take the average of the three ratios (0.729) as this I/O pair's association degree in the natural language part of the user manual.

- Tables like Fig. 3b provide structured ways to relate an input variable and an output variable. We therefore count the number of tables in which an I/O pair co-appears, and then divide it by the total number of tables the user manual has as an implication of how the pair of I/O variables may be structurally associated together. This calculation leads to a $\frac{1}{107}$=0.009 degree of association between "curb length" and "subcatchment" in the tabular part of the user manual.
- We combine the natural language part and the tabular part by taking the average of the above two measures. Thus, the association of "curb length" and "subcatchment" in the user manual is $\frac{0.729+0.009}{2}$=0.369.

Our rationale is to estimate how easy a user would find a pair of I/O variables being related in the user manual. By employing WordNet's lemmatizer (wordnet.princeton.edu) to convert words into the inflected roots (e.g., "conduits" to "conduit"), we rank SWMM's I/O pairs based on the degrees of association. Table 1 lists the ten top-ranked pairs and shows their associations in the natural

Table 1. I/O associations based on variable co-occurrences in SWMM's user manual.

Rank	Input variable	Output variable	textual part	tabular part	user manual
1	rain barrel	runoff	1.000	0.000	0.500
2	conduit	hours flooded	1.000	0.000	0.500
3	conduit	peak depth	1.000	0.000	0.500
4	conduit	peak runoff	1.000	0.000	0.500
5	aquifer	runoff	1.000	0.000	0.500
6	rainfall	hours flooded	1.000	0.000	0.500
7	outlet	flow routing	0.952	0.000	0.476
8	wet step	runoff	0.889	0.000	0.444
9	node invert elevation	depth	0.889	0.000	0.444
10	dynamic wave	flow	0.861	0.009	0.435

language part, the tabular part, and the user manual as a whole. More complete results can be found in our online data [23].

3.2   Association Rule Mining in User Forum

While SWMM's user manual is written by one scientific software developer, a forum like Open SWMM [21] records the questions, discussions, and interactions of thousands of SWMM users. The typical topics include how to install, configure, and run the software. The experience of running the software leads some users to post their frustrations about executions producing no result, their doubts about the validity of the generated results, and their dissatisfactions about the execution process. Sometimes, others respond to these questions to clarify confusions, offer diagnostic helps, or provide answers. A sample Open SWMM post and two replies [25] are shown in Fig. 4 where the concern regarding the number of threads that the user would choose to run SWMM was communicated.

We adapt *association rule mining* [2] for discovering patterns in the user forum data, which represents a step toward automating the construction of metamorphic relations [16]. Association rule mining was originally developed to identify products in large-scale transaction data recorded in supermarkets. For example, an association rule {diaper} ⇒ {beer} would indicate that customers who purchase diapers are also likely to purchase beer. In this example, diaper and beer are called *antecedent* and *consequent* respectively. Apriori [2] is among the most well-known algorithms to mine associate rules from a database containing various transactions (e.g., collections of items bought by customers).

It is therefore critical to define *transactions* in the context of user forums for algorithms like Apriori to work. As different users have different viewpoints and use different vocabularies, their posts shall be treated as different transactions. In addition, posts at different times reflect the user's evolving views, possibly influenced by the thread of discussions. Based on these observations, we deem a distinct forum user's post at a single time as a transaction, much like a customer's purchase at a given time being considered as a transaction in market basket analysis. As a result, Fig. 4 contains three transactions.

Q User   15-Feb-2018
I am simulating 3 urban catchments (totally 15.7 Km2). There are 273,000 stormwater pipes and the period of simulation is 18 years (rainfall data). When I only run the area for 2 months period (6 minutes time step) program shows up to 150 hours for running.
···
A Commenter
···
Finaly, are you using the maximum number of threads?
Q User
Many thanks for your useful comments. Yes, I have 237K pipes. I changed the threads from 1 to 4 and the simulation time has changed significantly. Now I am running for 10 years rainfall within 10 minutes.
···

Fig. 4. Sample Open SWMM post and two replies.

Algorithm 1: Generate {I} ⇒ {O} Association Rules

Input: a set of variables V manually identified from
   user manual, a user forum U
Output: an unordered list L of input-output associations

1 **Pre-processing:**
2 $U \leftarrow to\_lower\_case(U)$;
3 **for** *(each $v \in V$) ∧ (each $u \in U$)* **do**
4 | $VN \leftarrow \{v.name\} \cup \{v.alias\}$;
5 | **while** $VN \neq \varnothing$ **do**
6 | | $ln \leftarrow longest\_name(VN)$;
7 | | **if** *string_match(u, ln) ≥ δ* **then**
8 | | | substitute u with ln in U;
9 | | | break;
10 | | **else**
11 | | | $VN \leftarrow VN \setminus \{ln\}$;
12 | | **end**
13 | **end**
14 **end**
15 $U \leftarrow preserve\_and\_unify\_variable\_name(U)$;
16 $T \leftarrow split(U)$;
17 **Processing:**
18 $L \leftarrow Apriori\_algorithm(remove\_punctuation(T), min\_support)$;
19 **Post-processing:**
20 $L \leftarrow (L.antecedent \cap V.input) \wedge (L.consequent \cap V.output)$;

Fig. 5. Mining association rules from a user forum.

The raw posts shown in Fig. 4, however, must be processed to make the transactions amenable to association rule mining. Algorithm 1 of Fig. 5 shows our procedure to generate I/O associations. The pre-processing (lines 1–16) is tailored for user forum data U. Upon converting U into lower cases, Algorithm 1 sorts each variable's name identified from the user manual V based on length. If a term $u \in U$ matches the longest variable name $ln \in V$, then a match is found and u is replaced by ln (lines 8–9); otherwise, the next longest variable name is examined (line 11 continued with the while-loop back to line 5). This ensures that "wet weather time step" is recognized before "time step" is recognized.

We currently employ Levenshtein distance [10] and its fuzzywuzzy Python library (github.com/seatgeek/fuzzywuzzy) to implement *string_match* at line 7 of Fig. 5, and the threshold $\delta=0.85$ is determined heuristically by a small-scale pilot trial based on SWMM. Lines 15–16 show that pre-processing is completed with preserving and unifying the variable names in U, followed by splitting U into transactions. Once transactions are prepared, Algorithm 1 invokes Apriori to mine association rules where punctuations are removed from T. The post-processing of line 20 is to ensure that each rule's antecedent contains the input variable, and the consequent contains the output variable. We rank the mined association rules by Algorithm 1 via two metrics [2]: first with *support* that indicates how frequently the antecedent and consequence (i.e., the I/O variables) co-appear in the transactions, and then with *confidence* that determines the relative amount of the given consequence across all alternatives for a given antecedent. Table 2 lists the ten top-ranked association rules mined from the Open SWMM posts and their support and confidence values. More complete results of association rule mining can be found in our online data [23].

Table 2. I/O association rules mined from the Open SWMM posts based on a total of 15,958 transactions.

Rank	Association rule		Support	Confidence
1	{upstream}	⇒ {flow}	0.029	0.533
2	{downstream}	⇒ {flow}	0.027	0.532
3	{weir}	⇒ {flow}	0.022	0.543
4	{rain barrel}	⇒ {runoff}	0.018	0.518
5	{surcharge}	⇒ {flow}	0.011	0.539
6	{surface area}	⇒ {storage}	0.010	0.645
7	{previous area}	⇒ {total precipitation}	0.009	0.593
8	{depression storage}	⇒ {infiltration}	0.008	0.554
9	{wet step}	⇒ {runoff}	0.008	0.506
10	{shape curve}	⇒ {runoff}	0.006	0.933

3.3 Comparing Ranked Lists

The inputs and outputs identified from the user manual (Sect. 3.1) represent comprehensive yet static information, whereas the association rules mined from the actual software usage data of a user forum (Sect. 3.2) help uncover the dynamic regularities of inputs and outputs. To compare these two ranked lists of I/O associations, we adopt Kendall's τ which is a correlation measure for ordinal data [9]. The τ value ranges from -1 to 1 where values close to 1 indicate strong agreement between two rankings and values close to -1 indicate strong disagreement. We use the SciPy Python library [28] to calculate τ, and the scipy.stats.kendalltau() function implements the following measure:

$$\tau = \frac{P - Q}{\sqrt{(P + Q + T) * (P + Q + U)}} \tag{1}$$

where P is the number of concordant pairs, Q is the number of discordant pairs, T is the number of ties only in the first ranking, and U is the number of ties only in the second ranking.

In our analysis, we first identify the *overlapped* I/O pairs from two ranked lists, and then compute Kendall's τ for only the overlapped pairs. Figure 6 illustrates our calculation of two ranked lists, A and B, both having four pairs.

However, only three pairs are shared which we keep in A' and B'. The ranking of the three remaining pairs is preserved from the original list. In A' and B', "<weir, flow>" and "<area, storage>" are concordant, because the former is ranked higher than the latter in both lists. Similarly, "<aquifer, storage>" and "<area, storage>" are also concordant. The discordant comes from "<weir, flow>" and "<aquifer, storage>" because their relative rankings are different in A' and B'. Using equation (1), the Kendall's τ between A' and B' is: $\frac{2-1}{\sqrt{(2+1+0)*(2+1+0)}}=0.33$.

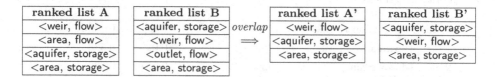

ranked list A	ranked list B		ranked list A'	ranked list B'
<weir, flow>	<aquifer, storage>	*overlap*	<weir, flow>	<aquifer, storage>
<area, flow>	<weir, flow>	$\Longrightarrow$	<aquifer, storage>	<weir, flow>
<aquifer, storage>	<outlet, flow>		<area, storage>	<area, storage>
<area, storage>	<area, storage>			

Fig. 6. Illustration of selecting the overlapped pairs and then calculating Kendall's τ.

The results of comparing the 200 top-ranked I/O associations are shown in Fig. 7. The number of overlapped I/O pairs increases in a linear fashion, and approximately a quarter (e.g., 40 out of 160) I/O variables are associated in both SWMM's user manual and in the Open SWMM forum. Among the overlapped I/O associations, the Kendall's τ correlation remains positive. This shows that the concordant pairs outnumber the discordant ones, which implies the degree of I/O associations is reasonably consistent between the user manual produced by the scientific software development team and the forum posts among the end users themselves. While we will make some qualitative observations of concordant and discordant I/O variables in Sect. 4, we next discuss some of the important aspects of our study that one shall take into account when interpreting our findings.

3.4   Threats to Validity

A threat to construct validity is how we define the degree of association between an input variable and an output variable. In particular, we use different measures for the two different data sources. As the user manual is written by somebody who is familiar with the scientific software, we quantify the I/O associations based on how coupled the two variables are within the textual part and the tabular part. From the thousands of end users' posts, we mine association rules, aiming to discover: "Forum users who mentioned an input variable also mentioned an output variable". We believe such measures account for the static and authoritative natures of the user manual, and the dynamic and idiosyncratic natures of the user forum.

An internal validity threat is our manual identification of SWMM's I/O variables from the user manual. Although an almost perfect inter-rater agreement

(Cohen's κ=0.87) was achieved on a randomly chosen sample, our manual effort may have false positives and false negatives. Another threat relates to the parameter values that we chose in association rule mining: δ=0.85 (line 7 of Algorithm 1) and $min\_support$=3 (line 18 of Algorithm 1). The former is determined by a small-scale SWMM pilot trial, and the latter is informed by a prior association rule study in software engineering [33].

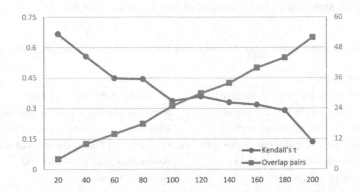

Fig. 7. Kendall's τ and the size of overlapped pairs between the I/O associations from the user manual (cf. Table 1) and the I/O associations from the user forum (cf. Table 2): x-axis represents the number of top-ranked I/O associations, left y-axis represents Kendall's τ values, and right y-axis represents the number of overlapped I/O pairs on which Kendall's τ is computed.

Several factors affect our study's external validity. Our results may not generalize to other user forums of SWMM and other scientific software systems. As for conclusion validity and reliability, we believe we would obtain the same results if we repeated the study. In fact, we publish all our analysis data in the institution's digital preservation repository [23] to facilitate reproducibility.

4   Discussion

While the Kendall's τ of Fig. 7 shows positive correlations, we share some observations of the I/O pairs in the two ranked lists. A few I/O variables are ranked high in both lists, e.g., <rain barrel, runoff> is number one in Table 1 and number four in Table 2. We observe that the input variables are often necessary and yet end users may encounter some barriers of setting up the proper values. SWMM's user manual provides prototypical values, e.g., "...single family home rain barrels range in height from 24 to 36 in. (600 to 900 mm)" [26]. Another necessary and oftentimes misused input variable is "date" which the user manual specifies the permissible formats. However, different countries have different date conventions, making the concrete values from the user forum valuable for metamorphic testing, especially for selecting source test cases (cf. Figure 1).

Some I/O variables have associations stronger in the user forum than in the user manual. For instance, <shape curve, runoff> ranks tenth in Table 2 and 6055th in the user manual results. A closer look shows that the use of "shape curve" in the implementation became deprecated after version 5.0.015[1], and the variable "storage curve" should have been used. This indicates that association rules mined from user forum posts may suggest problematic, and even deprecated variables. As a result, the user manual of scientific software shall be updated to better stay in sync with the evolution of the implementation.

Table 3. Comparing User Manual and User Forum of Scientific Software

	User manual	User forum
who	Written by the development team	Organized for & By end users
&	Focusing on the specific software	Software playing a part in meeting goals
why	Usage norms of the software	Idiosyncratic uses of the software
what	Features & Capabilities of the software	Questions & Dissatisfactions of software
&	Comprehensive intro to I/O	Attentions paid to partial I/O
how	Prototypical I/O demos	Actual I/O values
	Updated periodically & Authoritatively	Growing continuously & Organically

Not only are some higher-ranked I/O associations from the user forum indicative of deprecation, they also reveal frequently used features of the scientific software. For example, <snowmelt, runoff> is the twelfth-ranked association rule mined from Open SWMM, but ranks 98th in the user manual's results. This shows that the user manual's descriptions tend to be comprehensive, making core parameters like "snowmelt" less prominent. In contrast, end users commonly discuss the important variables, as "Snowmelt parameters are climatic variables that apply across the entire study area when simulating snowfall and snowmelt" [26]. Interestingly, the association rules mined from the forum posts can depict the simulation capabilities used, potentially suggesting requirements and their evolution of the scientific software [11].

5   Conclusions

I/O associations are integral to metamorphic testing which has helped to address some scientific software testing challenges [8]. This paper reports our analysis of the user manual and user forum of EPA's SWMM in order to quantify I/O associations. Our results show a positive correlation of the identified I/O pairs, and further reveal the differences between the two data sources. Table 3 highlights the complementary aspects, which could assist in choosing the proper data to

---

[1] https://www.epa.gov/sites/production/files/2020-03/epaswmm5_updates.txt   Last accessed: April 2021.

support scientific software's metamorphic testing, requirements engineering [11], software traceability [20], and other tasks.

Our future work includes developing automated and accurate ways to classify I/O variables, exploring associations beyond a single input variable and a single output variable, and instrumenting metamorphic testing with source test cases from the user manual and user forum. Our goal is to better support scientists in improving testing practices and software quality.

Acknowledgments. *We thank the EPA SWMM team, especially Michelle Simon, for the research collaborations. Funding is provided in part by the U.S. National Science Foundation Critical Resilient Interdependent Infrastructure Systems and Processes (CRISP 2.0) Award to Dr. Omar I. Abdul-Aziz (NSF CMMI Award #1832680).*

References

1. Abdul-Aziz, O.I., Al-Amin, S.: Climate, land use and hydrologic sensitivities of stormwater quantity and quality in a complex coastal-urban watershed. Urban Water J. **13**(3), 302–320 (2016)
2. Agrawal, R., Srikant, R.: Fast algorithms for mining association rules in large databases. In: International Conference on Very Large Data Bases, pp. 487–499 (1994)
3. Al-Amin, S., Abdul-Aziz, O.I.: Challenges in mechanistic and empirical modeling of stormwater: review and perspectives. Irrig. Drain. **62**(S2), 20–28 (2013)
4. Chen, T.Y., Feng, J., Tse, T.H.: Metamorphic testing of programs on partial differential equations: a case study. In: International Computer Software and Applications Conference, pp. 327–333 (2002)
5. Fleiss, J.L., Cohen, J.: The equivalence of weighted kappa and the intraclass correlation coefficient as measures of reliability. Educ. Psychol. Meas. **33**(3), 613–619 (1973)
6. Huq, E., Abdul-Aziz, O.I.: Climate and land cover change impacts on stormwater runoff in large-scale coastal-urban environments. Sci. Total Environ. **778**, 146017 (2021)
7. Kamble, S., Jin, X., Niu, N., Simon, M.: A novel coupling pattern in computational science and engineering software. In: International Workshop on Software Engineering for Science, pp. 9–12 (2017)
8. Kanewala, U., Chen, T.Y.: Metamorphic testing: a simple yet effective approach for testing scientific software. Comput. Sci. Eng. **21**(1), 66–72 (2019)
9. Kendall, M.G.: The treatment of ties in ranking problems. Biometrika **33**(3), 239–251 (1945)
10. Levenshtein, V.: Binary codes capable of correcting deletions, insertions and reversals. Sov. Phys. Dokladay **10**(8), 707–710 (1966)
11. Li, Y., Guzman, E., Tsiamoura, K., Schneider, F., Bruegge, B.: Automated requirements extraction for scientific software. In: International Conference on Computational Science, pp. 582–591 (2015)
12. Lin, X., Simon, M., Niu, N.: Exploratory metamorphic testing for scientific software. Comput. Sci. Eng. **22**(2), 78–87 (2020)
13. Lin, X., Simon, M., Niu, N.: Hierarchical metamorphic relations for testing scientific software. In: International Workshop on Software Engineering for Science, pp. 1–8 (2018)

14. Lin, X., Simon, M., Niu, N.: Releasing scientific software in GitHub: a case study on SWMM2PEST. In: International Workshop on Software Engineering for Science, pp. 47–50 (2019)
15. Lin, X., Simon, M., Niu, N.: Scientific software testing goes serverless: creating and invoking metamorphic functions. IEEE Softw. **38**(1), 61–67 (2021)
16. Lin, X., Simon, M., Peng, Z., Niu, N.: Discovering metamorphic relations for scientific software from user forums. Comput. Sci. Eng. **23**(2), 65–72 (2021)
17. Murphy, C., Kaiser, G.E., Hu, L., Wu, L.: Properties of machine learning applications for use in metamorphic testing. In: International Conference on Software Engineering & Knowledge Engineering, pp. 867–872 (2008)
18. Nguyen-Hoan, L., Flint, S., Sankaranarayana, R.: A survey of scientific software development. In: International Symposium on Empirical Software Engineering and Measurement, pp. 1–10 (2010)
19. Niu, N., Koshoffer, A., Newman, L., Khatwani, C., Samarasinghe, C., Savolainen, J.: Advancing repeated research in requirements engineering: a theoretical replication of viewpoint merging. In: International Requirements Engineering Conference, pp. 186–195 (2016)
20. Niu, N., Wang, W., Gupta, A.: Gray links in the use of requirements traceability. In: International Symposium on Foundations of Software Engineering, pp. 384–395 (2016)
21. Open SWMM. SWMM Knowledge Base. https://www.openswmm.org. Accessed Apr 2021
22. Pawlik, A., Segal, J., Petre, M.: Documentation practices in scientific software development. In: International Workshop on Cooperative and Human Aspects of Software Engineering, pp. 113–119 (2012)
23. Peng, Z., Lin, X., Niu, N.: Data of SWMM I/O Associations. https://doi.org/10.7945/0mn5-p763. Accessed Apr 2021
24. Peng, Z., Lin, X., Niu, N.: Unit tests of scientific software: a study on SWMM. In: Krzhizhanovskaya, V.V., et al. (eds.) ICCS 2020, Part VII. LNCS, vol. 12143, pp. 413–427. Springer, Cham (2020). https://doi.org/10.1007/978-3-030-50436-6_30
25. Rashetnia, S.: Long simulation time for very large models. https://www.openswmm.org/Topic/11289/long-simulation-time-for-very-large-models. Accessed Apr 2021
26. Rossman, L.A.: Storm Water Management Model User's Manual Version 5.1. https://www.epa.gov/sites/production/files/2019-02/documents/epaswmm5_1_manual_master_8-2-15.pdf. Accessed Apr 2021
27. Sanders, R., Kelly, D.: Dealing with risk in scientific software development. IEEE Softw. **25**(4), 21–28 (2008)
28. SciPy. Scientific Computing Tools for Python. https://docs.scipy.org/doc/scipy/reference/generated/scipy.stats.kendalltau.html. Accessed Apr 2021
29. United States Environmental Protection Agency. Agency-wide Quality System Documents. https://www.epa.gov/quality/agency-wide-quality-system-documents. Accessed Apr 2021
30. United States Environmental Protection Agency. Storm Water Management Model (SWMM). https://www.epa.gov/water-research/storm-water-management-model-swmm. Accessed Apr 2021
31. United States Geological Survey. Review and Approval of Scientific Software for Release (IM OSQI 2019–01). https://www.usgs.gov/about/organization/science-support/survey-manual/im-osqi-2019-01-review-and-approval-scientific. Accessed Apr 2021

32. Vilkomir, S.A., Swain, W.T., Poore, J.H., Clarno, K.T.: Modeling input space for testing scientific computational software: a case study. In: Bubak, M., van Albada, G.D., Dongarra, J., Sloot, P.M.A. (eds.) ICCS 2008, Part III. LNCS, vol. 5103, pp. 291–300. Springer, Heidelberg (2008). https://doi.org/10.1007/978-3-540-69389-5_34

33. Wang, W., et al.: Complementarity in requirements tracing. IEEE Trans. Cybern. **50**(4), 1395–1404 (2020)

Understanding Equity, Diversity and Inclusion Challenges Within the Research Software Community

Neil P. Chue Hong[1]([✉])(iD), Jeremy Cohen[2](iD), and Caroline Jay[3](iD)

[1] Software Sustainability Institute and EPCC, University of Edinburgh, Edinburgh, UK
`N.ChueHong@epcc.ed.ac.uk`
[2] Department of Computing, Imperial College London, London, UK
`jeremy.cohen@imperial.ac.uk`
[3] Department of Computer Science, University of Manchester, Manchester, UK
`Caroline.Jay@manchester.ac.uk`

Abstract. Research software – specialist software used to support or undertake research – is of huge importance to researchers. It contributes to significant advances in the wider world and requires collaboration between people with diverse skills and backgrounds. Analysis of recent survey data provides evidence for a lack of diversity in the Research Software Engineer community. We identify interventions which could address challenges in the wider research software community and highlight areas where the community is becoming more diverse. There are also lessons that are applicable, more generally, to the field of software development around recruitment from other disciplines and the importance of welcoming communities.

Keywords: Research software · Software engineering · Research software engineering · Diversity · EDI

1 Introduction

Developing specialist research software to support computational science is an especially challenging process. Unlike more traditional software engineering tasks, researchers or, increasingly, RSEs (Research Software Engineers) who write this software need an understanding of the underlying scientific challenge being addressed.

The methods used in research favour a hypothesis-driven approach. This creates a different working environment from the wider software engineering industry, where software is built to meet a client's specification. As noted by Hettrick et al.,

> *"To be effective, software development in research should be approached, not as a one-off transaction, but as a partnership between researcher and software expert."* [3]

© Springer Nature Switzerland AG 2021
M. Paszynski et al. (Eds.): ICCS 2021, LNCS 12747, pp. 390–403, 2021.
https://doi.org/10.1007/978-3-030-77980-1_30

The "partnership between researcher and software expert" mentioned here highlights a need for varied skills and good communication. It is clear that much computational research requires diversity of skills and experience, working in partnership, that spans software engineering and science. Researchers are more likely to be working with others who have different technical expertise, use different technical terminology, and may be communicating in a tertiary language.

Diversity in project teams, and workplaces in general, is important and it is widely accepted that there are many benefits from ensuring diversity within teams and communities. For example, there is evidence to show that diversity in terms of knowledge or skills can be beneficial [19] and that gender diversity within groups can result in higher quality scientific outputs [4]. However, there are cases where diversity can raise challenges, for example with reduced feelings of well-being amongst members in highly-diverse teams [31]. Diversity in the context of teams in industry and research is an area that has been the subject of extensive research and there are some contradictory results from the many studies undertaken. Stahl et al. [26] look at a number of different studies of cultural diversity concluding that there are both benefits and drawbacks. Whether benefits can be realised while minimising more challenging aspects will depend on effective process management. In the context of software engineering, Capretz and Ahmed [5] looked at how different personality traits relate to suitability for different roles in software projects. Ultimately they conclude that diversity in terms of personalities and skills is important in helping with problem solving tasks involved in building and maintaining software.

The different types of diversity highlighted so far, including diversity of skills and knowledge, culture, personality and gender are, of course, just some of the many different aspects that lead to diversity amongst a group of individuals. Others, among a huge range, include ethnicity, disability and age.

Research Software Engineers are, at present, much more likely to come from a research background than software engineering professionals working in other fields. Nonetheless, we know from surveys within the RSE community that a similar diversity crisis to that identified in other fields exists. In this paper, we examine current problems with diversity in research software engineering and consider potential causes. Our aim is to present a better understanding of existing diversity within the RSE community in order to develop insights and recommendations to address current issues. We achieve this through an empirical analysis of existing open data collected and made available through large-scale international surveys of research software engineers undertaken by the RSE community. We also review related work from allied areas which suggests ways of addressing the challenges identified.

As pointed out by Mathieson in [20], based on 2018 ONS data, just one in eight of almost 340,000 software development professionals in the UK are women. This example demonstrates that there is a long way to go to address the lack of diversity in the wider software engineering domain, as well as the more focused domain of research software engineers. Professional software engineers can gain their skills through a variety of different routes and they may come from a wide range of different disciplines. For example, they may have developed their skills through a degree programme or vocational training in computing or a related area. Alternatively, they may come from a completely different disciplinary background and have re-trained through one of a large number of "coding schools" that offer training, often via intensive courses, in software development skills. The wide array of routes into the field creates an expectation of a diverse field which makes it all the more surprising that this is not the case.

We highlight three core contributions this paper provides to understanding the importance of diversity among RSEs working in computational and data science:

- Providing an evidence-based analysis that demonstrates the current problems with diversity in the research software engineering community.
- Demonstrating that there is already extensive "domain mobility" for RSEs and that a lack of such mobility is therefore not likely to be a cause for a lack of diversity within the field.
- Offering four general recommendations that we believe can form a basis to support addressing the lack of diversity in research software engineering.

In Sect. 2 we examine the International RSE Survey results from 2018 and undertake further analysis on this data. Section 3 highlights the three areas that we see as both helping to explain and provide the basis for addressing the lack of diversity amongst RSEs, while discussion and conclusions are provided in Sect. 4.

2   Surveying RSE Diversity

To understand the diversity challenges facing those embarking on careers as Research Software Engineers, we require a better understanding of the landscape as a whole. The Software Sustainability Institute coordinates an ongoing series of surveys of the RSE community. In the most recent survey [24], from 2018, participants were asked a number of socio-demographic questions.

Table 1. Comparison of 2018 demographics for research software engineers, academics, software developers and general working population in the United Kingdom.

Percentage of role in UK who are:	RSEs [24]	Academics [15]	Softwaredevelopers [25]	All UK workers [25]
Gender (female)	14	46	14	48
Ethnicity (BAME/Mixed)	5	15	21	12
Report disability	6	4	10	13

Our reanalysis primarily focuses on UK data, as this is where the authors are located, but the numbers from other countries in the survey are broadly similar. Table 1 compares RSEs with UK Higher Education Statistics Authority (HESA) data [15], and a study by the British Computer Society (BCS) of software development professionals in the UK [25] based on 2018 Office for National Statistics data. Limitations in the RSE data currently available to us mean that we focus on aspects of gender, ethnicity and disability, as opposed to other types of diversity.

This comparison indicates a gender diversity gap, which we might expect given the percentage of women working as software developers is also 14%. But this could be better: software is a fundamental part of all research, and 46% of academic staff are female. The ethnicity data indicates a greater problem. Of the respondents who declared their ethnicity, there were only 6 non-white and 5 mixed race RSEs – 5%. These figures are significantly lower than the 28% BAME students studying computer science [30] and 10% of BAME physicists [17]. Although we should not aggregate non-white ethnic categories together, as they will experience different challenges and biases, it is striking that RSEs do

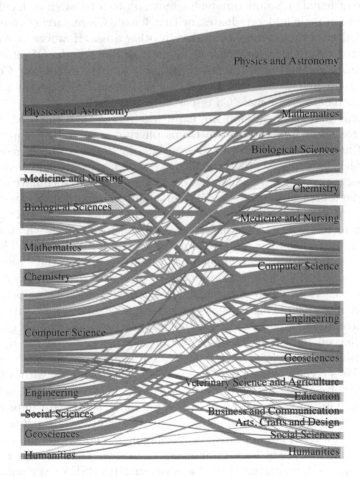

Fig. 1. Flow of UK RSEs from academic discipline studied to domains they now work in, categorised by gender: 79.8% male (green)/14.3% female (purple)/5.9% other responses (grey). N = 203. (Color figure online)

not fit the general profile of those working in the IT industry, where there is greater ethnic diversity.

An important area of research is understanding if diversity can be improved if RSEs are drawn from a wider set of degree backgrounds. Figure 1 shows where RSEs are working based on their academic degree. Only gender diversity was considered, given the low ethnic diversity and number of respondents reporting a disability. Gender balance is relatively uniform across fields rather than mirroring the gender balance in that field. Over half of RSEs have a first degree in Physics and Astronomy or Computer Science. In the UK, 17% of CS undergraduates were female compared with 41% of physical sciences undergraduates [16], suggesting that RSEs come from the "computational" subset of a subject. However, within computer science research within the UK just under 23% of academics and researchers are female [14]. This perhaps suggests that a larger percentage of female CS undergraduates move on to a research or faculty position in CS than male undergraduates, or that female CS researchers are moving into the field after undergraduate studies in other areas. However, it could also be attributed to other factors such as industry hiring trends. Of interest here is that, while over half of RSEs have a Physics and Astronomy or Computer Science-focused first degree, the percentage of female RSEs is still somewhat below the percentage of female CS research/faculty staff.

The areas that RSEs support are more widespread than the fields they are currently recruited from. There are teams offering general RSE support at a number of institutions and many RSEs working in the biosciences, geosciences, and medicine. This suggests that a challenge research software engineering has is encouraging more people with first degrees in these areas (which have more balanced gender representation) to become RSEs. This is important because domain knowledge can be especially valuable when undertaking RSE projects. Things to consider include the wording of job adverts, which can affect who applies for them [1]—can we encourage more candidates from other disciplines? There may be lessons for the software development field more generally—there are a significant number of RSEs whose background is in biological sciences or geosciences, as well as some from social sciences and the humanities, showing people with "non-traditional" degree backgrounds seek careers as software practitioners.

One encouraging aspect of Fig. 1 is there appear to be many lines going from one discipline on the left hand side to a different discipline on the right hand side, representing an RSE working in an area different from the one where they trained. This suggests that there is scope to improve diversity within the RSE community through attracting individuals from more diverse domains into the RSE space. This provides an opportunity for more immediate improvements in the diversity of the RSE community than the multi-year timeline that we might expect for efforts to improve diversity amongst undergraduate cohorts. To investigate this movement between domains, which we call *"domain mobility"*, we undertook further analysis of the most recent 2018 RSE survey data.

Table 2 shows an example of "domain mobility" for a series of domains where the number of survey respondents was >20. We define *partial mobility* as RSEs

Table 2. Example of domain mobility across a selection of disciplines from the 2018 RSE survey data where $N > 20$.

Discipline	N	Partially mobile (%)	Fully mobile (%)
Biological Sciences	95	44.21	8.42
Chemistry	22	40.91	22.73
Computer Science	241	51.45	22.41
Electrical & Electronic engineering	36	41.67	38.89
Geography & Environmental sciences	46	52.17	6.52
Mathematics	69	47.83	47.83
Physics and Astronomy	261	31.80	19.92

working in one or more domains outside the domain of their highest degree, while also still undertaking work in the domain of their degree. *Full mobility* applies a further filter to partial mobility by excluding RSEs who still work in the domain of their highest degree. So, in the case of Mathematics, where the figures for partial and full mobility are the same, none of the survey respondents who said their highest degree was in Mathematics undertake RSE work in the domain of Mathematics. It should be noted that while domain mobility is a positive concept, and the ability to work across a wide variety of different domains is a great benefit, we do not have any data on the reasons for these movements between domains. The movements may be individuals wanting to undertake RSE work in, and learn about, a different domain to their original area of study. However, there are also likely to be cases of movement between domains being made out of necessity due to a lack of job opportunities in an individual's domain of choice, for example.

2.1   Code and Data Availability

The Jupyter Notebooks used to perform the analysis are available from [8]. International RSE data can be obtained from [24], apart from some gender data which have not been publicly published at the time of writing.

3   Improving EDI in the Research Software Community

In this section we investigate three areas that represent both challenges and opportunities for improving equity, diversity and inclusion (EDI) within the research software community. Through the material in this section, we hope to highlight possible reasons for the lack of diversity in the research software space while also providing some thoughts and guidance on how these can be addressed and how the community can work together to help improve a wide range of aspects of diversity. The material here is aimed at both developers of research software, and at team leaders, technical managers and individuals in other management and community leadership roles. For individuals who write software, we aim to offer thoughts and advice that can help improve understanding of the benefits of diversity and recognise situations that can contribute to a lack of diversity. For managers and team leaders, we hope to improve recognition of opportunities to address diversity challenges, both locally within teams and in the wider community in the context of events and activities.

3.1   Safety in Similarity?

Diversity, whether in terms of gender, ethnicity, skills or other characteristics, is important in bringing different perspectives, ideas and experiences to a community. Diverse teams can lead to higher quality science [4] and improved technology business performance [22].

However, as highlighted by Merritt [21] in the context of hiring staff, individuals want to work with people who are like them and Lang and Liu [19] point out contradictory evidence for the benefits of diversity within teams. They highlight work by Byrne in the 1971 book "*The Attraction Paradigm*" that suggests certain similarities within teams can be helpful in ensuring effective working environments. However, they do also highlight a series of other work that supports the idea of diversity within teams and groups being beneficial. Van der Zee, et al. [31] highlight previous work that suggests that we tend to have a positive response to similarity while the opposite is true for dissimilarity. They also point out other previous work suggesting that we have an attraction to people who share similar attitudes and values to our own since this makes communication with them easier.

A very large study of over 20 years of scientific papers [12] shows that paper co-authors are more likely to be of similar ethnicity. However, where papers are the result of collaborations between authors at different locations, it was shown that this can result in them being published in journals with high impact factors and receiving more citations.

The above examples demonstrate that there is a perceived "safety" in being around people who have similar interests, backgrounds and/or values. We see that this is likely to work both ways and that in the case of a community, such as the RSE community or groups of software practitioners more generally, individuals choose whether or not to engage partly based on whether they see people like themselves within that community. Of course, while individuals may seek out others who they feel are similar or who they have things in common with, diversity can be extremely important and hugely valuable in providing different views, ideas, attitudes and perspectives. Ultimately this can lead to important benefits, even if there may be learning experiences to be had along the way. These include, for example, getting more used to working with people who may not look at every opportunity, challenge or research problem in the same way. Many of us will have had experiences that support this when attending community events and workshops which often involve breakout group discussions or problem-solving tasks.

Awareness is an important step here and this begins with **educating team leaders, community managers and event organisers to be more aware of situations where there is a lack of diversity and the benefits of addressing this**. Developing a stronger understanding of the benefits of diversity, but also the challenges that can arise in ensuring that diverse groups can collaborate effectively, is an important step towards addressing the lack of diversity in the scientific software community. This is especially important for individuals in leadership or organisational roles. Various guides, for example the Hopper Conference Diversity Guide [28] and NumFOCUS DISCOVER Cookbook [23], provide detailed information on approaches for helping to ensure diversity at conferences and events. There is also a role for RSE group leaders in encouraging and enabling the RSEs in their teams to collaborate with other institutions, as this diversity might lead to higher impact of their work.

3.2   Increasing Equity, Diversity and Inclusion at Events

The events that we consider in this section include everything from large conferences and workshops to small local community events with only a small number of attendees. Large events often have dedicated organising teams and potentially co-chairs specifically dedicated to areas such as EDI. This is unlikely to be the case for a small community event with perhaps only a few 10 s of participants. Nonetheless, being able to widen participation at events requires you to start with an understanding of the community the event is targeted at and an awareness of the particular aspects of diversity that you aim to improve [9]. One suggestion for improving equity and inclusion at conferences, from work looking at 30 conferences in the conservation and ecology domain, is to write up and promote details of the actions and processes followed to support improving EDI [29]. This work demonstrates that concerns around ensuring diversity at conferences and events are not specific to the research software community. The sort of open approach espoused through supporting and promoting EDI activities could be beneficial across many, if not all, domains. It would help to offer a demonstration of a conference's efforts to support aspects of EDI and also provide evidence for the wider community to help identify what works and what is less successful. This should help to avoid repeating less successful approaches to addressing diversity and inclusion concerns across different domains.

Ensuring that event committees, speakers and panel members reflect the diversity that organisers would like to see among an event's attendees has been effective in some fields. This builds on some of the ideas highlighted in Sect. 3.1. Research that looked at presenters across 21 meetings of the American Association of Physical Anthropologists found events that had either female, or both female and male organisers, resulted in a much higher percentage of women as first authors of presented papers and posters [18]. However a study by Bano and Zowghi of six software engineering conference series [2] found that having a female conference chair or program committee chair *did not* significantly affect the number of keynote speakers that were female, or the make up of the program committee. This work did not, however, look at all conference speakers.

While the figures shown in Sect. 2 show a lack of diversity within the RSE community, there is a perception, when attending RSE workshops and events, of gender and ethnic diversity being significantly better than the headline survey figures suggest. RSE events generally have a significant number of female speakers, workshop organisers and community leaders. We see with the Collaborations Workshop series, an event that focuses on general research software practice, a positive change in the diversity balance. It has had a policy of ensuring diversity of keynote speakers, and in 2020, 55% of steering committee members were female, 42% of all speakers were female, 34% of attendees reported their gender as female, and 20% of attendees reported their ethnicity as non-white or mixed.

Nonetheless, the challenge remains of attracting more individuals from a wider range of backgrounds to get involved with RSE. A concrete action that can be taken here is to work to **increase the diversity amongst organisers, speakers and sponsors at RSE events and be more open about the approaches taken to support this**. The latter part of this recommendation should help organisers of other events, both within the research software community and beyond, to learn from and build on efforts to improve EDI. Ultimately this should help to accelerate the process of improving diversity throughout the research community.

3.3 An Inclusive Culture and Safe Space

An inclusive culture is an important foundation of diversity [6]. Research teams that include RSEs frequently have only one or two individuals in RSE roles. Having valuable but different skills, and different career aims, can make RSEs feel like the "odd one out" within a research group or team. As a result, RSE communities of practice have developed in recent years to support individuals in this field. Possibly due to previous perceived marginalisation, they tend to be open, welcoming communities. These communities provide a great opportunity to meet, network and collaborate with others who understand the challenges RSEs face in their day-to-day work.

At conferences and events, one way of formally promoting inclusion is having a clear and well-publicised *Code of Conduct*. The importance of a Code of Conduct and the challenges it can help to address and provide guidance on are highlighted in [10]. While still not commonplace at traditional academic conferences [11], it is unusual to find an RSE conference or workshop that does not have a Code of Conduct – learning from the experience of the open source community. To ensure best practice and clarity of the message provided, many events choose to build on widely accepted, open codes of conduct such as the template provided by the Geek Feminism wiki [13], and also have a diversity statement.

In the context of workplaces and research teams, it is also important that individuals feel that they fit in, both within the workplace environment itself and with the people based there. Cheryan et al. [7] call this "ambient belonging". They undertook a series of studies within a computer science context looking at gender-based perspectives on the influence of different environments. This work showed that environments that had aspects that made them fit with computer science stereotypes reduced the interest of women participating in the field. Changing the environments to make them appear less like something that would be associated with computer science increased interest in the field. This provides an example of the sort of potentially small, but nonetheless significant, changes that can help to develop an inclusive culture, a welcoming environment and, ultimately, help to improve diversity. We consider that there may also be opportunities to help improve other aspects of diversity through identifying and making similarly small environmental or organisational changes in different contexts, for example to improve the experience for people with disabilities.

The wide ranging use of codes of conduct at research software events and the recognition of the importance of communities in helping to provide individuals with a place where they fit in is significant. It suggests that the research software community is developing events that are explicit in their desire to be open and diverse, with clear statements on acceptable behaviour. We see that progress is being made in developing an inclusive culture and providing environments through events and activities that potential participants feel are open and welcoming. It is of vital importance that this continues. There are two concrete recommendations from this analysis. Firstly, **within RSE groups, both team leaders and members should be aware of the importance of their working environment and how potentially small environmental changes may affect other team members' feelings of inclusion and belonging**. Secondly, in the context of community events and activities, **event organisers should look to highlight their support for diversity and inclusion through the use of a diversity statement and provide a clear code of conduct highlighting acceptable behaviour**.

The Software Sustainability Institute runs a fellowship programme recognising the diverse roles and skills of those working to promote research software practice. An evaluation of the programme showed it plays an important role in supporting communities of best practice and skills transfer, and that a significant benefit is the way it has raised the profile of software in research, and those people who develop and advocate for it [27]. This has had positive effects for those who may previously have considered themselves as 'outsiders' in the role, or lacked confidence. This is exemplified by a comment from a female respondent:

> *"Despite getting a PhD partially from a computer science programme, I could see that my skills and knowledge were always at least to some extent dismissed or doubted [...] since being elected a SSI fellow I most definitely observed a significant drop in mansplaining... I have little doubt that the SSI fellowship was a significant [reason] I got my current position as (Head of Division at a Supercomputing Center)."* [27]

4 Discussion and Conclusions

Our research has shown that the gender diversity of RSEs is similar to the field of software engineering, but does not reflect the gender balance of the academic research workplace. This is particularly true where RSEs are working in domains such as the biosciences, geosciences and medicine. Ethnic diversity among RSEs is worse than in the wider software industry, but it is unclear why. The number of RSEs with disabilities reflects the academic research workplace, and is poor in comparison to the IT profession. Given that 19% of the UK working age population report a disability, there is much to be done to provide an equitable workplace for RSEs with disabilities. Differences in workplace culture, environment or incentives may be factors, and further research is required. However, general interventions to improve diversity appear to be increasing gender and ethnic diversity at events.

We have highlighted the importance of communication and collaboration between individuals building software to support research and the computational scientists and researchers that they collaborate with. We have also shown that diversity among all parties in these collaborations can lead to better communication, a wider range of ideas and perspectives, and, ultimately more effective collaborations that produce higher quality outputs. The three areas discussed in Sect. 3 as key opportunities for improving equity, diversity and inclusion within the field of research software can be summarised as follows:

- **Safety in similarity?** The perceived safety of looking to collaborate and work with individuals who have similar interests, backgrounds or values is widely recognised. While there is research that suggests there can be benefits to this approach, there is also extensive work highlighting the benefits of diversity. However, this is something that can be approached from two different perspectives and individuals may be more likely to join and engage with a community if they see people like themselves within that community.
- **Increasing diversity at events:** It is important to highlight and promote actions being take to improve EDI in the context of events. It is also important to ensure that both organising groups and speakers reflect the level of diversity that an event's organisers would like to see among its participants.
- **An inclusive culture and safe space:** This is a very important aspect of both participating in a community, and being part of an RSE group, for everyone involved. A well-publicised Code of Conduct and diversity statement with clearly defined processes supporting them are key elements in helping to ensure this within the context of events. In RSE groups, leaders and team members need to be aware of the importance of an inclusive culture and working environment.

Poor diversity in the community of research software developers is likely a result of the low levels of diversity in the disciplines that currently form the major path towards an RSE role or career. We have identified scope to widen the range of areas from which RSEs are recruited and it is hoped that this can be achieved by better use of language in the way that RSE roles are advertised and in the way that RSE is promoted, more generally. This may also have relevance for professional software practitioners looking to get involved in the research community, as it is clear that there are many people with "non-traditional" degrees seeking careers developing software. There is a role for professional bodies, such as the British Computer Society, Association for Computing Machinery and IEEE Computer Society, to support and embrace these career paths and help improve the gender diversity at undergraduate and postgraduate level.

Nonetheless, we feel that despite the knowledge gained from our extended analysis of existing survey data, and the pre-existing material that we have referenced, this is a hugely complex but very important area. It would benefit extensively from additional evidence that could be gathered through a range of further empirical studies. Some of the areas that may be considered most important for such studies in the short term include:

- Exploring why levels of diversity among RSEs are lower than in many of the areas of study and research that already feed into RSE careers, such as Computer Science and Physics and Astronomy. What are the levels of diversity in other career paths for these subjects? What influences the career choices that individuals make and are there specific aspects that steer them away from an RSE career?
- Trialing approaches for increasing diversity and inclusion at events, including workshops and conferences. Gathering statistics on the relative success of these approaches and their contribution to improving diversity within the computational science developer community.
- Looking at opportunities to further increase "domain mobility" through taking advantage of existing courses/training material or offering new information that can help individuals move more practically between working in different domains. Look at take-up of such opportunities across different domains and use this as a basis for longer-term analysis on how diversity in the RSE community changes over time.

Once individuals become RSEs and engage with the RSE community, the RSE community meets many of the requirements to ensure that it welcomes and supports diversity. The open, inclusive, nature of the community is of great importance here and, using gender diversity as an example, there are already several women in highly-visible leadership roles.

Developing software for science and the wider computational research domain is challenging. The relatively new field of research software engineering encompasses many of the individuals undertaking these software development tasks, regardless of whether their official job title considers them to be a researcher, professor, software engineer or RSE. It is clear that research software engineering, and research software in general, shares many diversity challenges with the wider software development field but we feel there are many opportunities to address this. This begins with increasing awareness of the different aspects raised in this paper and using this as a basis to develop the environment and opportunities to help build equity, diversity and inclusion within the community. We hope that the research software community continues to grow, and become more diverse by sharing and learning from other practitioners.

Acknowledgement. We thank Simon Hettrick, James Graham and Rob Haines for their input, and Olivier Philippe who originally processed the international RSE survey data. NCH and CJ acknowledge support from EPSRC/BBSRC/ESRC/NERC/AHRC/STFC/MRC grant EP/S021779/1 for the UK Software Sustainability Institute. JC acknowledges support from EPSRC grant EP/R025460/1.

References

1. Apa, A.: Driving Gender Equality in the Tech Industry: Breaking down unconscious bias (June 2018). https://womeninhpc.org/hpc/driving-gender-equality
2. Bano, M., Zowghi, D.: Gender disparity in the governance of software engineering conferences. 2019 IEEE/ACM 2nd International Workshop on Gender Equality in Software Engineering (GE) (May 2019). https://doi.org/10.1109/ge.2019.00016
3. Brett, A., et al.: Research Software Engineers: State of the Nation Report **2017**, (2017). https://doi.org/10.5281/zenodo.495360
4. Campbell, L.G., Mehtani, S., Dozier, M.E., Rinehart, J.: Gender-heterogeneous working groups produce higher quality science. PLoS ONE **8**(10), (2013). https://doi.org/10.1371/journal.pone.0079147
5. Capretz, L.F., Ahmed, F.: Why do we need personality diversity in software engineering? ACM SIGSOFT Soft. Eng. Notes **35**(2), 1–11 (2010). https://doi.org/10.1145/1734103.1734111
6. Chamorro-Premuzic, T.: How to design a diversity intervention that actually works (June 2019). https://www.forbes.com/sites/tomaspremuzic/2019/06/20/how-to-design-a-diversity-intervention-that-actually-works/
7. Cheryan, S., Plaut, V.C., Davies, P.G., Steele, C.M.: Ambient belonging: how stereotypical cues impact gender participation in computer science. J. Pers. Soc. Psychol. **97**(6), 1045–1060 (2009). https://doi.org/10.1037/a0016239
8. Chue Hong, N., Cohen, J., Jay, C.: RSE Diversity Analysis Notebooks (2021). https://doi.org/10.5281/zenodo.4662166
9. Collis, T.: Improving diversity at HPC conferences and events In: West, J., Rivera, L., Chue Hong, N. (eds.) (2016). http://www.hpc-diversity.ac.uk/best-practice-guide/access-best-practice
10. Favaro, B., et al.: Your science conference should have a code of conduct. Front. Mar. Sci. **3**, (2016). https://doi.org/10.3389/fmars.2016.00103
11. Foxx, A.J., Barak, R.S., Lichtenberger, T.M., Richardson, L.K., Rodgers, A.J., Webb Williams, E.: Evaluating the prevalence and quality of conference codes of conduct. Proc. Nat. Acad. Sci. **116**(30), 14931–14936 (2019). https://doi.org/10.1073/pnas.1819409116
12. Freeman, R.B., Huang, W.: Collaborating with people like me: Ethnic coauthorship within the United States. J. Labor Econ. **33**(S1), S289–S318 (2015). https://doi.org/10.1086/678973
13. Gardiner, M., Aurora, V., Smith, S., Benjamin, D.: Geek Feminism Wiki: Conference anti-harassment/Policy. https://geekfeminism.wikia.org/wiki/Conference_anti-harassment/Policy. Accessed 10 Aug 2020
14. Graham, H., Raeside, R., Maclean, G.: Understanding the status of underrepresented groups in the information and communication technologies: A report to the engineering and physical sciences research council (July 2017). https://epsrc.ukri.org/newsevents/pubs/napierdiversityreport/

15. HESA: HE staff by activity standard occupational classification, atypical marker, contract marker, mode of employment, terms of employment, contract levels, sex, age group, disability, ethnicity and academic year. https://www.hesa.ac.uk/data-and-analysis/staff/table-3

16. HESA: HE student enrolments by subject area and sex. https://www.hesa.ac.uk/data-and-analysis/sfr247/figure-14

17. Institute of Physics: What does a physicist look like? (March 2016). http://www.iop.org/publications/iop/2016/file_67244.pdf

18. Isbell, L.A., Young, T.P., Harcourt, A.H.: Stag parties linger: Continued gender bias in a female-rich scientific discipline. PLoS ONE **7**(11), (2012). https://doi.org/10.1371/journal.pone.0049682

19. Liang, T., Liu, C., Lin, T., Lin, B.: Effect of team diversity on software project performance. Ind. Manage. Data Syst. **107**(5), 636–653 (2007). https://doi.org/10.1108/02635570710750408

20. Mathieson, S.A.: How diversity spurs creativity in software development (December 2019). https://www.computerweekly.com/feature/How-diversity-spurs-creativity-in-software-development. Accessed 26 Jan 2021

21. Merritt, C.: Why do we keep hiring ourselves? (September 2018). https://womeninhpc.org/diversity-and-inclusion/why-do-we-keep-hiring-ourselves

22. NCWIT: What is the impact of gender diversity on technology business performance? (2014). https://www.ncwit.org/sites/default/files/resources/impactgenderdiversitytechbusinessperformance_print.pdf

23. NumFOCUS: NumFOCUS DISCOVER Cookbook (Diverse & Inclusive Spaces and Conferences: Overall Vision and Essential Resources) (2019). https://discover-cookbook.numfocus.org/. Accessed 13 Aug 2020

24. Philippe, O., et al.: softwaresaved/international-survey: Public release for 2018 results (March 2019). https://doi.org/10.5281/ZENODO.2585783

25. Society, B.C.: BCS Diversity Report 2020: ONS Analysis (June 2020). https://www.bcs.org/media/5766/diversity-report-2020-part2.pdf

26. Stahl, G.K., Maznevski, M.L., Voigt, A., Jonsen, K.: Unraveling the effects of cultural diversity in teams: a meta-analysis of research on multicultural work groups. J. Int. Bus. Stud. **41**(4), 690–709 (2009). https://doi.org/10.1057/jibs.2009.85

27. Sufi, S., Jay, C.: Raising the status of software in research: a survey-based evaluation of the software sustainability institute fellowship programme [version 1; peer review: 3 approved with reservations]. F1000Research **7**(1599) (2018). https://doi.org/10.12688/f1000research.16231.1

28. The Hopper Fund: The Hopper Conference Diversity Guide: Do Better at Conference Diversity. https://conference.hopper.org.nz/. Accessed 13 Aug 2020

29. Tulloch, A.I.T.: Improving sex and gender identity equity and inclusion at conservation and ecology conferences. Nature Ecol. Evol. **4**(10), 1311–1320 (2020). https://doi.org/10.1038/s41559-020-1255-x

30. Universities UK: Patterns and trends in UK higher education 2018 (September 2018). https://www.universitiesuk.ac.uk/facts-and-stats/data-and-analysis/Documents/patterns-and-trends-in-uk-higher-education-2018.pdf

31. Van Der Zee, K., Atsma, N., Brodbeck, F.: The influence of social identity and personality on outcomes of cultural diversity in teams. J. Cross Cult. Psychol. **35**(3), 283–303 (2004). https://doi.org/10.1177/0022022104264123

Solving Problems with Uncertainty

The Necessity and Difficulty of Navigating Uncertainty to Develop an Individual-Level Computational Model

Alexander J. Freund[ID] and Philippe J. Giabbanelli[(✉)][ID]

Department of Computer Science and Software Engineering,
Miami University, Oxford, OH 45056, USA
{freundaj,giabbapj}@miamioh.edu
https://www.dachb.com

Abstract. The design of an individual-level computational model requires modelers to deal with uncertainty by making assumptions on causal mechanisms (when they are insufficiently characterized in a problem domain) or feature values (when available data does not cover all features that need to be initialized in the model). The simplifications and judgments that modelers make to construct a model are not commonly reported or rely on evasive justifications such as 'for the sake of simplicity', which adds another layer of uncertainty. In this paper, we present the first framework to transparently and systematically investigate which factors should be included in a model, where assumptions will be needed, and what level of uncertainty will be produced. We demonstrate that it is computationally prohibitive (i.e. NP-Hard) to create a model that supports a set of interventions while minimizing uncertainty. Since heuristics are necessary, we formally specify and evaluate two common strategies that emphasize different aspects of a model, such as building the 'simplest' model in number of rules or actively avoiding uncertainty.

Keywords: Agent-based model · Causal map · Graph algorithms · Information fusion

1 Introduction

The design of an individual-level model (e.g., Agent-Based Model) is a common activity in computational science. In computational *social* science, such models can serve to explain social phenomena or safely test interventions within an artificial society before selecting the most promising ones for real-world pilot testing [7,13,14]. In an individual-level model, the simulated entities have their own *features* (e.g., age, gender, beliefs and values) which need to be initialized at

Alexander J. Freund thanks the Department of Computer Science & Software Engineering and the Graduate School at Miami University for research funding. Both authors thank Ketra Rice and Nisha Nataraj for providing a case study.

© Springer Nature Switzerland AG 2021
M. Paszynski et al. (Eds.): ICCS 2021, LNCS 12747, pp. 407–421, 2021.
https://doi.org/10.1007/978-3-030-77980-1_31

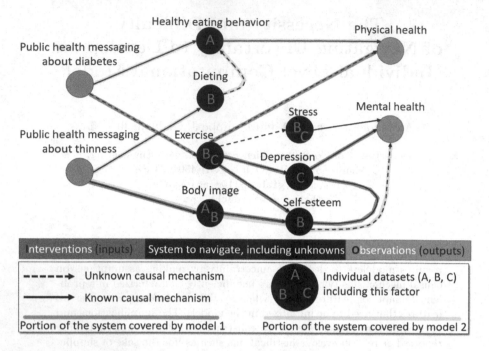

Fig. 1. A systems map (nodes and directed edges) of a problem space can be used to identify what salient factors and rules will go into a model based on available data and a systematic handling of uncertainties. (Color figure online)

the start of the simulation (i.e. given a *baseline value*). As the simulation unfolds, the entities' behaviors and some features will be updated based on a set of *rules* that can take into account the entity's features as well as the features of simulated peers or the environment. By specifying the rules, a modeler expresses how to build reactive entities that either continue to engage in an existing pattern of behavior or adopt a new one by reacting to socio-environmental stimuli. The recommended guidance is that a "model should be embedded in existing theories and make use of whatever data are available. [Its] assumptions need to be clearly articulated, supported by the existing theories and justified by whatever information is available." [1]

In contrast to the next steps which are more standardized (e.g., implementation in an object-oriented language, verification and validation) [3,10,11], the design of a model and the extent to which it should use information or theories is particularly subject to ad-hoc practices, hence to variations across modeling teams. To illustrate these variations, consider Fig. 1 which exemplifies the problem space (i.e., set of relevant factors and interrelationships) in the case of food-related behaviors. For a model to be 'fit for purpose' (i.e. adequate) [34], the rules should connect the interventions required by model users to observable model outcomes. In this situation, two teams could produce and justify very different models. One team with access to dataset *B* could create model 1 (Fig. 1, red highlights), which focuses on exercise and also touches on depression, body image, and self-esteem. This model will require making 3 assumptions: one to compensate for the lack of baseline data to initialize the entities' depression level, and two for unknown rules relating public messaging to exercise, or exercise to physical health. Another team with access to dataset A could create model 2 (Fig. 1, green highlights) which is more focused on eating and also requires three assumptions. A team with access to *both* datasets A and B could further reduce the number of assumptions necessary, as they could pass on representing dieting or the link from self-esteem to mental health while keeping the model fit for purpose. The design of an individual-based model is thus heavily impacted by how a team handles *parameter uncertainty* either in unknown causal mechanisms or in unknown feature values. However, the "different types of simplifications and scientific judgments that have to be made" [2] are not commonly reported, in part since they are not required by standardized documentations [16] (e.g., ODD, ODD+D). As essential aspects of model building are shaped by important yet unknown decisions, we face the additional problem of *structural uncertainty* [2].

Although the need to lower structural uncertainty in model building has long been established and mentioned in methodological guidance [35], this task has been hampered by the lack of a clear picture on the problem space (i.e. the map shown in Fig. 1). In other words, it's simple to *state* that 'modelers should explain how they systematically navigate the uncertainties in the problem space based on available data and assumptions', but it's difficult for modelers to follow a *systematic method* to handle a problem space that has not been precisely mapped. Indeed, the creation of a comprehensive map of a problem space[1] (known as a 'systems map') is often beyond the scope of creating one model, which usually only involves a brief literature review [1] and/or consulting subject matter experts. However, the growth of Participatory Modeling (PM) studies using techniques such as Fuzzy Cognitive Mapping or Causal Mapping [29,36] has resulted in an abundance of systems maps across topic areas, which can thus be used by modelers to create a model operating within any subset of the map. To appreciate the coverage of systems maps, consider examples from health [23] such as the Foresight Obesity Map [19] (over 100 factors and 300 links) or our map on mental and physical well-being in relation to body weight [5] (269 links). Mid-sized maps are even more common, and may remain sufficient for many modeling applications, such as health technology adoption [24] (52 factors and 105 connections) or radiotherapy [27] (66 links). A plethora of maps has also been developed to study ecology and sustainability [8,15,20] or social challenges [18]. By clearly summarizing the salient constructs within the problem space, such maps present an opportunity to shift from an ad-hoc model building approach to a systematic and transparent one, thus decreasing structural uncertainty. Specifically, the use of a map and accompanying datasets allows to *systematically* answer three essential interrelated questions for model building:

---

[1] A structure of the problem space may also be called 'domain model' [33] or 'conceptual model'. To avoid confusion on the multiple uses of 'model', we reserve this term for the *simulation* model, that is, the operational model that was obtained after making design decisions within the problem space.

- *Which factors* should we include?
- Where in the problem space is it necessary to make *assumptions*?
- What is the *resulting uncertainty* of the model?

In this paper, we propose a framework to express the task of creating a model as a matter of navigating a systems map given a set of datasets. Using this framework, the strategies used by modelers to answer the three questions above are made both transparent and systematic using graph algorithms. Through this framework, we note that creating the perfect model is an NP-Hard problem, hence there is no perfect strategy to automatically generate a model: a heuristic needs to be chosen by modelers based on measures that they explicitly wish to favor. Our contributions are as follows:

(1) We present the first mathematical framework to express model design choices based on data availability and their effects on uncertainty.
(2) We explain why a perfect model cannot be automatically created (NP-Hard problem), hence emphasizing the need for heuristics.
(3) We demonstrate how common heuristics used by modelers can be formulated in this framework, using a guiding example from a model of suicide.

The remainder of this paper is organized as follows. In Sect. 3, we introduce our framework formally and exemplify its elements using a model for suicide prevention. In Sect. 4, we explain why the model creation problem is NP-Hard and demonstrate the effects of common heuristics. Finally, we discuss the potential of shifting the process of model building from an ad-hoc unspecified approach to the use of transparent heuristics within a systems map.

2   Framework

Intuitively, modelers have access to a systems map describing the problem space as well as at least one dataset. For the model to be adequate, end users need to see the effects of simulated interventions, which requires each intervention to eventually impact at least one observable outcome through a chain of rules. The task of creating a model thus requires maintaining paths from interventions to observable outcomes through the problem space. Modelers use a *strategy* regarding which paths to take, in part based on data. Paths may travel through nodes for which modelers do not have data, thus they would need to make assumptions about the simulated entities' feature values for these nodes. Paths will also go through edges, which represent causal mechanisms that would be turned into simulation rules. Some of these edges may be intuitively understood (e.g., 'more trauma leads to more suicide ideation') but not sufficiently characterized to write model rules, thus leading to additional assumptions (e.g., every unit of trauma leads to an increase p in suicide ideation). A modeling strategy is thus an algorithm that identifies a subset of a systems map based on available data and makes assumptions to address unknown nodes and edges.

The main elements of the framework are listed in Table 1 and will now be explained along with their formal notation. The problem space is represented by

Table 1. Main elements of the framework

Notation	Meaning
$\mathbb{G} = (V, E)$	Problem space represented as a map $\mathbb{G}$, consisting of labeled nodes $\mathbb{V}$ and directed connections $\mathbb{E}$
I	Interventions (or 'inputs') needed by model users
O	Observations (or 'outputs') that model users require to quantify the effects of their interventions
$w(e)$	Uncertainty value of a causal connection in the problem space
$\mathbb{D}$	Set of all data sources available to modelers
$Sources(v)$	Set of data sources that include the node v; in other words, data sources that modelers can use to initialize v
S	A model design strategy. Given the map $\mathbb{G}$ and data sources $\mathbb{D}$, it specifies which subset of the map to use in a model, which data source will be involved, and what level of uncertainty will arise. For a model to be adequate, a valid strategy must ensure that each intervention from I has an observable effect in O

a directed, labelled, weighted causal graph $\mathbb{G} = (\mathbb{V}, \mathbb{E})$. The nodes $\mathbb{V}$ represent factors in the problem space and potential candidates for the entities' features, such as age, depression, or suicide ideation. To characterize the task of model building, we need to identify nodes that play particular roles. A subset of the nodes $I \subset \mathbb{V}$ represents the intervention nodes (e.g., public health campaigns, economic interventions), also known as inputs. Similarly, the subset $O \subset \mathbb{V}$ represents the outcome nodes (e.g., number of suicide attempts, prevalence of suicide ideation), also known as outputs. For a model to be viable, the interventions of interest need to eventually affect the outcomes; otherwise, the model does not offer support to examine the consequence of the intervention. For example, the evaluation of a suicide prevention package may measure the impact of economic interventions through a reduction in suicide attempts.

The edges $\mathbb{E}$ stand for causal connections and have an associated value (i.e., edge weight) denoted $w(e) \mapsto \mathbb{R}+, e \in \mathbb{E}$. This value denotes the uncertainty associated with the edge. We encode the value as a positive real number rather than a boolean because systems maps *may* provide fine grained information on the *amount* of uncertainty, which can thus be expressed without needing to amend our framework. For example, Fuzzy Cognitive Maps can specify uncertainty [17] by measuring the extent to which there is disagreement about the causal strength of an edge (via entropy) among participants [12] or by checking in a corpus whether there is enough supporting evidence for each proposed connection [22,28]. In the case of Fuzzy Grey Cognitive Maps, each edge has a Grey uncertainty [26]. In a coarse categorization, such as by asking subject-matter experts whether a factor impacts another [25], uncertainty would either be 1 (present; maximal) or 0 (absent; minimal).

Information on causal mechanisms is held at the level of the systems map via $w(e)$ because it represents a fact about the system, independently of any data source selected by modelers. This is reflective of the fact that a systems map serves as a synthesis of the evidence base [5, 19, 24, 27]. For example, a map may stipulate that abusing or neglecting children has an impact on their risk for suicidal ideation, or that a suicide attempt can lead to death. Data sources may help to understand how these general mechanisms work in a specific population, but the mechanisms exist irrespective of a population. In contrast, information on the entities' features depend on the data sources used, thus allowing to capture how traits are expressed in specific populations. The collection of **data** sources available to modelers is denoted by $\mathbb{D} = \{D_1, \cdots, D_n\}$. Each data source may hold information on some of the nodes. We denote the set of data sources for a node $v \in \mathbb{V}$ by $Sources(v)$. When $Sources(v) = \emptyset$, modelers have no data regarding this specific node hence its uncertainty is maximal and its inclusion in a model would come at the cost of making an assumption.

Modelers can employ one of several strategies $\mathbb{S}$ to design a model. As shown in the next section, examples may include finding the simplest set of rules from each intervention to an outcome, or finding rules that avoid uncertainty whenever possible. A strategy $S \in \mathbb{S}$ is thus a function:

$$S: \quad \underbrace{(\mathbb{G}, \mathbb{D})}_{\text{given map and datasets}} \quad \mapsto \underbrace{(G = (V \subseteq \mathbb{V}, E \subseteq \mathbb{E})}_{\text{selects a map subset}}, \underbrace{D \subseteq \mathbb{D}}_{\text{dataset used}}, \quad \underbrace{\mathbb{R}}_{\text{uncertainty cost}})$$

In line with the earlier explanations in this section, a strategy is *valid* if and only if there is a path from every intervention node to at least one observable node. Formally, this condition is enforced by checking the following:

$$S \text{ is valid} \iff \forall i \in I, \exists o \in O \text{ s.t. } (i, v_1), \ldots, (v_n, o) \in E$$

A core aim for modelers is to design a valid strategy while minimizing the uncertainty cost. The design of a model can thus be operationalized through this framework as a discrete optimization problem in a graph given a set of datasets.

3   The Necessity and Design of Heuristics

3.1   The Impossible Quest for Perfection in Model Design

Intuitively, an ideal model design is one that satisfies all needs of the end users while making the least number of assumptions. Formally, that would be an optimal strategy $S^* \in \mathbb{S}$ such that S^* is valid and its associated uncertainty cost is minimal among all valid strategies. However, finding the best strategy may not be feasible in practice as shown in the theorem below.

Theorem 1. *Identifying the best strategy S^* is an NP-Hard problem.*

Proof. To compute the optimal strategy S^*, we are given intervention and outcome nodes. The aim is to select nodes and edges in between (i.e., the 'inner part' of the network) such that the sum of selected nodes' and edges' weights is minimal, while maintaining connectivity. Minimizing this inner cost while providing connectivity is known as the *Minimum Spanning Tree with Inner nodes cost* problem (MSTI). The MSTI problem is NP-Hard as a special case of the Connected Dominated Set problem [9, 21], hence our problem is also NP-Hard.

Alternatively, our problem can be related to the node weighted Steiner tree in which 'terminals' must be connected and the cost function is the sum of the nodes' weights selected to provide this connectivity. Considering the special case in which there is a single outcome node and multiple intervention nodes, then they collectively form the 'terminals' used in a node weighted Steiner tree. Further consider that there is no uncertainty on any edge, hence leaving only the uncertainty on the nodes. Then, the special case is equivalent to minimizing a node weighted Steiner tree, which is an NP-Hard problem [4].

3.2   Imperfect yet Practical: Two Common Model Heuristics

The previous section established that building model that is fit-for-purpose and minimizes uncertainty is an NP-Hard problem. However, modelers routinely build models, which implies that they use heuristics. In this section, we show how two such heuristics can be expressed in our framework, thus making the model building process more transparent and systematic. An example for both heuristics is provided in Fig. 2.

One strategy employed by modelers is to 'Keep It Simple Stupid' (KISS) in which "one *only* tries a more complex model if simpler ones turn out to be inadequate" [6]. In other words, the model design is justified by 'the sake of simplicity'. Since each intervention must be measured via an observation, a translation of KISS into a process would be to find the *simplest* way of connecting each intervention to one observation (possibly the same). When that strategy is expressed algorithmically, it is equivalent to using the *shortest path* from each intervention to one observation. Algorithm 1 formalizes this strategy by using a Breadth-First Search to generate each shortest path. We make two observations:

(1) Focusing on the shortest number of rules will not take the locations of unknowns into consideration. A *slightly* more complex model in number of rules may have gone through a path better supported by data, thus producing a model that needs fewer assumptions.

(2) *Independently* finding a path for each intervention can produce a model that is simple for each intervention, but overall much more complex than strictly necessary. For instance, taking a short detour for one intervention may have allowed it to share the rest of the journey toward an observation using the path of another intervention. Sharing paths or finding 'synergies' may thus help to keep the whole model simpler and potentially lower its uncertainty.

A second approach captures the preference of modelers who actively avoid creating poorly understood rules. Their preference is not to keep the model 'simple': rather, they seek a more robust model in which rules can be backed by evidence as much as possible. From an algorithmic standpoint, this consists of generating paths from every intervention node to the observation nodes and selecting the one with the least overall uncertainty. This selected path will thus use a sequence of factors with well-understood relationships to lead from an intervention node to an observation node.

Algorithm 1: Generate all shortest paths using Breadth-First Search

Input: List of edges in causal map, List of intervention nodes, List of observation nodes

1 paths = $\{\emptyset\}$, finalPaths = $\emptyset$ // create empty list of lists and a map

2 **foreach** *intervention* **do**

3 | paths ← paths $\cup\{\{intervention\}\}$

4 **while** $\exists intervention \notin finalPaths$ // continue until we have a path for each intervention

5 **do**

6 | newPaths = $\{\emptyset\}$ // for each current path, look for next steps

7 | **foreach** *path* $i_1, ..., i_n \in paths$ **do**

8 | | targets ← $\{t|(i_n, t) \in edges\}$

9 | | **foreach** *target* $\in targets$ // for each possible next step, save the potential path

10 | | **do**

11 | | | newPaths ← path $\cup$ target

12 | **foreach** *path* $i_1, ..., i_n \in newPaths$ **do**

13 | | **if** $i_1 \notin finalPaths$ // ignore path if we already have a finalPath for its root

14 | | **then**

15 | | | **if** $i_n \in observations$ // if the path is complete, save to finalPaths

16 | | | **then**

17 | | | | finalPaths(i_1) ← path

18 | | | **else**

19 | | | | paths ← paths $\cup$ path // otherwise keep path for next iteration

---

This strategy may lead to long and awkward tangents. For example, instead of creating few rules with limited empirical or theoretical backing (e.g., after-school programs reduce school problems thus reducing suicide ideation), this strategy may result in creating *many* rules: after-school programs reduce school problems thus reducing parental frustration, which means parents may be less likely to cope with frustration through substance use, hence children are less exposed to unhealthy coping strategies, thus they can deal better with their own issues and overall are less likely to engage in suicidal thoughts. Each of these rules may be backed by stronger evidence, but the collective chain of rules produces a meandering model that can appear less plausible overall.

As modelers may avoid *both* arbitrarily long chains of rules *and* models whose rules lack evidence, a strategy would need to discourage both uncertainty and long paths. Algorithm 2 implements this approach by assigning a unit cost of 1 to all known edges and a penalty value to unknown edges.

Algorithm 2: Generate all minimal paths using Dijkstra's algorithm

---

Input: List of edges in causal map, List of intervention nodes, List of observation nodes, Penalty

1 finalPaths $\leftarrow \emptyset$
2 **foreach** *intervention* **do**
3 | visitedNodes $\leftarrow \emptyset$
4 | paths $\leftarrow \{\{intervention, 0\}\}$ // each node in a path is a pair (name, cost)
5 | bestTarget $\leftarrow \{0, 0\}$
6 | **while** $bestTarget_1 \notin observations$ **do**
7 | | bestTarget $\leftarrow \{\emptyset, \infty\}$ // track current most-optimal edge to add
8 | | bestPath $\leftarrow \{\emptyset\}$
9 | | **foreach** *path* $\in$ *paths* **do**
10 | | | traversal $\leftarrow \emptyset$ // track current traversal through path
11 | | | **foreach** *node* $\in$ *path* **do**
12 | | | | traversal $\leftarrow$ traversal $\cup$ node
13 | | | | targets $\leftarrow \{t | (node_1, t) \in edges\}$
14 | | | | **foreach** *target* $\in$ *targets* **do**
15 | | | | | **if** *target* $\notin$ *visitedNodes* // ignore already-visited nodes
16 | | | | | **then**
17 | | | | | | **if** $(node_1, target)_{weight} \in \,]-1, 1[$ **then**
18 | | | | | | | cost $\leftarrow node_2 + 1.0$ // if edge weight is known, marginal cost = 1
19 | | | | | | **else**
20 | | | | | | | cost $\leftarrow node_2 + penalty$ // otherwise, marginal cost = penalty
21 | | | | | | **if** $cost < bestTarget_2$ **then**
22 | | | | | | | bestTarget $\leftarrow$ target
23 | | | | | | | bestPath $\leftarrow$ traversal
24 | | newPath $\leftarrow$ bestPath $\cup$ bestTarget
25 | | **if** *bestTarget* $\in$ *observations*   // if target $\in$ observations, this root is finished
26 | | **then**
27 | | | finalPaths(intervention) $\leftarrow$ newPath
28 | | **else**
29 | | | paths $\leftarrow$ paths $\cup$ newPath   // otherwise, save to current paths and continue
30 | | | visitedNodes $\leftarrow$ visitedNodes $\cup$ bestTarget

---

Both algorithms are exemplified in Fig. 2. They produce the same paths for interventions $I1$ and $I2$ but differ on $I3$. In this example, Algorithm 2 is better: it uses 10 evidence-based rules and makes five assumptions, whereas Algorithm 1 needs more rules (11) and also makes more assumptions (six).

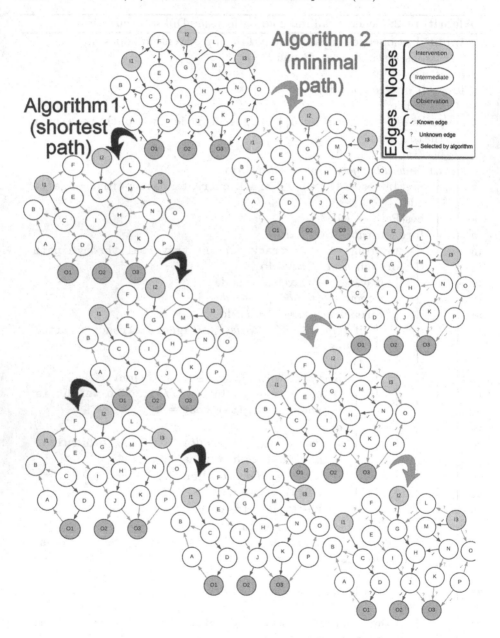

Fig. 2. The same map (top) being processed by both algorithms: shortest paths (left) and minimal paths (right). *This high resolution figure can be zoomed in using a digital copy of this article.*

3.3   Differences in Practice: A Sample Case Study

Algorithms 1 and 2 are *conceptually* different as they correspond to different preferences from modelers: the simplest set of rules to connect each intervention (Algorithm 1) or an emphasis on evidence-based rules (Algorithm 2). As exemplified by Fig. 2 on the previous page, these algorithms can produce different models in terms of number of rules or assumptions. Two items remain, in order to assess the impact of these different model design strategies. First, we need to examine how the number of rules or assumptions differ when these strategies are used *in practice* rather than in an idealized environment. Second, we need to evaluate the possible *choices for datasets* regarding the concept nodes, which also benefits from a case study rather than an artificially constructed situation.

Our brief case study is a model of suicide. Data is openly accessible at https://osf.io/7nxp4/, which include (i) the systems map together with the uncertainty level for each edge (3–'big map with weights') and (ii) which one(s) of four datasets can be used for each concept node (5–'Nodes data availability'). Our algorithms are implemented in a Jupyter Notebook for Python 3, which also contains the results (6–'Common modeling strategies'). The systems map is composed of 361 concept nodes and 946 causal edges. As detailed in the Notebook, there are 5 intervention nodes and 3 outcome nodes.

We performed a parameter sweep to assess how the penalty in Algorithm 2 impacts its two target outcomes: the percentage of unknown edges and the average size of the model. Results in Fig. 3 show that the only difference is encountered if the cost of unknowns is less than 1, that is, less than using a *known* edge. Specifically, the only values of penalty that had an effect on results were in the range $0.2 \leq penalty \leq 0.6$. However, creating evidence-based rules cannot be more expensive than making assumptions for unknown edges. The penalty of unknown would thus be always greater than 1. Consequently, when the penalty is within a *valid range*, then the results are always the same.

Using Algorithm 1, the model requires 3 assumptions on edges and 4 assumptions on nodes. The five paths created for each intervention had no overlap. Two datasets can be interchangeably used as they each support the same number of nodes. With Algorithm 2, *results are identical.* Despite valuing different aspects of a model and having a large problem space, the two design strategies result in the same model. Either strategy has room for improvement as the resulting model requires making many assumptions (7) and is composed of parts that may be

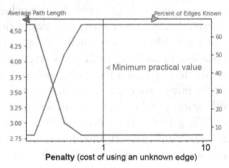

Fig. 3. Evaluation of the penalty parameter in Algorithm 2 with respect to unknown edges and model size.

locally optimal but miss an opportunity for global savings (disjoint paths across all interventions).

4   Discussion and Conclusion

Several frameworks such as the Characterisation and Parameterisation (CAP) framework [31] and its revised version [30] have been proposed to track how computational individual-level models are designed. Such frameworks provide valuable guidance to help standardize the description of modeling studies (c.f. the application of the CAP framework to 11 studies in [32]) by listing the broad types of data or methods involved at each stage. However, there is currently no framework to explain how modelers choose what factors to keep and which rules to make, based on the goal of the model and available data. These choices are important for the transparency and replicability of modeling studies, which often face a high level of structural uncertainty. In addition, these choices have consequences on the robustness and computational costs of models: inefficient model building strategies that result in making a very large number of assumptions may require extensive sensitivity analyses, which come at a high cost. In this paper, we propose the first formal framework to express model-building strategies. We demonstrated that the perfect strategy is NP-Hard, hence modelers will employ heuristics that emphasize specific aspects. We stress this point, as it means that a perfect model cannot be built automatically and instead modelers need to clearly state which measures they value, then map these preferences onto an algorithm. We showed how two common strategies can be systematically expressed in an algorithmic manner and demonstrated that they *can* result in different models, although differences may not be manifest *in practice*.

We noted that both of these common model-building strategies miss several opportunities to decrease the uncertainty of the resulting model. Creating the model with the least number of rules to keep it 'simple' may result in high uncertainty compared to having a tolerance for slightly more rules as long as they are evidence-based. Most importantly, when models are designed to support multiple interventions, it would be beneficial to aim for synergies by identifying common mechanisms across interventions rather than operationalizing each one independently. The design and evaluation of algorithms leveraging these opportunities would be a worthwile investment for future studies. Such tools can help modelers in shifting from the currently time consuming and ad-hoc practice of model design into a more efficient and systematic approach.

As our framework is the first to tackle complex practices, it comes with simplifications. When simplifications prevent a direct use of the framework by a modeling team, they become limitations and changes are necessary. Although we separately report uncertainty on causal mechanisms (i.e. on edges in the problem space) and concept nodes (i.e. insufficient data), our framework is limiting in capturing the cost of uncertainty on nodes. We focused on using *one* dataset to maximize node coverage, but modelers can use *multiple* datasets as long as the individuals captured in these datasets can be accurately linked. For example, consider that two datasets have nothing in common: one covers depression and stress, while the other contains information on poverty and bullying. If a simulated entity was independently given baseline values on depression and poverty, then it would ignore the correlation between these features and model uncertainty grows. Conversely, if the two datasets have shared features that strongly characterize individuals (e.g., age, gender, income, ethnicity) then we may preserve more (but not all) of the real-world dependencies, hence lowering model uncertainty. A possible extension of our framework would thus address how the cost of uncertainty is impacted by the ability to link datasets.

References

1. Abdou, M., Hamill, L., Gilbert, N.: Designing and building an agent-based model. In: Heppenstall, A., Crooks, A., See, L., Batty, M. (eds.) Agent-Based Models of Geographical Systems, pp. 141–165. Springer, Dordrecht. https://doi.org/10.1007/978-90-481-8927-4_8

2. Bojke, L., et al.: Characterizing structural uncertainty in decision analytic models: a review and application of methods. Value Health 12(5), 739–749 (2009)

3. Carley, K.M.: Social-behavioral simulation: Key challenges. Social-Behavioral Modeling for Complex Systems (2019)

4. Demaine, E.D., Hajiaghayi, M., Klein, P.N.: Node-weighted steiner tree and group steiner tree in planar graphs. ACM Trans. Algorithms (TALG) 10(3), 1–20 (2014)

5. Drasic, L., Giabbanelli, P.J.: Exploring the interactions between physical well-being, and obesity. Can. J. Diab. 39, S12–S13 (2015)

6. Edmonds, B., Moss, S.: From KISS to KIDS – an anti-simplistic modelling approach. In: Davidsson, P., Logan, B., Takadama, K. (eds.) MABS 2004. LNCS (LNAI), vol. 3415, pp. 130–144. Springer, Heidelberg (2005). https://doi.org/10.1007/978-3-540-32243-6_11

7. Epstein, J.M.: Why model? J. Artif. Soc. Soc. Simul. 11(4), 12 (2008)

8. Firmansyah, H.S., et al.: Identifying the components and interrelationships of smart cities in indonesia: Supporting policymaking via fuzzy cognitive systems. IEEE Access 7, 46136–46151 (2019)

9. Fleischer, R., et al.: Approximating spanning trees with inner nodes cost. In: Sixth International Conference on Parallel and Distributed Computing Applications and Technologies (PDCAT 2005), pp. 660–664. IEEE (2005)

10. Giabbanelli, P.J., Jackson, P.J.: Using visual analytics to support the integration of expert knowledge in the design of medical models and simulations. Procedia Comput. Sci. 51, 755–764 (2015)

11. Giabbanelli, P.J., et al.: Ideal, best, and emerging practices in creating artificial societies. In: 2019 Spring Simulation Conference (SpringSim), pp. 1–12. IEEE (2019)

12. Giabbanelli, P., Torsney-Weir, T., Mago, V.: A fuzzy cognitive map of the psychosocial determinants of obesity. Appl. Soft Comput. 12(12), 3711–3724 (2012)

13. Gilbert, N., Doran, J.: Simulating Societies: the Computer Simulation of Social Phenomena. Routledge, New York (2018)

14. Gilbert, N., Stoneman, P.: Researching Social Life. Sage, Los Angeles (2015)

15. Gray, S., et al.: Assessing (social-ecological) systems thinking by evaluating cognitive maps. Sustainability 11(20), 5753 (2019)

16. Grimm, V., et al.: The odd protocol for describing agent-based and other simulation models: a second update to improve clarity, replication, and structural realism. J. Artif. Soc. Social Simul. 23(2), 7 (2020)

17. Lavin, E.A., Giabbanelli, P.J.: Analyzing and simplifying model uncertainty in fuzzy cognitive maps. In: 2017 Winter Simulation Conference (WSC), pp. 1868–1879. IEEE (2017)

18. Mago, V.K., et al.: Analyzing the impact of social factors on homelessness: a fuzzy cognitive map approach. BMC Med. Inf. Decis. Making **13**(1), 94 (2013)
19. McPherson, K., Marsh, T., Brown, M.: Foresight report on obesity. Lancet **370**(9601), 1755 (2007)
20. Mourhir, A.: Scoping review of the potentials of fuzzy cognitive maps as a modeling approach for integrated environmental assessment and management. Environ. Model. Softw. **135**, 104891 (2020)
21. Peng, C., Tan, Y., Zhu, H.: On computing the backbone tree in large networks. In: 2008 IEEE Systems and Information Engineering Design Symposium, pp. 118–122. IEEE (2008)
22. Pillutla, V.S., Giabbanelli, P.J.: Iterative generation of insight from text collections through mutually reinforcing visualizations and fuzzy cognitive maps. Appl. Soft Comput. **76**, 459–472 (2019)
23. de Pinho, H.: Generation of Systems Maps: Mapping Complex Systems of Population Health. Systems Science and Population Health, pp. 61–76, Oxford University Press, New York (2017)
24. Rahimi, N., Jetter, A.J., Weber, C.M., Wild, K.: Soft data analytics with fuzzy cognitive maps: modeling health technology adoption by elderly women. In: Giabbanelli, P.J., Mago, V.K., Papageorgiou, E.I. (eds.) Advanced Data Analytics in Health. SIST, vol. 93, pp. 59–74. Springer, Cham (2018). https://doi.org/10.1007/978-3-319-77911-9_4
25. Reddy, T., Giabbanelli, P.J., Mago, V.K.: The artificial facilitator: guiding participants in developing causal maps using voice-activated technologies. In: Schmorrow, D.D., Fidopiastis, C.M. (eds.) HCII 2019. LNCS (LNAI), vol. 11580, pp. 111–129. Springer, Cham (2019). https://doi.org/10.1007/978-3-030-22419-6_9
26. Salmeron, J.L.: Modelling grey uncertainty with fuzzy grey cognitive maps. Exp. Syst. Appl. **37**(12), 7581–7588 (2010)
27. Salmeron, J.L., Papageorgiou, E.I.: A fuzzy grey cognitive maps-based decision support system for radiotherapy treatment planning. Knowl.-Based Syst. **30**, 151–160 (2012)
28. Sandhu, M., Giabbanelli, P.J., Mago, V.K.: From social media to expert reports: the impact of source selection on automatically validating complex conceptual models of obesity. In: Meiselwitz, G. (ed.) HCII 2019. LNCS, vol. 11578, pp. 434–452. Springer, Cham (2019). https://doi.org/10.1007/978-3-030-21902-4_31
29. Schmitt-Olabisi, L., McNall, M., Porter, W., Zhao, J.: Innovations in Collaborative Modeling. MSU Press, East Lansing (2020)
30. Smajgl, A., Barreteau, O.: Framing options for characterising and parameterising human agents in empirical abm. Environ. Model. Softw. **93**, 29–41 (2017)
31. Smajgl, A., et al.: Empirical characterisation of agent behaviours in socio-ecological systems. Environ. Model. Softw. **26**(7), 837–844 (2011)
32. Smajgl, A., Barreteau, O.: Empirical Agent-Based Modelling-Challenges and Solutions, vol. 1. Springer, New York (2014) https://doi.org/10.1007/978-1-4614-6134_0
33. Stepney, S., Polack, F.A.: Discovery phase patterns: building the domain model. In: Engineering Simulations as Scientific Instruments: A Pattern Language, pp. 117–145. Springer, Cham (2018) https://doi.org/10.1007/978-3-030-01938-9_5
34. Swarup, S.: Adequacy: what makes a simulation good enough? In: 2019 Spring Simulation Conference (SpringSim), pp. 1–12. IEEE (2019)

35. Treasury, H.: The aqua book: guidance on producing quality analysis for government (2015)
36. Voinov, A., et al.: Tools and methods in participatory modeling: selecting the right tool for the job. Environ. Model. Softw. **109**, 232–255 (2018)

Predicting Soccer Results Through Sentiment Analysis: A Graph Theory Approach

Clarissa Miranda-Peña$^{(\boxtimes)}$, Hector G. Ceballos, Laura Hervert-Escobar, and Miguel Gonzalez-Mendoza

Tecnologico de Monterrey, Escuela de Ingenieria y Ciencias, Ave. Eugenio Garza Sada 2501, 64849 Monterrey, NL, Mexico
A01400214@itesm.mx

Abstract. More than four out of 10 sports fans consider themselves soccer fans, making the game the world's most popular sport. Sports are season based and constantly changing over time, as well, statistics vary according to the sport and league. Understanding sports communities in Social Networks and identifying fan's expertise is a key indicator for soccer prediction. This research proposes a Machine Learning Model using polarity on a dataset of 3,000 tweets taken during the last game week on English Premier League season 19/20. The end goal is to achieve a flexible mechanism, which automatizes the process of gathering the corpus of tweets before a match, and classifies its sentiment to find the probability of a winning game by evaluating the network centrality.

Keywords: Graph theory · Machine learning · Sentiment analysis · Social networks · Sports analytics

1 Introduction

Most of today's literature on Machine Learning and Soccer talks about engineering the best indicators based on match statistics. Also, current research tries to figure out the best existing features to build models that could predict results before a game. However, retrieving data for a set of repeatable events is a difficult task to accomplish, as well, changes on the team, staff, management, and many other factors could have happened. Based on Graph Theory, Social Networks can be seen as a set of interconnected users with a weighted influence on its edges. Evaluating the spread influence of fans can serve as a metric for identifying fans' intensity. In order to decouple league, team, and even sports information, it is proposed a Sentiment Analysis Model which scores polarity on opinions made by soccer fans on Twitter.

1.1 Review on Social Network Analysis: Spread Influence

Some research studies, as the one developed by Yan, [16] evaluate the influence of users, represented as nodes, on other entities under the Social Network

© Springer Nature Switzerland AG 2021
M. Paszynski et al. (Eds.): ICCS 2021, LNCS 12747, pp. 422–435, 2021.
https://doi.org/10.1007/978-3-030-77980-1_32

Analysis, this is performed by calculating a value for each eigenvector by scoring the weight and the importance of the nodes it is connected to. This paper also adds betweenness centrality, which obtains the shortest paths and finds the most repetitive nodes, so that the most influential elements in the network are identified.

Riquelme [11] proposes two new centralization measures for evaluating networks. The model graph is a compound of labels representing the resistance of the actors to be influenced, and the weight of the edges is the power of influence from one actor to another.

The Eq. 1 of the **Node activation** is:

$$\sum_{j \in F_t(X)} W_{ij} \geq f(i) \tag{1}$$

The activation occurs when the sum of the weight of activated nodes connected to i, in the set of $F_t(X)$ is greater or equal to i's resistance denoted as $f(i)$.

The Eq. 2 of the **Spread of Influence X** is:

$$F(X) = \bigcup_{t=0}^{k} F_t(X) = F_0(X) \cup ... \cup F_k(X) \tag{2}$$

where t denotes the current spread level of X, and X is an initial activation set.

The first measure considered is called Linear Threshold Centrality and represents how much an actor i can spread his influence within a network, this by convincing his immediate neighbors.

The Eq. 3 of the **Linear Threshold Centrality** is:

$$LTR(i) = \frac{|F(\{i\} \cup neighbors(i))|}{n} \tag{3}$$

The second measure is Linear Threshold Centralization, this defines how centralized the network is, by finding a k-core which is the maximal subgraph $C(G)$ such that every vertex has a degree at least k.

The Eq. 4 of the **Linear Threshold Centralization** is:

$$LTC(G) = \frac{|F(C(\hat{G}))|}{n} \tag{4}$$

This relation shows that elements outside the core are easier to be influenced.

Kim [6] proposed a formula to address opportunity based on satisfying the fan's requirements. Korean National Football team's comments on the match against Uzbekistan on FIFA World Cup 2018 qualifications were ranked using TF-IDF, which reflects the relevance a word has in the document. After that, a clustering algorithm, such as K-Means, was implemented for topic modeling, once the topic was known, it was assigned a satisfaction value given by the Delphi Method.

The Eq. 5 of the **Delphi satisfaction expression** is:

$$TS_i = \frac{\sum_{j=1}^{j_i} CS_{i,j}}{J_i} \tag{5}$$

$CS_{i,j}$ satisfaction level of the j-th post in the i-th topic, TS_i average satisfaction for the i-th topic, and J_i total number of post in the i-th topic.

1.2   Review on Sentiment Analysis

Schumaker [13] applies sentiment analysis based on a combination of 8 models using either polarity, such as positive, negative, and neutral, and tone such as the objective, subjective, and neutral. This research has an odds-based approach that gathers an odds-maker's match balance sheet on demand of the wagers. The sentiment is calculated by normalizing a specific data model against tweets for a particular club and match.

The Eq. 6 of the **Normalize polarity** is:

$$max(\frac{\sum Tweets|Model_n,Club_1,Match_m}{\sum Club_1,Match_m}, \frac{\sum Tweets|Model_n,Club_2,Match_m}{\sum Club_2,Match_m}) \qquad (6)$$

When models tested with negative polarity were higher, they could predict a potential loss, whereas models of positive polarity as a possible win.

In contrast, Dharmarajan [2] applied the Multinomial Naive Bayes Algorithm into two main classifiers. The first one is oriented towards an objective tone, this model is trained with a self-made dataset of well-trusted sources, and the second one is a subjectivity classifier that can either label text as positive or negative. This last one achieved 79,50% accuracy over 32,000 instances, while the first one obtained 77,45% when trained with 86,000 records.

Ljajic [9] proposes a sentiment score by quantifying the logarithmic difference of terms in positive and negative sports comments. Again, sentiment classification is seen as a supervised task that requires creating a domain-specific dictionary and assigning a tag as positive or negative for each of the terms. The author proposed the principle of logarithmic proportion TF-IDF as a labeling mechanism.

The Eq. 7 of the **Polarity compute using TF-IDF** is:

$$tfidf_p = (1 + tf_p) * log_{10}\left(\frac{N_p}{N_{t,p}}\right) \qquad (7)$$

Where $tfidf_p$ is the polarity of the term in positive comments, tf_p is the term frequency in positive comments, N_p is the number of positive documents, and $N_{t,p}$ is the number of positive documents with term t. The same procedure will be followed for negative ratio $tfidf_n$, where the larger term will be set as a tag.

A methodology, for setting terms as stop words, is also concluded on this research, by finding boundaries due to the logarithmic difference of the terms, on the paper boundaries were set when accuracy stopped improving.

The Eq. 8 of the **Logarithmic difference of term** is:

$$DifLog_t = log_{10}\left(\frac{tfidf_p + 0.001}{tfidf_n + 0.001}\right) \qquad (8)$$

During World Cup 2018, Talha [14] constructed a database containing 38,371,358 tweets and 7,876,519 unique users, 9 different machine learning models were trained with the 48 matches on the group phase, and tested to predict round 16, and so on. The features considered for this model are detailed information about the user (number of followers, location, likes count, tweets counts, etc.) and the tweet (is it a retweet, reply to a user, retweet count, like count, etc.), the highest accuracy obtained was 81.25% when using a Multilayer Perceptron algorithm with 30,000 epochs.

Jai-Andaloussi [4] aims to summarize highlights in soccer events by analyzing tweets and scoring text sentiment they recommend the deep learning method implemented in Stanford NLP which categorizes comments from 0 being very negative to 4 being very positive. However, as the intention is to obtain the most relevant tweets, the moving-threshold burst detection technique is used.

The Eq. 9 of the **Moving threshold** is:

$$MT_i = \alpha * (mean_i + x * std_i) \tag{9}$$

Given l as the length of the sliding window at time t_i, $N(l_1)$ to $N(l_i)$ where N is the number of tweets, the mean and standard deviation at the time i can be calculated. α is the relaxation parameter, and x is a constant between 1.5 and 2.0. A highlight is defined as $N(l_i) > MT_i$.

2 Methodology

Soccer is constantly changing over time, in order to make this a real-time problem, a framework for gathering recent tweets was built. The methodology is summarized in two key components: first tweets are preprocessed for scoring sentiment polarity, and second, they are evaluated as a Social Network problem by applying graph theory.

2.1 Gathering and Preprocessing Data

The data is obtained through the Twitter's Standard Search API. The queries were performed in the last match week on Premier League's season 19/20 and were limited to the English language. The data was processed into a Dataframe with the next remaining fields as shown in Table 1. Twitter's documentation on standard operations shows that appending a string happy face ":)" on the query represents a positive attitude, while ":(" represents a negative attitude. The maximum number of tweets retrieved from a request is 100, to aid the evaluation process, three types of queries were performed: first adding the happy face, second adding the sad face, and finally a neutral request without a face.

In the end, a total of 30 JSON requests with a maximum count of 100 tweets were available for study, Table 2 indicates the keywords placed in each fixture query. However, not all fixture requests accomplished these 100 tweets as shown in Table 3.

2.2 Data Cleaning

The data cleaning process performed drop of duplicates with a count of one, and drop off empty tweets, the empty tweets were filtered after removing user

Table 1. Dataset fields

	Field	Description
1	Season	A YYYY representation of the match season
2	Weekgame	The number of the current week match
3	Home_team	A three-letter code abbreviating the home team
4	Away_team	A three-letter code abbreviating the away team
5	Favorite_count	The count of favorites in the tweet
6	Lang	A two-letter code abbreviating the language
7	Retweet_count	The count of retweets in the tweet
8	Retweeted	True or false if the tweet is a retweet
9	Text	The text of the tweet
10	Followers_count	The count of followers from the user
11	Verified	True or false if the account is verified

Table 2. Queries

	Match	Keywords
1	Arsenal vs Watford	#ARSFC @Arsenal #WatfordFC @WatfordFC #ARSWAT
2	Burnley vs Brighton	#BURFC @BurnleyOfficial #BHAFC @OfficialBHAFC #BURBHA
3	Chelsea vs Wolves	#CFC @ChelseaFC #WWFC @Wolves #CHEWOL
4	Crystal Palace vs Tottenham	#CPFC @CPFC #THFC @SpursOfficial #CRYTOT
5	Everton vs Bournemouth	#EFC @Everton #AFCB @afcbournemouth #EVEBOU
6	Leicester vs Manchester United	#LCFC @LCFC #MUFC @ManUtd #LEIMUN
7	Manchester City vs Norwich	#MCFC @ManCity @NorwichCityFC #MCINOR
8	Newcastle vs Liverpool	#NUFC @NUFC #LFC @LFC #NEWLIV
9	Southampton vs Sheffield Utd	#SaintsFC @SouthamptonFC #SUFC @SheffieldUnited
10	West Ham vs Aston Villa	@WestHam #AVFC @AVFCOfficial #WHUAVL

Table 3. Requests

	Match	# of *tweets*
1	Arsenalvs Watford	*247*
2	Burnley vs Brighton	*121*
3	Chelsea vs Wolves	*235*
4	Crystal Palace vs Tottenham	*156*
5	Everton vs Bournemouth	*143*
6	Leicester vs Manchester United	*256*
7	Manchester City vs Norwich	*198*
8	Newcastle vs Liverpool	*268*
9	Southampton vs Sheffield Utd	*128*
10	West Ham vs Aston Villa	*175*

mentions and ended up with no text to analyze, this returned a count of 11 empty tweets. At first, the library langdetect was used with a threshold of 50% probability for the English language, however, in practice, the language detection accuracy drops drastically on shorter tweets, so multilingual tweets were kept. The length of the dataset finished with a total of 1,915 tweets.

2.3 Data Engineering

In order to make the most of the available resources, extra variables were included, two of them relate to text transformations.

- pre_label: integer field that pre classifies the tweet according to the search query, for positive tweets gives 1, negative -1 and 0 if neutral.
- support: integer field that pre represents the support to a given team if it appears on the tweet a mention or hashtag to the home team returns 1 when away team returns -1 and 0 if both appearances happened.
- no_mentions: string field as a version of the tweet without mentions and removing anything that is not plain text.
- with_emojis: string field as a version of the tweet, this sophisticated text transformation keeps mentions, removes links, and uses the emoji library to encode emojis into text, also a regular expression matches happy and sad faces representation and replace happy faces by the word good and sad faces with the word bad.

Classifying polarity was possible by using available resources, such as the Open Source Library Stanza [10], previously named Stanford NLP. Stanza is a language-agnostic processing pipeline that groups together tokenization, lemmatization, part-of-speech tagging, dependency parsing, and named entity recognition. Stanza has a built-in model for Sentiment Analysis [5], this model is trained as a one-layer Convolutional Neural Network using word2vec which are the resulting vectors when applying bag-of-words on 100 billion articles on Google News.

Since it is possible to provide text previously tokenized to Stanza's pipeline, it was preferable to create tokens using NLTK Tweet Tokenizer as it applies regular expressions to maintain mentions. A mention identifies users with the prefix @, while Stanza Tokenizer split those characters underperforming entity recognition.

Figure 1 shows some word clouds comparing the results given on the assumption of the pre_label tag against Stanza's model evaluation.

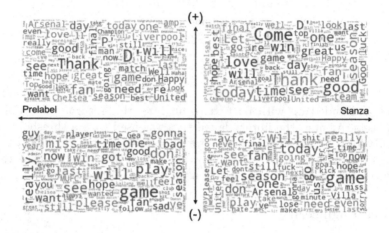

Fig. 1. Word clouds comparison by polarity and model

By applying Stanza's classification it is possible to measure the magnitude of the polarity tags. Now, phrases such as love, good luck, hope, took relevance on the positive tags, while on the negative tags curse words and negations took precedence.

After classifying polarity in tweets from a Machine Learning perspective, two new fields were created a modified support $m\_support$ and modified sentiment $m\_sentiment$. These fields were the result of matching a regular expression that identifies suggested scores of form 0:0 or 0-0 since a result with a goal difference greater than zero indicates clear favoritism to one of the adversarial teams.

Figure 2 shows relevance on the proposed characteristics, this is measured by the amount of neutral support and sentiment that was able to be classified as positive or negative, and as a side of the home team or away team.

2.4 Graph Theory

Popular teams such as Arsenal, Liverpool, Manchester United, etc., have higher rates of tweets, making it difficult to choose a favorite when comparing against its opponent. This section translates the imbalance of favoritism into a graph analysis.

A simple graph [7] has the form $G = (V, E)$ where V is a set of n vertexes and E is a set of n edges. An edge is a link between two vertexes, so an edge E_k is associated with an unordered pair of the vertex (V_i, V_j).

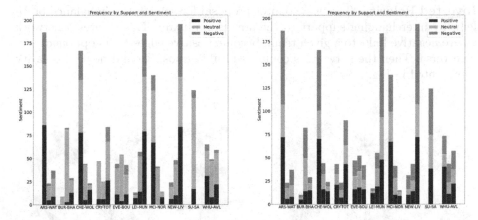

Fig. 2. On the left the frequency of the pre_label polarity by team support, on the right the frequency of the modified polarity by team modified support

Here the vertex are the users and each edge represents a tweet to a tagged user, the tagged user is the team which is mentioned on the tweet. The final tuple looks like: $(fan, team, edge_k)$. Two edges were added to the graph when a tweet mentioned both of the teams. Also, it was preferable to choose a multigraph representation, since it is possible to have multiple edges of a fan to the same team.

Edge's Weight. Setting a singular value for each edge's weight will miss out on tweets having likes or being retweeted, as well, it would not solve the imbalance problem, since the sum of all edges values from the team's vertex to the fans will be equal to the frequency of the fans of a given team.

The Eq. 10 of the **Tweet's weight imbalance** is:

$$E_k(U_i, team) = c - \frac{U_i(likes) + U_i(retweets)}{\sum_{i=0}^{k} support(team) + \frac{\sum_{i=0}^{k} support(match|neutral)}{2}} \quad (10)$$

An edge between the user and the team represents the distance the tweet has with the team, whenever a tweet is more retweeted or has a larger amount of likes, it means it is more reachable to the audience of a team, subtracting from a constant c, the relative number of likes and retweets to the support of the team, means reducing the distance between the fan and the team. By calculating the relative influence in a network as the sum of interactions over the frequency of support in a team, a team with fewer followers will represent a greater reach to its network, rather than the reach-in networks with larger amounts of fans, this way the class imbalance problem could be resolved. Neutral support was split in two and added to the frequency of each of the teams as seen in Fig. 3 where light blue lines are neutral, navy are the positive tweets and green negatives.

Inverted Polarity. This is a counter proposal for solving the imbalance problem, by interchanging support to the adversarial team. This creates a network where negative links to a given team, become positive edges to its opponent and vice versa. Then the network is composed only of positive and neutral polarity represented in Fig. 4.

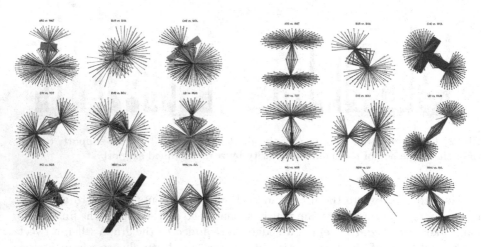

Fig. 3. Graph using tweet's weight imbalance

Fig. 4. Graph using tweet's inverted polarity

3   Results

3.1   Evaluation

A way for evaluating network entities is through indexing centrality [17], this metric indicates the influence of the vertex in the network. Degree centrality is discarded, since it counts the number of links to a node, and as mentioned earlier there is a clear imbalance between the number of fans, so it might present misleading results. Betweenness centrality is not taken into consideration neither, this measure gives precedence to mediation nodes that connect the network, here users that mentioned both of the teams have the highest scores. Closeness centrality is the selected measure for comparing independence and efficiency of communication in an entity [8].

Closeness Centrality. It is computed as the reciprocal of the average of shortest-path distance from an agent A_u to all other agents. The Eq. 11 of the **Closeness centrality** is:

$$C(u) = \frac{n-1}{\sum_{i=1}^{n} d(i,u)} \tag{11}$$

Current-Flow Closeness Centrality. It is based on information spreading efficiently like an electrical current. Edges are now resistors $r_e = 1/w(e)$ and each vertex has a voltage $v(u)$. The Eq. 12 of the **Current-flow closeness centrality** is:

$$C(u) = \frac{n}{\sum_{i=1}^{n} v(u)-v(i)} \tag{12}$$

This represents the ratio n to the sum of effective resistances between u and other vertexes quoted [3]. It is also equivalent to the information centrality which considers all path weights, not only the shortest ones, and instead computes its average from the originated vertex. The information in a path is the inverse of the length of a path [1]. The Eq. 13 of the **Information centrality** is:

$$\bar{I}_u = \frac{n}{\sum_{i=1}^{n} \frac{1}{I_{ui}}} \tag{13}$$

Harmonic Centrality. Applies harmonic mean to overcome outweighs from infinite distances, and it is computed as the sum of the reciprocal of the shortest path distances. The normalized harmonic centrality can reach up to 1 as the maximum connected vertex. Lower values occur when used on an unconnected graph representing the reduced capability of communication in the network [12]. The Eq. 14 of the **Harmonic centrality** is:

$$C(u) = \sum_{i=1}^{n} \frac{1}{d(i,u)} \tag{14}$$

3.2   Testing

For evaluation purposes, the study was extended to a set of 54 matches starting at week 38 from season 2019 up to week 8 from current season 2020. In total 7,833 tweets were analyzed. Besides a graph considering the three polarity links, three subgraphs, one for each polarity, were built. Based on centrality measures two cases were considered:

Case 1. Applying current-flow closeness centrality as a comparison measure inter-team, since low resistance will show efficiency in the way a team communicates to its fans. The difference between the current-flow closeness index on the home team and away team will reflect the favorite team given its communication effectiveness. The Eq. 15 of the **Inter-team closeness** is:

$$diff\_closeness = \|closeness(home) - closeness(away)\| \tag{15}$$

Case 2. Applying harmonic centrality as the leading polarity intra-team, to know which polarity has a better representation of the fan's sentiment towards a team. Communication is more difficult when having fewer connections. For each subgraph, the less fluctuated harmonic centrality given a polarity against the harmonic centrality considering all three polarities will support a good communication capability. The Eq. 16 of the **Intra-team closeness** is:

$$closeness(\tfrac{team}{polarity}) = \|closeness(team) - closeness(polarity)\| \tag{16}$$

Support Vector Machines were used for classifying a match as a win, draw, or lose at home. These models were trained with different centrality indexes and

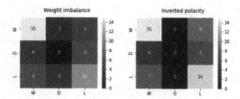

Fig. 5. Confusion matrix comparison

Table 4. Classification report when selecting features from tweet's weight imbalance.

Classification report				
	Precision	Recall	F1-score	Support
−1	0.60	0.65	0.63	23
0	0.00	0.00	0.00	10
1	0.50	0.52	0.51	21
Accuracy avg/total	0.48	0.48	0.48	54

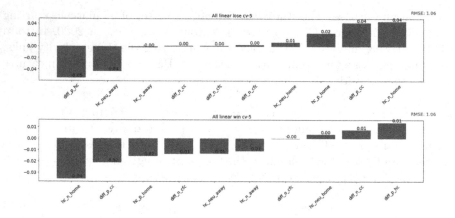

Fig. 6. SVM's selection on tweet's weight imbalance

Table 5. Classification report when selecting features from tweet's inverted polarity.

Classification report				
	Precision	Recall	F1-score	Support
−1	0.58	0.65	0.61	23
0	0.00	0.00	0.00	10
1	0.50	0.67	0.57	21
Accuracy avg/total	0.54	0.54	0.54	54

evaluated with five-fold cross-validation. During the pipeline different k best features were applied testing ANOVA.

4 Discussion

Table 5 shows higher values on the recall metric than Table 4, this metric references to all events classified correctly, while the precision metric focus on the predicted positive to be correctly [15]. As seen in Fig. 5 although the inverted polarity model has a better recall, the weight imbalance model did not loose precision and gave more diversity to the prediction model by attempting to guess draw matches.

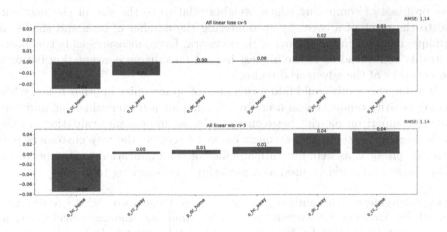

Fig. 7. SVM's selection on the inverted polarity

About the metrics used, Fig. 6 plots feature weights, and validates the inter-team centrality as the difference on the normalized harmonic closeness centrality, while the intra-team measure is given by singular polarities of a team. Figure 7 confirms the current-flow closeness centrality as a non-significant measure.

Compared to the work done by Schumaker [13] with the highest accuracy of 50.49%, the current model has a slightly better performance with 54% accuracy.

However, this is only a sample of all Premier League games, so there is a huge area of opportunity for testing future games.

Finally, this study brought relevant candidate features retrieved from fan expertise in Social Media which can be interchanged with statistics and odds information, for boosting accuracy in soccer prediction before a game.

5 Conclusion

Football as any other sport is unpredictable, modeling draws is a very difficult task, as well classifying a match previous to its start. Probabilities closer to the end of the game might give more accurate results, but it is also true that a match can be flipped in the last five minutes.

This study gave satisfactory results from the fact that it does not consider statistics at all, instead, it uses as a historical database the knowledge from fans' comments, for scoring polarity to a team and then generates a prediction previous to the start of the game.

This novelty presents a prediction mechanism that can be decoupled from football league's and even sports, whenever comparing a team's sports with scores of the form 0-0, future steps would consider testing this method in other sports. Also as far as our knowledge, it presents a unique methodology for mining Sports sentiment in Social Networks by engineering centrality measures to be considered as candidate features in order to train a Machine Learning model. The proposal of computing edges' weights relative to the size of the network, and to the reach of a tweet by encountering the number of likes and shares, is a unique mechanism for balancing the network. Even, measures as betweenness centrality could lead us to the highest impact fans driven conversation between the networks of the adversarial teams.

Beyond, this study could find a rich area of opportunity in other fields, when scoring polarity tendencies on users by applying intra-team evaluation, and generating comparison metrics between users by an inter-team evaluation. In the marketing sector, the first statement could be seen as the way customers perceive the product, as well, measure an efficient communication to the spectator, while the second could be used as a powerful benchmarking tool.

Acknowledgment. The authors are grateful to Tecnologico de Monterrey, who through its Academic Scholarship Program for Graduate Students provided technical and financial support for the development of this research. In addition, we are grateful to CONACyT for the financial support awarded through the National Scholarship for PNPC and SNI programs designed for promoting quality research and close the existing gap between industry and academia.

References

1. Amrit, C., ter Maat, J.: Understanding Information Centrality Metric: A Simulation Approach (December 2018)

2. Dharmarajan, K., Abuthaheer, F., Abirami, K.: Sentiment analysis on social media. vol. 6, pp. 210–217 (2019)
3. Huan, L., Richard, P., Liren, S., Yuhao, Y., Zhongzhi, Z.: Current flow group closeness centrality for complex networks?. pp. 961–971 (2019)
4. Jai-Andaloussi, S., El Mourabit, I., Madrane, N., Chaouni, S.B., Sekkaki, A.: Soccer events summarization by using sentiment analysis. In: Proceedings - 2015 International Conference on Computational Science and Computational Intelligence, CSCI 2015, (September 2018), pp. 398–403 (2016)
5. Kim, Y.: Convolutional Neural Networks for Sentence Classification. CoRR, abs/1408.5 (2014)
6. Kim, Y.S., Kim, M.: A wisdom of crowds: social media mining for soccer match analysis. IEEE Access **7**, 52634–52639 (2019)
7. Kumar, R., Pattnaik, P.K.: Graph Theory. Laxmi Publications Pvt Ltd (2018)
8. Liotta, G., Tamassia, R., Tollis, I.G.: Graph algorithms and applications 4. World Scientific (2006)
9. Ljajić, A., Ljajić, E., Spalević, P., Arsić, B., Vučković, D.: Sentiment analysis of textual comments in field of sport Sentiment analysis of textual comments in field of sport (November 2015)
10. Qi, P., Zhang, Y., Zhang, Y., Bolton, J., Manning, C.D.: Stanza: A python natural language processing toolkit for many human languages. arXiv preprint arXiv:2003.07082 (2020)
11. Riquelme, F., Gonzalez-Cantergiani, P., Molinero, X., Serna, M.: Centrality measure in social networks based on linear threshold model. Knowl. Based Syst. **140**(January), 92 102 (2018)
12. Rochat, Y.: Closeness centrality extended to unconnected graphs: the harmonic centrality index (2009)
13. Schumaker, R.P., Jarmoszko, A.T., Labedz, C.S.: Predicting wins and spread in the Premier League using a sentiment analysis of twitter. Decis. Support Syst. **88**, 76–84 (2016)
14. Talha, A., Simsek, M., Belenli, I.: The wisdom of the silent crowd: predicting the match results of world cup 2018 through twitter. Int. J. Comput. Appl. **182**(27), 40–45 (2018)
15. Lee, W.M.: Python Machine Learning. Wiley, Indianapolis (2019)
16. Yan, G., Watanabe, N.M., Shapiro, S.L., Naraine, M.L., Hull, K.: Unfolding the Twitter scene of the 2017 UEFA champions league final: social media networks and power dynamics. Eur. Sport Manage. Quart. **19**(4), 419–436 (2019)
17. Zhang, J., Luo, Y.: Degree Centrality, Betweenness Centrality, and Closeness Centrality in Social Network, vol. 132, pp. 300–303 (2017)

Advantages of Interval Modification of NURBS Curves in Modeling Uncertain Boundary Shape in Boundary Value Problems

Marta Kapturczak[(✉)] [iD], Eugeniusz Zieniuk[iD], and Andrzej Kużelewski[iD]

Institute of Computer Science, University of Bialystok, Białystok, Poland
{mkapturczak,ezieniuk,akuzel}@ii.uwb.edu.pl

Abstract. In this paper, the advantages of interval modification of NURBS curves for modeling uncertainly defined boundary shapes in boundary value problems, are presented. The different interval techniques for modeling the uncertainty of linear as well as curvilinear shapes are considered. The uncertainty of the boundary shape is defined using interval coordinates of control points. The knots and weights in the proposed interval modification of NURBS curves are defined exactly. Such a definition allows for modification of the uncertainly defined shape without any change of interval values. The interval NURBS curves are compared with other interval techniques. The correctness of modeling the shape uncertainty is confirmed by solving the problem using the interval parametric integral equations system method. Such solutions (obtained using a program implemented by authors) confirm the advantages of using interval NURBS curves for modeling the boundary shape uncertainty. The shape approximation is improved using less number of interval input data and the obtained solutions are correct and less over-estimated.

Keywords: NURBS curves · Modeling uncertainty · Interval arithmetic · Boundary value problems · PIES

1 Introduction

All kinds of shapes can be modeled using computer graphics curves. Nowadays, the application of even a complex mathematical model to determine these curves, using computer techniques, is a very effective approach. This allows for more accurate and realistic modeling of any of the object shapes. Using analytical methods would be very troublesome and time-consuming.

The NURBS curves [1] are more and more frequently used in the boundary problems [2,3]. These curves increase the accuracy of modeling the shape even with a small number of points. Additional parameters that increase the modeling possibilities are point weights and the knots vector. The weights determine the influence of the point on the curve and enable correct modeling of a circle or an ellipse. The knots allow to obtain corners and to change the degree of the curve.

© Springer Nature Switzerland AG 2021
M. Paszynski et al. (Eds.): ICCS 2021, LNCS 12747, pp. 436–443, 2021.
https://doi.org/10.1007/978-3-030-77980-1_33

The advantages of NURBS curves in modeling exactly defined problems [4, 5] motivated the authors to verify them in modeling uncertainly defined problems. In this paper, the boundary shape uncertainty in the boundary problems is modeled by NURBS curves using interval arithmetic [6, 7]. For this purpose, the control points' coordinates are defined using interval numbers. Consideration of the uncertainty (e.g., measurement errors) is a better approximation of reality.

The interval NURBS curves are compared with interval linear segments and interval Bézier curves to emphasize their advantages in modeling the boundary shape uncertainty. The impact of such modeling on the interval solutions of the problem is also analyzed. The mentioned modeling methods with the strategy of its inclusion into the mathematical formalism of the interval parametric integral equation system (interval PIES) [8] are presented below.

2 Modeling the Boundary Shape Uncertainty

Direct application of classical or directed interval arithmetic [6, 7] in modeling boundary problems with any, uncertainly defined boundary shape is troublesome even with linear segments. A detailed description of the arising problems is presented in [9]. Among others, there is a lack of continuity between boundary segments (unrealistic problems are considered). Modeling the shape in different quadrants of the Cartesian coordinate system gives different results. Therefore, the authors proposed a modification of directed interval arithmetic by shifting arithmetic operators to the positive semi-axis as follows (for multiplication):

$$
\boldsymbol{x} \cdot \boldsymbol{y} = \begin{cases} \boldsymbol{x}_s \cdot \boldsymbol{y}_s - \boldsymbol{x}_s \cdot \boldsymbol{y}_m - \boldsymbol{x}_m \cdot \boldsymbol{y}_s + \boldsymbol{x}_m \cdot \boldsymbol{y}_m & \text{for } \boldsymbol{x} \leq 0, \boldsymbol{y} \leq 0 \\ \boldsymbol{x}_s \cdot \boldsymbol{y} - \boldsymbol{x}_m \cdot \boldsymbol{y} & \text{for } \boldsymbol{x} > 0, \boldsymbol{y} \leq 0 \\ \boldsymbol{x} \cdot \boldsymbol{y}_s - \boldsymbol{x} \cdot \boldsymbol{y}_m & \text{for } \boldsymbol{x} \leq 0, \boldsymbol{y} > 0 \\ \boldsymbol{x} \cdot \boldsymbol{y} & \text{for } \boldsymbol{x} > 0, \boldsymbol{y} > 0 \end{cases}, \quad (1)
$$

where $(\cdot)$ is an interval multiplication and for any $\boldsymbol{a} = [\underline{a}, \overline{a}]$ can be defined

$$
\boldsymbol{a}_s = \boldsymbol{a} + a_m \text{ and } a_m = \begin{cases} |\overline{a}| & \text{for } \overline{a} > \underline{a} \\ |\underline{a}| & \text{for } \overline{a} < \underline{a} \end{cases}, \text{ where } \begin{cases} a > 0 \rightarrow \underline{a} > 0 \text{ and } \overline{a} > 0 \\ a \leq 0 \rightarrow \underline{a} < 0 \text{ or } \overline{a} < 0 \end{cases}.
$$

Significant advantages of exactly defined NURBS curves [1] in PIES are presented in [4, 5]. Therefore in this paper, for modeling uncertainly defined boundary shape, it is decided to verify the effectiveness of its interval modification:

$$
\boldsymbol{S}_m(s) = \frac{\sum\limits_{i=0}^{n} w_i \boldsymbol{P}_i N_i^k(s)}{\sum\limits_{i=0}^{n} w_i N_i^k(s)} \quad \text{dla} \quad t_k \leq s \leq t_{n+1}, \quad (2)
$$

where $\boldsymbol{P}_i (i = 0, 1, \ldots, n)$ are the interval control points, $w_i (i = 0, 1, \ldots, n)$ are exactly defined weights corresponding to points, and the base function $N_i^k(s)$ of k degree is exactly defined as normalized B-spline blending function [1]. Its definition requires also exactly defined elements of the knot vector.

The advantage of such an interval modification of NURBS curves is the possibility to change the uncertainly defined boundary shape using only exactly defined knots and weights (without changing the interval coordinates of control points). In Fig. 1 the examples of such kind of shape modifications are presented.

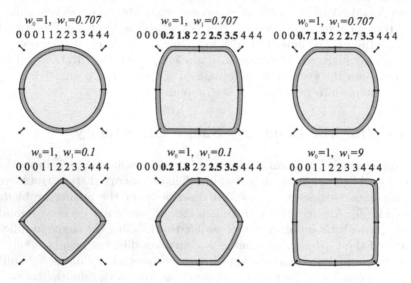

Fig. 1. Modification of interval shape using exactly defined weights and knots.

So, interval NURBS curves make modification much easier. Additionally, using the second-degree curve, a fewer amount of interval input data can be used what means fewer calculations on these numbers. This significantly reduces overestimation and improves obtained interval solutions.

3 Inclusion of Interval Curves into Interval PIES Method

The effectiveness of the PIES method and the accuracy of its solutions have been confirmed for exactly defined problems [10,11]. Therefore, in this paper, to obtain solutions on the boundary (of uncertainly defined two-dimensional problem modeled by Laplace's equation), the interval PIES method [8] is proposed:

$$\frac{1}{2}u_l(s_1) = \sum_{j=1}^{n} \int_{\widehat{s}_{j-1}}^{\widehat{s}_j} \left\{ U_{lj}^*(s_1,s)p_j(s) - P_{lj}^*(s_1,s)u_j(s) \right\} J_j(s)ds, \qquad (3)$$

where $\widehat{s}_{l-1} \leq s_1 \leq \widehat{s}_l$, $\widehat{s}_{j-1} \leq s \leq \widehat{s}_j$ are defined exactly in a parametric coordinate system and correspond to the beginning and the end of the segment of the interval curve S_m (where m = j, l).

The functions $p_j(s)$, $u_j(s)$ are parametric boundary functions on individual segments S_j of the interval boundary. One of these will be given as boundary conditions, while the other will be searched for as a result of the numerical solution of the interval PIES. In this paper, to analyze only the influence of the boundary shape uncertainty, the boundary conditions will be defined exactly.

To include the uncertainly defined boundary shape in PIES, the kernels should be modified. Hence, they will be defined as following interval functions $U_{lj}^*(s_1, s) = [\underline{U}_{lj}^*(s_1, s), \overline{U}_{lj}^*(s_1, s)]$, $P_{lj}^*(s_1, s) = [\underline{P}_{lj}^*(s_1, s), \overline{P}_{lj}^*(s_1, s)]$:

$$U_{lj}^*(s_1, s) = \frac{1}{2\pi} \ln \frac{1}{[\eta_1^2 + \eta_2^2]^{0.5}}, \quad P_{lj}^*(s_1, s) = \frac{1}{2\pi} \frac{\eta_1 n_1(s) + \eta_2 n_2(s)}{\eta_1^2 + \eta_2^2}, \qquad (4)$$

where $n_1(s) = [\underline{n}_1(s), \overline{n}_1(s)]$, $n_2(s) = [\underline{n}_2(s), \overline{n}_2(s)]$ are the interval components of $n(s)$ - the normal vector to the interval segment S_j. The kernels analytically include the boundary shape uncertainty into its mathematical formalism. Such shape is defined as relation between interval segments S_m ($m = l, j = 1, 2, ..., n$):

$$\eta_1 = S_l^{(1)}(s_1) - S_j^{(1)}(s_1), \eta_2 = S_l^{(2)}(s_1) - S_j^{(2)}(s_1). \qquad (5)$$

The uncertainty of the boundary shape should be also included in the Jacobian $J_j(s) = [\underline{J}_j(s), \overline{J}_j(s)]$ for the segment of the interval curve $S_j(s)$.

The PIES numerical solution does not require classical discretization, unlike the boundary integral equation (BIE). To include the boundary uncertainty directly in functions (4) the interval segments will be defined by interval NURBS curves (2). The interval Bézier curves [12, 13] of the second and third-degree are used for comparison:

$$S_m(s) = (1 - s)^2 P_0 + 2s(1 - s) P_1 + s^2 P_2, \qquad (6)$$

$$S_m(s) = (1 - s)^3 P_0 + 3s(1 - s)^2 P_1 + 3(1 - s)s^2 P_2 + s^3 P_3, \qquad (7)$$

where $m = l, j$. The second-degree curve (6) depends on three interval points: approximating (P_1) and interpolating (P_0, P_2) and the third-degree curve (7) on four points respectively: approximating (P_1, P_2) and interpolating (P_0, P_3).

4   Comparison of Interval PIES Solutions

The shape of the first example is modeled using interval linear segments (Fig. 2a) and using a second-degree interval NURBS curve (Fig. 2b). The Dirichlet boundary conditions $u = 0.5(x^2 + y^2)$ are defined. The analytical solution [14] of the problem with error obtained by total differential method [15] is defined as:

$$u_a = \frac{x^3 - 3xy^2}{2a} + \frac{2a^2}{27}, \quad \Delta u_a = \left| \frac{3xy^2 - x^3}{2a^2} + \frac{4a}{27} \right| |\Delta a|, \qquad (8)$$

where the height of the triangle is uncertainly defined as $a = [\underline{a}, \overline{a}] = [2.9, 3.1]$, then $a = 0.5(\overline{a} + \underline{a})$ and $\Delta a = 0.5|\overline{a} - \underline{a}|$. The analytical interval solution will be

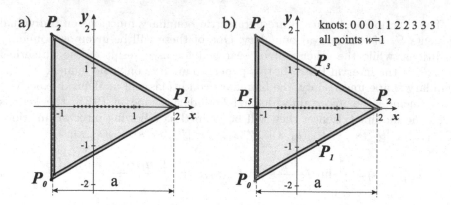

Fig. 2. Boundary shape obtained using interval a) linear segments, b) NURBS curves.

Table 1. Comparison of interval PIES solutions in domain (example from Fig. 2).

$y = 0$	Analytical		Interval PIES		Interval PIES	
x	Total differential		NURBS		Linear	
−0.4	0.611	0.701	0.612	0.702	0.612	0.702
−0.1	0.622	0.711	0.623	0.712	0.623	0.712
0.2	0.624	0.712	0.624	0.713	0.624	0.713
0.5	0.644	0.731	0.645	0.732	0.645	0.732
0.8	0.710	0.794	0.711	0.794	0.711	0.794
1.1	0.851	0.926	0.852	0.927	0.852	0.927

defined as: $u_a = [u_a - \Delta u_a, u_a + \Delta u_a]$. The interval PIES solutions with both modeling methods and the interval analytical solutions are presented in Table 1.

The interval PIES solutions with linear segments are almost equal to those with the interval NURBS curves (the average relative error is $3 \cdot 10^{-7}\%$). The average relative error of solutions in comparison to interval analytical ones is 0.1%. Obtained solutions are correct and almost without overestimation.

The correctness of the algorithm has been confirmed, so to emphasize the advantages of the strategy the problem with elliptical domain is also considered. The shape is modeled using the second-degree interval NURBS curve (Fig. 3a) with double knots (0 0 0 1 1 2 2 3 3 4 4 4) and using interval Bézier curves of second (Fig. 3b) [13] and third degree (Fig. 3c). The Dirichlet boundary conditions $u = 0.5(x^2 + y^2)$ are defined and exact analytical solution is [14]:

$$u_a = \frac{x^2 + y^2}{2} - \frac{a^2 b^2 \left(\frac{x^2}{a^2} + \frac{y^2}{b^2} - 1\right)}{a^2 + b^2}, \tag{9}$$

where semi-major axis a and semi-minor axis b of the ellipse are defined as $a = [\underline{a}, \overline{a}] = [1.95, 2.05]$ and $b = [\underline{b}, \overline{b}] = [0.9, 1.1]$. Therefore, analogically to the

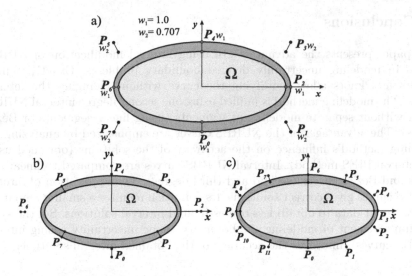

Fig. 3. Uncertainly defined elliptical shape of the boundary modeled by interval curves a) NURBS II degree, b) Bézier II degree, c) Bézier III degree.

previous example, interval analytical solutions (with error Δu_a obtained using the total differential method [15]) are presented as $\boldsymbol{u}_a = [u_a - \Delta u_a, u_a + \Delta u_a]$.

The average of the lower and upper bound relative error of interval PIES solutions in comparison with analytical ones are presented in Fig. 4. The solutions obtained using the second-degree NURBS curves (8 interval points) and third-degree Bézier curves (12 interval points) are almost equal (maximum error 0.8%). The maximum error after application of second-degree Bézier curves is about 1.8%. Therefore, the interval NURBS curves are not only easy to model and modify but also the obtained results are correct and less overestimated.

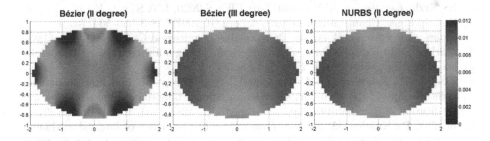

Fig. 4. Average relative error of interval PIES solutions in domain Fig. 3.

5   Conclusions

This paper presents the advantages of using interval modification of NURBS curves in modeling uncertainly defined boundary problems. Exactly defined weights and knots allow modifying the curve without changing the interval points. The modeling method is unified using one second-degree interval NURBS curve, without separate modeling of segments (using linear segments or Bézier curves). The advantages of the NURBS curves are emphasized by analyzing the modeling method's influence on the accuracy of the solutions (obtained using the interval PIES method). Interval NURBS curves are compared to linear segments and Bézier curves (second and third-degree). The application of interval NURBS curves gives correct solutions. Its definition requires a smaller amount of interval input data to obtain less overestimated interval solutions. So, the accuracy improvement of modeling the boundary shape uncertainty (using interval NURBS curves), improves the accuracy of the obtained interval solutions.

References

1. Piegl, L., Tiller, W.: The NURBS Book, 2nd edn. Springer, Heidelberg (1997). https://doi.org/10.1007/978-3-642-59223-2
2. Hughes, T.J.R., Evans, J.A., Reali, A.: Finite element and NURBS approximations of eigenvalue, boundary-value, and initial-value problems. Comput. Meth. Appl. Mech. Eng. **272**, 290–320 (2014)
3. Wang, Q., Zhou, W., Cheng, Y., Ma, G., Chang, X., Chen, E.: NE-IIBEFM for problems with body forces: a seamless integration of the boundary type meshfree method and the NURBS boundary in CAD. Adv. Eng. Softw. **118**, 1–17 (2018)
4. Zieniuk, E., Kapturczak, M.: Modeling the shape of boundary using NURBS curves directly in modified boundary integral equations for Laplace's equation. Comput. Appl. Math. **37**(4), 4835–4855 (2018)
5. Kapturczak, M., Zieniuk, E., Kużelewski, A.: NURBS curves in parametric integral equations system for modeling and solving boundary value problems in elasticity. In: Krzhizhanovskaya, V.V., et al. (eds.) ICCS 2020. LNCS, vol. 12138, pp. 116–123. Springer, Cham (2020). https://doi.org/10.1007/978-3-030-50417-5_9
6. Moore, R.E.: Interval Analysis. Prentice-Hall, Englewood Cliffs, New York (1966)
7. Markov, S.M.: Extended interval arithmetic involving infinite intervals. Mathematica Balkanica (New Ser.) **6**, 269–304 (1992)
8. Zieniuk, E., Kapturczak, M., Kużelewski, A.: Concept of modeling uncertainly defined shape of the boundary in two-dimensional boundary value problems and verification of its reliability. Appl. Math. Model. **40**(23–24), 10274–10285 (2016)
9. Zieniuk, E., Kużelewski, A., Kapturczak, M.: The influence of interval arithmetic on the shape of uncertainly defined domains modelled by closed curves. Comput. Appl. Math. **37**(2), 1027–1046 (2016). https://doi.org/10.1007/s40314-016-0382-0
10. Zieniuk, E., Szerszen, K., Kapturczak, M.: A numerical approach to the determination of 3D stokes flow in polygonal domains using PIES. In: Wyrzykowski, R., Dongarra, J., Karczewski, K., Waśniewski, J. (eds.) PPAM 2011. LNCS, vol. 7203, pp. 112–121. Springer, Heidelberg (2012). https://doi.org/10.1007/978-3-642-31464-3_12

11. Kużelewski, A., Zieniuk, E., Kapturczak, M.: Acceleration of integration in parametric integral equations system using CUDA. Comput. Struct. **152**, 113–124 (2015)
12. Hosaka, M.: Modeling of Curves and Surfaces in CAD/CAM. Computer Graphics - Systems and Applications. Springer, Berlin (1992). https://doi.org/10.1007/978-3-642-76598-8
13. Rababah, A.: The best uniform quadratic approximation of circular arcs with high accuracy. Open Math. **14**(1), 118–127 (2016)
14. Hromadka II, T.V.: The Complex Variable Boundary Element Method in Engineering Analysis. Springer, New York (1987). https://doi.org/10.1007/978-1-4612-4660-2
15. Peng, F.Y., Ma, J.Y., Wang, W., Duan, X.Y., Sun, P.P., Yan, R.: Total differential methods based universal post processing algorithm considering geometric error for multi-axis NC machine tool. Int. J. Mach. Tools Manuf. **70**, 53–62 (2013)

Introducing Uncertainty into Explainable AI Methods

Szymon Bobek$^{(\boxtimes)}$ and Grzegorz J. Nalepa

Jagiellonian Human-Centered Artificial Intelligence Laboratory (JAHCAI) and Institute of Applied Computer Science, Jagiellonian University, 31-007 Kraków, Poland
{szymon.bobek,grzegorz.j.nalepa}@uj.edu.pl

Abstract. Learning from uncertain or incomplete data is one of the major challenges in building artificial intelligence systems. However, the research in this area is more focused on the impact of uncertainty on the algorithms performance or robustness, rather than on human understanding of the model and the explainability of the system. In this paper we present our work in the field of knowledge discovery from uncertain data and show its potential usage for the purpose of improving system interpretability by generating Local Uncertain Explanations (LUX) for machine learning models. We present a method that allows to propagate uncertainty of data into the explanation model, providing more insight into the certainty of the decision making process and certainty of explanations of these decisions. We demonstrate the method on synthetic, reproducible dataset and compare it to the most popular explanation frameworks.

Keywords: Machine learning · Rules · Uncertainty · Decision trees · Explainability

1 Introduction

Introducing uncertainty into the machine learning (ML) process is an important research topic in the field of knowledge discovery across different areas of applications. It gained special importance in pervasive and mobile systems, where contextual information is delivered by different, possibly distributed providers. It is also an important field of study in the area of data analysis of sensor data, e.g., from industrial machines. Such sensors may not be always available, and produce uncertain, vague or ambiguous information (e.g., noisy readings, missing values, anomalous events). In an intelligent systems that use such information for knowledge discovery and decision support, capability of learning and reasoning under different types of uncertainty is a fundamental requirement.

A lot of research was devoted to providing robust methods for handling uncertainty in machine learning algorithms starting from Quinlan's C4.5 algorithm [12] for dealing with missing values in training sets and ending on more recent advances on uncertainty management in knowledge discovery from data streams [9]. Most of this research focuses primarily on the efficiency of algorithms in terms of accuracy or resources. However, due to the UE GDPR regulations, the understanding of the model becomes one of the fundamental requirement for every artificial intelligence system [8].

© Springer Nature Switzerland AG 2021
M. Paszynski et al. (Eds.): ICCS 2021, LNCS 12747, pp. 444–457, 2021.
https://doi.org/10.1007/978-3-030-77980-1_34

Explainability is not a new concept in the field of artificial intelligence [15]. However, it has been most extensively developed over the last decade due to the huge successes of black-box ML models such as deep neural networks in sensitive application contexts like medicine, industry 4.0 etc. Although a variety methods were developed over the years to support explainability of ML models such as Lime [13], Shap [11], Anchor [14] the quality of their explanations depends highly on the quality of the model predictions (e.g., accuracy). Yet, the information on the accuracy is not transferred in an explicit way to the explanation itself, leaving the final assessment of the explanation quality to the user.

In this paper we present an extension to our work in the area of semi-automatic knowledge discovery from data streams with uncertain or missing class labels [2–4], that aims at exploiting the semantic knowledge representation and uncertainty handling for the purpose of explainability improvement. We based our method on decision tree generation algorithm that uses modified information gain split criterion, which takes into account uncertainty of data. We show how our research can be used to increase explainability and understandability (intelligibility) of intelligent systems. In particular we show how uncertain decision trees can be used to build models that approximate arbitrary machine learning model locally and provide rule-based explanation for each decision made by the ML model.

In our approach we focus on a robust model-agnostic solution. First of all, our goal was to provide a method for building models which will be human understandable. Secondly, we wanted to create an algorithm that will not only inform the user about possible uncertainty in decision process, but could also inform other system components (or an expert) about the impact that the uncertainty may have on the model performance. The former could be used in user-centric solutions to trigger mediation with the user and request human assistance in upgrading or modifying the model.

The rest of this paper is organized as follows. The algorithm for building uncertain decision trees is presented in Sect. 2. In Sect. 3 we discuss the interpretability of models generated with our algorithm. Application of these models to generating local uncertain explanations is given in Sect. 4. We demonstrate our solution and present a comparison with existing explainability frameworks in Sect. 5. Finally, a short summary of our work is presented in Sect. 6.

2   Uncertain Decision Trees

In this section we describe the underlying mechanism that allows to build LUX models. The mechanism is based on the uID3 decision tree generation algorithm proposed by us in [2] and extended here for the purpose of explanation generation.

The uID3 algorithm is based on a heuristic that uses the modified information gain split criterion that includes uncertainty of training data into the calculation. This allows to apply it to a variety of algorithms that are based on it, such as classic ID3 algorithm, or more complex, incremental versions such as VFDT [7] or CVFDT [10].

The classic information gain formula for the attribute A and a training set X is defined as follows:

$$Gain(A) = H(X) - \sum_{v \in Domain(A)} \frac{|X_v|}{|X|} H(X_v) \tag{1}$$

Where X_v is a subset of X, such that for every $x \in X$ value of $A = v$. The entropy for the training set X is defined as follows:

$$H(X) = - \sum_{v \in Domain(C)} p(v) \log_2 p(v) \tag{2}$$

Where $Domain(C)$ is a set of all classes in X and $p(v)$ is a ratio of the number of elements of class v to all the elements in X.

In case of the uncertain data, the $p(v)$ from the Eq. (2) has to be defined as a probability of observing element of class v in the dataset X. This probability will be denoted further as capital $P_{total}(C = v)$. Similarly, a fraction $\frac{|X_v|}{|X|}$ from Eq. (1) has to be redefined as a probability of observing value v of attribute A in the dataset X. This probability will be referred later as $P_{total}(A = v)$ and is defined as a probability of observing a value v_j^i of an attribute A_i in the set X_t that contains k training instances. This can be defined as follows:

$$P_{total}(A_i = v_j^i) = \frac{1}{k} \sum_{X_t \ni P_{j=1...n}} P_j(A_i = v_j^i) \tag{3}$$

Similarly $P_{total}(C = v_j)$ can be defined, which represents probability of observing class label in a set. Having that, the uncertain information gain measure can be defined as shown in the Eq. (4).

$$Gain^U(A) = H^U(X) - \sum_{v \in Domain(A)} P_{total}(A = v) H^U(X_v) \tag{4}$$

Where the H^U is the uncertain entropy measure defined as:

$$H^U(X) = - \sum_{v \in Domain(C)} P_{total}(C = v) \log_2 P_{total}(C = v) \tag{5}$$

Uncertain information gain and entropy represented by the Eq. (4) and (5) are used to build decision tree. The complete procedure of generating the uncertain tree is presented in the Algorithm 1. In such a tree, every branch connecting two nodes contains statistics about the accuracy of data used to grow the subtree, as shown in Fig. 1a.

Data used to generate this tree is a special format of ARFF (the format of the WEKA ML tool), called uARFF. The example of such file used to generate the tree from Fig. 1a is presented in Listing 1.1. It presents a dataset for classifier using certain parameters of a machine to predict its malfunctions. Some of the data is uncertain, and denoted as an alternative divided by semicolon, with probabilities or confidence in square brackets.

Algorithm 1: uID3 algorithm to grow a decision tree from uncertain data

---

Input: data X; set of attributes A
Output: uncertain decision tree uT
if Homogeneous(X) **then**
    return MajorityClass(X)
end if
$R \leftarrow$ Best split using $Gain^U(X)$
split X into subsets X_i according to Domain(R);
for each i **do**
    if $X_i \neq \emptyset$ **then**
        $uT_i \leftarrow$ uID3(X_i, A)
    else
        uT_i is a leaf labeled with MajorityClass(X)
    end if
end for
return a root R of the decision tree

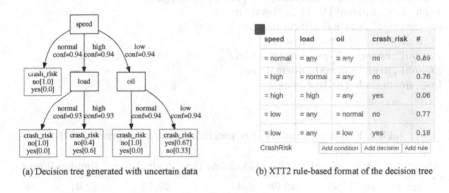

(a) Decision tree generated with uncertain data   (b) XTT2 rule-based format of the decision tree

Fig. 1. Decision tree and a corresponding decision table (Color figure online)

In [2], we presented an evaluation of this method on highly distorted dataset for predicting human emotional condition based on the physiological readings. As shown in Table 1, our method was not worse than approaches that included only most probable class. However, the main advantage was the ability to quantify decision accuracy not only by the statistics in leaves but also by the certainty of data used to generate the tree. We argue that this helps in increasing transparency of both learning and decision making, improving interpretability of the model. This topic will be discussed in the next section.

Listing 1.1. uARFF file format

```
@relation  machine.symbolic

@attribute  speed  {high, normal, low}
@attribute  temperature  {high, normal, low}
@attribute  load  {high, normal}
@attribute  oil  {low, normal}
@attribute  crash_risk  {yes, no}

@data
high[0.5]; low[0.3]; normal[0.2], high, high, low, yes
high, high, high, low, yes
normal[0.2]; high[0.7]; low[0.1], high, high, normal, no
low, normal, high, normal, no
low, low, normal, normal[0.3], no
low, low, normal, low, yes
normal, low, normal, low, no
high, normal, high, normal[0.6]; low[0.4], yes
high, low, normal[0.4], normal, no
low, normal, normal, normal, no
high, normal, normal, low, no
normal, normal, high, low[0.4], no
normal, high, normal, normal, no
low, normal, high, low, yes
```

Table 1. Evaluation summary of an uncertain decision tree generator [2]

Algorithm	uID3	ZeroR	J48	HoeffdingTree	NaiveBayes	RandomForest
Accuracy	49.71	21	48.96	46.42	49.55	42.91

3   Interpretability of Uncertain Decision Trees

The learning algorithm presented in previous section does not improve drastically the accuracy of the classification [2,4]. However, the additional information that is stored in the model may be used to give user a deeper insight into the decision and learning processes. It provides more compact, and efficient way of encoding uncertain knowledge than more sophisticated methods.

Specifically, the $P_{total}(A = v)$ is included for each branch. Such information is useful while translating decision tree into rule-based knowledge representation. In such a translation uncertain branches can be verified by the user or skipped, keeping the size of the knowledge model small.

The translation from the uncertain decision tree into rule representation is straightforward. Every branch of the tree is considered one rule. Additionally, the information about the branch uncertainty is translated as a certainty factor [6] of a particular rule.

The total uncertainty of the branch is calculated as a product of all the probabilities associated with edges that form the branch. Let assume that there is a branch B, which includes n attributes $A_1, A_2, \ldots, A_n$, and n edges associated with the values of these attributes. Therefore, the total branch uncertainty is defined as follows:

$$U(B) = \prod P_{total}(A_i = v_i) \cdot Acc(B) \tag{6}$$

Where Acc is the accuracy of the classification of the branch B denoted by the leaf node.

An example of the rule set for the uncertain decision tree presented in Fig. 1a is given in Fig. 1b. Our XTT2 rule representation [6] uses certainty factors algebra for representing uncertainty. This allows for direct modification of confidence levels of rules by the user as it does not require any knowledge in the area of probability theory.

In the Fig. 1b the first rule is a translation of the first branch of the tree that was given in Fig. 1a, according to Eq. (6)[1]. Probabilities are expressed by a number $p \in [0; 1]$, while certainty factors algebra operates on the range $[-1; 1]$. Therefore a simple transformation from probability space to certainty factors space was given: $cf = p \cdot 2 - 1$. It is worth noting, that the probability theory has different foundations than certainty factors algebra, and such transformation is performed by us only to capture a simple intuition for what the value of probability may mean in the certainty factors space. This transformation does not have any formal basis though.

Such a notation allows the user to verify potentially incorrect rules (e.g., rule three in Fig. 1b) by deleting, adding or modifying the certainty factors. Additionally, it allows the system itself to identify the parts of the model that may be corrupted by uncertain data as the knowledge about the uncertain dataset is retained in the model.

4 Local Uncertain Explanations (LUX)

The uncertain decision tree defined previously can be used as stand alone model that solves ML classification tasks. However, in order to exploit its properties, the knowledge about certainty of readings or class labels are required. This knowledge is not always available in real-life setting, but can be obtained as an output from other machine learning models, as the predictions generated by such models are in most of the cases returned with some level of certainty that can be considered an approximation of probability.

Such an observation makes our uID3 algorithm perfectly fit the requirements of the local explanation models, which aim at building simple (and therefore interpretable) model on a fraction of data, which forms a local neighborhood of the instance in consideration. In this section the detailed information on how such an explanation can be generates will be given.

---

[1] The model was created with HWeD editor we developed, available on-line on https://heartdroid.re/hwed/.

4.1 Building the Local Model

The main goal of the local, interpretable model L is to approximate the model M in surrounding to some instance $x^{(i)} \in X$ in order to provide explanation to the decision of the model M. This assumes that locally, the decision boundary of the original model M is simple enough to be approximated with simpler, yet interpretable model L. Under this assumption we get that $L(x^{(i)}) = M(x^{(i)})$ holds in a neighborhood N of $x^{(i)}$ and hence the impact of the features of L are similar in M as well. Such an impact may be considered a simple explanation of a decision of the model M for instance $x^{(i)}$.

For the sake of the discussion let us assume that we want to provide an explanation of a decision of any model M that can be trained on dataset X. An example of such a setting was given in Fig. 2, where two different models (SVM and XGBoost) were trained on the same training set. The figure shows decision boundaries for both of the classifiers and the instance (marked red) for which we would like to obtain an explanation.

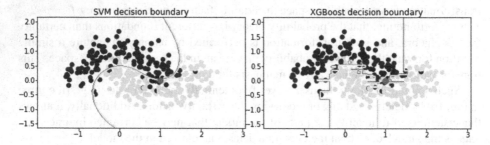

Fig. 2. Dataset with two models trained along with their decision boundaries. The red dot represents an instance for which an explanation is needed. (Color figure online)

The main goal of the explanation mechanism is not to provide a correct solution to the classification problem, but explain the decision of the model M. Hence, the local model L should approximate the model M even in cases where the latter one is wrong. This leads to the conclusion that the approximated model should be trained with target

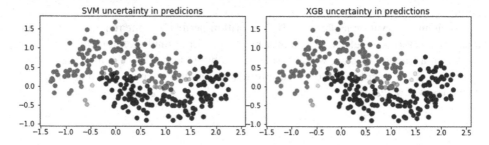

Fig. 3. Uncertainty of a prediction for two models. High transparency level depicts low uncertainty. (Color figure online)

labels acquired directly from the model M. However, the predictions from the model M are uncertain, as the model itself is an approximation of an unknown function. Figure 3 shows the uncertainty of a prediction of a SVM and XGBoost classifiers. Such an uncertainty should not only be taken into consideration while training local approximation model L but also should be transferred to the final explanations. This will allow for better assessment of the quality of the explanation by an expert.

We exploit the fact that the local model L is trained with uncertain labels to use uID3 mechanism. In order do build the local model, we first select a neighborhood N of a training instance $x^{(i)}$ in consideration. The neighborhood is created in a stratified way in order to assure existence of both positive and negative training examples. Hence, the neighborhood N of size K is defined in Eq. (7).

$$N(x^{(i)}, K) = \left\{ x^{(k)} \in X : d(x^{(i)}, x^{(k)}) \leq D_i^{(K)} \right\} \tag{7}$$

Where $D_i^{(K)}$ is K-th element from a tuple D_i defined for all m instances from a training set as:

$$D_i = \left\{ d(x^{(i)}, x^{(1)}), d(x^{(i)}, x^{(2)}), \dots, d(x^{(i)}, x^{(m)}) \right\}$$

Where D is sorted in ascending order, and $d(x^{(i)}, x^{(j)})$ is a distance between instances i-th and j-th. Figure 4 depicts the neighborhood for the instance $x^{(i)} = (1.0, 0.0)$ (marked red).

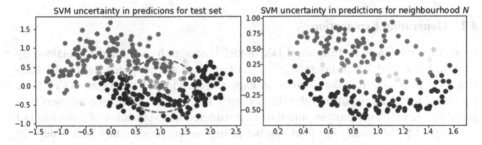

Fig. 4. Neighborhood of a point being explained, with transparency level reflecting uncertainty of the prediction. Right figure provides the full dataset, while the left figure only the neighborhood selection. (Color figure online)

Finally, the uID3 algorithm is run on the training dataset formed by $N(x^{(i)}, K)$ and the LUX model is created. Figure 5a presents a fragment of training set obtained from neighborhood of a point $x^{(i)} = (1.0, 0.0)$. Figure 5b on the other hand depicts the uncertain decision tree obtained by running uID3 algorithm on the dataset. In the next section we discuss how the final explanation is obtained from the LUX model and how it can be combined with external systems to increase its intelligibility.

```
@relation lux

@attribute x1 @REAL
@attribute x2 @REAL
@attribute class {1,0}

@data
0.94,0.01,1[0.48]
0.87,-0.04,1[0.64]
1.02,-0.16,1[0.78]
1.14,0.08,1[0.37]
1.01,-0.21,1[0.83]
1.10,-0.19,1[0.81]
0.80,-0.13,1[0.81]
0.91,-0.23,1[0.87]
0.77,-0.12,1[0.83]
1.01,-0.28,1[0.89]
0.97,-0.28,1[0.89]
...
```

(a) Training set

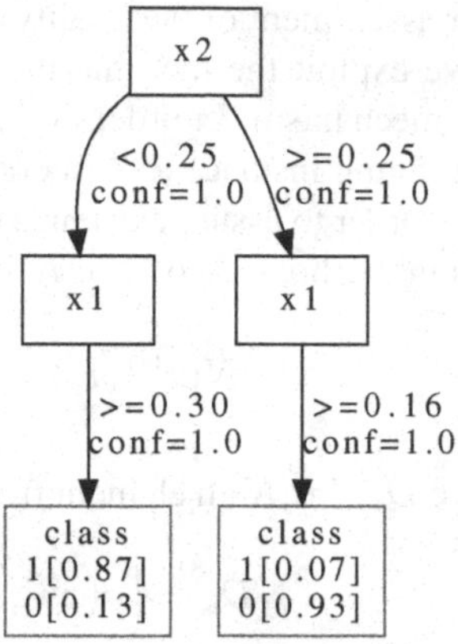

(b) LUX model

Fig. 5. Training set obtained by sampling neighborhood of a point $x^{(i)}$ (a) and LUX model (b) generated form the data using uID3 algorithm. (Color figure online)

4.2 Generating Explanation

Generating explanations from the LUX model is straightforward and is obtained by feeding LUX model with instance $x^{(i)}$. The branch that is activated during the classification forms the rule that defines an explanation. Figure 6 shows an XTT2 table generated for SVM classifier with uID3 toolkit[2]. The first rule is the one that was triggered by the inference process, and thus is considered an explanation of a decision of the original SVM model. The float number in the last column marked with # denotes the certainty of the rule.

x1 ▾	x2 ▾	class ▾	#
>= 0.30	< 0.25	set 1	0,8 ●
>= 0.16	>= 0.25	set 0	0,9 ●

tree   Add condition | Add decision | Add rule

Fig. 6. XTT2 table generated as an explanation to the LUX model given in Fig. 5b (Color figure online)

---

[2] See: https://github.com/sbobek/udt.

It is worth noting that the explanation is a valid XTT2 table, that can be combined with the larger rule-based system and process with HeaRTDroid inference engine [5]. This is especially useful in settings where there exists some kind of domain knowledge that cannot be detected on such a small fraction of data, defined with N. Such a knowledge can be encoded with rules and serve as additional guards of correctness of the explanation. For instance if a LUX system forms an explanation rule that based on two attributes *temperature* and *pressure* that a water *boils*, one can easily develop additional rules that will put more strict constraints on both *pressure* and *temperature*, as there exist a law of physics that forbids some of the values coexist.

The next section provides more insight into the validity of the explanation generated by the LUX in comparison to most popular frameworks such as Lime, Shap, Anchor.

5   Evaluation

In this section we present a comparison of LUX explanations and selected explanation mechanism. We will focus on two aspects of the explanation: qualitative and quantitative. In qualitative explanation we would like to emphasize the way the explanation is presented to the user, how much information it presents to the user, and how this information can be used in the system to perform more advanced reasoning. In quantitative comparison we focus on measurable aspect of explanations as a whole, such as consistency and stability. For that purpose we will use our InXAI toolkit[3].

5.1   Qualitative Comparison

Lime. Lime presents its explanation in a visual form given in Fig. 7. Negative (blue) bars indicate class 0, while values (orange) indicate class 1. Values of the bars represent the importance of each feature in making the prediction. The way to interpret the importance is by applying them to the prediction probabilities obtained from the original model. For example, if we remove the variable x2, we expect the classifier to predict class 0 with probability $0.52 - 0.14 = 0.38$.

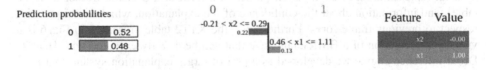

Fig. 7. Lime explanation for instance $x^{(i)} = (1.0, -0.0)$ (Color figure online)

This information itself brings an insight on the performance of the main model M and its prediction confidence, however it does not include any information on the confidence of the explanation itself. Although the results are available also as numerical values, they are not formalized in any form of logical rules, nor executable model to provide further inference.

---

[3] See https://github.com/sbobek/inxai.

Shap. This framework provides a lot more interactive methods for visualizing explanations, however the quality of the explanation is not included within. The explanation presented in Fig. 8 shows how features contribute to push the model output from the base value (the average model output over the training dataset) to the model output. Features pushing the prediction higher are shown in red, those pushing the prediction lower are in blue. Similarly to Lime, there is no information on the confidence of the explanation. The raw values are returned in the same way as in the case of Lime.

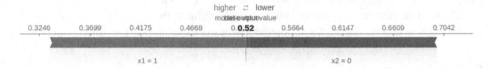

Fig. 8. Shap explanation for instance $x^{(i)} = (1.0, -0.0)$ (Color figure online)

Anchor. Opposite to the former two, Anchor produces only a textual representation of an explanation. It is given to the user in a form of *anchor* which define an area in the space of features that is dominated by samples of a given class. The example of an explanation generated with Anchor is given in Listing 1.2.

Listing 1.2. Anchor eplanation for instance $x^{(i)} = (1.0, -0.0)$

```
Anchor: x2 <= 0.29 AND x1 > -0.17
Precision: 0.96
Coverage: 0.45
```

Although the explanation is presented as a rule and it can be executed within a framework to obtain prediction, it is not possible to export the rule to a format acceptable by rule-based systems in a straightforward way.

LUX. In comparison to the former explanation mechanism, our solution provides both: a visual representation of explanation (either in a form of a decision tree or a XTT2 table), and information about the confidence of the explanation, which is unique with respect to previous frameworks. Furthermore, the XTT2 table presented in Fig. 6 is a visual representation of a HMR+ language that can be directly executed with HeaRT-Droid inference engine we developed as a part of larger explanation system that integrates also domain and expert knowledge [5].

5.2   Quantitative Comparison

In this section we focus on comparison of the frameworks in more quantitative way, providing means of assessing their quality in an automate way. For this purpose the InXAI framework will be used. We excluded Anchor from this comparison, as it does not provide feature importance information which is required to compute appropriate statistics.

Consistency measures how explanations generated for predictions of different ML models are similar to each other. Therefore, it is more related to stability of ML models with respect to decision making rather than to explanation mechanisms directly. Figure 9 (left) presents Consistency measures obtained for the dataset presented in Fig. 2 and SVM and XGBoost classifiers.

It can be observed that the overall consistency of LUX is better than in other frameworks, meaning that the explanations are not that much sensitive to different models. However, the spread in the values of consistency is high, meaning that there exists regions in dataset, where different models yield different explanations. This reflects the points that are located near decision boundary.

Stability (or robustness) assures generation of similar explanations for similar input. To obtain a numerical value to this property, modified notion of Lipschitz continuity has been proposed in [1]. Figure 9 (right) presents Consistency measures obtained for the dataset presented in Fig. 2 and SVM classifier.

Similarly as in case of the consistency, the Stability of LUX is generally better, although the spread is much larger than in two remaining frameworks. It means that there are regions in the dataset that yield different explanation for neighborhood points. This is caused by regions that contains points of different classes that are mixed. It can also be the case of too large neighborhood selected for the LUX model.

To summarize, both qualitative and quantitative evaluation shows the advantage of LUX explanations in cases where the interpretability and accountability of the explanation is crucial. Although the stability and consistency measures are better on average in LUX, the large spread of these values will be a subject of further investigation, especially in consideration to neighborhood selection for LUX training.

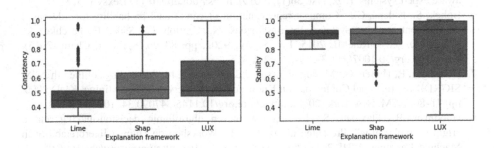

Fig. 9. Consistency and stability plots for explanation frameworks (Color figure online)

6 Summary

In this paper we presented LUX, an algorithm for building decision trees from uncertain data and using it for generating local uncertain explanation of an arbitrary machine learning model. Such an approach transfers uncertainty from the machine learning model to the explanation in an explicit way, which helps in assessing the quality of the explanation. We demonstrated our solution on an exemplary dataset and compared it with the most popular frameworks.

Currently, our toolkit is implemented as a hybrid solution that integrates Java and Python and is available online along with the dataset for reproducing experiments presented in the paper[4].

For the future work, we plan to test different approaches for selecting neighborhood for local classifier in order to improve stability and consistency of the model. In particular we would like to test similarity kernels and use values obtained from them to weight samples from neighborhood according to their applicability to the explanation.

Acknowledgements. The paper is funded from the XPM project funded by the National Science Centre, Poland under CHIST-ERA programme (NCN UMO-2020/02/Y/ST6/00070).

References

1. Alvarez-Melis, D., Jaakkola, T.S.: On the robustness of interpretability methods (2018)
2. Bobek, S., Misiak, P.: Uncertain decision tree classifier for mobile context-aware computing. In: Rutkowski, L., Scherer, R., Korytkowski, M., Pedrycz, W., Tadeusiewicz, R., Zurada, J.M. (eds.) ICAISC 2018. LNCS (LNAI), vol. 10842, pp. 276–287. Springer, Cham (2018). https://doi.org/10.1007/978-3-319-91262-2_25
3. Bobek, S., Nalepa, G.J.: Uncertain context data management in dynamic mobile environments. Fut. Gener. Comput. Syst. **66**, 110–124 (2017). https://doi.org/10.1016/j.future.2016.06.007
4. Bobek, S., Nalepa, G.J.: Uncertainty handling in rule-based mobile context-aware systems. Pervasive Mob. Comput. **39**, 159–179 (2017). https://doi.org/10.1016/j.pmcj.2016.09.004
5. Bobek, S., Nalepa, G.J., Ślażyński, M.: HeaRTDroid - rule engine for mobile and context-aware expert systems. Exp. Syst. **36**(1), (2019). https://doi.org/10.1111/exsy.12328
6. Bobek, S., Nalepa, G.J.: Compact representation of conditional probability for rule-based mobile context-aware systems. In: Bassiliades, N., Gottlob, G., Sadri, F., Paschke, A., Roman, D. (eds.) RuleML 2015. LNCS, vol. 9202, pp. 83–96. Springer, Cham (2015). https://doi.org/10.1007/978-3-319-21542-6_6
7. Domingos, P., Hulten, G.: Mining high-speed data streams. In: Proceedings of the 6th ACM SIGKDD International Conference on Knowledge Discovery and Data Mining, KDD 2000, pp. 71–80. ACM, New York (2000). https://doi.org/10.1145/347090.347107
8. Goodman, B., Flaxman, S.: EU regulations on algorithmic decision-making and a "right to explanation". Presented at 2016 ICML Workshop on Human Interpretability in Machine Learning, WHI 2016, New York, NY (2016). http://arxiv.org/abs/1606.08813, arxiv:1606.08813Comment
9. Goyal, N., Jain, S.K.: A comparative study of different frequent pattern mining algorithm for uncertain data: a survey. In: 2016 International Conference on Computing, Communication and Automation (ICCCA), pp. 183–187 (April 2016). https://doi.org/10.1109/CCAA.2016.7813714
10. Hulten, G., Spencer, L., Domingos, P.: Mining time-changing data streams. In: Proceedings of the 7th ACM SIGKDD International Conference on Knowledge Discovery and Data Mining, KDD 2001, pp. 97–106. ACM, New York (2001). https://doi.org/10.1145/502512.502529

---

[4] See: https://github.com/sbobek/lux.

11. Lundberg, S.M., Lee, S.I.: A unified approach to interpreting model predictions. In: Proceedings of the 31st International Conference on Neural Information Processing Systems, NIPS 2017, pp. 4768–4777. Curran Associates Inc. (2017)
12. Quinlan, J.R.: C4.5: Programs for Machine Learning. Morgan Kaufmann Publishers Inc., San Francisco (1993)
13. Ribeiro, M.T., Singh, S., Guestrin, C.: "Why should i trust you?": explaining the predictions of any classifier. In: Proceedings of the 22nd ACM SIGKDD International Conference on Knowledge Discovery and Data Mining, KDD 2016, pp. 1135–1144. Association for Computing Machinery, New York (2016). https://doi.org/10.1145/2939672.2939778
14. Ribeiro, M.T., Singh, S., Guestrin, C.: Anchors: high-precision model-agnostic explanations. In: 32nd AAAI Conference on Artificial Intelligence. AAAI Publications (2018)
15. Schank, R.C.: Explanation: a first pass. In: Kolodner, J.L., Riesbeck, C.K. (eds.) Experience, Memory, and Reasoning, pp. 139–165. Lawrence Erlbaum Associates, Hillsdale (1986)

New Rank-Reversal Free Approach to Handle Interval Data in MCDA Problems

Andrii Shekhovtsov[ID], Bartłomiej Kizielewicz[ID], and Wojciech Sałabun[✉][ID]

Faculty of Computer Science and Information Technology, Department of Artificial Intelligence and Applied Mathematics, Research Team on Intelligent Decision Support Systems, West Pomeranian University of Technology in Szczecin, ul. Żołnierska 49, 71-210 Szczecin, Poland
wojciech.salabun@zut.edu.pl

Abstract. In many real-life decision-making problems, decisions have to be based on partially incomplete of uncertain data. Since classical MCDA methods were created to be used with numerical data, they are often unable to process incomplete or uncertain data. There are several ways to handle uncertainty and incompleteness in the data, i.e., interval numbers, fuzzy numbers, and their generalizations. New methods are developed, and classical methods are modified to work with incomplete and uncertain data. In this paper, we propose an extension of the SPOTIS method, which is a new rank-reversal free MCDA method. Our extension allows for applying this method to decision problems with missing or uncertain data. The proposed approach is compared in two study cases with other MCDA methods: COMET and TOPSIS. Obtained rankings would be analyzed using rank correlation coefficients.

Keywords: MCDA · COMET · SPOTIS · TOPSIS · Uncertainty · Interval numbers

1 Introduction

There are many complex problems which require handling a relatively significant number of opposing criteria to evaluate decision alternatives. Classical multi-criteria decision problem consists of three elements: a set of criteria, a set of the alternatives and criteria weights. For that kind of problems, Multi-Criteria Decision-Analysis (MCDA) methods help support decision-maker in the decision process. Applying the MCDA methods to the decision problem allows determining the most reliable solution for this particular decision problem [8].

The complete dataset about alternatives should be collected to use the MCDA method to solve a particular decision problem. However, in many real-life cases, we faced with uncertain or incomplete data. This problem could appear in different cases, for example, when we collect data from various sources, or when some values in the data just not provided [22,26]. There are several methods, which decision-makers could apply to handle uncertain data, e.g., interval numbers [21], fuzzy numbers [5] and their generalizations [6,23]. Besides, if a single

© Springer Nature Switzerland AG 2021
M. Paszynski et al. (Eds.): ICCS 2021, LNCS 12747, pp. 458–472, 2021.
https://doi.org/10.1007/978-3-030-77980-1_35

criterion attribute is missing for a particular alternative, we have to consider all possible values from the domain [24].

The other problem to cope with in the decision-making process is a rank reversal paradox [1,25]. It is a phenomenon of reversing alternative's order in ranking when the set of alternatives is changed, e.g., alternative A_1 which was better than A_2 in the initial ranking could be worse than A_2 in the ranking calculated after adding a new alternative. The most MCDA methods, such as Technique for Order of Preference by Similarity to Ideal Solution (TOPSIS), Preference Ranking Organization Method for Enrichment Evaluation (PROMETHEE), VlseKriterijumska OptimizacijaI Kompromisno Resenje (VIKOR) are susceptible to this paradox [3,7,11]. Classical MCDA methods are modified in order to eliminate rank reversal paradox in them. However, there are also new methods created which were designed to eliminate rank reversal paradox in them, e.g., Ranking of Alternatives through Functional mapping of criterion sub-intervals into a Single Interval (RAFSI) [27], Characteristic Object METhod (COMET) [14] and Stable Preference Ordering Towards Ideal Solution method (SPOTIS) [4].

SPOTIS is a new MCDA method which aims to eliminate rank reversal paradox by design [4]. It is a simple method that uses reference objects to evaluate final preferences, similarly to COMET and TOPSIS methods. Unlike classic MCDA methods, such as TOPSIS, VIKOR and PROMETHEE, SPOTIS method requires criteria bounds to be defined. Using criteria bounds as reference objects allows distributing alternatives linearly between ideal positive and negative solutions. Thus SPOTIS method stays completely rank reversal free [4].

In this paper, we propose extending the SPOTIS method, which allows applying this method in the decision problems with incomplete or uncertain data using interval values. The proposed approach is based on using monotonic criteria, i.e., each criterion is profit or cost type. Moreover, we compare the proposed approach with two other MCDA methods that also use the reference objects' concept, i.e., COMET and TOPSIS. In order to compare these three methods, we present two numerical study cases. The final preferences are determined based on interval number comparison according to priority degree. Finally, the rankings are compared using ranking similarity coefficients and literature reference results.

The rest of the paper is organized as follows: In Sect. 2, basic preliminary concepts on selected MCDA methods are presented. Section 3 introduces the proposed approach. In Sect. 4, we present and discuss two study cases that show the efficiency of the proposed approach. In Sect. 5, we present the summary and conclusions.

2 Preliminaries

2.1 TOPSIS

The Technique of Order Preference Similarity (TOPSIS) method measures the distance of alternatives from the reference elements, respectively, positive and negative ideal solution (PIS and NIS). This method was widely presented in

[2,12]. The TOPSIS method is a simple MCDA technique used in many practical problems. Thanks to its simplicity of use, it is widely used in solving multi-criteria problems. Below we present its algorithm [2]. We assume, that we have decision matrix with m alternatives and n criteria is represented as $X = (x_{ij})_{m \times n}$.

Step 1. Calculate the normalized decision matrix. The normalized values r_{ij} calculated according to Eq. (1) for profit criteria and (2) for cost criteria. We use this normalization method because [13,20] shows that it performs better than the classical vector normalization. However, we can also use any other normalization method.

$$r_{ij} = \frac{x_{ij} - min_j(x_{ij})}{max_j(x_{ij}) - min_j(x_{ij})} \tag{1}$$

$$r_{ij} = \frac{max_j(x_{ij}) - x_{ij}}{max_j(x_{ij}) - min_j(x_{ij})} \tag{2}$$

Step 2. Calculate the weighted normalized decision matrix v_{ij} according to Eq. (3).

$$v_{ij} = w_i \cdot r_{ij} \tag{3}$$

Step 3. Calculate Positive Ideal Solution (PIS) and Negative Ideal Solution (NIS) vectors. PIS is defined as maximum values for each criteria (4) and NIS as minimum values (5). We do not need to split criteria into profit and cost here, because in step 1 we use normalization which turns cost criteria into profit criteria.

$$v_j^+ = \{v_1^+, v_2^+, \cdots, v_n^+\} = \{max_j(v_{ij})\} \tag{4}$$

$$v_j^- = \{v_1^-, v_2^-, \cdots, v_n^-\} = \{min_j(v_{ij})\} \tag{5}$$

Step 4. Calculate distance from PIS and NIS for each alternative. As shows Eqs. (6) and (7).

$$D_i^+ = \sqrt{\sum_{j=1}^{n}(v_{ij} - v_j^+)^2} \tag{6}$$

$$D_i^- = \sqrt{\sum_{j=1}^{n}(v_{ij} - v_j^-)^2} \tag{7}$$

Step 5. Calculate each alternative's score according to Eq. (8). This value is always between 0 and 1, and the alternatives which got values closer to 1 are better.

$$C_i = \frac{D_i^-}{D_i^- + D_i^+} \tag{8}$$

2.2   The COMET Method

The Characteristic Objects METhod (COMET) is based on fuzzy logic and triangular fuzzy sets. The COMET method's accuracy was verified in previous works [15–17]. The formal notation of the COMET must be recalled based on [14]. Figure 1 presents the flowchart of the COMET method as summarizing.

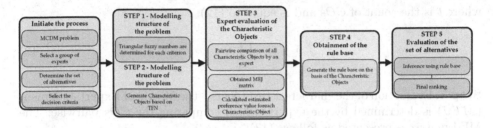

Fig. 1. The procedure of the COMET method

Step 0. Initiate the process – it is a preparatory stage, which aims to identify the problem to be further analysed clearly. In the beginning, it is necessary to define the purpose of the research and determine the specificity of the MCDA problem. We should then select an expert or a group of experts whose task will be to select decision alternatives and criteria for their evaluation. After selecting a group of alternatives, a set of criteria that should be taken into account in the further analysis should also be selected.

Step 1. Definition of the space of the problem – the dimensionality of the problem is determined by the expert, which selecting r criteria, $C_1, C_2, \ldots, C_r$. For each criterion C_i, e.g., $\{\tilde{C}_{i1}, \tilde{C}_{i2}, \ldots, \tilde{C}_{ic_i}\}$ (9) a set of fuzzy numbers is carefully selected:

$$C_1 = \left\{\tilde{C}_{11}, \tilde{C}_{12}, \ldots, \tilde{C}_{1c_1}\right\}$$
$$C_2 = \left\{\tilde{C}_{21}, \tilde{C}_{22}, \ldots, \tilde{C}_{2c_2}\right\}$$
$$\ldots$$
$$C_r = \left\{\tilde{C}_{r1}, \tilde{C}_{r2}, \ldots, \tilde{C}_{rc_r}\right\} \tag{9}$$

where $c_1, c_2, \ldots, c_r$ are the cardinality for all criteria.

Step 2. Generation of the characteristic objects – the characteristic objects (CO) are obtained with the usage of the Cartesian product of the fuzzy numbers' cores of all the criteria (10):

$$CO = \langle C(C_1) \times C(C_2) \times \cdots \times C(C_r) \rangle \tag{10}$$

As a result, an ordered set of all CO is obtained (11):

$$CO_1 = \langle C(\tilde{C}_{11}), C(\tilde{C}_{21}), \ldots, C(\tilde{C}_{r1}) \rangle$$
$$CO_2 = \langle C(\tilde{C}_{11}), C(\tilde{C}_{21}), \ldots, C(\tilde{C}_{r1}) \rangle$$
$$\cdots$$
$$CO_t = \langle C(\tilde{C}_{1c_1}), C(\tilde{C}_{2c_2}), \ldots, C(\tilde{C}_{rc_r}) \rangle$$

(11)

where t is the count of COs and is equal to (12):

$$t = \prod_{i=1}^{r} c_i \tag{12}$$

Step 3. Evaluation of the characteristic objects – the Matrix of Expert Judgment (MEJ) is determined by the expert, which comparing the COs pairwise. The MEJ matrix is presented as follows (13):

$$MEJ = \begin{pmatrix} \alpha_{11} & \alpha_{12} & \cdots & \alpha_{1t} \\ \alpha_{21} & \alpha_{22} & \cdots & \alpha_{2t} \\ \cdots & \cdots & \cdots & \cdots \\ \alpha_{t1} & \alpha_{t2} & \cdots & \alpha_{tt} \end{pmatrix} \tag{13}$$

where α_{ij} is the result of comparing CO_i and CO_j by the expert. The function f_{exp} express the individual judgement function of the expert. It is a representation of the knowledge of the selected expert, whose preferences can be presented as (14):

$$\alpha_{ij} = \begin{cases} 0.0, f_{\exp}(CO_i) < f_{\exp}(CO_j) \\ 0.5, f_{\exp}(CO_i) = f_{\exp}(CO_j) \\ 1.0, f_{\exp}(CO_i) > f_{exp}(CO_j) \end{cases} \tag{14}$$

The number of query is equal $p = \frac{t(t-1)}{2}$ because for each element α_{ij} we can observe that $\alpha_{ji} = 1 - \alpha_{ij}$. After the MEJ matrix is constructed, a vertical vector of the Summed Judgments (SJ) is obtained by using moudus ponens tautology as follows (15):

$$SJ_i = \sum_{j=1}^{t} \alpha_{ij} \tag{15}$$

Finally, the values of preference are estimated for each characteristic object, and a vertical vector P is obtained. The $i - th$ row includes the estimated value of preference for CO_i.

Step 4. The rule base—each characteristic object and its value of preference is converted to a fuzzy rule as (16):

$$IF \quad C\left(\tilde{C}_{1i}\right) \quad AND \quad C\left(\tilde{C}_{2i}\right) \quad AND \quad \ldots \quad THEN \quad P_i \tag{16}$$

In this way, a complete fuzzy rule base is obtained, which will then be used to infer alternatives' evaluation.

Step 5. Inference and the final ranking – each alternative is represented as a set of values, e.g., $A_i = \{\alpha_{i1}, \alpha_{i2}, \alpha_{ri}\}$. This set is addressed to the criteria $C_1, C_2, \ldots, C_r$. Mamdani's fuzzy inference technique is used to calculate the preference of the $i - th$ decision variant. The constant rule base guarantees that the determined results are unequivocal, and it makes the COMET completely rank reversal free.

2.3 SPOTIS

Stable Preference Ordering Towards Ideal Solution (SPOTIS) is a new method for multi-criteria decision support [4]. The authors of this method aim to create a new method free of the Rank Reversal problem (the phenomenon of reversing the ranking when changing the number of alternatives in the input data). This method uses the concept of reference objects. Unlike other MCDA methods such as TOPSIS and VIKOR, which creates reference objects based on decision matrix, SPOTIS requires defining the data boundaries. Using data borders to define Ideal Positive and Ideal Negative Solution allows for a linear distribution of alternatives between IPR and INR and avoids ranking reversals.

To apply this method, data boundaries should be defined. For each criterion C_j the maximum S_j^{max} and minimum S_j^{min} bounds should be selected. Ideal Positive Solution S_j^* is defined as $S_j^* = S_j^{max}$ for profit criterion and as $S_j^* = S_j^{min}$ for cost criterion. Decision matrix is defined as $X = (x_{ij})_{m \times n}$, where x_{ij} is attribute value of the i-th alternative for j-th criterion.

Step 1. Calculation of the normalized distances to Ideal Positive Solution (17).

$$d_{ij}(A_i, S_j^*) = \frac{|S_{ij} - S_j^*|}{|S_j^{max} - S_j^{min}|} \tag{17}$$

Step 2. Calculation of weighted normalized distances $d(A_i, S^*) \in [0, 1]$, according to (18).

$$d(A_i, S^*) = \sum_{j=1}^{N} w_j d_{ij}(A_i, S_j^*) \tag{18}$$

Final ranking should be determined based on $d(A_i, S^*)$ values. Better alternatives have smaller values of $d(A_i, S^*)$.

This method has an alternative algorithm which is described in [4]. We describe and use this version because it is easier to understand, and both versions give identical results.

3 The Proposed Approach

An interval number is a set of real numbers with the property that any number that lies between two numbers included in the set is also included in the set. The interval of numbers between a^L and b^R, including a^L and b^R, is denoted $[a^L, b^R]$.

The two numbers a^L and b^R are called the endpoints of the interval. Interval numbers are used when an attribute has an indefinite or uncertain value. This entails analysing all the values from a given interval. Only one of the interval's values is the unknown real value.

The SPOTIS method is designed to solve problems with crisp numbers. The values of the decision attributes will be converted to interval numbers, which will be noted as (19):

$$a_j = [\alpha_j^L, \alpha_j^R] \tag{19}$$

where j means number of criterion. Each real number can be written as a degenerate interval numbers, i.e., $\alpha_j^L = \alpha_j^R$. Let suppose there is no attribute value for an individual alternative in a given set of alternatives. In that case, the smallest and the biggest value in the criterion should be taken respectively. Each alternative will be written as an interval data set (20).

$$A = \{[\alpha_1^L, \alpha_1^R], [\alpha_2^L, \alpha_2^R], ..., [\alpha_n^L, \alpha_n^R]\} \tag{20}$$

Note here that when evaluating an alternative with at least one attribute given in terms of an interval number that is not degenerate, the assessment result will always be returned as an interval number in the proposed approach. Also, for monotonic decision criteria, the lowest and the highest evaluation value will always be on the interval boundaries. Therefore, to calculate the resulting evaluation interval, it suffices to determine the set of alternatives A', which will arise as the Cartesian product of all interval boundaries of the form (21):

$$A' = \{\{\alpha_1^L, \alpha_1^R\} \times \{\alpha_2^L, \alpha_2^R\} \times ... \times \{\alpha_n^L, \alpha_n^R\}\} \tag{21}$$

The set of alternatives A' contains exactly 2^n crisp alternatives which must be calculated by using SPOTIS algorithm. The final ranking's left boundary is the lowest preference value determined from the set A', and the right boundary is the highest value.

4   Study Cases

In order to demonstrate the proposed approach, we have chosen two MCDA problems with interval data from recent studies, which are presenting in the Subsect. 4.1 and 4.2. Both topics deal with the assessment of electric vehicles, which is motivated by phasing out diesel and petrol engines in Europe. Many of these vehicles' parameters are of an interval nature, demonstrating the superiority of the proposed approach.

4.1   Assessment of Electric Bikes

The first problem is the choice of the best electric bicycles for city transport. There is currently an increasing tendency to look for more sustainable transport solutions, especially in highly congested urban areas. It seems that electric bicycles can be a good option, as they allow more benefits than combustion cars.

Because of missing data in the manufacturer's specifications, we should apply interval MCDA methods to handle incompleteness in the data.

The alternatives used for this study case are adopted from [18]. Criteria used for the analysis are presented in the Table 1, where we can also find characteristic values, which are needed to be defined to use COMET and SPOTIS methods. Characteristic values are also needed to create intervals instead of missed data.

Table 1. The selected criteria C_1–C_8 and their characteristic values [18].

C_i	Name	Unit	Low	Medium	High
C_1	Battery capacity	Ah	4	9	15
C_2	Charging time	hours	3	5	8
C_3	Number of gears	units	1	7	21
C_4	Engine power	W	250	350	500
C_5	Maximum speed	km/h	20	27	35
C_6	Range	km	20	60	100
C_7	Weight	kg	10	20	25
C_8	Price	USD	300	2500	6300

Table 2 presents chosen alternatives from the original study. Alternatives A_1–A_8 contains several interval attribute values and alternatives A_9–A_{13} contains only real values. This selection of alternatives shows how the proposed approach works when only part of the alternatives have interval data.

The structured COMET approach was used in the original paper to solve the considered MCDA problem. In this study, we also use COMET, but with the monolithic approach [22]. This is because we assume that the structure of the problem is unknown to the decision-maker. We have used stochastic optimization methods to obtain preference values for CO from preference values from the alternatives [10].

Table 3 presents raw preference values calculated for each alternative by using three MCDA methods. For SPOTIS method, smaller values means better alternative, for other methods bigger values means better alternative. However, preference data alone is not sufficient for determining the rankings, as these intervals overlap to some extent, which poses a problem in unambiguously assessing the final ranking.

We apply the approach described in [9] to obtain ranking values from interval data. Rankings from Table 3 are calculated in two steps: build comparison matrix for interval values using $P(a \geq b)$ and then rank sums for each row of this matrix. This ensures obtained the most likely ranking.

Table 2. The performance table of the alternatives A_1–A_{13}.

A_i	Name	C_1	C_2	C_3	C_4	C_5	C_6	C_7	C_8
A_1	Aceshin	8	[4, 6]	21	250	30	40	22.2	730
A_2	ANCHEER Plus	8	5	21	250	25	[25, 50]	23	615
A_3	Carrera Crossfuze	11	[6, 7]	9	400	25	80	20.3	2300
A_4	ECOTRIC	12	[5, 8]	7	500	32	55	24.9	999
A_5	Emu Crossbar	14.5	[6, 8]	7	250	25	[55, 100]	23	1560
A_6	Kemanner	8	[4, 6]	21	250	25	[35, 70]	20	[615, 700]
A_7	Merax 26" Aluminum	8.8	[5, 6]	7	350	32	[35, 45]	22	690
A_8	Rattan	10.4	[4, 5]	7	350	32	50	23.5	740
A_9	Desiknio Pinion Classic	7	3	6	250	24.8	80	15.7	6135
A_{10}	e-Joe Gadis	11	5	7	350	32	72	24.9	1699
A_{11}	California Bicycle S	8	4	1	250	32	56	22.6	2499
A_{12}	Coboc ONE Soho	9.6	3	1	250	24.8	88	13.1	5520
A_{13}	Gazelle CityZen C8 HM	11	3.5	8	350	32	94	23.1	2999

Table 3. Considered alternatives and their results A_1–A_{13}.

A_i	Preference				Ranking			
	Ref	SPOTIS	COMET	TOPSIS	Ref	SPOTIS	COMET	TOPSIS
A_1	[0.4414 0.4804]	[0.4756 0.5256]	[0.4439 0.5030]	[0.4727 0.5050]	9	5	9	6
A_2	[0.3752 0.4693]	[0.5309 0.5700]	[0.4015 0.4380]	[0.4207 0.4421]	10	9	10	12
A_3	[0.4802 0.4918]	[0.4875 0.5125]	[0.4853 0.5007]	[0.4565 0.4765]	8	7	8	7
A_4	[0.5308 0.5686]	[0.4056 0.4806]	[0.5615 0.5959]	[0.5228 0.5759]	5	2	5	2
A_5	[0.4219 0.6119]	[0.5111 0.6314]	[0.4284 0.5954]	[0.3970 0.4782]	6	11	6	10
A_6	[0.4116 0.5959]	[0.4496 0.5561]	[0.4356 0.5619]	[0.4288 0.4982]	7	6	7	8
A_7	[0.5264 0.5778]	[0.5020 0.5426]	[0.5301 0.6368]	[0.4738 0.5036]	4	8	4	5
A_8	[0.5800 0.6056]	[0.4646 0.4896]	[0.5991 0.6580]	[0.5168 0.5344]	2	3	2	4
A_9	0.3945	0.5950	0.3603	0.4071	12	12	12	13
A_{10}	0.5555	0.4800	0.6262	0.5248	3	4	1	3
A_{11}	0.3669	0.5991	0.3736	0.4344	13	13	11	11
A_{12}	0.4016	0.5497	0.2609	0.4499	11	10	13	9
A_{13}	0.6204	0.4140	0.6046	0.5768	1	1	3	1

The rankings obtained by the methods used are quite different. Alternative A_{13} has the first position in the reference ranking and the ranking obtained with interval SPOTIS and TOPSIS method. The COMET method placed this alternative in the third position in the ranking. Alternative A_{10} has first position in monolithic COMET ranking, but has lower values in other rankings. Alternative A_8 has the second position in the reference ranking, where only COMET return the same position. It should be noted that both SPOTIS and TOPSIS have different ranked this alternative, but in the case of TOPSIS, the position is more distant. In order to comprehensively assess the similarity of the obtained rankings, r_w and WS values will be determined [19].

Table 4 presents r_w ranking correlation coefficient values. These values point out that ranking obtained monolithic COMET method is strongly correlated with reference ranking. Other two rankings, obtained with interval SPOTIS and interval TOPSIS methods have a high correlation between themselves, but the quite good correlation with reference ranking.

Table 4. Comparison of rankings using r_w coefficient.

r_w	Ref	SPOTIS	COMET	TOPSIS
Ref	1.0000	0.7975	0.9560	0.8615
SPOTIS	0.7975	1.0000	0.7316	0.9317
COMET	0.9560	0.7316	1.0000	0.8144
TOPSIS	0.8615	0.9317	0.8144	1.0000

Table 5 shows WS ranking similarity coefficient values. According to calculated values, ranking obtained with interval SPOTIS method is strongly correlated with the reference ranking. Ranking obtained using monolithic COMET approach also has a quite strong correlation with reference ranking. The last ranking obtained with interval TOPSIS method also has a good correlation with the reference.

Table 5. Comparison of rankings using WS coefficient.

WS	Ref	SPOTIS	COMET	TOPSIS
Ref	1.0000	0.9111	0.8915	0.9240
SPOTIS	0.8904	1.0000	0.7929	0.9696
COMET	0.8915	0.7735	1.0000	0.8157
TOPSIS	0.9044	0.9650	0.7957	1.0000

4.2 Assessment of Electric Vans

An ecological footprint in the urban environment is made by urban freight transport. This problem has become the key challenge for all groups involved in freight transport in urban areas. Therefore electric vans should be considered as an alternative for combustion vehicles. The second study case is on assessing electric vans and the data with reference ranking for this investigation is taken from [26]. In the original study, the authors use PROMETHEE II and Fuzzy TOPSIS methods to rank electric vans for city logistic. In this work, we would use chosen alternatives from the original paper in order to demonstrate how efficient the proposed methods are.

Table 6 presents criteria description and characteristic values calculated from complete decision matrix. Characteristic values are necessary to determine when using COMET and SPOTIS methods. Criteria C_6, C_7 and C_9 are cost type criteria and should be minimized. Other criteria are profit type.

Table 6. The selected criteria C_1–C_9 and their characteristic values.

C_i	Name	Unit	Low	Medium	High
C_1	Carrying capacity	kg	340.00	1770.00	3200.00
C_2	Max velocity	km/h	40.00	95.00	150.00
C_3	Travel range	km	100.00	250.00	400.0
C_4	Engine power	kW	9.00	104.50	200.00
C_5	Engine torque	Nm	80.00	490.00	900.00
C_6	Battery charging time 100%	h	2.00	7.00	12.00
C_7	Battery charging time 80%	min	10.00	95.00	180.00
C_8	Battery capacity	kWh	2.70	61.35	120.00
C_9	Price	thous. USD	12.90	81.45	150.00

Table 7 presents alternatives chosen from original work with criteria attributes values. Alternatives A_5, A_6 and A_9 have only real number attributes, and the other alternatives have missed data which are substituted with intervals based on characteristic values from Table 6.

Table 7. Considered alternatives and their results A_1–A_{10}.

A_i	Name	C_1	C_2	C_3	C_4	C_5	C_6	C_7	C_8	C_9
A_1	Berlingo Electric	695	110	170	49	200	7.5	30	22.5	[12.9, 150.0]
A_2	Boulder Delivery Truck	2700	104	160	80	[80.0, 900.0]	8	[10.0, 180.0]	80	100.0
A_3	Ecomile	935	80	120	28	[80.0, 900.0]	8	[10.0, 180.0]	14.4	51.5
A_4	Electric Delivery Van 1000	830	40	118	14	98	8	120	2.7	[12.9, 150.0]
A_5	EVI MD	3000	96	145	200	610	10	120	99	120.0
A_6	e-NV200+	705	120	170	80	270	4	30	24	25.0
A_7	Kangoo Maxi Z.E	650	130	170	44	226	8	[10.0, 180.0]	22	22.0
A_8	Mercedes-Benz Sprinter E-CELL	1200	80	135	100	220	2	[10.0, 180.0]	35.2	[12.9, 150.0]
A_9	Minicab-MiEV Truck	350	100	110	30	196	4.5	15	10.5	12.9
A_{10}	Peugeot eBipper	350	100	100	30	[80.0, 900.0]	3	[10.0, 180.0]	20	60.0

In the Table 8 calculated preference values are presented. Reference ranking is a ranking obtained with Fuzzy TOPSIS from the original work [26]. Next columns show preferences obtained with the proposed interval SPOTIS method, monolithic COMET, and interval TOPSIS method.

For obtained the rankings we use the same methodology as for Sect. 4.1, and rankings from Table 8 are calculated in two steps: build comparison matrix for interval values using $P(a \geq b)$ and then rank sums for each row of this matrix. In this problem, obtained rankings are quite similar, according to the Table 8. Alternative A_5 has the first position in all four rankings. Alternative A_6 has the third position in the reference ranking and the second position in other rankings. Rank positions for alternative A_2 are also similar: second position in the reference ranking and third position in other rankings.

Table 9 contains r_w ranking correlation coefficient values. Despite very similar first positions in rankings, correlations between reference ranking and other rankings are quite low. Other rankings have quite strong correlations between themselves.

Table 8. Vans preferences and rankings

A_i	Preference			Ranking			
	SPOTIS	COMET	TOPSIS	Ref	SPOTIS	COMET	TOPSIS
A_1	[0.5721 0.6833]	[0.2390 0.4164]	[0.4370 0.5103]	7	8	8	5
A_2	[0.3997 0.6220]	[0.3144 0.7213]	[0.4791 0.6431]	2	3	3	3
A_3	[0.5604 0.7827]	[0.1028 0.4328]	[0.2875 0.4709]	4	9	9	9
A_4	[0.7742 0.8853]	[0.0235 0.0974]	[0.1667 0.3137]	5	10	10	10
A_5	0.4635	0.5802	0.5737	1	1	1	1
A_6	0.5036	0.5476	0.5668	3	2	2	2
A_7	[0.5534 0.6645]	[0.2616 0.4571]	[0.4549 0.5227]	6	6	6	4
A_8	[0.4772 0.6994]	[0.2100 0.6026]	[0.3960 0.5588]	8	4	5	6
A_9	0.5977	0.3794	0.4497	10	5	4	7
A_{10}	[0.5152 0.7375]	[0.1589 0.5359]	[0.3509 0.4976]	9	7	7	8

Table 9. Comparison of rankings using r_w coefficient for vans.

r_w	Ref	SPOTIS	COMET	TOPSIS
Ref	1.0000	0.5592	0.5405	0.6672
SPOTIS	0.5592	1.0000	0.9857	0.8766
COMET	0.5405	0.9857	1.0000	0.8645
TOPSIS	0.6672	0.8766	0.8645	1.0000

Finally, Table 10 shows WS ranking similarity coefficient values calculated for obtained rankings. This values point that correlation is quite strong, because first positions in the ranking is more important when WS similarity coefficient is calculated. As we can see, the results obtained show that the interval SPOTIS method provides solutions comparable to other methods, while being very simple to apply.

Table 10. Comparison of rankings using WS coefficient for vans.

WS	Ref	SPOTIS	COMET	TOPSIS
Ref	1.0000	0.8630	0.8634	0.8570
SPOTIS	0.8731	1.0000	0.9833	0.9574
COMET	0.8647	0.9833	1.0000	0.9533
TOPSIS	0.9051	0.9510	0.9528	1.0000

5   Conclusions

In this paper, we present a way to extend SPOTIS method to work with interval numbers which allow handling uncertainty and incompleteness in decision problems. We also compare the proposed approach with two other interval methods, COMET and TOPSIS to show how it performs in real-life decision problems.

The main contribution is providing an extension of the SPOTIS method, which is performed comparably with other interval methods. The main advantage of the SPOTIS method is its simplicity. The SPOTIS method consists of two simple steps, and the only additional requirement is defining criteria bounds. The study cases confirm it performs as good as COMET and TOPSIS methods, but it is much simpler to use than COMET and is also rank-reversal free, unlike TOPSIS. In order to compare the performance of these methods, the priority degree approach was used to build rankings. Then, the rankings were compared using rank correlation coefficients.

The future works may include

- developing the more complex approach which would be possible to apply to any criteria types without limitations,
- research of possibility using other number generalizations instead of interval numbers,
- comparing the proposed approach with other MCDA methods.

Acknowledgments. The work was supported by the National Science Centre, Decision number UMO-2016/23/N/HS4/01931.

References

1. Aires, R.F.F., Ferreira, L.: The rank reversal problem in multi-criteria decision making: a literature review. Pesqui. Oper. **38**(2), 331–362 (2018)
2. Behzadian, M., Otaghsara, S.K., Yazdani, M., Ignatius, J.: A state-of the-art survey of TOPSIS applications. Exp. Syst. Appl. **39**(17), 13051–13069 (2012)
3. Ceballos, B., Pelta, D.A., Lamata, M.T.: Rank reversal and the VIKOR method: an empirical evaluation. Int. J. Inf. Technol. Decis. Making **17**(02), 513–525 (2018)
4. Dezert, J., Tchamova, A., Han, D., Tacnet, J.M.: The SPOTIS rank reversal free method for multi-criteria decision-making support. In: 2020 IEEE 23rd International Conference on Information Fusion (FUSION), pp. 1–8. IEEE (2020)

5. Dubois, D., Prade, H.: Fuzzy numbers: an overview. In: Readings in Fuzzy Sets for Intelligent Systems, pp. 112–148. Elsevier (1993)
6. Faizi, S., Rashid, T., Sałabun, W., Zafar, S., Wątróbski, J.: Decision making with uncertainty using hesitant fuzzy sets. Int. J. Fuzzy Syst. **20**(1), 93–103 (2018)
7. García-Cascales, M.S., Lamata, M.T.: On rank reversal and TOPSIS method. Math. Comput. Model. **56**(5–6), 123–132 (2012)
8. Figueira, J.É., Greco, S., Ehrogott, M.: Multiple Criteria Decision Analysis: State of the Art Surveys. ISORMS, vol. 78. Springer, New York (2005). https://doi.org/10.1007/b100605
9. Gu, Y., Zhang, S., Zhang, M.: Interval number comparison and decision making based on priority degree. In: Cao, B.-Y., Wang, P.-Z., Liu, Z.-L., Zhong, Y.-B. (eds.) International Conference on Oriental Thinking and Fuzzy Logic. AISC, vol. 443, pp. 197–205. Springer, Cham (2016). https://doi.org/10.1007/978-3-319-30874-6_19
10. Kizielewicz, B., Sałabun, W.: A new approach to identifying a multi-criteria decision model based on stochastic optimization techniques. Symmetry **12**(9), 1551 (2020)
11. Mareschal, B., De Smet, Y., Nemery, P.: Rank reversal in the PROMETHEE II method: some new results. In: 2008 IEEE International Conference on Industrial Engineering and Engineering Management, pp. 959–963. IEEE (2008)
12. Papathanasiou, J., Ploskas, N.: Multiple Criteria Decision Aid. SOIA, vol. 136. Springer, Cham (2018). https://doi.org/10.1007/978-3-319-91648-4
13. Sałabun, W.: The mean error estimation of TOPSIS method using a fuzzy reference models. J. Theor. Appl. Comput. Sci. **7**(3), 40–50 (2013)
14. Sałabun, W.: The characteristic objects method. a new distance-based approach to multicriteria decision-making problems. J. Multi-Criteria Decis. Anal. **22**(1–2), 37–50 (2015)
15. Sałabun, W., Karczmarczyk, A.: Using the COMET method in the sustainable city transport problem: an empirical study of the electric powered cars. Procedia Comput. Sci. **126**, 2248–2260 (2018)
16. Sałabun, W., Karczmarczyk, A., Wątróbski, J.: Decision-making using the hesitant fuzzy sets comet method: an empirical study of the electric city buses selection. In: 2018 IEEE Symposium Series on Computational Intelligence (SSCI), pp. 1485–1492. IEEE (2018)
17. Sałabun, W., Karczmarczyk, A., Wątróbski, J., Jankowski, J.: Handling data uncertainty in decision making with COMET. In: 2018 IEEE Symposium Series on Computational Intelligence (SSCI), pp. 1478–1484. IEEE (2018)
18. Sałabun, W., Palczewski, K., Wątróbski, J.: Multicriteria approach to sustainable transport evaluation under incomplete knowledge: electric bikes case study. Sustainability **11**(12), 3314 (2019)
19. Sałabun, W., Urbaniak, K.: A new coefficient of rankings similarity in decision-making problems. In: Krzhizhanovskaya, V.V., et al. (eds.) ICCS 2020. LNCS, vol. 12138, pp. 632–645. Springer, Cham (2020). https://doi.org/10.1007/978-3-030-50417-5_47
20. Sałabun, W., Wątróbski, J., Shekhovtsov, A.: Are MCDA methods benchmarkable? A comparative study of TOPSIS, VIKOR, COPRAS, and PROMETHEE II methods. Symmetry **12**(9), 1549 (2020)
21. Sengupta, A., Pal, T.K.: On comparing interval numbers. Eur. J. Oper. Res. **127**(1), 28–43 (2000)
22. Shekhovtsov, A., Kołodziejczyk, J., Sałabun, W.: Fuzzy model identification using monolithic and structured approaches in decision problems with partially incomplete data. Symmetry **12**(9), 1541 (2020)

23. Torra, V.: Hesitant fuzzy sets. Int. J. Intell. Syst. **25**(6), 529–539 (2010)
24. Utkin, L.V., Augustin, T.: Decision making under incomplete data using the imprecise Dirichlet model. Int. J. Approx. Reason. **44**(3), 322–338 (2007)
25. Wang, Y.M., Luo, Y.: On rank reversal in decision analysis. Math. Comput. Model. **49**(5–6), 1221–1229 (2009)
26. Wątróbski, J., Małecki, K., Kijewska, K., Iwan, S., Karczmarczyk, A., Thompson, R.G.: Multi-criteria analysis of electric vans for city logistics. Sustainability **9**(8), 1453 (2017)
27. Žižović, M., Pamučar, D., Albijanić, M., Chatterjee, P., Pribićević, I.: Eliminating rank reversal problem using a new multi-attribute model-the RAFSI method. Mathematics **8**(6), 1015 (2020)

Vector and Triangular Representations of Project Estimation Uncertainty: Effect of Gender on Usability

Dorota Kuchta⊙, Jerzy Grobelny⊙, Rafał Michalski(✉)⊙, and Jan Schneider⊙

Faculty of Computer Science and Management, Wroclaw University of Science and Technology,
27 Wybrzeże Wyspiańskiego st, 50-370 Wrocław, Poland
{dorota.kuchta,jerzy.grobelny,rafal.michalski,
jan.schneider}@pwr.edu.pl

Abstract. The paper proposes a new visualisation in the form of vectors of not-fully-known quantitative features. The proposal is put in the context of project defining and planning and the importance of visualisation for decision making. The new approach is empirically compared with the already known visualisation utilizing membership functions of triangular fuzzy numbers. The designed and conducted experiment was aimed at evaluating the usability of the new approach according to ISO 9241–11. Overall 76 subjects performed 72 experimental conditions designed to assess the effectiveness of uncertainty conveyance. Efficiency and satisfaction were examined by participants subjective assessment of appropriate statements. The experiment results show that the proposed visualisation may constitute a significant alternative to the known, triangle-based visualisation. The paper emphasizes potential advantages for the proposed representation for project management and in other areas.

Keywords: Fuzzy number visualisation · Fuzzy number vector representation · Visual processing · Project uncertainty · Usability

1 Introduction

Project estimating is a crucial element of project planning [1]. It involves providing quantitative estimates of various parameters of the project: e.g. cost and duration of individual activities or the necessary amount of resources needed. The problem is that in the stage of project planning those parameters are often not completely known. This is natural, as project is per definition a unique endeavour [2] and at least some of its elements are performed for the first time in the given circumstances. The uniqueness is especially acute for innovative or R&D projects.

The reason for the incomplete knowledge in the stage of project planning is either the lack of information (e.g. it will be known only in the future how many persons will be necessary to perform a task) or ambiguity of available information (in numerous cases the customer is unable to communicate clearly what they expect [3]). The "not knowing for sure, due to lack of information or ambiguous information" [4] is one of possible

© Springer Nature Switzerland AG 2021
M. Paszynski et al. (Eds.): ICCS 2021, LNCS 12747, pp. 473–485, 2021.
https://doi.org/10.1007/978-3-030-77980-1_36

definitions of uncertainty. This means that dealing with the problem of not being able to provide exact estimations in the stage of project planning is part of project uncertainty management [5].

Project plan plays a crucial role in project-related decision making [2]. It is on the basis of project plan that resources are assigned, or even such critical decisions taken as acceptance or rejection of projects. Uncertain information complicates the decision-making process. That is why it is important to search for ways of supporting decision makers in this uneasy but extremely important process.

One possibility of providing such support is visualisation [6]. It is used in various areas and supports decision makers in analysing the current situation and drawing conclusions. As one possible representation of uncertainty are fuzzy numbers (according to [7], fuzziness is a consequence of uncertainty), their graphical representation (the graphs representing their membership function) is a visualisation of uncertainty and should support decision makers in analysing projects in their planning stage.

It is true that fuzzy numbers [8], together with the graphical representations of their membership functions, are widely used in the literature to represent uncertainty in the planning stage of projects (e.g. [9]). However, we hypothesize that graphs representing the membership functions of fuzzy numbers may not always be the optimal way of visualising uncertainty and that the same information can be conveyed in an alternative way, more attractive and useful from the point of view of at least some decision makers.

The objective of this paper is thus to propose a new, vector-based method of visualizing uncertainty linked to project estimation, a method which would convey the same information as triangular fuzzy numbers but possibly in a more attractive and appealing form, and to compare the two uncertainty visualisations from the point of view of their potential users.

To achieve the assumed goal, empirical research was designed and carried out. It was focused on identifying the basic features of the new uncertainty representation proposition in comparison with the traditional form of triangular fuzzy numbers. The experimental study was conducted in the perspective of the usability definition proposed in the ISO 9241–11 standard [10]. According to this norm, the key usability assessment dimensions of any information conveying system (e.g. computer program interface) include effectiveness (to what extent the system meets user's needs), efficiency (what resources are necessary to meet those needs), and user satisfaction (related to using the system). In this paper, the effectiveness of conveying information about uncertainty was tested be means of an objective indicator. We measured the accuracy of identifying information presented by means of vectors and triangles in relation to their textual description. Efficiency and user satisfaction for both representation methods were identified subjectively by means of appropriate survey scales measuring participants' preferences. The questions about interpretation easiness of both graphical representations allowed for their assessment in terms of the efficiency dimension. Participants' subjective opinions on the attractiveness of vector and triangular visualisations refer to the satisfaction component of the uncertainty conveyance usability. The last aspect of the usability has been explored in recent decades in the area of human-computer interaction. As it was shown many years ago, an attractive message significantly influences the objective results in other usability dimensions (e.g., [11–13]).

The outline of the paper is as follows: In Sect. 2 a literature review on the present usage of visualisation in project uncertainty management is performed. In Sect. 3 the two representations of estimation uncertainty are shortly described: that using the graph of a triangular fuzzy number membership function and the new one, based on vectors. Sect. 4 includes all details about the experimental study, whose results are presented in Sect. 5. We finish the paper by discussing the results and summing up the findings.

2 Visualisation in Project Uncertainty Management: State of Art

In order to analyse the state of art with respect to the use of visualisation in project uncertainty management, we assumed first of all that project risk is a form of uncertainty [14]. The search string 'TITLE (project AND (visualisation OR visual)) AND (uncertainty OR risk)' was applied to scientific literature bases ScienceDirect and Scopus. The results can be summarised as follows:

The primordial role of uncertainty communication to project managers is underlined. At the same time it is stated that visualisation can play a significant role in conveying uncertainty information, and shown that appropriate visualisation methods can improve the communication process in project uncertainty management [15].

Visualisation is used to represent the following aspects of projects: probability distributions of estimated project parameters [15, 16], resource flexibility [17], interdependencies between projects [18], project initial data generally [19], project portfolio information [20], project constraints [21]. Other identified papers treat specific projects or problems related to project control, which is not the object of our considerations here.

Visualisation techniques used to represent project uncertainty are graphs, maps, tables, grids, boxplots, violin plots, strip charts, tree diagrams and stacked bar charts [15]. No vectors have been used in this context so far.

As mentioned above, fuzziness is widely used to model uncertainty of project estimates [9]. Here the only visualisation technique used are graphs representing the membership functions. This will be presented in the next section, along with a new visualisation proposal.

3 Membership Functions Versus Vectors – Two Uncertainty Visualisation Approaches

Let us suppose that a project parameter P is not known exactly in the stage of project planning. The only information which is given is that the most possible value of the parameter is $\hat{p}$, that the value of the parameter will be included in the interval $\left[\underline{p}, \overline{p}\right]$, $\hat{p} \in \left[\underline{p}, \overline{p}\right]$ and that the further a value lies from $\hat{p}$, the less its possibility degree is, and the changes are linear. Obviously, the possibility degree of numbers outside $\left[\underline{p}, \overline{p}\right]$ is 0. Let us mention in this place that possibility degree is not the same as probability. The discussion about the relationship of the two notions has been subject to a vast research [22], but here let us limit ourselves to the statement that possibility can be determined more subjectively and it expresses the subjective feeling of an (or a group of) expert(s) about the possibility of occurrence of the given crisp number in the role of the actual value of P.

Our knowledge about parameter P can be represented first of all as a graph of the membership function of the triangular fuzzy number determined by the parameters $\underline{p}, \hat{p}, \overline{p}$ [8]. This fuzzy number will be denoted as $\tilde{P} = \left(\underline{p}, \hat{p}, \overline{p}\right)$. Its membership function μ_P is defined on the set $\mathfrak{R}$ of real numbers and represents the possibility degrees of the respective real numbers. It is defined as follows:

$$\mu_P(x) = \begin{cases} 0 \text{ for } x \leq \underline{p} \text{ or } x \geq \overline{p} \\ \frac{x-\underline{p}}{\hat{p}-\underline{p}} \text{ for } x \in \left(\underline{p}, \hat{p}\right) \\ \frac{\overline{p}-x}{\overline{p}-\hat{p}} \text{ for } x \in \left[\hat{p}, \overline{p}\right) \end{cases} \tag{1}$$

Its representation can be seen in Fig. 1 (for P represented by numbers 2, 4, 5).

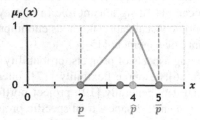

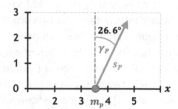

Fig. 1. Membership function-based visualisation of parameter P determined by numbers 2, 4, 5.

Fig. 2. Vector visualisation of parameter P determined by numbers 2, 4, 5.

The interpretation of Fig. 1 is as follows: value 4 is the most possible value of the unknown parameter P and the width of interval $[2, 5]$, thus 3, represents the indeterminacy degree linked to the estimation of P. The wider the support of the triangle, the less is known about the actual value of the parameter being estimated.

Here we propose an alternative representation of the same information about parameter P. The information will be represented as vector whose end will point to the most possible value and whose length will indicate the indeterminacy degree. The vector $\overrightarrow{P}\left(\underline{p}, \hat{p}, \overline{p}\right) = \{m_p, s_p, \gamma_P\}$ will be defined by:

- its beginning: point with coordinates $\left(m_p, 0\right)$, where $m_p = \left(\overline{p}+\underline{p}\right)\big/2$
- its length $s_P = \overline{p} - \underline{p}$
- the angle γ_P between the line $x = m_p$ defined as $\gamma_P = arctan(\hat{p} - m_p)$, where positive values of γ_P stand for the inclination to the right and negative – to the left.

The vector representation for the considered example of parameter P is given in Fig. 2. The interpretation of Fig. 2 is as follows: the inclination of the vector to the right from the line $x = m_p = 3, 5$ indicates the distance of the most possible value (here 4) from the mean value 3.5: the inclination is to the right because here the most possible value is higher than the middle value. Inclinations to the left correspond to most possible values lower than the middle values. The length of the vector (here 3) shows the indeterminacy degree linked to the estimated parameter.

We hypothesize that vectors may act as clock hands and be more appealing to some recipients of the information about the estimated parameter than triangles. The inclinations show changes of the most possible values with respect to the centre points of the possible range of the parameter: both the magnitude and the direction of those changes. The lengthening of the clock hand indicates that our knowledge about the parameter decreases, the shortening shows the opposite direction.

The two representations were compared in an experiment described in next sections.

4 Method

4.1 Subjects

Overall, 76 volunteer students of Wroclaw University of Science and Technology (Poland) took part in the experiments. There were 37 (48.7%) female and 39 (51.3%) male participants. Their age ranged from 21 to 29 years, with the mean of 22.5 and standard deviation equal 1.1 years.

4.2 Experimental Task and Measures

Factors and their levels were chosen so that it would be possible to verify the effectiveness of both triangular and vector representations. We included both clear-cut conditions where it was easy to check the correctness of answers, and a number of variants that served as noise. That is, an unambiguous answer was not available. The subjects were to assess whether the textual information about two features of the examined visualisations (the indeterminacy and the most possible value) match fuzzy number-based and vector-based graphical representations of unknown parameters being estimated.

Independent Variables. We examined two representations, that is, the vector and triangular ones that were described in detail in Sect. 3. The knowledge about the investigated not-fully-known parameters differed in two aspects: (1) the indeterminacy, which was examined on two levels (low and high), and (2) the most possible value (MPV) which varied on three levels (small, medium, big). The factors and their levels are graphically shown in Fig. 3.

Dependent Measures. Two types of dependent variables were employed. For determining the effectiveness of both visualisation types, we examined the number of perfectly correct responses and the quantity of entirely false selections. Subjects were presented with a following statement: "The description fully corresponds to the graphics". They were to assess on a five-point Likert scale (1 – "I do not agree", 2 – "I rather disagree", 3 – "Hard to tell", 4 – "I rather agree", 5 – "I agree") to what extent the description of uncertainty and the most possible value match the visual representation. A sample experimental task is demonstrated in Fig. 4.

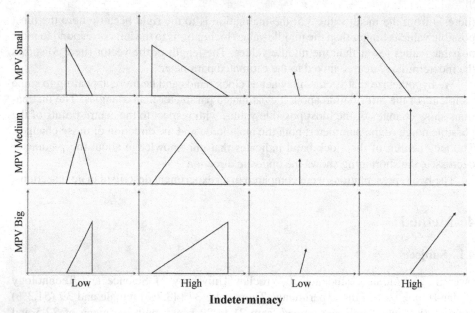

Fig. 3. Factors and their levels examined in the current study: visual representation (vectors, triangles), indeterminacy (low, high), the most possible value (MPV: small, medium, big).

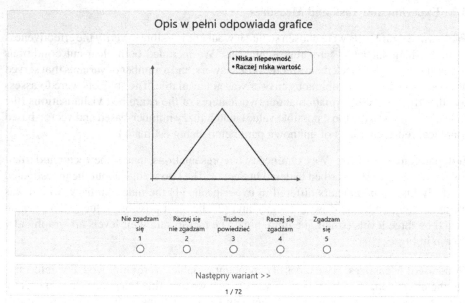

Fig. 4. Sample screen shot of an experimental task. Subjects assessed on a 5-point Likert scale to what degree the description of indeterminacy and the most possible value match the visual representation.

The participants' preferences towards the investigated stimuli were examined by asking them questions regarding both representations, once all effectiveness tasks were completed. Subjects were presented with four statements and instructed to specify their degree of agreement or disagreement on a seven-point Likert scale (1 – "I strongly disagree", 2 – "I disagree", 3 – "I rather disagree", 4 – "I do not have an opinion", 5 – "I rather agree", 6 – "I agree", 7 – "I strongly agree"). The questions were as follows: "Triangular/Vector representation was easy to interpret", and "Triangular/Vector representation was more attractive to me". One of the four questions is illustrated in Fig. 5.

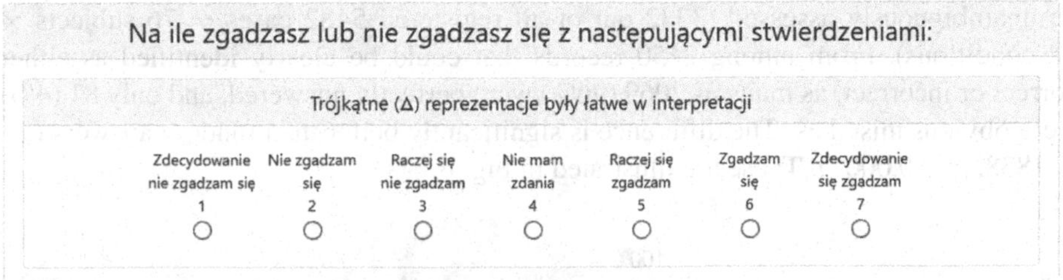

Fig. 5. Sample screen shot of one of the questions related to the subjects' preferences. Participants assessed on a 7-point Likert scale to what extent they agree or disagree with one of the statements on the easiness of interpretation and attractiveness of representations.

4.3 Experimental Design

A combination of independent variables' levels resulted in 72 experimental conditions. There were 6 not-fully-known parameters differing in the indeterminacy degree (2 levels) and the most possible value (3 levels). The information about all these parameters was prepared in two graphical versions, that is triangular and vector. Each of the 12 graphical variants could be displayed with corresponding 6 different descriptions varying in the same way as the investigated 6 parameters. A within-subject design was applied, which means that every participant examined all 72 experimental conditions.

4.4 Procedure

The experiments were conducted entirely online. Students received information about the possibility of participating in the study. They were provided with the hyperlink to a slideshow including voice recorded explanation of the experiment in the context of project estimation uncertainty. On the last slide, the subjects were asked to click the button that opened the React.js-based supporting software in the default local web browser. The experimental software was freely available on the Internet. Due to the web page structure, participants were asked to use devices having the screen larger than 10 inches in diagonal. In the application, subjects had to read and accept the informed consent for taking part in the examination and provide their basic data such as gender and age. Next, the main part of the study took place, that is, they performed the evaluation of all 72 conditions presented in a random order, followed by assessing four questions about their subjective opinions on the unknown parameter information visualisations (see Fig. 4 and 5). In the final thank-you page, they had an opportunity to input free-text comments before sending the data to the server.

5 Results

Generally, the obtained results prove that both graphical representations were to a large extent effective, despite a significant number of experimental conditions that could not be unambiguously assessed (3332 out of all registered 5482 cases = 76 subjects × 72 conditions). From among 2150 records that could be clearly identified as either correct or incorrect, as many as 2069 (96%) were perfectly answered, and only 81 (4%) were obvious mistakes. The difference is significantly better than random answers (χ^2 = 1838, $p < 0.0001$). These are illustrated in Fig. 6.

Fig. 6. Fully correct answers and clear-cut mistakes as percentages of unambiguous cases.

5.1 Effectiveness

Detailed results regarding the 2150 unambiguous records, in relation to the two types of unknown parameter representations, are presented in Fig. 7. They show that the number of the correct responses and errors were comparable both in the triangular and vector variants, with a tiny advantage in favour of the vector representation. To formally verify the differences, in this section we use typical Chi-Square (χ^2) test based in frequencies. Such an analysis showed that the influence of the graphical representation on the number of correct answers was not statistically meaningful ($\chi^2 = 0.22$, $p = 0.64$).

Since the sample was almost perfectly balanced in terms of gender, we examined if there were any discrepancies in this regard. In Fig. 8, one may notice that females more often provided correct answers than men. This effect was statistically significant ($\chi^2 = 13.2$, $p = 0.0003$). Correspondingly, there were fewer clear-cut mistakes registered for women than for male participants however, the difference was not statistically meaningful ($\chi^2 = 1.49$, $p = 0.22$). This interactive factor could be the reason of almost identical general effectiveness.

Due to the significant gender impact on the number of correct and incorrect responses, we checked whether this effect influenced the results of visualisation effectiveness. The

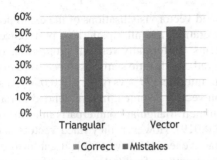
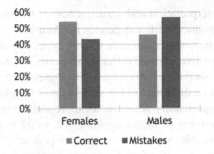

Fig. 7. Triangular and vector percentages of correct answers and mistakes.

Fig. 8. Females and males as percentages of correct answers and mistakes.

results of this analysis are illustrated in Fig. 9. The data show that there were decidedly more correct responses among women than men both for triangular ($\chi^2 = 6.55$, $p = 0.0105$) and vector representations ($\chi^2 = 6.60$, $p = 0.0102$), which is consistent with the results from Fig. 8. There were no differences between triangular and vector variants – both sexes performed equally (for females: $\chi^2 = 0.72$, $p = 0.79$, and males: $\chi^2 = 0.67$, $p = 0.8$).

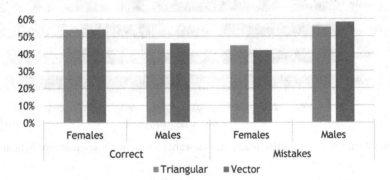

Fig. 9. Triangular and vector interaction with gender. Correct answers and errors.

As far as errors are concerned, women made more mistakes while assessing triangular than vector representations. Men, in turn, committed more errors for vector than triangular visualisations. However, since the general number of incorrect answers was small, none of these differences were statistically significant on the level of 0.1. Similarly, males made more obvious mistakes than females both for triangular and vector representations, but in both cases the differences were statistically irrelevant (for triangular: $\chi^2 = 0.42$, $p = 0.52$, and vector: $\chi^2 = 1.14$, $p = 0.29$).

5.2 Preferences

The outcomes of the analysis of the subjects' preferences expressed after performing the 72 experimental tasks are illustrated in Fig. 10. Participants rated the easiness of interpretation and attractiveness of triangular and vector visualisations of the examined parameters. If the gender is taken into consideration, a similar pattern of preferences emerges in responses to both types of statements. To formally verify if proportions of positive answers differ significantly, we employed classic ratio statistics. It occurred that male participants regarded triangles as easier to interpret (77% vs 65%, $p = 0.0509$) and more attractive (74% vs 57%, $p = 0.014$) than vectors. Women, in turn, tended to rate significantly better vector representations than their triangular counterparts in terms of easiness of interpretation (73% vs 57%, $p = 0.019$). However, they rated vectors and triangles equally from the perspective of attractiveness (65% vs 65%). It can also be observed that females were more decisive as the number of neutral responses from men was twice as big (6 vs 12).

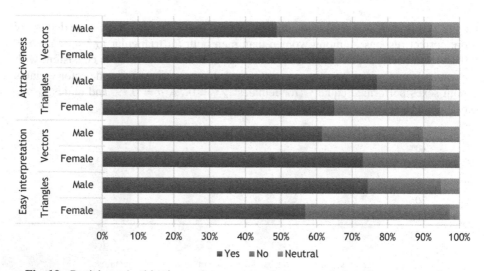

Fig. 10. Participants' subjective preferences towards vector and triangular representations.

6 Discussion and Conclusion

The paper presents a new concept of applying a vector representation to project parameters' uncertainty. The performed experimental studies allowed to assess this idea in confrontation with the classical approach based on triangular fuzzy numbers. The research was designed and conducted in view of the usability concept specified in three dimensions: effectiveness, efficiency, and satisfaction (ISO 9241–11). The level of the number of correct identifications of the examined graphic messages was adopted as the measure of effectiveness. Efficiency and satisfaction were estimated by examining the appropriately subjective participants' preferences towards the easiness of interpretation, and attractiveness of the investigated stimuli. As there was a comparable number of men and women in the sample, we analysed the differences between both genders.

The obtained results allow us to evaluate the proposed vector representation of uncertainty as a promising alternative to the classic triangular approach. First of all, both forms of representation are generally equally effective. They correctly convey the designed messages. The vector representation is significantly better in relation to errors made by women, but worse as far as for men's mistakes are concerned. The effectiveness of both graphic solutions is differentiated by gender as well. For women, the information conveyed by vectors is significantly easier to interpret than in the triangular variants, while for men the opposite is true. The subjects' satisfaction measured by the subjective assessment of attractiveness is slightly higher for vectors (but not statistically significantly) in the case of women and considerably higher for triangles in the group of men.

The comparison of men and women overall performance showed a significantly greater number of correct answers and a lower number of errors among women than among men. This outcome is interesting, especially in light of relatively scarce empirical studies on the gender heterogeneity in various aspects. In a review of studies in this area, Vanston and Strother [23] presented the results showing significant differences in visual information processing systems by women and men, both at the level of eye physiology and neural mechanisms. The results of performing numerous visual tests discussed in [24] suggest generally higher efficiency of men (especially in the group under 30) in various types of visual message processing. However, some experimental studies indicate that women, though on the average slower in performing visually guided tasks, are superior to males when the accuracy is taken into account (e.g., [25–27]). It seems that our findings support this hypothesis, all the more that subjects had unlimited time to perform the experimental tasks.

The new uncertainty representation proposed here, along with the traditional, triangle based one, may find a wide application in defining and planning projects. In those processes a countless set of parameters and project quantitative features have to be given, even though most of them cannot possibly be known exactly yet. Thus, the two representations can be used alternatively, according to each user preferences, to represent incomplete knowledge about duration, cost, the available and the needed number of resource units, the risk occurrence probability and consequences etc.

It is important that our proposal makes it possible to adapt the visualisation method to the user: instead of just one, traditional visualisation it will be possible to offer to project managers and project teams a choice. It is particularly important in the context of this study results, which show a significant gender influence on the graphical representations usability in conveying uncertainty.

The two visualisation methods can be of use in other than project management areas such as strategical and tactical management of organizations. Everywhere, where there is incomplete knowledge with respect to some important quantitative features, it might be useful to provide an appropriate form of visualisation to facilitate making decisions. Of course, both the vector and triangular forms can be used in any visual communication system that provides information of this type, e.g., multimedia presentations or printed materials.

While drawing conclusions from this article results, one should be aware that the research is preliminary in nature and exhibits a number of limitations. The sample size was moderate and included almost exclusively young students having similar characteristics. Thus, more experiments are needed to validate the two graphical representations and determine for which recipient groups which one should be selected. Conducting similar experiments in the environment of people professionally involved in project management seems particularly interesting. The two visualisations should be applied to real-world projects. In agilely managed projects, for instance, visualisation has already been being widely used [28]. Project teams are accustomed to analyse, discuss and make decisions on the basis of various types of graphics, therefore they might be open to testing new approaches in this regard. Triangles and vectors could visualize, e.g., task effort estimation for Scrum sprints. The application of triangle-based representation for this purpose has been already positively validated in practice [29]. A comparison with the vector-based representation should be the next step.

In this paper, we considered only static situations, which is acceptable in the context of project planning. During the project realization, however, the situation is dynamic and the most important goal is the identification of trends, in the project course. In our opinion, the visualisation akin to clock hands could be much more appealing in indicating the trends than triangles. But, of course, this is merely a hypothesis which need to be verified.

The results have shown that both representations of the incomplete knowledge about a parameter are generally accepted and understandable. Different groups of users may differ with respect to miscellaneous mistakes committed or personal preferences, but the results of the experiment show clearly that the new, vector-based representation is efficient in conveying the information and has a chance to be accepted and preferred by a large group of users.

Acknowledgement. The work of Jerzy Grobelny and Rafał Michalski was partially financially supported by the Polish National Science Centre grant no. 2019/35/B/HS4/02892: *Intelligent approaches for facility layout problems in management, production, and logistics.* The work of Dorota Kuchta and Jan Schneider was partially financially supported by the Polish National Science Centre grant no. 2017/27/B/HS4/01881: *Selected methods supporting project management, taking into consideration various stakeholder groups and using type-2 fuzzy numbers.*

References

1. Project Management Institute. Project Management Body of Knowledge (PMBOK® Guide). A Guid to Proj Manag Body Knowl (PMBOK® Guid - Fourth Ed (2008). https://doi.org/10.1007/s13398-014-0173-7.2

2. Kerzner, H.: A systems approach to planning scheduling and controlling. New York (2017). https://doi.org/10.1016/j.drudis.2010.11.015

3. Yen, J., Lee, J.: Fuzzy logic as a basis for specifying imprecise requirements (1993). https://doi.org/10.1109/FUZZY.1993.327535

4. Grote, G.: Promoting safety by increasing uncertainty—implications for risk management. Saf. Sci. **71**, 71–79 (2015)

5. Cleden, D.: Managing project uncertainty. Manag. Proj. Uncertain (2017). https://doi.org/10.4324/9781315249896

6. Burnay, C., Dargam, F., Zarate, P.: Special issue: data visualization for decision-making: an important issue. Oper. Res. Int. J. **19**(4), 853–855 (2019). https://doi.org/10.1007/s12351-019-00530-z

7. Hawer, S., Schönmann, A., Reinhart, G.: Guideline for the classification and modelling of uncertainty and fuzziness. Procedia CIRP **67**, 52–57 (2018)

8. Zimmermann, H.J.: Fuzzy set theory. Wiley Interdiscip. Rev. Comput. Stat. (2010). https://doi.org/10.1002/wics.82
9. Masmoudi, M., Haït, A.: Project scheduling under uncertainty using fuzzy modelling and solving techniques. Eng. Appl. Artif. Intell. **26**, 135–149 (2013)
10. Bevan, N., Carter, J., Harker, S.: Iso 9241-11 revised: what have we learnt about usability since 1998? In: Kurosu, M. (ed.) Human-Computer Interaction: Design and Evaluation. LNCS, vol. 9169, pp. 143–151. Springer, Cham (2015). https://doi.org/10.1007/978-3-319-20901-2_13
11. Tractinsky, N., Katz, A., Ikar, D.: What is beautiful is usable. Interact. Comput. **13**, 127–145 (2000)
12. Hassenzahl, M.: The interplay of beauty, goodness, and usability in interactive products. Hum.-Comput. Interact. **19**, 319–349 (2004)
13. Lavie, T., Tractinsky, N.: Assessing dimensions of perceived visual aesthetics of web sites. Int. J. Hum. Comput. Stud. **60**, 269–298 (2004)
14. Carvalho, M.: Risk and uncertainty in projects management: literature review and conceptual framework RECEBIDO. GEPROS Gestão da Produção, Operações e Sist **12**, 93–120 (2017)
15. Dikmen, I., Hartmann, T.: Seeing the risk picture: visualization of project risk information. EG-ICE 2020 Work. In: Intell. Comput. Eng. Proc. (2020)
16. Jaber, K., Sharif, B., Liu, C.: An empirical study on the effect of 3D visualization for project tasks and resources. J. Syst. Softw. **115**, 1–17 (2016)
17. Lima, R., Tereso, A., Faria, J.: Project management under uncertainty: resource flexibility visualization in the schedule. Procedia Compu. Sci. **164**, 381–388 (2019)
18. Killen, C.P.: Visualizations of project interdependencies for portfolio decision making: evaluation through decision experiments. In: Decision Science Institute Annual Conference (2013)
19. Kolychev, V.D., Rumyantsev, V.P.: Visual models' system of project management. Sci. Vis. **6**(3), 14–54 (2014)
20. da Silva, C.G., et al.: An improved visualization-based approach for project portfolio selection. Comput. Hum. Behav. **73**, 685–696 (2017)
21. van der Hoorn, B.: The project-space model: visualising the enablers and constraints for a given project. Int. J. Proj. Manag. **34**, 173–186 (2016)
22. Drakopoulos, J.A.: Probabilities, possibilities, and fuzzy sets. Fuzzy Sets Syst. **75**, 1–15 (1995)
23. Vanston, J.E., Strother, L.: Sex differences in the human visual system. J. Neurosci. Res. **95**, 617–625 (2017)
24. Shaqiri, A., et al.: Sex-related differences in vision are heterogeneous. Sci. Rep. **8**, 1–10 (2018)
25. Ives, J.C., Kroll, W.P., Bultman, L.L.: Rapid movement kinematic and electromyographic control characteristics in males and females. Res. Q. Exerc. Sport **64**, 274–283 (1993)
26. Rohr, L.E.: Gender-specific movement strategies using a computer-pointing task. J. Mot. Behav. **38**, 431–437 (2006)
27. Michalski, R.: The effects of panel location, target size, and gender on efficiency in simple direct manipulation tasks. Hum. Comput. Interact. (2008). https://doi.org/10.5772/6311
28. Wysocki, R.K., Kaikini, S., Sneed, R.: Effective project management : traditional, agile, extreme (2014)
29. Rola, P.M., Kuchta, D.: Application of fuzzy sets to the expert estimation of Scrum-based projects. Symmetry-Basel **11**, 1–22 (2019)

The Use of Type-2 Fuzzy Sets to Assess Delays in the Implementation of the Daily Operation Plan for the Operating Theatre

Barbara Gładysz[1]($\boxtimes$) ⓘ, Anna Skowrońska-Szmer[1] ⓘ, and Wojciech Nowak[2] ⓘ

[1] Faculty of Computer Science and Management, Wrocław University of Science and Technology, Wybrzeże Wyspiańskiego 27, 50-370 Wrocław, Poland
{barbara.gladysz,anna.skowronska-szmer}@pwr.edu.pl
[2] EURNET POLSKA LLC, Piotra Michalowskiego 15/2, 51-637 Wroclaw, Poland

Abstract. In the paper we present a critical time analysis of the project, in which there is a risk of delay in commencing project activities. We assume that activity times are type-2 fuzzy numbers. When experts estimate shapes of membership functions of times of activities, they take into account both situations when particular activities of the project start on time and situations when they start with a delay. We also suggest a method of a sensitivity analysis of these delays to meeting the project deadline. We present a case study in which the critical tome analysis was used to analyse processes implemented in the operating ward of a selected hospital in the South of Poland. Data for the empirical study was collected in the operating theatre of this hospital. This made it possible to identify non-procedural activities at the operating ward that have a significant impact on the duration of the entire operating process. In the hospital selected for testing implementation of the daily plan of surgeries was at risk every day. The research shows that the expected delay in performing the typical daily plan - two surgeries in one operating room – could be about 1 h. That may result in significant costs of overtime. Additionally, the consequence may also include extension of the queue of patients waiting for their surgeries. We show that elimination of occurrence of surgery activity delays allows for execution of the typical daily plan of surgeries within a working day in the studied hospital.

Keywords: Operating theatre · Project critical time · Scheduling · Type-2 fuzzy number

1 Introduction

The literature has numerous proposals as to the models of optimization of schedules on operating wards in hospitals. A review of the methods may be found in the work by Gür and Eren [1]. Here, we will present models in which authors used the fuzzy sets theory to define the schedules. Al-Refaie et al. [2] propose multiple-period fuzzy optimization models for scheduling and sequencing of patients in operating theatres. In their model, the surgery time is deterministic. The goals are fuzzy, the criterion functions

© Springer Nature Switzerland AG 2021
M. Paszynski et al. (Eds.): ICCS 2021, LNCS 12747, pp. 486–499, 2021.
https://doi.org/10.1007/978-3-030-77980-1_37

are: minimization of undertime and overtime and maximization of patients' satisfaction. Dexter et al. [3] proposes an algorithm to schedule add-on elective cases that maximizes operating room suite utilization. Surgery times are given as real numbers. In the model they use the fuzzy constrain theory. Nasiri et al. [4] consider three-criterion surgery scheduling problem. The criteria are maximizing the number of surgeries that can be done using given fixed resources, minimizing the total fixed costs and overtime costs and minimizing the cost of completion time. They assume stochastic times of surgeries and fuzzy constraints for resources and overtime. Rachuba and Werners [5] propose algorithms of determining robust schedules where they assume stochastic surgery times and randomly arriving emergency demand. For target functions they assume a degree of satisfaction of patients with a short time of waiting for their surgeries, minimization of rejected requests and minimization of overtime. Gül et al. [6] for the time analysis of flow in an emergency room apply the fuzzy CPM and fuzzy PERT. Times of activities are described here as type-1 fuzzy numbers. Nazif and Makis [7] solve the problem of determining times for commencing particular surgeries in the operating ward, taking into account availability of resources. The scheduling issue is modelled as a fuzzy flexible flow shop scheduling problem, assuming that times of activities are type-1 fuzzy variables. To determine the schedules they use simulation and heuristics methods (Ant colony optimization algorithms). Lahijanian [8] proposes a mixed-integer programming model for scheduling operating theatres. Surgery times are given as triangular type-1 fuzzy numbers. The target function is to minimize the total weighted time. Wang and Xu [9] use the hybrid intelligent algorithm to determine the schedule of surgeries with limited resources. Times of surgeries are given as type-1 fuzzy variables. As the criterion function they assumed minimization of overtime and undertime costs. Behmanesh, Zandieh and Hadji Molana [10] as target functions adopt minimization of makespain, Behmanesh and Zandieh [11] used minimization of makespain and minimization of the number of unserved patients. Times of surgeries are type-1 fuzzy numbers. The optimal solution is determined with the Fuzzy Pareto Envelope-based Selection Ant System algorithm.

In this work we present a method of time analysis of a project, in which there is a risk of delay in starting project activities. We also suggest a method of analysing sensitivity of these delays to meeting the project deadline. We assume that activity times are given as type-2 fuzzy numbers. When experts estimate shapes of membership functions of times of activities, they take into account both situations when particular activities of the project start on time, and situations when they start with a delay. Therefore, this method of estimating fuzzy times of particular phases of a surgery indirectly takes into account availability and readiness of human resources. We are also presenting a case study in which we implement the critical time analysis method to the time analysis of the process of surgery in an operating ward of a hospital in Poland.

The paper is organized as follows. Next Section presents main terms from the fuzzy theory. In Sect. 3 we present a method of time analysis of a project, in which there is a risk of delay in starting project activities. We assume that activity times are type-2 fuzzy numbers. We also suggest a method of sensitivity analysis of these delays to meeting the project deadline. In Sect. 4 we present a case study in which our method was used to analyse processes implemented in the operating ward of a hospital in Poland. Section 5 and Sect. 6 are Discussion and Conclusions.

2 Basic Notions

In this section we are presenting a basic notion from the fuzzy sets theory.

An interval number $\overline{A}$ is a closed interval $\overline{A} = [\underline{a}, \overline{a}] = \{x \in \mathfrak{R} : \underline{a} \leq x \leq \overline{a}\}$. Values $\underline{a} = -\infty$ and $\overline{a} = +\infty$ are allowed. The interval number $\overline{A} = [\underline{a}, \overline{a}]$, means unknown realization x which may take values from the interval $[\underline{a}, \overline{a}]$.

Let $\overline{A} = [\underline{a}, \overline{a}]$ and $\overline{B} = [\underline{b}, \overline{b}]$ be two interval numbers. The sum of the two interval numbers $\overline{A} = [\underline{a}, \overline{a}]$ and $\overline{B} = [\underline{b}, \overline{b}]$ is the interval number of the form $\overline{A} + \overline{B} = [\underline{a} + \underline{b}, \overline{a} + \overline{b}]$. Maximum of two interval numbers $\overline{A} = [\underline{a}, \overline{a}]$ and $\overline{B} = [\underline{b}, \overline{b}]$ is the interval number of the form $max\{\overline{A}, \overline{B}\} = [max\{\underline{a}, \underline{b}\}, max\{\overline{a}, \overline{b}\}]$ [12]. The degree to which the number $\overline{A}$ is greater than the number $\overline{B}$ is defined as follows:

$$degree(\overline{A} \geq \overline{B}) = \left|\left\{x : x \in \overline{A} \text{ and } \bigwedge y \epsilon \overline{B}, x \geq y\right\}\right|/|\overline{A}| \tag{1}$$

In 1965 Zadeh proposed his concept of possibility theory [13] We will present the basic notions of this theory. First, we will present the concept of a fuzzy number (type-1 fuzzy number). Let $\tilde{X}$ be a single valued variable whose value is not precisely known. The membership for $\tilde{X}$ is a normal, quasi concave and upper semi continuous function $\mu_X : \mathcal{R} \to [0, 1]$, see [14, 15]. The value $\mu_X(x)$ for $x \in \mathcal{R}$ denotes the possibility of the event that the fuzzy variable $\tilde{X}$ takes the value of x. We denote this as follows $\mu(x) = Pos(\tilde{X} = x)$. For a given fuzzy number $\tilde{X}$ and a given λ, the λ-level is defined to be the closed interval $[\tilde{X}]_\lambda = \{x : \mu(x) \geq \lambda\} = [\underline{x}(\lambda), \overline{x}(\lambda)]$.

An interval type-1 fuzzy number $\tilde{X}$ is called an $L-R$ fuzzy number if its membership function takes the form of [14]:

$$\mu_X(x) = \begin{cases} L\left(\frac{m-x}{\alpha}\right) & for \quad x < \underline{m} \\ \mu_m & for \quad \underline{m} \leq x \leq \overline{m} \\ R\left(\frac{x-\overline{m}}{\beta}\right) & for \quad x > \overline{m} \end{cases} \tag{2}$$

where: $L(x), R(x)$ - continuous non-increasing functions x; $\alpha, \beta > 0$.

Functions $L(x). R(x)$ are called shape functions of a fuzzy number. The most commonly used shape functions are: $max\{0, 1 - x^p\}$ and $exp(-x^p), x \in [0, +\infty), p \geq 1$. An interval-valued fuzzy number for which $L(x) = R(x) = max\{0, 1 - x^p\}$ and $p = 1$ is called a trapezoid fuzzy number, which we denote as $(\underline{x}, \underline{m}, \overline{m}, \overline{x})$. A trapezoid fuzzy number for which $\underline{m} = \overline{m} = m$ is called a triangular fuzzy number. A type-2 fuzzy set (T2FS) $\tilde{A} \in \mathcal{F}_2(X)$ is an ordered pair $\tilde{A} = \{(x, u), J_x, f_x(u)/x \in X; u \in J_x \subseteq [0, 1]$, where $\tilde{A}$ represents uncertainty around the word A, J_x the *primary membership* function of x, u is the domain of uncertainty, and $\mathcal{F}_2(X)$ is a class of type-2 fuzzy sets [16, 17]:

$$\tilde{A} : X \to [0, 1]$$

$$\tilde{A} = \int_{x \in X} \int_{u \in J_x} \frac{1}{(x, u)}, J_{x \in [0,1]} \tag{3}$$

An interval type-2 fuzzy number is a simplification of a T2FS. Its secondary membership function is assumed to be 1:

$$\tilde{A} = \int_{x \in X} \int_{u \in J_x} 1/(x, u) = \int_{x \in X} \left[\int_{u \in J_x} 1/u \right]/x, \tag{4}$$

where x, u are primary and secondary variables, and $f_x(u)/u = 1$ is the secondary membership function.

The footprint of uncertainty of the interval type-2 fuzzy number $\tilde{A}$ is bound by two functions: an upper membership function and lower membership function. For trapezoid type-2 fuzzy number we will use the following notion $\tilde{X} = \left(\left(\underline{x}^U, \underline{m}^U, \overline{m}^U, \overline{x}^U; \mu_m^U \right), \left(\underline{x}^L, \underline{m}^L, \overline{m}^L, \overline{x}^L; \mu_m^L \right) \right)$, where both upper and lower memberships functions have the shapes given by formula (2). When upper and lower membership functions are the same (equal to each other) type-2 fuzzy number is type-1 fuzzy number.

In arithmetic of fuzzy numbers we will apply Zadeh's extension principle [13] $\mu_Z(z) = \sup\limits_{z=f(x_1,\dots,x_n)} min\{\mu_{X_1}(x_1), \dots, \mu_{X_n}(x_n)\}$ extended to type-2 fuzzy numbers [18]:

$$\mu_Z(z) = \left(\left(\sup\limits_{z=f(x_1,\dots,x_n)} min\{\mu_{X_1}^U(x_1), \dots, \mu_{X_n}^U(x_n)\} \right), \left(\sup\limits_{z=f(x_1,\dots,x_n)} min\{\mu_{X_1}^L(x_1), \dots, \mu_{X_n}^L(x_n)\} \right) \right) \tag{5}$$

The interval possibility that the realization of type-2 fuzzy number $\tilde{X}$ will be greater or equal to the realization of type-2 fuzzy number $\tilde{Y}$ is equal [19]:

$$\overline{Pos\left(\tilde{X} \geq \tilde{Y} \right)} = \left[sup_{x \geq y}\left(min\left(\mu_X^L(x), \mu_Y^L(y) \right) \right), sup_{x \geq y}\left(min\left(\mu_X^U(x), \mu_Y^U(y) \right) \right) \right] \tag{6}$$

Those index is an extension of index of relations of majority of those proposed by Dubois and Prade [14].

The interval expected value of type-2 fuzzy variable is [19]:

$$\overline{E(\tilde{X})} = \left[min\left\{ \int_0^{\mu_m^L} \frac{1}{2}\left(\underline{x}^L(\lambda) + \overline{x}^L(\lambda) \right)d\lambda, \int_0^{\mu_m^U} \frac{1}{2}\left(\underline{x}^U(\lambda) + \overline{x}^U(\lambda) \right)d\lambda \right\}, max\left\{ \begin{array}{l} \int_0^{\mu_m^L} \frac{1}{2}\left(\underline{x}^L(\lambda) + \overline{x}^L(\lambda) \right)d\lambda, \\ \int_0^{\mu_m^U} \frac{1}{2}\left(\underline{x}^U(\lambda) + \overline{x}^U(\lambda) \right)d\lambda \end{array} \right\} \right] \tag{7}$$

3 Method of Time Analysis of the Project

In this section we present a method of time analysis of a project, in which there is a risk of delay in starting project activities. We also suggest a method of the sensitivity analysis of these delays to meeting the project deadline.

Let a project be represented as an acyclic network $G(N, \mathcal{A}, \tilde{T})$, where $N = \{1, \dots, n\}$ is the set of nodes (events), $\mathcal{A} \subset N \times N$ is the set of arcs (activities), and $\tilde{T} : \mathcal{A} \rightarrow \mathcal{F}^+$ – a function representing the type-2 fuzzy durations of these activities. For each activities $(i, j) \in \mathcal{A}$, the experts determine the optimistic duration t_{ij}^{opt}, the most possible duration t_{ij} and the pessimistic one t_{ij}^{pes} for stable (normal, most typical, scenarios) conditions (circumstances) for the realization of a project. The experts also judge which activities could start later because of poor organizational reason and determine the optimistic duration τ_{ij}^{opt}, the most possible duration τ_{ij} and the pessimistic one τ_{ij}^{pes} for such conditions and give the possibility μ_{ij} of occurrence of the most possible lateness τ_{ij}. Based on that data we estimate the type-2 fuzzy membership function of time of realization of those activities. Upper membership function represents a possibility that time of realization of activities will be realized in standard conditions, namely in conditions which may entail a lateness in commencement of activity realization (i, j). Lower membership function defines a possibility of activity realization (i, j) in optimal conditions, namely in conditions where an activity commences on time.

Times of realization of activities for which there is no risk that activities will not be commenced on time we estimate with trapezoid type-2 fuzzy numbers in a form of:

$$\tilde{T}_{ij} = \left(\left(t_{ij}^{opt}, t_{ij}, t_{ij} t_{ij}^{pes}; 1 \right), \left(\tau_{ij}^{opt}, \tau_{ij}, \tau_{ij}, \tau_{ij}^{pes}; \mu_{ij} \right) \right) \tag{8}$$

Let assume that we have due date d for our project. We will now present a critical time analysis method where we will apply arithmetic based on the Zadeh's extension principle (4) and modified interval index of majority relation (5)[1].

Algorithm 1

Step 1. Number nodes of network (events of project) $i \in N$ ascendingly starting with the initial node: $i = 1, 2, \ldots, n$.

Step 2. Set $\tilde{T}_1 = 0$.

Step 3. For $i = 2, \ldots, n$:

Find the earliest time of the event i implying Zadeh's extension principle (5):

$$\tilde{T}_i = \max_{(j,i) \in P_i} \tilde{T}_j + \tilde{T}_{ji}$$

where P_i – the set of predecessors of i.

Step 4. Find the critical time $\tilde{T}^{crit} = \tilde{T}_n$.

Step 5. Find using equations (6) interval possibility that due date d is greater or equal than critical time $\overline{Pos(d \geq \tilde{T}^{crit})}$.

Step 6. Using the Zadeh's extension principle (5) and equation (7), find project lateness $\tilde{L} = \tilde{T}_{crit} - d$ and the interval expected value of the project lateness $\overline{E(\tilde{L})}$.

---

[1] In order to find the sum and the maximum of interval fuzzy numbers one may apply the rule of dividing a part [0, 1] of the value of membership functions into intervals and to find approximate value of the maximum, applying interval arithmetic for those intervals.

We will now propose the method of analysing the impact of particular delays of activities on the total project time.

Sensitivity Analysis

For a given activity $(i, j) \in \mathcal{A}$, for which there is a risk of lateness of its commencement, assume that this lateness will not occur, so the time of realization of that activity is in a form of:

$$\tilde{T}_{ij} = \left(\left(t_{ij}^{opt}, m_{ij}, m_{ij}, t_{ij}^{pes}; 1 \right), \left(t_{ij}^{opt}, m_{ij}, m_{ij} t_{ij}^{pes}; 1 \right) \right) \tag{9}$$

Step 1. Find the critical time $\tilde{T}_{crit}$ of the project using Algorithm 1.

Step 2. Find using Eq. (5) interval possibility that due date d is greater or equal than critical time $Pos_{Int}\left(d \geq \tilde{T}_{crit} \right)$.

Step 3. Using the Zadeh's extension principle (5) and Eq. (7), find project lateness

$\tilde{L} = \tilde{T}_{crit} - d$ and the interval expected value of project lateness $E\left(\tilde{L} \right)$.

It should be emphasized that in the arithmetic applied in the algorithm based on the Zadeh's extension principle the maximum value (lower, upper) of the membership function of the critical time is equal to the minimum value of membership functions of particular times of the project activities. Instead of the Zadeh's arithmetic, one may apply another arithmetic for type-2 fuzzy numbers, e.g. arithmetic proposed by Hu et al. [20]. Then, for projects composed of numerous activities the maximum value (lower, upper) of the membership function of the project critical time will equal approximately 1.

4   Case Study

In this section we will apply Algorithm 1 to analyse the time of a surgery in an operating ward. The research was conducted in a large hospital in the south of Poland. There, the operating theater is a separate department and consists of 10 operating theaters. In the subject unit of a selected hospital, the main problem regarded untimely realization of the operating schedule. As shown in the analysis, that was caused with occurring of latenesses in commencement of subsequent activities of the process occurring in the operating ward.

In the selected hospital work in the operating ward was planned one day ahead. The daily schedule of surgeries is established based on the queue of patients awaiting surgeries and expert knowledge of a person authorized to prepare the schedule. It contains the sequence of surgeries and numbers of operating theatres assigned to given surgeries. It does not define the expected duration of the surgery as it is difficult to determine. However, the person preparing the schedule consults experts (surgeon - operator) in order to determine estimated duration of a given surgery. This knowledge to a large extent facilitates planning of work in each operating theatre.

The operating ward is a specific organizational unit of a hospital whose structure is very complex and at the same time dependent on numerous aspects of work of the whole hospital. The analysed process of performing surgeries composes of many activities that may be divided into three groups: pre-surgery activities, the surgery and post-surgery activities. They take into account both the place of performing the surgery and the required medical personnel taking part in a given activity. Medical personnel creates a team. The team composes of various professional groups, among others those are nurses, anaesthetists or operators of various specializations, and supporting personnel such as e.g. cleaner's staff. For the process to run smoothly it is necessary that all people involved in the process cooperate closely. It is also necessary to coordinate efficiently and plan work of the whole team. Each lateness in realization of an activity results in extension of the duration of the whole process, and latenesses in realization of particular processes have consequences in a form of failure to realize the daily schedule of surgeries. Correct planning of that schedule has very large impact on problems with realization of the planned surgeries. It has to take into account the estimated duration of each surgery. That duration is defined based on the type of surgery based on expert knowledge of the ward head and operators.

The subject hospital operating theatres in the operating ward it function from Monday to Friday from 7:25 am to 3 pm. Additionally, shall that be the case, one operating theatre is available also on Sundays and bank holidays, as well as on weekdays afternoon and night (from 3 pm to 7:25 am). All that accounts for theoretical availability of the operating ward calculated to 303 h and 20 min within one week. The procedures assume that in each operating theatre within one working day there should be three surgeries. If a planned surgery is not realized on the planned date and time, it has adverse consequences for the whole hospital. A person that should already be after the surgery returns to the original ward, is planned to another date which results in postponing of next surgeries and, in consequence, drastically extends the queue. It has adverse impact both on patients and on functioning of the hospital itself. The former still have health issues, their health is not improving, what is more – they lose their confidence in public health care. And for the hospital this situation generates additional costs related with return of the same patient to their wards, as well as reputational damage.

The person responsible for planning of the schedule must very precisely both select types of surgeries to a given theatre, and specify the estimated duration of each of them. Information on duration of each surgery is key so that the schedule of work in one theatre of an operating ward is prepared in the most reliable way. Improper selection of surgeries may lead to overtime of the operating ward personnel. For the hospital this is not a demanded situation due to economic reasons.

According to the initial analysis of the situation in the operating ward, duration of the process in a given operating theatre was the most significant factor influencing timeliness of work of the whole ward. Therefore, one needed to determine durations of particular activities composing the whole surgery process. That process was defined as one surgery with pre-surgery activities and post-surgery activities. In order to determine durations of particular activities of the process there were studies conducted in the operating ward including observation of a working day of the whole medical personnel.

There was a snapshot taken of the working day using an original form. It included all procedural activities that are performed in the process occurring in an operating ward. Description of all activities are presented in Table 1. Those are so called procedural

activities, namely those which are described in work procedures applicable in the operating ward of a given hospital. These activities are mandatory and the manner of their
realization is precisely defined.

Table 1. Network's structure and activities' times.

Activity	Predecessor	Description of activities	Duration [min]
A	-------	Taking over the patient to the operating ward by operating room nurses and taking him to the waiting room	((2, 3, 3, 27; 1), (2, 3, 3, 6; 0.93))
D	-------	Preparing the operating room by operating room nurses	((5, 7, 7, 18; 1), (5, 7, 7, 18; 1))
E	-------	Preparing instruments in the operating room by the operating room nurse	((6, 7, 7, 38; 1), (6, 7, 7, 38; 1))
B	A	Preparing the patient by the anaesthesiologist team	((9, 10, 10, 46; 1), (9, 10, 10, 25; 0.92))
C	A	Preparing the patient by the nurse in the preparatory room	((6, 7, 7, 19;1), (6, 7, 7, 19; 1))
F	B, C, D	Taking in the patient to the operating room (time between the patient's readiness in the waiting room and taking the patient to the operating room)	((2, 2, 2, 28; 1), (2, 2, 2, 7; 0.89))
G	E, F	Anaesthesia	((4, 5, 5, 24; 1), (4, 5, 5, 24; 1))
H	G	Performing the surgery (duration is counted from cutting the patient until suturing)	((54, 95, 95, 405; 1), (54, 95, 95,375; 0.88))
I	H	Filling in documentation by the anaesthesiologist	((5, 7, 7, 17; 1), (5, 7, 7, 17; 1))
J	G	Filling in documentation by the operating room nurse	((6, 8, 8, 12; 1), (6, 8, 8, 12; 1))
K	H	Postoperative activities with the patient performed in the operating room	((4, 5, 5, 44; 1), (4, 5, 5, 44; 1))
L	K	Taking the patient to the recovery room	((2, 2, 2, 17; 1), (2, 2, 2, 17; 1))
M	K	Postoperative activities performed by operating room nurses in the operating room	((5, 5, 5, 31; 1), (5, 5, 5, 31; 1))
N	H	Preparing instruments for sterilization by operating room nurses	((4, 5, 5, 26; 1), (4, 5, 5, 26; 1))
O	I, J, L, M, N	Cleaning of the operating room by the cleaning personnel	((6, 7, 7, 57; 1), (6, 7, 7, 35; 0.91))

During the research it appeared that duration of the whole process is significantly impacted by latenesses in commencement of some activities procedural. Additional activities have been identified. Those are not procedural activities, they do not contribute any added value for the process, only significantly extend its duration. They were classified, it was determined commencement of which procedural activities is disturbed that way (see Table 2). Table 2 includes also example reasons for latenesses. During the studies duration of procedural and non-procedural activities were registered.

Table 2. Lateness – "non-productive" observed in an operating room of an operating ward.

Activity	Description of the "waiting" time of activity	Employee performing the activity
A	Patient waiting to be taken in the waiting room reason: e.g. lack of patient's documents	Operating room nurses
B	Waiting for the first anaesthesiologist reason: occurs only before the first surgery as anaesthetists have the morning briefing at 7:30 am	The anaesthesiologist team
F	'Prepared' patient's waiting to be taken in the operating room (time between completion of anaesthetic preparation and taking the patient in the operating room), change of surgeries in the daily plan, related with e.g. change of instruments reason: e.g. unprepared room, unprepared instruments, no instructions	Operating room nurses
H	Waiting for the operator reason: e.g. the operator is in the ward, fills in documentation	Operator
O	Waiting for the cleaning service reason: e.g. cleaning rooms as they are located, not according to priorities	Cleaning service

It needs to be emphasized that it was possible to identify latenesses thanks to observation conducted by the researchers when taking the snapshot of a working day in an operating ward. If data on duration of surgeries was taken from the data bases serviced in that ward, it would not be possible to observe their occurrence. As emphasized by experts from the subject hospital, one should aim at minimizing durations of those latenesses but their total elimination is not possible, among other due to financial policy of the hospital and the state.

There was a network of activities prepared for a process occurring in one selected operating theatre. Its correctness was consulted with experts (operators, manager and director of the operating ward, and theatre nurses).

The subject of this case study is to check whether and how elimination of the reasons for latenesses will impact the total time of the process of surgeries performed during the day in the operating ward.

Based on the time analysis of 107 surgeries performed in the operating ward, fuzzy times were approximated $\tilde{T}_{ij} = \left(\left(t_{ij}^{opt}, t_{ij}, t_{ij}^{pes}; 1 \right), \left(\tau_{ij}^{opt}, \tau_{ij}, \tau_{ij}^{pes}; \mu_{ij} \right) \right)$ of particular activities, $(i, j) \in \mathcal{A} = \{A, B, C, \ldots, O\}$. For activities $(i, j) = A, B, F, H, O$ it was adopted that $t_{ij}^{opt}, t_{ij}, t_{ij}^{pes}$ are correspondingly: the shortest time, the most possible time and the longest time of realization of activity (i, j) together with the time of non-procedural activity corresponding to that activity. And times $\tau_{ij}^{opt}, \tau_{ij}, \tau_{ij}^{pes}$ are correspondingly: the shortest time, the most possible time and the longest time of realization of a procedural activity, whereas $\mu_{ij} = 1 - p_{ij}$, where is p_{ij} is the frequency of occurrence of latenesses in activities (i, j). For other activities $(i, j) = C, D, E, G, I, J, K, L, M, N$, for which there were no latenesses observed, it was assumed that $\tau_{ij}^{opt} = t_{ij}^{opt}$ tis the shortest, $\tau_{ij} = t_{ij}-$ is the most possible and $\tau_{ij}^{pes} = t_{ij}^{pes}$ – the longest time of realization of activity (i, j) and $\mu_{ij} = 1$.

Let us now perform a time analysis for a working day (from 7:25 am to 3 pm) in one operating theatre, assuming that there are two surgeries planned in that theatre. In the analysed hospital, in one operating theatre, one to three surgeries are performed during a working day. Most often, in about 50% of cases, these are two surgeries. Therefore, we will carry out our time analysis for a daily plan of 2 surgeries. We will determine in line with the Algorithm 1 the critical time of the process of 2 surgeries and a possibility that we will perform two surgeries by 3 pm (working time $d = 455$ min). Let us assume that all non-procedural activities may (but do not have to) occur. Then, the critical time for 2 surgeries is equal to $\tilde{T}_{crit} = ((172, 264, 264, 1324; 1)(172, 263.3, 387.7, 1094; 0.88))$ and the possibility that we will finish all surgeries by 3 pm ranges from 0.88 to 1 (see Table 3).

We will now perform a sensitivity analysis of the critical time of 2 surgeries with the assumption that we will eliminate particular latenesses, see Table 3. The possibility of performing 2 surgeries during a business day ranges from 0.88 to 1. The largest will be when we eliminate lateness of the activity *Performing the surgery* – according to Eq. (1) $degree([0.89, 1] \geq [0.88, 1]) = 0.09$. Therefore, the bottleneck here are non-procedural activities causing lateness of commencement of the procedural activity *Performing the surgery*, such as waiting for the operator caused by the fact that the operator is in the ward, fills in documentation, etc. When we eliminate any other one non-procedural activity, then the possibility of completing 2 surgeries by 3 pm ranges from 0.88 to 1. If we eliminate all non-procedural activities, i.e. all activities will commence on time, then the expected possibility of performing 2 surgeries during a working day will be equal to 1.

5 Discussion

The conducted analysis of critical time of two surgeries (a typical daily surgery plan) showed that with the current organization of work in the subject operating ward the expected delay goes from half an hour to around one hour (from 36 to 51 min) (see Table 3). In the case there is no delay in the activity *Performing the surgery* (activity H), the time reserve is the lowest and amounts to 36 min. Elimination of delay in commencing this activity may reduce the expected time of delay to around ½ h. Elimination of reasons for delay of each remaining activity of the surgery process could result in reduction of

Table. 3. Critical time and possibility of realization 2 surgeries on time, and expected lateness.

Type of lateness	Critical time	Possibility of finish on time	Expected lateness
Possible each lateness – current status	((172, 264, 264, 1324; 1), (172, 263.3, 287.7, 1094; 0.88))	[0.88, 1]	[−55, 51]
No lateness in commencement of activity A	((172, 264, 264, 1282; 1), (172, 263.1, 288.1, 1094; 0.88))	[0.88, 1]	[−55, 41]
No lateness in commencement of activity B	((172, 264, 264, 1282; 1), (172, 263.1, 290.0, 1094; 0.88)	[0.88, 1]	[−55, 41]
No lateness in commencement of activity F	((172, 264, 264, 1282; 1), (172, 263.2, 288.8, 1094; 0.88))	[0.88, 1]	[−55, 41]
No lateness in commencement of activity H	((172, 264, 264, 1264; 1), (172, 254.3, 346.5, 1094; 0.89))	[0.89, 1]	[−38, 36]
No lateness in commencement of activity O	((172, 264, 264, 1280; 1), (172, 263.1, 292.6, 1094; 0.88))	[0.88, 1]	[−54, 40]
No lateness	((172, 264, 264, 1094; 1), (172, 264, 264, 1094; 1))	1	− 7

the working day by around 10 min. It needs to be emphasized that as a result of the analysis it was proven that there is a possibility to perform two surgeries within less than 7 h and 35 min. Optimistic time reserve is equal to around 1 h. Such a situation will occur e.g. if the performed surgeries are not complicated surgeries, i.e. do not require large outlay of time or human resources or/and if particular activities of the surgery process start on time. If in the surgery process there are no non-procedural activities, then the possibility to perform 2 surgeries within one working day is equal to 1, and the expected time reserve equals 7 min.

In the considered process of surgery, in the selected hospital in Poland, the bottleneck is the delay in starting the activity Performing the surgery. The reason for this, namely not commencing the activity *Performing the surgery* on time, is among others lack of doctors in Poland. Currently, the Polish government prepared a draft bill on possibility to employ doctors from outside of the European Union on more advantageous conditions than at present. Other types of delays occurring in the operating theatre are also caused by lack of medical personnel, including anaesthetists and operating room nurses. The reasons for occurrence of these delays, and thus significant costs incurred by the hospital related to overtime worked by employees, also include mistakes in the organization of work in the operating theatre.

As a result of the lateness analysis described in the article, in the studied hospital several corrective activities were introduced, the main purpose of which is to minimize latenesses. Due to them, the daily surgery plan is not implemented. The most important improvements concerned work of the cleaning staff and of anaesthetists. For the first employee group, to reduce work delays it was enough to reorganize the work system. It was indicated in the procedures that rooms in which surgeries were to be performed on a given day should be cleaned first. Anaesthetists who assist with the first surgery on a given day have been released from the obligation to attend the morning briefing. Implementation of the recommendations improved implementation of the plan of surgeries, reducing the number of surgeries not performed by 16%. It should be noted that the introduced improvements did not require additional financial outlays, but only reorganization of work and improvement of existing procedures.

Let's compare the method presented in this article with the classical PERT method [21]. Let's assume as the activities time evaluation (i, j): the optimistic duration t_{ij}^{opt}, the most possible duration t_{ij} and the pessimistic one t_{ij}^{pes} for stable (normal, most typical, scenarios) conditions (circumstances) for the realization of the project. If the delay of the activity (i, j) is possible, we assume $t_{ij}^{pes} + \tau_{ij}^{pes}$ as the pessimistic time. Critical time distributions of two surgeries and their characteristics for these cases are given in Table 4. In PERT method, lateness time interval estimates are significantly wider than time interval estimates calculated according to the method used in this article. It is due to the fact that, among others, in PERT method, it is assumed that project's critical time has normal distribution with variance equal to the sum of variances of activities times on critical path. Whereas in the case of project time analysis presented in this article, critical time variance is lower than the sum of critical activities times, (see [22]). In an actual case, the decision maker should establish which of the methods (presented in this article, PERT) is more adequate to conditions of the surgery in the operating theatre.

Table 4. Critical time, probability of finish on time and confidence interval for lateness (0.95).

Type of lateness	Critical time	Probability of finish on time	Confidence interval for lateness
Possible each lateness	$N(398.6, 77.4)$	0.95	$(-274.4, 23.8)$
No lateness	$N(329.67, 76.1)$	0.77	$(-207.9, 95.3)$

Let's analyze the study case presented in Santibáñez et al. [23]. The authors analyze the time spent by the patient in British Columbia Cancer Agency's ambulatory care. The simulation technique is used to determine the total time spent by the patient in the clinic. They assume that patient wait times at the various stages of the process (waits for exam room, consults and discharge) have normal distributions. Therefore, total patient wait time in process for an ACU appointment has normal distribution with variance equal to the sum of variances of the wait times in particular points of the process in the ACU. Expected time of total patient wait time (min.) in Clinic of Medical Oncology equals 20.1, and a confidence interval (at confidence level 0.95) equals (19.7, 20.5), whereas, expected actual patient wait time equals 16.8, and confidence interval (at confidence level 0.95) equals (13.7, 20.0). Therefore, also in this case, interval estimate of patient wait time is significantly wider than the actual wait time. It is due to the fact that, among others, normal distribution of patient wait time at the various stages of the process was used.

6 Conclusions

As a result of our study we came up with a proposal of time analysis of a project implemented in conditions of risk associated with delayed start of activities. We assumed that activity times are type-2 fuzzy numbers. When experts estimate shapes of membership functions of times of activities, they take into account both situations when particular activities of the project start on time and situations when they start with a delay. Therefore, when estimating shapes of membership functions, experts should take into account availability of the personnel which is, among others, the condition for timely realization of particular stages of the project. There was also suggested a method of a sensitivity analysis of these delays to meeting the project completion deadline. At the end we presented a case study in which we analysed a typical daily plan of surgeries in the operating ward of a hospital in Poland. In this case, when estimating the shape of membership functions, experts should take into account availability of the personnel, the qualifications of the medical staff, as well as medical factors. As a result of the surgical process time analysis and the sensitivity analysis, we identified a certain type of delay that significantly increases the time of the surgery process. In this particular case, the largest delay is caused by waiting for the operator due to the fact that e.g. the operator is in the ward, fills in documentation. If in the surgery process there are no non-procedural activities (no delay in starting any procedural activity), then the expected time reserve is equal to 7 min.

Acknowledgment. This work was partially supported by Narodowe Centrum Nauki (Poland), under Grant 2017/27/B/HS4/01881: "Selected methods supporting project management, taking into consideration various stakeholder groups and using type-2 fuzzy numbers" and partially supported by the Polish Ministry of Science and Higher Education.

References

1. Gür, S., Eren, T.: Application of operational research techniques in operating room scheduling problems: Literature Overview. J. Healthcare Eng. **2018**(5341394), 15 (2018)
2. Al-Refaie, A., Judeh, M., Li, M.-H.: Optimal fuzzy scheduling and sequencing of multiple-period operating room. Artif. Intell. Eng. Des. Anal. Manuf. **32**(1), 108–121 (2018)
3. Dexter, F., Macario, A., Traub, R.D.: Which algorithm for scheduling add-on elective cases maximizes operating room utilization? Use of bin packing algorithms and fuzzy constraints in operating room management. Anesthesiolog **91**(5), 1491–2150 (1999)
4. Nasiri, M.M., Shakouhi, F., Jolai, F.: A fuzzy robust stochastic mathematical programming approach for multi-objective scheduling of the surgical cases. Opsearch **56**(3), 890–910 (2019). https://doi.org/10.1007/s12597-019-00379-y
5. Rachuba, S., Werners, B.: A Fuzzy multi-criteria approach for robust operating room schedules. Ann. Oper. Res. **251**(1–2), 325–351 (2017)

6. Gül, M., Güneri, A.F., Güneş, G.: Project management in healthcare: a case study for patient flow evaluation in an emergency room using fuzzy CPM and fuzzy PERT. Sigma J. Eng. Natiral Sci. **8**(1), 41–51 (2017)
7. Nazif, H., Makis, V.: Operating room surgery scheduling with fuzzy surgery durations using a metaheuristic approach. Adv. Oper. Res. **2018**(8637598), 8 (2018)
8. Lahijanian, B., Zarandi, M.F., Farahani, F.V.: Proposing a model for operating" room scheduling based on fuzzy surgical duration. In: Annual Conference of the North American Fuzzy Information Processing Society (NAFIPS), IEEE, El Paso, TX, USA (2016)
9. Wang, D., Xu, J.: A fuzzy multi-objective optimizing scheduling for operation room in hospital. In: IEEE International Conference on Industrial Engineering and Engineering Management, pp. 614–618 (2008)
10. Behmanesh, R., Zendieh, M., Hadji Molana, S.M.: The surgical case scheduling problem with fuzzy duration time: an ant system algorithm. Scientia Iranica E **26**(3), 1824–1841 (2018)
11. Behmanesh, R., Zendieh, M.: Surgical case scheduling problem with fuzzy surgery time: an advances bi-objective ant system approach. Knowl.-Based Syst. **186**, 104913 (2019)
12. Moore, R.E.: Interval Analysis. Prentice-Hall, Englewood Cliff (1966)
13. Zadeh, L.A.: Fuzzy sets. Inf. Control **8**, 338–353 (1965)
14. Dubois, D., Prade, H.: Operations on fuzzy numbers. Int. J. Syst. Sci. **9**(6), 613–626 (1978)
15. Zadeh, L.A.: Fuzzy sets as a basis of theory of possibility. Fuzzy Sets Syst. **1**, 3–28 (1978)
16. Mendel, J.: Uncertain Rule-Based Fuzzy Systems. Springer, Cham (2001). https://doi.org/10.1007/978-3-319-51370-6
17. Mendel, J., Wu, D.: Linguistic summarization using IF–THEN rules and interval type-2 fuzzy sets. IEEE Trans. Fuzzy Syst. **19**(1), 136–151 (2010)
18. Dinagar, D.S.; Anabalagan, A.A.: New type -2 fuzzy number arithmetic using extension principle. In: IEEE- International Conference on Advances in Engineering, Science and Management (ICAEMS-2012), pp. 30–31 (2012)
19. Gładysz, B.: Type-2 fuzzy numbers in models of the duration of a project affected by risk. J. Decis. Syst. p. 8 (2020)
20. Hu, J., Zhang, Y., Chen, X., Liu, Y.: Multi-criteria decision making method based on possibility degree of interval type-2 fuzzy number. Knowl.-Based Syst. **43**, 21–29 (2013)
21. Malcolm, D.G., Roseboom, J.H., Clark, C.E., Fazar, W.: Application of a technique for research and development program evaluation. Oper. Res. **7**, 646–669 (1959)
22. Carlsson, C., Fullér, R.: On possibilistic mean value and variance of fuzzy numbers. Fuzzy Sets Syst. **122**, 315–326 (2001)
23. Santibáñez, P.S., Chow, V.S., French, J., Puterman, M.L., Tyldesley, S.: Reducing patient wait times and improving resource utilization at British Columbia cancer agency's ambulatory care unit through simulation. Health Care Manag. Sci. **12**, 392–407 (2009)

Linguistic Summaries Using Interval-Valued Fuzzy Representation of Imprecise Information - An Innovative Tool for Detecting Outliers

Agnieszka Duraj$^{(\boxtimes)}$ (iD) and Piotr S. Szczepaniak (iD)

Institute of Information Technology, Łódź University of Technology,
ul. Wólczańska 215, 90-924 Lodz, Poland
`agnieszka.duraj@p.lodz.pl`

Abstract. The practice of textual and numerical information processing often involves the need to analyze and test a database for the presence of items that differ substantially from other records. Such items, referred to as outliers, can be successfully detected using linguistic summaries. In this paper, we extend this approach by the use of non-monotonic quantifiers and interval-valued fuzzy sets. The results obtained by this innovative method confirm its usefulness for outlier detection, which is of significant practical relevance for database analysis applications.

Keywords: Intelligent data analysis · Outlier detection · Linguistic summaries · Fuzzy handling of linguistic uncertainty · Interval-valued fuzzy sets · Non-monotonic quantifiers · Data mining · Knowledge discovery

1 Introduction

Fuzzy sets (also known as uncertain sets) and fuzzy logic have been jointly used to deal with the uncertainty related to human perception and classification [1,2]. This methodology enables the modeling of uncertainty features. It allows one to represent, in an integrated way, the uncertainty of qualitative terms based on quantitative aspects of various phenomena. Linguistic summarization based on fuzzy concepts has been found to be a useful method for the qualitative analysis of databases, including outlier detection [3].

The concept of an outlier, as related to data analysis, has been widely discussed in the literature, and consequently, many definitions of this term have been proposed, e.g. [4–6]. Information outliers represent information granules that are unexpected, occur rarely, or exhibit abnormal characteristics. Outlier detection is a crucial step in many data mining applications, as the presence of outliers in a data set may translate into serious errors. When using computational approaches, the representation of information hidden in the records of big databases must be extracted in a human-friendly form. In data mining

© Springer Nature Switzerland AG 2021
M. Paszynski et al. (Eds.): ICCS 2021, LNCS 12747, pp. 500–513, 2021.
https://doi.org/10.1007/978-3-030-77980-1_38

and knowledge discovery tasks, outliers are understood as a degree of deviation from a specified information pattern. The present paper examines databases that contain both numeric and textual records.

In general, there are two approaches to detecting outliers: they may be either eliminated at the stage of data preparation [7,8] or saved [9]. In the latter case, the unique objects are defined as small clusters that are dissimilar to the other data. The basis for the approach presented is Yager's [10–12] idea of linguistic summaries and the selected extensions developed by Kacprzyk and Zadrozny [13–17]. This paper provides a new insight into how this idea may be applied to the problem of outlier detection. The innovative approach presented involves the use of two concepts: non-monotonic quantifiers and interval-valued fuzzy sets, both applicable to outlier detection problems arising in practice. Specifically, we examine the usefulness of non-monotonic quantifiers for outlier detection methods based on linguistic summarization of database contents. The present study endeavors to show that non-monotonic interval-valued fuzzy sets can provide a more significant value of the degree of truth of a linguistic summary.

The content of this work is split into several sections. In Sect. 2, a brief survey of related literature is presented. Basic definitions of non-monotonic interval-valued fuzzy quantifiers are given in Sect. 3. In the next section, the concept of a linguistic summary and the way of its generation is explained. The practical rules for determining the degree of truth for monotonic and non-monotonic quantifiers are demonstrated in Sect. 4. Section 5 begins by introducing the concept of an outlier and presents the procedure of outlier detection using linguistic summaries. Section 6 demonstrates the practical application of the concept of outlier detection using linguistic summaries and two types of quantifiers, namely interval-valued monotonic and non-monotonic ones. Finally, our conclusions are drawn in Sect. 7.

2 Related Works

The past decade has seen the rapid development of outlier detection methods. Often designed for a particular type of data, these methods have been used in many varied applications, for example, medical research, where the outlier is defined as an anomaly or pathogen [18–20].

Outlier detection is also often used in public monitoring systems [21], climate change research [22], computer networks—for identifying hacker attacks [4,23], banking—for detecting fraud and fraudulent transactions on credit cards [24], or in manufacturing—for detection of defects [25]. The classic approach to outlier detection is based on distance or density. There are also other methods dedicated to specific types of data.

As observed by Kacprzyk and Zadrożny [17,26], there is an ongoing trend towards natural language-based knowledge discovery systems. Many researchers have demonstrated the use of linguistic summaries in decision-making processes [15–17,27–29]. In [30,31], linguistic summaries based on classic and interval-valued fuzzy sets were successfully applied to outlier detection tasks. As demonstrated in [32,33], outlier detection can also be performed using monotonic quantifiers.

Non-monotonicity is closely related to default conclusions. Many papers have presented the usefulness of this kind of formalism for developing natural language-based systems, e.g. [34–37]. The new aspect of this work lies in the use of non-monotonic interval-valued fuzzy quantifiers. We demonstrate that linguistic summaries supported by non-monotonic interval-valued fuzzy quantifiers may prove even more useful than previous outlier detection solutions.

3   Non-monotonic Quantifiers and Interval-Valued Fuzzy Implementation

The concept of a linguistic variable, introduced by Zadeh [38,39], enables the description of complex and ill-defined phenomena, which are difficult to specify using quantitative methods.

The concepts used in the natural language, such as *less than, almost half, about, hardly, few*, can be interpreted as mathematically fuzzy linguistic concepts determining the number of objects that meet a given criterion. Note that relative quantifiers are defined on the interval of real numbers $[0; 1]$. They describe the relationship of objects that meet the summary feature for all objects in the analyzed data set. Absolute quantifiers are defined on a set of non-negative real numbers. They describe the exact number of objects that meet the summary feature.

A linguistic quantifier being a determination of the cardinality is then a fuzzy set or a single value of the linguistic variable describing the cardinality of objects that meet specific characteristics.

In practical solutions, monotone quantifiers are defined as classic fuzzy sets or interval fuzzy sets. For example, the linguistic variable $Q = \{few\}$ can be a trapezoidal or triangular fuzzy membership function. It can also be designated as a function in the form of a fuzzy interval set.

Obviously, not all quantifiers of practical significance meet the condition of monotonicity [28]. The *few* and *very few* quantifiers are of particular importance in the context of detecting abnormal objects.

The quantifiers should be normal and convex. Normal—because the height of the fuzzy set representing the quantifiers is equal to 1. Convex—because for any $\lambda \in [0, 1]$, $\mu_Q(\lambda x_1 + (\lambda - 1)x_2) \geq \min(\mu_Q(x_1) + \mu_Q(x_2))$, where Q is a chosen, relevant quantifier, e.g. *fe"*, and $x_1, x_2 \in X$ are the objects considered.

We use the $L-R$ fuzzy number to model the quantifiers with the membership function, where $L, R : [0, 1] \longrightarrow [0, 1]$ non-decreasing shape functions and $L(0) = R(0) = 0$, $L(1) = R(1) = 1$. If the term *few* is a non-monotonic quantifier, it can be defined as a membership function in the form of (1).

$$
\mu_Q(r) = \begin{cases} L(\frac{r-a}{b-a}) & r \in [a, b] \\ 1 & r \in [b, c] \\ 0 & otherwise \\ R(\frac{d-r}{d-c}) & r \in [c, d] \end{cases} \tag{1}
$$

The function (1) can be written as a combination of functions L and R defined by Eqs. (2) and (3).

$$\mu_{Q_L}(r) = \begin{cases} 0 & r < a \\ L(\frac{r-a}{b-a}) & r \in [a,b] \\ 1 & r > b \end{cases} \tag{2}$$

$$\mu_{Q_R}(r) = \begin{cases} 0 & r < c \\ R(\frac{r-c}{d-c}) & r \in [c,d] \\ 1 & r > d \end{cases} \tag{3}$$

In interval-valued fuzzy sets, the degree of membership to the set is defined as an interval of real numbers $[0; 1]$. Thus, we obtain two membership functions: lower membership function $\underline{\mu}$, which determines the minimum degree of membership of an element, and upper membership function $\overline{\mu}$, which determines the maximum degree of the membership.

Having defined the linguistic variable Q as a non-monotonic interval-valued fuzzy set, we make similar changes for Eqs. (1) as (4) where $\mu_{\underline{Q}}(x)$ is calculated as (5) and $\mu_{\overline{Q}}(x)$ as (6).

$$\mu_Q(x) - [\mu_{\underline{Q}}(x), \mu_{\overline{Q}}(x)] \tag{4}$$

According to the definition of interval-valued fuzzy sets, we can define a non-monotonic quantifier Q (1) as interval-valued fuzzy set as (5) and (6).

$$\mu_{\underline{Q}}(r) = \begin{cases} L(\frac{r-\underline{a}}{\underline{b}-\underline{a}}) & r \in [\underline{a},\underline{b}] \\ 1 & r \in [\underline{b},\underline{c}] \\ 0 & otherwise \\ R(\frac{r-\underline{d}}{\underline{d}-\underline{c}}) & r \in [\underline{c},\underline{d}] \end{cases} \tag{5}$$

$$\mu_{\overline{Q}}(r) = \begin{cases} L(\frac{r-\overline{a}}{\overline{b}-\overline{a}}) & r \in [\overline{a},\overline{b}] \\ 1 & r \in [\overline{b},\overline{c}] \\ 0 & otherwise \\ R(\frac{\overline{d}-r}{\overline{d}-\overline{c}}) & r \in [\overline{c},\overline{d}] \end{cases} \tag{6}$$

Additionally, it is known that $\mu_Q(x)$ as Q non-monotonic quantifiers can be written as a combination of functions (2) and (3) as (7) and (8), where $\mu_{\underline{Q_L}}$, $\mu_{\overline{Q_L}}$, $\mu_{\underline{Q_R}}$ and $\mu_{\overline{Q_R}}$ are determined by (9), (10), (11) and (12).

$$\mu_{Q_L}(x) = [\mu_{\underline{Q_L}}(x), \mu_{\overline{Q_L}}(x)] \tag{7}$$

$$\mu_{Q_R}(x) = [\mu_{\underline{Q_R}}(x), \mu_{\overline{Q_R}}(x)] \tag{8}$$

$$\mu_{\underline{Q_L}}(r) = \begin{cases} 0 & r < \underline{a} \\ L(\frac{r-\underline{a}}{\underline{b}-\underline{a}}) & r \in [\underline{a},\underline{b}] \\ 1 & r > \underline{b} \end{cases} \tag{9}$$

$$\mu_{\overline{Q_L}}(r) = \begin{cases} 0 & r < \overline{a} \\ L(\frac{r-\overline{a}}{\overline{b}-\overline{a}}) & r \in [\overline{a}, \overline{b}] \\ 1 & r > \underline{b} \end{cases} \tag{10}$$

$$\mu_{\underline{Q_R}}(r) = \begin{cases} 0 & r < \underline{c} \\ R(\frac{r-\underline{c}}{\underline{d}-\underline{c}}) & r \in [\underline{c}, \underline{d}] \\ 1 & r > \underline{d} \end{cases} \tag{11}$$

$$\mu_{\overline{Q_R}}(r) = \begin{cases} 0 & r < \overline{c} \\ R(\frac{r-\overline{c}}{\overline{d}-\overline{c}}) & r \in [\overline{c}, \overline{d}] \\ 1 & r > \overline{d} \end{cases} \tag{12}$$

In the following sections, the above definitions of non-monotonic quantifiers are used.

4   Linguistic Summary

The collection of linguistic variables, referred to as linguistic quantifiers, is the expert knowledge used in linguistic summaries. This linguistic summary in a strictly structured form, expressed in a natural (or close to natural) language, is generated on the basis of the information contained in the information system and expert knowledge in the particular field. The definition of a linguistic summary is given by Def. 1.

Def. 1 *Yager's linguistic summary*
Yager's linguistic quantifier is in the form of ordered four elements
$< Q; P; S; T >$
where:
Q - a linguistic quantifier, or quantity in agreement, which is a fuzzy determination of amount. Quantifier Q determines how many records in the analyzed database fulfill the required condition - have the characteristic S;
P - the subject of summary; the actual objects stored in the database;
S - the summarizer, i.e., the feature by which the database is scanned;
T - the degree of truth; it determines the extent to which the result of the summary, expressed in a natural language is true.

According to the definition of linguistic summaries, we get the response in the natural language of the form:
Q objects being P are (have a feature) S [the degree of truth of this statement is [T], or in short:
Q P are/have the property S [T].

Generating natural language responses as Yager's summaries consists of creating all possible expressions for the predefined quantifiers and summarizers of the analyzed set of objects. The value of the degree of truth for each summary is determined according to formula $T = \mu_Q(r)$, where r is defined in (13).

The value r is determined for each attribute $a_i \in A$. We determine the membership function $\mu_A(a_i)$, thus defining how well attribute a_i matches characteristic S.

$$r = \frac{\sum_{i=1}^{n}(\mu_R(a_i) \cdot \mu_S(b_i))}{\sum_{i=1}^{n} \mu_R(a_i)} \qquad (13)$$

For linguistic summaries that apply interval-valued fuzzy sets, we obtain the value of the degree of truth in the form of an interval, not a number. It is possible to introduce interval-valued fuzzy sets for sets of linguistic variables Q or features R, S. We then get T in the form (14).

$$T = [\underline{T}, \overline{T}] = [\underline{\mu}_Q(r), \overline{\mu}_Q(r)] \qquad (14)$$

If features R or S are defined as an interval-valued fuzzy set, then $r = [\underline{r}, \overline{r}]$ is defined as follows (15).

$$[\underline{r}, \overline{r}] = [\frac{\sum_{i=1}^{n}(\underline{\mu}_R(a_i) \cdot \underline{\mu}_S(b_i))}{\sum_{i=1}^{n} \underline{\mu}_R(a_i)}, \frac{\sum_{i=1}^{n}(\overline{\mu}_R(a_i) \cdot \overline{\mu}_S(b_i))}{\sum_{i=1}^{n} \overline{\mu}_R(a_i)}] \qquad (15)$$

The application of non-monotonic linguistic quantifiers to linguistic summaries affects the manner of calculating the degree of truth. For linguistic summaries with the defined interval-valued non-monotonic quantifiers, the degree of truth is determined as (16).

$$T = [(\mu_{Q_L}(r) - \mu_{Q_R}(r)), (\mu_{\overline{Q_L}}(r) - \mu_{\overline{Q_R}}(r))] \qquad (16)$$

The membership function for the classifier is selected by the user, i.e. the expert. For non-monotonic quantifiers obtained in the linguistic summary, the value of the degree of truth could be higher. The summary becomes more reliable, which is important for detecting exceptions.

5   Detection of Outliers

An outlier is treated as a single element or a very small group of objects which, in comparison with other objects in the database, differ in the values of the analyzed feature. Let us define the concept of an outlier using a linguistic summary.

Def. 2 *Let:*
$X = \{x_1, x_2, \ldots, x_N\}$ *for $N \in \mathbb{N}$ be a finite, non-empty set of objects;*
S be a finite, non-empty set of attributes (features) of the set of objects X,
$S = \{s_1, s_2, \ldots, s_n\}$;
Q be non-monotonic interval-valued quantifiers defining the requested low cardinality.
A collection of objects (the subjects of a linguistic summary) are called
outliers if Q objects having the feature S is a true statement in the
sense of interval-valued fuzzy logic. If the linguistic summary of Q objects
in the P is/has S, $[\underline{T}, \overline{T}]$ has $[\underline{T}, \overline{T}] > 0$ (therefore, it is true in the sense of fuzzy
logic) which means that outliers have been found.

The procedure for detecting outliers using linguistic summaries according to Def. 2 begins with defining a set of linguistic values $Q = \{Q_1, Q_2, ..., Q_n\}$. For example, $Q_1 = very\ few$, $Q_2 = few$, $Q_3 = many$, $Q_4 = almost\ all$. The next step is to calculate the value of r according to the procedure for generating the linguistic summary described in Sect. 4.

As for detecting outliers using linguistic summaries, according to Def. 1, the most important elements are linguistic variables defining cardinality as *very few, few, little, almost none*, etc. If for the linguistic variable Q_i (e.g. $Q_1 = very\ few$, $Q_2 = few$) defined according to Def. 1, the value of a measure $[\underline{T}, \overline{T}] > 0$, then the resulting sentence is true in the sense of Zadeh's fuzzy logic, and thus, according to Def. 2, outliers have been detected.

In the practical applications [3, 30, 32, 40], the authors took into account a maximum of two variables characterizing the outliers. In such a case, four responses are possible.

The same assumptions should be made when the set of linguistic variables is determined by interval-values fuzzy sets. According to the definition of linguistic summary, in which the set of linguistic variables Q is defined as interval-valued fuzzy sets, we obtain the degree of truth for each generated sentence in the form of (14) for monotonic quantifiers and (16) for non-monotonic quantifiers. Outliers are found if the interval-valued fuzzy T contains values greater than 0. Then, four responses are possible, as shown in Table 1.

Table 1. Types of responses of the system based on interval-valued fuzzy sets, with two quantifiers Q_1 and Q_2: *very few* and *few*.

Response	Degree of truth	Result
$Q_1\ P$ is (has) $S\ [\underline{T}, \overline{T}]$	$[\underline{T}, \overline{T}] = [0, 0]$	Outliers were found
$Q_2\ P$ is (has) $S\ [\underline{T}, \overline{T}]$	$[\underline{T}, \overline{T}] > [0, 0]$	for the linguistic variable Q_2
$Q_1\ P$ is (has) $S\ [\underline{T}, \overline{T}]$	$[\underline{T}, \overline{T}] > [0, 0]$	Outliers were found
$Q_2\ P$ is (has) $S\ [\underline{T}, \overline{T}]$	$[\underline{T}, \overline{T}] = [0, 0]$	for the linguistic variable Q_1
$Q_1\ P$ is (has) $S\ [\underline{T}, \overline{T}]$	$[\underline{T}, \overline{T}] > [0, 0]$	Outliers were found
$Q_2\ P$ is (has) $S\ [\underline{T}, \overline{T}]$	$[\underline{T}, \overline{T}] > [0, 0]$	for the linguistic variables Q_1 and Q_2
$Q_1\ P$ is (has) $S\ [\underline{T}, \overline{T}]$	$[\underline{T}, \overline{T}] = [0, 0]$	Outliers were not found
$Q_2\ P$ is (has) $S\ [\underline{T}, \overline{T}]$	$[\underline{T}, \overline{T}] = [0, 0]$	

6 Practical Examples

The dataset used for the present analysis is composed of publicly available data from Statistics Poland [41]. It is a collection of 20 attributes on the basis of which we may reason about the financial liquidity of enterprises. The attributes are: company size, short-term liabilities, long-term liabilities, company assets, number of employees, financial liquidity ratio, bankruptcy risk, etc. The novelty of the present approach and its practical evaluation lies in the introduction of non-monotonic quantifiers based on interval-valued fuzzy sets.

The method of detecting outliers using linguistic summaries is demonstrated using two queries.

Query 1:

How many enterprises with a high current liquidity ratio are in the group with a high risk of bankruptcy?

Query 2:

How many enterprises with low profitability are in the high-risk group?

A bankruptcy risk score is a number that indicates whether a company or an individual has a high probability of becoming insolvent. There is no single, universally agreed-upon index of measurement [42, 43]. For example, the Altman Z-score [44] relies on five financial factors: profitability, leverage, liquidity, solvency and activity. In the banking sector, it is common practice to employ various bankruptcy risk scoring methodologies as a tool for assessing people's creditworthiness [45]. Below, we present a linguistically motivated approach to bankruptcy risk estimation. The exact definition of linguistic expressions such as low, medium, high etc. depends on the policy adopted by a given financial institution. A similar linguistic approach can be applied to the liquidity, profitability, leverage, solvency, and activity ratios used in the Altman Z-score. The analysis of bankruptcy risk involves two key steps: definition of uncertainty levels and estimation of the impact of uncertainty. One also needs a bridge between the qualitative and quantitative analysis. This bridge is provided by the fuzzy sets theory where the shapes of membership functions and their parameters are defined by the users or domain experts.

The linguistic variables describing the risk of bankruptcy are expressed as *low, medium* and *high* values. The current liquidity ratio of the company is expressed as *very low, low, medium*, or *high*. All values of bankruptcy risk *(low, medium, high)* are determined as trapezoidal membership functions:

$Tr[a, b, c, d]$;
$Tr_{low}[0, 0, 0, 2, 0.4]$;
$Tr_{medium}[0.3, 0.5, 0.7, 0.9]$;
$Tr_{high}[0.6, 0.8, 1, 1]$.

The membership functions of the current liquidity indicator are defined in a similar way.

The value of the coefficient r_{query1} is calculated as (17), where cls is a current liquidity indicator and $risk$ is the risk of bankruptcy for Query 1 and (18) for Query 2. The definition of monotonic quantifiers is not change for Query 2.

$$r_{query1} = \frac{\sum_{i=1}^{n}(\mu_{risk}(a_i) \cdot \mu_{cli}(b_i))}{\sum_{i=1}^{n} \mu_{risk}(a_i)} = 0.28 \tag{17}$$

$$r_{query2} = \frac{\sum_{i=1}^{n}(\mu_{risk}(a_i) \cdot \mu_{c}(b_i))}{\sum_{i=1}^{n} \mu_{risk}(a_i)} = 0.34 \tag{18}$$

6.1 Quantifier Q as a Monotonic Interval-Valued Fuzzy Set

Let us now consider a set of linguistic variables defined by interval-valued fuzzy sets. For the variable Q_1 there are two membership functions, namely $\underline{Q_1}$ and $\overline{Q_1}$, defined by the trapezoidal membership functions. Figure 1 shows a graphical representation of the membership function of a set of linguistic variables $Q_1 = very\ few$ and $Q_2 = few$ defined by interval-valued fuzzy sets.

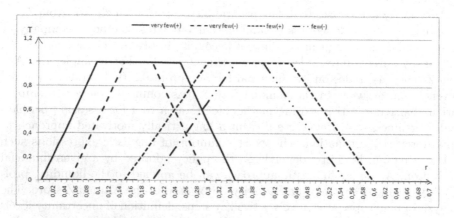

Fig. 1. Graphical presentation of the membership function of linguistic variables *very few* and *few* defined by interval-valued fuzzy sets.

The coefficient r is calculated according to Eqs. (17) and (18) for Query 1 and Query 2, respectively. The value of the degree of truth for Q defined as an interval-valued fuzzy set is calculated as (14).

The obtained linguistic summaries have the following values of measure T.

Query 1:

Very few enterprises with a high current liquidity ratio are in the group with a high risk of bankruptcy $T[0.2; 0.67]$.

Few enterprises with a high current liquidity ratio are in the group with a high risk of bankruptcy $T[0.53; 0.86]$.

Many enterprises with a high current liquidity ratio are in the group with a high risk of bankruptcy $T[0; 0]$.

Almost all enterprises with a high current liquidity ratio are in the group with a high risk of bankruptcy $T[0; 0]$

For the linguistic variable $Q_1 = very\ few$ and $Q_2 = few$, defined in accordance with Def. 2, the sentences are true in the sense of Zadeh's logic. Thus, the outliers have been found. A graphical interpretation of the determined degree of truth for the variable Q defined as an interval-valued fuzzy set is given in Fig. 2.

Query 2:

Very few enterprises with low profitability are in the high-risk group $T[0.0; 0.1]$.

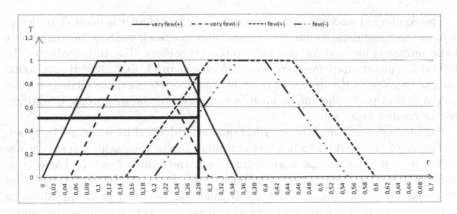

Fig. 2. Graphical presentation of the determined degree of truth for $Q_1 = very\ few$ and $Q_2 = few$.

Few enterprises with low profitability are in the high-risk group $T[0.93; 1]$.
Many enterprises with low profitability are in the high-risk group $T[0; 0]$.
Almost all enterprises with low profitability are in the high-risk group $T[0; 0]$.
Outliers were found for Query 2 for linguistic variables Q_1 and Q_2.

6.2   Quantifier Q as Non-monotonic Interval-Valued Fuzzy Set

For quantifiers Q defined as non-monotonic interval-valued fuzzy sets, the relevant membership functions are defined according to (9), (10), (11), (12). Consequently, the linguistic variable $Q_1 = very\ few$ is defined using four membership functions $\underline{\mu_Q}_{L1}$, $\underline{\mu_Q}_{R1}$, $\overline{\mu_Q}_{L1}$, $\overline{\mu_Q}_{R1}$ and $r = 0.28$. The degree of true T is determined using the Eq. (16).

The generated linguistic summaries are:

Query 1:

Very few enterprises with a high current liquidity ratio are in the group of a high risk of bankruptcy $T[0.2; 0.7]$.

Few enterprises with a high current liquidity ratio are in the group of a high risk of bankruptcy $T[0.53; 0.86]$.

Many enterprises with a high current liquidity ratio are in the group of a high risk of bankruptcy $T[0; 0]$.

Almost all enterprises with a high current liquidity ratio are in the group of a high risk of bankruptcy $T[0; 0]$

Query 2:

Very few enterprises with low profitability are in the high-risk group $T[0.0; 0.1]$.

Few enterprises with low profitability are in the high-risk group $T[0.95; 1]$.

Many enterprises with low profitability are in the high-risk group $T[0; 0]$.

Almost all enterprises with low profitability are in the high-risk group $T[0; 0]$.

The conducted research and experiments confirm that it is possible to detect outliers using linguistic summaries. In addition, the work verified the functioning of the proposed method for non-monotonic quantifiers. The functioning of the method for monotonic classifiers was also shown in [3,30,32,33]. It was found that the increase of the degree of truth for non-monotonic quantifiers as normal fuzzy sets is of particular importance as compared to the monotonic quantifiers used as normal fuzzy sets.

For the first query and the *very few* quantifier defined by a normal fuzzy set, the degree of truth equal to 0.2 was obtained. In the case where *very few* was defined as a non-monotonic normal fuzzy set, the result was 0.7. There is no doubt that outliers were detected in both cases. It should be emphasized that the increase in T for Q non-monotonic normal fuzzy set results in the validation of our research. We have received a greater degree of truth. For the second query, no outliers were detected for the quantifier $Q=very few$ defined as a normal fuzzy set. However, after using the non-monotonic quantifier, the value of truth was 0.1. Thus, the degree of truth increased.

The results of the degree of truth for the monotonic and non-monotonic quantifiers *very few* and *few* are given in Table 2. In the case of interval-valued fuzzy sets, the degree of truth obtained for monotonic and non-monotonic quantifiers was similar or the same. An increase in the degree of truth was observed for both Query 1 and Query 2.

Table 2. The results of the degree of truth for the monotonic and non-monotonic quantifiers *very few* and *few*.

Query 1	Monotonic	Non-monotonic
Very few (normal fuzzy set)	0.2	0.7
Few (normal fuzzy set)	0.86	0.86
Very few (interval-valued fuzzy set)	[0.2;0.67]	[0.2;0.7]
Few (interval-valued fuzzy set)	[0.53;0.86]	[0.55;0.86]
Query 2		
Very few (normal fuzzy set)	0.0	0.1
Few (normal fuzzy set)	1	1
Very few (interval-valued fuzzy set)	[0.0;0.1]	[0.0;0.1]
Few (interval-valued fuzzy set)	[0.93;1]	[0.95;1]

7   Conclusion

This paper has proposed a novel method for outlier detection. The solution presented provides the user with human-understandable natural language responses, in the form of fuzzy numerical values given as linguistic variables. The response generated by the system concerns a linguistic variable, which constitutes a linguistic specification of the records found, e.g., *about a half, not many, a lot, almost all,* etc.

As demonstrated by practical examples, the application of non-monotonic interval-fuzzy sets, which characterize the least numerous groups of objects (*very few, few*), thus corresponding to the definition of an outlier, improves the reliability of the results, as it leads to an increase in the degree of truth of a linguistic summary. This proves the usefulness of our method for outlier detection.

References

1. Shareef, D.M.A.M., Aminifar, S.A.: Uncertainty handling in big data using fuzzy logic-literature review (2021)
2. Ross, T.J., et al.: Fuzzy Logic with Engineering Applications, vol. 2. Wiley, Hoboken (2004)
3. Duraj, A., Szczepaniak, P.S.: Information outliers and their detection. In: Burgin, M., Hofkirchner, W. (eds.) Information Studies and the Quest for Transdisciplinarity, vol. 9, pp. 413–437, Chapter 15. World Scientific Publishing Company (2017)
4. Hawkins, D.M.: Identification of Outliers. Monographs on Statistics and Applied Probability, vol. 11. Springer, Heidelberg (1980). https://doi.org/10.1007/978-94-015-3994-4
5. Hawkins, S., He, H., Williams, G., Baxter, R.: Outlier detection using replicator neural networks. In: Kambayashi, Y., Winiwarter, W., Arikawa, M. (eds.) DaWaK 2002. LNCS, vol. 2454, pp. 170–180. Springer, Heidelberg (2002). https://doi.org/10.1007/3-540-46145-0_17
6. Barnett, V., Lewis, T.: Outliers in Statistical Data, vol. 3. Wiley, New York (1994)
7. Guevara, J., Canu, S., Hirata, R.: Support measure data description for group anomaly detection. In: ODDx3 Workshop on Outlier Definition, Detection, and Description at the 21st ACM SIGKDD International Conference On Knowledge Discovery And Data Mining (KDD 2015) (2015)
8. Xiong, L., Póczos, B., Schneider, J., Connolly, A., Vander Plas, J.: Hierarchical probabilistic models for group anomaly detection. In: International Conference on Artificial Intelligence and Statistics 2011, pp. 789–797. Springer (2011)
9. Jayakumar, G., Thomas, B.J.: A new procedure of clustering based on multivariate outlier detection. J. Data Sci. **11**(1), 69–84 (2013)
10. Yager, R.R.: A new approach to the summarization of data. Inf. Sci. **28**(1), 69–86 (1982)
11. Yager, R.R.: Linguistic summaries as a tool for database discovery. In: FQAS, pp. 17–22 (1994)
12. Yager, R.: Linguistic summaries as a tool for databases discovery. In: Workshop on Fuzzy Databases System and Information Retrieval (1995)
13. Kacprzyk, J., Wilbik, A., Zadrozny, S.: Linguistic summaries of time series via a quantifier based aggregation using the sugeno integral. In: 2006 IEEE International Conference on Fuzzy Systems, pp. 713–719. IEEE (2006)
14. Kacprzyk, J., Wilbik, A., Zadrożny, S.: Linguistic summarization of time series using a fuzzy quantifier driven aggregation. Fuzzy Sets Syst. **159**(12), 1485–1499 (2008)
15. Kacprzyk, J., Yager, R.R., Zadrozny, S.: Fuzzy linguistic summaries of databases for an efficient business data analysis and decision support. In: Abramowicz, W., Zurada, J. (eds.) Knowledge Discovery for Business Information Systems. SECS, vol. 600, pp. 129–152. Springer, Boston (2002). https://doi.org/10.1007/0-306-46991-X_6

16. Kacprzyk, J., Zadrożny, S.: Linguistic database summaries and their protoforms: towards natural language based knowledge discovery tools. Inf. Sci. **173**(4), 281–304 (2005)
17. Kacprzyk, J., Wilbik, A., Zadrożny, S.: An approach to the linguistic summarization of time series using a fuzzy quantifier driven aggregation. Int. J. Intell. Syst. **25**(5), 411–439 (2010)
18. Ng, R.: Outlier detection in personalized medicine. In: Proceedings of the ACM SIGKDD Workshop on Outlier Detection and Description, p. 7 ACM (2013)
19. Aggarwal, C.C.: Toward exploratory test-instance-centered diagnosis in high-dimensional classification. IEEE Trans. Knowl. Data Eng. **19**(8), 1001–1015 (2007)
20. Cramer, J.A., Shah, S.S., Battaglia, T.M., Banerji, S.N., Obando, L.A., Booksh, K.S.: Outlier detection in chemical data by fractal analysis. J. Chemom. **18**(7–8), 317–326 (2004)
21. Knorr, E.M., Ng, R.T., Tucakov, V.: Distance-based outliers: algorithms and applications. VLDB J.-Int. J. Very Large Data Bases **8**(3–4), 237–253 (2000)
22. Angiulli, F., Pizzuti, C.: Fast outlier detection in high dimensional spaces. In: Elomaa, T., Mannila, H., Toivonen, H. (eds.) PKDD 2002. LNCS, vol. 2431, pp. 15–27. Springer, Heidelberg (2002). https://doi.org/10.1007/3-540-45681-3_2
23. Giatrakos, N., Kotidis, Y., Deligiannakis, A., Vassalos, V., Theodoridis, Y.: In-network approximate computation of outliers with quality guarantees. Inf. Syst. **38**(8), 1285–1308 (2013)
24. Last, M., Kandel, A.: Automated detection of outliers in real-world data. In: Proceedings of the Second International Conference on Intelligent Technologies, pp. 292–301 (2001)
25. Guo, Q., Wu, K., Li, W.: Fault forecast and diagnosis of steam turbine based on fuzzy rough set theory. In: Second International Conference on Innovative Computing, Information and Control 2007. ICICIC 2007, p. 501. IEEE (2007)
26. Kacprzyk, J., Zadrozny, S.: Protoforms of linguistic database summaries as a human consistent tool for using natural language in data mining. Int. J. Softw. Sci. Comput. Intell. (IJSSCI) **1**(1), 100–111 (2009)
27. Kacprzyk, J., Yager, R.R.: Linguistic summaries of data using fuzzy logic. Int. J. General Syst. **30**(2), 133–154 (2001)
28. Wilbik, A., Keller, J.M.: A fuzzy measure similarity between sets of linguistic summaries. IEEE Trans. Fuzzy Syst. **21**(1), 183–189 (2013)
29. Boriah, S., Chandola, V., Kumar, V.: Similarity measures for categorical data: a comparative evaluation. Red **30**(2), 3 (2008)
30. Duraj, A., Niewiadomski, A., Szczepaniak, P.S.: Outlier detection using linguistically quantified statements. Int. J. Intell. Syst. **33**(9), 1858–1868 (2018)
31. Duraj, A., Niewiadomski, A., Szczepaniak, P.S.: Detection of outlier information by the use of linguistic summaries based on classic and interval-valued fuzzy sets. Int. J. Intell. Syst. **34**(3), 415–438 (2019)
32. Duraj, A.: Outlier detection in medical data using linguistic summaries. In: 2017 IEEE International Conference on INnovations in Intelligent SysTems and Applications (INISTA), pp. 385–390. IEEE (2017)
33. Duraj, A., Szczepaniak, P.S., Ochelska-Mierzejewska, J.: Detection of outlier information using linguistic summarization. In: Flexible Query Answering Systems 2015. AISC, vol. 400, pp. 101–113. Springer, Cham (2016). https://doi.org/10.1007/978-3-319-26154-6_8
34. van Benthem, J., Ter Meulen, A.: Handbook of Logic and Language. Elsevier, Amsterdam (1996)

35. Benferhat, S., Dubois, D., Prade, H.: Nonmonotonic reasoning, conditional objects and possibility theory. Artif. Intell. **92**(1–2), 259–276 (1997)
36. Giordano, L., Gliozzi, V., Olivetti, N., Pozzato, G.L.: A non-monotonic description logic for reasoning about typicality. Artif. Intell. **195**, 165–202 (2013)
37. Schulz, K., Van Rooij, R.: Pragmatic meaning and non-monotonic reasoning: the case of exhaustive interpretation. Linguist. Philos. **29**(2), 205–250 (2006). https://doi.org/10.1007/s10988-005-3760-4
38. Zadeh, L.A.: Fuzzy sets. Inf. Control **8**(3), 338–353 (1965)
39. Zadeh, L.A.: The concept of a linguistic variable and its application to approximate reasoning-iii. Inf. Sci. **9**(1), 43–80 (1975)
40. Niewiadomski, A., Duraj, A.: Detecting and recognizing outliers in datasets via linguistic information and type-2 fuzzy logic. Int. J. Fuzzy Syst. **23**(3), 878–889 (2020). https://doi.org/10.1007/s40815-020-00919-5
41. Databases: Statistic Poland. https://stat.gov.pl/en/databases/
42. Arora, N., Kaur, P.D.: A Bolasso based consistent feature selection enabled random forest classification algorithm: an application to credit risk assessment. Appl. Soft Comput. **86**, 105936 (2020)
43. Kaur, S.: Comparative analysis of bankruptcy prediction models: An Indian perspective. CABELL'S DIRECTORY, USA 19
44. Altman, E.I., Iwanicz-Drozdowska, M., Laitinen, E.K., Suvas, A.: Financial distress prediction in an international context: a review and empirical analysis of Altman's Z-score model. J. Int. Financ. Manag. Account. **28**(2), 131–171 (2017)
45. Greco, S., Matarazzo, B., Slowinski, R.: A new rough set approach to evaluation of bankruptcy risk. In: Zopounidis, C. (ed.) Operational Tools in the Management of Financial Risks, pp. 121–136. Springer, Boston (1998). https://doi.org/10.1007/978-1-4615-5495-0_8

Combining Heterogeneous Indicators by Adopting Adaptive MCDA: Dealing with Uncertainty

Salvatore F. Pileggi[1,2(✉)]

[1] School of Information, Systems and Modelling (ISM), University of Technology Sydney, Sydney, Australia
SalvatoreFlavio.Pileggi@uts.edu.au
[2] Centre on Persuasive Systems for Wise Adaptive Living (PERSWADE), University of Technology Sydney, Sydney, Australia

Abstract. Adaptive MCDA systematically supports the dynamic combination of heterogeneous indicators to assess overall performance. The method is completely generic and is currently adopted to undertake a number of studies in the area of sustainability. The intrinsic heterogeneity characterizing this kind of analysis leads to a number of biases, which need to be properly considered and understood to correctly interpret computational results in context. While on one side the method provides a comprehensive data-driven analysis framework, on the other side it introduces a number of uncertainties that are object of discussion in this paper. Uncertainty is approached holistically, meaning we address all uncertainty aspects introduced by the computational method to deal with the different biases. As extensively discussed in the paper, by identifying the uncertainty associated with the different phases of the process and by providing metrics to measure it, the interpretation of results can be considered more consistent, transparent and, therefore, reliable.

Keywords: MCDA · Uncertainty · Sustainability

1 Introduction

There is an intrinsic relationship between decision theory [29] and uncertainty. Indeed, uncertainty often characterises typical decision-making scenarios in different disciplines [34]. Intuitively, more uncertainty in a given context results in a more difficult decision-making process in that context. Depending on the extent in which decision theory is applied (e.g. under ignorance, risk), minimising the uncertainty by adopting the different techniques to inform decision-making may become a key factor for the whole decision-making process.

Such considerations evidently also apply to Multi-Criteria Decision Analysis (MCDA) [30], whose methods are often integrated with explicit mechanisms or models [8] to deal with uncertainty (e.g. [14]). The relevance of the different uncertainty factors often suggests the combined use of MCDA and probablistic

© Springer Nature Switzerland AG 2021
M. Paszynski et al. (Eds.): ICCS 2021, LNCS 12747, pp. 514–525, 2021.
https://doi.org/10.1007/978-3-030-77980-1_39

approaches [10,33]. In general terms, the different applications of MCDA deal with different kinds of uncertainty. Concrete examples are, among others, in the field of energy planning [6], waste-water infrastructure planning [35], assessment of strategic options [23], healthcare [13,19], transport infrastructure appraisal [5], sustainability assessment [7], life-cycle assessment [22] and marine conservation [9].

In this paper we discuss the uncertainty associated with *Adaptive MCDA* [21], which systematically supports the dynamic combination of heterogeneous indicators to assess overall performance. Such a method is completely generic and is currently adopted to undertake a number of studies in the area of sustainability (e.g. [20]). The intrinsic heterogeneity characterizing this kind of analysis leads to a number of biases, which need to be properly considered and understood to correctly interpret computational results in context. While on one side the method provides a comprehensive data-driven analysis framework, on the other side it introduces a number of uncertainties that are object of discussion in this paper. Uncertainty is approached holistically, meaning we address all uncertainty aspects introduced by the computational method to deal with the different biases. As extensively discussed in the paper, by identifying the uncertainty associated with the different phases of the process and by providing metrics to measure it, the interpretation of results can be considered more consistent, transparent and, therefore, reliable.

Previous Work. *Adaptive MCDA* is described in [21], while an application on global sustainable development adopting such a method is proposed in [20]. The proposed contribution is strongly related to previous work as (i) the uncertainty analysis provided refers to *Adaptive MCDA* only and (ii) the case study on sustainable development is used as practical example for uncertainty analysis.

Aims and Scope. This paper focuses on uncertainty analysis in *Adaptive MCDA*. Such a topic is not addressed in previous work. Analysis and considerations in the paper apply only to *Adaptive MCDA*, while a more generic uncertainty analysis along the different MCDA techniques is out of the scope of this paper.

Structure of the Paper. The introductory part of the paper follows with Sect. 2, which briefly addresses MCDA. The core part of the paper (Sect. 3) deals with the uncertainty analysis in *Adaptive MCDA*, looking at a concrete case study. As usual, the conclusions section provides a concise summary of the contribution both with a brief outline of possible future work.

2   MCDA and Adaptive MCDA

MCDA is a consolidated concept within decision science, where MCDA-based techniques aim to provide a more comprehensive decision framework to contrast decisions based merely on intuition [25]. As many problems can be structured as multi-criteria decision problems [11,18], MCDA proliferated in the past decades by defining a relevant number of different approaches, methods and techniques.

While MCDA can be holistically considered as a completely generic app-roach, applications within the different domains resulted in a number of more specific and fine-grained methods, as reported by different contributions in lit-erature (e.g. [16] in the area of sustainable and renewable energy and [17] for transportation systems). In [15], the authors reviewed the different techniques and their application within the different disciplines, while an overview from a software perspective is proposed in [32].

Adaptive MCDA is a relatively simple technique that, overall, aims to make the weighting step as simple as possible by adapting computations to available data. Additionally, the method is expected to provide a rich analysis frame-work by combining multiple assessment metrics and visualizations in presence of heterogeneity. This paper discusses uncertainty factors, metrics and mitigations related to *Adaptive MCDA*, with emphasis on quantitative aspects and their relationship with qualitative ones.

A more holistic discussion of uncertainty along the different MCDA tech-niques could be complex and very articulated. It is out of the scope of this paper.

3   Uncertainty Analysis in Adaptive MCDA

In this section we discuss the implications of *Adaptive MCDA* in terms of uncer-tainty. We have identified two main kinds of uncertainty: one of them is associ-ated with the need to weight the considered criteria, while the other one results from the adaptive mechanism for parameter tuning to mitigate the numerical differences among the different indicators. The two categories and the respective metrics will be object of a separate discussion in the next subsections.

The experiments reported belong to the previously mentioned case study in the field of sustainable global development. By adopting multiple configurations that reflect different design decisions for the target case study, we point out the meaning of the uncertainty in context and its quantification according to the proposed metrics. The practical impact of such an uncertainty on final results and interpretations can vary very much depending on the extent and the intent of the considered case study. As the reference method allows customization and largely relies on interpretations in context, understanding uncertainty becomes a critical step for a correct computation set-up and result interpretation. More concretely, the use case object of analysis includes six different indicators: *Temperature Anomaly* [2], *Life Expectancy* [4,26], *GDP x capita* [3], *People living in extreme Poverty* [24,28], *People living in Democracy* [27] and *Terrorist attacks* [1].

3.1   Uncertainty Associated with Case Study Design: Indicator Selection and Weighting

The meaning of the weights associated with the different criteria may vary very much from case to case. Generally speaking, the weight set reflects some kind of importance or relevance related to the indicators framework in a given context.

However, while indicators themselves are somehow objective, yet not always perfect, measures, weighting indicators in a specific context can be a much more volatile and subjective concept. Indeed, depending on the intent and extent of a given study, weights can be simply not relevant (i.e. - "neutral" computations that assume the same weight for all indicators are considered acceptable), can be understood as static parameters estimated, measured or anyway known a priori, as well as they can be considered as variables or their estimation may even be the main purpose of the study (e.g. typical of Participatory Modelling [12,31]).

As weighting plays a very critical and key role to establish holistically the performance of the target system and it can rarely be considered completely objective, we assume an intrinsic uncertainty which reflects the potential variability of weights. *Adaptive MCDA* always proposes final computations taking into account of such an uncertainty. Indeed, the result of a computation $\alpha(t)$ at a given time i, that assumes a weight set from users, is integrated with extreme computations, $\gamma(t)$ and $\beta(t)$, that adopt the weight sets that correspond, respectively, to the best and the worst possible performance at the time i. Thus, at the generic time i, it is always $\beta(t_i) \leq \alpha(t_i) \leq \gamma(t_i)$. The *Uncertainty Range* is defined as $Range(t_i) = \{\beta(t_i), \gamma(t_i)\}$, while the resulting uncertainty Δ is measured as $\Delta(t_i) = |\beta(t_i) - \gamma(t_i)|$.

Looking at the six selected indicators, we consider different combinations of indicators to define the different configurations of the experiment (Table 1).

Table 1. Configuration of the experiment to measure the uncertainty (Δ) associated with weighting.

UC	Temp.	Life Exp.	GDP	Pov.	Dem.	Terr.	Range	$\Delta(t_{MAX})$
$UC\_1.1$	✓	✓	✓	x	x	x	$-1450/1000$	**2450**
$UC\_1.2$	x	x	x	✓	✓	✓	$-7000/2550$	**9550**
$UC\_1.3$	x	x	✓	✓	✓	x	$0/1075$	**1075**

For each configuration we have computed the *Uncertainty Range* and Δ. Both metrics are reported in Table 1. As shown, even considering the same number of indicators (3 indicators), the corresponding uncertainty varies notably for the different cases: the second configuration presents a significantly high uncertainty that is almost 5 times higher than in the first configuration and almost 10 times higher than the one associated with the last configuration.

In order to fully understand the computation results also considering the associated uncertainty as previously defined, the method's output (Fig. 1) is always proposed in context.

The first two configurations (Fig. 1a and 1b) present classical ranges, meaning that, depending on the weight set adopted, the system may have positive or negative performance. However, the Δ value is much higher for the second configuration that is, therefore, characterised by an higher level of uncertainty. The third configuration (Fig. 1c) proposes a completely different pattern, as the

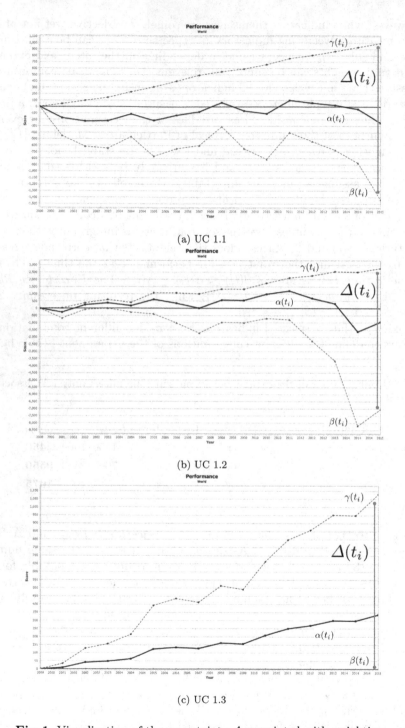

(a) UC 1.1

(b) UC 1.2

(c) UC 1.3

Fig. 1. Visualization of the uncertainty Δ associated with weighting.

performance of the system can only be positive. Additionally, this use case is associated with a significantly lower uncertainty.

Impact on Results and Interpretations. Weights play a key role in final computations and, therefore, in the whole decision analysis process. Weighting doesn't necessarily raise uncertainty, as well as the potential impact of uncertainty on results and interpretations may vary significantly from case to case. The method discussed in this paper is based on the contextual interpretation of the results computed. The results are always proposed both with extreme computations (Fig. 1) in order to provide a clear understanding of the range of possible results as the function of the chosen weights set. The uncertainty associated with the weighting process may also be understood as a possible driver factor to select target criteria, as the minimization of the uncertainty can be, at least in theory, a way to have a more "objective" and transparent analysis.

3.2 Numerical Bias

An additional uncertainty is inducted by the *Adaptive MCDA* algorithm used to mitigate numerical differences existing among heterogeneous indicators. Such an adaptive algorithm is not always adopted as it is a user choice. It can be very useful any time the target indicator framework presents strong differences in scale among the different indicators. If not properly addressed by the computational method, this *numerical bias* can make certain criteria as non-relevant in the assessment of overall performance regardless of the weights associated.

Adaptive MCDA adopts a metric, which we refer to as *distance*, to estimate the accuracy of the algorithm, as lower values correspond to higher accuracy. This metric is associated with the *neutral computation*, that is a reference computation which assumes the same weight $w_i = \hat{w}$ for the i target indicators. $\hat{w}$ is normally the average value over the allowed values for weighting (e.g. $\hat{w} = 5$ for the range [0,10]). More concretely, it measures the distance between the neutral computation function and the x-axis. Some visualizations assuming $\hat{w} = 5$ are reported in Fig. 3.

We adopt *distance* to assess the uncertainty introduced by the adaptive parameter tuning. In this case, uncertainty is mostly synonymous with precision. Indeed, an ideal parameter tuning assumes *distance* $= 0$, while in fact such a value is normally not null. It introduces an uncertainty in the computation which is estimated by *distance*, which measures the distance between the current parameters tuning for computation and the ideal one.

The experiment proposed consists in the analysis of the four different configurations as reported in Table 2. These configurations include all available indicators. The time frame for the analysis is 2000–2015 (16 points). The configurations differ from each other on the number of points considered (Fig. 2).

Table 2. Summary of results for uncertainty associated with parameter tuning.

UC	Temp.	Life Exp.	GDP	Pov.	Dem.	Terr.	Period	#Points	*Distance*
UC_2.1	✓	✓	✓	✓	✓	✓	2000–2015	2	≈**10**
UC_2.2	✓	✓	✓	✓	✓	✓	2000–2015	3	≈**100**
UC_2.3	✓	✓	✓	✓	✓	✓	2000–2015	5	≈**650**
UC_2.4	✓	✓	✓	✓	✓	✓	2000–2015	16	≈**2000**

In general terms, considering more points contributes to have a more fine grained analysis by providing a clearer and more detailed understanding of trends. Reducing the number of points considered affects the analysis of trends. The proper number of points to consider depends first of all on data availability and can be normally considered very specific of a given case study. We assume that many studies consider all available data so we expect a relatively high number of points.

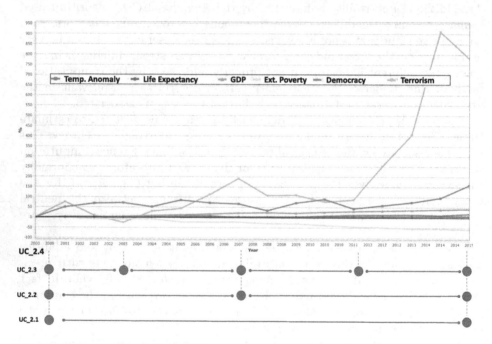

Fig. 2. Configuration of the experiment to measure the uncertainty associated with parameter tuning.

Average values for *distance* resulting by computing the different configurations are reported in Table 2, as well as a visualization is reported in Fig. 3. As expected, a lower number of points is associated with a lower uncertainty, which indeed increases with the amount of data considered.

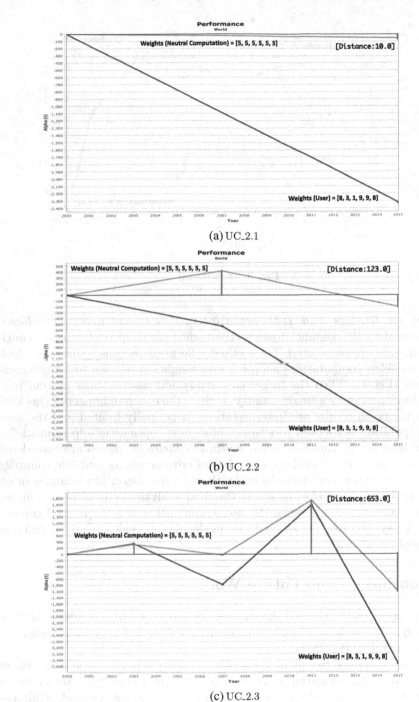

(a) UC_2.1

(b) UC_2.2

(c) UC_2.3

Fig. 3. Visualization of the uncertainty (*distance*) introduced by the parameter tuning algorithm.

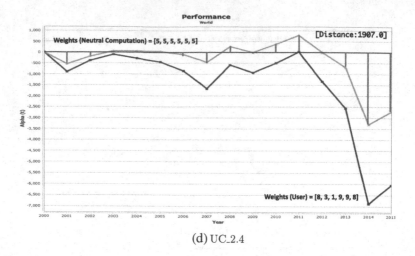

(d) UC_2.4

Fig. 3. (*continued*)

Impact on Results and Interpretations. This second uncertainty factor is introduced by the computational method when the adaptive features to mitigate numerical biases are used. The method reflects an analysis framework which always considers computations adopting user weights in relation to neutral computations (Fig. 3). When the target case study addresses a single system (e.g. global development, or a single country or city), the concrete impact of this kind of uncertainty on results and interpretations is normally limited and the neutral computation is adopted just to establish how "optimistic"/"pessimistic" a given weights set is. However, when multiple systems/scenarios are considered (e.g. some comparison based on a number of criteria among multiple countries or cities), the uncertainty introduced by the adaptive algorithm becomes much more relevant. In such kind of study, the final analysis needs to be conducted looking at the distance between the user computation and the neutral computation. Indeed absolute values could be misleading as scales could be different because of the algorithm.

4   Conclusions and Future Work

In this paper we have analysed the uncertainty associated with *Adaptive MCDA*, a method to systematically and dynamically combine heterogeneous indicators to assess overall performance.

We have identified two main kinds of uncertainty related to the weighting of criteria and to the mitigation of numerical bias. The former uncertainty factor is a direct consequence of the weights relevance within the method, while the latter is introduced by the adaptive features of the method.

As extensively discussed, such an uncertainty can be measured and computational results are always proposed in the context of the metrics associated.

Uncertainty can be understood in two possible ways: on one side, a clear understanding and quantification of uncertainty makes the analysis framework richer, more accurate and transparent; on the other side, the minimization of uncertainty can be considered as a valuable driver factor to design reasonable case studies in terms of amount of data and heterogeneity.

Future work is still in the field of sustainability and aims at a more fine-grained analysis. More concretely, *Adaptive MCDA* will be used to measure expected country resilience to situations of pandemic (e.g. COVID-19). Because of the notable heterogeneity of criteria, we expect uncertainty to be even more relevant than in the cases approached so far (global development).

References

1. Global terrorism database (GTD). University of Maryland
2. Temperature anomalies - Met Office Hadley Centre. https://www.metoffice.gov.uk/hadobs/hadcrut4/index.html. Accessed 11 June 2020
3. World Development Indicators, The World Bank. http://data.worldbank.org/data-catalog/world-development-indicators
4. Zijdeman, R., Ribeira da Silva, F.: Life Expectancy at Birth (Total), ISH Data Collection, V. 1 (2005). https://hdl.handle.net/10622/LKYT53
5. Ambrasaite, I., Barfod, M.B., Salling, K.B.: MCDA and risk analysis in transport infrastructure appraisals: the Rail Baltica case. Procedia-Soc. Behav. Sci. **20**, 944–953 (2011)
6. Diakoulaki, D., Antunes, C.H., Gomes Martins, A.: MCDA and energy planning. In: Multiple Criteria Decision Analysis: State of the Art Surveys. ISORMS, vol. 78, pp. 859–890. Springer, New York (2005). https://doi.org/10.1007/0-387-23081-5_21
7. Dorini, G., Kapelan, Z., Azapagic, A.: Managing uncertainty in multiple-criteria decision making related to sustainability assessment. Clean Technol. Environ. Policy **13**(1), 133–139 (2011)
8. Durbach, I.N., Stewart, T.J.: Modeling uncertainty in multi-criteria decision analysis. Eur. J. Oper. Res. **223**(1), 1–14 (2012)
9. Estévez, R.A., Gelcich, S.: Participative multi-criteria decision analysis in marine management and conservation: research progress and the challenge of integrating value judgments and uncertainty. Mar. Policy **61**, 1–7 (2015)
10. Fenton, N., Neil, M.: Making decisions: using Bayesian nets and MCDA. Knowl.-Based Syst. **14**(7), 307–325 (2001)
11. Franco, L.A., Montibeller, G.: Problem structuring for multicriteria decision analysis interventions. In: Wiley Encyclopedia of Operations Research and Management Science (2010)
12. Gray, S., et al.: Combining participatory modelling and citizen science to support volunteer conservation action. Biol. Conserv. **208**, 76–86 (2017)
13. Hansen, P., Devlin, N.: Multi-criteria decision analysis (MCDA) in healthcare decision-making. In: Oxford Research Encyclopedia of Economics and Finance (2019)
14. Hyde, K., Maier, H.R., Colby, C.: Incorporating uncertainty in the PROMETHEE MCDA method. J. Multi-Criteria Decis. Anal. **12**(4–5), 245–259 (2003)

15. Mardani, A., Jusoh, A., Nor, K., Khalifah, Z., Zakwan, N., Valipour, A.: Multiple criteria decision-making techniques and their applications-a review of the literature from 2000 to 2014. Econ. Res.-Ekonomska istraživanja **28**(1), 516–571 (2015)
16. Mardani, A., Jusoh, A., Zavadskas, E.K., Cavallaro, F., Khalifah, Z.: Sustainable and renewable energy: an overview of the application of multiple criteria decision making techniques and approaches. Sustainability **7**(10), 13947–13984 (2015)
17. Mardani, A., Zavadskas, E.K., Khalifah, Z., Jusoh, A., Nor, K.M.: Multiple criteria decision-making techniques in transportation systems: a systematic review of the state of the art literature. Transport **31**(3), 359–385 (2016)
18. Marttunen, M., Lienert, J., Belton, V.: Structuring problems for multi-criteria decision analysis in practice: a literature review of method combinations. Eur. J. Oper. Res. **263**(1), 1–17 (2017)
19. Mühlbacher, A.C., Kaczynski, A.: Making good decisions in healthcare with multi-criteria decision analysis: the use, current research and future development of MCDA. Appl. Health Econ. Health Policy **14**(1), 29–40 (2016)
20. Pileggi, S.F.: Life before COVID-19: how was the World actually performing? Qual. Quant. https://doi.org/10.1007/s11135-020-01091-6
21. Pileggi, S.F.: Is the world becoming a better or a worse place? A data-driven analysis. Sustainability **12**(1), 1–24 (2019)
22. Prado, V., Rogers, K., Seager, T.P.: Integration of MCDA tools in valuation of comparative life cycle assessment. In: Life Cycle Assessment Handbook: A Guide for Environmentally Sustainable Products, pp. 413–432 (2012)
23. Ram, C., Montibeller, G., Morton, A.: Extending the use of scenario planning and MCDA for the evaluation of strategic options. J. Oper. Res. Soc. **62**(5), 817–829 (2011)
24. Ravallion, M.: The Economics of Poverty: History, Measurement, and Policy. Oxford University Press, Oxford (2015)
25. Rew, L.: Intuition in decision-making. Image J. Nurs. Scholarsh. **20**(3), 150–154 (1988)
26. Riley, J.C.: Estimates of regional and global life expectancy, 1800–2001. Popul. Dev. Rev. **31**(3), 537–543 (2005)
27. Roser, M.: Democracy. Our World in Data (2013) https://ourworldindata.org/democracy
28. Roser, M., Ortiz-Ospina, E.: Global extreme poverty. Our World in Data (2013). https://ourworldindata.org/extreme-poverty
29. Steele, K., Stefánsson, H.O.: Decision theory (2015)
30. Stewart, T.J.: Dealing with uncertainties in MCDA. In: Multiple Criteria Decision Analysis: State of the Art Surveys. ISORMS, vol. 78, pp. 445–466. Springer, New York (2005). https://doi.org/10.1007/0-387-23081-5_11
31. Videira, N., Antunes, P., Santos, R., Lopes, R.: A participatory modelling approach to support integrated sustainability assessment processes. Syst. Res. Behav. Sci. **27**(4), 446–460 (2010)
32. Weistroffer, H.R., Li, Y.: Multiple criteria decision analysis software. In: Greco, S., Ehrgott, M., Figueira, J.R. (eds.) Multiple Criteria Decision Analysis. ISORMS, vol. 233, pp. 1301–1341. Springer, New York (2016). https://doi.org/10.1007/978-1-4939-3094-4_29
33. Yang, A., Huang, G., Qin, X., Fan, Y.: Evaluation of remedial options for a benzene-contaminated site through a simulation-based fuzzy-MCDA approach. J. Hazard. Mater. **213**, 421–433 (2012)

34. Zarghami, M., Szidarovszky, F.: Mcda problems under uncertainty. In: Multicriteria Analysis, pp. 113–147. Springer, Heidelberg (2011). https://doi.org/10.1007/978-3-642-17937-2_7
35. Zheng, J., Egger, C., Lienert, J.: A scenario-based MCDA framework for wastewater infrastructure planning under uncertainty. J. Environ. Manag. **183**, 895–908 (2016)

Solutions and Challenges in Computing FBSDEs with Large Jumps for Dam and Reservoir System Operation

Hidekazu Yoshioka(✉) 🆔

Graduate School of Natural Science and Technology, Shimane University,
Nishikawatsu-cho 1060, Matsue 690-8504, Japan
yoshih@life.shimane-u.ac.jp

Abstract. Optimal control of Lévy jump-driven stochastic differential equations plays a central role in management of resource and environment. Problems involving large Lévy jumps are still challenging due to their mathematical and computational complexities. We focus on numerical control of a real-scale dam and reservoir system from the viewpoint of forward-backward stochastic differential equations (FBSDEs): a new mathematical tool in this research area. The problem itself is simple but unique, and involves key challenges common to stochastic systems driven by large Lévy jumps. We firstly present an exactly-solvable linear-quadratic problem and numerically analyze convergence of different numerical schemes. Then, a more realistic problem with a hard constraint of state variables and a more complex objective function is analyzed, demonstrating that the relatively simple schemes perform well.

Keywords: Forward-backward stochastic differential equations · Resource and environment · Tempered stable subordinator · Stochastic maximum principle · Least-squares monte-carlo

1 Introduction

Uncertainties are ubiquitous in mathematical modeling and control for environmental and resource management. Stochastic differential equations (SDEs) have been principal tools for efficiently as well as rigorously describing stochastic dynamics of environments and resources [1–3]. Stochastic control based on Markovian feedback policy [4] is a well-established concept implementable in applications because it enables decision-makers to make decisions based on system observations.

Operation of dam and reservoir systems has been a major problem involving management of both resource and environment [5]. A dam and reservoir system consists of a reservoir created in a river to receive and store stochastic inflow discharge and an associated dam as a hydraulic structure to control outflow discharge [6]. Each dam and reservoir system has different operation goal depending on its construction purpose; however, it has a common principle that there exist some targeted reservoir water volume

© Springer Nature Switzerland AG 2021
M. Paszynski et al. (Eds.): ICCS 2021, LNCS 12747, pp. 526–539, 2021.
https://doi.org/10.1007/978-3-030-77980-1_40

and outflow discharge. In this way, operation of a dam and reservoir system is understood as a stochastic control problem having a target state.

Existing stochastic control models of dam and reservoir systems are based on dynamic programming approaches where finding an optimal control reduces to solving an optimality equation of a degenerate elliptic or parabolic type [6–8]. This methodology works only if the optimality equation is solvable analytically or its dimension is relatively low, one or two in most cases, such that a common numerical method like a finite difference scheme is implementable [9]. Such cases are too simple from an engineering viewpoint [10]. Furthermore, the previous study suggested that the inflow discharge follows an SDE driven by both small and large Lévy jumps [11], leading to an optimality equation of an integro-differential type having a singular integral kernel that is not necessarily easy to numerically discretize.

To tackle the above-mentioned issue in the stochastic control of dam and reservoir systems, we introduce forward-backward stochastic differential equations (FBSDEs) [4, 12] as a new mathematical tool in this research area. FBSDEs are often equivalent to the optimality equations of the degenerate elliptic and parabolic types [12, 13] but are more suited to higher-dimensional problems. This is because they can be implemented using a Monte-Carlo method that can mitigate or even defeat the curse of dimensionality [14–16]. Our FBSDEs are based on a stochastic maximum principle [4] and fully couple forward and backward processes, both of which are driven by a Lévy jump process having infinite activities. To the best of our knowledge, FBSDEs, especially those driven by jumps, have not been investigated in dam and reservoir control problems. In addition, studies focusing on numerical computation of jump-driven FBSDEs are still rare except for purely theoretical ones [17, 18].

The objective of this paper is thus set to be formulation and analysis of new jump-driven FBSDEs for stochastic control of dam and reservoir systems. The model proposed in this paper is simple but unique and oriented to engineering applications. A simplified linear-quadratic problem that is solvable analytically but still non-trivial is firstly derived and analyzed numerically using different least-squares Monte-Carlo methods. We use the exact discretization formula [19] to efficiently simulate the inflow discharge process driven by Lévy jumps having infinite activities. A more realistic case having constrained state variables and a more complex objective function is then numerically analyzed with a least-squares Monte-Carlo method. Through the numerical experiments conducted in this paper, we discuss difficulties and remaining challenges in modeling and computation of dam and reservoir systems using FBSDEs.

2 Stochastic Process Model

2.1 Stochastic Differential Equations

We consider a continuous-time operation problem of a dam and reservoir system having the three state variables: the inflow discharge $(I_t)_{t\geq0}$ as a non-negative variable having the range $\Omega_I = [0, +\infty)$, the water volume $(V_t)_{t\geq0}$ of a reservoir having the range $\Omega_V = [\underline{V}, \overline{V}]$ with constants $\underline{V} < \overline{V}$, and the outflow discharge $(O_t)_{t\geq0}$ having the range $\Omega_O = [\underline{O}, \overline{O}]$ with constants $\underline{O} < \overline{O}$. Typically, we have $\underline{V} = \underline{O} = 0$. Assume that the inflow is uncontrollable while the outflow is indirectly controllable by tuning its acceleration $(a_t)_{t \geq 0}$ having the range $A = [-\overline{a}, \overline{a}]$ with $\overline{a} > 0$. The state and control variables must be constrained in the corresponding ranges for any $t \geq 0$. Mathematically, the constants $\underline{O}, \overline{O}, \underline{V}, \overline{V}, \overline{a}$ are not necessarily bounded. We consider both bounded and unbounded cases in this paper. Clearly, the former is more realistic.

We consider the problem in a complete probability space as in the usual setting [4]. The stochastic system dynamics we consider are formulated as follows:

$$dI_t = \rho(\underline{I} - I_t)dt + \int_0^\infty zN(dz, dt), \tag{1}$$

$$dO_t = a_t dt, \tag{2}$$

$$dV_t = (I_t - O_t)dt, \tag{3}$$

where (1), (2), and (3) describe inflow, outflow, and storage processes, respectively. Here, $\rho > 0$ is the inverse of the correlation time of the inflow, N is a Poisson random measure of a subordinator type having only positive jumps with the Lévy measure $v(dz)$. Based on a recent identification result for a real river [11], set.

$$v(dz) = \rho a z^{-(1+\alpha)} e^{-bz} dz, \quad z > 0 \tag{4}$$

with positive constants $a, b,$ and $\alpha \in (0, 1)$. This Lévy measure is of the infinite activities type since $\int_0^\infty v(dz) = +\infty$. Its first-order moment $M_1 = \int_0^\infty zv(dz)$ is bounded and is given as $M_1 = \rho ab^{1-\alpha}\Gamma(1 - \alpha) > 0$ with a Gamma function Γ. The inflow discharge is more intermittent as α gets closer to 0; in real cases α is close to 0.5 [11].

We assume that the system (1)–(3) is equipped with a deterministic initial condition (I_0, O_0, V_0) belonging to the space $\Omega_I \times \Omega_O \times \Omega_V$. Furthermore, a natural filtration generated by the Poisson random measure N is denoted as $(F_t)_{t \geq 0}$. We consider a Markovian setting, and the control a_t is assumed to be progressively measurable with respect to the filtration F_t at each $t \geq 0$. The control should be chosen so that the following constraint is satisfied as explained above: $O_t \in \Omega_O$ and $V_t \in \Omega_V$ with $a_t \in A$ for each $t \geq 0$. The earlier studies considered that the outflow discharge is directly controllable [6–8]; however, instantaneously adjusting the outflow discharge is not only technically difficult but also triggers catastrophic failures of reservoir walls [20]. We therefore consider problems without such an impulsive adjustment.

For simplicity of the analysis, we assume that the system (1)–(3) with each control $(a_t)_{t \geq 0}$ has a unique path-wise solution that is right-continuous and has left limits. This assumption itself is interesting because our problem is a of state-constrained problem: a non-classical problem requiring a careful treatment at boundaries [6, 8], but its rigorous mathematical analysis is beyond the scope of this paper.

2.2 Objective Function

The objective function to optimize the process $(a_t)_{t \geq 0}$ is formulated. We focus on a discounted case whose long-run limit is an infinite horizon case. Assume that the time horizon of the optimization problem is $\Omega_T = [0, T]$ with a terminal time $T > 0$. Our objective function ϕ is a functional of the initial condition and the process $(a_t)_{t\geq 0}$:

$$\phi(I_0, O_0, V_0; a_{(\cdot)}) \equiv \mathbb{E}^{0,i,o,v}\left[\int_0^T f(I_s, O_s, V_s, a_s)ds\right]$$

$$= \mathbb{E}^{0,i,o,v}\left[\int_0^T e^{-\delta s}\left(\frac{w_1}{2}a_s^2 + \frac{w_2}{2}(I_s - O_s)^2 + \frac{w_3}{2}\left(V_s - \hat{V}\right)^2 + \frac{w_4}{2}\max\left\{\hat{O} - O_s, 0\right\}^2\right)ds\right].$$

$$(5)$$

Here, $\mathbb{E}^{t,i,o,v}$ is the conditional expectation using the information $(t, I_t, O_t, V_t) = (t, i, o, v)$, $\delta > 0$ is the discount rate, $\hat{V} \in \Omega_V$ is the target water volume, and $\hat{O} \in \Omega_O$ is the threshold discharge, and $w_1, w_2, w_3, w_4 \geq 0$ are weighting coefficients satisfying $w_1 w_2 w_3 w_4 \neq 0$. The parameters in (5) are allowed to be time-dependent if necessary.

In the expectation of (5), the first term penalizes sudden changes of the outflow discharge; the second term penalizes the deviation from the run-of-river condition ($I_t = O_t$) not only for sustainable operation of the reservoir with smaller impacts against the downstream river but also for continuous hydropower generation [21]; the third term penalizes deviation of the water volume from the prescribed target value $\hat{V}$; and the fourth term penalizes outflow discharges smaller than the prescribed threshold value $\hat{O}$ below which the downstream environment is severely threatened [6]. The discount rate δ is the inverse of the effective time length of the decision-making, meaning that the decision-maker, an operator of the dam and reservoir system, controls the system considering future events of the time δ^{-1} ahead in the mean. The objective function is simple but is oriented to concurrent management of resource and environment considering multiple objectives using a scalarization technique. It reduces to be quadratic if $w_4 = 0$.

We introduce the dynamic counterpart of (5) to optimize the process $(a_t)_{t\geq 0}$ using a maximum principle approach based on FBSDEs:

$$\hat{\phi}(t, I_t, O_t, V_t; a_{(\cdot)}) = \mathbb{E}^{t,i,o,v}\left[-\int_t^T e^{\delta t}f(I_s, O_s, V_s, a_s)ds\right] \qquad (6)$$

and set its dynamically optimized counterpart, the value function, as

$$\Phi(t, i, o, v) = \sup_{a_{(\cdot)}} \hat{\phi}(t, I_t, O_t, V_t; a_{(\cdot)}) \qquad (7)$$

Clearly, (6) is equivalent to (5) when $t = 0$. Taking the minus sign in (6) is simply to follow the existing argument of the stochastic maximum principle [4]. A maximizing control of (7) (and also minimizing (5)) is called an optimal control and is denoted as $(a_t^*)_{t\geq 0}$. We explore Markovian optimal feedback controls, with an abuse of notations, of the form $a_t^* = a_t(t, I_t, O_t, V_t)$. This is a dynamic and thus adaptive control based on the observation process $(t, I_t, O_t, V_t)_{t\geq 0}$ up to the current time.

2.3 Stochastic Maximum Principle

Stochastic maximum principle approach [4] reduces the optimization problem (6) to initial and terminal value problems of FBSDEs. It is a standard machinery for analyzing stochastic control problems in finance, economics, insurance, and related research fields. Our problem does not fall into them, but the approach is still applicable. Remarkable advantages of the approach based on FBSDEs over that using the dynamic programming principle is that the former can naturally handle high-dimensional problems and that both optimal controls and the controlled processes are derived simultaneously without any postprocessing. On the other hand, its disadvantage is that the FBSDEs are computationally inefficient for low-dimensional problems having one or two state variables. Another advantage of using FBSDEs is the capability to manage non-Markovian problems. Among these advantages, our modeling and computation benefit from the capability to manage high-dimensional problems and the characteristic that is able to simulate both the forward and backward processes simultaneously.

In our case, if we neglect the constraint on the state variables for simplicity, a lengthy but straightforward calculation [e.g., Theorem 5.4 of 4] with (6) leads to the following backward system of adjoint equations to be coupled with (1)–(3):

$$
dp_t^{(I)} = \left\{ w_2(I_t - O_t) + (\rho + \delta)p_t^{(I)} - p_t^{(V)} \right\} dt + \int_0^\infty \theta^{(I)}(t, z)\{N(dz, dt) - v(dz)dt\},
$$
(8)

$$
dp_t^{(O)} = \left\{ \begin{array}{c} w_2(O_t - I_t) + \delta p_t^{(O)} + p_t^{(V)} \\ -w_4 \max\left\{ \hat{O} - O_t, 0 \right\} \end{array} \right\} dt + \int_0^\infty \theta^{(O)}(t, z)\{N(dz, dt) - v(dz)dt\},
$$
(9)

$$
dp_t^{(V)} = \left\{ w_3\left(V_t - \hat{V} \right) + \delta p_t^{(V)} \right\} dt + \int_0^\infty \theta^{(V)}(t, z)\{N(dz, dt) - v(dz)dt\}
$$
(10)

with the terminal condition $p_T^{(I)} = p_T^{(O)} = p_T^{(V)} = 0$. The triplet $\left(p_t^{(I)}, p_t^{(O)}, p_t^{(V)} \right)_{t \geq 0}$ is the adjoint variables to find an ("candidate of", see also **Remark 1**) optimal control as

$$
a_t^*(t, I_t, O_t, V_t) = \min\left\{ \overline{a}, \max\left\{ -\overline{a}, p_t^{(O)} \right\} \right\} \text{ if } O_t \in \left(\underline{O}, \overline{O} \right) \text{ and } V_t \in \left(\underline{V}, \overline{V} \right),
$$
(11)

otherwise it is replaced by a_t with the smallest $|a_t|$ to guarantee the constraints $O_t \in \Omega_O$ and $V_t \in \Omega_V$.

The right-hand side of (11) should be understood as a Markovian variable determined by (t, I_t, O_t, V_t) at each t. The triplet $\left(\theta^{(I)}(t, \cdot), \theta^{(O)}(t, \cdot), \theta^{(V)}(t, \cdot) \right)_{t \geq 0}$ represents the predictable stochastic fields such that jump integral terms (8)–(10) exist. Consequently, the FBSDEs consist of the forward system (1)–(3), the backward system (8)–(10), the control law (11), and the corresponding initial and terminal conditions. The integrated system is a jump-driven fully-coupled FBSDEs not well-studied previously.

Finally, we introduce Markovian representations of (8)–(10) that are employed in numerical discretization of the FBSDEs:

$$
p_t^{(I)} = \mathbb{E}^{t, I_t, O_t, V_t}\left[\int_t^T \left\{ -w_2(I_s - O_s) - (\rho + \delta)p_s^{(I)} + p_s^{(V)} \right\} ds \right],
$$
(12)

$$
p_t^{(O)} = \mathbb{E}^{t, I_t, O_t, V_t}\left[\int_t^T \left\{ -w_2(O_s - I_s) + w_4 \max\left\{ \hat{O} - O_s, 0 \right\} - \delta p_s^{(O)} - p_s^{(V)} \right\} ds \right],
$$
(13)

$$p_t^{(V)} = \mathbb{E}^{t,I_t,O_t,V_t}\left[\int_t^T \left\{-w_3\left(V_s - \hat{V}\right) - \delta p_s^{(V)}\right\}ds\right] \tag{14}$$

Remark 1. Recall that we have neglected the state constraint in the derivation of the backward system (8)–(10). Incorporating a state-constraint to the stochastic maximum principle results in FBSDEs having state constraints in both forward and backward systems [22]. In our case, neglecting the constraint implies that the FBSDEs give only sub-optimal controls that are inferior to optimal ones. Nevertheless, we numerically show that the feasible controls based on the presented FBSDEs serve well.

2.4 An Exact Solution

The derived FBSDEs are analytically solvable under the unconstrained case because it then reduces to a linear-quadratic problem. The following proposition states that a solution to the backward system (8)–(10) is derived as a function of the (controlled) forward process (t, I_t, O_t, V_t). This analytical solution is used to verifying our numerical schemes. Hereafter, A_t' represents $\frac{dA_t}{dt}$ etc.

Proposition 1. *Assume $w_4 = 0$ and Ω_O, Ω_V, $A = \mathbb{R}$. Then the backward system (8)–(10) admits a unique square-integrable solution of the following parametric form*

$$p_t^{(I)} = A_t I_t + B_t O_t + C_t V_t + D_t, \tag{15}$$

$$p_t^{(O)} = B_t I_t + F_t O_t + G_t V_t + H_t, \tag{16}$$

$$p_t^{(V)} = C_t I_t + G_t O_t + L_t V_t + P_t, \tag{17}$$

with the following backward system of Riccati type ODEs for $t < T$:

$$A_t' = w_2 + (2\rho + \delta)A_t - 2C_t - w_1^{-1}B_t^2, \ A_T = 0, \tag{18}$$

$$B_t' = -w_2 + (\rho + \delta)B_t + C_t - G_t - w_1^{-1}B_t F_t, \ B_T = 0, \tag{19}$$

$$C_t' = (\rho + \delta)C_t - L_t - w_1^{-1}B_t G_t, \ C_T = 0, \tag{20}$$

$$D_t' = -(\rho \underline{I} + M_1)A_t + (\rho + \delta)D_t - P_t - w_1^{-1}B_t H_t, \ D_T = 0, \tag{21}$$

$$F_t' = w_2 + \delta F_t + 2G_t - w_1^{-1}F_t^2, \ F_T = 0, \tag{22}$$

$$G_t' = \delta G_t + L_t - w_1^{-1}F_t G_t, \ G_T = 0, \tag{23}$$

$$H_t' = -(\rho \underline{I} + M_1)B_t + \delta H_t + P_t - w_1^{-1}F_t H_t, \ H_T = 0, \tag{24}$$

$$L'_t = w_3 + \delta L_t - w_1^{-1} G_t^2, \quad L_T = 0, \tag{25}$$

$$P'_t = -(\rho \underline{L} + M_1) C_t - w_3 \hat{V} + \delta P_t - w_1^{-1} G_t H_t, \quad P_T = 0 \tag{26}$$

The proof of **Proposition 1** is omitted since it uses a direct substitution of (15)–(17) into the FBSDEs. The uniqueness and optimality follow from proofs similar to the existing ones [Theorem 5.4 of 4, Theorem 3.1.1 of 12].

3 Numerical Experiments

3.1 Discretization

Numerical computation of the FBSDEs consists of explicit discretization of the forward system (1)–(3) using (11) and an explicit or semi-implicit discretization of the backward system (8)–(10). We use a Monte-Carlo method with $n \gg 1$ sample paths and temporal resolution $\Delta t = k^{-1} T$ with $k \in \mathbb{N}$. The SDEs (2)–(3) are discretized with a classical forward-Euler scheme, while the SDE (1) is with the exact sampling scheme [19] free from temporal discretization errors: $I_{j\Delta t} = Y_{j\Delta t} + \underline{I}$ and at each time step

$$Y_{(j+1)\Delta t} = e^{-\rho \Delta} Y_{j\Delta t} + \eta_0 + \sum_{l=1}^{\overline{N}(\Delta t)} \eta_l (j = 0, 1, 2, \ldots), \quad Y_0 = I_0 - \underline{I} \tag{27}$$

Here, η_0 is a tempered stable variable with the exponent b based on the stable one:

$$\left(\frac{a(1 - e^{-\alpha \rho \Delta t}) \Gamma(1 - \alpha)}{\alpha \cos(V)} \right)^{\frac{1}{\alpha}} \sin(\alpha(V + \pi/2)) \left(\frac{\cos(V - \alpha(V + \pi/2))}{e} \right)^{\frac{1-\alpha}{\alpha}}, \tag{28}$$

where V follows a uniform distribution in $(0, 1)$, e follows an exponential distribution with the intensity 1, $\overline{N}(\Delta t)$ is a Poisson process with the intensity $-a(1 - e^{-\alpha \lambda \Delta t}) \Gamma(-\alpha) b^{\alpha}$, and each η_l ($l = 1, 2, 3, \ldots$) are independent random variables generated by the probability density function.

$$p(\eta) = \frac{1}{(1 - e^{\alpha \lambda \Delta t}) \Gamma(-\alpha) b^{\alpha}} \eta^{-(1+\alpha)} \left(e^{-b\eta} - e^{-be^{\lambda \Delta t} \eta} \right), \quad \eta > 0. \tag{29}$$

The independent random variables appearing the above-presented formula can be easily generated using a common rejection sampling method. Excellent theoretical performance of the scheme compared with classical Euler-Maruyama type scheme has been demonstrated in Kawai and Masuda [19]; especially, the scheme works stably for arbitrary Δt without blowing up. This stability is important in engineering applications of the proposed model because it is not always possible to choose sufficiently small Δt under limited computational resources. This is the reason why we do not apply the classical Euler-Maruyama scheme to (1). To the best of the author's knowledge, this special scheme has not been used for computing FBSDEs.

The backward system (8)–(10) is discretized explicitly or semi-implicitly. Both schemes are represented in a unified manner as follows:

$$p_{j\Delta t}^{(I)} = E^{j\Delta t, I_{j\Delta t}, O_{j\Delta t}, V_{j\Delta t}} \left[p_{(j+1)\Delta t}^{(I)} + \Delta t \left\{ -w_2(I_{j\Delta t} - O_{j\Delta t}) - (\delta + \rho)p_{(j+S)\Delta t}^{(I)} + p_{(j+1)\Delta t}^{(V)} \right\} \right],$$
(30)

$$p_{j\Delta t}^{(O)} = E^{j\Delta t, I_{j\Delta t}, O_{j\Delta t}, V_{j\Delta t}} \left[p_{(j+1)\Delta t}^{(O)} + \Delta t \left\{ \begin{array}{c} -w_2(O_{j\Delta t} - I_{j\Delta t}) + w_4 \max\left\{ \hat{O} - O_{j\Delta t}, 0 \right\} \\ -\delta p_{(j+S)\Delta t}^{(O)} - p_{(j+1)\Delta t}^{(V)} \end{array} \right\} \right],$$
(31)

$$p_{j\Delta t}^{(V)} = E^{j\Delta t, I_{j\Delta t}, O_{j\Delta t}, V_{j\Delta t}} \left[p_{(j+1)\Delta t}^{(V)} + \Delta t \left\{ -w_3\left(V_{j\Delta t} - \hat{V} \right) - \delta p_{(j+S)\Delta t}^{(V)} \right\} \right],$$
(32)

where $S = 0$ corresponds to the semi-implicit scheme while $S = 1$ to the explicit scheme since we are dealing with a time-backward system. Implicit and semi-implicit numerical schemes usually require some iterative evaluation of the system of the form (30)–(32); however, the adjoint variables in the conditional expectations of (30)–(32) are of the affine form and we do not need such an iteration. For example, in the semi-implicit scheme, (32) can be rewritten to directly find $p_{j\Delta t}^{(V)}$:

$$p_{j\Delta t}^{(V)} = (1 + \delta\Delta t)^{-1} E^{j\Delta t, I_{j\Delta t}, O_{j\Delta t}, V_{j\Delta t}} \left[p_{(j+1)\Delta t}^{(V)} - \Delta t w_3 \left(V_{j\Delta t} - \hat{V} \right) \right]$$
(33)

Each conditional expectation in (30)–(32) must be evaluated numerically for implementing the schemes. We employ a least-squares Monte-Carlo method [14] using monomial and piece-wise smooth basis functions. The method itself is quite standard in numerical computation of FBSDEs as explained in Chassagneux et al. [14] but the admissible set S of basis are specialized for the proposed model:

$$S = \left\{ 1, i^{\lambda_i} o^{\lambda_o} v^{\lambda_v}, i^{\lambda_i} \max\left\{ \hat{O} - o, 0 \right\}^{\lambda_o} v^{\lambda_v} \middle| \lambda_i \lambda_o \lambda_v \neq 0, \lambda_i, \lambda_o, \lambda_v = 0, 1, 2, \dots \right\}$$
(34)

This is a collection of constant, monomial, and modified monomial functions considering the functional form of the fourth term of (5) that is inherited in the driver of (9).

After discretizing the conditional expectations of (30)–(32), the multiplication coefficient of each base is computed using a least-square procedure [14] with a classical conjugate gradient method. Other numerical solvers for inverting linear systems can be equally used as well if preferred. The optimal control (11) at each time step is implemented using the corresponding basis representation of $p_{j\Delta t}^{(O)}$.

The forward and backward systems are computed in an alternating manner using a Picard algorithm [Chapter 4 of 14]. The convergence criterion of the Picard iteration in this paper is the following: the candidate of numerical solution obtained at the i th Picard iteration ($i \in \mathbb{N}$) is a numerical solution if the updated increment of $p_0^{(O)}$, which is a deterministic value because we are assuming a deterministic initial condition, between the i th and the $(i - 1)$ th iteration becomes smaller than a threshold value $\varepsilon(= 10^{-6})$. The candidate of numerical solution at the 0 th iteration is the initial guess of the iteration.

The iteration procedure starts from simulating forward processes using the initial guesses $p_t^{(I)} \equiv 0$, $p_t^{(O)} \equiv 0$, $p_t^{(V)} \equiv 0$. In our computation below, the iteration is terminated at most 200 steps. It requires less than 10 steps for the unbounded cases.

3.2 Unconstrained Case

We use the following parameter values of an existing dam and reservoir system in Japan [11]: $\overline{V} = 6 \times 10^7$ (m³), $\underline{I} = 0$ (m³/s), $\alpha = 0.5$ (–), $\rho = 0.696$ (1/day), $a = 0.195$ (m$^{1.5}$/s$^{0.5}$), $b = 0.007$ (s/m³), $I_0 = O_0 = \underline{I}$ (m³/s), $V_0 = \overline{V}$ (m³). The parameter values for the objective function are determined so that each penalization is balanced: $\delta = 0.1$ (1/day) assuming a decision-maker having a daily perspective, $w_1 = 7,000$, $w_2 = 10$, $w_3 = 1.2 \times 10^{-3}$, $w_4 = 0$, $\hat{V} = 0.5\overline{V}$ (m³), $\hat{O} = 5$ (m³/s). These values are used unless otherwise specified. Statistical moments of the modelled inflow are as follows: average 5.130 (m³/s), standard deviation 17.71 (m³/s), skewness 12.84 (–), and kurtosis 259.1 (–), agreeing well with the observation [11]. Set $T = 30$ (day) for the unconstrained case and $T = 120$ (day) for the constrained case. Set Ω_O, Ω_V, $A = \mathbb{R}$ in the unconstrained case.

We firstly numerically analyze the unconstrained case having the exact solution of **Proposition 1**. The coefficients of the adjoint variables were obtained using a forward-Euler method with a sufficiently fine time resolution: 0.0005 (day). This numerical solution is considered as a reference solution with which performance of the explicit and semi-implicit schemes is discussed. Here, we focus on the adjoint variable $p_t^{(O)}$, namely its coefficients B_t, F_t, G_t, H_t, because they directly determine the optimal control (11). The basis employed for the computation here are $1, i, o, v$. This is the exact choice by **Proposition 1**. Hence, we can analyze statistical and temporal errors.

Figure 1 compares these coefficients for the reference solution, numerical solution with the explicit scheme, and numerical solution with the semi-implicit scheme. The total number of sample paths is $n = 10,000$ and the time increment is 1/24 (day). Both schemes capture characteristics of the reference solution despite n is not large, especially the semi-implicit scheme performs better with smaller bias. Sharp transitions of the coefficients are captured in the numerical solutions despite they have remarkably varied sizes. Both schemes have oscillatory coefficients near the initial time $t = 0$, which is considered due to using a deterministic initial condition with which least-squares Monte-Carlo methods become more ill-conditioned near $t = 0$. One may be able to avoid this issue by using some probabilistic initial condition or by adding some regularization term to the least-squares procedure.

Considering the superior performance of the semi-implicit scheme, only this scheme is analyzed in the rest of this paper. Figure 2 compares the coefficient F_t among the reference solution (black curve) and numerical solutions (unfilled circles) of the semi-implicit scheme for different computational resolution: $n = 2,500$ (red), $n = 10,000$ (blue), and $n = 40,000$ (green), where the time increment depends on n as $\Delta t = \sqrt{n}/100$ based on the basic error analysis result of FBSDEs with Monte-Carlo method [e.g., Chapter 14 of 4]. Increasing n gives less oscillatory numerical solutions closer to the reference one. The spiky oscillation visible for smaller n is considered due to the lack of total number of samples to be used in the least-squares Monte-Carlo method. Similar

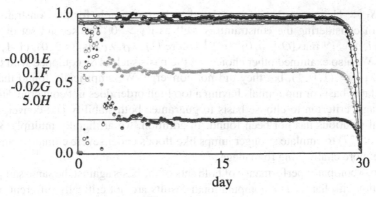

Fig. 1. Comparison of the coefficients B_t (black), F_t (red), G_t (blue), and H_t (green). Curve represent reference solution; unfilled circles (◯) numerical solution with the explicit scheme; filled circles (●) numerical solution with the semi-implicit scheme. (Color figure online)

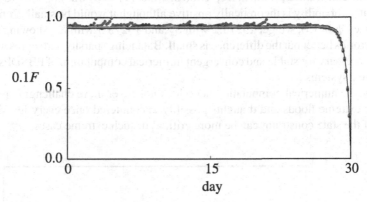

Fig. 2. Comparison of the coefficient F_t among the reference solution (black curve) and numerical solutions (unfilled circles) of the semi-implicit scheme for different computational resolution: $N = 2,500$ (red), $N = 10,000$ (blue), and $N = 40,000$ (green). (Color figure online)

problems would be encountered in simulating FBSDEs driven by large jumps but have not been reported so far. Sampling large jumps are insufficient for the small n because they are rare. The l^2-errors of the coefficient F for $n = 2,500$, $n = 10,000$, and $n = 40,000$ in $(0, T)$ are 0.2204, 0.0992, and 0.0496, respectively, suggesting an almost first-order convergence of the scheme in Δt.

3.3 Constrained Case

We compare the impacts of basis in computing more realistic cases having the state constraint. We choose $w_4 = 5$ to consider a case that is not a linear-quadratic type, and set $w_1 = 3,000$ and $w_3 = 1.2 \times 10^{-4}$ to allow for larger acceleration of the outflow discharge. The semi-implicit scheme is employed here, and the two sets of basis functions are considered for approximating the adjoint variables. The first set of basis is $S_1 = \{1, i, o, v\}$ that is considered to be too simple for the constrained case because of considering the constraint as well as $w_4 > 0$. The second set of basis is $S_2 = \{1, i, o, v, i^{\lambda_I} \max\{O - o, 0\}^{\lambda_O} v^{\lambda_V}\}$ where $(\lambda_I, \lambda_O, \lambda_V)$ is $(0, 2, 0)$, $(1, 1, 0)$, and $(0, 1, 1)$. We also examined other choices of the basis such as S_2 equipped with the base $(\lambda_I, \lambda_O, \lambda_V) = (1, 0, 1)$, but they did not converge. We empirically found that using too correlated basis or monomials having a too high order does not converge. However, at this stage, criterion to choose basis to guarantee both stability and convergence of numerical solutions has not been found. In addition, we artificially multiply 3 by the last term of (27) to emulate stronger jumps like floods under severe climate changes that would be more challenging to manage.

Figure 3 compares performance of both sets of the basis against the same sample path of the inflow discharge. The computational results are not critically different between the two sets of the basis, but there is a visible difference between the controlled water volumes near $t = T$. The fact that the set S_1, which is a too simple to approximate the adjoint variables, reasonably works suggests usefulness of sub-optimal controls in analyzing the constrained problem. Depletion of water ($V_t = 0$) was not observed in both cases, but its probability is theoretically positive although it would be small. Concerning the optimized ϕ of (5), we get 15278.2 with S_2 and 15278.5 with S_1, showing that the former performs better but the difference is small. Balancing sparsity and representability would be necessary for stable and convergent numerical computation of FBSDEs of dam and reservoir systems.

Finally, the numerical computation here does not cover more challenging problems to manage extreme floods and draughts possibly encountered once every few decades. Impacts of the state constraint can be more critical in such extreme cases.

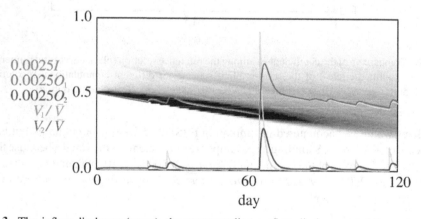

Fig. 3. The inflow discharge (green), the corresponding outflow discharges, and the water volumes. Colors of the legends correspond to the colored curves. The sub-scripts 1 and 2 represent the set of basis S_1 and S_2. The black and white contour plots are proportional to the probability density of the water volume with S_2; S_1 gives a similar result. (Color figure online)

4   Conclusion

We analyzed a new jump-driven stochastic control problem of dam and reservoir systems receiving stochastic inflow. The FBSDEs to find the optimal acceleration of outflow discharge were derived based on a maximum principle. The FBSDEs were solvable analytically in the unbounded state. The exact solution itself is useful because it serves as a benchmark for evaluating accuracy of numerical schemes and can also be used as an initial guess of iteration schemes for solving extended problems in future.

For the unconstrained case, the least-squares Monte-Carlo methods generated reasonable numerical solutions except near the initial time $t = 0$ at which the deterministic initial condition was specified. The computational results suggested that using too complicated basis do not converge, suggesting importance of analyzing mathematical structure of the problem, especially regularity of solutions to the FBSDEs. Indeed, the computational results of the second case where the state variables are constrained suggested that using basis considering regularity of the coefficients in the objective function works well. Although the true solution of the FBSDEs in this case was not found, we numerically demonstrated that the controlled outflow tracks the inflow while effectively following the targeted states. Solving the forward SDEs via parallel computing, which was not used here, is a valuable option for more efficient computation.

Our contribution in this paper is only a starting point for modeling and control of dam and reservoir systems based on FBSDEs, and there remain a number of challenges. Firstly, the full well-posedness of the FBSDEs considering the constraints of the adjoint state variables, as implied in **Remark 1**, should be addressed theoretically. We expect that this issue is resolved by a singular control approach [23]. For an engineering implementation, we must construct approximation sequences of singular control variables, but this is an unresolved issue in general. We can somehow manage this issue if a closed-form solution to the corresponding FBSDEs is found.

Secondly, not only water quantity dynamics but also water quality dynamics are important because both critically affect environmental and ecological conditions of the reservoir and further its downstream river [24]. Other factors affecting the operation goal, such as flood mitigation, should also be considered as well when necessary. Adding these factors to the problem as new state variables is straightforward, but the size of the system and thus computational cost increase. Establishment of a massive computational environment is necessary to numerically investigate such extended problems. The forward SDEs can be simulated in a parallel manner, while the backward SDEs are not necessarily so unless the special basis having disjoint domains are employed [25]. Exploring the existence issue of such useful basis for jump-driven FBSDEs is interesting both from mathematical and engineering standpoints. To construct sparse as well as well-functioning basis is also important. We are currently addressing this issue based on a regularized least-squares Monte-Carlo approach.

Acknowledgements. Kurita Water and Environment Foundation 19B018 and 20K004 and a grant from MLIT Japan for surveys of the landlocked *Ayu* sweetfish and management of seaweeds in Lake Shinji support this research. The author thanks all the members of mathematical analysis study group in Shimane University for their valuable comments on this research.

References

1. Jónsdóttir, G.M., Milano, F.: Stochastic modeling of tidal generation for transient stability analysis: a case study based on the all-island Irish transmission system. Electr. Power Syst. Res. **189**, 106673 (2020). https://doi.org/10.1016/j.epsr.2020.106673
2. Wang, Y., Hu, J., Pan, H., Failler, P.: Ecosystem-based fisheries management in the pearl river delta: applying a computable general equilibrium model. Mar. Policy **112**, 103784 (2020). https://doi.org/10.1016/j.marpol.2019.103784
3. Zavala-Yoe, R., Iqbal, H.M., Ramirez-Mendoza, R.A.: Understanding the evolution of pollutants via hierarchical complexity of space-time deterministic and stochastic dynamical systems. Sci. Total Environ. **710**, 136245 (2020). https://doi.org/10.1016/j.scitotenv.2019.136245
4. Øksendal, B., Sulem, A.: Applied Stochastic Control of Jump Diffusions. Springer, Cham (2019). https://doi.org/10.1007/978-3-030-02781-0
5. Bertoni, F., Castelletti, A., Giuliani, M., Reed, P.M.: Discovering dependencies, trade-offs, and robustness in joint dam design and operation: an ex-post assessment of the Kariba dam. Earth's Future **7**(12), 1367–1390 (2019). https://doi.org/10.1029/2019EF001235
6. Yoshioka, H.: Stochastic control of dam discharges. Wiley StatsRef: Statistics Reference Online. 0.1002/9781118445112.stat08365. (In press)
7. Picarelli, A., Vargiolu, T.: Optimal management of pumped hydroelectric production with state constrained optimal control. J. Econ. Dyn. Contr. **126**, 103940 (2020). https://doi.org/10.1016/j.jedc.2020.103940
8. Yoshioka, H., Yoshioka, Y.: Regime switching constrained viscosity solutions approach for controlling dam-reservoir systems. Comput. Math. Appl. **80**(9), 2057–2072 (2020). https://doi.org/10.1016/j.camwa.2020.09.005
9. Lesmana, D.C., Wang, S.: An upwind finite difference method for a nonlinear black-scholes equation governing European option valuation under transaction costs. Appl. Math. Comput. **219**(16), 8811–8828 (2013). https://doi.org/10.1016/j.amc.2012.12.077
10. Janga Reddy, M., Nagesh Kumar, D.: Evolutionary algorithms, swarm intelligence methods, and their applications in water resources engineering: a state-of-the-art review. H2Open J. **3**(1), 135–188 (2020). https://doi.org/10.2166/h2oj.2020.128
11. Yoshioka, H., Yoshioka, Y.: Tempered stable Ornstein–Uhlenbeck model for river discharge time series with its application to dissolved silicon load analysis. In: IOP Conference Series: Earth and Environmental Science, vol. 691, p. 012012 (2021). https://doi.org/10.1088/1755-1315/691/1/012012
12. Delong, Ł: Backward Stochastic Differential Equations with Jumps and Their Actuarial and Financial Applications. Springer, London (2020)
13. Wu, Z., Yu, Z.: Dynamic programming principle for one kind of stochastic recursive optimal control problem and Hamilton–Jacobi–Bellman equation. SIAM J. Contr. Optim. **47**(5), 2616–2641 (2008). https://doi.org/10.1137/060671917
14. Chassagneux, J.F., Chotai, H., Muûls, M.: A Forward-Backward SDEs Approach to Pricing in Carbon Markets. Springer, Cham (2017)
15. Chau, K.W., Tang, J., Oosterlee, C.W.: An SGBM-XVA demonstrator: a scalable python tool for pricing XVA. J. Math. Ind. **10**(1), 1–19 (2020). https://doi.org/10.1186/s13362-020-000 73-5
16. Fujii, M., Takahashi, A., Takahashi, M.: Asymptotic expansion as prior knowledge in deep learning method for high dimensional BSDEs. Asia-Pacific Finan. Markets **26**(3), 391–408 (2019). https://doi.org/10.1007/s10690-019-09271-7
17. Khedher, A., Vanmaele, M.: Discretisation of FBSDEs driven by càdlàg martingales. J. Math. Anal. Appl. **435**(1), 508–531 (2016). https://doi.org/10.1016/j.jmaa.2015.10.022

18. Madan, D., Pistorius, M., Stadje, M.: Convergence of BSΔEs driven by random walks to BSDEs: The case of (in) finite activity jumps with general driver. Stoch. Process. Appl. **126**(5), 1553–1584 (2016). https://doi.org/10.1016/j.spa.2015.11.013

19. Kawai, R., Masuda, H.: Exact discrete sampling of finite variation tempered stable Ornstein-Uhlenbeck processes. Monte Carlo Method Appl. **17**(3), 279–300 (2011). https://doi.org/10.1515/mcma.2011.012l

20. Song, K., Wang, F., Yi, Q., Lu, S.: Landslide deformation behavior influenced by water level fluctuations of the three gorges reservoir (China). Eng. Geol. **247**, 58–68 (2018). https://doi.org/10.1016/j.enggeo.2018.10.020

21. Abdelhady, H.U., Imam, Y.E., Shawwash, Z.: Ghanem, A: Parallelized Bi-level optimization model with continuous search domain for selection of run-of-river hydropower projects. Renew. Energ. **167**, 116–131 (2021). https://doi.org/10.1016/j.renene.2020.11.055

22. Ji, S., Zhou, X.Y.: A maximum principle for stochastic optimal control with terminal state constraints, and its applications. Commun. Inform. Syst. **6**(4), 321–338 (2006). https://projecteuclid.org/euclid.cis/1183729000

23. Hu, Y., Øksendal, B., Sulem, A.: Singular mean-field control games. Stoch. Analy. Appl. **35**(5), 823–851 (2017). https://doi.org/10.1080/07362994.2017.1325745

24. Zhang, P., Yang, Z., Cai, L., Qiao, Y., Chen, X., Chang, J.: Effects of upstream and downstream dam operation on the spawning habitat suitability of Coreius guichenoti in the middle reach of the Jinsha River. Ecol. Eng. **120**, 198–208 (2018). https://doi.org/10.1016/j.ecoleng.2018.06.002

25. Chau, K.W., Oosterlee, C.W.: Stochastic grid bundling method for backward stochastic differential equations. Int. J. Comput. Math. **96**(11), 2272–2301 (2019). https://doi.org/10.1080/00207160.2019.1658868

Optimization of Resources Allocation in High Performance Computing Under Utilization Uncertainty

Victor Toporkov⬤, Dmitry Yemelyanov(✉)⬤, and Maksim Grigorenko

National Research University "MPEI", Ul. Krasnokazarmennaya, 14, Moscow 111250, Russia
{ToporkovVV,YemelyanovDM,GrigorenkoMO}@mpei.ru

Abstract. In this work, we study resources co-allocation approaches for a dependable execution of parallel jobs in high performance computing systems with heterogeneous hosts. Complex computing systems often operate under conditions of the resources availability uncertainty caused by job-flow execution features, local operations, and other static and dynamic utilization events. At the same time, there is a high demand for reliable computational services ensuring an adequate quality of service level. Thus, it is necessary to maintain a trade-off between the available scheduling services (for example, guaranteed resources reservations) and the overall resources usage efficiency. The proposed solution can optimize resources allocation and reservation procedure for parallel jobs' execution considering static and dynamic features of the resources' utilization by using the resources availability as a target criterion.

Keywords: Computing · Grid · Resource · Scheduling · Uncertainty · Dynamic · Availability · Probability · Job · Allocation · Optimization

1 Introduction

Today, Grid and cloud computing systems are used universally. Due to their commercial reach and low entry threshold, they attract users with different technical skills, who solve a wide range of computational tasks (time- and volume-wise) and require different quality of service.

It usually takes certain economic costs to build and manage the necessary computing infrastructure, including the purchase and installation of equipment, the provision of power supply, and user support. Thus, when a budget for job performance is limited, it becomes important to allocate suitable resources efficiently in accordance with both technical specification and a constraint on the total cost [1–6].

The system's resources may include computational nodes, storage devices, data communication links, software, etc. Each resource has a set of characteristics, their values determine its suitability for performing a specific job. Generally, computational nodes have the widest set of characteristics. For example, a virtual machine is the main computing resource in the commonly used CloudSim simulator [2, 3], its characteristics

© Springer Nature Switzerland AG 2021
M. Paszynski et al. (Eds.): ICCS 2021, LNCS 12747, pp. 540–553, 2021.
https://doi.org/10.1007/978-3-030-77980-1_41

include overall performance, number of CPU cores, size of RAM and disk memory, bandwidth limit of the data link.

It is worth mentioning the dynamic utilization issue of available resources and computational nodes at time. High performance and distributed computing systems (HPDCS) are the dynamic systems, in which the following processes take place: execution of parallel jobs from multiple users, utilization with local jobs, maintenance works, a physical shutdown of nodes (both scheduled and unscheduled). To procure the reliability and dependability of such systems, an advance allocation mechanism is used [5–8]. This mechanism allows one to pre-allocate resources for a specific job and, thereby, prevents possible contention between jobs. Thus, a utilization schedule for each resource can be obtained: a list of utilization intervals (allocated time, scheduled maintenance, and outages) and downtime periods. Downtime periods can be used to perform other jobs and to allocate the resources for the execution of user jobs.

The problem of scheduling and co-allocating resources for executing parallel jobs in a distributed computing system with non-dedicated resources is stated as follows.

- The set R of the computing system resources, as a rule, is heterogeneous and includes resources r_i of several types with different sets of characteristics C_i. The values of these characteristics for the resources of the same type may also differ. Among the most important characteristics of a resource, one can single out its performance, which affects the execution time of a job, as well as the cost required to allocate the resource. Besides, at any specific time, some subsets of the resources may be unavailable for a user job. Therefore, available resources, as a rule, are represented in classical models as a set of slots - intervals of availability of each resource [5–8].
- Resources co-allocation for a parallel job execution typically requires selection (allocation) of a set of resources with types and characteristics defined by the user who is running the job. The resource request for the job execution includes the number of concurrently required resources n, the minimum suitable values of the characteristics Ch_i, the volume of task V (the number of calculations/instructions) or the ordered resource allocation time T, as well as the total execution budget C [1–8].

However, as a rule, the structure and specifics of submitted jobs in HPDCS imply some uncertainty, primarily in the execution time and load of the allocated resources. So, users can only roughly estimate the execution time of their jobs, while special expert systems for predicting the execution time of user programs or the level of resource load (based on the use of machine learning, statistics, and big data) present the results in the form of probabilities of outcomes [4, 9–14].

In this paper, we propose proactive algorithm for resources allocation and reservation in heterogeneous market-based computing environments considering static and dynamic resources availability uncertainties. The uncertainties are formalized with the availability probability functions as a natural way of statistical and machine learning predictions presentation. The novelty of the proposed solution is in general knapsack-based resources selection procedure performing resources availability maximization according to the parallel job requirements.

The paper is organized as follows. Section 2 presents a brief overview of works, related to the jobs execution uncertainties and probabilities in parallel computing environments. Section 3 presents a formal model of the resources' utilization and a general procedure for the dynamic resources' allocation optimization. Additional details are provided for the subset selection and time scan algorithms. Section 4 provides details about the simulation experiment setup, simulation results and analysis. Section 5 summarizes the paper and describes further research topics.

2 Related Works

Existing systems of distributed computing usually perform resources allocation and distribution based on deterministic models of the resource scheduling [1–3, 5–7]. As a result, the expected efficiency and accuracy of such scheduling methods are reduced due to unforeseen resource events (failures, maintenance works), inaccurate estimates and predictions of the jobs' characteristics and execution times. Late job completion time requires rescheduling of all the subsequent jobs or shutting down the job with possible loss of results. Early release of resources also requires rescheduling to minimize the resource downtime. For example, according to an existing approach [8], a scheduler may double the user's runtime estimates to improve the efficiency of the job flow.

Many such systems rely on the reactive approach [4, 9, 10], when an actual state of the computing environment is analyzed, and the appropriate migration and re-scheduling decisions are made on the fly. However, these rescheduling and migration operations incur additional time, cost, and network losses. Thus, proactive algorithms, which concentrate on the resource utilization predictions and advanced resources allocations may improve the overall resources usage efficiency.

In [4] a simple uncertainty-based scheduling approach is proposed for a workflow job execution. Concepts of deadline, budget and execution surety are defined to choose the Pareto optimal set of the schedules, satisfying the user requirements. The task execution surety parameter is provided for each available resource by their owners/administrators.

Paper [11] discusses the problem of scheduling a flow of sequential jobs with the execution time uncertainties. Different resources allocation strategies are studied to minimize the total execution time based on the runtime probabilities of the queued jobs. The jobs' execution times are modeled as self-similar heavy-tail processes. In [12] a single-machine scheduling model is proposed to minimize a total flowtime of jobs with processing times characterized by normally distributed random variables. In [13], a set of distinct availability states is defined to model resource behavior and probabilities state transitions.

In [14] we studied the problem of a static resource co-allocation for a parallel job execution in a computing environment with utilization uncertainties. Similarly to [4] we used concepts of the execution time deadline, cost limit (budget) and the probability of a successful execution as a target optimization criterion. We used a knapsack-based algorithm to maximize an aggregate availability probability of a set of simultaneously allocated resources with the corresponding time and cost constraints. However, in this static scenario, the resources' availability probabilities are modeled as simple normally distributed estimates at the given static moment of time.

Current paper extends [14] by studying the dynamic variation of a resources co-allocation problem during some scheduling interval: when the parallel job may be executed at any time inside the given interval. For this purpose, for each independent resource we use heavy-tail distribution to model utilization uncertainties caused by inaccurate estimates in other jobs' execution runtimes. Additional scheduling optimization methods are proposed and analyzed to handle the emerging time scan problem.

3   Resource Selection Algorithm

3.1   Resources Utilization Model

We consider a set R of heterogeneous computing nodes with different performance p_i and price c_i characteristics.

The probabilities (predictions) of the resource's availability and utilization for the whole scheduling interval L are provided as input data. Dynamic job execution uncertainties are modeled as a sequence of *allocation, occupation* (actual execution) and *release* events with the *occupation* probability $P_o \leq 1$. Global (static) resources utilization uncertainties, such as maintenance works or network failures, are modeled as a continuous *occupation* events with $P_o << 1$ during the whole considered scheduling interval.

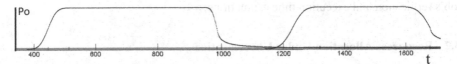

Fig. 1. Example of a resource utilization probability schedule.

Figure 1 shows an example of a single resource occupation probability P_o schedule. With two jobs already assigned to the resource, there are two resources allocation events (with expected times of allocation at 445 and 1230 time units), two resources occupation events (starting at 513 and 1319 time units) and two resources release events (expected release times are 986 and 1676 time units respectively). Gray translucent bar at the bottom of the diagram represents a sum of global utilization events with a total resource occupation probability $P_o = 0.05$.

A detailed analysis of the main utilization characteristics of real HPDCS systems was made to design and simulate an adequate resources utilization model. As the basis for modeling the availability and utilization probability of computational nodes, the log files of the ForHLR II supercomputer from the Karlsruhe Institute of Technology in Germany were taken for the analysis [15, 16]. The available files contain information on the execution of jobs from June 2016 to January 2018.

After carrying out many experiments, the normal distribution on a logarithmic scale (lognormal) was chosen as the most suitable for modeling the jobs' length and size characteristics. The main parameters of the distribution (mathematical expectation and variance) were selected experimentally to achieve an acceptable accuracy. As a result, the

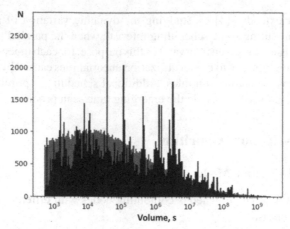

Fig. 2. Job size distributions of real (black) and simulated (blue) job-flows. (Color figure online)

generated distribution by form largely replicates the original one (Fig. 2). More formal comparison gives 0.14 value by the Kolmogorov - Smirnov test.

Thus, the resources *allocation* events are modeled by random variables with a normal distribution. Resources *release* events are modeled with lognormal distribution and expose heavy tails [11, 16]. Expected allocation and release times are derived from the job's replication and execution time estimations.

3.2 Resources Allocation Under Uncertainties

To execute a parallel job a set of simultaneously idle nodes (a window) should be allocated ensuring user requirements from the resource request. The resource request usually specifies number n of nodes required simultaneously, their minimum applicable performance p, job's total computational volume V and a maximum available resources allocation budget C.

These parameters constitute a formal generalization for resource requests common among distributed computing systems and simulators [2, 5, 7].

Common allocation and release times for all the window resources ensure the possibility of inter-node communications during the whole job execution. In this way, the occupation and availability probabilities should be estimated for each resource during the scheduling interval L. For the job scheduling, values $P_a^{r_i}(t; t + T)$ may be derived, representing a probability that resource r_i will be available for the whole job execution interval T starting at time t.

When a set of n resources is required for a job execution for a period T, the total window availability P_a^w during the expected job execution interval can be estimated as a product of availability probabilities of each independent window nodes:

$$P_a^w(t) = \prod_i^n P_a^{r_i}(t; t + T). \tag{1}$$

If any of the window nodes will be occupied during the expected job execution interval T, the whole parallel job will be postponed or even aborted. Therefore, a common resources allocation problem is a maximization of a total resources' availability probability.

Based on the model above we consider the following job resources allocation problem in heterogeneous computing environment with non-dedicated resources and utilization uncertainties: during a scheduling interval L allocate a set of n nodes with performance $p_i \geq p$ for a time T, with common allocation and release times and a restriction C on the total allocation cost. As a target optimization criterion, we assume maximization of a whole window availability probability $P_a^w(1)$.

The solution for this problem may be divided into two sub-problems.

1. Static sub-problem. Given the time t_k and values $P_a^{r_i}(t_k; t_k + T)$ of the resources' availability for the following period T, allocate a subset of n resources according to the job requirements with the maximum probability $P_a^w(t_k)$.
2. Dynamic generalization. Perform time scan and execute the first sub-problem for each time moment $t_k \in [0; L]$. The solution is then obtained as a maximum from all the intermediate solutions: $P_a^w = \max_{t_k} P_a^w(t_k)$.

Thus, further in this paper we study different approaches for these two sub-problems implementation.

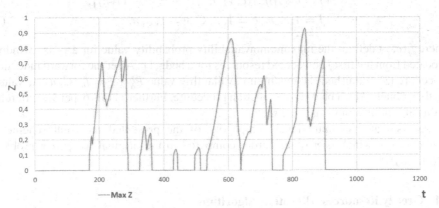

Fig. 3. An example of $\max P_a^w(t)$ function for a parallel job resources allocation.

As an example, Fig. 3 shows maximum values of function $Z = P_a^w(t)$ obtained for a parallel job on the interval $[0; 1200]$ with the maximum availability probability reaching 0.93 at $t^{\max} = 834$.

3.3 Near-Optimal Resources Allocation

Let us discuss in more details the procedure which allocates an optimal (according to the probability criterion P_a^w) subset of n resources at some static time moment t_k.

We consider the following total resources availability criterion $P_a^w = \prod_i^n P_a^{r_i}$, where $P_a^{r_i} = P_i$ is an availability probability of a single resource r_i on the interval $[t_k; t_k + T]$.

In this way we can state the following problem of an n-size window subset allocation out of m available nodes in the system:

$$P_a^w = \prod_{j=1}^{m} x_j P_j, \tag{2}$$

with the following restrictions:

$$\sum_{j=1}^{m} x_j c_j \leq C$$

$$\sum_{j=1}^{m} x_j = n,$$

$$x_j \in \{0, 1\}, j = 1..m,$$

where c_j is total cost required to allocate resource r_j for a time T, x_j - is a decision variable determining whether to allocate resource r_j ($x_j = 1$) or not ($x_j = 0$) for the current window.

In [14] based on a classical 0–1 *Knapsack* problem solution we proposed the following dynamic programming recurrent scheme to solve problem (2):

$$f_j(c, v) = \max\{f_{j-1}(c, v), f_{j-1}(c - c_j, v - 1) * P_j\},$$
$$j = 1, .., m, c = 1, .., C, v = 1, .., n, \tag{3}$$

where $f_j(c, v)$ defines the maximum availability probability value for a v-size window allocated from the first j considered resources for a budget c. After the forward induction procedure (3) is finished the maximum availability value $P_{a\ max}^w = f_m(C, n)$. x_j values are then obtained by a backward induction procedure. Further in this paper we will refer to this algorithm simply as *Knapsack*.

An estimated computational complexity of the presented recurrent scheme is $O(m * n * C)$, which is n times harder compared to the original *Knapsack* problem ($O(m * C)$).

3.4 Greedy Resources Allocation Algorithm

Another approach for the static subset allocation sub-problem is to use more computationally efficient greedy algorithms. We outline four main greedy algorithms to solve the problem (2).

1. *MaxP* selects first n nodes providing maximum availability probability P_j values. This algorithm does not consider total usage cost limit and may provide infeasible solutions. Nevertheless, *MaxP* can be used to determine the best possible availability options and estimate a budget required to obtain them.

2. An opposite approach *MinC* selects first n nodes providing minimum usage cost c_j or an empty list in case it exceeds a total cost limit C. In this way, *MinC* does not perform any availability optimization, but always provides feasible solutions when it is possible. Besides, *MinC* outlines a lower bound on a budget required to obtain a feasible solution.

3. Third option is to use a weight function to regularize nodes in an appropriate manner. *MaxP/C* uses $w_j = P_j/c_j$ as a weight function and selects first n nodes providing maximum w_j values. Such an approach does not guarantee feasible solutions but performs some availability optimization by implementing a compromise solution between *MaxP* and *MaxC*.

4. Finally, we consider a joint approach *GreedyJnt* for a more efficient greedy-based resources allocation. The algorithm consists of three stages.

    a. Obtain *MaxP* solution and return it if the constraint on a total usage cost is met.
    b. Else, obtain *MaxP/C* solution and return it if the constraint on a total usage cost is met.
    c. Else, obtain *MinC* solution and return it if the constraint on a total usage cost is met.

This combined algorithm is designed to perform the best possible greedy optimization considering restrictions on total resources allocation size and cost.

Estimated computational complexity for the greedy resources' allocation step is $O(m * \log m)$.

3.5 Time Scan Optimization

Dynamic generalization of the static resources' allocation problem requires a full-time scan performed over all the considered scheduling interval L. In general, this leads to a significant increase in the computational cost of the dynamic scheduling algorithm (especially, when a full knapsack-based optimization should be performed for all time moments $t_k \in [0; L]$).

To optimize the performance of the proposed resources allocation procedure during the time scan we consider a computational method which performs search for the maximum from a set of starting time points. Assuming, that the functions $P_a^{r_i}(t)$ for each resource are continuous in time (see Fig. 1), then their product $P_a^w(t)$ will be continuous as well. This means that certain computational algorithms are applicable for $P_a^w(t)$ function study and the extrema search. Figure 2 shows an example of $P_a^w(t)$ function calculated by the resources allocation algorithm after scanning all time points if [0; 1200] interval.

A general procedure for $\max P_a^w(t)$ search optimization during the scheduling interval L can be presented as follows.

1. A set of starting time points is allocated on the interval L. Their particular locations can be given as 1) uniform, 2) randomized, 3) a combination of options 1 and 2.
2. At each starting time point t_i^s the value of $P_a^w(t_i^s)$ is calculated by the static resources' allocation algorithm (*Knapsack* or *GreedyJnt*) based on actual resources state at t_i^s.

3. The gradient value is determined for each starting point by calculating and comparing neighbor values $P_a^w(t_i^s + 1)$ and $P_a^w(t_i^s - 1)$ with $P_a^w(t_i^s)$.
4. From each starting point t_i^s an incremental movement is performed in the direction of increasing the gradient by the sequential calculation of $P_a^w(t_i^s \pm \delta * k) = P_a^w(t_i^s, k)$, where k is a step number. The movement is stopped if the maximum is reached (when $P_a^w(t_i^s, k) < P_a^w(t_i^s, k - 1)$) and, thus, can be found on the interval $[t_i^s \pm \delta * (k - 1); t_i^s \pm \delta * k]$. Besides, the search movement stops if any other starting points t_{i+1}^s or t_{i-1}^s are reached. In this case, the search will be continued independently, starting from the corresponding points.

It should be noted that the above optimization procedure does not guarantee an exact solution: scenarios of finding local maxima or missing abrupt function changes are possible. Improving the accuracy is possible by increasing the set of starting points and by decreasing the search step length δ. On the other hand, the performance of this procedure is significantly increased compared to the full-time scan: the calculation of function $P_a^w(t)$ is performed on a limited set of time points, guaranteed to be smaller than the whole interval L.

4 Simulation Study

4.1 Simulation Environment

We performed a series of simulations to study optimization properties of the proposed dynamic resources allocation approaches. An experiment was prepared as follows using a custom distributed environment simulator [5, 6, 14]. For our purpose, it implements a heterogeneous resource domain model: nodes have different usage costs and performance levels. A space-shared resources allocation policy simulates a local queuing system (like in CloudSim [2, 3]) and, thus, each node can process only one task at any given simulation time. Additionally, each node supports a list of active global and local job utilization events.

Global static uncertainty events represent resources failure or shutdown susceptibility and keep a constant occupation probability during the whole scheduling interval L. Static utilization is generated for each resource based on a random variable P_o of occupancy probability with a normal. System-wide global-load parameter defines a standard deviation for P_o and is used to set an average global utilization for the whole computing environment. Thus, for example, when global load $= 0.05$, about 68% of the resources on average have global occupancy probability $P_o^{rj} < 0.05$. More detailed study of a static resources' allocation problem under global utilization uncertainties was provided in [14].

Dynamic job-based utilization uncertainty is generated based on a preliminary job-flow scheduling simulation. For each resource, a list of single-node jobs is generated with random jobs submit times, lengths, start time and finish time uncertainty estimations. The jobs are ordered by their submit time and are scheduled in advance starting either at the submit time, or after the previous job is finished. During this scheduling, a chain of the resource *allocation, occupation* and *release* events is generated for each job. Corresponding expected times and standard deviations are defined by the job length

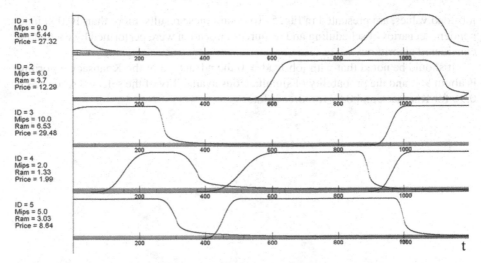

ID = 1
Mips = 9.0
Ram = 5.44
Price = 27.32

ID = 2
Mips = 6.0
Ram = 3.7
Price = 12.29

ID = 3
Mips = 10.0
Ram = 6.53
Price = 29.48

ID = 4
Mips = 2.0
Ram = 1.33
Price = 1.99

ID = 5
Mips = 5.0
Ram = 3.03
Price = 8.64

Fig. 4. An example of static and dynamic load generated for system resources.

and uncertainty parameters. More details regarding the simulated job-flow properties provided in Sect. 3.1. A total length of jobs generated for each resource is determined by a system wide job-load parameter. For example, when job-load = 0.1, a total length of locally generated jobs constitutes nearly 10% of the considered scheduling interval L.

Figure 1 shows a single resource utilization schedule with global and dynamic utilization events generated based on the procedures described above. Figure 4 shows an example of global and dynamic utilization uncertainties generated for a subset of the system resources in simulator [14].

4.2 Dynamic Resources Allocation

To solve the dynamic resources allocation problem for a parallel job, it is necessary to consider the available resources' schedule and utilization events which change over time Thus, the scheduling problem requires allocation of a set of suitable resources not at some static moment t_k, but during a given time interval.

Since the computational complexity and working time of the algorithms under consideration increase in proportion to the size of the considered scheduling interval, the following parameters of the scheduling problem were chosen to minimize the simulation time. It is required to maximize the probability of simultaneous availability of 6 concurrently available nodes to perform a job with a volume of 200 computational units on a time interval [0; 800] in a computing environment that includes 64 heterogeneous computational nodes. Initial load of computational nodes with global events global load = 0.05. The dynamic load of the computing system changed during the simulation within the limits of job-load ∈ [0; 1].

The obtained results indicating the availability of the resources selected by the Knapsack (Sect. 3.3) and GreedyJnt (Sect. 3.4) algorithms depending on the dynamic load job-load values, are presented in Fig. 5. To obtain these results, more than 10,000 independent scenarios of scheduling and resources allocation were performed by each of the considered algorithms.

It should be noted that with job-load $= 0$ the advantage of the Knapsack algorithm is about 9%, and the probability of simultaneous availability of the selected resources is 0.96 for Knapsack and 0.87 for GreedyJnt.

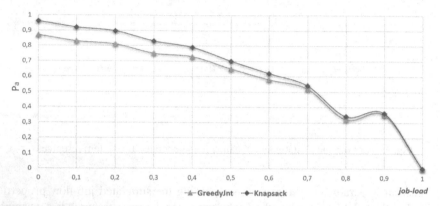

Fig. 5. Simulation results: P_a^w resources availability obtained by Knapsack and GreedyJnt algorithms depending on the resources utilization level.

With an increase in the dynamic load of the system (*job-load* > 0), the highest achievable probability P_a^w of simultaneous resource availability, as expected, sensibly monotonically decreases. The local maximum at *job-load* $= 0.9$ is explained by the fact that under conditions of extra high dynamic load, the number of experiments in which it was possible to find six concurrently available resources, turned out to be statistically insignificant (about 10 results). At the same time, when the value of job-load $= 1$ (full initial utilization of the system) was reached, a suitable set of resources ($P_a^w = 0$) was not found in any of the experiments at any time instant $t_k \in [0; 800]$.

Also, *Knapsack* provided a higher availability probability P_a^w of the required set of resources at all the considered values of the dynamic load (*job-load* < 1) in comparison to *GreedyJnt*. However, the relative advantage decreases from about 9% to almost 0% as the *job-load* increases. This decrease in relative efficiency is explained by a decrease in the dimensionality and variability of problem (2) with an increase of the resources load. For example, when *job-load* $= 0$ all 64 resources are available at every instant with a probability of at least 0.95 (due to global-load $= 0.05$). Then as the *job-load* increases, many resources fall out of consideration due to a high probability of being utilized by other jobs (see Fig. 4). Thus, for large *job-load* values, the static algorithms often solved the degenerated problem of selecting a set of 6 concurrently available resources from 6 resources in the system that remained unloaded.

4.3 Time Scan Optimization

The time and accuracy characteristics of the proposed time scan optimization procedure (Sect. 3.5) were studied based on a resource's allocation problem in computing environment with dynamically changing utilization level. Figure 4 presents an example of a utilization schedule generated for a few computational nodes in the simulation environment [14].

To obtain reliable results, we performed more 1000 independent simulation scenarios of resources allocation for a single parallel job. The computing environment consisted of 64 heterogeneous computing nodes of varying cost and performance characteristics with dynamically changing occupation function $P_o(t)$: *job-load* $= 0.5$, *global-load* $= 0.05$. The job scheduling problem required allocation of six nodes for 200 units of time on the interval $L \in [0; 800]$. The target optimization criterion P_a^w is a simultaneous availability of the selected resources. As an additional criterion, a total algorithm working times was measured.

The time scan optimization procedure described in Sect. 3.5 was implemented with a different number of the starting points: $\{1, 5, 10, 20, 50, 100\}$.

Table 1 shows the relative results in terms of working time (performance acceleration) and accuracy in comparison with the full scan approach.

Table 1. Algorithms' efficiency comparison in terms of accuracy and performance (working time acceleration) relative to the *Knapsack* full time scan implementation

Algorithm	Optimization accuracy	Time acceleration
Full scan (*Knapsack*)	1	1
1 starting point (*Knapsack*)	0,8	65
5 starting points (*Knapsack*)	0,93	17
10 starting points (*Knapsack*)	0,96	10,5
20 starting points (*Knapsack*)	0,97	8,3
50 starting points (*Knapsack*)	0,99	6,8
100 starting points (*Knapsack*)	0,99	3,7
Full scan (*GreedyJnt*)	0,953	143

As expected, with an increase in the number of starting points the accuracy of the approximate procedure tends to 1 (i.e., to the optimal solution obtained with a full scan search). Already with 50 starting points (on an interval of 801 points) the accuracy of the optimized solution reaches 99%, while the calculation time is accelerated by almost 7 times.

On the other hand, full scan procedure with *GreedyJnt* algorithm achieves 95% accuracy with a 143× speedup! Thus, it is advisable to apply *Knapsack* with this time scan optimization technique if it is necessary to achieve a high accuracy in the presence of the light computation time restrictions. In this case, it is possible to speed up the work time by about an order of magnitude. With tighter time constraints, additional speedup can be achieved by using a greedy counterpart. In addition, the time scan optimization is applicable to *GreedyJnt* algorithm as well: for example, running *GreedyJnt* algorithm from 50 starting points allows you to speed up the computation time by 1000 times, while the solution accuracy will decrease only to 94%.

5 Conclusion and Future Work

In this work, we presented procedure for a reliable resources' allocation in high performance computing systems with heterogeneous hosts considering utilization uncertainty. The uncertainties are formalized with probability functions as a natural way of statistical and machine learning predictions representation. The proposed solution uses an availability criterion to optimize resources allocation under static and dynamic utilization features. *Knapsack*-based and greedy algorithms were implemented and compared in a dynamic procedure performing optimized time scan over a specified scheduling interval. Both approaches were able to successfully optimize availability of the selected resources.

We considered several types of static and dynamic job-based resources utilization events with different load levels.

The simulation study addressed two main criteria: optimization efficiency and algorithms working time. *Knapsack*-based solution advantage over the greedy approach by the resources availability criterion at average reaches 5% but requires nearly 100 times more time for the calculations. Considering a relatively high computation complexity of the *Knapsack*-based solution, several optimization options were proposed to provide 99% accuracy 10 times faster or almost 94% accuracy 1000 times faster.

In our further work, we will refine the resource utilization model to simulate different types of global and local utilization events closer to real systems.

References

1. Lee, Y.C., Wang, C., Zomaya, A.Y., Zhou, B.B.: Profit-driven scheduling for cloud services with data access awareness. J. Parallel Distrib. Comput. **72**(4), 591–602 (2012)
2. Calheiros, R.N., Ranjan, R., Beloglazov, A., De Rose, C.A.F., Buyya, R.: CloudSim: a toolkit for modeling and simulation of cloud computing environments and evaluation of resource provisioning algorithms. J. Softw. Pract. Experience **41**(1), 23–50 (2011)
3. Samimi, P., Teimouri, Y., Mukhtar, M.: A combinatorial double auction resource allocation model in cloud computing. J. Inf. Sci. **357**, 201–216 (2016)
4. Sample, N., Keyani, P., Wiederhold, G.: Scheduling under uncertainty: planning for the ubiquitous grid. In: Arbab, F., Talcott, C. (eds.) Coordination Models and Languages. LNCS, vol. 2315, pp. 300–316. Springer, Heidelberg (2002). https://doi.org/10.1007/3-540-46000-4_28
5. Toporkov, V., Yemelyanov, D.: Optimization of resources selection for jobs scheduling in heterogeneous distributed computing environments. In: Shi, Y., et al. (eds.) Computational Science – ICCS 2018. LNCS, vol. 10861, pp. 574–583. Springer, Cham (2018). https://doi.org/10.1007/978-3-319-93701-4_45
6. Toporkov, V., Yemelyanov, D., Toporkova, A.: Coordinated global and private job-flow scheduling in Grid virtual organizations. Simul. Model. Pract. Theory **107**, 102228 (2021)

7. Jackson, D., Snell, Q., Clement, M.: Core Algorithms of the Maui Scheduler. In: Feitelson, D.G., Rudolph, L. (eds.) JSSPP 2001. LNCS, vol. 2221, pp. 87–102. Springer, Heidelberg (2001). https://doi.org/10.1007/3-540-45540-X_6
8. Tsafrir, D., Etsion, Y., Feitelson, D.G.: Backfilling using system-generated predictions rather than user runtime estimates. IEEE Trans. Parallel Distrib. Syst. **18**(6), 789–803 (2007)
9. Tchernykh, A., Schwiegelsohn, U., El-ghazali, T., Babenko, M.: Towards understanding uncertainty in cloud computing with risks of confidentiality, integrity, and availability. J. Comput. Sci. **36**, 100581 (2019)
10. Chaari, T., Chaabane, S., Aissani, N., Trentesaux, D.: Scheduling under uncertainty: survey and research directions. In: 2014 International Conference on Advanced Logistics and Transport (ICALT), pp. 229–234 (2014)
11. Ramírez-Velarde, R., Tchernykh, A., Barba-Jimenez, C., Hirales-Carbajal, A., Nolazco-Flores, J.: Adaptive resource allocation with job runtime uncertainty. J. Grid Comput. **15**(4), 415–434 (2017)
12. Wu, C.W., Brown, K.N., Beck, J.C.: Scheduling with uncertain durations: modeling beta-robust scheduling with constraints. J. Comput. Oper. Res. **36**(8), 2348–2356 (2009)
13. Rood, B., Lewis, M.J.: Grid resource availability prediction-based scheduling and task replication. J. Grid Comput. **7**, 479–500 (2009)
14. Toporkov, V., Yemelyanov, D.: Availability-based resources allocation algorithms in distributed computing. In: Voevodin, V., Sobolev, S. (eds.) Supercomputing. CCIS, vol. 1331, pp. 551–562. Springer, Cham (2020). https://doi.org/10.1007/978-3-030-64616-5_47
15. https://www.cse.huji.ac.il/labs/parallel/workload/ (2021)
16. Feitelson, D.G.: Workload Modeling for Computer Systems Performance Evaluation, pp. 501–540. Cambridge university press, New York (2015)

A Comparison of the Richardson Extrapolation and the Approximation Error Estimation on the Ensemble of Numerical Solutions

Aleksey K. Alekseev[(⊠)] [iD], Alexander E. Bondarev[iD], and Artem E. Kuvshinnikov[iD]

Keldysh Institute of Applied Mathematics, RAS, Moscow, Russia
bond@keldysh.ru

Abstract. The epistemic uncertainty quantification concerning the estimation of the approximation error using the differences between numerical solutions treated in the Inverse Problem statement is addressed and compared with the Richardson extrapolation. The Inverse Problem is posed in the variational statement with the zero order Tikhonov regularization. The ensemble of numerical results, obtained by the OpenFOAM solvers for the inviscid compressible flow with a shock wave is analyzed. The approximation errors, obtained by the Richardson extrapolation and the Inverse Problem are compared with the exact error, computed as the difference of numerical solutions and the analytical solution. The Inverse problem based approach is demonstrated to be an inexpensive alternative to the Richardson extrapolation.

Keywords: Richardson extrapolation · Approximation error · Ensemble of numerical solutions · Euler equations · OpenFOAM · Inverse problem

1 Introduction

The estimation of the approximation error that is a subject of the epistemic uncertainty quantification is the main element for the verification of numerical calculations. The standards [1, 2] recommend the Richardson extrapolation (RE) as one of the main tools for the verification of solutions and codes in the Computational Fluid Dynamics. RE provides the pointwise approximation of the approximation error, unfortunately, at the cost of the extremely high computational burden [3–5]. There exist some computationally inexpensive approaches for the approximation error norm estimation, for example [6]. However, these approaches do not provide the pointwise information on the error. Thus, the need for a computationally inexpensive a posteriori estimation of the pointwise approximation error exists. For this reason we consider herein the computationally inexpensive approach to a posteriori error estimation [7] that is based on the ensemble of numerical solutions obtained by different algorithms. The approximation error is estimated using the differences of solutions at every grid node that are treated by the Inverse Ill-posed Problem (IP) stated in the variational statement with the Tikhonov zero order regularization [8, 9]. The results of the numerical tests for compressible Euler equations are provided that demonstrate both the estimated error and the exact error (obtained by a

© Springer Nature Switzerland AG 2021
M. Paszynski et al. (Eds.): ICCS 2021, LNCS 12747, pp. 554–566, 2021.
https://doi.org/10.1007/978-3-030-77980-1_42

comparison of numerical solution with the exact analytical one). The paper [7] analyzed the pointwise error by comparison with the etalon solutions [10]. In the present paper we compare the error computed by Inverse problem with the exact error (engendered by the analytic solutions) and the results by the Richard-son extrapolation.

2 The Richardson Extrapolation for Flows with Discontinuities

We consider the numerical solution u_h obtained by some CFD solver, the exact (unknown) solution $\tilde{u}$, the approximation error $\Delta u = u_h - \tilde{u}$. The Richardson extrapolation (RE) applies two numerical solutions obtained for consequently refined grids for the pointwise (m is the number of the coarse grid point) estimation of exact solution and error:

$$
\begin{aligned}
u_m^{(1)} &= \tilde{u}_m + C_m h_1^\alpha, \\
u_m^{(2)} &= \tilde{u}_m + C_m h_2^\alpha.
\end{aligned}
\tag{1}
$$

This equation enables to estimate the approximation error as $\Delta u_m^{(1)} \approx C_m h_1^\alpha$. To apply RE the convergence order α should be *a priori* known and the solutions should belong to the asymptotic range of the convergence (the upper order terms should be small and may be neglected). To ensure the sequence of solutions to belong the asymptotic range one should use several additional levels of mesh refinement that caused an additional computational cost. The traditional domain for the Richardson extrapolation corresponds to the elliptic and parabolic problems with smooth solutions. The behavior of Richardson extrapolation error estimates for simulations of solutions with jumps, such as shock and contact lines for fluid mechanics, is known to be problematic [4, 5]. It is caused by the fact that for CFD problems of inviscid compressible fluid containing shock waves and contact discontinuities the error order is essentially spatially local and depends on the type of flow structure [3, 4, 11, 12]. So, it is necessary to extend RE for the additional estimation of the local order of convergence that is performed by [3, 5].

The pointwise results of numerical computation for three consequent meshes of different steps (to avoid the interpolation issue, the steps corresponds consequent doubling of the number of grid nodes: $h_q \sim (1/2)^{q-1}$, $q = 1, 2, 3$) may be presented as:

$$
\begin{aligned}
u_m^{(1)} &= \tilde{u}_m + C_m h_1^{\alpha_m}, \\
u_m^{(2)} &= \tilde{u}_m + C_m h_2^{\alpha_m}, \\
u_m^{(3)} &= \tilde{u}_m + C_m h_3^{\alpha_m}.
\end{aligned}
\tag{2}
$$

This system (generalized Richardson extrapolation (GRE), [3]) may be resolved regarding $\tilde{u}_m, C_m, \alpha_m$ by several methods described by [3–5] if C_m is independent on h and higher order terms may be neglected, that is, the solution is in the asymptotic range. This approach requires the use of several sequentially refined grids. The number of such grids can increase if the results for the coarse grid fall outside the asymptotic range. Unfortunately, in the contrast to the standard RE, the estimation of α_m is the ill-posed problem [5] and requires a regularization in order to obtain the stable results. The approximation error on the rough grid in the frame of GRE may be estimated as

$$
\Delta u_m^{(1)} \approx C_m h_1^{\alpha_m}.
\tag{3}
$$

It should be mentioned that the accuracy of RE (and GRE) for the error estimation remains unresolved quantitatively that excludes estimates by computable inequalities of the form $\left| \Delta u_m^{(1)} \right| \leq C$. So, the Richardson extrapolation provides the pointwise approximation for the error field at the cost of the extremely high computational burden, requires a regularization (in its generalized form) and does not provide the mathematically rigorous estimates in the form of the inequality.

3   The Relation of Approximation Error and the Distances Between Numerical Solutions

Let's consider the approximation error estimation using the distances between numerical solutions treated using the Inverse problem in accordance with [7]. We analyze an ensemble of numerical solutions $u_m^{(i)}$ ($i = 1...n$), obtained by n different numerical algorithms (different solvers) on the same grid. Herein, we apply certain vectorization, so m is the grid point number ($m = 1,...,L$). We note the projection of the exact solution $\tilde{u}$ onto the grid as $\tilde{u}_{h,m}$ and the approximation error for i-th solution as $\Delta u_m^{(i)}$ $\left(u_m^{(i)} = \tilde{u}_m + \Delta u_m^{(i)} \right)$. Since the differences of numerical solutions $d_{ij,m} = u_m^{(i)} - u_m^{(j)} = \tilde{u}_{h,m} + \Delta u_m^{(i)} - \tilde{u}_{h,m} - \Delta u_m^{(j)} = \Delta u_m^{(i)} - \Delta u_m^{(j)}$ are equal to the differences of approximation errors one may get $N = n \cdot (n-1)/2$ independent equations relating unknown approximation errors and computable differences of numerical solutions

$$D_{ij} \Delta u_m^{(j)} = f_{i,m}. \tag{4}$$

Herein, D_{ij} is a rectangular $N \times n$ matrix, $f_{i,m}$ is a vectorized form of the differences $d_{ij,m}$, the summation over a repeating index is implied.

Formally, the approximation error may be expressed as

$$\Delta u_m^{(j)} = (D_{ij})^{-1} f_{i,m}. \tag{5}$$

In the considered case ($n = 4$) the Eq. (4) has the form

$$\begin{pmatrix} 1 & -1 & 0 & 0 \\ 1 & 0 & -1 & 0 \\ 1 & 0 & 0 & -1 \\ 0 & 1 & -1 & 0 \\ 0 & 1 & 0 & -1 \\ 0 & 0 & 1 & -1 \end{pmatrix} \begin{pmatrix} \Delta u_m^{(1)} \\ \Delta u_m^{(2)} \\ \Delta u_m^{(3)} \\ \Delta u_m^{(4)} \end{pmatrix} = \begin{pmatrix} f_{1,m} \\ f_{2,m} \\ f_{3,m} \\ f_{4,m} \\ f_{5,m} \\ f_{6,m} \end{pmatrix} = \begin{pmatrix} u_m^{(1)} - u_m^{(2)} \\ u_m^{(1)} - u_m^{(3)} \\ u_m^{(1)} - u_m^{(4)} \\ u_m^{(2)} - u_m^{(3)} \\ u_m^{(2)} - u_m^{(4)} \\ u_m^{(3)} - u_m^{(4)} \end{pmatrix}. \tag{6}$$

At the first glance the system is overdetermined. However, the solution of system (4) is invariant relatively a simultaneous shift transformation of all terms $u_m^{(j)} = \tilde{u}_m^{(j)} + b$ (and corresponding errors $\Delta u_m^{(j)} = \Delta \tilde{u}_m^{(j)} + b$) for any $b \in (-\infty, \infty)$ due to the usage of the difference of solutions as the input data. For this reason, the problem of approximation error estimation from the difference of solutions is really underdetermined and therefore ill-posed. We solve the system of Eqs. (6) by the method considered in following section.

4 The Estimation of Approximation Error Using Regularized Inverse Problem

In general, a regularization [8, 9] is necessary in order to obtain the steady and bounded solution of the ill-posed problems. Herein we apply the zero order Tikhonov regularization in order to obtain solutions with the minimum shift error $|b|$. The minimal L_2 norm of $\Delta u^{(j)}$ restricts the absolute value of b, since:

$$\min_{b_m}(\delta(b_m)) = \min_{b_m} \sum_j^n (\Delta u_m^{(j)})^2/2 = \min_{b_m} \sum_j^n (\Delta \tilde{u}_m^{(j)} + b_m)^2/2. \tag{7}$$

This expression is used as the regularizing term in variational statement of the Inverse Problem.

One may see that

$$\Delta\delta(b_m) = \sum_j^n (\Delta \tilde{u}_m^{(j)} + b_m)\Delta b_m, \tag{8}$$

and the minimum occurs at b_m that equals the mean error (with the opposite sign):

$$b_m = -\frac{1}{n} \sum_j^n \Delta \tilde{u}_m^{(j)} = -\Delta \bar{u}_m. \tag{9}$$

So, the expression (7) may be treated as the minimum of the deviation of the exact error from the mean $\Delta u^{(j)} = \Delta \tilde{u}^{(j)} - \Delta \bar{u}$ (the exact error dispersion). The minimality of δ ensures the boundedness of the shift error b_m. Thus, the accuracy of the error $\Delta u^{(j)}$ estimation in considered approach is restricted by the mean error value.

We pose the Inverse Problem for $\Delta u^{(j)}$ estimation in the variational statement [9] that implies the minimization of the following functional:

$$\varepsilon_m(\Delta \bar{u}) = 1/2(D_{ij}\Delta u_m^{(j)} - f_{i,m}) \cdot (D_{ik}\Delta u_m^{(k)} - f_{i,m}) + \alpha/2(\Delta u_m^{(j)} E_{jk} \Delta u_m^{(k)}). \tag{10}$$

The first term of (10) is the discrepancy of the predictions and observations, the second term is the zero order Tikhonov regularization (α is the regularization parameter, E_{jk} is the unite matrix). We apply the gradient based (steepest descent) iterations (k is the number of the iteration) for the minimization of the functional:

$$\Delta u_m^{(j),k+1} = \Delta u_m^{(j),k} - \tau \nabla \varepsilon_m. \tag{11}$$

The gradient is obtained in the present work by the direct numerical differentiation, the iterations terminate at certain small value of the functional $\varepsilon \leq \varepsilon_*$ ($\varepsilon_* = 10^{-8}$ was used). The obtained solution depends on the choice of the regularization parameter α. Without regularization ($\alpha = 0$) $|\Delta u^{(j)}(\alpha)|$ is not bounded and is not acceptable for this reason. The limit $\alpha \to \infty$ is not acceptable also, since $|\Delta u^{(j)}(\alpha)| \to 0$. A range of the regularization parameter α exists where the weak dependence of the solution on α is manifested. In this range, the solution $\Delta u^{(j)}(\alpha)$ is close to the exact one $\Delta \tilde{u}^{(j)}$ and is considered as the regularized solution [8]. By the rearranging Eq. (10) one may obtain

$$\varepsilon_m^{(1)}(\Delta \vec{u}) = 1/2(\Delta u_m^{(j)} E_{jk} \Delta u_m^{(k)}) + 1/(2\alpha)(D_{ij}\Delta u_m^{(j)} - f_{i,m}) \cdot (D_{ik}\Delta u_m^{(k)} - f_{i,m}) \quad (12)$$

and

$$\varepsilon_m^{(1)}(\Delta \vec{u}) = \varepsilon_m(\Delta \vec{u})/\alpha. \quad (13)$$

This expression enables the estimation of the mean local error in the form of inequality:

$$\sum_k \left\| \Delta u_m^{(k)} \right\|^2 \leq 2\varepsilon_m^{(1)}. \quad (14)$$

The corresponding estimation of the global error norm has the appearance

$$\sum_k \left\| \Delta u^{(k)} \right\|^2 \leq 2/M \cdot \sum_{m=1}^{M} \varepsilon_m^{(1)}. \quad (15)$$

So, in contrast to the Richardson extrapolation, the Inverse Problem based approach enables the estimation of the averaged error in the form of strict inequalities.

5 The Test Problem

The estimation of the approximation error for the problems containing discontinuities is a challenging task. In CFD problems the errors arising at shock waves and contact discontinuities are of the significant magnitude, demonstrate the oscillating behavior and are specified by nonstandard order of the convergence. For this reason our attention in the present paper is focused on the errors engendered by the shock waves. The similar topics arising at the contact discontinuities and the shock interferences are reserved for the future works. The test problem is governed by the two dimensional compressible Euler equations describing a shock wave. The following flowfield parameters: density, velocity components along x and y axes, pressure ($\{\rho, v_x, v_y, p\}$) and the Mach number (M) are used in the analysis. We estimate the approximation error by minimizing the functional (7) using Expression (11) for parameters $u_m^{(i)}$ that correspond the flowfield variables $\{\rho, v_x, v_y, p\}$ at every grid point. The flowfield around a plate at the angle of attack $\alpha = 6°$ and $\alpha = 20°$ in the uniform supersonic flow ($M = 2$ and $M = 4$) of ideal gas is analyzed. The approximation error is estimated using generalized Richardson extrapolation [3, 5] and the Inverse problem based statement. The results are compared with the exact error obtained by the subtraction of the numerical solution and the projection of the analytic solution on the computational grid. At $\alpha = 6°$ and $M = 2$ we obtain a relatively weak shock wave, while at $\alpha = 20°$ and $M = 4$ the shock wave is strong. On the left boundary ("*inlet*") and on the upper boundary ("*top*"), the inflow parameters are set for Mach number $M = 2$, $M = 4$ an corresponding angles. On the right boundary ("*outlet*") the zero gradient condition for the gas dynamic functions is specified. On the plate surface, the condition of zero normal gradient is posed for the pressure and the temperature, and the condition "*slip*" is posed for the speed, corresponding to the non-penetration. The parameters of the OpenFOAM package are the same as in [7].

6 OpenFOAM Solvers

The solvers from the OpenFOAM software package [13] that were used are the following:

- *rhoCentralFoam* (marked as rCF), which is based on a central-upwind scheme [14, 15].
- *sonicFoam* (sF), which is based on the PISO algorithm [16].
- *pisoCentralFoam* (pCF) [17], which combines the Kurganov-Tadmor scheme [14] and the PISO algorithm [16].
- *QGDFoam* (QGDF), which implements the quasi-gas dynamic equations [18].

These solvers are of the second approximation order, while they are based on the algorithms of the quite different ideas and inner structure.

7 Numerical Results

The exact errors are obtained by the comparison of the numerical solution with the analytic one for the shock wave (Rankine-Hugoniot relations). The relative exact errors in L_1 and L_2 norms are presented in Tables 1, 2, 3, 4, 5 and 6 for $\alpha = 20°$, $M = 4$ and three grids (20000, 80000 and 320000 nodes). For QGDF code the coefficient $\beta = 0.1$ is used that controls the artificial viscosity.

Table 1. The relative errors in L_1 norm ($M = 4$, $\alpha = 20°$, 20000 nodes).

	rCF	pCF	sF	QGDF
M	0.001295	0.001489	0.002267	0.01155
p	0.010745	0.011226	0.025567	0.018022
ρ	0.005817	0.006573	0.015622	0.008541

Table 2. The relative errors in L_1 norm ($M = 4$, $\alpha = 20°$, 80000 nodes).

	rCF	pCF	sF	QGDF
M	0.000701	0.000816	0.001205	0.000611
p	0.005512	0.005964	0.013733	0.009629
ρ	0.003086	0.003601	0.008663	0.004549

Tables 1, 2, 3, 4, 5 and 6 demonstrate the order of convergence about 1.0 in L_1 norm and about 0.5 in L_2 that is far from the nominal (second) order of considered algorithms and corresponds results by [3, 4, 11, 12].

We estimate the approximation error using the generalized Richardson extrapolation [3, 5] on three consequent grids containing 20000, 80000 and 320000 nodes (with doubling number of nodes along both directions at every refinement).

Table 3. The relative errors in L_1 norm (M = 4, α = 20°, 320000 nodes).

	rCF	pCF	sF	QGDF
M	0.000381	0.000439	0.000678	0.000439
p	0.002944	0.003217	0.007621	0.003217
ρ	0.001654	0.001938	0.005043	0.001938

Table 4. The relative errors in L_2 norm (M = 4, α = 20°, 20000 nodes).

	rCF	pCF	sF	QGDF
Ma	0.013951	0.015138	0.013520	0.009920
p	0.080503	0.079493	0.143794	0.098895
ρ	0.047797	0.047157	0.092308	0.055613

Table 5. The relative errors in L_2 norm (M = 4, α = 20°, 80000 nodes)

	rCF	pCF	sF	QGDF
M	0.010127	0.011202	0.009191	0.007125
p	0.055094	0.056591	0.104672	0.070171
ρ	0.033143	0.033612	0.066215	0.039410

Table 6. The relative errors in L_2 norm (M = 4, α = 20°, 320000 nodes)

	rCF	pCF	sF	QGDF
M	0.007527	0.008263	0.006610	0.008263
p	0.039382	0.040675	0.081605	0.040675
ρ	0.024024	0.024262	0.051199	0.024262

In the inverse problem statement, we minimize the functional (7) for each flow parameter from the set $\{\rho, v_x, v_y, p\}$ separately at every grid point.

The Figs. 1, 2, 3, 4, and 5 present the pieces of vectorized grid function of density error obtained by the Inverse Problem in comparison with the results of the generalized Richardson extrapolation and the exact error. The index along abscissa axis $i = N_x(k_x - 1) + m_y$ is defined by indexes along $X(k_x)$ and $Y(m_y)$. The periodical jump of solution variables corresponds to the transition through the shock wave. One may see that the error at the shock wave is under resolved by both (RE and IP) methods. This behavior is expected since the error at a shock tends to be singular at the mesh refinement.

The impact of the shock wave intensity on the quality of the error estimation may be observed from Figs. 1 and 2 that demonstrate the dependence of both the IP-based and GRE estimation quality on the strength of the shock wave (for $M = 2$ and $M = 4$ correspondingly). The Fig. 1 presents the piece of vectorized grid function of density error (computed by rCF) for $\alpha = 6°$, $M = 2$. The Fig. 2 presents the piece of vectorized grid function of density error (computed by rCF) for $\alpha = 20°$, $M = 4$. For the small approximation error (small Mach number and deflection angle, Fig. 1) the generalized Richardson extrapolation outperforms the Inverse problem based results despite some instability past shock wave. These results are expectable, since the Richardson extrapolation is known to well behave for the rather regular solutions (weak shock waves in our case). For the relatively great approximation errors the IP-based results are rather smoothed and shifted in comparison with the exact error. This is caused by the using the set of solutions having slightly shifted position of the shock waves. Nevertheless the total quality of the IP-based error estimate improves and may compete with the results obtained by GRE (which suffer from instabilities).

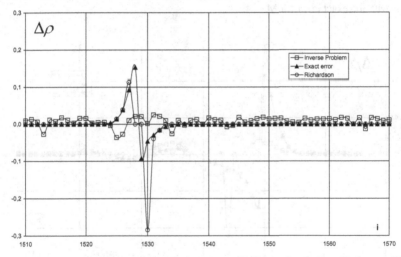

Fig. 1. The comparison of the vectorized density error (rCF), estimated by the Inverse Problem and GRE with the exact error for M = 2.

The Figs. 3, 4 and 5 provide the density errors ($\alpha = 20°$, $M = 4$) for pCF, QGDF, and sF correspondingly.

The results significantly depend on the choice of the solver. On the above numerical tests the sF solver provides most error over all set of codes (see Tables 1, 2, 3, 4, 5 and 6).

The quality of *a posteriori* error estimate may be described by the effectivity index [19] that equals the relation of the estimated error norm to the exact error norm:

$$I_{eff,k} = \left\| \Delta \vec{\rho}^{(k)} \right\|_{L_2} / \left\| \Delta \vec{\tilde{\rho}}^{(k)} \right\|_{L_2} \tag{16}$$

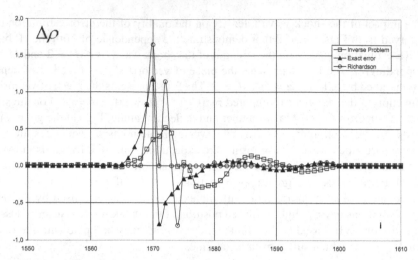

Fig. 2. The comparison of the vectorized density error (rCF), estimated by the Inverse Problem and GRE with the exact error for M = 4.

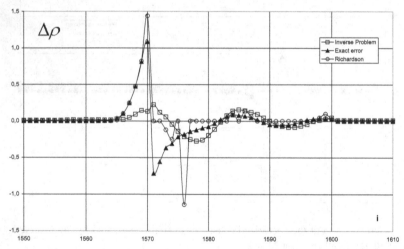

Fig. 3. The comparison of the vectorized density error (pCF), estimated by the Inverse Problem and GRE with the exact error for M = 4.

The vectors $\Delta\vec{\rho}^{(k)}$, $\Delta\vec{\tilde{\rho}}^{(k)} \in R^M$ (M is the number of grid nodes) in this relation denote the grid functions. Thus, the norms, herein, imply averaging of pointwise errors over the total flowfield. To provide the reliability of the error estimation, this index should be greater the unit. On the other hand, the estimation should be not too pessimistic, so the value of the effectivity index should be not too great. According [19], the range $1 \leq I_{eff} \leq 3$ is acceptable for the finite elements in the domain of elliptic equations. However, for the present discontinuous solutions these values are problem dependent. The upper bound may by corrected using the tolerance of the valuable functionals and the

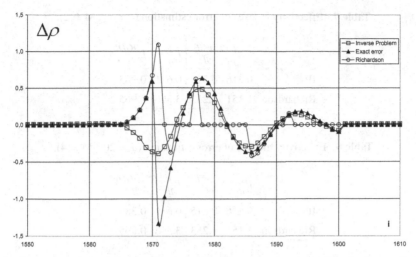

Fig. 4. The comparison of the vectorized density error (QGD), estimated by the Inverse Problem and GRE with the exact error for $M = 4$.

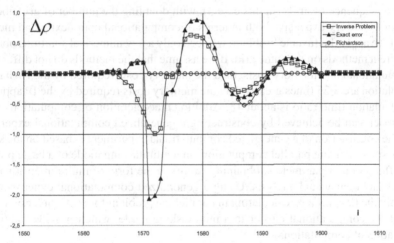

Fig. 5. The comparison of the vectorized density error (sF), estimated by the Inverse Problem and GRE with the exact error for $M = 4$.

Cauchy–Bunyakovsky–Schwarz inequality [6]. The down boundary may be corrected using certain safety coefficient. The corresponding values of the effectivity index are provided in the Tables 7 and 8 for IP-based statement and the Richardson extrapolation.

Tables 7 and 8 present the effectivity index for $M = 2$ and $M = 4$. One may see that the reliability of the Richardson extrapolation decreases as the intensity of the shock waves increases (Mach number and flow deflection angle rise). On contrary, the reliability of the IP-based estimated increases. In general, from the standpoint of the global error estimation the GRE provides more reliable results.

Table 7. Effectivity indexes of error estimation ($\alpha = 6°$, $M = 2$).

	I_{eff}^{rCF}	I_{eff}^{pCF}	I_{eff}^{sF}	I_{eff}^{QGDF}
IP	0.316	0.315	0.631	0.385
Richardson	1.151	1.253	3.366	0.965

Table 8. Effectivity indexes of error estimation ($\alpha = 20°$, $M = 4$).

	I_{eff}^{rCF}	I_{eff}^{pCF}	I_{eff}^{sF}	I_{eff}^{QGDF}
IP	0.316	0.315	0.631	0.385
Richardson	1.151	1.253	3.366	0.965

Especially important is the question of comparing the computational costs for the GRE and IP methods for a given test problem. Since the Richardson extrapolation requires a sequence of grids (in this case 3), with doubling the number of grid nodes, it turns out to be very expensive both in terms of computational complexity and memory costs. If we apply the Inverse Problem, we need only a few numerical solutions obtained by different methods on the same grid. If we assume that the methods do not differ much in computational and memory costs, then the memory costs required by the Richardson extrapolation are 5.25 times greater than the memory costs required by the IP approach. The calculation time ratio is about 18. Additional acceleration of computations in the IP approach can be achieved by constructing a generalized computational experiment [22]. The construction of a generalized computational experiment is based on the simultaneous solution using parallel computations in a multitasking mode of a basic problem with different input parameters, obtaining results in the form of multidimensional data volumes and their visual analysis. Using a generalized computational experiment, we can apply the IP approach, calculating in parallel the problem for each solver on its own node of the computational cluster in a multitasking mode, which provides additional acceleration of computations.

8 Discussion

In the paper [7] the Inverse Problem based approach was used for the supersonic axisymmetric flows around cones. The comparison with the etalon (high precision solution by [10]) was presented. In [20] the flow modes obtained by the crossing shocks (Edney-I and Edney-VI patterns by [21]) are analyzed. Herein, the comparison for the flat flow is performed for the Inverse Problem based errors, exact error (obtained by the comparison with the analytic solutions) and the results of the Richardson approximation. Formally, the Inverse Problem based approach is less accurate if compared with the Richardson extrapolation due to the presence of the unremovable error, proportional to the mean error over the ensemble of solutions. However, this statement is valid only for the highly smooth solutions. For the above considered problems with shock waves the generalized Richardson extrapolation should be used that is the highly unstable, that deteriorates the results. In most cases, the GRE demonstrates highly nonsmooth solutions that may approximate the part of the exact error (usually, before the shock wave) with a relatively high resolution, while another part of the exact error (past the shock wave) is poorly approximated. The comparison of the results obtained by GRE and IP with the exact error demonstrates the high smoothing properties of the pointwise IP error estimation and the visible shift of the error location. Since the IP-based error estimate are polluted by the mean error over the ensemble (9), the best results are obtained for the less accurate solutions (herein, for sF, see Fig. 5).

In general, numerical tests demonstrate the accuracy of the error estimates obtained using generalized Richardson extrapolation to be superior in the comparison with the Inverse Problem based results for the weak shocks and comparable for the strong shocks. In contrast to the generalized Richardson extrapolation, the considered IP-based postprocessor is much more computationally inexpensive, since it uses only single grid computations. Additionally, it possesses some natural parallelism, since different solvers may be independently computed by different nodes of the cluster, which fits into the concept of constructing a generalized computational experiment [22].

9 Conclusion

The numerical tests demonstrate the feasibility for the estimation of the point-wise approximation error via the Inverse Problem treating of the ensemble of numerical solutions obtained using the four solvers from the OpenFOAM software package for the two-dimensional inviscid flow pattern engendered by the oblique shock wave. The Inverse Problem based estimation of the point-wise approximation error using the differences of numerical solutions as the input data provides the accuracy comparable with the generalized Richardson extrapolation, however, it is much more computationally inexpensive.

Acknowledgments. This work was supported by grant of RSF № 18-11-00215.

References

1. Guide for the Verification and Validation of Computational Fluid Dynamics Simulations. American Institute of Aeronautics and Astronautics, AIAA-G-077-1998 (1998)
2. Standard for Verification and Validation in Computational Fluid Dynamics and Heat Transfer, ASME V&V 20-2009 (2009)
3. Roy, C.: Grid convergence error analysis for mixed-order numerical schemes. AIAA J. **41**(4), 595–604 (2003). https://doi.org/10.2514/2.2013
4. Banks, J.W., Aslam, T.D.: Richardson extrapolation for linearly degenerate discontinuities. J. Sci. Comput. **57**, 1–15 (2013). https://doi.org/10.1007/s10915-013-9693-0
5. Alekseev, A.K., Bondarev, A.E.: On some features of Richardson extrapolation for compressible inviscid flows. Mathematica Montisnigri **XL**, 42–54 (2017)

6. Alekseev, A.K., Bondarev, A.E., Kuvshinnikov, A.E.: On uncertainty quantification via the ensemble of independent numerical solutions. J. Comput. Sci. **42**, 10114 (2020). https://doi.org/10.1016/j.jocs.2020.101114

7. Alekseev, A.K., Bondarev, A.E., Kuvshinnikov, A.E.: A posteriori error estimation via differences of numerical solutions. In: Krzhizhanovskaya, V.V., et al. (eds.) ICCS 2020. LNCS, vol. 12143, pp. 508–519. Springer, Cham (2020). https://doi.org/10.1007/978-3-030-50436-6_37

8. Tikhonov, A.N., Arsenin, V.Y.: Solutions of Ill-Posed Problems. Winston and Sons, Washington DC (1977)

9. Alifanov, O.M., Artyukhin, E.A., Rumyantsev S.V.: Extreme Methods for Solving Ill-Posed Problems with Applications to Inverse Heat Transfer Problems. Begell House (1995)

10. Babenko, K.I.,Voskresenskii, G.P., Lyubimov, A.N., Rusanov, V.V.: Three-Dimensional Ideal Gas Flow Past Smooth Bodies. Nauka, Moscow (1964). (in Russian)

11. Carpenter, M.H., Casper, J.H.: Accuracy of shock capturing in two spatial dimensions. AIAA J. **37**(9), 1072–1079 (1999). https://doi.org/10.2514/2.835

12. Godunov, S.K., Manuzina, Y., Nazareva, M.A.: Experimental analysis of convergence of the numerical solution to a generalized solution in fluid dynamics. Comput. Math. Math. Phys. **51**, 88–95 (2011). https://doi.org/10.1134/S0965542511010088

13. OpenFOAM. http://www.openfoam.org. Accessed 25 Jan 2021

14. Kurganov, A., Tadmor, E.: New high-resolution central schemes for nonlinear conservation laws and convection-diffusion equations. J. Comput. Phys. **160**(1), 241–282 (2000). https://doi.org/10.1006/jcph.2000.6459

15. Greenshields, C., Wellerr, H., Gasparini, L., Reese, J.: Implementation of semi-discrete, non-staggered central schemes in a colocated, polyhedral, finite volume framework, for high-speed viscous flows. Int. J. Numer. Meth. Fluids **63**(1), 1–21 (2010). https://doi.org/10.1002/fld.2069

16. Issa, R.: Solution of the implicit discretized fluid flow equations by operator splitting. J. Comput. Phys. **62**(1), 40–65 (1986). https://doi.org/10.1016/0021-9991(86)90099-9

17. Kraposhin, M., Bovtrikova, A., Strijhak, S.: Adaptation of Kurganov-Tadmor numerical scheme for applying in combination with the PISO method in numerical simulation of flows in a wide range of Mach numbers. Procedia Comput. Sci. **66**, 43–52 (2015). https://doi.org/10.1016/j.procs.2015.11.007

18. Kraposhin, M.V., Smirnova, E.V., Elizarova, T.G., Istomina, M.A.: Development of a new OpenFOAM solver using regularized gas dynamic equations. Comput. Fluids **166**, 163–175 (2018). https://doi.org/10.1016/j.compfluid.2018.02.010

19. Repin, S.I.: A Posteriori Estimates for Partial Differential Equations, vol. 4. Walter de Gruyter (2008). https://doi.org/10.1515/9783110203042

20. Alekseev A., Bondarev A.: The estimation of approximation error using the inverse problem and the set of numerical solutions. arXiv:2101.00381v1 (2021)

21. Edney, B.: Effects of shock impingement on the heat transfer around blunt bodies. AIAA J. **6**(1), 15–21 (1968). https://doi.org/10.2514/3.4435

22. Bondarev A.E.: On the construction of the generalized numerical experiment in fluid dynamics. Mathematica Montisnigri **XLII**, 52–64 (2018)

Predicted Distribution Density Estimation for Streaming Data

Piotr Kulczycki[1,2(✉)] and Tomasz Rybotycki[2]

[1] Faculty of Physics and Applied Computer Science, AGH University
of Science and Technology, Kraków, Poland
kulczycki@agh.edu.pl, kulczycki@ibspan.waw.pl
[2] Systems Research Institute, Polish Academy of Sciences, Warsaw, Poland

Abstract. Recent growth in interest concerning streaming data has been forced by the expansion of systems successively providing current measurements and information, which enables their ongoing, consecutive analysis. The subject of this research is the determination of a density function characterizing potentially changeable distribution of streaming data. Stationary and nonstationary conditions, as well as both appearing alternately, are allowed. Within the distribution-free procedure investigated here, when the data stream becomes nonstationary, the procedure begins to be supported by a forecasting apparatus. Atypical elements are also detected, after which the meaning of those connected with new tendencies strengthens, while diminishing elements weaken. The final result is an effective procedure, ready for use without studies and laborious research.

Keywords: Streaming data · Distribution density · Nonparametric estimation · Distribution-free procedure · Prediction · Atypical element (outlier)

1 Introduction

Technological progress within the scope of numerical techniques has enabled the comprehensive analysis and exploration of data with different natures. Recently interest in a specific type, characterized by successive and unlimited inflow of sequential elements, named streaming data, has grown. In current practice, data of this type may be nonstationary (evolving in time), therefore, their characteristics are variable, which additionally makes all analysis considerably more difficult. Frequently the character of streaming data undergoes changes from stationary to nonstationary and *vice-versa*, implying further research challenges. Moreover, the nature of permanently incoming, often unverified data causes that they may also contain atypical elements, mostly as a result of errors of different kinds. Their automatic removal may, however, result in the elimination of valuable information about newly forming tendencies. Finally, effective analysis of streaming data fulfilling requirements of contemporary applications needs a range of significant factors, frequently absent in classic problems with an assumed finite dataset size, to be taken into account. This makes the analysis of streaming data

© Springer Nature Switzerland AG 2021
M. Paszynski et al. (Eds.): ICCS 2021, LNCS 12747, pp. 567–580, 2021.
https://doi.org/10.1007/978-3-030-77980-1_43

extremely valuable from the applicational point of view, but also demanding from a research perspective.

The subject of this paper is the synthesis of a procedure enabling the determination of distribution density of streaming data, both stationary and nonstationary, also with these both cases appearing alternately. The mathematical apparatus is based on the procedures of contemporary data analysis and mathematical statistics, allowing calculation of density without any assumption concerning the specific type of distribution. The particular elements, applied later during the creation of the procedure, will be presented in Sect. 2. Successively, the nonparametric method of kernel estimators, procedure for atypical (rare) elements detection, the statistical test of stationarity, and elements of forecasting theory, will be presented in the subsequent four sections of this chapter. The concept of the procedure developed here for the predicted estimation of distribution density of streaming data is the subject of Sect. 3. This procedure is modular in nature. In the succeeding four sections, the concepts of fixing the size of a reservoir, the outdatedness of its elements, introduction of forecasting methods, and detection of atypical elements strengthening the importance of those connected with newly arising tendencies and the weakening associated with disappearing trends, have been elaborated. In consequence, a procedure for determining the current distribution density of streaming data will be created, while in the case of discovery of nonstationarity, procedures for adaptation and forecasting are activated in order to effectively match to the changing environment. The calculation complexity of all algorithms used are linear and quadratic; the whole cycle of the calculations is enclosed within few seconds. The memory requirements do not exceed the typical capabilities of contemporary computer systems. The final conclusions and numerical results of the designed method will be briefly described in Sect. 4.

The estimation of distribution density of streaming data is a current topic being studied and various methods have been applied, e.g. histogram [19] or wavelets [22], however, concepts based on kernel estimators dominate (see [5] for a rich bibliography) consisting of a proper selection of incoming elements [20], specialized clustering [9, 24], local approach [7], and using calculational intelligence methods, e.g. self-organizing maps [8]. Fundamental information concerning streaming data can be found in the recent books [3, 21].

2 Mathematical Preliminaries

2.1 Nonparametric Estimation, Kernel Estimators

Consider a set consisting of the m elements being the n-dimensional vectors with continuous attributes:

$$x_1, x_2, \ldots, x_m \in \mathbb{R}^n. \tag{1}$$

The kernel estimator $\hat{f} : \mathbb{R}^n \to [0, \infty)$ of the density of a dataset (1) distribution, is defined then as [11, 23]:

$$\hat{f}(x) = \frac{1}{m} \sum_{i=1}^{m} K(x, x_i, h), \tag{2}$$

where after separation into coordinates

$$x = \begin{bmatrix} x_1 \\ x_2 \\ \vdots \\ x_n \end{bmatrix}, \quad x_i = \begin{bmatrix} x_{i,1} \\ x_{i,2} \\ \vdots \\ x_{i,n} \end{bmatrix} \quad \text{for } i = 1, 2, \ldots, m, \quad h = \begin{bmatrix} h_1 \\ h_2 \\ \vdots \\ h_n \end{bmatrix}, \tag{3}$$

while the positive constants h_j are the so-called smoothing parameters; the kernel K will be used here in the product form:

$$K(x, x_i, h) = \prod_{j=1}^{n} \frac{1}{h_j} K_j \left(\frac{x_j - x_{i,j}}{h_j} \right), \tag{4}$$

whereas the one-dimensional kernels $K_j : \mathbb{R} \to [0, \infty)$, for $j = 1, 2, \ldots, n$, are measurable with unit integral $\int_{\mathbb{R}} K_j(y) \, dy = 1$, symmetrical, and non-increasing for $[0, \infty)$; (in consequence: non-decreasing for $(-\infty, 0]$). For the needs of further considerations, the definition (2) will be generalized to the weighted form:

$$\hat{f}(x) = \frac{1}{\sum_{i=1}^{m} w_i} \sum_{i=1}^{m} w_i K(x, x_i, h), \tag{5}$$

where the introduced parameters $w_i \geq 0$ are not all equal to 0. Assuming $w_i \equiv 1$, one simply obtains the formula (2). The kernel estimator allows us to calculate the density on the basis of the dataset (1) without any arbitrary assumption concerning the type of its distribution.

Generally, the selection of the kernel K_j form is practically meaningless and the user should, above all, take into account the properties of the desired estimator or/and computational aspects, beneficial for the application being worked out. In the following, the normal (Gauss) kernel

$$K_j(x) = \frac{1}{\sqrt{2\pi}} \exp\left(-\frac{x^2}{2} \right) \tag{6}$$

will be applied, as generally used.

The fixing of the smoothing parameter h_j has significant meaning for quality of estimation. Fortunately, many suitable procedures for calculating its optimal value have been worked out. In particular, for simple unimodal distributions and in the preliminary phase of investigation, the normal concept is suggested. Then

$$h_j = \left(\frac{8\sqrt{\pi}}{3} \frac{W(K)}{U(K)^2} \frac{1}{m} \right)^{1/5} \hat{\sigma}_j, \tag{7}$$

where $W(K) = \int_{-\infty}^{\infty} K(y)^2 dy$ and $U(K) = \int_{-\infty}^{\infty} y^2 K(y) dy$. For the normal kernel (6) one has $W(K) = 1/2\sqrt{\pi}$ and $U(K) = 1$. The standard deviation estimator $\hat{\sigma}_j$, occurring above, can be calculated for the dataset (1) from the classic formula, potentially extended for the weighted form (5) as follows:

$$\hat{\sigma}_j^2 = \frac{1}{m-1} \sum_{i=1}^{m} w_i x_{i,j}^2 - \frac{1}{m(m-1)} \left(\sum_{i=1}^{m} w_i x_{i,j} \right)^2. \tag{8}$$

In other situations we propose testing the plug-in method [11, Sect. 3.1.5; 23, Sect. 3.6.1], where its degree should be equal to the number of separated factors (modes), but in practice not greater than 3; the value 2 can be treated as a standard. A generalization of this method to the weighted form should be made similarly to the formula (8).

In practice various modifications, generalizations and fitting properties of the estimator to specific realities can be applied, e.g. other algorithms for fixing the smoothing parameter, its adaptation, or boundary of the function $\hat{f}$ support. The procedure worked out in this paper has no limits in this range besides requirements regarding time and memory as well as excessive complexity of interpretation, which should be individually considered. The classic textbooks on kernel estimators constitute the monographs [11, 23]. The effective determination of distribution density enables comprehensive data analysis [12, 13] and various valuable applications [14, 15].

2.2 Detection of Atypical Elements (Outliers)

The determination of distribution density enables effective detection of atypical elements [2], which are understood here in the sense of rare occurrence. Unlike distance methods, one can then find atypical observations not only on the peripheries of the population, but in the case of multimodal distributions with wide-spreading segments, also those lying in between these segments, even if they are close to the 'center' of the set.

Consider the dataset (1) containing elements representative of the considered population. Based on the material from Sect. 2.2, the kernel estimator (5) can be calculated. Then, consider also the set of its values for elements of the dataset (1), therefore

$$\hat{f}_{-1}(x_1), \hat{f}_{-2}(x_2), \ldots, \hat{f}_{-m}(x_m), \tag{9}$$

where $\hat{f}_{-i}$ means the kernel estimator $\hat{f}$ calculated excluding the i-th element of the dataset. Next, define the number $r \in (0, 1)$ determining the sensitivity of the procedure for identifying atypical elements. This number will simply determine the assumed proportion of atypical elements in relation to the total population; therefore, the ratio of the number of atypical elements to the sum of atypical and typical elements. In practice

$$r = 0.01, 0.05, 0.1 \tag{10}$$

is the most often used. Next, for the set (23) one can calculate the positional estimator for the quantile of the degree r given by the formula

$$\hat{q}_r = \begin{cases} s_1 & \text{for} \quad mr < 0.5 \\ (0.5 + i - mr)s_i + (0.5 - i + mr)s_{i+1} & \text{for} \quad mr \geq 0.5 \end{cases}, \tag{11}$$

where $i = \underline{int}(mr + 0.5)$, while $\underline{int}$ denotes an integral part of a number, and s_i is the i-th value in size of the set (9) after being sorted; thus

$$\{s_1, s_2, \ldots, s_m\} = \left\{ \hat{f}_{-1}(x_1), \hat{f}_{-2}(x_2), \ldots, \hat{f}_{-m}(x_m) \right\}. \tag{12}$$

with $s_1 \leq s_2 \leq \ldots \leq s_m$. Generally, there are no special recommendations concerning the choice of the sorting algorithm used for specifying set (12). However, let us interpret the definition (11), taking into account the values (10). So, it is enough to sort only the $i + 1$ smallest values in the set (9), therefore, about 1–10% of its size. One can apply a simple algorithm that subsequently finds the $i + 1$ smallest elements of the set (9).

Finally, if for a given tested element $\tilde{x} \in \mathbb{R}^n$, the condition $\hat{f}(\tilde{x}) \leq \hat{q}_r$ is fulfilled, then this element should be considered atypical; for the opposite $\hat{f}(\tilde{x}) > \hat{q}_r$ it is typical.

The details of the above method can be found in the paper [16]. A review of various methods of outlier detection is given in the monograph [2].

2.3 Testing of Stationarity, KPSS Test

Let the real time series $\{X_t\}_{t=1,2,\dots}$ be given. The stationarity of the stochastic process, from which this series originate, will be verified using the KPSS test [18]. The hypothesis being tested here is the stationarity, with respect to the alternative hypothesis that the process is nonstationary. Generally, the KPSS test is applied in two options: without considering the trend and assuming its presence. Here, the first of them will be used — in the investigated procedure, each trend will be treated as a nonstationary factor.

The test statistics, calculated on the basis of T values $X_1, X_2, \dots, X_T$ takes the form

$$KPSS = \frac{\sum_{t=1}^{T} S_t^2}{\hat{\sigma}_c^2}, \tag{13}$$

where S_t denotes the partial sum of the residuals of mean-square approximation of the series by a constant function (the optimal value here is equal to the arithmetic mean), i.e.

$$S_t = \sum_{l=1}^{t} R_l \tag{14}$$

$$R_l = X_l - \overline{X} \quad \text{for} \quad l = 1, 2, \dots, t \tag{15}$$

$$\overline{X} = \frac{1}{T} \sum_{l=1}^{T} X_l, \tag{16}$$

and $\hat{\sigma}_c$ means the consistent estimator of a standard deviation, given by the formulas

$$\hat{\sigma}_c^2 = T \sum_{l=1}^{T} R_l^2 + 2T \sum_{s=1}^{L} W(s, L) \sum_{z=s+1}^{T} R_z R_{z-s} \tag{17}$$

$$W(s, L) = 1 - \frac{s}{L+1} \tag{18}$$

$$L = int\left[4 \cdot (0.01T)^{\frac{1}{4}}\right], \tag{19}$$

where int means rounding to an integer. To avoid $0/0$, define $KPSS = 0$ for $T = 1$.

The critical set takes the right-hand form, while the critical values for critical levels equal respectively

$$
\begin{array}{lcccc}
\text{critical level} & 0.1 & 0.05 & 0.025 & 0.01 \\
\text{critical value} & 0.347 & 0.463 & 0.574 & 0.739
\end{array}. \tag{20}
$$

Based on the fuzzy approach, a quantity with values from the interval [0, 1] will now be defined, characterizing the 'degree of nonstationarity' of the data stream under research. Namely, the function KPSS will be subjected to a linear transformation and then covered by the sigmoid function $sgm : \mathbb{R} \to (0, 1)$ given as

$$sgm(x) = \frac{1}{1 + e^{-x}}. \tag{21}$$

After determining the transformation parameters one obtains

$$sgmKPSS = sgm(0.995KPSS - 2.932). \tag{22}$$

The coefficients of the linear transformation, appearing in the formula (22) were fixed heuristically such that the biggest, used in practice, critical value 0.739 is transformed into the highest critical level 0.1, and the smallest critical value 0.347 into the level lower by 30%. The last value has been fixed through the inspiration of automatic control practice, in particular the Ziegler-Nichols method of tuning PID controllers [6]. Namely, the integral quality index L_1 was minimized in a response to the unique step in the time series $\{X_t\}_{t=1,2,...}$. Such a value generally seems to be the most favorable (Sect. 4). Using the classic automatic control language, one then obtains a course without or with small over-regulations.

For purposes of the procedure investigated here, we fixed by the same method $T = 600$. Its increase results in sensitivity improving, however, at the expense in a slower reaction; a reduction brings opposite effects. Naturally, in the initial t steps when $t < 600$ we should employ as many elements as we have; therefore

$$T = \begin{cases} t & \text{when} & t < 600 \\ 600 & \text{when} & t \geq 600 \end{cases}. \tag{23}$$

For simple unimodal distributions, the value $T = 600$ can be reduced to 500.

In the multidimensional case

$$sgmKPSS = \max_{i=1,2,...,n} sgmKPSS_i, \tag{24}$$

where $sgmKPSS_i$ denotes the quantity $sgmKPSS$ given by the formula (22) for the i-th continuous attribute. The maximum norm assumed in the formula (24) allow the strongest, among particular attributes, nonstationarity to be identified. Note that using smooth functions in the above formulas will result in relatively mild fluctuations in time of the estimated density. For further considerations recall also that $0 < sgmKPSS < 1$.

2.4   Forecasting, Exponential Smoothing

If a nonstationarity is detected, the possibility appears of identification of a potential trend of the changes that have occurred, and regarding in the algorithm the values related with it. In this paper the exponential smoothing forecasting method [10] will be applied with the assumption of linear form of the trend. This method enables effective updating of the prediction model after receiving the subsequent value of the time series $\{X_t\}_{t=1,2,...}$.

The identified trend is assumed in the form $a_2 t + a_1$; denote the coefficients existing here in the form of a line vector, additionally denoting they dependence on t:

$$A_t = [a_{1,t}, a_{2,t}]. \tag{25}$$

The prognosis calculated at the moment t, with the anticipation $p \in \mathbb{N} \backslash \{0\}$ is given by:

$$X_t^p = A_t \begin{bmatrix} 1 \\ p \end{bmatrix}, \tag{26}$$

and then $a_{2,t}$ characterizes the velocity of changes.

In order to determine the matrix A_t, first define the following matrixes:

$$L = \begin{bmatrix} 1 & 0 \\ 1 & 1 \end{bmatrix} \tag{27}$$

$$B_1 = \begin{bmatrix} 1 & 0 \\ 0 & 0 \end{bmatrix}, \quad B_{t+1} = B_t + v^t \begin{bmatrix} 1 & -t \\ -t & t^2 \end{bmatrix} \quad \text{for} \quad t = 1, 2, \ldots \tag{28}$$

$$b_1 = \begin{bmatrix} X_1 \\ 0 \end{bmatrix}, \quad b_{t+1} = vL^{-1}b_t + \begin{bmatrix} 1 \\ 0 \end{bmatrix} X_{t+1} \quad \text{for} \quad t = 1, 2, \ldots, \tag{29}$$

where the parameter $v \in [0, 1]$ determines the intensity of adaptation of the forecasting model, fitting it to the changing reality. The possible increase in its value reduces the speed of reaction for forecasting errors, while decrease intensifies this reaction but threatens instability. The parameter v value will be determined in the following.

On the basis of the values successively obtained in the examined time series $X_1, X_2, \ldots$ one can calculate the matrixes (28) and (29), and finally

$$A_t = B_t^{-1} b_t \quad \text{for } t = 1, 2, \ldots . \tag{30}$$

Its second element $a_{2,t}$ will be used in the next chapter for the procedure designed there.

Detailed information on the exponential smoothing method can be found in the monograph [10] and the classic textbook [1].

3 Procedure for Predicted Distribution Density Estimation

The distribution density of streaming data will be determined using the moving window concept. Assume three parameters m_{min}, m, $m_0 \in \mathbb{N} \backslash \{0\}$ such that $m_{min} \leq m \leq m_0$. They represent a minimal, current and standard (in practice also maximal) number of elements, on the basis of which the kernel estimator $\hat{f}$ will be calculated. A reservoir consisting of m_0 last elements of the data stream under research will be created and successively updated. The elements of the reservoir are stored with the order of currency, from the newest x_1 to the oldest x_{m_0}.

The parameters m_{min}, m_0 are constant, while m changes depending on the current behavior of the data stream (see Sect. 3.1). They are assigned weights resulting from the outdatedness with intensity depending on the nature of the stream under research (Sect. 3.2). Following its characteristics, the procedure will be supported by forecasting methods (Sect. 3.3). Atypical elements are accordingly amplified or reduced depending on whether it represents new or diminishing tendencies (Sect. 3.4). Each of the above concepts reduces the estimation error, while these gains are independent and cumulative (Sect. 4).

3.1 Variable Reservoir Size

The reservoir size m, on the basis of which the kernel estimator is calculated, has a fundamental meaning for the quality of estimation. In the stationary case, when the characteristics of the data stream do not change, the higher value of this parameter gives more accurate results. However in the case of nonstationarity, smaller values of m enable us to more effectively keep up with changes.

We have assumed the following heuristic evaluation concerning the accuracy of the basic one-dimensional ($n = 1$) estimator:

$$
\begin{aligned}
m &= 100 \quad - \text{ acceptable quality} \\
m &= 1,000 \quad - \text{ good quality} \\
m &= 5,000 \quad - \text{ very good quality.}
\end{aligned}
\tag{31}
$$

The accuracies obtained experimentally for exemplary one-, two-, and three-modals distributions are shown in Table 1. Of course, all intermediate values as well as outside of the above range are possible. Enlarging the data dimension by one, requires about 4-fold increase in the size m to maintain quality.

Table 1. Accuracy of estimation of the exemplary distributions one-modal N(0, 1) with formula (7), two-modals 60% N(0, 1) + 40% N(5, 1) with plug-in of degree 2, three-modals 30% N(-5, 1) + 40% N(0, 1) + 30%N (5, 1) with plug-in of degree 3, for L^1, L^2 and *sup* norms.

Accuracy	One-modal	Two-modals	Three-modals
$m = 50$	0.184, 0.300, 0.080	0.246, 0.340, 0.056	0.254, 0.350, 0.044
$m = 100$	0.141, 0.224, 0.062	0.187, 0.264, 0.046	0.205, 0.283, 0.037
$m = 1,000$	0.060, 0.098, 0.030	0.076, 0.112, 0.023	0.091, 0.127, 0.018
$m = 5,000$	0.032, 0.054, 0.018	0.041, 0.062, 0.013	0.050, 0.071, 0.011
$m = 10,000$	0.024, 0.041, 0.014	0.031, 0.047, 0.011	0.038, 0.055, 0.009

Therefore let m_0 constitute the assumed reservoir size for the conditions of stationarity, as well as m_{min} its minimal permissible level. Then define the value on the basis of which the kernel estimator $\hat{f}$ will be calculated as

$$
m = \begin{cases}
m_{min} & \text{when} & m_* < m_{min} \\
m_* & \text{when} & m_{min} \leq m_* \leq m_0 , \\
m_0 & \text{when} & m_* > m_0
\end{cases}
\tag{32}
$$

while

$$
m_* = int(1.1 m_0 (1 - sgmKPSS)),
\tag{33}
$$

where $sgmKPSS$ is given by the formula (22) substituting (13)–(19) and (21)–(24). Therefore, if one is dealing with a stationary process, then $sgmKPSS \cong 0$ and in consequence $m \cong m_0$. In turn, in the case of distinct nonstationarity $sgmKPSS \cong 1$, and then

$m \cong m_{min}$. The intermediate values consecutively fluctuate in a continuous manner, as the term *sgmKPSS* successively changes its value. Multiplication of the parameter m_0 by 1.1 (i.e. increase by 10%) in the formula (33) and then restriction m to m_0 in (33) eliminates possible fluctuations of m near m_0, resulting from the "tail" of the KPSS test statistics. To justify the level of 10%, see the determination of the parameters of the linear transformation in the equality (22). Note also that for purposes of the KPSS test (and only here) described in Sect. 2.4, the elements should be provided in the opposite order, from the oldest with the index 1 to the newest with m_0.

Using the classic automatic control methods, in the basic one-dimensional case $n = 1$, the value $m_0 = 1,000$ has been obtained as a standard (compare the formula (31) and Table 1). In the case of complex, significantly multimodal distributions, it can be increased by 100 for each additional mode.

The parameter m_{min} value should be dependent on the biggest speed of changes. In particular, we propose

$$m_{min} = int\left(\frac{m_0}{10}\right), \tag{34}$$

The value $m_{min} = 100$ can be treated as a standard. Such a value enables an effective tracking of changes not faster than $0.01 \, \hat{\sigma}$ per step. For slower changes, the bottom boundary by m_{min} will be simply inactive. For faster alternations $m_{min} = 50$ is possible, however, runs can excessive fluctuating in time. Further decreasing of this value is not recommended (compare the formula (31) and Table 1).

3.2 Outdatedness

Particular elements used to calculate the kernel estimator will undergo outdatedness over time. This function will be performed by appropriate definition of values of the coefficients w_t, introduced in the definition (5). The linear formula will be applied

$$w_i^* = 2\left[1 - \frac{\alpha(i-1)}{m}\right] \quad \text{for} \quad i = 1, 2, \ldots, m, \tag{35}$$

where $\alpha \in [0, 1]$ specifies the intensity of outdatedness. In particular $\alpha = 0$ means its absence; all the reservoir elements then have the same weight. In contrast, if $\alpha = 1$, the weights successively decreased from 2 for the newest element with the index 1, to $2/m$ for the oldest with the index m, with the step $2/m$.

In the case of stationarity, it is worth assuming the value $\alpha = 0$, successively growing it as nonstationarity increases, to the maximum permissible value 1. As a natural consequence, it has been accepted that

$$\alpha = sgmKPSS, \tag{36}$$

where *sgmKPSS* is given by the formula (22) substituting (13)–(19) and (21)–(24).

Finally, to take account of the above outdatedness procedure, for the purposes of constructing the estimator (5) one should assume $w_i = w_i^*$ for $i = 1, 2, \ldots, m$, where w_i^* are given by the formulas (35) and (36).

3.3 Prediction

In the case when nonstationarity of the data stream under research results from a formed trend, it is worthwhile suitably introducing elements of forecasting methods, described in Sect. 2.4, to the model.

For each new reservoir element x_i, sequentially, from the moment of its receiving, one builds a forecasting model, following the material presented in Sect. 2.4, for which the consecutive quantities $\hat{f}(x_i)$, where $\hat{f}$ is the kernel estimator calculated on the basis of Sect. 3.1 are treated as successive values of the observed time series. Thanks to this we have the vector (25) and in particular its second component $a_{2,t}$, which for the element x_i can be naturally denoted as $a_{2,t,i}$. Note also that the forecasting model is assigned to the specific element x_i and when its index i changes over time within the reservoir, this model moves with it for $i = 1, 2, \ldots, m_0$.

Now define the function representing changes of the kernel estimator (5). Let, therefore, for the fixed t, the function $g_t : \mathbb{R}^n \to \mathbb{R}$ be given by the formula

$$g_t(x) = \frac{1}{m} \sum_{i=1}^{m} a_{2,t,i} K(x, x_i, h), \qquad (37)$$

where $a_{2,t,i}$ is the second element of the vector A_t (25), at the moment t, for the element x_i; the function K remains unchanged (4), while the parameter h value is the same as in the estimator $\hat{f}$ calculated on the base of Sect. 3.1.

Introduce now the coefficients

$$w_i^{**} = 1 + \beta_i sgmKPSS \quad \text{for } i = 1, 2, \ldots, m , \qquad (38)$$

where $sgmKPSS$ is given by the formula (22) substituting (13)–(19) and (21)-(23), while $\beta_i \in [-1, 1]$. The presence in the above dependence (38) of the factor $sgmKPSS$ causes that in the case when the data stream is stationary, the coefficients w_i^{**} are close to 1, while in the nonstationary case the influence of the parameters β_i is manifested accordingly. Define their values as

$$\beta_i = \beta_0 \cdot \frac{g_t(x_i)}{\overline{g}_t} \quad \text{for } i = 1, 2, \ldots, m , \qquad (39)$$

where

$$\overline{g}_1 = 1 , \quad \overline{g}_t = \max_{i=1,2,\ldots,m_t} |g_t(x_i)| \quad \text{for } t = 2, 3, \ldots , \qquad (40)$$

m_t denotes the size of the reservoir in the moment t and the constant $\beta_0 \in [0, 1]$ indicates the intensity of the forecasting function constructed herein. For the stationary conditions the value $\beta_0 = 0$ is natural. In the case of nonstationarity, initially consider $\beta_0 = 0.5$ as a standard. Generally the values from the range $[1/3, 2/3]$ are satisfactory, while for the slow changes smaller values are preferable (also because of the function $\hat{f}$ fluidity over time) and for fast − bigger. For the nonstationarity, values smaller than $1/3$ result in too weak prediction, larger than $2/3$ seem to be somewhat extreme (in particular, for $\beta_0 = 1$ some kernels could be removed, which unintentionally reduces a sample size assumed in Sect. 3.1). Finally we propose

$$\beta_0 = \frac{2}{3} sgmKPSS. \qquad (41)$$

Note that the condition $\beta_i \in [-1, 1]$ is fulfilled only with accuracy of determining the maximum of the function g_t only on the finite set $\{x_i\}$ as assumed in the formula (40). It has no meaning from the applicational point of view, because these parameters are multiplied in the dependence (38) by $sgmKPSS$, which is strictly less than 1, what in practice ensures the meaningful inequality $w_i^{**} \geq 0$.

The parameter v introduced in the formulas (28) and (29), defining the intensity of adaptation of the forecasting model, can be determined by the natural dependence:

$$v = 1 - \frac{1}{m}. \tag{42}$$

The intensity of adaptation of the forecasting model is therefore proportional to information provided by every new element of the data stream with the current reservoir size m.

Finally, if the prediction is used without the outdatedness procedure, it should be assumed that $w_i = w_i^{**}$, where w_i^{**} was defined above by the formulas (37)–(42), while if the both concepts are implemented, then $w_i = w_i^* \cdot w_i^{**}$, for $i = 1, 2, \ldots, m$.

3.4 Atypical (Rare) Elements

By calculating the distribution density, one can easily detect, separately in every moment t, atypical elements in the sense of rarely occurring. As previously, introduce the coefficients

$$w_i^{***} = 1 + \gamma_i sgmKPSS \quad \text{for } i = 1, 2, \ldots, m, \tag{43}$$

where $sgmKPSS$ is given by the formula (22) substituting (13)–(19), (21) and (23), and moreover $\gamma_i \in [-1, 1]$ are defined as

$$\gamma_i = \begin{cases} \frac{g_t(x_i)}{\bar{g}_t} & \text{when} \quad x_i \text{ is atypical element} \\ 0 & \text{when} \quad x_i \text{ is typical element} \end{cases} \quad \text{for } i = 1, 2, \ldots, m, \tag{44}$$

while g_t and $\bar{g}_t$ were specified in the previous Sect. 3.3; see formulas (37) and (40). Thus, in the case of stationarity, the values of the coefficients w_i^{***} will be close to 1, while the data stream is nonstationary, the more amplified will be atypical elements which represents a rising tendency (thanks to $\gamma_i > 0$), and reduced recessive elements (due to $\gamma_i < 0$).

The procedure presented in Sect. 2.3. can be used to identify atypical elements. Based on suggestions from the formula (10), the value of the parameter r, determining the procedure sensitivity, will be assumed as

$$r = 0.01 + 0.09 \, sgmKPSS. \tag{45}$$

In the stationary case, a few (around 1%) atypical elements are specified, with an indication which are connected with new trends ($\gamma_i > 0$) and which with diminishing ($\gamma_i < 0$). It may be a valuable suggestion in the fundamental analysis of the results obtained. In turn, in conditions of strong nonstationarity, when $sgmKPSS \cong 1$, almost 10% of elements are recognized as atypical, which introduces an additional forecasting factor, as the importance of elements of increasing significance grows ($\gamma_i > 0$) and of decreasing meaning shrinks ($\gamma_i < 0$), which generally improves estimation quality. (If the results are given in the graphical form, it is worthwhile to mark on the graph the atypical elements by different color whose with positive γ_i values and other with negative.)

Finally, if the procedure described in Sects. 3.2–3.4 are used, then the coefficients introduced in the definition (5) should be taken in the form

$$w_i = w_i^* \cdot w_i^{**} \cdot w_i^{***} \quad \text{for } i = 1, 2, \ldots, m \ . \tag{46}$$

If any of these procedures, outdatedness, prediction or detection of atypical elements, should be omitted, then the appropriate element w_i^*, w_i^{**} or w_i^{***} should be removed from the above formula. For clarity of interpretation, each of them varies in the same range from 0 to 2. All of them change continuously, which smooths fluctuations of the density $\hat{f}$.

4   Conclusion, Additional Aspects, and Numerical Evaluation

This paper investigates the concept of calculation of the current distribution density of the streaming data. The function $\hat{f}$ is defined by the formula (5), whereas standard quantities are associated with kernel estimators are presented in Sect. 2.2, while the determination of the reservoir size and the construction of the coefficients w_i are given in the particular sections of Sect. 3.

In the first step (to avoid zeros in the denominator in the formula (8)) it is arbitrarily assumed that $h = 1$. If in the initial steps, the number of the elements received is insufficient to fill the reservoir of the size obtained in Sect. 3.1 or the average $\bar{g}_t$ from the formula (40), then this size should be reduced naturally to the number we have, similarly to the formula (23). Up to the step $t = m_{min}$, the results obtained are only indicative and do not give ground for further analysis at the assumed accuracy level. Only after the moment $t = m_0$ does the procedure work under appropriate sufficient stabilized conditions.

Similarly, during an increase in the parameter m value, one should add to the reservoir only new elements, even if they come slower than the m grow speed.

The procedure investigated has been comprehensively verified using both, simulated and real, data streams. For the basic illustration, consider a single continuous attribute, when the testing stochastic process is given in the form

$$X_t = p\,t + 0.6\,N(0,\ 1) + 0.4\,N(5,\ 1) \quad \text{for } \ t = 1, 2, \ldots \ , \tag{47}$$

where $p\,t$ represents a deterministic trend, while

$$p = \begin{cases} 0.000,5 & \text{when} & t < 2{,}000 \\ 0.01 & \text{when} & 2{,}000 \le t < 5{,}000 \\ 0.005 & \text{when} & 5{,}000 \le t < 8{,}000 \\ +1 & \text{when} & t = 8{,}000 \\ 0 & \text{when} & 8{,}000 < t \le 10{,}000 \end{cases} \tag{48}$$

whereas $+1$ denotes a unit step. Therefore during the initial period with very slow changes, a consolidation of the algorithm occurs, and then the data stream increases firstly very fast and then with medium speed, and finally, after a unit step, the process becomes stationary. Such changes in dynamics pose a big challenge for the worked out method.

Three performance indexes were used; averaged over time differences between the real density resulting from the formulas (47) and (48) and the estimator, in the senses of the L^1, L^2, sup norms. The presented results were obtained on the basis of 20 averaged runs. Each calculation cycle was performed in a few seconds.

In the stationary case the results were close (with accuracy to 1%) to those obtained simply on the basis of the last m_0 elements; the KPSS test correctly classified the stationarity of the data stream. The above simple strategy of the last m_0 elements was generally treated as the reference. The introduction of the variable m, as indicated in Sect. 3.1, resulted in a decrease in the value of these indexes of 56%, 67% and 64%, respectively for particular indexes. The addition of outdatedness (Sect. 3.2) improved the indexes by further 23%, 24% and 9%, while the addition of forecasting (Sect. 3.3) reduces their values by 21%, 30% and 16%, and finally, adding atypical elements detection (Sect. 3.4) by further 3%, 2% and 1%. In total, all four factors (Sects. 3.1–3.4) improved quality by about 63%, 83% and 65%.

Atypical elements detection does not introduce significant improvement of indexes. It works in those areas, in which the distribution density values are small and is also their influence of numerical indicators. This factor, however, captures even insignificant changes but often very important in the fundamental analysis of datasets, and also extraordinary errors and situations, not covered by the above research scheme. Note that forecasting as well as atypical elements detection work on characteristics which were already significantly improved by the modification of the reservoir size and outdatedness.

Similar results were obtained for multidimensional cases, also in the presence of categorical attributes [4], and with a noise correlated in time. Detailed experimental studies are the subject of the comprehensive paper [17], where a comparative analysis with the other methods quoted at the end of Sect. 1, is also contained. Generally, the procedure presented here gives much better results, where the more clear and ambiguous formed trend is present.

Future researches will lead to the substitution of the removal of the oldest elements of the reservoir by sampling with probabilities dependent on the current size of the reservoir, outdatedness, prognosis and atypical elements detection presented in the successive sections of Sect. 3 of this paper. It prevents the complete omission of phenomena manifested by elements older than the current reservoir size as it is in the case of the moving window method applied here. Thanks to forecasting, this goal can be achieved without, both qualitative and quantitative, deterioration of a quality.

References

1. Abraham, B., Ledolter, J.: Statistical Methods for Forecasting, Wiley (2005)
2. Aggarwal, C.C.: Outlier Analysis. Springer, Heidelberg (2013). https://doi.org/10.1007/978-3-319-47578-3

3. Aggarwal, C.C. (ed.): Data Streams. Models and Algorithms. Springer US, Boston, MA (2007). https://doi.org/10.1007/978-0-387-47534-9
4. Agresti, A.: Categorical Data Analysis. Wiley (2002)
5. Amiri, A., Dabo-Niang, S.: Density estimation over spatio-temporal data streams. Econometrics Stat. **5**, 148–170 (2018)
6. Bequette, B.W.: Process Control: Modeling, Design, and Simulation. Prentice Hall (2010)
7. Boedihardjo, A.P., Lu, C.-T., Chen, F.: Fast adaptive kernel density estimator for data streams. Knowl. Inf. Syst. **42**(2), 285–317 (2013). https://doi.org/10.1007/s10115-013-0712-0
8. Cao, Y., He, H., Man, H.: SOMKE: kernel density estimation over data streams by sequences of self-organizing maps. IEEE Trans. Neural Netw. Learn. Syst. **23**, 1254–1268 (2012)
9. Heinz, C., Seeger, B.: Cluster kernels: resource-aware kernel density estimators over streaming data. IEEE Trans. Knowl. Data Eng. **20**, 880–893 (2008)
10. Hyndman, R., Koehler, A., Ord, K., Snyder, R.: Forecasting with Exponential Smoothing: The State Space Approach. Springer, Berlin (2008). https://doi.org/10.1007/978-3-540-719 18-2
11. Kulczycki, P.: Estymatory jadrowe w analizie systemowej. WNT (2005)
12. Kulczycki, P.: Methodically unified procedures for outlier detection, clustering and classification. In: Arai, K., Bhatia, R., Kapoor, S. (eds.) Proceedings of the Future Technologies Conference (FTC) 2019, FTC 2019. Advances in Intelligent Systems and Computing, San Francisco, CA USA, vol 1069, pp. 460–474. Springer, Cham (2020).https://doi.org/10.1007/978-3-030-32520-6_35
13. Kulczycki, P., Franus, K.: Methodically unified procedures for a conditional approach to outlier detection, clustering, and classification. Inf. Sci. **560**, 504–527 (2021). https://doi.org/10.1016/j.ins.2020.08.122
14. Kulczycki, P., Kacprzyk, J., Kóczy, L.T., Mesiar, R., Wisniewski, R. (eds.): ITSRCP 2018. AISC, vol. 945. Springer, Cham (2020). https://doi.org/10.1007/978-3-030-18058-4
15. Kulczycki, P., Korbicz, J., Kacprzyk, J. (eds.): Automatic Control, Robotics, and Information Processing. SSDC, vol. 296. Springer, Cham (2021). https://doi.org/10.1007/978-3-030-485 87-0
16. Kulczycki, P., Kruszewski, D.: Identification of atypical elements by transforming task to supervised form with fuzzy and intuitionistic fuzzy evaluations. Appl. Soft Comput. **60**, 623–633 (2017)
17. Kulczycki, P., Rybotycki, T.: Predicted Kernel Estimator for Data Stream (2021, in press)
18. Kwiatkowski, D., Phillips, P.C.B., Schmidt, P., Shin, Y.: Testing the null hypothesis of stationarity against the alternative of a unit root. J. Econometrics **54**, 159–178 (1992)
19. Muthukrishnan, S., Strauss, M., Zheng, X.: Workload-optimal histograms on streams. In: Brodal, G.S., Leonardi, S. (eds.) ESA 2005. LNCS, vol. 3669, pp. 734–745. Springer, Heidelberg (2005). https://doi.org/10.1007/11561071_65
20. Qahtan, A., Wang, S., Zhang, X.: KDE-track: an efficient dynamic density estimator for data streams. IEEE Trans. Knowl. Data Eng. **29**, 642–655 (2017)
21. Rutkowski, L., Jaworski, M., Duda, P.: Stream Data Mining: Algorithms and Their Probabilistic Properties. Springer, Heidelberg (2019). https://doi.org/10.1007/978-3-030-139 62-9
22. Trevino, E.S.G., Hameed, M.Z., Barria, J.A.: Data stream evolution diagnosis using recurs. ACM Trans. Knowl. Discov. Data **12**(14), 28 (2018)
23. Wand, M.P., Jones, M.C.: Kernel Smoothing. Chapman and Hall (1995)
24. Zhou, Z., Matterson, D.S.: Predicting Ambulance Demand: A Spatio-Temporal Kernel Approach. arXiv:1507.00364 (2015)

LSTM Processing of Experimental Time Series with Varied Quality

Krzysztof Podlaski[1]([✉])[iD], Michał Durka[1], Tomasz Gwizdałła[1][iD],
Alicja Miniak-Górecka[1][iD], Krzysztof Fortuniak[2][iD], and Włodzimierz Pawlak[2][iD]

[1] Faculty of Physics and Applied Informatics, University of Lodz, Lodz, Poland
{krzysztof.podlaski,tomasz.gwizdalla,alicja.miniak}@uni.lodz.pl
[2] Faculty of Geographical Sciences, University of Lodz, Lodz, Poland
{krzysztof.fortuniak,wlodzimierz.pawlak}@geo.uni.lodz.pl

Abstract. Automatic processing and verification of data obtained in experiments have an essential role in modern science. In the paper, we discuss the assessment of data obtained in meteorological measurements conducted in Biebrza National Park in Poland. The data is essential for understanding the complex environmental processes, such as global warming. The measurements of CO2 flux brings a vast amount of data but suffer from drawbacks like high uncertainty. Part of the data has a high-level of credibility while, others are not reliable. The method of automatic evaluation of data with varied quality is proposed. We use LSTM networks with a weighted square mean error loss function. This approach allows incorporating the information on data reliability in the training process.

Keywords: lstm · Neural networks · Time series · Prediction · co2 flux

1 Introduction

Verification and prediction of real data are important and challenging. Many experiments conducted every day produce new raw data that has to be assessed and analyzed. The vast amount of new data acquired every day puts high demand on automatic methods of data processing. Moreover, some values obtained in the experiments have higher credibility than the others. Therefore, machine learning systems have to be sensitive to such issues. In the literature, prediction and forecasting are connected with the use of neural networks [4,14,20]. Here we discuss Long Short Term Memory (LSTM) networks as the ones that suit well for the task of analysis and prediction of time series [8,18].

In the paper, we consider the data acquired in a continuous meteorological experiment. Some records in the dataset have higher quality (are more reliable) than others. The prediction of time series with varied quality is similar to the classification of imbalanced datasets [19]. The methods used in classification cannot be directly applied to time series prediction but can be used as suggestions.

© Springer Nature Switzerland AG 2021
M. Paszynski et al. (Eds.): ICCS 2021, LNCS 12747, pp. 581–593, 2021.
https://doi.org/10.1007/978-3-030-77980-1_44

We propose a modification of the usual loss function in order to incorporate the issue of data credibility. The paper combines the computer and meteorological sciences. The importance of research conducted in the area of global warming is hard to neglect. The paper follows the idea of using computer science methods for a better future.

Understanding the complex environmental processes, such as global warming, belongs to the most challenging, both for cognitive reasons and social consequences. It is a truism to say that reliable data are of fundamental meaning for this purpose. The standard geophysical data provided by national and international services (such as WMO) are generally continuous and high-quality but do not include all parameters needed. Therefore, they must be supplemented by the results of experiments that often use very sophisticated measurement techniques. These experiments allow us to gain a vast amount of unique data but may suffer from various drawbacks, including a large number of unreliable or missing values. The missing values introduce high uncertainty when the data is analyzed from a long-term perspective and must be replaced by the most likely one. The short gaps in data sets might be filled up with simple interpolation methods, but the longer ones should be completed with adequately modeled values. This paper focuses on the time series of CO_2 flux collected in the wetlands of Biebrza National Park, northeastern Poland [9,10]. The measured CO_2 flux represents the net exchange of this greenhouse gas between the surface and the atmosphere, which is vital to understand the role of such ecosystems in the global carbon cycle. The eddy-covariance method used, although considered to be the most adequate in measurements of surface-atmosphere exchange in the whole ecosystem scale [2,3], results in a large number of non-randomly distributed missing data due to required quality control and sensitivity to unfavorable weather conditions. To evaluate the total uptake or emission of CO_2 by the ecosystem in an annual or multi-year perspective, these gaps must be filled up, taking into account the sensitivity of the CO_2 flux to changing hydrometeorological conditions [2]. Otherwise, the results could be biased towards the flux recorded in fine weather conditions. Various approaches to the gap-filling procedure have been suggested [2,7,10,15,16], but the problem has still not been fully standardized by the eddy-covariance community. Therefore, we propose using methods known from automatic prediction and verification of time series of CO_2 flux data, which can improve gap-filling methods and consequently reduce uncertainty in assessing the carbon balance in terrestrial ecosystems.

2   The Dataset

In the paper, we use raw, real data obtained during measurements conducted at the Biebrza National Park's wetlands in northeastern Poland. The measurement site was located in the middle basin of Biebrza valley (53°35′30.8″N, 22°53′32.4″E, 109m a.s.l.) in a large flat area covered by patches of reeds, high sedges, and rushes, very typical of wetlands of the Biebrza National Park. The measurement period used in this analysis covers the years 2013–2017. The open-path eddy-covariance system consisted of fast-response sensors: CO_2/H_2O gas

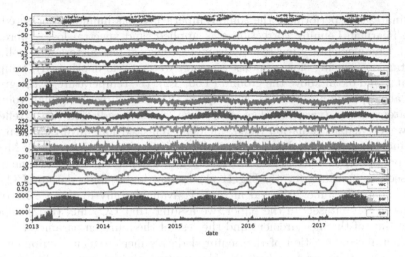

(a) Years 2013-17

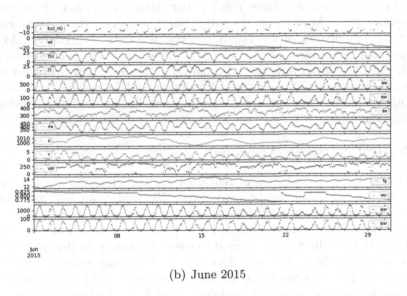

(b) June 2015

Fig. 1. Parameters measured during geographical experiments in Biebrza National park.

analyzer (Li7500, Li-cor Inc., USA) and a sonic anemometer (81000, R. M. Young, USA) with the middle of the path at 3.7 m above ground level [9,10]. The CO_2 fluxes were calculated for 1-hour block averaging (and auxiliary on a 5-min basis) with the aid of EddyPro 6.0 (Li-cor Inc., USA) software to ensure compatibility with other studies. In the flux calculations, covariance was maximized within a ± 2s window, a double rotation of the natural wind coordinates was performed, the sonic temperature was humidity corrected, the Webb-Pearman-Leuning correction was applied, and spectral losses were corrected. Complementary fourteen hydrometeorological parameters were collected simultaneously using slow response sensors: water table depth (wd), temperature at 2 and 0.5 m above the ground (T50, T2), the temperature of the ground (Tg), atmospheric pressure (p), wind speed (v), wind direction (vdir), volumetric water content (vwc), incident shortwave (visible) radiation (isw), reflected shortwave (visible) radiation (rsw), incident/reflected longwave radiation (ilw, rlw), incident/reflected photosynthetically active radiation (ipar, rpar). Most of these parameters show a seasonal behavior and a dependence on the time of the day (see Fig. 1). Moreover, the Fig. 1b shows that high-quality data on CO_2 flux is very sparse in June 2015 and in total is available only in about 25% of the records in the dataset. In the paper, we assume that CO_2 flux (fco2) depends on the time of the measurement and the rest of the fourteen parameters.

A detailed description of the acquired data's measurement equipment and postprocessing can be found in [9]. The postprocessing data quality included three stationarity tests and the friction velocity threshold criterion. In the result, the data was divided into three classes that define the quality of the result, High-quality (HQ), Medium-quality (MQ), and Low-quality (LQ). These groups describe the credibility of the raw data. The HQ data (flagged by EddyPro and accepted by all three additional tests) passed more rigorous criteria than usually used in eddy-covariance data analysis. The MQ data (accepted by at least one of the three additional tests) are similar to those usually analyzed. The remaining data were classified as LQ, and they may be burdened with substantial measurement errors.

3   Neural Network Approach

The verification and prediction of real data are two areas of data processing. These approaches are similar in many ways. We build, train, and validate the prediction model with historical data. We can use the system to predict new values that will be obtained in the future. Usually, the real result has higher credibility and any differences are used to assess the model. In that way, we verify the model with historical data. If we do not observe any external effects that should result in model modification, we can assume some level of correctness of the model. Such an approved prediction models can be used for verification purposes. The differences between the predictions made by the approved model and the data obtained from the measurements can suggest low-quality measurements. Researchers use artificial neural networks for gap-filling in meteorological for some time [6,16,17]. First, we have to build the model and validate it against the historical dataset.

All changes in time in a meteorological system preserve continuity. Therefore, we can treat the measurements as a time series. The changes between two consecutive measurements should express a level of smoothness. In the paper, we plan to follow the presented approach and assess the measurement data using Long Short Term Memory (LSTM) neural network [11,13].

3.1 Neural Networks

The neural network is a technique based on the neural structures of living animals. A feed-forward artificial network is built from neurons grouped in layers. Each layer receives signals only from the preceding layer. The information flows sequentially from layer to layer. The first layer is usually called the Input Layer. The last layer is denoted as the Output Layer. All layers in between are called hidden ones. Each neuron receives information from neurons in the preceding layer. Each input signal is assigned with a weight. A neuron counts a weighted sum of inputs, adds a bias b, and applies an appropriate activation function. All neurons in a layer have the same activation function, and all of them have individual weights and bias. The networks can be trained in a supervised approach using an optimization method (for example, SGD, Adam, AdaGrad, RMSProp) [12]. The training goal is to find a set of network parameters that minimize a given loss function. Usually, the loss function measures the distance between an expected answer and the one obtained from the network.

LSTM neural networks proved to be adequate for the prediction of time-series data. In short, we can perceive the LSTM cell as a subsystem with a set of fully connected sub-layers and gates. Input signal contains a time series of signals that are fed to four fully connected sub-layers one by one. The sub-layers' signals are joined with three types of gates (forget, input, and output). The so-called long-term state represents the memory of the cell. Forget gate controls how long-term state should incorporate a new signal from a given series. An activation function accompanied each LSTM cell. The LSTM cell processes inputs row by row using the embedded memory feature. Each row contains a single event in a time series processed. Therefore, the cell size depends only on the size of the information stored in one row.

In the classical approach, all time series provided during training have the same credibility. Therefore, all are treated in the same way. We know that the experiment's data has a varied quality and needs to be treated in a special way. In classification approaches, a few solutions to an imbalance in class representation are proposed [19]. For example, each training input is assigned to a weight used during training. We can enforce the underrepresented class signals to appear more often in the training process. In the prediction approach, we can either change the ratio of occurrences of the high-quality signals during training or change the loss function to take the quality into account. In the paper, we use the latter approach and define the weighted mean square error as a loss function. Similar solutions to our approach can be found in [5].

3.2 Input Data

The dataset contains fourteen meteorological parameters measured simultaneously, the date and time of the measurements, and registered CO_2 flux. We decide not to limit ourselves to a subset of experimental data and assume that the CO_2 flux depends on all fourteen meteorological parameters (described in the previous section). It should be mentioned that the authors of the paper [9] limit themselves to nine measured parameters. Additionally, as neural networks treat better signals in the range $(-1, 1)$, all parameters are normalized to this range. We prepare input data representing a set of $n + 1$ consecutive occurrences of measured parameters for our experiments. That means the input of the network has a shape of an $(n + 1) \times m$ matrix, where m denotes the size of a single set of simultaneous meteorological parameters accompanied by the measurement time, n denotes the number of consecutive predecessors of the actual measurement. We can interpret $n + 1$ as the size of the time window used to predict the value of CO_2 flux. At each time step t_i, we have a vector of input parameters $\mathbf{v}_i$. Each vector $\mathbf{v}_i$ contains the year, month, day, hour, minute of measurement, as well as all fourteen meteorological parameters. In the preprocessing phase, we create a set of input matrices of the form:

$$I_i = \begin{pmatrix} \mathbf{v}_i \\ \mathbf{v}_{i-1} \\ \ldots \\ \mathbf{v}_{i-n} \end{pmatrix} \tag{1}$$

The input matrix I_i has a size $(n + 1) \times m$, with $n + 1$ rows of size m. Each row contains consecutive measurement vectors $\mathbf{v}_i$. The matrix I_i is accompanied by the expected output that contains the value of registered CO_2 flux at the moment t_i (fco2$_i$). The values fco2$_i$ are the results the model will be trained to predict.

3.3   LSTM Model

LSTM model uses an LSTM layer as well as dense ones (fully connected) as presented in Fig. 2. The model contains an LSTM layer of size s and one dense layer with $2s$ neurons. Both layers use tanh activation functions. The output layer is discussed in the next subsection.

3.4   Output Layer and Loss Function

During training, the neural network use loss function and appropriate optimization method to derive modifications of the actual network parameters. In the dataset, we have three types of measurements connected to their quality. High-quality CO_2 flux values, the most reliable data (HQ) accounted for about 25% of all records, Medium-quality (MQ) around (35%), and the rest is a Low-quality (LQ) (40%). Therefore, during the training, the HQ data need to have the highest priority. The designed network has two neurons in the output layer. The first one y_{fco} predicts the value of CO_2 flux the other y_q expected quality. The expected results, used for training and testing has the same structure:

$$\hat{y} = [\hat{y}_{fco}, \hat{y}_q], \qquad \text{where: } \hat{y}_q = \begin{cases} 0 & \text{for result in LQ} \\ 1 & \text{for result in MQ}. \\ 2 & \text{for result in HQ} \end{cases} \tag{2}$$

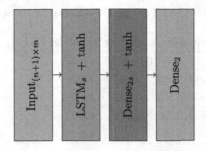

Fig. 2. LSTM neural network model, the subscripts denote size of the Layer

The results with high quality have higher reliability than the ones for other classes. Thus, we should enforce the network to take the data credibility into account during the training process. We propose to incorporate quality into the loss function. Our self-defined loss function has the form:

$$\mathrm{loss}(\mathbf{y}, \hat{\mathbf{y}}) = \mathrm{mean}\left((y_{\mathrm{fco}} - \hat{y}_{\mathrm{fco}})^2 \cdot 2^{\hat{y}_q}\right), \tag{3}$$

where $\mathbf{y}, \hat{\mathbf{y}}$ denote the batch of predicted and expected results, respectively. The presented loss function is a weighted version of the mean standard error function (WMSE), where HQ results have a weight of 4, the weight of MQ results equals 2, while LQ results have the weight factor set to 1. In this manner, we do not totally neglect the results with low quality but assign them with much lower priority.

4 The Numerical Experiment

We conducted our numerical experiments using the Tensorflow library [1] in python on a computer with AMD Ryzen 9 3900X processor and 32 GB of RAM. In all models, we use the Stochastic Gradient Descent (SGD) optimization algorithm. The dataset was split into two disjoint sets, a training set consisting of 30 766 elements (80% of the total number of measurements) and a test set with 7 666 elements (20%). The ratio 80/20 between training and test sets is typically used in classification and prediction tasks that use neural networks.

The different models have been tested with a few selections of network layer the size parameter s, size of series time window n (Fig. 2). The number of trainable parameters for each tested model is presented in Table 1. As mentioned before, the number of trainable parameters in the LSTM layer depends on the row's size in an input signal. It does not depend on the number of rows in the input (the input represents a time series, and each row is a single event). Therefore, the increase of n, number of elements in a time series, has no impact on the number of trainable parameters. There are no rules on how to set network layer sizes for data to be processed. We have tried to keep our neural model small. The s parameter used had values $10, 15, 20$. The parameter n represents the size

Table 1. The size of the LSTM models trained in numerical experiment.

Trainable parameters for model LSTM$_{n,s}$			
	$s = 10$	$s = 15$	$s = 20$
$n = 3$	1502	2702	4202
$n = 4$	1502	2702	4202
$n = 5$	1502	2702	4202
$n = 6$	1502	2702	4202

of the time series of measurements taken into account in a single input for our model. The measurements in the dataset in consideration were measured hourly. Thus, the parameter $n = 6$ means we use for prediction the actual results and values measured in the previous six hours. As most of the parameters have a visible daily routine (Table 1), we have decided to experiment with $n \in (2, 3, 4, 5, 6)$. This expresses the continuity and causality property of the measured parameters in a short time frame from two to six hours. For timeframes longer than six hours, many unmeasured parameters can have non-negligible impact.

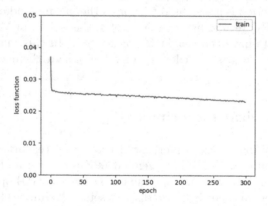

Fig. 3. Training process for LSTM network with $n = 6$ and $s = 20$.

All trained networks achieve stability after around 50 epochs, and the loss function decreases at a stable slow pace. No vital difference has been observed when trained for 300 epochs. For example, we present the training process of the LSTM network with $n = 6$ and $s = 20$ in Fig. 3. Therefore, for most of the cases, we have trained the models for 120 epochs.

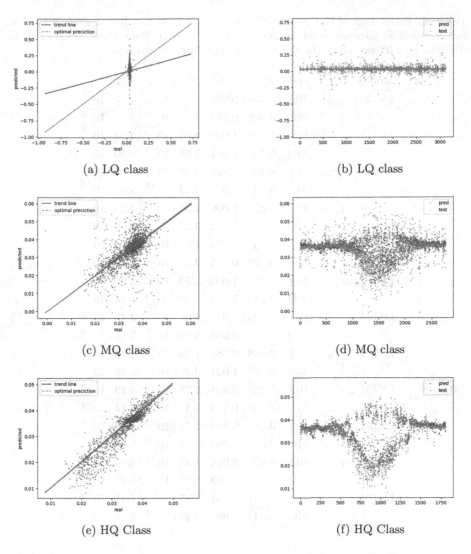

Fig. 4. Prediction given from LSTM network with $s = 20$, $n = 6$ after 120 epochs. On both axes we present normalized predicted CO_2 flux compared to expected(real) CO_2 flux.

The models are trained to focus on HQ results. Therefore, for a detailed assessment of the training process, we measure each class's effectiveness independently. We present a comparison of the results obtained from a neural network with the expected value of CO_2 flux for each class. The assessment was conducted on the test dataset that was not used in training. As shown in Fig. 4, the results predicted from $LSTM_{n=6,s=20}$ network give results with a high level of agreement with the expected (real) values for HQ and MQ classes. The results

Table 2. Estimation of the quality of prediction of CO_2 flux for considered models. For each model parameters Mean Squared Error (SME), Root Mean Squared Error (RSME) and R^2 as well as the slope of the trend line are presented.

Model	class	R^2	slope	MSE	RMSE
$LSTM_{n=3,s=10}$	HQ	0.798	0.926	$2.12 \cdot 10^{-5}$	$4.60 \cdot 10^{-3}$
	MQ	0.499	0.973	$3.58 \cdot 10^{-5}$	$5.98 \cdot 10^{-3}$
$LSTM_{n=3,s=15}$	HQ	0.819	1.006	$2.06 \cdot 10^{-5}$	$4.54 \cdot 10^{-3}$
	MQ	0.529	1.011	$3.96 \cdot 10^{-5}$	$6.29 \cdot 10^{-3}$
$LSTM_{n=3,s=20}$	HQ	0.794	0.978	$2.10 \cdot 10^{-5}$	$4.58 \cdot 10^{-3}$
	MQ	0.542	0.941	$3.52 \cdot 10^{-5}$	$5.94 \cdot 10^{-3}$
$LSTM_{n=4,s=10}$	HQ	0.824	1.006	$1.93 \cdot 10^{-5}$	$4.43 \cdot 10^{-3}$
	MQ	0.562	0.972	$3.24 \cdot 10^{-5}$	$5.69 \cdot 10^{-3}$
$LSTM_{n=4,s=15}$	HQ	0.857	0.978	$1.81 \cdot 10^{-5}$	$4.26 \cdot 10^{-3}$
	MQ	0.536	0.975	$3.37 \cdot 10^{-5}$	$5.81 \cdot 10^{-3}$
$LSTM_{n=4,s=20}$	HQ	0.848	1.006	$1.59 \cdot 10^{-5}$	$3.99 \cdot 10^{-3}$
	MQ	0.584	1.002	$3.29 \cdot 10^{-5}$	$5.74 \cdot 10^{-3}$
$LSTM_{n=5,s=10}$	HQ	0.824	0.975	$1.96 \cdot 10^{-5}$	$4.43 \cdot 10^{-3}$
	MQ	0.549	0.967	$3.31 \cdot 10^{-5}$	$5.75 \cdot 10^{-3}$
$LSTM_{n=5,s=20}$	HQ	0.836	0.986	$1.76 \cdot 10^{-5}$	$4.19 \cdot 10^{-3}$
	MQ	0.569	1.027	$3.36 \cdot 10^{-5}$	$5.79 \cdot 10^{-3}$
$LSTM_{n=5,s=15}$	HQ	0.835	0.961	$1.76 \cdot 10^{-5}$	$4.19 \cdot 10^{-3}$
	MQ	0.553	0.949	$3.41 \cdot 10^{-5}$	$5.84 \cdot 10^{-3}$
$LSTM_{n=6,s=10}$	HQ	0.847	0.965	$1.72 \cdot 10^{-5}$	$4.14 \cdot 10^{-3}$
	MQ	0.561	0.954	$3.24 \cdot 10^{-5}$	$5.69 \cdot 10^{-3}$
$LSTM_{n=6,s=15}$	HQ	0.823	0.976	$1.73 \cdot 10^{-5}$	$4.16 \cdot 10^{-3}$
	MQ	0.541	0.963	$3.33 \cdot 10^{-5}$	$5.77 \cdot 10^{-3}$
$LSTM_{n=6,s=20}$	HQ	0.804	0.981	$1.85 \cdot 10^{-5}$	$4.31 \cdot 10^{-3}$
	MQ	0.543	0.983	$3.46 \cdot 10^{-5}$	$5.88 \cdot 10^{-3}$

for the lowest quality data are not similar to the expected ones. As was mentioned before, these measurements have very low reliability. In Figs. 4a, 4c, 4e we present the dependence of predicted CO_2 flux on the real value, this plot should be as near as possible to a line. In the figures, we draw the orange lines that represents the optimal prediction (a line with slope equal to 1) and the trend line based on data points (in blue). We can see that the HQ data is predicted with high accuracy. The MQ data is not so well concentrated around the optimal prediction line as HQ one is. The last low-quality class is not predicted properly, but as was mentioned before, this was expected, as this data has low credibility. Figures 4b, 4d, 4f compare the behaviour of predictions and expected values for a test set in time. As we can see again, the prediction for HQ class well agrees with the expected values.

The optimal prediction would result in a simple relation: $\hat{y} = 1 * y$ for all $\hat{y}$ that belong to the class in consideration. Therefore, we derive linear regression estimators for every class and all models: R^2, slope, MSE (Mean Squared Error), RMSE (Root Mean Square error) to qualify the prediction. The estimator values are presented in Table 2.

Table 2 presents that most of the LSTM network models have similar quality. It seems that the LSTM with a window length set to 4 gives the best prediction. However, the differences between the qualities of the models are small. The models that have been trained can be used for additional assessment of the quality of experimental data. As we can see in the Fig. 4f, the model predicts more often lower values of high quality CO_2 flux than measured. Therefore, we analyze the residuum of prediction (i.e., the difference between prediction and a real value) Fig. 5. We can see that the shape of the histogram is not symmetric. As a result, the model predicts more often values lower than the real experimental data. It can be understood as extreme values of CO_2 flux are rare and hard to predict.

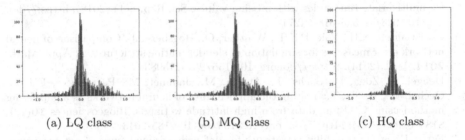

|(a) LQ class|(b) MQ class|(c) HQ class|

Fig. 5. The histograms of residuum for prediction using LSTM network with $s = 20$, $n = 6$ after 120 epochs on test set. On the horizontal axis we can see the value of residuum on the vertical one the number of occurrences.

5   Conclusions

In the paper, we present LSTM neural network usage for CO_2 flux prediction in the meteorological experiment. The data have three different levels of credibility and have to be treated differently during training. We propose to use a weighted means squared error as a loss function during the training process. The method leads to good results with a high level of agreement with the expected values for high-quality results. The network also correctly assesses the quality of the record in consideration. The prepared LSTM network can be used for the automatic verification of experimental raw data. Additionally, we can use the method for gap filling to repair the records with flows. In the future, the usage of the method for gap filling will be analyzed in detail.

Acknowledgements. Funding for this research was partly provided (data collection) by the National Science Centre, Poland under the project UMO-2015/17/B/ST10/02187. The site was established in 2012 under project UMO-2011/01/B/ST10/07550 founded by National Science Centre, Poland. The authors thank the authorities of the Biebrza National Park for allowing continuous measurements in the area of the Park.

References

1. Abadi, M., et al.: Tensorflow: a system for large-scale machine learning. In: Proceedings of the 12th USENIX Conference on Operating Systems Design and Implementation, OSDI 201, pp. 2265–2836. USENIX Association, USA (2016)
2. Aubinet, M., Vesala, T., Papale, D. (eds.): Eddy Covariance: A Practical Guide to Measurement and Data Analysis. Springer, Netherlands (2012). https://doi.org/10.1007/978-94-007-2351-1
3. Baldocchi, D.D.: Assessing the eddy covariance technique for evaluating carbon dioxide exchange rates of ecosystems: past, present and future. Global Change Biol. **9**(4), 479–492 (2003). https://doi.org/10.1046/j.1365-2486.2003.00629.x
4. Che, Z., Purushotham, S., Cho, K., Sontag, D., Liu, Y.: Recurrent neural networks for multivariate time series with missing values. Sci. Rep. **8**(1) (2018). https://doi.org/10.1038/s41598-018-24271-9
5. Christiansen, N.H., Voie, P.E.T., Winther, O., Høgsberg, J.: Comparison of neural network error measures for simulation of slender marine structures. J. Appl. Math. **2014**, 1–11 (2014). https://doi.org/10.1155/2014/759834
6. Dengel, S., Zona, D., Sachs, T., Aurela, M., Jammet, M., Parmentier, F.J.W., Oechel, W., Vesala, T.: Testing the applicability of neural networks as a gap-filling method using CH4 flux data from high latitude wetlands. Biogeosciences **10**(12), 8185–8200 (2013). https://doi.org/10.5194/bg-10-8185-2013
7. Falge, E., et al.: Gap filling strategies for defensible annual sums of net ecosystem exchange. Agric. Forest Meteorol. **107**(1), 43–69 (2001). https://doi.org/10.1016/s0168-1923(00)00225-2
8. Fathalla, A., Salah, A., Li, K., Li, K., Francesco, P.: Deep end-to-end learning for price prediction of second-hand items. Knowl. Inf. Syst. **62**(12), 4541–4568 (2020). https://doi.org/10.1007/s10115-020-01495-8
9. Fortuniak, K., Pawlak, W., Bednorz, L., Grygoruk, M., Siedlecki, M., Zielinski, M.: Methane and carbon dioxide fluxes of a temperate mire in central Europe. Agri. Forest Meteorolog. **232**, 306–318 (2017)
10. Fortuniak, K., Pawlak, W., Siedlecki, M., Chambers, S., Bednorz, L.: Temperate mire fluctuations from carbon sink to carbon source following changes in water table. Sci. Total Environ. **756** (2021). https://doi.org/10.1016/j.scitotenv.2020.144071
11. Gers, F.A., Eck, D., Schmidhuber, J.: Applying LSTM to time series predictable through time-window approaches. In: Dorffner, G., Bischof, H., Hornik, K. (eds.) ICANN 2001. LNCS, vol. 2130, pp. 669–676. Springer, Heidelberg (2001). https://doi.org/10.1007/3-540-44668-0_93
12. Goodfellow, I., Bengio, J., Courville, A., Bach, F.: Deep Learning. MIT Press Ltd (2016)
13. Hochreiter, S., Schmidhuber, J.: Long short-term memory. Neural Comput. **9**(8), 1735–1780 (1997). https://doi.org/10.1162/neco.1997.9.8.1735

14. Ke, J., Zheng, H., Yang, H., Chen, X.M.: Short-term forecasting of passenger demand under on-demand ride services: a spatio-temporal deep learning approach. Transp. Res. Part C Emerging Technol. **85**, 591–608 (2017). https://doi.org/10.1016/j.trc.2017.10.016

15. Kim, Y., et al.: Gap-filling approaches for eddy covariance methane fluxes: a comparison of three machine learning algorithms and a traditional method with principal component analysis. Global Change Biol. **26**(3), 1499–1518 (2019). https://doi.org/10.1111/gcb.14845

16. Moffat, A.M., et al.: Comprehensive comparison of gap-filling techniques for eddy covariance net carbon fluxes. Agric. Forest Meteorol. **147**(3–4), 209–232 (2007). https://doi.org/10.1016/j.agrformet.2007.08.011

17. Papale, D.: Data gap filling. In: Eddy Covariance, pp. 159–172. Springer, Netherlands (2011). https://doi.org/10.1007/978-94-007-2351-1_6

18. Selvin, S., Vinayakumar, R., Gopalakrishnan, E.A., Menon, V.K., Soman, K.P.: Stock price prediction using LSTM, RNN and CNN-sliding window model. In: 2017 International Conference on Advances in Computing, Communications and Informatics (ICACCI). IEEE (2017). https://doi.org/10.1109/icacci.2017.8126078

19. Spelmen, V.S., Porkodi, R.: A review on handling imbalanced data. In: 2018 International Conference on Current Trends towards Converging Technologies (ICCTCT). IEEE (2018). https://doi.org/10.1109/icctct.2018.8551020

20. Sutskever, I., Vinyals, O., Le, Q.V.: Sequence to sequence learning with neural networks. In: Proceedings of the 27th International Conference on Neural Information Processing Systems - Volume 2, pp. 3104–3112. MIT Press, Cambridge (2014)

Sampling Method for the Robust Single Machine Scheduling with Uncertain Parameters

Paweł Rajba[✉][iD]

Institute of Computer Science, University of Wrocław,
Joliot-Curie 15, 50-383 Wrocław, Poland
pawel@cs.uni.wroc.pl

Abstract. Many real problems are defined in an uncertain environment where different parameters such as processing times, setup times, release dates or due dates are not known at the time of determining the solution. As using deterministic approach very often provides solutions with poor performance, several approaches have been developed to embrace the uncertainty and the most of the methods are based on: stochastic modeling using random variables, fuzzy modeling or bound form where values are taken from a specific interval. In the paper we consider a single machine scheduling problem with uncertain parameters modeled by random variables with normal distribution. We apply the sampling method which we investigate as an extension to the tabu search algorithm. Sampling provides very promising results and it is also a very universal method which can be easily adapted to many other optimization algorithms, not only tabu search. Conducted computational experiments confirm that results obtained by the proposed method are much more robust than the ones obtained using the deterministic approach.

Keywords: Single machine scheduling · Uncertain parameters · Stochastic scheduling · Normal distribution · Tabu search · Sampling method

1 Introduction

Research on optimization problems for the last decades has been primarily focusing on deterministic models where we assume that problem parameters are specific and well defined. Unfortunately in many production processes we can observe different levels of uncertainty what has a direct impact on their smooth execution. For instance in many businesses delivering goods with no delays has a direct financial consequences. Unfortunately, it is not easy to meet this requirement as the transportation time depends on many external factors like weather conditions, traffic jams, driver's condition and many others. Moreover, solving such problems effectively requires very often a thorough knowledge of the process or production system.

© Springer Nature Switzerland AG 2021
M. Paszynski et al. (Eds.): ICCS 2021, LNCS 12747, pp. 594–607, 2021.
https://doi.org/10.1007/978-3-030-77980-1_45

Depending on the nature of the problem and the level of our knowledge about the measured parameters, uncertainty can be modeled in different ways. For instance, approximated values can be taken in case the variance is very small, we can use ranges of values or fuzzy representation in case we have limited understanding of the parameters' variation, finally, we can leverage random variables with specific probabilistic distributions and this approach we investigated in our paper. In literature scheduling based on random variables with probabilistic distributions is recognized as *stochastic scheduling*. Over the recent decades many different problems and their variants were investigated, during this time also many good reviews have been introduced. Basics of stochastic scheduling one can find in Pinedo [19] and more extensive reviews dedicated to methods solving scheduling problems in stochastic models are presented in Cai et al. [8], Dean [10] and Vondrák [26].

There are different ways how randomness is considered and key ones are: uncertain problem parameters and machine breakdowns. The single machine scheduling problem where different problem parameters like processing times or due dates are uncertain is also approached in different ways. One way is to develop a scheduling policy. Rothkopf in [22] introduced WSEPT (weighted shortest expected processing time, ordering jobs with nonincreasing ratio $w_j/E[\tilde{p}_j]$) rule proved to be optimal for single machine scheduling with identical release dates in [23]. The approach is still being investigated and recently in [29] an optimal policy EWCT (the expected weighted completion time) for single-machine scheduling with random resource arrival times has been introduced. Earlier Cai in [7] showed that for single-machine scheduling with processing times modeled by random variables with exponential distribution and cost functions and due dates with any distribution, the WSEPT policy (weighted shortest expected processing time first) is optimal. More one can find in [13,15] or [24]. The another way to solve the problem is to, instead of developing a scheduling policy, hire metaheuristics like tabu or simulated annealing. This approach was investigated in Bożejko et al. [1–4] and Rajba et al. [20] where effective methods were proposed for single machine scheduling problem where parameters are modelled with random variables with the normal distribution. In [1,2] and [20] additionally Erlang distribution was investigated and those papers cover $\sum w_i U_i$ and $\sum w_i T_i$ problem variants. The main goal of [3] and [4] was to introduce techniques to shorten the computational time (i.e. elimination criteria and random blocks) keeping the robustness of the determined solutions on a good level. An interesting approach has been also presented in Urgo et al. [25] where a variant of stochastic single machine scheduling problem is considered with release times and processing times as uncertain parameters and solved using the classic branch-and-bound method. The other dimension is to define the appropriate stochastic objective function (what is related to stochastic dominance, see [12,16,19]). For the most of discussed research so far the goal is to minimize the average (expected) value, the variance, or some combination of those. It is worth to indicate that there is also another significant area of robust scheduling where to goal is to minimize and control the worst-case scenario (see [9,14,27] and [28]), however usually in those scenarios uncertainty is modelled by the bounded form instead of the stochastic approach.

Random machine breakdowns is considered in the literature mainly in the following two variants: preemptive-resume mode where partially done work is continued after repair (see [5,18]) and preemptive-repeat mode where partially done work is discarded and job needs to be started again (see [6,11]). However, as in this paper we focus on uncertain parameters, we conclude machine breakdowns with this short introduction.

In this paper we consider a single machine scheduling problem with due dates in two variants where either job execution times or due dates are uncertain and modeled by independent variables with the normal distribution. We investigate the sampling method which is an extension to the tabu search algorithm and offers a probabilistic approach of finding solutions. To the best of author's knowledge, this technique was introduced in [21] for the first time and it wasn't studied in the scheduling literature. In this paper we introduce the following novel contribution:

- We apply the sampling method to the considered single machine scheduling problem with uncertain processing times and due dates,
- We introduce 3 optimization rules which makes samples more tailored and significantly decreases required samples' sizes keeping the robustness on a good level,
- We conduct an extensive experimental evaluation of the proposed method.

As verified in the computational experiments by applying the proposed method we obtain much more robust solutions than the ones obtained in the classic deterministic approach, moreover, the experiments also confirmed that applying optimization rules significantly reduces the samples' sizes.

The rest of the paper is structured as follows: in Sect. 2 we describe a classic deterministic version of the problem, then in Sect. 3 we introduce a randomized variant of the one. In Sect. 4 we present the method and optimization rules what is the main contribution of the paper, and in Sect. 5 a summary of computational experiments is described. Finally, in Sect. 6 conclusions and future directions close the paper.

2   Deterministic Scheduling Problem

Let $\mathcal{J} = \{1, 2, \ldots, n\}$ be a set of jobs to be executed on a single machine with conditions that (1) at any given moment a machine can execute exactly one job and (2) all jobs must be executed without preemption. For each $i \in \mathcal{J}$ we define p_i as a *processing time*, d_i as a *due date* and w_i as a cost for a delay.

Let Π be the set of all permutations of the set $\mathcal{J}$. For each permutation $\pi \in \Pi$ we define

$$C_{\pi(i)} = \sum_{j=1}^{i} p_{\pi(j)}$$

as a completion time of a job $\pi(i)$.

We investigate the following ways of calculating cost function:

- sum of weights for tardy jobs,
- the total weighted tardiness.

Therefore we introduce the delay indicator

$$U_{\pi(i)} = \begin{cases} 0 & \text{for} \quad C_{\pi(i)} \leqslant d_{\pi(i)}, \\ 1 & \text{for} \quad C_{\pi(i)} > d_{\pi(i)}. \end{cases}$$

and cost factor

$$T_{\pi(i)} = \begin{cases} 0 & \text{for} \quad C_{\pi(i)} \leqslant d_{\pi(i)}, \\ C_{\pi(i)} - d_{\pi(i)} & \text{for} \quad C_{\pi(i)} > d_{\pi(i)}. \end{cases}$$

Then, the cost function for the permutation π is either

$$\sum_{i=1}^{n} w_{\pi(i)} U_{\pi(i)}. \tag{1}$$

or

$$\sum_{i=1}^{n} w_{\pi(i)} T_{\pi(i)}. \tag{2}$$

Finally, the goal is to find a permutation $\pi^* \in \Pi$ which minimizes either

$$W(\pi^*) = \min_{\pi \in \Pi} \left(\sum_{i=1}^{n} w_{\pi(i)} U_{\pi(i)} \right).$$

or

$$W(\pi^*) = \min_{\pi \in \Pi} \left(\sum_{i=1}^{n} w_{\pi(i)} T_{\pi(i)} \right).$$

(depending on the considered variant).

3 Probabilistic Model

In this section we introduce the randomization of the problem described in Sect. 2. We investigate two variants: (a) uncertain processing times and (b) uncertain due dates.

In order to simplify the further considerations we assume w.l.o.g. that at any moment the considered solution is the natural permutation, i.e. $\pi = (1, 2, \ldots, n)$.

3.1   Random Processing Times

Random processing times are represented by random variables with the normal distribution $\tilde{p}_i \sim N(p_i, c \cdot p_i)$ ($i \in \mathcal{J}$, c determine the disturbance level and will be defined later) while due dates d_i and weights w_i are deterministic. Then, completion times $\tilde{C}_i$ are random variables

$$\tilde{C}_i \sim N\left(p_1 + p_2 \ldots + p_i, c \cdot \sqrt{p_1^2 + \ldots + p_i^2}\right). \tag{3}$$

Furthermore, the delay's indicators are random variables

$$\tilde{U}_i = \begin{cases} 0 & \text{for} \quad \tilde{C}_i \leqslant d_i, \\ 1 & \text{for} \quad \tilde{C}_i > d_i, \end{cases} \tag{4}$$

and the cost's factors are random variables

$$\tilde{T}_i = \begin{cases} 0 & \text{for} \quad \tilde{C}_i \leqslant d_i, \\ \tilde{C}_i - d_i & \text{for} \quad \tilde{C}_i > d_i. \end{cases} \tag{5}$$

For each permutation $\pi \in \Pi$ the cost in the random model is defined as a random variable:

$$\widetilde{W}(\pi) = \sum_{i=1}^{n} w_i \tilde{U}_i, \tag{6}$$

or

$$\widetilde{W}(\pi) = \sum_{i=1}^{n} w_i \tilde{T}_i. \tag{7}$$

3.2   Random Due Dates

Random due dates are represented by random variables with the normal distribution $\tilde{d}_i \sim N(d_i, c \cdot d_i)$ ($i \in \mathcal{J}$, c determine the disturbance level and will be defined later) while processing times p_i and weights w_i are deterministic. Delay's indicators are random variables

$$\tilde{U}_i = \begin{cases} 0 & \text{dla} \quad C_i \leqslant \tilde{d}_i, \\ 1 & \text{dla} \quad C_i > \tilde{d}_i, \end{cases} \tag{8}$$

and the cost's factors are random variables

$$\tilde{U}_i = \begin{cases} 0 & \text{dla} \quad C_i \leqslant \tilde{d}_i, \\ C_i - \tilde{d}_i & \text{dla} \quad C_i > \tilde{d}_i. \end{cases} \tag{9}$$

Cost functions are the same as for the variant with random processing times.

4   Sampling

We have introduced sampling for the first time in [21] for the flowshop problem. In this paper we apply the same idea, but with additional optimizations and tailored for the single machine scheduling problem.

Let's first recall that since permutations' costs defined in (6) and (7) are random variables, we need some way to be able to compare different solutions, so in each tabu search algorithm's iteration when we are testing different candidate solutions from the neighbourhood, we are able to find the best one and improve the current global best solution. The sampling method is a way how to make this comparison and the main idea is as follows.

Let us consider a problem instance $\{(\tilde{p}_i, w_i, d_i)\}$ (or $\{(p_i, w_i, \tilde{d}_i)\}$, respectively for uncertain due dates variant) and the examined candidate solution, a permutation π. As $\{\tilde{p}_i\}$ ($\{\tilde{d}_i\}$, respectively) are defined as random variables and we don't know the actual values that may come, the main idea of sampling is to generate samples of disturbed data based on $\{\tilde{p}_i\}$ ($\{\tilde{d}_i\}$, respectively) and simulate the potential different scenarios evaluating those disturbed candidate solutions. More formally we can describe the method as follows.

Algorithm 1: Sampling overview

1: Generate l vectors $\{(\overline{p}_i^k)\} = \{(\overline{p}_1^k, \ldots, \overline{p}_n^k)\}$ based on $\{\tilde{p}_i\}$ what gives l deterministic instances $\{(\overline{p}_i^k, w_i, d_i)\}$, $i \in \{1, \ldots, n\}$, $k \in \{1, \ldots, l\}$.
2: For each deterministic instance $(\overline{p}_i^k, w_i, d_i)$ a cost is calculated based on the candidate solution π. By that we obtain a sample $\{W_1^\pi, \ldots, W_l^\pi\}$.
3: We calculate a mean $\overline{x}$ and a standard deviation s from the sample which are used in the comparison criteria W by tabusearch.

The above listing is applicable for the random processing times variant. We can easily obtain a version for the random due dates by generating and using samples for $\tilde{d}_i$ instead of $\tilde{p}_i$.

In the basic scheme we use the fixed size of a sample (parameter l) and we investigate values dependent on the jobs' number: $0.25n, 0.5n, \ldots, 2n$.

Let's first specify the formula for criteria W. We investigated two main options: either only mean or some kind of combination of the mean and the standard deviation. During the analysis it turned out that the comparison criteria efficiency depends on the considered problem variant and even though for the most analyzed variants the best was $W = \overline{x}$, for instance for the one the best comparison criteria was $W = 50 \cdot \overline{x} + s$. However, as the differences were very small, for the simplicity reasons we assumed everywhere

$$W = \overline{x}.$$

Next, we wanted to learn something about the sample $\{W_1^\pi, \ldots, W_l^\pi\}$. Unfortunately it turned out that the distribution of the samples are not representing the normal distribution (according to the Shapiro-Wilk test), so we calculated how many different values are produced taking the large sample size (we took $10 \cdot n$).

Table 1. Number of different values in samples obtained in Step 2 in Algorithm 1

N	Factor	$w_i U_i, \tilde{p}_i$ Mean	StdDev	$w_i U_i, \tilde{d}_i$ Mean	StdDev	$w_i T_i, \tilde{p}_i$ Mean	StdDev	$w_i T_i, \tilde{d}_i$ Mean	StdDev
40	0.05	1,9	1,0	4,4	2,3	36,8	2,1	37,0	1,6
	0.10	2,6	1,4	8,0	2,5	37,0	2,1	38,1	1,1
	0.15	3,4	1,7	11,2	2,4	37,2	2,0	39,2	0,7
	0.20	3,9	1,7	14,1	2,3	37,2	2,0	39,7	0,4
	0.25	4,7	1,8	16,7	2,2	37,2	2,0	39,9	0,3
	0.30	5,3	1,9	18,6	2,1	37,3	1,9	39,9	0,2
50	0.05	1,9	1,1	5,4	2,5	45,9	2,5	46,6	1,8
	0.10	2,7	1,6	10,4	2,8	46,2	2,5	48,5	1,0
	0.15	3,5	1,8	14,8	2,7	46,3	2,5	49,5	0,5
	0.20	4,2	2,0	18,4	2,5	46,4	2,4	49,8	0,3
	0.25	4,9	2,0	21,3	2,3	46,5	2,4	49,9	0,3
	0.30	5,6	2,1	23,7	2,2	46,6	2,3	49,9	0,2
100	0.05	3,6	2,6	13,0	4,8	90,7	4,4	92,5	3,3
	0.10	5,8	4,3	21,3	4,0	91,5	4,4	97,5	1,6
	0.15	7,4	4,6	27,9	3,3	92,0	4,5	99,5	0,6
	0.20	9,0	5,0	33,5	2,9	92,2	4,6	99,8	0,4
	0.25	9,9	4,7	38,4	2,6	92,3	4,6	99,8	0,4
	0.30	11,5	4,7	42,3	2,4	92,6	4,5	99,9	0,3

Looking at Table 1 we can make a few observations. First, what is the most significant, the standard deviation from the number of different values is very small among all considered cases and it varies from below 1 to 5 (however majority of values are around 1–2). Second, the average value for a specific problem variant (i.e. defined by the number of jobs, the random parameter and the cost criteria) is also quite stable. It is interesting that the average value is around number of jobs for the $\sum w_i T_i$ criteria and it is much smaller for the $\sum w_i U_i$ criteria. Moreover, for that criteria it is still much smaller both for random processing times and for random due dates.

We use those observations to define the first optimization rule to reduce the size of samples keeping the robustness coefficient on a good level.

Optimization 1. *Based on the number of different values in samples presented in Table 1 we state that for the problem variant with the cost criteria $\sum w_i T_i$, the size of the sample S is enough and there is no need to generate new samples when $|S| = n$.*

Unfortunately, based on the initial evaluation, for the cost variant $\sum w_i U_i$ the values are to small define the similar rule that brings any positive contribution.

Another approach for generating not too big samples is looking into the confidence intervals for average. Along with generating successive values we can analyze how confidence intervals change for sequence of samples. As we don't know the sample $W_1^{\pi}, \ldots, W_l^{\pi}$ distribution we apply the following variant of the theory

$$\overline{x} - \mu_\alpha \frac{s}{\sqrt{l}} < m < \overline{x} + \mu_\alpha \frac{s}{\sqrt{l}}$$

where l is a sample size (at least 30), $\overline{x}$ is the sample mean, s is the sample standard deviation and μ_α is the value of random variable $N(0, 1)$ under the condition:

$$\Phi(\mu_\alpha) = 1 - \frac{\alpha}{2}$$

what, assuming the standard significance level $\alpha = 5\%$, provides $\mu_\alpha = 1,96$.

Now we are ready to introduce the second optimization rule.

Optimization 2. *Let $S_1, S_2, \ldots, S_k$ ($k > n/2$) be a sequence of samples where $S_i = S_{i-1} \cup \{new\ element\}$ and $CI_1, CI_2, \ldots, CI_k$ be the sequence of the confidence intervals obtained from $S_1, S_2, \ldots$. Let $len(CI)$ denote the length of the confidence interval CI. We state that the size of the sample $|S| = k$ is enough and there is no need to generate new samples when the lengths of the last $n/2$ confidence intervals are more less the same, i.e.:*

$$\sum_{i=k,k-1,\ldots,k-n/2} |len(CI_i) - len(CI_{i-1})| \lessapprox 3$$

Of course the values $n/2$ and 3 are arbitrary and they are based on some initial evaluation of different values.

Making an initial evaluation of confidence intervals for average we observed that the later iteration is, the smaller confidence intervals are. This observation triggered to introduce one more optimization rule which is very simple, but quite strong and by default it reduces the total sum of samples' sizes by half.

Optimization 3. *Let i be the iteration number in the tabu search algorithm execution and let's assume tabu search is executing n iterations. Then in the i-th iteration the sample size is determined by the following formula:*

$$|S| = (n - i + 1) \cdot 2;$$

Obviously the above formula can be easily adapted for any number of the tabu search iterations.

All the optimization rules are applied into the tabu search as follows:

- Optimization 3 is defining the upper bound for the sample size and provides the guarantee on the overall execution time.
- If the any of the rules defined by Optimization 1 and Optimization 2 is fulfilled, we stop generating more items in the sample before Optimization 3 rule holds. We can't estimate at which stage those rules are fulfilled as they are depended on the actual samples' values. However, the initial investigation shows that applying those makes a significant reduce in the total sum of samples' sizes.

5   Robustness of the Solutions

In this section we present the results of the robustness property comparison between the tabu search method with and without the sampling method applied. All the tests are executed using a modified version of tabu search method described in [1]. The algorithm has been configured with the following parameters:

- $\pi = (1, 2, \ldots, n)$ is an initial permutation,
- n is the length of tabu list and
- n is the number of algorithm iterations,

where n is the tasks number.

Both methods have been tested on instances from OR-Library ([17]) where there are 125 examples for $n = 40$, 50 and 100 (in total 375 examples). For each example and each parameter $c = 0.05$, 0.1, 0.15, 0.2, 0.25 and 0.3 (expressing 6 levels of data disturbance) 100 randomly disturbed instances were generated according to the normal distribution defined in Sect. 3 (in total 600 disturbed instances per example). The full description of the method for disturbed data generation can be found in [3].

All the presented results in this section are calculated as the relative coefficient according to the following formula:

$$\delta = \frac{W - W^*}{W^*} \cdot 100\% \tag{10}$$

which expresses by what percentage the investigated solution W is worse than the reference (best known) solution W^*. Details of calculating robustness of the investigated methods can also be found in [3].

A classic version of the algorithm we denote by $\mathcal{AD}$, the one with applied sampling with the fixed sample size by $\mathcal{AP}^F$ and the one with applied sampling with the sample sized based on optimization rules by $\mathcal{AP}^O$.

5.1   Results

In Tables 2 and 3 we present a complete summary of results for cost criteria $\sum w_i U_i$ and in Tables 4 and 5 we present a complete summary of results for cost criteria $\sum w_i T_i$. Values from columns $\mathcal{AD}$, $\mathcal{AP}^F$ and $\mathcal{AP}^O$ in all tables represent a relative distance between solutions established by a respective algorithm and the best known solution. The distance is based on (10) and it is the average of all solutions calculated for the disturbed data on a respective disturbance level expressed by the parameter c. For $\mathcal{AP}^F$ values are broken down for different sample sizes (which is based on number of jobs) and we can observe how those values are changing depending on the sample size. For the cost criteria $\sum w_i U_i$ the highlighted column ($0.75n$) represents in the author's opinion the best choice when it comes to balance between the obtained robustness coefficient value and the sample size. The version of the algorithm with optimizations applied is presented only for random p_i and cost criteria $\sum w_i T_i$ as only for this variant optimizations brought the expected improvements.

Looking at the results we can quickly conclude that by applying sampling method we obtain significantly more robust solutions than in the deterministic approach. Column *IF (2n)* represents how much relatively presented approach is better and actually the level of improvement depends on the problem variant. For random p_i improvements are enormous and they start from ca. 400% and they reaches values over 7000% for the cost criteria $\sum w_i U_i$ and they start from ca. 900% and reaches over 32000% for the cost criteria $\sum w_i T_i$. For random d_i those values are smaller, but still showing great improvements.

We can also notice that by applying Optimizations 1, 2 and 3 we obtain a better ratio between the sample size and solution robustness. As in this approach there is no longer a fixed sample size, the actual sample size has been calculated during performing the tests and on average the sample size was $0,78n$ with very small deviations. Comparing values presented in the column $\mathcal{AP}^O$ to similar variant in the $\mathcal{AP}^F$ version $(0,75n)$ we observe that for all cases either values are similar or version with optimizations provides much better robustness keeping more less the same samples' size on average.

In general results follow the expectations, i.e. the bigger disturbance factor, the worse robustness coefficient, the bigger sample size, the better robust coefficient (to some degree), however, there are some exceptions from those rules what we plan to investigate further in the future research.

Finally, in Table 6 we present aggregated on all disturbance levels the percentage for how many instances $\mathcal{AP}^F$ gives not worse solution than $\mathcal{AD}$ assuming

Table 2. Results for random p_i and $\sum w_i U_i$ cost criteria. Values are the relative errors between algorithm being compared to the best known solution

N	Factor	$\mathcal{AD}$	$\mathcal{AP}^F$								IF (2n)
			0,25n	0,5n	**0,75n**	1n	1,25n	1,5n	1,75n	2n	
40	0,05	58,0	28,1	24,7	**13,2**	13,4	12,2	12,1	11,8	12,0	382%
	0,1	172,7	68,2	33,5	**22,6**	20,6	18,7	17,6	16,6	16,9	920%
	0,15	505,0	281,3	128,9	**51,6**	48,0	52,7	50,3	51,9	52,2	867%
	0,2	827,7	320,0	228,0	**93,8**	86,8	78,8	79,0	69,0	71,6	1055%
	0,25	1213,6	781,8	197,4	**119,0**	153,9	133,1	132	98,7	97,2	1148%
	0,3	1299,4	578,4	473,8	**156,8**	162,2	163,2	145,7	164,8	110,3	1078%
50	0,05	70,7	21,2	17,4	**16,5**	15,7	15,4	14,3	13,4	12,6	461%
	0,1	610,0	82,3	105,3	**76,2**	56,1	54,8	39	39,8	29,0	2002%
	0,15	553,7	171,9	100,1	**81,4**	62,7	42,9	40,8	34,9	21,4	2484%
	0,2	2064,7	529,8	390,9	**181,0**	192,4	121,7	96,6	104	97,8	2011%
	0,25	2248,8	394,3	271,0	**200,3**	126,3	117,0	114,0	92,3	114,2	1868%
	0,3	1752,5	315,2	202,4	**111,4**	110,3	108,8	59,2	81,1	53,2	3195%
100	0,05	546,4	186,2	160,8	**110,8**	150,1	151,9	100,2	58,8	55,6	882%
	0,1	717,1	74,4	44,1	**33,5**	37,0	12,6	38,9	22,3	14,2	4932%
	0,15	1585,8	252,0	79,7	**58,4**	58,5	57,3	82,7	68,4	42,3	3648%
	0,2	1670,8	247,6	102,9	**120,5**	54,7	56,5	48,8	51,4	38,9	4199%
	0,25	1551,4	165,0	98,2	**70,6**	69,4	51,2	40,3	38,2	36,6	4133%
	0,3	2199,2	186,2	74,5	**48,9**	48,5	37,2	31,8	33,9	29,4	7377%

Table 3. Results for random d_i and $\sum w_i U_i$ cost criteria. Values are the relative errors between algorithm being compared to the best known solution

N	Factor	$\mathcal{AD}$	$\mathcal{AP}^F$								IF (2n)
			0,25n	0,5n	0,75n	1n	1,25n	1,5n	1,75n	2n	
40	0,05	1191,6	451,2	297,0	**197,2**	200,6	200,8	191,1	197,3	207,2	475%
	0,1	2891,4	1062,7	881,5	**709,1**	785,1	734,1	645,6	684,4	653,9	342%
	0,15	4460,0	1757,4	1193	**1152,6**	1128,3	1109	1132,8	1171,1	1095,6	307%
	0,2	2953,8	1480,0	1209,8	**1073,5**	1034,5	1055,4	1045,7	1016,3	1012,6	191%
	0,25	2457,6	1304,8	1153,3	**1129,5**	1100,4	1100,1	1111,4	1103,4	1087,4	126%
	0,3	1391,3	874,9	777,9	**732,8**	731,9	718,8	716,7	714,8	712,3	95%
50	0,05	3045,5	873,6	586,3	**512,9**	487,1	566,6	563,6	494,7	583,2	422%
	0,1	1128,6	472,6	370,5	**361,4**	323,7	317,9	338,6	321,5	323,2	249
	0,15	3387,1	1210,3	1112,1	**1067,2**	1000,9	955,5	928,4	918,9	963,1	251%
	0,2	2217,6	1171,6	943,7	**943,0**	908,5	892,4	876,5	858,3	877,1	152%
	0,25	2462,2	1501,2	1328,2	**1247,6**	1225,1	1195	1178,3	1181,6	1190,2	106%
	0,3	1758,1	1171,4	1036,6	**1002,8**	980,0	989,8	965,6	977,5	980,5	79%
100	0,05	3898,6	640,9	404,4	**373,9**	340,4	261,0	256,0	308,5	261,6	1390%
	0,1	1671,4	464,3	424	**379,8**	369,8	355,1	351,0	338,3	337,0	395%
	0,15	1537,7	576,4	530,6	**495,8**	477,5	485,9	497,1	504,7	473,5	224%
	0,2	1445,3	729,8	688,3	**676,7**	663,3	670,3	641,9	644,7	641,0	125%
	0,25	1180,5	735,7	700,8	**674,2**	664,6	659,2	650,5	652,4	640,3	84%
	0,3	784,6	548,8	525,5	**506,9**	503,2	503,7	499,7	497,0	494,1	58%

Table 4. Results for random p_i and $\sum w_i T_i$ cost criteria. Values are the relative errors between algorithm being compared to the best known solution

N	Factor	$\mathcal{AD}$	$\mathcal{AP}^O$	$\mathcal{AP}^F$								IF (2n)
			0,78n	0,25n	0,5n	0,75n	1n	1,25n	1,5n	1,75n	2n	
40	0,05	452,8	28,3	58,5	57,1	37,4	28,1	22,2	23,2	23,9	21,7	1982%
	0,1	851	31,3	250,6	163	133	51,2	121,8	42,2	37,5	35,2	2320%
	0,15	1532,6	49,2	571	226,8	94,9	101,1	70,9	79,3	49,9	52,2	2836%
	0,2	2012	98,3	762	311,2	126,1	108,8	97,6	105,2	86,3	68,5	2836%
	0,25	4693,6	900,9	5013,9	2073,4	1793,9	1737,6	1662,8	1628,9	456,4	458,4	923%
	0,3	5238,1	1014,7	6297,6	1440,9	1065,3	1044,7	983	932,2	280,5	279,8	1772%
50	0,05	373,1	10,4	49,6	45,1	21	10	9,7	9,9	9,8	9,5	3831%
	0,1	990,6	17,7	123,2	63,7	32,4	21,1	19,4	16,9	16,6	15,8	6186%
	0,15	2666	192,9	501,2	368,9	237,7	101,3	100,4	95,3	100,7	98,1	2617%
	0,2	4524	87,5	1068,9	484,4	148,4	92,3	89,6	83,1	74,9	73,9	6018
	0,25	12048,7	528,2	1921,3	1334,8	855,2	553,5	498	476,7	480,6	382	3053%
	0,3	7585,5	115,8	1099,6	633,6	283,2	113,5	154,5	119,1	117,7	44,5	16931%
100	0,05	830,9	6,9	19	7,5	7,1	7,7	7,5	7,5	6,9	7,3	11345%
	0,1	2452,8	13,7	39,5	13,3	12	8,9	10,8	5,3	8,8	6,4	37991%
	0,15	4095	62,1	222,2	56	63,3	70,6	46,2	31,7	40,1	42,2	9613%
	0,2	12436,8	78,8	336,3	78,3	144,9	67,3	59,6	64,9	42,3	38,5	32173%
	0,25	9104,2	114	305,4	161,2	135	80,5	85,4	82,6	75,6	59,4	15214%
	0,3	11229,8	93,3	501,7	167,2	145,7	129,3	123,6	113,3	86,5	87,3	12756%

Table 5. Results for random d_i and $\sum w_i T_i$ cost criteria. Values are the relative errors between algorithm being compared to the best known solution (all values for $\mathcal{AD}$ and $\mathcal{AP}^F$ should be multiplied by 10^3)

N	Factor	$\mathcal{AD}$ ($\cdot 10^3$)	$\mathcal{AP}^F$ ($\cdot 10^3$)								IF (2n)
			$0,25n$	$0,5n$	$0,75n$	$1n$	$1,25n$	$1,5n$	$1,75n$	$2n$	
40	0,05	10,5	5,4	4,2	1,8	1,9	1,8	1,8	1,7	1,5	595%
	0,1	26,4	11	7,5	7,0	7,1	6,5	6,6	6,4	6,2	325%
	0,15	35,2	18,8	15,1	14,7	14	13,6	13,2	13,5	13,2	165%
	0,2	56,2	32,6	26,4	25,5	24,6	24,1	24,3	23,7	23,1	142%
	0,25	51,9	32	28,5	27,8	27,8	26,8	27,1	27	26,4	96%
	0,3	21,1	15,5	14,2	13,3	13,3	13	13,1	12,9	13	62%
50	0,05	10,3	1,9	1,4	1,1	0,9	1,0	0,8	0,8	0,8	1164%
	0,1	68,2	23,9	18	15,6	16,6	14,7	15,3	15,8	14,4	372%
	0,15	55,1	27,7	22	21	19,8	19,8	19,5	19,6	19,8	178%
	0,2	103,6	54	46,3	42,8	42,6	42,3	43	42,3	42,1	146%
	0,25	221,2	136,3	121,4	116,3	116,4	115,8	114,1	114,2	114,7	92%
	0,3	25,8	16,5	14,4	14	13,7	13,8	13,6	13,5	13,6	89%
100	0,05	26	3,4	2,8	2,4	2,1	2,1	2	1,9	1,8	1319%
	0,1	97,4	28,6	24,2	22,6	21,8	21,8	21,3	20,8	20,1	384%
	0,15	576,1	283,7	270,6	254,1	246,2	244,9	241,7	242,4	241,1	138%
	0,2	2345,7	1506,1	1402	1332,5	1301,1	1308,8	1302,7	1296,5	1288,4	82%
	0,25	657,5	454,3	425,6	400,8	395,2	392,5	394,9	395,1	390,8	68%
	0,3	37,7	29,6	27,5	26,2	25,8	25,9	25,6	25,5	25,5	47%

that the samples size $S = 2n$. We can quickly see that for all problem variants (except the one for random p_i and cost criteria $\sum w_i T_i$) all results are almost 95% and higher and the bigger n is, the better percentage we get. Moreover, for random d_i and cost criteria $\sum w_i U_i$ all values are very close or even equal to 100%. For random p_i and cost criteria $\sum w_i T_i$, even if a little smaller, we also get very strong result where all results are above 74%. This another perspective also shows predominance of the proposed sampling method.

Table 6. The percentage for how many instances $\mathcal{AP}^F$ gives not worse solution than $\mathcal{AD}$ assuming $S = 2n$

n	p_i		d_i	
	$\sum w_i U_i$	$\sum w_i T_i$	$\sum w_i U_i$	$\sum w_i T_i$
40	94,7%	77,9%	98,8%	94,9%
50	96,8%	74,3%	99,2%	97,7%
100	99,9%	80,0%	100,0%	99,9%

6   Conclusions

In this paper we proposed the sampling method with a set of optimizations which can be applied to tabu search and other similar methods in order to improve the robustness of solutions calculated in an uncertain environment modeled by random variables with the normal distribution. Based on the performed computational experiments we can conclude that the proposed method provides substantially more robust solutions than the ones obtained by the deterministic approach and the proposed optimization rules reduces the samples' sizes generated during the algorithm's execution

As there are several ways how the described method can be further investigated, we can see the following ways for continuation. Obviously, as there are several results which don't follow the expected trends, we plan to investigate the topic further and understand better the nature of those exceptions and by that, hopefully, improve the method and make it more tailored where applicable. There is also a question how strong is that method comparing to other methods solving the same problem based on stochastic and fuzzy description. The other area of investigation might be also to verify how much the input data distribution is important in the final results as the obtained samples doesn't reflect the distribution of the input data. Please note that this might be both advantage and disadvantage depending on the properties we would like to obtain in the end. Finally, this method is very universal and can be applied to many other types of optimization problems, so we plan also to follow this direction.

References

1. Bożejko, W., Rajba, P., Wodecki, M.: Stable scheduling with random processing times. In: Klempous, R., Nikodem, J., Jacak, W., Chaczko, Z. (eds.) Advanced Methods and Applications in Computational Intelligence, vol. 6, pp. 61–77. Springer, Heidelberg (2014). https://doi.org/10.1007/978-3-319-01436-4_4
2. Bożejko, W., Rajba, P., Wodecki, M.: Stable scheduling of single machine with probabilistic parameters. Bull. Polish Acad. Sci. Techn. Sci. 65(2), 219–231 (2017)
3. Bożejko, W., Rajba, P., Wodecki, M.: Robustness of the uncertain single machine total weighted tardiness problem with elimination criteria applied. In: Zamojski, W., Mazurkiewicz, J., Sugier, J., Walkowiak, T., Kacprzyk, J. (eds.) DepCoS-RELCOMEX 2018. AISC, vol. 761, pp. 94–103. Springer, Cham (2019). https://doi.org/10.1007/978-3-319-91446-6_10
4. Bożejko, W., Rajba, P., Wodecki, M.: Robust single machine scheduling with random blocks in an uncertain environment. In: Krzhizhanovskaya, V.V., et al. (eds.) Robust Single Machine Scheduling with Random Blocks in an Uncertain Environment. LNCS, vol. 12143, pp. 529–538. Springer, Cham (2020). https://doi.org/10.1007/978-3-030-50436-6_39
5. Cai, X., Zhou, S.: Stochastic scheduling on parallel machines subject to random breakdowns to minimize expected costs for earliness and tardy jobs. Oper. Res. 47(3), 422–437 (1999)
6. Cai, X., Sun, X., Zhou, X.: Stochastic scheduling subject to machine breakdowns: the preemptive-repeat model with discounted reward and other criteria. Naval Res. Logistics (NRL) 51(6), 800–817 (2004)
7. Cai, X., Zhou, X.: Single-machine scheduling with exponential processing times and general stochastic cost functions. J. Global Optim. 31(2), 317–332 (2005)
8. Cai, X., Wu, X., Zhou, X.: Optimal Stochastic Scheduling, vol. 4. Springer, New York (2014)

9. Daniels, R.L., Carrillo, J.E.: β-Robust scheduling for single-machine systems with uncertain processing times. IIE Trans. **29**(11), 977–985 (1997)
10. Dean, B.C.: Approximation algorithms for stochastic scheduling problems (Doctoral dissertation, Massachusetts Institute of Technology) (2005)
11. Glazebrook, K.D.: Optimal scheduling of tasks when service is subject to disruption: the preempt-repeat case. Math. Methods Oper. Res. **61**(1), 147–169 (2005)
12. Hadar, J., Russell, W.R.: Rules for ordering uncertain prospects. Am. Econ. Rev. **59**(1), 25–34 (1969)
13. Jang, W., Klein, C.M.: Minimizing the expected number of tardy jobs when processing times are normally distributed. Oper. Res. Lett. **30**(2), 100–106 (2002)
14. Kuo, C.Y., Lin, F.J.: Relative robustness for single-machine scheduling problem with processing time uncertainty. J. Chin. Inst. Ind. Eng. **19**(5), 59–67 (2002)
15. Leus, R., Herroelen, W.: The complexity of machine scheduling for stability with a single disrupted job. Oper. Res. Lett. **33**(2), 151–156 (2005)
16. Levy, H.: Stochastic dominance and expected utility: survey and analysis. Manage. Sci. **38**(4), 555–593 (1992)
17. OR-Library. http://www.brunel.ac.uk/~mastjjb/jeb/info.html. Accessed 11 May 2020
18. Pinedo, M., Rammouz, E.: A note on stochastic scheduling on a single machine subject to breakdown and repair. Probability Eng. Inf. Sci. **2**(1), 41–49 (1988)
19. Pinedo, M.L.: Scheduling: Theory, Algorithms, and Systems. Springer, New York (2016). https://doi.org/10.1007/978-3-319-26580-3
20. Rajba, P., Wodecki, M.: Stability of scheduling with random processing times on one machine. Applicationes Mathematicae **2**(39), 169–183 (2012)
21. Rajba, P., Wodecki, M.: Sampling method for the flow shop with uncertain parameters. In: Saeed, K., Homenda, W., Chaki, R. (eds.) CISIM 2017. LNCS, vol. 10244, pp. 580–591. Springer, Cham (2017). https://doi.org/10.1007/978-3-319-59105-6_50
22. Rothkopf, M.H.: Scheduling independent tasks on parallel processors. Manage. Sci. **12**(5), 437–447 (1966)
23. Rothkopf, M.H.: Scheduling with random service times. Manage. Sci. **12**(9), 707–713 (1966)
24. Soroush, H.M.: Scheduling stochastic jobs on a single machine to minimize weighted number of tardy jobs. Kuwait J. Sci. **40**(1), 123–147 (2013)
25. Urgo, M., Váncza, J.: A branch-and-bound approach for the single machine maximum lateness stochastic scheduling problem to minimize the value-at-risk. Flexible Serv. Manuf. J. **31**, 472–496 (2019)
26. Vondrák, J.: Probabilistic methods in combinatorial and stochastic optimization (Doctoral dissertation, Massachusetts Institute of Technology) (2005)
27. Yang, J., Yu, G.: On the robust single machine scheduling problem. J. Comb. Optim. **6**(1), 17–33 (2002)
28. Yue, F., Song, S., Jia, P., Wu, G., Zhao, H.: Robust single machine scheduling problem with uncertain job due dates for industrial mass production. J. Syst. Eng. Electron. **31**(2), 350–358 (2020)
29. Zhang, L., Lin, Y., Xiao, Y., Zhang, X.: Stochastic single-machine scheduling with random resource arrival times. Int. J. Mach. Learn. Cybern. **9**(7), 1101–1107 (2018)

Teaching Computational Science

Biophysical Modeling of Excitable Cells - A New Approach to Undergraduate Computational Biology Curriculum Development

Sorinel A. Oprisan(✉)

Department of Physics and Astronomy, College of Charleston, Charleston, SC, USA
oprisans@cofc.edu
http://oprisans.people.cofc.edu/bmec.html/

Abstract. As part of a broader effort of developing a comprehensive neuroscience curriculum, we implemented an interdisciplinary, one-semester, upper-level course called Biophysical Modeling of Excitable Cells (BMEC). The course exposes undergraduate students to broad areas of computational biology. It focuses on computational neuroscience (CNS), develops scientific literacy, promotes teamwork between biology, psychology, physics, and mathematics-oriented undergraduate students. This course also provides pedagogical experience for senior Ph.D. students from the Neuroscience Department at the Medical University of South Carolina (MUSC). BMEC is a three contact hours per week lecture-based course that includes a set of computer-based activities designed to gradually increase the undergraduates' ability to apply mathematics and computational concepts to solving biologically-relevant problems. The class brings together two different groups of students with very dissimilar and complementary backgrounds, i.e., biology or psychology and physics or mathematics oriented. The teamwork allows students with more substantial biology or psychology background to explain to physics or mathematics students the biological implications and instill realism into the computer modeling project they completed for this class. Simultaneously, students with substantial physics and mathematics backgrounds can apply techniques learned in specialized mathematics, physics, or computer science classes to generate mathematical hypotheses and implement them in computer codes.

Keywords: Computational neuroscience · Undergraduate education · Interdisciplinary curricula

1 Introduction

Neuroscience is an interdisciplinary endeavor that requires biology and psychology knowledge and challenges undergraduates to cross the boundaries of their primary major [21]. Computational neuroscience (CNS) requires a further fundamental understanding of physics, mathematics, and computer science [2].

© Springer Nature Switzerland AG 2021
M. Paszynski et al. (Eds.): ICCS 2021, LNCS 12747, pp. 611–625, 2021.
https://doi.org/10.1007/978-3-030-77980-1_46

Nowadays, neuroscience graduate programs require at least a basic knowledge of computer algorithms used in a wide range of experimental techniques, from electrophysiological measurements, such as data acquisition, storage, spike sorting, and pattern recognition, to elaborate and computationally-intensive data analysis in functional brain imaging [7]. The students who attend a primarily undergraduate institution (PUI) and aspire to a spot in a neuroscience graduate program are often at a disadvantage regarding the depth of their experience when competing against those coming from research universities with dedicated state-of-the-art facilities and established neuroscience programs. Of all sub-fields of neuroscience research, CNS is the only one that requires the least infrastructure and investment into specialized equipment. Luckily, CNS only requires readily available computers, e.g., a fully functional Linux-powered Raspberry Pi costs only \$35, and freely available software packages, such as NEURON, Genesis, or XPP [13].

Eric L. Schwartz coined the computational neuroscience (CNS) name in the mid-1980s to characterize vastly disparate computational research approaches to understanding how neurons process information and how the brain functions. [10].

In addition to contributing to students' personal and professional development [8], undergraduate research opportunities in CNS enhance their chances of successful and more productive completion of a graduate degree [16]. However, gaining research experience in CNS may be especially difficult due to (1) the lack of CNS-trained faculty at PUIs and (2) the breadth of knowledge required from undergraduates. National Institute of Health's (NIH) Blueprint initiative supports five large and well-established graduate programs in neuroscience that include some undergraduate CNS components [26]. Early and engaging neuroscience education is critical in recruiting and retaining undergraduate students to CNS [9]. While some undergraduates will pursue graduate studies, their first exposure to neuroscience will still occur in undergraduate classes. Moreover, every neuroscience graduate program has a computational component, and, therefore, there is a great need for bottom-up funding and growth of CNS starting at PUIs.

At the College of Charleston (CofC), we developed and implemented an innovative interdisciplinary curriculum in CNS consisting of an introductory research rotation class offered to incoming first-year students as a First-Year Experience (see http://fye.cofc.edu/) followed by an upper-level Biophysical Modeling of Excitable Cells (BMEC) class [19,20]. The BMEC course is an enriching interdisciplinary learning experience that seemingly fits into the existing Neuroscience Minor and Biomedical Physics Minor at the CofC.

The primary goal of the BMEC course is to provide CofC undergraduates with an overview of the potential applications of CNS, offer a glimpse at the mathematical and computational techniques used in CNS, and analyze in-depth a few computational implementations of single-cell conductance-based models (see http://oprisans.people.cofc.edu/bmec.html/ for a complete list of computational models developed for this class). Another goal of the course is to foster the collaboration between our undergraduate institution (CofC) and the Neuroscience Department at Medical University of South Carolina (MUSC) through the direct involvement of graduate students in teaching undergraduate classes. We aim to contribute to both undergraduate and graduate students learning.

2   Our Curriculum Deployment Approach

1. Identify Your Target Audience. In our case, with a successful and fast-growing Neuroscience Minor (developed and staffed by Biology and Psychology Departments) and a Biomedical Physics Minor (developed and staffed by Department of Physics and Astronomy), it became apparent that one crucial learning experience was missing, i.e., a computationally-oriented component of these two programs. After running our CNS class for a few years as a special topic course (2006–2010) and analyzing the enrollment data, we concluded that our CNS course's target audience is the biology, psychology, physics, and mathematics undergraduates with a strong interest in computational biology.

2. Secure Administrative Support. With the administrative support of the two chairpersons (Biology and Physics), the Director of Neuroscience Minor, and the coordinators of Biomedical Physics Minor, we adjusted the prerequisites and the content of the course to serve both biology or psychology majors, presumably exploring a Neuroscience Minor option, and physics or mathematics majors, likely interested in Biomedical Physics. The BMEC course is maximizing the use of faculty who serves both Biology and Physics majors and two interdisciplinary minors (Neuroscience and Biomedical Physics).

3. Effective Academic Advising. One mechanism we used is the inclusion of two faculty members from the Physics and Astronomy Department in the Neuroscience Steering Committee. This ensures that biology and psychology students were aware of the BMEC course offered by the Physics and Astronomy Department. Simultaneously, physics instructors who were members of the Neuroscience Steering Committee make sure that physics majors are aware of biology and psychology prerequisites for the BMEC course and encourage them to consider a Neuroscience Minor pathway. The goal is to create the mindset, very early on, that mathematically-oriented physics students who would like to consider a career in life sciences need to enroll in biology and psychology classes. The earlier and more pervasive the academic advising is, the more effective it will be in breaking down the natural compartmentalization along discipline and department lines. In addition to academic advising, we use posters and short 10-min talks delivered during the open enrollment period in targeted biology, physics, and neuroscience classes. The talks were pitched to the appropriate courses and majors to inform the students about the new curriculum.

4. Course Cross-listing. The BMEC course is cross-listed with the Department of Physics and Astronomy as PHYS 396 and the Department of Biology as BIOL 396. This course serves as one of the mandatory electives for the Neuroscience Minor at the CofC. It is also a core required class for the Biomedical Physics Minor.

3 Designing the Computational Neuroscience Curriculum for Undergraduates

3.1 Aims

BMEC curriculum was developed and implemented with three aims: 1) to provide an interdisciplinary undergraduate experience in CNS, 2) to promote interdisciplinary and computational thinking, and 3) create an environment that fosters pedagogical awareness in Ph.D. students from the MUSC early in their careers.

The main goal of the BMEC course is to provide a learning environment where undergraduates can apply their specialized major-related knowledge to solving problems outside their disciplines. For example, physics majors are well-versed in manipulating analytically and numerically differential equations for diffusion, thermal conduction, etc. Still, they rarely find themselves in an undergraduate class that requires solving diffusion equations across the cell membrane, using electrostatic interactions among different ionic species to determine the spatial distribution of charges in a living cell, or using finite difference schemes to solve action potential propagation along an axon. The same is true for biology students who know all the details of biochemical mechanisms involved in transmembrane transport but rarely make the connection with the mathematical description of such processes as they learn it in a traditional physics class [18].

3.2 Prerequisites

The current prerequisites for our BMEC course are unique, and this course introduced a new interdisciplinary curriculum design model. We adopted the model of a single course with two cross-listed sections and prerequisites both in Biology and Physics.

The Physics/Mathematics/Computer Science section of the BMEC course, i.e., PHYS 396, requires calculus-based physics classes through electricity (Physics 111 and 112) and introductory-level biology classes (BIOL 111 Introduction to Cell and Molecular Biology and BIOL 112 Evolution, Form, and Function of Organisms).

The Biology/Psychology/Neuroscience Minor section of the BMEC course, i.e., BIOL 396, requires algebra-based physics classes through electricity (PHYS 101 and PHYS 102) and advanced biology classes (BIOL 211 Biodiversity, Ecology, and Conservation Biology and BIOL 212 Genetics).

There is no formal computer literacy requirement for the BMEC course since the focus is on using dedicated computational tools to model biological processes and not the implementation of computer codes *per se*. Furthermore, to reduce the apprehension related to computer modeling, the course starts with a user-friendly set of computer activities from the Neuron in Action package [25]. After the students are comfortable manipulating modeling parameters (cell radius, the density of ionic channels, half-activation potentials, etc.), the course gradually introduces them to the actual computer code that runs user-friendly interface of Neuron in Action package [13].

3.3 Class Resources

Like any computational biology course, BMEC requires extra resources.

Computers. Our CNS laboratory has 20 dedicated laptops which can be loaned by students enrolled in BMEC at the beginning of the semester. The Department of Physics and Astronomy also offers free access to all students to 20 desktop computers with all the necessary software.

Undergraduate Teaching Assistant. One teaching assistant (TA) is recruited from amongst the students who completed the BMEC class. The TA's primary duties are to help the instructor during the hands-on computational activities so that the computational questions and issues raised by students are promptly solved. The TA holds office hours to help students with home-work assignments and computer codes, and grade assignments. The TA also ensures that all hardware and software issues related to BMEC class are solved expeditiously, updates the software for all laptops serving the class, and coordinates with Information Technology Department regarding required updates for all computational packages used in class and installed on departmental computers.

Teaching Training Fellowships for Ph.D. Students from MUSC. The ongoing collaboration between the CofC and MUSC extends beyond the research and co-mentoring undergraduates students. We involve senior Ph.D. students from MUSC as Teaching Training Fellows (TTFs) in our undergraduate classes. Although teaching is not required of Ph.D. graduate students at MUSC, studies showed that teaching experience has beneficial effects on enhancing oral and written communication skills and improving research performance [6]. The TTF program between the CofC and MUSC was designed to avoid usual pitfalls, i.e., unstructured pedagogical approach to teaching [1] and lack of precise and quantifiable learning goals and objectives [14].

We selected for the BMEC course only highly-motivated senior Ph.D. graduate students, usually from the labs of the instructor's collaborators, based on first-hand interaction during our collaborative research projects with MUSC. Opportunities are created for increasing pedagogical awareness of TTFs through discussions with other instructors and feedback from undergraduate students and the course instructor about the effectiveness of the teaching methods used in their guest lectures.

The purpose of TTF is to set aside protected time for Ph.D. students interested in pursuing a teaching career such that they can focus on pedagogy, develop teaching competencies, and help graduate students decide their career path beyond the Ph.D. The prospective TTF's research adviser agrees to release the senior Ph.D. student, who is in the final stages of the thesis preparation, from some of the research duties and consign a contract with a CofC instructor with whom the TTF will work. Although the TTF does not directly receive a stipend for teaching at the CofC, the research adviser's lab at MUSC receives compensatory funds from a CofC grant to help substitute the Ph.D. student with other undergraduate or graduate students.

The TTF conducts review sessions with undergraduates, prepares and presents two-three guest lectures, and attends the graduate-level Teaching Techniques class at MUSC. Many office hours go beyond the class topics and sometimes cross into personal advice from TTF to undergraduate students regarding effective learning strategies, what it takes to get admitted into a graduate program, what it takes to be a successful graduate student, etc. The undergraduates do not feel intimidated by the TTF and tend to ask questions more openly. At the same time, the undergraduates know from the research lab visits at MUSC conducted by the TTF and the lectures taught by the TTF towards the end of the semester in our BMEC class that the TTF masters the neuroscience research techniques and respect him/her as a looked-up professional and potential role model.

4   Biophysical Modeling of Excitable Cells Course Logistics

Teaching computational biology at the undergraduate level is challenging primarily because of the required breadth of knowledge. BMEC is a one-semester, upper-level, undergraduate course for students interested in exploring computational approaches to biologically-relevant questions. The course is a three contact hours per week class.

Gaining hands-on experience is crucial in computational biology [23], and we opted for a mix of lectures and hands-on computational activities during a three contact-hour class. Depending on the week's topic, at least one out of the three one-hour class periods is allocated to hands-on computational activities. The three credits/three contact hours BMEC course is offered every fall semester.

BMEC is a single-instructor course and consists of standard two-hour lectures per week and one-hour hands-on computer activities weekly. Computational activities are selected from the Neuron in Action package during the first half of the semester and implemented in XPP during the second half of the semester. Class discussions focus on primary literature and database models. The students' workload consists of weekly reading and homework assignments, one term paper, and one end-of-semester computational project.

4.1 Delivering BMEC Content

Lectures and Hands-On Computational Activities. The course content is delivered primarily through lectures focused on the major points of the assigned readings. To increase class participation and get a feeling of the students' grasp on some tricky concepts, we use a classroom response system (iClickers) for anonymous polling. The small class size (10–16 students) helps students feel more comfortable speaking up, although iClickers is the preferred method to scale-up and involve larger classes [24].

The purpose of computational activities is to familiarize undergraduate students with the user-friendly interface provided by Neurons in Action software [17] and guide them through a series of tutorials modeling different aspects of electric activity of excitable cells, e.g., passive properties of membrane bilayers, effect of voltage-gated ionic channels on electric activity of excitable cells, action potential propagation, synaptic coupling, etc.

During the second part of the semester, the undergraduate students are introduced to the computer programming behind the friendly graphic user interface. Since computer programming experience is neither required nor the main focus of the BMEC course, we decided to use a simple, straightforward, yet powerful scripting language called XPP developed at Pittsburg University by Dr. Bard Ermentrout (see http://www.math.pitt.edu/~bard/xpp/xpp.html). Extensive tutorials also support this cross-platform software package [5].

After the first half of the semester, the students become comfortable linking biological concepts, such as transmembrane ionic currents induced by protein conformation changes, with the corresponding mathematical representation. For example, all students in BMEC are familiar with the mathematical description of the leak current, which has the general form of Ohm's law $I_L = g_L \times (V - E_L)$, where g_L is the leak current conductance, V is the membrane potential, and E_L is the reversal potential of the leak current. In XPP, the implementation of the above current is straightforward:

$$\text{par } gl = 0.1, el = -50$$
$$il = gl * (v - el)$$
$$\ldots$$

The reserved word **par** in the above computer code signals that what follows is a list of parameters. Understanding XPP scripts is straightforward and gives undergraduate students the necessary confidence that CNS is not (only) about computer programming. After practicing with XPP during the second half of the semester, the students become confident that they can implement any conductance-based mathematical model.

Guest Lecturers and Lab Visits. The close ties with the MUSC and TTF Ph.D. graduate students' involvement offer CofC undergraduates the opportunity of class visits to observe wet electrophysiological or behavior experiments carried out at MUSC. We also monitor all steps of *in vitro* experiments using mice prefrontal slices from glass electrode and slice preparation, cell identification, to spike sorting. In a different lab visit at MUSC, we observed *in vivo* multielectrode recordings from mice performing a maze task. After completing the BMEC course, our undergraduates have the opportunity of gaining additional first-hand research experience and apply what they learned in the classroom by working in MUSC labs through the Summer Undergraduate Research Program (SURP) initiative.

To further strengthen the ties between the CofC and MUSC departments, experts from different fields are invited to give guest lectures. We usually do not have more than one guest speaker per academic year. In some particular academic years, we incorporated more electrophysiology background information by asking a faculty member from the Biology Department at the CofC to present this specific neuroscience-related topic. In other years we had MUSC guest speakers from Biomedical Imaging Center at MUSC and focused more on brain networks and behavioral-level neuroscience with connectomics emphasis. To secure a guest speaker, we usually send the invitation to targeted faculty at least three-four months before the beginning of the semester.

4.2   Online Course Management System

Class materials are organized and distributed via a secure wiki page. A screened wiki webpage summarizing materials from our CNS course is available at http:// oprisans.people.cofc.edu/bmec.html/. The instructor edits and maintains a few webpages linked to the main wiki, e.g., the syllabus, the list of recommended computational projects, the list of recommended essays, the criteria for evaluating the presentations, the office hour schedule, the end-of-semester student presentation schedule, etc.

Course's wiki webpage contains pages created and managed by each student or group of students. Each student must store all files related to his/her wiki webpage in an individual folder to avoid mixing and overwriting files from other users. A typical student webpage contains a detailed description of the end-of-semester computational project. It is organized as a research paper, i.e., it must have a title, abstract, introduction, method, and materials, results, conclusions, acknowledgments, references. The PowerPoint presentation and the final written report for the end-of-semester computational project are also linked to the student's wiki page. All references used during the project are stored as pdf files in the related student's folder to allow every student enrolled in the BMEC course quick access to the particular project's primary literature.

4.3   Course Topics

Course topics include (1) exploration of physics basis of (bio)electrical signals, such as electric potential, electric current, Ohm's law, Kirchhoff rules, an RC time constant, (2) search for experimental data and computational models in open databases, such as CRCNS - Collaborative Research in

Computational Neuroscience (https://crcns.org/), ModelDB (https://senselab. med.yale.edu/modeldb/), and BioModels (https://www.ebi.ac.uk/biomodels/), (3) fit experimental data to analytic functions and calibrate ionic currents based on empirical data, (4) use freely available (NEURON, XPP, etc.) and proprietary (Matlab, Mathematica, etc.) computational tools to integrate model equations, and (5) relate computational predictions and results to behavioral and clinical data.

In BMEC, we cover four main topics: (1) electrical properties of excitable cells, e.g., capacitance, time constant, axial and trans-membrane electric resistances, length constant, and leak currents, (2) ionic currents, e.g., Ohmic and voltage-gated ionic channels, Kirchhoff rules, and Hodgkin-Huxley equations, (3) synaptic coupling, e.g., neurotransmitter release, calcium-sensing, GABA- and NMDA-gated ionic channels and long-term potentiation, and (4) multi-compartment models and neural networks (see http://oprisans.people.cofc.edu/ bmec.html/ for weekly topics, objectives, assignments, etc.).

4.4   Textbooks

Due to the heterogeneity of students' backgrounds and the breadth of required knowledge, no single textbook can cover both the biological foundations and the mathematical/computational topics for an undergraduate-level CNS class. We used a combination of textbooks (available at CofC library) and review papers to cover the foundations of CNS:

1. Principles of Neural Science, E.R. Kandel, J.H. Schwartz, T.M. Jessell, 2005.
2. Computational Cell Biology: An Introduction to Computer Modeling in Molecular Cell Biology. Chris Fall, Eric Marland, John Tyson, and John Wagner (editors). Springer-Verlag. New York, NY. 2002.
3. Ionic Channels of Excitable Membranes, Third Edition. Bertil Hille. Sinauer Associates. Sunderland, MA. 2001.

Additional assigned readings are listed on our BMEC website.

4.5   Assignments

Reading Assignments. Primary literature is employed in three ways: (1) inclusion of experiments and models in all lectures, (2) individual written essays focused on a literature review of a specific computational model, and (3) end-of-semester computational projects. Through active reading assignments from the CNS field's primary literature, we reinforce concepts presented in class and guide students on integrating real data into computational models of excitable cells. The reading assignment also expose students to current research and teach them the importance of staying abreast of new developments in the rapidly changing field of CNS [22].

Literature Review Essay. All students are required to summarize assigned primary literature articles in a short review paper, targeting non-science peers, called essay assignment. Through these assignments, the students became aware of the difficulties of conveying technical information to a broader audience and gain experience in science writing and communication. As opposed to the regular class reading assignments, the literature review essay is a comprehensive overview focused on the novelty and uniqueness of a particular computer model applied to a biologically-relevant question. Each literature review essay must provide a historical overview emphasizing a broader understanding of the interdisciplinary and often sinuous track towards the "right" answer. In addition to introducing the specific subject of the essay, the literature review essay details the design, results, and interpretation of computer simulations and their relevance to biology.

Computational Project. Further student engagement in CNS comes in the form of an end-of-semester computational project. A list of projects, together with minimal bibliography and computer codes, is provided at the beginning of the semester. However, the students are free to select any other CNS-related topics (see http://oprisans.people.cofc.edu/bmec.html/).

At the end of the first month of the semester-long course, every group of (maximum three) students must select a computational topic and submit a brief one-page proposal containing a title, an abstract, and some references for their intended computational project.

The project groups must have (at least) one biology/psychology/neuroscience minor member from the BIOL 396 section of the class and one from the mathematics/physics/computer science PHYS 396 section of the course. Course evaluations showed that students with stronger biology/psychology backgrounds and weaker physics/mathematical backgrounds benefit from the mixed teams' arrangement to explore quantitative and computational solutions. Simultaneously, students with stronger physics/mathematics backgrounds had the opportunity to understand better the hypothesis and assumptions made in tackling biological problems using mathematical and computational models.

The midterm checkpoint of the computational project provides feedback to students regarding the consistency of their numerical simulations, the content, and the style of their presentation. For the end-of-semester computational project, the midterm requirement is a working computer code and a strategy for conducting numerical simulations for the rest of the term, e.g., computational resource allocation, time management, etc.

During the second part of the semester, the students summarize their numerical simulations, design a 10–15 minutes PowerPoint presentation and an accompanying, detailed paper summarizing their findings. The emphasis of the end-of-semester accompanying paper is on the clarity of writing, the relationship between a biologically-relevant question and the computational model they developed and tested, judicious selection of simulation parameters and their biological relevance, creativity regarding data mining and visualization of results, and the critical overview of shortcomings and possible improvement of the project.

At least a week before the end-of-semester presentation, each student is required to meet with the instructor to assess both the content and the format of the intended 10-minutes class presentation. This meeting also allows the instructor to evaluate student fluency with the material presented and individual contribution to the team project. Many students reported that this was the first, or one of the only few, presentations required in any of their classes. Thus, BMEC also helps develop oral communication skills, which is a crucial feature of a liberal arts education.

While both the essay assignments and the end-of-semester computational projects develop and sharpen students' critical evaluation of the primary literature, teach students how to identify strengths and weaknesses, and present the results logically, they serve different purposes. The literature review essay's primary focus is to engage undergraduates in scientific thinking, critique what they have read, and understand what questions have been left unanswered or what new problems have emerged from the primary literature they covered. Critical evaluation of the primary literature during essay writing reflects their assessment of results' reliability, replicability, methods, and conclusions reached by the study's authors. The end-of-semester computational project also requires a literature review. Still, it focuses on a single computational model and, more importantly, requires a working computer implementation of the model under investigation to (1) replicate the data published in the literature and (2) suggest new implementations and new parameter ranges explored.

4.6 Learning Objectives Assessment

Reading Quizzes. The reading quizzes consist of multiple-choice questions that require 5–10 minutes of class time. One purpose of the reading quizzes is to check that the students read the assigned material and are familiar with the new vocabulary before starting the lecture. An equally important purpose is to signal to the instructor where the most misconceptions are so that the lecture's emphasis can be tuned to the student's actual needs. There is at least one reading quiz per chapter with the option of delivering them electronically via the college-wide course management system.

Homework. There is usually one homework assignment per week (see http://oprisans.people.cofc.edu/bmec.html/ for a detailed list of tasks). The homework assignments are primarily quantitative and require computational proficiency in preparing charts, data visualizations, and statistical analysis.

Literature Review Essays and End-of-Semester Projects. Assessing course outcomes is crucial to determine what areas need improvements. We adopted and modified a rubric model of assessment to help students with the essay preparation and the end-of-semester computational project. For the essay rubric, the emphasis is on understanding the literature's scientific background and tune writing and communication skills to technical aspects of the paper. The rubric also makes several explicit aspects of scientific content and methodologies, such as identifying each experiment's aim in the assigned literature, stating and justifying the hypothesis, discussing the results, and identifying obvious short-comings and possible suggestions for improvement. Critical evaluation of the primary literature is also required. Students must identify strengths and weaknesses and suggest experiments that would logically follow from the paper's results or fill in the article's gaps.

For the literature review essay, the students were required to organize multiple sources, emphasize the flow and progression from one experiment to the next, and identify open questions. Throughout the entire BMEC course, we caution our students that "all models are wrong, but some are useful" (George E. P. Box) and encourage students to consider both the limited scope of a given study and the totality of the evidence presented.

For the end-of-semester class presentations of the computational project, the project's team will cover the background literature and their numerical simulation results. Other students are randomly called upon to discuss or critique the hypothesis, assumptions, and limitations of the work presented by their colleagues.

Assessment of Teaching Training Fellows. In our experience, the TTF program benefits both our undergraduates, a fact that we attributed to the beneficial role of peer coaching [11], and the Ph.D. students. TTF students are encouraged to focus on multiple aspects of pedagogy, from general teaching style to specific teaching strategies. The lecture's roster faculty serves as a teaching mentor for the TTF and helps him/her develop lecture plans and assignments appropriate for undergraduates.

By observing how various instructional methods engage undergraduates in the course, TTF students can incorporate effective teaching strategies into their lectures. Of particular importance is the early exposure and adoption of modern classroom technology by TTFs. In our BMEC classroom, we use as many modern teaching technology tools as possible, from the Peer Instruction [15] with classroom response systems (iClickers) [3] to dynamic PowerPoint presentations that integrate multimedia content [4].

Our goal is to guide TTF students towards identifying effective teaching strategies that match both their personality and the intended audience, help them design the assignments given to undergraduates at the appropriate level of difficulty, design qualitative questions and computational problems that require synthesis and manipulation of the material to make sure students understood the key concepts, and convey the message that the instructor must always have a much deeper understanding of the topic than he/she may be teaching. To evaluate improved pedagogical skills of the TTF students, we rely on undergraduates' ratings on the clarity of learning objectives, the material presented, and qualitative feedback on the most and least effective components of the lecture from both students and course directors.

5 Discussion

We implemented at the CofC a computational neuroscience (CNS) program that fuses neuroscience with more math-oriented topics. It afforded new courses to be developed which have served a multitude of purposes, i.e., growth of both Neuroscience and Biomedical Physics Minors, introduced freshmen students to computational biology through our first-year experience, empower biology students to search for computational solutions of biologically-relevant questions, and open a new realm of discovery for mathematically-oriented and computer science savvy student.

The Biophysical Modeling of Excitable Cells course is unique at the CofC for several reasons. It bridges (1) multiple disciplines (physics, mathematics, computer science, psychology, and biology), (2) numerous programs on campus (Neuroscience Minor, Biomedical Physics Minor, and DATA Science), and (3) the neuroscience research at MUSC with undergraduate teaching at the CofC.

I the absence of a TA and TTF, this class would be challenging to implement with an enrollment greater than 15–20 students. One of the limitations is the number of end-of-semester presentations, which could be reduced by creating larger teams. It is more challenging to coordinate students' schedules to run numerical simulations for larger groups. Suppose a TA or TTF is not available during hands-on sessions. In that case, the instructor must adopt a synchronous, step-by-step approach such that any computation roadblock is removed for all students before moving forward.

This course promotes pedagogical awareness through its TA and TTF programs and helps prepare the next generation of neuroscience instructors. Through the strong ties we have with research labs at MUSC and the constant interaction among our students and the TTF, we created steady steam of students who successfully pursue Ph.D. degrees in (computational) neuroscience. Among other approaches, this is one of the most effective ways of preparing the next generation of computational biologists [12].

Acknowledgments. This work was supported by a Research & Development grant form the CofC and an award from the South Carolina Space Grant Consortium.

References

1. Austin, A.: Preparing the next generation of faculty: graduate school as socialization to the academic career. J. High. Educ. **73**, 94–122 (2002). https://doi.org/10.1353/jhe.2002.0001
2. Bialek, W., Botstein, D.: Introductory science and mathematics education for 21st-century biologists. Science **303**, 788–790 (2004). https://doi.org/10.1126/science.1095480
3. Caldwell, J.E.: Clickers in the large classroom: current research and best-practice tips. CBE Life Sci. Educ. **6**(1), 9–12 (2007). https://doi.org/10.1187/cbe.06-12-0205
4. Chien, Y., Smith, M.L.: Powerpoint: is it an answer to interactive classrooms? Int. J. Instr. Media **35**(3), 271 (2008)

5. Ermentrout, B.: Simulating, Analyzing, and Animating Dynamical Systems: A Guide to XPPAUT for Researchers and Students. SIAM (2002). https://doi.org/10.1137/1.9780898718195
6. Feldon, D., et al.: Graduate students' teaching experiences improve their methodological research skills. Science **333**, 1037–1039 (2011). https://doi.org/10.1126/science.1204109
7. Holley, K.A.: The longitudinal career experiences of interdisciplinary neuroscience PhD recipients. J. High. Educ. **89**(1), 106–127 (2018). https://doi.org/10.1080/00221546.2017.1341755
8. Hunter, A.B., Laursen, S., Seymour, E.: Becoming a scientist: the role of undergraduate research in students? Cognitive, personal, and professional development. Sci. Educ. **91**, 36–74 (2006). https://doi.org/10.1002/sce.20173
9. Hurd, M., Vincent, D.: Functional magnetic resonance imaging (fMRI): a brief exercise for an undergraduate laboratory course. J. Undergrad. Neurosci. Educ. **5**, A22–A27 (2006)
10. Kaplan, D.: Explanation and description in computational neuroscience. Synthese **183**(3), 339–373 (2011). https://doi.org/10.1007/s11229-011-9970-0
11. Knight, J.: A primer on instructional coaching. Principal Leadersh. **5**, 17–20 (2005)
12. Kozeracki, C., Carey, M., Colicelli, J., Levis-Fitzgerald, M.: An intensive primary-literature-based teaching program directly benefits undergraduate science majors and facilitates their transition to doctoral programs. Life Sci. Educ. **5**, 340–347 (2006). https://doi.org/10.1187/cbe.06-02-0144
13. Latimer, B., Bergin, D., Guntu, V., Schulz, D., Nair, S.: Open source software tools for teaching neuroscience. J. Undergrad. Neurosci. Educ. **16**(3), A197–A202 (2018)
14. Luft, J., Kurdziel, J., Roehrig, G., Turner, J.: Growing a garden without water: graduate teaching assistants in introductory science laboratories at a doctoral/research university. J. Res. Sci. Teach. **41**, 211–233 (2004). https://doi.org/10.1002/tea.20004
15. Mazur, E.: Peer Instruction: A User's Manual. Series in Educational Innovation. Prentice Hall, Hoboken (1997)
16. Miller, J., Martineau, L., Clark, R.: Technology infusion and higher education: changing teaching and learning. Innov. High. Educ. **24**, 227–241 (2000). https://doi.org/10.1023/B:IHIE.0000047412.64840.1c
17. Moore, J., Stuart, A.: Neurons in Action Version 2: Tutorials and Simulations Using NEURON. Sinauer Associates, Sunderland, MA, USA (2007)
18. Muir, G.: Mission-driven, manageable and meaningful assessment of an undergraduate neuroscience program. J. Undergrad. Neurosci. Educ. **13**(3), A198–A2015 (2015)
19. Oprisan, S.: Teaching computational neuroscience at a liberal arts and sciences undergraduate college. Society for Neuroscience, Washington, DC (2011)
20. Oprisan, S.: Introducing computational neuroscience concepts and research projects to undergraduates. Society for Neuroscience, New Orleans, LA (2012)
21. Ramirez, J.J.: Undergraduate neuroscience education: meeting the challenges of the 21st century. Neurosci. Lett. **739**, 135418 (2020). https://doi.org/10.1016/j.neulet.2020.135418
22. Salomon, D., Martin-Harris, L., Mullen, B., Odegaard, B., Zvinyatskovskiy, A., Chandler, S.: Brain literate: making neuroscience accessible to a wider audience of undergraduates. J. Undergrad. Neurosci. Educ. **13**(3), A64–A73 (2015)
23. Schultheiss, S.: Ten simple rules for providing a scientific web resource. PLoS Comput. Biol. **7**(5), e1001126 (2011). https://doi.org/10.1371/journal.pcbi.1001126

24. Stanley, D.: Can technology improve large class learning? The case of an upper-division business core class. J. Educ. Bus. **88**, 265–270 (2013). https://doi.org/10.1080/08832323.2012.692735
25. Stuart, A.: Teaching neurophysiology to undergraduates using neurons in action. J. Undergrad. Neurosci. Educ. **8**(1), A32–A36 (2009)
26. Wiertelak, E., Hardwick, J., Kerchner, M., Parfitt, K., Ramirez, J.: The new blueprints: undergraduate neuroscience education in the twenty-first century. J. Undergrad. Neurosci. Educ. **16**(3), A244–A251 (2018)

Increasing the Impact of Teacher Presence in Online Lectures

David Iclanzan(✉) and Zoltán Kátai

Faculty of Technical and Human Sciences, Sapientia Hungarian University
of Transylvania, Târgu-Mureş, Romania
{iclanzan,zoltan_katai}@ms.sapientia.ro

Abstract. We present a freely available, easy to use system for promoting teacher presence during slide-supported online lectures, meant to aid effective learning and reduce students' sense of isolation. The core idea is to overlay the teacher's body directly onto the slide and move it and scale it dynamically according to the currently presented content. Our implementation runs entirely locally in the browser and uses machine learning and chroma keying techniques to segment and project only the instructor's body onto the presentation. Students not only see the face of the teacher but they also perceive as the teacher, with his/her gaze and hand gestures, directs their attention to the areas of the slides being analyzed.

We include an evaluation of the system by using it for online teaching programming courses for 134 students from 10 different study programs. The gathered feedback in terms of attention benefit, student satisfaction, and perceived learning, strongly endorse the usefulness and potential of enhanced teacher presence in general, and our web application in particular.

Keywords: Teacher presence · Teaching aid tools · Online teaching

1 Introduction

In recent decades, there have been a number of predictions about an increase in demand for online courses. For example, the 2009 Chronicle of Higher Education research report, "The College of 2020", predicted that students would be increasingly interested in online courses [27]. However, it is quite certain that even the boldest predictions did not imply the actual need generated by the COVID-19 pandemic, which in addition, required an almost instant shift.

In most countries, video conferencing tools (Zoom, Microsoft Teams, Google Meet, etc.) provided a quick solution, enabling webinar-like courses. The teaching environments often include PowerPoint slides [4] which are preferred and used by many CS instructors too [22]. The video conferencing tools have a screen sharing feature which allows the teacher to use the same PowerPoint presentation as for

This study was partially supported by the Sapientia Foundation Institute for Scientific Research.

M. Paszynski et al. (Eds.): ICCS 2021, LNCS 12747, pp. 626–639, 2021.
https://doi.org/10.1007/978-3-030-77980-1_47

face-to-face lectures. Common learning settings are: i) students listen to the teacher's explanation while their eyes are fixed on the slides; ii) the teacher turns on its camera and its video appears next to the slides in a separate smaller window; iii) instruction is presented as full-screen lecture slides with a small embedded video overlay of the instructor speaking (usually, the teacher's face on the slide has a fixed size and position).

A major concern with online courses is a sense of isolation that can hinder students' ability to learn [8]. When social cues disappear, communication becomes more "task-oriented, cold, and less personal than face-to-face communication" [29, p. 461]. Accordingly, research in the field of online education highlights the key role the "strong teacher presence" plays in engaging learners in meaningful learning experiences [5,28]. This phenomenon was investigated mostly within the framework of asynchronous video lectures. It was suggested that including the instructor's face in the online lecture (via a small embedded video) will result in more effective learning because it has the potential to amplify the social cues coming from the teacher [7]. On the other hand, empirical evidence does not support (consequently and consistently) the validity of this social argument. According to Kizilcec, Papadopoulos, and Sritanyaratana [16] a possible reason could be that the resulting social cue is too weak to induce such positive social responses in learners, that could surpass the generated attention division between the two video inputs (teacher and slides).

In line with this suggestion we tried to increase the impact of teacher presence in synchronous video lectures by adding a new, platform-independent feature to video conferencing tools, in the form of a simple web application accessible from any modern internet browser. The application segments in real time the teacher's web camera feed and projects the teacher's body onto the slides. The component allows the instructor to change the position of the video image in the slide area and to zoom in or out, respectively. We expect that this new feature will promote a stronger teacher presence by supporting a "more alive" teacher-student and teacher-slide interactivity. The teacher's gaze and hand gestures engages and focuses students' attention onto the region in the slides that are currently discussed. The teacher may also choose to move his/her image to the region currently analyzed in the slide, or to use his/her scaled down image as a pointer. When it is deemed important for students to pay undivided attention to the teacher's explanation, the instructor may scale his/her image to obscure the slide. Since the communication occurs live, the increased "visual flexibility" of the teacher allows them to consider student feedback on how they relate to the content displayed on the slides. In this study, we investigated the effectiveness of this tool with respect to student satisfaction, increased attention and perceived learning.

2   Background

A number of research investigated the effectiveness of instructional videos in engaging students in meaningful online learning experiences. To facilitate a stronger teacher presence, many of these videos feature a picture-in-picture view of the instructor. However, in a recent study, Wang and Antonenko [30] emphasize that it is not clear how teacher presence influences learners' visual attention and what it contributes to learning and affect.

2.1 Split Attention Versus Social Cue

Several studies in multimedia learning research have examined the phenomenon of split attention resulting from multiple channel presentation [19,25]. According to Baddeley's Theory of Working Memory [3], separate processing units are employed for different input modalities: the so-called "visual-spatial sketchpad area" of working memory stores visual input and the "phonological loop area" stores auditory information. According to this model, the two visual inputs (the video of the instructor's face and the slide content) would compete with each other for visual-spatial cognitive resources while the instructor's narration is processed separately (although potentially supported by nonverbal information encoded in the instructor's presence; e.g., gestures and facial expressions).

Since all lecture-relevant information is encoded on the slides and in the narration, someone might consider the instructor's face as a source of unnecessary extra load as it could obstruct cognitive processing of relevant information and, consequently, hinder learning. On the other hand, according to Clark and Mayer [7], social cues from the instructor may enhance the learning process by triggering social responses in the learner [21] and promoting deeper engagement with the lecture content.

In line with these contradictory viewpoints, prior research on the effect of including the instructor's face in lecture videos provided mixed results. A considerable amount of experimental evidence supports Mayer's [18] image principle that adding a picture or video of the instructor to a multimedia instruction does not necessarily support learning [14]. On the other hand, there is also competing empirical evidence on learners' affective response to the instructor's face in video lectures. In a review of the social presence theory and its instructional design implications, Cui, Lockee, and Meng [8] refer to several studies which conclude that social presence is one of the most significant factors in improving learners' satisfaction and perceived learning.

A possible reason for these apparently contradictory findings is suggested by Homer, Plass, and Blake [14]. These authors report that learners who saw the speaker's face did not report a greater sense of social presence than those who did not see the speaker. In addition, as mentioned above [16] argues that this may happen when the generated social cue is too weak to induce positive social responses in learners. Three immediate indicators of a stronger teacher presence could be attention benefit, increased perceived learning and student satisfaction.

2.2 Visual Attention, Student Satisfaction, and Perceived Learning in Online Learning Environments

Several studies revealed that human eyes elicit strong attentional shifts in the direction of their gaze [1]. Since attention is a prerequisite of learning [26], displaying the instructor in an online learning environment could enhance the learning process because of the generated attentional cueing effect. For example, van Wermeskerken and van Gog [31] investigated learning settings that embodied video examples in which an instructor demonstrates how to perform a task. These authors conclude that the teacher's gaze may be a powerful cue for students because it may help them to switch their attention timely from the instructor to the task.

Another important component of the learning process is student satisfaction [2]. According to the Online Learning Consortium, student satisfaction is one of the defining elements of high-quality online learning. A widely used conceptual framework for evaluating learning environments is Kirkpatrick's four-level model [15]: reactions, learning, behavior, and results. A cornerstone of Kirkpatrick's model is that reaction affect, such as satisfaction, results in effective learning.

Perceived learning is also considered as an indicator of learning and a key element for course evaluation [32]. It is defined as a student's self-report of knowledge gain [24]. According to Alqurashi [2] it is important for teachers to evaluate how students perceive their learning to improve the quality of their online courses.

In a recent study, Wang and Antonenko [30] analyzed the impact of teacher presence on visual attention, perceived learning and student satisfaction in the context of online mathematics education. Participants were invited to watch two instructional videos, with the instructor either present or absent. These authors report that the teacher attracted considerable visual attention and teacher presence positively influenced participants' perceived learning and satisfaction.

We have proposed to test the effectiveness of the tool we designed from a similar perspective but within the framework of Computer Science education. A common particularity of mathematics and Computer Science topics is that the teacher often analyzes the slide content together with students.

3 Client-Side Web Application

With the proliferation of online meetings in general, and online classes in particular, video conferencing tools are frequently adding new features to better suit client needs and maximize market share. Most of these features expand the collaboration and coordination capabilities within these tools, but some of them are meant to offer avenues for more engaging presentations.

Recently, the Zoom platform started offering a new feature, still in beta (as of January 2021), in which users "can share presentations as Virtual Backgrounds for an immersive sharing experience". As the Google Meet platform used by our university does not yet have similar capabilities, we set out to develop a custom solution, to offer a better online learning experience for our students.

Fig. 1. Stages for obtaining an image-frame that contains the current slide overlaid with the image of the presenter.

The main requirements we established were i) ease of use, minimal learning curve; ii) flexibility - the presenter's projected image can be freely moved/scaled, customizable opacity, etc.; iii) privacy - no registration or data sharing; iv) out-of-the-box functionality - many teachers do not have administrator rights on the school-issued devices, therefore no installation should be required; v) platform independence - the solution should work on all major desktop operating systems and with all video conferencing tools that support screen sharing.

A client-side web application that runs locally in the browser can satisfy all above requirements. Therefore, we developed the application using JavaScript, relying primarily on p5.js - "a client-side library for creating graphic and inter-active experiences".

For combining a live web camera video feed of the presenter with a presentation (series of images), one must perform the steps presented in Fig. 1 many times per second. If this process takes too long and cannot be performed fast enough, the combined animation becomes choppy and unenjoyable.

In this time critical sequence, separating the presenter out from the web camera feed (segmentation) is the most computationally expensive one. Trying to maximize the out-of-the-box functionality, first we used a Machine Learning approach, namely, the BodyPix "person segmentation in the browser" neural network to extract the isolated image of the presenter. While the model works well, we found that the segmentation is not pixel-perfect. Usually, there are visually noticeable differences between two consecutive segmentations of a still standing person, which can lead to flickering edges, a known and so far unresolved issue[1].

High-quality video stream layering can be achieved relatively easily with chroma keying (a technique also refereed to as green-screen or colour-separation overlay), therefore we also implemented support for this approach. The green screen removal and combination of the segmented image and slides has been implemented using Seriously.js, a highly efficient real-time video compositor for the web.

The disadvantage of this technique is that it requires the purchase and instal-lation of a uniformly colored backdrop, hindering the ease-of-use and zero deploy-ment cost aspect of the system. Alternatively, software packages using hardware acceleration that can efficiently place a virtual green screen behind the user.

---

[1] https://github.com/tensorflow/tfjs/issues/3902.

Fig. 2. Snapshot from a lecture delivered through the Google Meet platform. (Color figure online)

However, their installation requires administrator rights and some configuration, again raising the entry point for the usage of the application.

A screenshot from a lecture presented with our web application can be seen in Fig. 2. The web application is available freely at the first author's university homepage[2]. We encourage the community to try it and use it for enhanced teacher presence.

4   Method

We designed an evaluation method to compare the following synchronous online learning settings: i) students are focused on the slide contents accompanied by teacher narration ("only narration"); ii) in addition, the video of the presenting teacher is displayed next to the slides ("teacher in a separate window"); iii) the teacher is projected onto the slides and moves dynamically using our tool ("teacher overlaid"). We developed a survey focusing on two main aspects: 1) students' general opinion in relation to the importance of nonverbal communication elements in online lectures; 2) students' feedback regarding the learning experience facilitated by the new tool.

Based on the above brief literature review, we addressed the following research questions: (RQ1.1) To what extent do students miss visual nonverbal communication elements in online lectures? (RQ1.2) To what extent do these elements contribute to students' perceived attention and learning? (RQ2.1) Are the learning settings that visualize the teacher more effective than "only narration", due to the generated social cues? (RQ2.2) Is the "teacher overlaid" condition more effective than the "teacher in a separate window", because of the stronger teacher presence effect? (RQ3.1) What is the level of satisfaction and general feedback regarding an actual lecture delivered in the new format? (RQ3.2) To what extent is the lecture in the new format more effective, compared to the "teacher in a separate window" setting, for perceived attention and learning?

---

[2] https://ms.sapientia.ro/~iclanzan/prezcam/.

4.1   Pilot Study

We scheduled a pilot study for the last teaching week of the first semester (school year 2020–2021), to test the usability of the system, evaluate the extent, if any, of added value for students, and to potentially unravel early, overlooked pitfalls and disadvantages. As we did not have access to a physical green screen, we used a virtual one via the XSplit VCam software.

During this week, all programming courses for the first and second year students attending the Sapientia Hungarian University of Transylvania were delivered with the help of the proposed system. After a brief introduction at the start of the class, the teacher navigated to the tool's website, dragged and dropped the lecture slides, started the camera feed, and then shared his full-screen browser window with the online audience. Because these presentations were not designed (or modified) to specifically suit the overlaid image of the presenter, the teacher resized and repositioned his image on the slides, whenever deemed it necessary. To collect feedback, the students were asked to respond to an online questionnaire right after the lecture.

Roughly 80% of students attending the lectures answered the questionnaire, resulting in 134 answers from 10 different study programs (Informatics, Computer Science, and various other engineering programs). Out of these, 93 (69.4%) responders studied in the first year and 18 (13.43%) answers came from female students.

4.2   Online Survey

The survey language was Hungarian and the responses were collected online through Google Forms. We tried to make the questionnaire as brief as possible, as students often report feeling overwhelmed during the pandemic and therefore tend to ignore complex surveys that require more than a few minutes to answer.

After (optionally) indicating their study program, year of study, and sex (questions 0.1–0.3), the students were asked to answer four general questions, pertinent to their experience of following online lectures during the pandemic (when on-premises teaching was not permitted).

Question 1.1 inquired on the extent of students that students had missed nonverbal communication elements, such as the teachers' *body language, gestures, facial expressions, eye contact*, during online lectures. The question explicitly indicated that students should consider all their attended subjects and report on the general impression. Questions 1.2 and 1.3 asked students to gauge the importance of the above-mentioned visual nonverbal communication elements *during online lectures*, in engaging and retaining their *attention* and facilitating their *understanding* of the presented concepts. Answers for questions 1.1–1.3 were indicated on a unipolar 5-point Likert scale. As unipolar scales allow responders to focus on the presence of a single characteristic, it can hopefully generate more clear and inclined responses.

Question 1.4 asked students to consider and rank (from most preferred to least preferred) the following three online lecture delivery methods: i) "only narration" ii) "teacher in a separate window"; iii) "teacher overlaid".

In the second part of the survey, students were asked to reflect on the experience of the new lecture format and compare it with the experience of the previous 13 lectures that were delivered with the "teacher in a separate window" method.

By answering questions 2.1–2.3, students provided feedback on i) how frequently the overlaid video of the teacher was distracting or disturbing (unipolar 5-point Likert scale) and optionally specify details why or when the projection was intrusive; ii) whether they would like to attend further lectures in this format ("Yes", "No", "Other" (free-text)); iii) and on the overall usefulness of having the video of the teacher overlaid onto the slides ("Distracting", "Neutral", "Useful").

Questions 2.4 and 2.5 inquired on the positive effect (if any) of the new lecture format, in comparison with the previous ones: *"The projection of the teacher on the presentation helped you"* i) "to FOCUS better? Were you able to pay more attention?"; ii) "to BETTER UNDERSTAND the material?". For both questions, the answers could be indicated using a unipolar 5-point Likert scale.

Lastly, students were asked to provide general feedback and suggestions for improvement through a free-text field.

5   Results and Discussion

5.1   Evaluation of the General Opinion (RQ1.1 and RQ1.2)

Student feedback with respect to question 1.1 (see Fig. 3) shows that more than 60% of the students miss a more prominent teacher presence in online lectures considerably or to a great extent (the most chosen option). Only 4 male students reported that they did not miss additional nonverbal communication cues coming from the lecturer. This result provides further support that the benefits of nonverbal communication (appearance, posture, limb movement, sight and facial expressions) improve the learning experience of students. As for the possible reasons, Cavanagh et al. [6] underline that visual cues such as body language can play a dominant role in transmitting emotional content [13] and contributes to the credibility of the teacher [12]. In addition, nonverbal communication elements

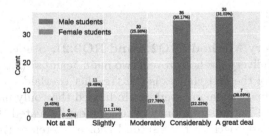

Fig. 3. Student feedback on how much they miss visual nonverbal communication elements, from the teacher.

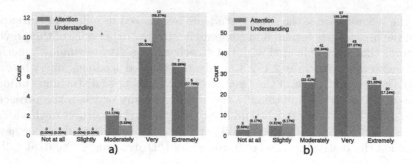

Fig. 4. Importance of visual nonverbal communication elements coming from the teacher, for engaging attention and facilitating a better understanding, as perceived by a) female; b) male students.

support teachers in manifesting more expressively their willingness to communicate and transmit valuable information which could have a great impact on student engagement and learning [20].

In line with the above, the vast majority of both female (Fig. 4a), respectively, male students (Fig. 4b) indicated that a stronger visual presence of the teacher would support them in being more focused during the online lectures and would be helpful in fostering a better understanding. Around 10% of male students saw very little value or no benefit at all in these factors.

With respect to possible gender differences, the [11] study concludes that (in accordance with studies of gender in virtual teams [17]) female students have higher dialogue in distance learning environments than males. Our findings also confirm that the subjective importance of the teacher's body language (gestures, facial expressions, eye contact) is assessed as more important by female students for engaging and retaining students' attention and fostering a better understanding (questions 1.2 and 1.3). For both questions they choose "considerably" or "to a great extent" in a greater proportion compared to their male counterparts: 79% vs. 73% for question 1.2 regarding the assessed attention benefit, and 94% vs. 54% for question 1.3 - understanding benefit. However, the sample size of female responders is small and therefore prone to a larger statistical error, hence further investigations are required to determine if the observed difference is significant.

Preferred Delivery Method (RQ2.1 and RQ2.2). Student rankings of the preferred lecture delivery methods reveal two main "camps" (Fig. 5). The biggest one, with 43.28% of the first choices is the "teacher overlaid" presentation approach. However, almost as many students indicated the "only narration" option as their first choice. The vast majority of responders indicated the "teacher in a separate window" approach as their second choice, therefore this option seems to offer a good compromise between the two camps.

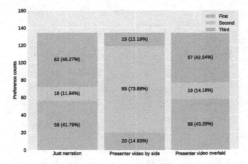

Fig. 5. Students rankings regarding the preferred explanation delivery mechanism during online lectures.

Firstly, we ranked the delivery mechanisms by applying the Ranked Choice Voting[3] method (also known as Instant Runoff Voting - IRV). The "teacher overlaid" method came out as the winner with 69 votes after redistribution. The "only narration" method totaled 65 votes.

Secondly, we applied a rank ordering weighting method [23] to obtain numerical scores. Associating rank positions I., II., and III. with the weights 0.64, 0.29, and 0.07 (rank exponent weights for $p = 2$), respectively, we obtained the following average numerical scores: 0.34 ("only narration"), 0.32 ("teacher in a separate window"), 0.35 ("teacher overlaid"). By applying the Wilcoxon signed-rank test we found that the "teacher overlaid" method was scored significantly higher ($p = 0.01$) and the "only narration" method marginally significantly higher ($p = 0.05$) than the "teacher in a separate window" one. This result, on the one hand, supports our expectation that our tool has the potential to significantly increase the impact of teacher presence. On the other hand, it also suggests that simply displaying the teacher' face next to the slides does not induce a social cue large enough to outweigh the extraneous processing needed to concomitantly follow the two visual inputs [3,16,18].

5.2   Lecture Evaluation and Feedback (RQ3.1 and RQ3.2)

Figure 6 depicts the students' feedback on three general aspects. The first one, a), evaluates how often students were distraught by the overlaid teacher video feed. As nobody responded "Always", the pie chart contains only four regions. Half the responders did not notice any disturbance, while one third was rarely distraught, 17% sometimes, while the remaining 3%, unfortunately, often. In the text feedback on why and when the overlay was distracting or disturbing, the consensus (27 times out of the 29 text responses) was that the overlaid teacher sometimes obstructed parts of the text or relevant source code. Two responses indicated that when the teacher repositions itself on a new slide, the flow is interrupted as it takes too much time.

---

[3] https://www.fairvote.org/rcv#how_rcv_works.

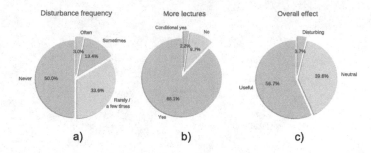

Fig. 6. Students' general feedback on three aspects: a) how often was the presenter's overlay distracting or disturbing; b) whether they would enjoy following other online lectures in this format; c) overall net effect of the technology (if any).

In spite of these occasional inconveniences, more than 88% of the students (see Fig. 6b)) enjoyed and would like to attend further lectures delivered in this format. Three students chose "Other" and in their text response specified a conditional yes: they would like to follow this format only if the obstruction of the content is completely eliminated. Almost 10% of the students would rather not attend online lectures in this format. Considering that 42% of the students indicated that they prefer the narration only delivery, and around 10% saw no value in enhanced teacher presence in general, this percentage is not surprising.

Almost 57% of the students found the overlay of the presenter overall positive and useful, almost 40% neutral, having its advantages but also disadvantages. 5 responders (3.7%) judged it distracting, having a mostly negative impact.

Through the optional free-text feedback and suggestion field, we received 38 entries. Many of these reiterated the observation that the teacher's projection should never cover content on the slides, the placement should be already taken into account when preparing the slides. Others suggested the use of wide aspect ratio slides that would fill more of the screens where students watch the lectures; 4:3 aspect ratio slides leave too many unused regions on the sides of the screen. A few responses mentioned dropped frames and a slight delay between the audio and video feed. Two students observed the occasional artefacts produced by the digital green screen and recommended that we invest in a physical one.

Some responses pointed out that as the presenter is already on the slides, a greater interaction with the content would be welcome. The teacher could point to and touch things, underline and emphasize content during the explanation, similarly to how it is done when the delivery happens on a blackboard in class.

5.3   Effect on Attention and Understanding

The results of the self-assessment, regarding the benefits of the "teacher overlaid" method with respect to attention and understanding are reported in Fig. 7.

The vast majority of students report that the enhanced teacher presence had helped them "considerably" or to a "great extent", in both aspects. Again, from the limited data it seems that the self-reported effect is greater in the case of

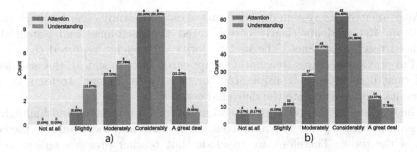

Fig. 7. Perceived benefits of the new online lecture format, as reported by a) female students; b) male students.

female students. 6 male students reported no improvement at all. As the scales were unipolar, it is possible that these students had a worse learning experience compared to the previous lectures, where the teacher's video was on the side.

As a last step, we compared the corresponding answers regarding attention and understanding benefits from the general opinion and lecture evaluation sections (question 1.2 vs. question 2.4 and question 1.3 vs. question 2.5). We coded the response categories of the 5-point Likert item with numerical scores from 1 to 5 [9]. Since we had to compare relatively highly correlated paired data derived from a medium-sized sample, following the guidelines from [9], we used the modification proposed by Pratt for the Wilcoxon signed-rank test. The test revealed significant differences in favour of general expectation ($p < 0.001$ in both cases). This result emphasizes that students anticipate more potential in nonverbal communication elements than what we have been able to exploit so far with the proposed system for the enhancement of teacher presence. Since generating strong teacher presence is a complex task, this finding emphasizes that it is important for teachers to approach this challenge with professional humility. Dockter [10] argues that online teachers' unfounded belief that they create and control their teaching presence can result in increased distance between teacher and students.

6   Conclusions

Prior research suggests that simply including the speaking instructor's face in online lectures does not necessarily result in more effective learning and reduce students' sense of isolation. In this paper we proposed a new, easy to use system meant to promote teacher presence in the form of a client-slide web application. The tool segments in real-time user's web camera feed, and by removing the background projects only the presenter onto the slides. The overlay is not static, the user has the flexibility to move and scale their body size or set transparency as needed. With his/her gaze and hand gestures, the teacher can focus students' attention, easily indicating the areas of the slides currently discussed.

We evaluated the system during a one-week pilot study. The feedback gathered from 134 students clearly corroborated the usefulness and potential of enhanced teacher presence. The vast majority of students reported that, compared to previous lectures (teacher' face appeared next to slides), the new delivery format helped them "considerably" or to a "great extent" to focus on the presentation and assimilate the delivered content.

The assessment also revealed several areas for improvement and highlighted that some students prefer to avoid attention division and focus only on the contents of the slides. Therefore, we conclude that teacher presence enhancement techniques must i) ensure that their benefits compensate and offsets the additional mental effort needed to follow multiple visual channels of information; ii) keep a balance and ensure that they are not intrusive for students who prefer to solely focus on the delivered content. Further work and improvements to the system will be made according to these guidelines.

References

1. Admoni, H., Bank, C., Tan, J., Toneva, M., Scassellati, B.: Robot gaze does not reflexively cue human attention. In: Proceedings of the Annual Meeting of the Cognitive Science Society, vol. 33 (2011)
2. Alqurashi, E.: Predicting student satisfaction and perceived learning within online learning environments. Distance Educ. **40**(1), 133–148 (2019)
3. Baddeley, A.: Working memory. Science **255**(5044), 556–559 (1992)
4. Baker, J.P., Goodboy, A.K., Bowman, N.D., Wright, A.A.: Does teaching with powerpoint increase students' learning? A meta-analysis. Comput. Educ. **126**, 376–387 (2018)
5. Boton, E.C., Gregory, S.: Minimizing attrition in online degree courses. J. Educators Online **12**(1), 62–90 (2015)
6. Cavanagh, M., Bower, M., Moloney, R., Sweller, N.: The effect over time of a video-based reflection system on preservice teachers' oral presentations. Aust. J. Teach. Educ. **39**(6), 1 (2014)
7. Clark, R.C., Mayer, R.E.: E-Learning and the Science of Instruction: Proven Guidelines for Consumers and Designers of Multimedia Learning. Wiley, Hoboken (2016)
8. Cui, G., Lockee, B., Meng, C.: Building modern online social presence: a review of social presence theory and its instructional design implications for future trends. Educ. Inf. Technol. **18**(4), 661–685 (2013)
9. Derrick, B., White, P.: Comparing two samples from an individual likert question. Int. J. Math. Stat. **18**(3), 1–13 (2017)
10. Dockter, J.: The problem of teaching presence in transactional theories of distance education. Comput. Compos. **40**, 73–86 (2016)
11. Ekwunife-Orakwue, K.C., Teng, T.L.: The impact of transactional distance dialogic interactions on student learning outcomes in online and blended environments. Comput. Educ. **78**, 414–427 (2014)
12. Goodboy, A.K., Martin, M.M., Bolkan, S.: The development and validation of the student communication satisfaction scale. Commun. Educ. **58**(3), 372–396 (2009)
13. Goodboy, A.K., Myers, S.A.: The effect of teacher confirmation on student communication and learning outcomes. Commun. Educ. **57**(2), 153–179 (2008)

14. Homer, B.D., Plass, J.L., Blake, L.: The effects of video on cognitive load and social presence in multimedia-learning. Comput. Hum. Behav. **24**(3), 786–797 (2008)
15. Kirkpatrick, J.D., Kirkpatrick, W.K.: Kirkpatrick's four levels of training evaluation. Association for Talent Development (2016)
16. Kizilcec, R.F., Papadopoulos, K., Sritanyaratana, L.: Showing face in video instruction: effects on information retention, visual attention, and affect. In: Proceedings of the SIGCHI Conference on Human Factors in Computing Systems, pp. 2095–2102 (2014)
17. Martins, L.L., Gilson, L.L., Maynard, M.T.: Virtual teams: what do we know and where do we go from here? J. Manag. **30**(6), 805–835 (2004)
18. Mayer, R.E.: Principles of multimedia learning based on social cues: personalization, voice, and image principles. In: The Cambridge Handbook of Multimedia Learning, pp. 201–212 (2005)
19. Mayer, R.E., Moreno, R.: A split-attention effect in multimedia learning: evidence for dual processing systems in working memory. J. Educ. Psychol. **90**(2), 312 (1998)
20. McCroskey, L., Richmond, V., McCroskey, J.: The scholarship of teaching and learning: contributions from the discipline of communication. Commun. Educ. **51**(4), 383–391 (2002)
21. Reeves, B., Nass, C.: How people treat computers, television, and new media like real people and places (1996)
22. Reuss, E.I., Signer, B., Norrie, M.C.: PowerPoint multimedia presentations in computer science education: what do users need? In: Holzinger, A. (ed.) USAB 2008. LNCS, vol. 5298, pp. 281–298. Springer, Heidelberg (2008). https://doi.org/10.1007/978-3-540-89350-9_20
23. Roszkowska, E.: Rank ordering criteria weighting methods-a comparative overview. Optimum. Studia Ekonomiczne **5**(65), 14–33 (2013)
24. Rovai, A.P.: Sense of community, perceived cognitive learning, and persistence in asynchronous learning networks. Internet High. Educ. **5**(4), 319–332 (2002)
25. Schmidt-Weigand, F., Kohnert, A., Glowalla, U.: A closer look at split visual attention in system-and self-paced instruction in multimedia learning. Learn. Instr. **20**(2), 100–110 (2010)
26. Schunk, D.H.: Learning Theories an Educational Perspective, 6th edn. Pearson, London, England (2012)
27. Van Der Werf, M., Sabatier, G.: The college of 2020: students. Technical report, Chronicle Research Services (2009)
28. Vincenzes, K.A., Drew, M.: Facilitating interactive relationships with students online: recommendations from counselor educators. Distance Learn. **14**(4), 13–22 (2017)
29. Walther, J.B., Anderson, J.F., Park, D.W.: Interpersonal effects in computer-mediated interaction: a meta-analysis of social and antisocial communication. Commun. Res. **21**(4), 460–487 (1994)
30. Wang, J., Antonenko, P.D.: Instructor presence in instructional video: effects on visual attention, recall, and perceived learning. Comput. Hum. Behav. **71**, 79–89 (2017)
31. van Wermeskerken, M., van Gog, T.: Seeing the instructor's face and gaze in demonstration video examples affects attention allocation but not learning. Comput. Educ. **113**, 98–107 (2017)
32. Wright, V.H., Sunal, C.S., Wilson, E.K.: Research on Enhancing the Interactivity of Online Learning. Information Age Publishing Inc, Charlotte (2006)

Model-Based Approach to Automated Provisioning of Collaborative Educational Services

Raul Llopis Gandia[1], Sławomir Zieliński[2], and Marek Konieczny[2]

[1] Universitat Politecnica de Valencia, Campus d'Alcoi, Alcoi, Spain
raulloga@epsa.upv.es

[2] AGH University of Science and Technology, Kraków, Poland
{slawek,marekko}@agh.edu.pl

Abstract. The purpose of the presented work was to ease the creation of new educational environments to be used by consortia of educational institutions. The proposed approach allows teachers to take advantage of technological means and shorten the time it takes to create new remote collaboration environments for their students, even if the teachers are not adept at using cloud services. To achieve that, we decided to leverage the Model Driven Architecture, and provide the teachers with convenient, high-level abstractions, by using which they are able to easily express their needs. The abstract models are used as inputs to an orchestrator, which takes care of provisioning the described services. We claim that such approach both reduces the time of virtual laboratory setup, and provides for more widespread use of cloud-based technologies in day-to-day teaching. The article discusses both the model-driven approach and the results obtained from implementing a working prototype, customized for IT trainings, deployed in the Małopolska Educational Cloud testbed.

Keywords: Model driven architecture · Cloud services provisioning · Collaborative education · STEM

1 Introduction

Cloud environments, though conceptually straightforward, are perceived as not easy to get started with, because they require significant effort and knowledge to be configured and used properly [9]. For that reason, they are not used in education as frequently as they could be. On the other hand, the proliferation of broadband Internet access in recent years resulted in significant reduction of technological barriers against integrating such services in courses' curricula.

The research presented in this paper has been partially supported by the funds of Polish Ministry of Science and Higher Education assigned to AGH University of Science and Technology.

M. Paszynski et al. (Eds.): ICCS 2021, LNCS 12747, pp. 640–653, 2021.
https://doi.org/10.1007/978-3-030-77980-1_48

As pointed out by the 2020 EDUCAUSE Horizon Report, the numbers of students grew in recent years, but "much of the growth has come from a significant increase in adult learners who are either returning for additional learning or seeking postsecondary credentialing" [4, p. 33]. Cloud based collaboration environments provide for addressing the needs of that group, especially when the respective courses are led according to distance or blended distance learning patterns, which are among the students' favourites.

As also stated in the report, by "forming innovative consortia, many smaller institutions have been able to avoid closure". However, as we observe from both from our practice, and from the opinions of people participating in the Małopolska Educational Cloud project, briefly described in Sect. 4.2, it is much easier to form a consortium than to provide added value to the learners. From our perspective, reluctance to exit one's comfort zone is a very important obstacle to wider adoption of modern educational tools. Even if the teachers use cloud services, they are not willing to change the tools they use, mainly because of the time it takes to set up a new environment. Tasks such as installation, configuration and – especially – granting privileges to the students, can be time consuming and error-prone. That discourages teachers from preparing short-lived environments, to be used, e.g., only to illustrate a single topic. Moreover, the typical unstructured way of sharing knowledge, which relies only on instructing people on how to use a particular tool or providing 'howto' documentation, does not result in increased willingness to experiment with new tools.

The approach presented in this paper assumes that cloud-based ICT tools are needed by multiple members of an educational community. We propose an MDA-based service orchestration system, capable of instantiating educational environments (compound services) on the resources possessed by the community, according to teachers' demands. The system relies both on resource pooling, and on pooling of expertise provided by the community members. By 'expertise' we understand both the knowledge possessed by IT staff about particular environment setup, and the knowledge of tools' applicability areas possessed by the teachers. We present a way of providing a structured pool of templates and propose a high level interface for describing educational environments, to make the deployment of new environments straightforward. Aside from hiding the technicalities from a teacher, the presented system automates the installation and configuration tasks, and reduces the setup times from several hours to minutes.

The structure of the paper is as follows. Section 2 surveys the important developments in the subject area. Section 3 describes the processing of user-defined models, which results in creating a ready-to-use service. Section 4 discusses the results of evaluating a proof of concept implementation of the system. Section 5 concludes the paper and points out the directions of future work.

2   Related Work

Since the introduction of MIT OpenCourseWare in 2001, many universities opened their curricula to online communities. The wide adoption of Massive

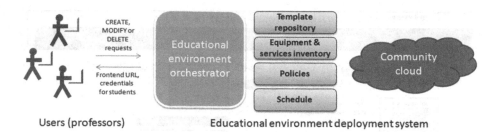

Fig. 2. Educational environment deployment system building blocks.

The template repository contains templates which simplify the conversions between Computation Independent Models (CIMs) and Platform Independent Models (PIMs). The templates include technical parameters set to values chosen by IT staff, and presented to the users in an easily understandable form, so that the teachers who request an educational environment are not overwhelmed with details. The inventory of equipment and services simply keeps track of computational nodes and existing instances of basic services, e.g., firewalls, available to the system. The policies repository contains declaratively specified administrative policies, and the schedule organizes the tasks to be performed by EEDS.

A CIM, according to the Model Driven Architecture standard specified by the Object Management Group - "only describes business concepts" [2, p. 13]. In EEDS case, the users are responsible for preparing a valid CIM that describes the educational environment they want to use during classes, and submit it to the orchestrator, which takes care of the instantiation process. It is safe to assume that the environment is composed of many services, especially given the typical class phases, which are preparation of tools and tasks, instruction, distribution of the tasks, collection of results, grading and providing feedback, and – finally – persisting the process outcomes. In a typical scenario, the class is expected to use at least an audiovisual connectivity service, a file storage, a learning management system, and a sort of virtual notebooks, which can be treated as separate services. Depending on the topic, additional services may be taken into account, including groupware and collaboration tools. Additional backend services, e.g., authentication, authorization and accounting service are also sure to be used, because the environment's use is both time and user constrained.

The main design guideline of EEDS was to provide an easy to use provisioning mechanism that would cover the details of instantiating complex, cloud-based collaboration environments. That was achieved by providing a common layer of abstraction, expressed in the form of CIMs. That is an important advantage from the point of view of educators who are not adept to rapidly changing contemporary cloud technologies. The CIMs, specified by the professors, undergo conversion into respective PIMs, which in turn are converted to a single or multiple Platform Specific Models (PSMs). Finally, a single PSM is selected for execution by administrative policies. The process is depicted in Fig. 3.

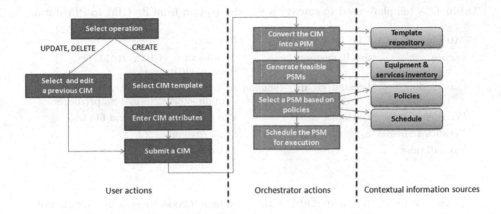

Fig. 3. Simplified workflow of CIM processing.

The process begins when an educator – using a web interface – specifies the model to be processed, as well as the operation to be performed by the EEDS orchestrator. There are three operations that can be selected: creation, modification or deletion of an environment. In the following considerations we focus on service creation. Once the data set describing attributes of a CIM is complete and valid, the request can be processed. Figure 4 contains a fragment of a web interface used to define a CIM.

Service Details

Lessons Type	Lesson topic	Service Size
● IT	○ Networking	● Small
○ Maths	○ Cybersecurity	○ Medium
○ Biology	○ Programming	○ Large
○ Physics	● Operating Systems	
	○ Neural Networks	

Fig. 4. CIM details specification interface.

Using a three-models structure is important, because it allows for handling multiple implementations of the same abstract service, as the PIM – though technical – is by definition technologically agnostic. The conversion process is carried out by using templates, which convert descriptive CIM attributes to technical PIM attributes. The resulting model is still abstract, and can be mapped to one of as many implementations as the particular system offers. The conversion rules are defined by the contents of the template repository. Table 1 shows an excerpt of such rules regarding VM configuration.

Table 1. A template used to convert a VM description from its CIM to PIM form.

Attribute	CIM form	PIM form
vmSize	Small/medium/large	Architecture, vCPUs, RAM, OS, storage, virtualization
Cooperation	Isolated/groups/ common	Admin and users' accounts, groups, data directories, filesystem privileges
Persistence	Datastore name	File transfer protocol and file URI
Network addresses	n/a	Static/dynamic
Firewall rules	n/a	Rules to be applied

In the example shown in Table 1 the educator needs to specify the size of a service (which should be related to the number of students participating in the online class), the mode of cooperation - isolated students, isolated groups or one large group, and a name of datastore to hold the results of the students' work. Those values are converted to more technical ones, but the rules of conversion are not stiff, rather they depend on the subject of the course. For example, different values regarding virtual machine size are assigned to a programming course, and different to a statistical analysis course. At the same stage of conversion, usernames are generated, the data store name converted into a file transfer protocol name and URLs for the files. Moreover, subnet IP address, gateway, DNS, and firewall rules are generated. Note that the requesting user does not need to deal with the network-related details, which – from our observation – is also an important hurdle for many teachers.

After converting the descriptive attributes to technical ones, all PSMs which describe the possible implementations of the educational environment, are generated. In case the equipment and services inventory contains multiple implementations of the requested services, multiple PSMs are generated. Again, the template repository is queried for converting the PIM attributes to attributes specific for the implementation (the respective template contains environment variable names, command line arguments, etc.).

Eventually, administrative policies are applied, and service schedules are consulted to select a single PSM. First of all, the screening policy decides which of the feasible PSMs are within the privileges of the requesting user. Then, the selection policy checks which of the PSMs can be run on the infrastructure at the specified time period, and whether or not it requires preemption of previously scheduled tasks. Finally, a single PSM is scheduled for execution.

In the event all the feasible PSMs do not comply with the policies, the process ends and the requesting user is properly notified by an e-mail message. In order to allow for reacting (e.g., by submitting a modified request), the EEDS defines the minimum time before execution threshold, after which the requests cannot be proceeded. That allows also for including human work in the environment preparation phase in the future.

Once selected, the PSM describing the compound service requested by a teacher is scheduled in a priority queue. After taking such a model out of the queue, the orchestrator generates code for the specific platform, execution of which results in provisioning the requested environment. After the requested usage time, the orchestrator takes care of persisting the results of students' work and destroys the environment.

4 Proof of Concept Implementation and Evaluation

This section describes a proof-of-concept implementation of EEDS[6] – a system that complies with the architecture described in Sect. 3.2, designed for teaching information technologies, but not limited to that area. We start the discussion with a brief description of the implementation, including technologies used in the process (Subsect. 4.1) to show that the architecture can be implemented using open technologies only. Then we describe the Małopolska Educational Cloud project (Subsect. 4.2), which may benefit from the service in the future, and which infrastructure was used during evaluation. We continue with describing the environment in which the evaluation was performed (4.3), and present some of the functional evaluation results (4.4) and processing time measurements (4.5).

4.1 Proof of Concept Implementation

The frontend of the EEDS prototype was implemented using the LAMP Open Source software package (Linux, Apache, MySQL, PHP) which makes up a web service stack to dynamically provide the webform and have control of the request reception system. The stack architecture is based on the Presentation-Business-Data architecture – it distinguishes between three independent modules implemented with different technologies but connected together.

We considered two options for representing the templates and policies: YAML and JSON. We chose YAML mainly because it is visually easier to understand and therefore facilitates the readability of the data (which leads to easier error detection). The readability is especially useful in defining policies, such as presented in the following excerpt, which defines a screening policy for a teacher:

```
maximums:
  teacher:
    small:  {users: 15 , groups: 5  , availability: 7 }
    medium: {users: 25 , groups: 8  , availability: 3 }
    large:  {users: 35 , groups: 11 , availability: 1 }
```

The prototype is capable of setting up educational environments using either Vagrant (to create a workflow to provision a virtual machine) or Docker (for deployment of containers). The design and architecture of the EEDS are open and allow integration of other environments. The prototype was configured to instantiate two educational environments, based on The Littlest JupyterHub (TLJH), and Antidote SelfMedicate, respectively.

---

[6] EEDS Repository: https://github.com/llopisga/cloud-orchestration.

TLJH was chosen as a platform for providing various types of lessons. The software was designed to handle 100 users on a single machine, so it is able to satisfy most of the educators' requests in that aspect. TLJH is a lightweight, Docker-based solution, capable of being adapted for arbitrarily chosen topics, presented in the form of Jupyter notebooks [1]. Figure 5 depicts a TLJH-based lesson interface.

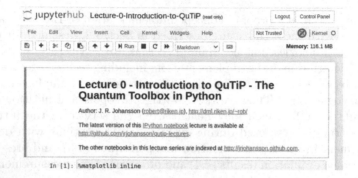

Fig. 5. TLJH service instantiated for a quantum physics lesson

Antidote (an open source project developed by NRE Labs) aims to facilitate learning network automation, and is provisioned by EEDS as a VM set up with Vagrant. The service splits the lesson window into two sides, containing the lesson text content and a Bash terminal, respectively (see Fig. 6). Antidote Selfmedicate allows to implement an identical service by spinning up a virtual machine with Vagrant, to which lessons can be added using YAML configuration files. The lessons may refer to different topics that involve the use of the terminal. Each Antidote-based environment is made up of four layers:

- infrastructure - virtual machine run on Virtualbox,
- MiniKube - a single-node cluster used for providing the lessons on demand,
- Antidote platform - for loading the lessons and providing web interface,
- curriculum - where the lessons are specified in YAML.

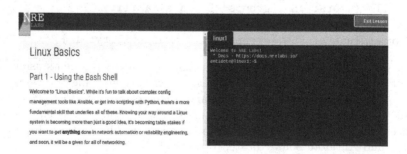

Fig. 6. Antidote service instantiated for a Linux basics lesson

4.2 Małopolska Educational Cloud

Małopolska Educational Cloud (MEC) is a community formed by 20 university departments, more than 120 high and vocational schools and several other institutions, including 11 pedagogical libraries and 7 teacher excellence centers. The community was formed to foster collaboration between universities and schools scattered over the Małopolska region (see Fig. 7). By participating in MEC activities, high school students learn more about particular universities and areas of study before they make decisions regarding their further education. MEC partners organize regular courses on various topics (e.g., information technologies, civil engineering, vocational English), each of which lasts at least one semester. The courses are led collaboratively by university and school teachers, the former being responsible mainly for online classes, the latter for offline activities. The rationale for implementing MEC and user activity analysis is discussed in [15,16] in more detail.

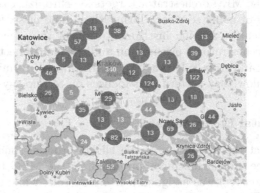

Fig. 7. MEC coverage as of Feb 2021; the numbers denote the quantities of network devices installed at respective locations.

From the beginning, MEC activities are split into:

- **didactic activities**, led by respective groups of institutions, and
- **infrastructural activities**, led by AGH University, the leader of the MEC community.

Didactic Activities. Each of the MEC high schools participates in at least two interest groups which organize online courses. Currently, MEC supports over 50 such groups. The groups – consisting of one university and up to five schools – organize online courses every semester. Typically, a group is established for a period of two semesters and involves about 100 high school students.

Infrastructure for Resource Pooling. MEC resources are offered to the participating schools according to either IaaS or SaaS paradigms. AGH University, the leader of the project, runs a private cloud hosting most of the services on its premises, providing participants with access to storage and computational resources. Additionally, users may leverage public clouds of their choice.

MEC implemented a dedicated overlay network, connecting all project participants. They use the overlay to access MEC services hosted in a private cloud. The list of MEC services includes: audiovisual connectivity, recording and storage, collaborative editing, etc., to be used during or after online classes. The outcomes of the online classes, as well as other materials prepared by or for students, are kept in a social media portal, which plays the role of a virtual notebook (see Fig. 1). MEC does not provide any unified LMS - the respective institutions use systems of their choices.

MEC is developing its own orchestrator of services, conceptually similar to the one presented in the article. However, a few differences are present. First, the goal of MEC orchestration is to provide an interface for moderating online classes led for people coming from many institutions. That is outside EEDS scope of interest. Second, the MEC orchestrator targets synchronizing the changes in the operational modes of many services, during the online class, while EEDS does not, as it would require specific instrumentation of the orchestrated services. Third, the MEC orchestrator is oriented on audiovisual connectivity as the most important service, and tries to leverage the hardware AV terminals capabilities. EEDS does not follow that pattern, but focuses on technology openness.

4.3   Experimental Environment

In order to conduct the evaluation of the EEDS proof-of-concept implementation, we used three virtual machines provided by the MEC IaaS service. One of the machines (master) was assigned a frontend role, while the others (workers) formed a small pool of servers which were used to instantiate the requested educational environments. Table 2 contains technical specifications of the VMs.

Table 2. Technical parameters of testbed VMs

VM	Master	Worker 1	Worker 2
vCPUs	2	4	4
RAM	4 GB	64 GB	64 GB
Storage	16 GB	16 GB	16 GB
Operating system	CentOS 8	CentOS 8	CentOS 8
Virtualization	IVT	IVT	IVT

The machines were accessed by the MEC VPN, and were accessible from the schools' internal networks by using the aforementioned MEC overlay network.

4.4   Functional Evaluation

At first, we conducted functional evaluation of the EEDS prototype. Among other features, we tested the orchestrator capability to detect policy breaches (note that policies are applied after the feasible PSMs set is generated). One of the tests that were conducted used four requests, described in Table 3. As expected, requests 1,2 and 4 were accepted and resulted in deployment of environments, while request 3 was rejected due to a policy breach (see Fig. 8 for rejection message contents). The characteristics of the respective environments generated by the successful requests 1, 2 and 4, are summarized in Table 4.

Table 3. Most important attributes specified in the initial test requests.

Attribute	Request 1	Request 2	Request 3	Request 4
Machine size	Small	Medium	Medium	Large
Course	IT	Biology	Maths	IT
Topic	Operating systems	Bioinformatics	Calculus	Networking
Users	8	14	26	14
Groups	4	Isolated	Common	Isolated

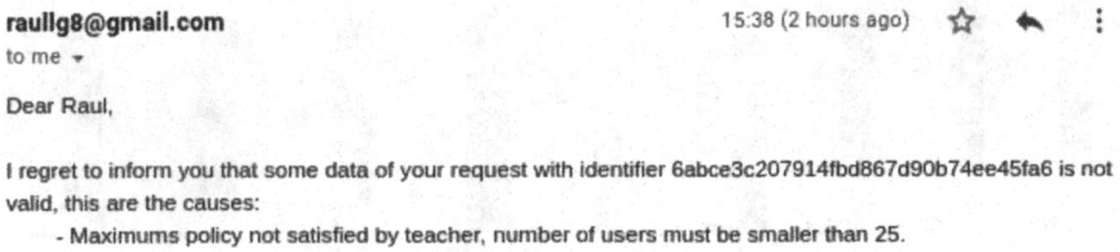

raullg8@gmail.com   15:38 (2 hours ago)   ☆ ↰ ⋮
to me ▾

Dear Raul,

I regret to inform you that some data of your request with identifier 6abce3c207914fbd867d90b74ee45fa6 is not valid, this are the causes:
    - Maximums policy not satisfied by teacher, number of users must be smaller than 25.

Fig. 8. Automatically generated message indicating a policy breach.

Table 4. Most important attributes of the platforms generated by requests 1, 2 and 4.

Attribute	Request 1	Request 2	Request 4
Dest. node	node02	node02	node01
Service	Antidote	Jupyter	Antidote
Virtualization	Vagrant	Docker	Vagrant
Resources	2 vCPUs, 8GB RAM	4 vCPUs, 16 GB RAM	8 vCPUs, 32 GB RAM

4.5   Processing Time Measurements

We measured deployment and lesson loading times for both kinds of environments the EEDS prototype was capable of creating, and for different machine sizes. The machine sizes and key characteristics are summarized in Table 5.

Not surprisingly, both the deployment and lesson loading times decreased with increased machine capabilities. As depicted in Fig. 9, the creation times of the Antidote machines were up to 15 min - most of the time was spent on downloading the needed packages from Internet repositories. The time it took to load a lesson (the Linux Basics lesson) was decreasing from slightly over a minute to about 30 s. The deployment of TLJH took about 5 min, and the respective lesson loading time was less than 10 s.

Table 5. Characteristics of environments instantiated by EEDS prototype.

Attribute	Antidote 1	Antidote 2	Antidote 3	Antidote 4	TLJH 1	TLJH 2	TLJH 3
Machine size	Small	Medium	Medium	Large	Small	Medium	Large
vCPUs	2	2	4	6	2	4	8
RAM	4 GB	8 GB	8 GB	12 GB	8 GB	16 GB	32 GB
Virt. provider	Virtualbox	Virtualbox	Virtualbox	Virtualbox	Docker	Docker	Docker

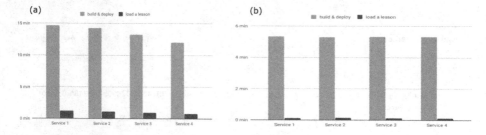

Fig. 9. Deployment and lesson load times for a) Antidote and b) TLJH services.

5   Conclusions

The described work proved that providing an MDA-based platform for deploying educational environments for online classes on arbitrary topics is feasible, and can be accomplished using open technologies only. The presented prototype of the EEDS system proved to be capable to instantiate the declaratively specified environments in a short time, measured in minutes. Therefore we claim that the main goals were achieved. Nonetheless, we see fields for improvement – which we set as our development goals – that could make the system usable on a larger scale. One of them is integrating the system with users' registry and an external authentication and authorization service to free the educators from the task of distributing user credentials. Second on the list is the integration with an LMS. The third goal is to provide the system with a monitoring service that would provide for reflecting upon the services' current and future use.

References

1. A gallery of interesting Jupyter Notebooks. https://github.com/jupyter/jupyter/wiki/A-gallery-of-interesting-Jupyter-Notebooks. Accessed 10 Feb 2021
2. OMG Model Driven Architecture (MDA) Guide rev. 2.0, OMG Document ormsc/2014-06-01 (2014). https://www.omg.org/cgi-bin/doc?ormsc/14-06-01.pdf. Accessed 10 Feb 2021
3. Auer, M.E., Villach, C.: A Toolkit to Facilitate the Development and Use of Educational Online Laboratories in Secondary Schools. Seattle, Washington (2015)
4. Brown, M., McCormack, M., Reeves, J., Brook, D.C., et al.: 2020 Educause Horizon Report Teaching and Learning. Technical report, EDUCAUSE (2020)
5. Cadenas, J.O., Sherratt, R.S., Howlett, D., Guy, C.G., Lundqvist, K.O.: Virtualization for cost-effective teaching of assembly language programming. IEEE Trans. Educ. **58**(4), 282–288 (2015)
6. Costa, R., Pérola, F., Felgueiras, C.: μLAB A remote laboratory to teach and learn the ATmega328p μC. In: 2020 IEEE Global Engineering Education Conference (EDUCON), pp. 12–13. IEEE (2020)
7. Demchenko, Y., Belloum, A., de Laat, C., Loomis, C., Wiktorski, T., Spekschoor, E.: Customisable data science educational environment: from competences management and curriculum design to virtual labs on-demand. In: 2017 IEEE International Conference on Cloud Computing Technology and Science (CloudCom), pp. 363–368. IEEE (2017)
8. Feisel, L.D., Rosa, A.J.: The role of the laboratory in undergraduate engineering education. J. Eng. Educ. **94**(1), 121–130 (2005)
9. Kim, B., Henke, G.: Easy-to-use cloud computing for teaching data science. J. Stat. Educ. **29**S1, S103–S11 (2021). https://doi.org/10.1080/10691898.2020.1860726
10. Lee, H.S., Kim, Y., Thomas, E.: Integrated educational project of theoretical, experimental, and computational analyses. In: ASEE Gulf-Southwest Section Annual Meeting 2018 Papers. American Society for Engineering Education (2019)
11. Lynch, T., Ghergulescu, I.: Review of virtual labs as the emerging technologies for teaching STEM subjects. In: INTED2017 Proceedings of the 11th International Technology, Education and Development Conference, Valencia Spain, 6–8 March, pp. 6082–6091 (2017)
12. Morales-Menendez, R., Ramírez-Mendoza, R.A., et al.: Virtual/remote labs for automation teaching: a cost effective approach. IFAC-PapersOnLine **52**(9), 266–271 (2019)
13. Perales, M., Pedraza, L., Moreno-Ger, P.: Work-in-progress: improving online higher education with virtual and remote labs. In: 2019 IEEE Global Engineering Education Conference (EDUCON), pp. 1136–1139. IEEE (2019)
14. Soceanu, A., Vasylenko, M., Gradinaru, A.: Improving cybersecurity skills using network security virtual labs. In: Proceedings of the International MultiConference of Engineers and Computer Scientists 2017, vol. II, IMECS (2017)
15. Zieliński, K., Czekierda, Ł., Malawski, F., Straś, R., Zieliński, S.: Recognizing value of educational collaboration between high schools and universities facilitated by modern ICT. J. Comput. Assist. Learn. **33**(6), 633–648 (2017)
16. Zygmunt, M., Konieczny, M., Zielinski, S.: Accuracy of statistical machine learning methods in identifying client behavior patterns at network edge. In: 2019 42nd International Conference on Telecommunications and Signal Processing (TSP), pp. 575–579. IEEE (2019)

A Collaborative Peer Review Process for Grading Coding Assignments

Pratik Nayak$^{(\boxtimes)}$, Fritz Göbel, and Hartwig Anzt

Steinbuch Center for Computing, Karlsruhe Institute of Technology,
Karlsruhe, Germany
pratik.nayak@kit.edu

Abstract. With software technology becoming one of the most important aspects of computational science, it is imperative that we train students in the use of software development tools and teach them to adhere to sustainable software development workflows. In this paper, we showcase how we employ a collaborative peer review workflow for the homework assignments of our course on Numerical Linear Algebra for High Performance Computing (HPC). In the workflow we employ, the students are required to operate with the git version control system, perform code reviews, realize unit tests, and plug into a continuous integration system. From the students' performance and feedback, we are optimistic that this workflow encourages the acceptance and usage of software development tools in academic software development.

Keywords: Peer review · Continuous integration · Collaborative learning · Sustainable software development

1 Introduction

With the digital revolution, much of the research, engineering, and production is no longer realized by the human workforce but automatized and controlled by computer programs. This radically changes the labor market and the skill-set wanted by employers. While a significant portion of the automate-able work will be handled by robots and computer programs in the future, other skills such as expertise in developing software and using the tools that enable sustainable software development will be important. Though there exist tutorials and talks discussing the use of development tools, experience tells us that only the practical use of the tools prepares the researchers for operating them in larger software projects. Against this background, we decided to expose graduate students to the practical use of sustainable software development tools by establishing a collaborative peer review concept for grading coding assignments.

In this paper, we elaborate on how we encourage the use of sustainable software development paradigms and enforce the usage of software development tools in a course on *Numerical Linear Algebra for High Performance Computing* offered at the Karlsruhe Institute of Technology. The course content includes

© Springer Nature Switzerland AG 2021
M. Paszynski et al. (Eds.): ICCS 2021, LNCS 12747, pp. 654–660, 2021.
https://doi.org/10.1007/978-3-030-77980-1_49

the design of efficient algorithms for basic linear algebra operations, direct and iterative linear solvers, hierarchical methods, preconditioners, etc., concentrating on massively parallel architectures such as multi-core CPUs and GPUs. By the end of the course, we expect the typical student to be able to use the techniques of parallel programming they have learned in the course and apply them to their thesis projects and their future coding endeavours. Additionally, we want to provide them with some knowledge about paradigms for sustainable software development and give them hands-on experience in using tools that are popular in the realization of software projects. We reiterate that the course does not focus on programming tools and software engineering paradigms but on the design of high performance computing algorithms, and the course content can be taught without touching the topic of sustainable software development. But we use our first-hand experience in large software efforts to propagate software sustainability paradigms and require the use of collaborative software development tools in the homework assignments.

Before detailing in Sect. 3 the peer review workflow we employ for the homework assignments, we provide in Sect. 2 some background about the tools we use for this. In Sect. 4 we discuss our experience with the approach, and conclude in Sect. 5.

2 Background

For sustainable software development, a set of tools has proven to ease software development and maintenance. While there exist tutorials, textbooks, and seminars on the use of those tools, we want to review some of them that we consider most important and thus include in our homework peer review workflow.

Version Control Systems (VCS). Version control systems are a popular way to manage codebases. They started as a snapshot capability, enabling the developer to roll-back the code history, thereby easing debugging and maintenance. They evolved as powerful platforms for collaborative software development. Some of the popular open-source version control systems are: Revision Control System (RCS), Concurrent Version Control (CVS), Subversion (SVN), Mercurial, and Git.

Figure 1 shows a typical version control workflow with the git version control system. In this workflow, a developer can create a feature branch from the main branch, add new functionality or change existing functionality, and create a merge request (MR) to merge the feature branch back to the main branch. The process where peers inspect and criticize the code changes and provide feedback to the developer is called a *code review*. This step may appear tedious, but is one of the most important components of sustainable software development. It is essential that both the feedback-giving peer and the developer take the code review seriously, and the developer acknowledges and adopts the reviewers' comments. Once all flaws and improvement suggestions are taken care of, the reviewers approve the merge request and the changes are merged into the main branch. The git version control system orchestrates the merge in case multiple developers want to merge to the main branch and enables the developers to retrieve new changes from the main branch into the feature branch.

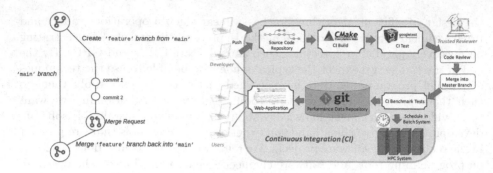

Fig. 1. A typical git workflow [1].　　**Fig. 2.** A software development workflow [8]

Continuous Integration. With the increasing complexity of the coding tool-chain, from different hardware platforms to the different build systems and the different compilers that need to be supported, it becomes increasingly challenging to ensure code robustness. Continuous integration (CI) aims to improve the reliability of a codebase [2,3,5] by automatically checking the code correctness after a change to the codebase has been pushed through the version control system. Specifically, the CI replicates the complete code usage process from source code cloning to the compilation of the code in a specified hardware-software environment. This is important in particular if the software is developed as a collaborative effort, and distinct developers are independently making changes to the codebase. The goal of CI is to ensure the permanent compile-ability of the codebase on the target system and to notify the developers about compilation issues. In addition to verifying whether the codebase compiles, CI systems usually also employ an automated testing workflow.

Automated Testing. Automated software testing pursues the idea of having an automatic mechanism that verifies the correctness of all functionality of a software ecosystem. Of course, this is a very complex goal, as there exists a large number of use cases and functionality combinations. Ultimately, testing all combinations, and all end-to-end user applications would be infeasible in terms of effort and time. Therefore, one typically uses a hierarchical testing strategy: On the lowest level, each basic functionality is accompanied by a *unit test*. This function-specific unit test verifies the correctness of the basic routine for all possible input and output scenarios. A failing unit test can easily point the developer to the part of the code that needs debugging.

As the correctness of all basic functionality (passing unit tests) does not guarantee the correctness of functionality combinations, unit tests are usually complemented by *integration tests* that form the second level of the testing hierarchy. These tests verify the correctness of different functionality combinations. Often, it is impossible to cover all combinations due to the sheer quantity of the possible permutations. Therefore, integration tests usually only cover the most popular functionality combinations.

Finally, the third level of automated testing is formed by *end-to-end tests*, which try to emulate a user's workflow. As it is again impossible to emulate all possibilities, end-to-end tests only emulate a few, typical use cases. For all levels of automated testing, there exist sophisticated tools such as GOOGLETEST [7] and the CATCH2 TEST FRAMEWORK [6] that provide frameworks.

Services in the Cloud. Gitlab and Github offer platforms that allow users to run Continuous Integration pipelines and automated tests on remote servers in the cloud. They also offer web interfaces for interactive and collaborative coding using ideas such as Merge Requests (MRs). Using these services removes the need to procure and run servers that provide version control systems, continuous integration, and automated testing.

3 Methodology

Numerical Linear Algebra for High Performance Computing (HPC). The course we offer at KIT is aimed at Masters students seeking to learn about computational numerical linear algebra methods suitable for computational science and its realization in HPC settings. To encourage the students to apply these techniques in larger projects, we require a final course project instead of an end-of-term examination. We offer some project ideas, particularly to enable them to contribute to GINKGO [9], but also encourage the students to come up with ideas or extend their thesis works. Additionally, the students have to complete several homework assignments where they are required to develop a parallel version of an algorithm, implement and run benchmarks on an HPC architecture, and analyze the algorithm performance in a report.

To prepare the students for collaborative coding, an exercises framework is provided in a version control system which handles the automatic compilation and testing of the code written by the students [10]. This allows the students to focus on the algorithms and parallel programming ideas rather than spending time on the build system. An additional advantage is the uniformization of the build and execution process for all students making it easier to debug and help the students.

The Peer Review Process. Figure 3 shows a schematic of this peer review process. Each student creates a fork of the exercise framework and a student-labeled sub-folder from the main folder in which they add their implementations. On the submission date, the CI is run (which compiles and runs their code with the unit tests) and the students create an MR for their exercise. The MRs are assigned to their peers in a round-robin fashion, and with the help of the Gitlab interface, they can provide feedback to their assigned peer. They are encouraged to follow code reviewing guidelines [4] while reviewing their peers' code. The review process is iterative, and the students are encouraged to update and enhance their code according to the feedback received. After approval by the reviewing peers, the students' codes are merged into the main repository to replicate an actual project workflow.

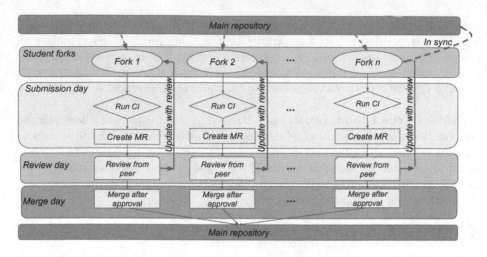

Fig. 3. The peer review process

Figure 4 shows a snippet of a review by Student 1 of the code written by Student 2. Student 1 is assigned to the MR opened by Student 2 and can provide comments on specific lines of the code based on a diff (the difference between the code in the target branch against the code added by the student).

Grading. In addition to the coding assignments, the students are required to work on a final course project. We split the final grade into three parts: 40% of the grade is made up by the project code and report, 30% is assigned for a short (approximately 10 min) presentation on the student's project, given at

Fig. 4. A typical peer review comment

the end of the term, and the remaining 30% of the grade is made up by the homework assignments. (70% of the points are required to pass the course.) Each homework assignment carries a maximum of 10 points. A performance analysis report makes up 4 of the 10 achievable points. Another 4 points are assigned for the code, where 2 points are awarded for functioning code, 1.5 points for the code quality, 2 points for the quality of the review a student provides, and 0.5 points for incorporating the review feedback that they receive from their peers.

4   Discussion

The motivation for establishing this peer review process for the homework assignments is that we are convinced that the practical experience in using software development tools, CI, and adhering to a peer review process is an essential skill for graduates entering the software-focused labor market.

In addition to equipping students with practical experience in using tools for sustainable software development, we observed that the framework allowed students to concentrate on the algorithms and their implementations rather than concerning themselves with tackling the issues from the build systems. We also observed a significant improvement in the code quality of most students throughout the course with students acknowledging the feedback received in the peer review process in the concurrent exercises, which has not been the case in previous versions of the course.

Table 1 shows the feedback we received from the students on the homework workflow. The questions are rated from 1 to 5 with 1 being the best rating meaning that it was very useful/very easy and 5 being the worst rating meaning that it was not useful at all/very difficult.

Table 1. Student feedback

Question	Avg rating (1–5)
How easy was it to use the framework?	2
How useful did you find the exercises instructions?	2
How easy was it to compile and run the code as provided?	2.3
How useful was the code review from your peer?	1.6
How easy was the reviewing process?	3.6
Would you like to see this type of frameworks in other courses?	1

5   Conclusion

With the wide and easy access to computing and the rise of Open-Source software, it is necessary that we train our students to be familiar with tools for sustainable software development. In this paper, we elaborate on how we introduced a collaborative peer review workflow for the homework assignments in a course on high performance computing. In the future, we plan to enhance the framework with automatic benchmarking of the coding assignments and test the workflow's viability for other courses.

Acknowledgements. The authors were supported by the "Impuls und Vernetzungs-fond of the Helmholtz Association" under grant VH-NG-1241. The authors would also like to thank Jan-Patrick Lehr of TU Darmstadt for his helpful discussions and perspectives on this subject.

References

1. Git cheat sheets. https://training.github.com/
2. GitLab CI/CD. https://docs.gitlab.com/ee/ci/
3. Jenkins CI. https://www.jenkins.io/index.html
4. The standard of code review. https://google.github.io/eng-practices/review/reviewer/standard.html
5. Travis CI - test and deploy your code with confidence. https://travis-ci.org/
6. Catch2 - testing framework. https://github.com/catchorg/Catch2
7. Googletest - testing framework. https://github.com/google/googletest
8. Anzt, H., et al.: Towards continuous benchmarking: an automated performance evaluation framework for high performance software. In: Proceedings of the Platform for Advanced Scientific Computing Conference, PASC 2019, pp. 1–11. Association for Computing Machinery, New York (June 2019). https://doi.org/10.1145/3324989.3325719
9. Anzt, H., et al.: Ginkgo: a high performance numerical linear algebra library. J. Open Source Softw. (August 2020). https://doi.org/10.21105/joss.02260
10. Nayak, P.: Exercises framework. https://github.com/pratikvn/nla4hpc-exercises-framework

How Do Teams of Novice Modelers Choose an Approach? An Iterated, Repeated Experiment in a First-Year Modeling Course

Philippe J. Giabbanelli[1]([✉])[iD] and Piper J. Jackson[2][iD]

[1] Department of Computer Science and Software Engineering,
Miami University, Oxford, OH 45056, USA
giabbapj@miamioh.edu
[2] Department of Computing Science, Thompson Rivers University,
Kamloops, BC V2C 0C8, Canada
pjackson@tru.ca
https://www.dachb.com

Abstract. There are a variety of factors that can influence the decision of which modeling technique to select for a problem being investigated, such as a modeler's familiarity with a technique, or the characteristics of the problem. We present a study which controls for modeler familiarity by studying novice modelers choosing between the only modeling techniques they have been introduced to: in this case, cellular automata and agent-based models. Undergraduates in introductory modeling courses in 2018 and 2019 were asked to consider a set of modeling problems, first on their own, and then collaboratively with a partner. They completed a questionnaire in which they characterized their modeling method, rated the factors that influenced their decision, and characterized the problem according to contrasting adjectives. Applying a decision tree algorithm to the responses, we discovered that one question (*Is the problem complex or simple?*) explained 72.72% of their choices. When asked to resolve a conflicting choice with their partners, we observed the repeated themes of mobility and decision-making in their explanation of which problem characteristics influence their resolution. This study provides both qualitative and quantitative insights into factors driving modeling choice among novice modelers. These insights are valuable for instructors teaching computational modeling, by identifying key factors shaping how students resolve conflict with different preferences and negotiate a mutually agreeable choice in the decision process in a team project environment.

Keywords: Agent-based model · Cellular automata · Education · First-year experience · Team-based modeling.

We thank faculty at Furman University (Kevin Treu, Paula Gabbert, Bryan Catron, Chris Healy, Andrea Tartaro, Chris Alvin) for many fruitful discussions on teaching a topic-based introductory course.

© Springer Nature Switzerland AG 2021
M. Paszynski et al. (Eds.): ICCS 2021, LNCS 12747, pp. 661–674, 2021.
https://doi.org/10.1007/978-3-030-77980-1_50

1   Introduction

For a given project, modelers can choose from a very large number of modeling techniques, both discrete (e.g., Agent-Based Modeling, Cellular Automata) or continuous (e.g., System Dynamics). *Ideally*, this choice would be driven by the availability of data, the project scope negotiated with the end-user, or the model performance given available computing resources [3]. However, all of these factors overlook an essential aspect of the model-building process: models are created *by modelers*. As most available literature provides little justification on the choice of a particular modeling approach, our previous study performed a survey of practitioners and found that the selection of methods primarily depended on factors related to the modelers, as 92% of respondents admitted that they chose the most familiar method [34]. However, 87% of respondents also declared that they chose a method based on the problem characteristics. This paradox suggests that a modeler may look at a problem and then decide on one method, convinced that it is the best fit, while a modeler with a different experience would make a case for another method in the same context. These two choices may have to be reconciled, since computational modeling is often constructed as a team-based discipline. In this paper, we examine the process through which novice modelers within the educational setting of a first year course negotiate the choice of a modeling approach.

A survey of 51 professionals in simulation education showed that 95.2% of programs on simulation education include a project, which involves a team in 92.1% of cases [20]. It is thus very common for students to be faced with the problem of reconciling multiple viewpoints when designing a model collaboratively, starting with the choice of a modeling approach. Although there has been research for several decades on how people behave when building a model, much of the focus has been on scenarios involving a single modeler [30,37] rather than collaborative settings. When a team is studied, the focus may be on the dialogues between subject-matter experts (also called 'domain experts') and modelers rather than among the modelers themselves [8]. Similarly to Peter Rittgen, we conducted experiments involving groups of students who were provided with a textual description of four problems and asked to model them by choosing among techniques [23]. While Rittgen's experiment involved ARIS-EPC, Petri Nets, UML, DEMO, our approach focused on the choice between two closely related techniques: Cellular Automata (CA) or Agent-Based Models (ABM). Our experiments were repeated over two classes and used a scaffolding technique starting with independently choosing and justifying a model, then reviewing the choices made by a student with different arguments, and finally working in pairs to reach a consensus.

Our three main contributions are as follows:

(1) Through experiments, we identify a *core set of three problems that lead to high divergence among students*. These problems can be used by instructors to intentionally maximize divergent thinking and practice skills in collaborative problem solving for an introductory course on computational science.

(2) By applying machine learning on experimental data, we show that the initial choice of agent-based models or cellular automata is primarily motivated by the perceived level of complexity of the problem.

(3) Through a thematic analysis of narratives by teams, we explain how students have a different perception of agent-environment interactions compared to interactions among agents. When a problem evokes interactions with the environment (e.g. through mobility), students choose Agent-Based Modeling; conversely, the absence of such interactions often justifies their use of Cellular Automata. In contrast, interactions among agents can lead to either option through the notion of networks or neighborhoods.

The remainder of this paper is organized as follows. In Sect. 2, we introduce the setting in which we taught the course, including the teaching philosophy, core material, and institutional factors. In Sect. 3, we explain how we designed experiments and which measurements were recorded at each step. Our results in Sect. 4 are subdivided to focus on problems leading to different choices, explaining individual choices, and examining the process of co-creation as a pair. Finally, we contextualize our findings in the broader domain of collaborative problem solving in computational science and discuss the potential for applications to other areas such as processes in organizations.

2 The Setting: Teaching Philosophy and Implementation

The course was offered at Furman University, which is a liberal arts institution located in South Carolina, USA. The Computer Science Department used topic-based introductory courses to introduce key concepts of computer science to both majors and non-majors. The motivation is to "contextualize computing in a real-world, interdisciplinary problem upfront, and show how a variety of computer science topics apply to solving problems in that context" [32].

In the same manner as personal preferences and experiences shape a modeler's actions, simulation education depends on the teaching philosophy of the instructor [29]. Our teaching philosophy for this course, titled 'CSC 105: Virtual Worlds', rests on the four objectives presented at ICCS2016 [11]:

① We provide an *overview of the field followed by three modeling techniques*, covering both individual- and aggregate-level models. Our implementation of this objective was similar to courses at peer institutions, such as 'Modeling and Simulation for the Sciences' at Wofford College [27]. Specifically, students were exposed to system dynamics as aggregate models, then to cellular automata and agent-based models as individual-level models. In contrast to higher-level classes such as 'Introduction to Computational Modeling and Data Analysis', we do not include topics such as Markov chains or coupling models (i.e. hybrid modeling) [28].

② We cover *one programming environment*, starting from basic syntax. We complemented conceptual lectures by using the widely adopted `NetLogo` in weekly hands-on labs involving paired programming. We also emphasizing best practices from a software engineering standpoint [33].

③ We practice through *interdisciplinary projects*, based on years of experience in interdisciplinary curricular activities [12]. This objective is also made necessary given that the course includes the general student population, who may not major in computer science. Projects provide opportunities for creative expression, which we see as a cornerstone of a student-centered approach [9].

④ We develop *critical thinking skills*. In line with departmental practices [31], we used scholarly readings. The emphasis was on identifying limitations and on contrasting studies rather than in original writing, which is covered in a separate first-year seminar course [2].

As part of the learning objectives for the course, students should be able to choose between Cellular Automata (CA) and Agent Based Models (ABMs) in a given application context, program their model in NetLogo, and analyze simulation results. At a high-level, CA and ABMs are similar as they are both *discrete*, *individual-level* modeling techniques [1]. Consequently, students have to identify an initial configuration for the individual entities (e.g., initial state for the *cells* and/or baseline values for the *agents*), provide update rules that are applied to these entities at discrete 'ticks', and decide when the simulation should end. In a CA, the update rules can change the *state* of each cell based on neighboring cells, time, or probabilities. Examples include biological models such as the spread of an infection within a body [18] and the growth of tumor [19], or geographical models such as forest fires [6] and land use [36]. An ABM optionally includes a CA, which may serve to represent a physical substrate such as a soil model. An ABM necessarily includes agents, which can interact with the space (e.g., animals foraging) and/or with each other. Agents may be equipped with elaborate anthropomorphic notions, such as manipulating others, making errors, or having a range of personalities [15]. As an ABM requires more data (for calibration and validation) and a deeper theoretical understanding than a CA (to craft rules), applied computational models in fields such as obesity have gradually shifted from CA in early research to ABMs as they gained maturity [13]. Given the introductory nature of the course, we focused on homogeneous and synchronous CA [26] (e.g., all cells are updated at the same time) and we did not cover the connection between ABM and geographical information systems [16].

3   Experiments and Measurements

The course was offered on two occasions (Fall 2018 and Spring 2019), which allowed for repeated measurements. Our experiment involved three consecutive parts (Fig. 1). First, students were given four problem statements and invited to create their own. In each of the five problems, a student independently chose to use either CA or ABM. Students had to argue for their choice and provide a complete design including a description of states, transitions, initial configuration, and condition to stop a simulation. In the second part, the instructor identified pairs of students who had made different modeling choices on at least some of the four shared problems. Each student received the anonymized submission from his or her pair-mate, such that they were unaware of each other's identity and hence unable to communicate directly. Each student reviewed the submission and sent it to the instructor, who then passed it onto the other student. Finally, the two students met each other and had to decide on *one* technique for each of their five problems. For problems in which they argued for different solutions in part 1, students had to explicitly write how they resolved the difference.

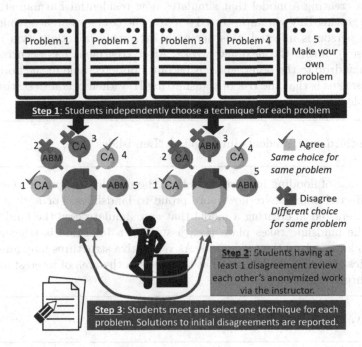

Fig. 1. Three steps process through which pairs of students resolve differences in their choice of ABM or CA.

In Fall 2018, the four problem statements started with "Spreading the flu in a classroom", described as follows:

A handful of hard-working students may make the selfish choice of coming to class when they know they have the flu, thus infecting others and causing problems throughout the community. This model seeks to capture how flu spreads in the classroom. We consider that students are either infectious, or not (e.g., susceptible, infected but not yet infectious, recovered). A student may cough/sneeze/talk, which (to really simplify) produces a cloud of infectious droplets invisible to the human eye. If a person walks through this cloud, the droplets can land in the mouth/nose or be inhaled into the lungs, and this person can get sick. Droplets survive in the air for a few hours.

The next problem statement was titled "Spreading HIV on campus":

The Centers for Disease Control and Prevention (CDC) announced last August that sexually transmissible diseases (STDs) in US reached record high. The Human Immunodeficiency Virus (HIV) is an STD. We are interested in creating a model that simulates how residential Furman students may be getting HIV via unprotected sex over the course of an academic year. Once a person is infected, this person is extremely infectious (via unprotected sex) during the next few weeks. After treatment has occurred, the virus may drop to the point where it becomes undetectable in the body, and the consensus is that the risk of transmitting HIV then becomes statistically negligible.

In the third case, students investigated "Landslides":

Because of flooding in North Carolina, the stability of some surfaces has been affected and they're now more prone to landslides. For a given land, we're interested in creating a model that can simulate how the land moves when the landslide takes place (i.e., assuming a landslide is triggered we want to know where the land goes). As vegetative structures may affect the dynamics of the landslide, note that the lands in the area of interest all have some shrubs and trees.

Finally, landslides were revisited in the last problem:

Assume the setting and goals are identical to problem (3) above. In addition, assume that landowners with peculiar hobbies also have bongos, oryxes, kudus, and lechwe grazing on the land. As they graze, they damage shrubs (but not trees) and thus remove some of the vegetative structures that hold the land together.

Problem statements were re-written in Spring 2019 to provide additional data, thus limiting the possibility that results are an artifact of our initial problem statements. The new problem 1 was "Peer-influence on smoking":

Smoking is partially driven by social norms, that is, whether peers smoke or endorse smoking. We want to model how students at Furman may choose non-smoking, vaping, or cigarette smoking. We are particularly interested in modelling peer influence on these choices within the academic community.

The second problem focused on "Animal migrations":

Animals migration are based on the availability of food, climate, and migration of other animals. For example, birds or grazing animals may want to be close to similar animals, but too many will result in lack of sufficient nutrients. We are particularly interested in modelling animal migrations in the Carolinas, where migrating species include many types of ducks, swans, snow geese, various other birds for fall coastal migrations (e.g., warblers, grosbeaks, tanagers, orioles, vireos), fall mountain migrations (e.g., hawks, eagles, falcons) or springtime mountain migrations (e.g., thrushes, flycatchers). As a model is a simplification, you are not expected to become an expert in bird ecology to answer this question!

The third problem examined "Shopping malls":

Retail sales in shopping malls are important to model: where do people go if we make changes in the mall? How can we promote traffic? How can we charge rent for a specific shop based on how much traffic it can get? To answer such questions, we want to model the influence of foot traffic, customers, location, and neighboring shops, on retail sales in a shopping mall. You can pick your favorite mall (e.g., Haywood mall) it it helps you to think of something concrete.

Finally, the fourth problem was about "Laughter":

Some people have a contagious laughter that will make others laugh too. We want to simulate how laughter may spread in a room full of people. Laughter originates from one person, who may or may not have a 'contagious laugh'. Then, others may laugh or not.

For all students, we collected information on their gender (male, female) and whether they liked students with whom they worked during paired-programming sessions (which does *not* include their pair-mate on the experiment). In Spring 2019, we also administered two questionnaires: one upon completion of step 1 to understand how students made their selection (e.g., were they confident? did they feel it was appropriate for the problem?), and the other upon completion of step 3 to characterize how they resolved differences (e.g., was it easy to come to an agreement? were they impacted by the strength of their partner's argument?).

4   Results

4.1   Which Problems Lead Modelers to Make Different Choices?

We were able to construct 8 pairs with at least one different modeling choice across the four problems (out of 9) in Fall 2018 and 4 such pairs (out of 7) in Spring 2019.

Table 1. Modeling choices of individuals and pairs across questions and semesters.

	Fall 2018 (9 pairs)				Spring 2019 (7 pairs)			
	Q1	Q2	Q3	Q4	Q1	Q2	Q3	Q4
Students using CA (%)	33.33	11.11	94.44	27.78	71.43	14.29	7.14	100
Students using ABM (%)	66.67	88.89	5.56	72.22	28.57	85.71	92.86	0.00
Pairs with differences	6/9	2/9	1/9	3/9	4/7	2/7	1/7	0/7
↪ *Number of differences*	4 had two diff., 4 had one				3 had two diff., 1 had one			
Pairs using CA (%)	44.4	0	100	11.1	85.7	0	0	100
Pairs using ABM (%)	55.6	100	0	88.9	14.3	100	100	0
Differences solved using CA	3	0	1	0	3	0	0	0
Differences solved using ABM	3	2	0	3	1	2	1	0

Our ability to create pairs that need to resolve differences depended on the extent to which each question led to a consensus. As shown in Table 1, the problem of "spreading the flu in a classroom" (Fall 2018 Q1) sharply divides students both as individuals and as pairs. The next questions leading to the most differences was "Peer-influence on smoking (Spring 2019 Q1) followed by our revisited landscape problem (Fall 2018 Q4). All other questions had less variations in modeling choices among individuals and no variation in pairs. This first result thus provides an experimentally established set of questions that can be used to promote differences in modeling choices.

4.2   Can We Explain How Modelers Choose?

We analyzed the rationale for the individual modeling choices based on the questionnaire administered in Spring 2019. For each of the five problems (four created by the instructor and their own), students characterized their choice of a modeling method (confidence in the choice they made? Was it easy to decide?), rated the extent to which five factors influenced their choice (complexity of implementation, preference for the modeling technique, appropriateness for problem, explanatory power, flexibility of model), and characterized the problem by choosing between pairs of adjectives (social/physical, complex/simple, rules/ideas, individual/group, spatial/relational, decisions/behaviors).

The data for analysis thus consists of 14 questions, answered by each respondent (n=11) for five problems thus forming 55 entries. We used the supervised machine learning approach of binary classification to explain the modeling choice (CA or ABM) as a function of the 14 questions. The baseline prediction accuracy achieved by a 0R rule (i.e. simply looking at how often students tend to choose CA or ABM without considering the questions) was 54.54%, thus any classification model able to truly explain choices based on the questions would have to outperform this baseline. Using a decision tree algorithm without the restriction of depth, we found that *one question suffices to correctly explain 72.72% of choices*: seeing a problem as simple or complex. Complex problems overwhelmingly resulted in ABMs (21 out of 26 times), simple problems in CA (28 out of

32 times), and problems that could not be characterized by either term were still ABM (6 out of 8 times). If the notion of simple versus complex was removed, then no combination of the remaining 13 questions had explanatory power as the accuracy fell to 52.72%, which is below the baseline.

We examined whether simple dynamics could predict whether the final modeling choice in a pair would follow the initial decision of one student or the other. Since gender is occasionally used as mediating variables in studies on teamwork, we analyze the final choice in a mixed pair would espouse the initial decisions of the male student. Due to a small number of mixed teams, we only noted this situation in 3 out of 5 pairs, which is close to parity. We also tested the theory that a team may be led by a more 'knowledgeable' student. However, in pairs with different final exam grades, differences were resolved in favor of the student with *the lowest grade* GPA in 7 out of 10 pairs. The problem of identifying the nature and impact of leadership in a team would require further studies and a larger sample size, thus allowing a more fine-grained analysis of whether demographics or personality factors play a role in driving the final decision of a modeling team.

4.3   How Do Modelers Co-Create Models?

In the survey administered upon resolution of modeling choices in a team, all students stated that resolving differences was very easy. To further characterize *how* differences were resolved, we examined the narratives provided by each pair on each resolution. In this section, we discuss the themes identified across narratives and briefly exemplify them through selected quotes in which students' *names were anonymized.*

Two themes were present across most narratives: mobility and decision-making processes. The need for an ABM was overwhelmingly motivated by the *perceived* need for entities to move over a space and/or engage in complex decision-making activities that require a cognitive framework. In contrast, the choice of a CA was justified by the perceived *absence* of these needs. For example, consider the flu problem. One group conceptualized students as "confined to one space within the classroom" and this absence of mobility resulted in a CA. Another group similarly debated the matter of movements:

> "*Simba claimed that [the flu] moved based on proximity because [it] cannot decide where it wants to go. Mufasa claimed that the flu was in the students who could walk around and make choices [...]. In the end, Mufasa agreed that the model is being used to show where the flu is and it is true that the flu cannot move so we agreed to explain this model as being a CA.*"   (Problem 1, Fall 2018)

In contrast, several other groups endorsed the hypothesis for the same problem that "agents can walk" thus leading to an ABM:

> "*Bernard believed that using Cellular Automata was the best approach and that we could have the students be stationary in an area, as if they*

were in sitting at desks. Bianca believed that using an Agent-Based modeling system was best because she has been in classroom environment that are hands-on more recently rather than lectures-based meaning the students would come in contact with others more often."

(Problem 1, Fall 2018)

An ABM is defined by interactions between agents and the environment as well as among agents. While the agents-environment interactions (e.g. through spatial mobility) were a recurring motivation for ABM, interactions among agents were less commonly discussed. In addition, the presence of these interactions was equally likely to motivate the use of a CA or an ABM, since both models include interactions (either through a neighborhood or via networks). For instance, consider the model of peer pressure over smoking. In one team, "the agents are interacting with other types of agents and also with the states of the environment, which are defining qualities of an ABM." However, for another team, these interactions can be handled by a CA:

"We believe that this situation is less about agent interaction and more about spatial orientation. The state of the smokers, non-smokers and vapers is more influenced by their neighbors than by individual decision making. Peer pressure often comes from people that are close to you, usually your friends. We will assume that friendship is stable and you will not randomly leave to find new friends. This implies that if you have many friends that smoke or vape it is more likely that you will smoke or vape. This is well illustrated by a Cellular Automata because while you interact with your own friends, they also interact with other friends who interact with other friends and so on and so forth."

(Problem 1, Spring 2019)

Through the narratives, we also notice that the resulting model and its justification is far from a one-sided triumph of one student's ideas over another. Indeed, students describe how features of the model are obtained *collaboratively*:

"We took things that Esmeralda's ABM could represent better and simplified them to fit the CA model. We used Quasimodo's base CA model and added those simplified elements for a more complete simulation, adding components like coughing and entering/leaving the classroom."

(Problem 1, Fall 2018)

"When we had peer reviewed each other's responses, we decided to try to combine each other's ideas (Mulan had the idea of cells being stable or unstable and Yao had thought of trying to show land movement through differences in elevation). This allowed us to be able to use both ideas to attempt a better model."

(Problem 3, Fall 2018)

5 Discussion

In order to improve the practice of computational modeling and simulation, it is essential to understand the process by which we construct models, from the initial stages of defining the problem, designing the model, implementing it in software and testing its behavior. Certainly, the hard-earned experience and well-developed expertise that guide the decisions of experienced modelers are an invaluable component of building successful models. However, novices provide an excellent opportunity to examine which other factors may be highly influential, especially from a fresh perspective that is not invested in a certain way of doing things. Our study employs a mixed methods approach to investigate many aspects of the initial decision process as experienced by two groups of new modelers. We found that, at least in the case of considering cellular automata versus agent-based models, the apparent complexity of the problem under consideration is the key determining factor in terms of which technique was chosen. We also identified common themes found in the justification of model choice in cases where there was disagreement about the ideal technique. For problems where mobility and/or decision-making were key aspects needing to be modeled, agent-based models were preferred over cellular automata.

The exercise described here provides a sample of how modeling choices can be studied in an instructional setting. One key component is the set of sample problems which are designed to provoke consideration and discussion within the frame of the modeling techniques being considered. Another important component is the kind of data gathered by the study. Demographic factors did not appear to be highly influential among our groups, but further experimentation could help to clarify this facet of collaborative modeling. This kind of exploration also prompts us as educators to consider what we need to be teaching alongside the technical aspects of model building. If problem complexity is a driving factor in decision making (as we have seen here), should we encourage the consideration of other aspects? Should we provide more guidance on how to consider complexity?

Thus the primary contributions of our work pertain to simulation education and the practice of computational science in the classroom. Our work can also be situated within the broader theme of education and training in Collaborative Problem Solving (CPS), which is often motivated by the fact that most professional work is accomplished by teams within organizations [14]. Oppl further highlights how human work in organizations presents several salient characteristics that are also found in our study context [21]. Organizational actors can reach a shared understanding by working on *shared conceptual models*. As stated by Oppl, "existing approaches in general assume that the contributing actors have existing modeling skills [but] actors operatively involved in a work process do not necessarily have these modeling skills" and the task of model creation cannot be left to a third-party expert since the active and direct involvement of actors in the modeling process is "beneficial for the collaborative construction of a shared understanding" [21]. Our work thus also contributes to a growing evidence base on the process of collaborative modeling and the negotiations that are involved [24].

A benefit of performing our experiments in one course at one institution is that we have a relatively consistent student population for analysis. However, this lack of diversity is a limitation when it comes to assessing the factors involved in shaping the co-creation of models by students of various levels, in different institutions, or involved in other curricula on computational modeling. In particular, the *experience* of students may be a mediated factor, since "conceptual modeling is often thought of as a skill that improves with experience" [35]. For example, studies on novice modelers by Powell and Willemain found several issues such as taking shortcuts [22,38]. More recent empirical studies on the creation of models by students have confirmed that experience leads to different patterns [17]. It would thus be of particular interest to complement our study of freshman (1st year, 1st semester) with follow-up examinations in courses focused on rising sophomore (2nd year, 2nd semester) or senior (4th year).

A second limitation of our approach is the reliance on a prepared in-person session requiring a joint decision. In other words, students had to reflect in detail on each other's proposed idea then met in-person with the objective of finding one modeling technique. Results may thus be different if there is less incentive to achieve a joint decision, which may affect the willingness of participants to engage in co-creation and find a consensus. Results could also be affected by a switch to a remote scenario, which introduces an element of technology-mediated collaboration (e.g. via Zoom, WebEx, and similar platforms). An asynchronous scenario may rely even more on technology, for example through software for distributed model negotiation such as COMA in which modelers can propose a model, support or challenge a proposal (by tracking arguments for/against), and view the latest agreed upon version [25].

Although our study has collected detailed qualitative and quantitative data on modeling choices and reconciliation, software provide additional opportunities to track the series of steps taken by modelers, for example via a replay function [7]. The time series of modeling steps can be of particular interest if the models are *structured* rather than provided as narratives, for instance by using flow diagrams to document transitions of states for both cells in a cellular automaton [10] or agents in an agent-based model. A structured graphical notation may be better aligned with the task of model creation [4] and it supports new analytical tasks. If modelers co-construct a model as an annotated flow diagram, then automated metrics from network analysis become available both as a means of analyzing the evolution of collaboratively created artifacts and as a feedback tool for students [5]. Future research may include the development of tools that support asynchronous collaborations and mine structures as they are generated to either offer guidance to students or inform the instructor.

References

1. Badham, J.: A compendium of modelling techniques (2010). http://i2s.anu.edu.au/sites/default/files/integration-insights/integration-insight_12.pdf

2. Barr, V., et al.: Computer science topics in first-and second-year seminar courses. In: Proceedings of the 2017 ACM SIGCSE Technical Symposium on Computer Science Education, pp. 643–644 (2017)
3. Brooks, R.J., Tobias, A.M.: Choosing the best model: level of detail, complexity, and model performance. Math. Comput. Model. **24**(4), 1–14 (1996)
4. Cachero, C., Meliá, S., Hermida, J.M.: Impact of model notations on the productivity of domain modelling: an empirical study. Inf. Softw. Technol. **108**, 78–87 (2019)
5. Chounta, I.A., et al.: When to say "enough is enough!" a study on the evolution of collaboratively created process models. In: Proceedings of the ACM on Human-Computer Interaction 1(CSCW), pp. 1–21 (2017)
6. Encinas, A.H., et al.: Simulation of forest fire fronts using cellular automata. Adv. Eng. Softw. **38**(6), 372–378 (2007)
7. Forster, S., Pinggera, J., Weber, B.: Toward an understanding of the collaborative process of process modeling. In: CAiSE Forum, pp. 98–105 (2013)
8. Frederiks, P.J., Van der Weide, T.P.: Information modeling: the process and the required competencies of its participants. Data Knowl. Eng. **58**(1), 4–20 (2006)
9. Giabbanelli, P.J.: Ingredients for student-centered learning in undergraduate computing science courses. In: Proceedings of the Seventeenth Western Canadian Conference on Computing Education, pp. 7–11 (2012)
10. Giabbanelli, P.J., Baniukiewicz, M.: Visual analytics to identify temporal patterns and variability in simulations from cellular automata. ACM Trans. Modeling Comput. Simul. (TOMACS) **29**(1), 1–26 (2019)
11. Giabbanelli, P.J., Mago, V.K.: Teaching computational modeling in the data science era. Procedia Comput. Sci. **80**, 1968–1977 (2016)
12. Giabbanelli, P.J., Reid, A.A., Dabbaghian, V.: Interdisciplinary teaching and learning in computing science: three years of experience in the mocssy program. In: Proceedings of the Seventeenth Western Canadian Conference on Computing Education, pp. 47–51 (2012)
13. Giabbanelli, P.J., Tison, B., Keith, J.: The application of modeling and simulation to public health: accessing the quality of agent-based models for obesity. Simul. Model. Practice Theory **108**, (2021)
14. Graesser, A.C., et al.: Advancing the science of collaborative problem solving. psychol. Sci. Public Interest **19**(2), 59–92 (2018)
15. Grantham, E.O., Giabbanelli, P.J.: Creating perceptual uncertainty in agent-based models with social interactions. In: 2020 Spring Simulation Conf. (SpringSim). IEEE/ACM (2020)
16. Heppenstall, A.J., Crooks, A.T., See, L.M., Batty, M.: Agent-based models of Geographical Systems. Springer Science & Business Media (2011)
17. Kavak, H., Padilla, J., Diallo, S., Barraco, A.: Modeling the modeler: an empirical study on how modelers learn to create simulations. In: 2020 Spring Simulation Conference (SpringSim), pp. 1–12. IEEE (2020)
18. Köster, T., Giabbanelli, P.J., Uhrmacher, A.M.: Performance and soundness of simulation: A case study based on a cellular automaton for in-body spread of hiv (2020)
19. Mallet, D.G., De Pillis, L.G.: A cellular automata model of tumor-immune system interactions. J. Theor. Biol. **239**(3), 334–350 (2006)
20. de Mesquita, M.A., da Silva, B.C., Tomotani, J.V.: Simulation education: a survey of faculty and practitioners. In: 2019 Winter Simulation Conference (WSC), pp. 3344–3355. IEEE (2019)

21. Oppl, S.: Supporting the collaborative construction of a shared understanding about work with a guided conceptual modeling technique. Group Decis. Negot. **26**(2), 247–283 (2017)
22. Powell, S.G., Willemain, T.R.: How novices formulate models. part i: qualitative insights and implications for teaching. J. Oper. Res. Soc. **58**(8), 983–995 (2007)
23. Rittgen, P.: Negotiating models. In: Krogstie, J., Opdahl, A., Sindre, G. (eds.) CAiSE 2007. LNCS, vol. 4495, pp. 561–573. Springer, Heidelberg (2007). https://doi.org/10.1007/978-3-540-72988-4_39
24. Rittgen, P.: Collaborative modeling-a design science approach. In: 2009 42nd Hawaii International Conference on System Sciences, pp. 1–10. IEEE (2009)
25. Rittgen, P.: Self-organization of interorganizational process design. Electron. Mark. **19**(4), 189 (2009)
26. Schönfisch, B., de Roos, A.: Synchronous and asynchronous updating in cellular automata. Biosystems **51**(3), 123–143 (1999)
27. Shiflet, A.B., Shiflet, G.W.: An introduction to agent-based modeling for undergraduates. Procedia Comput. Sci. **29**, 1392–1402 (2014)
28. Silvia, D., O'Shea, B., Danielak, B.: A learner-centered approach to teaching computational modeling, data analysis, and programming. In: Rodrigues, J.M.F., et al. (eds.) ICCS 2019. LNCS, vol. 11540, pp. 374–388. Springer, Cham (2019). https://doi.org/10.1007/978-3-030-22750-0_30
29. Smith, J.S., Alexopoulos, C., Henderson, S.G., Schruben, L.: Teaching undergraduate simulation—4 questions for 4 experienced instructors. In: 2017 Winter Simulation Conference (WSC), pp. 4264–4275. IEEE (2017)
30. Srinivasan, A., Te'eni, D.: Modeling as constrained problem solving: an empirical study of the data modeling process. Manage. Sci. **41**(3), 419–434 (1995)
31. Tartaro, A.: Scholarly articles in the introductory computer science classroom. J. Comput. Sci. Coll. **34**(2), 188–198 (2018)
32. Tartaro, A., Cottingham, H.: A problem-based, survey introduction to computer science for majors and non-majors. J. Comput. Sci. Coll. **30**(2), 164–170 (2014)
33. Vendome, C., Rao, D.M., Giabbanelli, P.J.: How do modelers code artificial societies? investigating practices and quality of netlogo codes from large repositories. In: 2020 Spring Simulation Conference (SpringSim). pp. 1–12. IEEE (2020)
34. Voinov, A., et al.: Tools and methods in participatory modeling: selecting the right tool for the job. Environ. Model. Softw. **109**, 232–255 (2018)
35. Wang, W., Brooks, R.J.: Empirical investigations of conceptual modeling and the modeling process. In: 2007 Winter Simulation Conference, pp. 762–770. IEEE (2007)
36. White, R., Engelen, G.: Cellular automata and fractal urban form: a cellular modelling approach to the evolution of urban land-use patterns. Environ. Plan A **25**(8), 1175–1199 (1993)
37. Willemain, T.R.: Insights on modeling from a dozen experts. Oper. Res. **42**(2), 213–222 (1994)
38. Willemain, T.R., Powell, S.G.: How novices formulate models. part ii: a quantitative description of behaviour. J. Oper. Res. Soc. **58**(10), 1271–1283 (2007)

Uncertainty Quantification
for Computational Models

Detection of Conditional Dependence Between Multiple Variables Using Multiinformation

Jan Mielniczuk[1,2]([✉]) [ID] and Paweł Teisseyre[1,2] [ID]

[1] Institute of Computer Science, Polish Academy of Sciences, Warsaw, Poland
{Jan.Mielniczuk,Pawel.Teisseyre}@ipipan.waw.pl
[2] Faculty of Mathematics and Information Sciences, Warsaw University
of Technology, Warsaw, Poland

Abstract. We consider a problem of detecting the conditional dependence between multiple discrete variables. This is a generalization of well-known and widely studied problem of testing the conditional independence between two variables given a third one. The issue is important in various applications. For example, in the context of supervised learning, such test can be used to verify model adequacy of the popular Naive Bayes classifier. In epidemiology, there is a need to verify whether the occurrences of multiple diseases are dependent. However, focusing solely on occurrences of diseases may be misleading, as one has to take into account the confounding variables (such as gender or age) and preferably consider the conditional dependencies between diseases given the confounding variables. To address the aforementioned problem, we propose to use conditional multiinformation (CMI), which is a measure derived from information theory. We prove some new properties of CMI. To account for the uncertainty associated with a given data sample, we propose a formal statistical test of conditional independence based on the empirical version of CMI. The main contribution of the work is determination of the asymptotic distribution of empirical CMI, which leads to construction of the asymptotic test for conditional independence. The asymptotic test is compared with the permutation test and the scaled chi squared test. Simulation experiments indicate that the asymptotic test achieves larger power than the competitive methods thus leading to more frequent detection of conditional dependencies when they occur. We apply the method to detect dependencies in medical data set MIMIC-III.

Keywords: Detection of conditional dependence · Conditional multiinformation · Information theory · Weighted chi squared distribution · Kullback-Leibler divergence

1 Introduction

Detecting conditional dependence is a fundamental problem in machine learning and statistics [7]. It has significant applications in several problems such as causal

© Springer Nature Switzerland AG 2021
M. Paszynski et al. (Eds.): ICCS 2021, LNCS 12747, pp. 677–690, 2021.
https://doi.org/10.1007/978-3-030-77980-1_51

inference [12], learning structure of Bayesian Networks [15,18] and feature selection [3]. Most of the research focuses on testing the conditional independence between two variables given a third one (which can be multi-dimensional), (see [11]). The existing methods are based on different approaches, such as kernel methods [22], information theoretic measures [14], permutation methods [2,19], generalized adversarial networks [1] and knockoffs [4].

In this work we investigate the more general problem of testing the conditional independence of multiple variables, i.e. we consider the null hypothesis H_0 of the form

$$P(X_1 = x_1, \ldots, X_d = x_d | Y = y) = \prod_{j=1}^{d} P(X_j = x_j | Y = y), \qquad (1)$$

where $X_1, \ldots, X_d, Y$ are discrete random variables. The above hypothesis reduces to a classical one for $d = 2$. Surprisingly, such generalization attracted much less attention despite its wide potential applicability. For example, in epidemiology there is often a need to test the independence of multiple diseases. However the task is challenging, as one should take into account the possible confounding variables, such as age, gender or race. The occurrences of diseases can be independent, but dependent when conditioning on a confounding variable. In this case, an important information will be missed when we focus on unconditional dependence. On the other hand, the diseases may be dependent but become independent when conditioning on a confounding variable. In the latter case, there is a risk of finding spurious dependences due to ignoring the latent conditioning variables. Therefore, one should rather focus on testing the conditional independence between diseases given confounding variable/variables. The problem of testing (1) appears naturally in the context of supervised learning and the Naive Bayes (NB) method which is one of the simplest and most popular classifiers. In NB method, it is assumed that all features are conditionally independent given a class variable. Using this assumption, it is possible to avoid the challenging estimation of the joint conditional probabilities and instead estimate the univariate conditional probabilities which is much easier. Usually, in practice, the NB method is used without verifying the assumption, which may lead to poor predictive performance of the classifier. Finally, testing (1) can be useful in multi-label classification where the goal is to predict binary output variables (labels) $Y_1, \ldots, Y_K$ using feature X. If the labels are conditionally independent given X, then classification models can be independently fitted for each label. Otherwise, it is necessary to use more complex approaches that take into account conditional dependencies among labels.

In this work we consider the case of discrete random variables. Then the problem of detecting conditional independence is equivalent to the problem of independence testing on each strata $Y = y$. However, performing the test for each strata separately will lead to multiple testing problem and lack of control of false discovery rate. Thus the need for a specialised test for (1).

To test the hypothesis (1) we propose to use conditional multiinformation (CMI), which is a measure derived from Information Theory. The conditional multiinformation reduces to conditional mutual information for $d = 2$. Although the latter attracts a great interest, the properties of conditional multiinformation and its usefulness to detect departures from (1) remain mainly unexplored. The present paper aims to fill the gap. We prove some new interesting theoretical properties of CMI. In particular we provide upper and lower bounds for it which are tight when $CMI = 0$. Most importantly, with the aim of reducing uncertainty concerning significance of positive value of sample CMI, we determine its asymptotic distribution (for both cases of H_0 and its alternative), which in particular allows to construct the asymptotic test for hypothesis (1). For $d = 2$, our result reduces to well known result for conditional mutual information, which states that the asymptotic distribution is chi squared with the number of degrees of freedom depending on the number of possible values of X_1, X_2, Y. For $d > 2$, the asymptotic distribution under null hypothesis is weighted sum of squared independent normally distributed variables. When $X_1, \ldots, X_d$ are conditionally dependent given Y, the asymptotic distribution is normal.

We compare the proposed asymptotic test with the permutation test as well as test based on the scaled chi-squared distribution. The advantage of asymptotic test over the permutation test is that we avoid generating permuted samples which is the main obstacle in applying permutation method.

2   Preliminaries

2.1   Conditional Multiinformation

We define first the main object of our interest here, namely conditional multiinformation. Let $(X_1, \ldots, X_d, Y)$ be $(d + 1)$-dimensional random variable such that any of its coordinates admits finite number of values and the corresponding mass function $P(X_1 = x_1, \ldots, X_d = x_d, Y = y)$ is denoted by $p(x_1, \ldots, x_d, y)$. Moreover, define $p(x_i) = P(X_i = x_i), p(y) = P(Y = y)$ and $p(x_i|y) = P(X_i = x_i|Y = y)$. Let $X = (X_1, \ldots, X_d)$ and consider the discrete distribution $P_{X,Y}^{ind}$ having mass function $p(y)p(x_1|y) \cdots p(x_d|y)$ which fulfills (1). It means that for any random variable having this distribution, X_is are conditionally independent given Y. The common approach to measure the strength of dependence for (X, Y) is to study a distance of its distribution from some distribution which satisfies the required type of independence and is moreover similar in a specified way to the distribution of (X, Y). Note that $P_{X,Y}^{ind}$ satisfies this requirement as it has the first d components conditionally independent given the last one and also has the same bivariate marginal distributions as $P_{X,Y}$ i.e. $P_{X_i,Y} = P_{X_i,Y}^{ind}$ for any $i = 1, \ldots, d$. Recall that for any two discrete distributions having the same support, Kullback-Leibler (K-L) divergence (relative entropy) is defined as

$$D_{KL}(P||Q) = \sum_i p(x_i) \log \left(p(x_i)/q(x_i) \right),$$

where $p(x_i)$ and $q(x_i)$ denote the corresponding probability mass functions [6]. K-L divergence plays a role of pseudo-distance between two distributions and is frequently used to describe their discrepancy.

We define now conditional multiinformation (aka conditional total correlation) as

$$CMI(X|Y) = CMI(X_1,\ldots,X_d|Y) = D_{KL}(P_{X,Y}||P_{X,Y}^{ind}) \qquad (2)$$

The term 'conditional' in the name CMI is explained by an equivalent definition of this quantity. Namely, note that it follows from definition (2) that

$$CMI(X|Y) = \sum_{x_1,\ldots,x_d,y} p(x_1,\ldots,x_d,y) \log \left(\frac{p(x_1,\ldots,x_d,y)}{p(x_1|y)\cdots p(x_d|y)p(y)} \right)$$

$$= \sum_y p(y) \sum_{x_1,\ldots,x_d} p(x_1,\ldots,x_d|y) \log \left(\frac{p(x_1,\ldots,x_d|y)}{p(x_1|y)\cdots p(x_d|y)} \right). \qquad (3)$$

Thus $CMI(X|Y) = E_{Y=y}(CMI(X|Y = y))$, where $CMI(X|Y = y)$ is Kullback-Leibler divergence between conditional distributions $P_{X|Y=y}$ and $P_{X_1|Y=y} \times \cdots \times P_{X_d|Y=y}$ and the expectation $E_{Y=y}$ is calculated with respect to the distribution of Y. Thus CMI is an *averaged* KL-divergence between these two conditional distributions. Note that, for $d = 2$, CMI reduces to the conditional mutual information for two variables. It follows from non-negativity of K-L divergence [6] that CMI is non-negative. Moreover, the same argument implies that

$$CMI(X|Y) = 0 \quad \Leftrightarrow \quad X_1 \perp X_2 \ldots \perp X_d|Y$$

as $CMI(X|Y) = 0$ entails that for any y in the support of Y distributions $P_{X|Y=y}$ and $P_{X_1|Y=y} \times \cdots \times P_{X_d|Y=y}$ coincide. We observe that $CMI(X|Y)$ can be de-constructed and represented as the combination of conditional entropies $H(X_i|Y)$ and $H(X|Y)$. Namely recalling that the conditional entropy of X given Y is defined as $H(X|Y) = -\sum_{x,y} p(x,y) \log p(x|y)$ we have from (3)

$$CMI(X|Y) = \sum_{i=1}^d H(X_i|Y) - H(X|Y) = \sum_{i=1}^d H(X_i,Y) - H(X,Y) - (d-1)H(Y).$$

$$(4)$$

We also remark that (2) can be written as

$$CMI(X|Y) = -\sum_y p(y) \sum_{x_1,\ldots,x_d} p(x_1|y)\cdots p(x_d|y) \log(p(x_1|y)\cdots p(x_d|y))$$

$$-(-\sum_{x,y} p(x,y) \log p(x,y)),$$

which yields interpretation of CMI as the change of entropy for the conditionally independent and conditionally dependent (X,Y). We also note that when Y is independent from $(X_1,\ldots,X_d)$ and thus conditioning can be omitted in (3), then definition of $CMI(X|Y)$ coincides with definition of unconditional multiinformation $MI(X)$ [17,21] which measure how much structure of dependence of X deviates from the unconditional dependence of its coordinates. $MI(X)$ is frequently applied to detect interactions in Genome Wide Association Studies (see

[5]). Let $\widehat{CMI}(X|Y)$ be defined as plug-in estimator of $CMI(X|Y)$ i.e. probabilities $p(x_1, \ldots, x_d)$ are replaced by fraction based on iid sample from $P_{X,Y}$ consisting of n observations. Properties of $\widehat{MI}(X)$ were studied by Studený in [16]. In the case when X is multivariate normal $N(0, \Sigma)$, where $\Sigma = (\sigma_{ij})$ is the covariance matrix, we have $MI(X) = 2^{-1} \sum_{i=1}^{d} \log \sigma_{ii}^2 - \log(|\Sigma|)$ and the properties of its sample counterpart has been studied in [13].

2.2 Properties of Conditional Multiinformation

Below we list some properties of conditional multiinformation. The first two are well-known but we find it useful to state them together with (iii)–(v).

Theorem 1. *Let $X = (X_1, \ldots, X_d)$. We have*
(i) For any $i < d$

$$CMI(X_1, \ldots X_{i+1}|Y) \geq CMI(X_1, \ldots X_i|Y)$$

(ii)

$$CMI(X|Y) = \sum_{i=2}^{d} MI(X_i; X_1, \ldots, X_{i-1}|Y),$$

where $MI(X_i; X_1, \ldots, X_{i-1}|Y)$ denotes the mutual information between X_i and $(X_1, \ldots, X_{i-1})$ given Y [6]. (iii) We have

$$CMI(X|Y) = \inf_{\tilde{X}_1, \ldots \tilde{X}_d} D_{KL}(P_{X|Y} || P_{\tilde{X}_1|Y} \times \cdots \times P_{\tilde{X}_p|Y}|Y),$$

where $(\tilde{X}_1, \ldots, \tilde{X}_d, Y)$ is any discrete random vector supported on $\mathcal{X}_1 \times \cdots \mathcal{X}_d \times \mathcal{Y}$ with distribution of Y equal to P_Y.
(iv) Let $P_{X,Y}^{ind}$ be a distribution with mass function $p(y)p(x_1|y) \cdots p(x_d|y)$. Then

$$CMI(X|Y) = D_{KL}(P_{Y|X} || P_{Y|X}^{ind}) + D_{KL}(P_X || P_X^{ind}) \qquad (5)$$

(v) We have

$$\frac{1}{2} \Big(\sum_{x_1, \ldots, x_d, y} |p(x_1, \ldots, x_d, y) - p(x_1|y) \cdots p(x_d|y)p(y)| \Big)^2 \leq CMI(X|Y) \leq \log(\chi^2 + 1),$$

where χ^2 index is defined as

$$\chi^2 = \sum_{x_1, \ldots, x_d, y} \frac{(p(x_1, \ldots, x_d, y) - p(x_1|y) \cdots p(x_d|y)p(y))^2}{p(x_1|y) \cdots p(x_d|y)p(y)}.$$

LHS and RHS equal 0 for the conditional independence case (1).

We prove part (v) here, the remaining proofs are relegated to the on-line supplement[1].

---

[1] https://github.com/teisseyrep/cmi.

Proof. Note that the RHS inequality in (v) follows from Jensen's inequality [6]

$$\sum_{x_1,\dots,x_d,y} p(x_1,\dots,x_d,y) \log\left(\frac{p(x_1,\dots,x_d,y)}{p(x_1|y)\cdots p(x_d|y)p(y)}\right) \le$$

$$\log\left(\sum_{x_1,\dots,x_d,y} \frac{p^2(x_1,\dots,x_d,y)}{p(x_1|y)\cdots p(x_d|y)}\right) = \log(\chi^2+1).$$

LHS is a direct consequence of Pinsker's inequality [20] and (2).

Note that (iii) may be interpreted as $P_{X_1|Y}\times\dots\times P_{X_d|Y}$ is the closest distribution consisting of conditionally independent coordinates given Y to $P_{X_1,\dots,X_d|Y}$.

3   Main Theoretical Result

Let $CMI(X|Y)$ and $\widehat{CMI}(X|Y)$, where $X = (X_1,\dots,X_d)$ be defined as before and assume that $p(x_1,\dots,x_d,y) > 0$ for all $(x_1,\dots,x_d,y) \in \mathcal{X}_1 \times \cdots \mathcal{X}_d \times \mathcal{Y}$. Obviously, even in the case when X_is are conditionally independent given Y $\widehat{CMI}(X|Y)$ will be strictly larger than 0 and we need to assess whether the deviations from 0 are due to random error caused by estimation of probabilities $p(x_1,\dots,x_d)$ or to the fact that $CMI(X|Y) > 0$. In order to account for this uncertainty, the distribution of $\widehat{CMI}(X|Y)$ under conditional independence is needed. We state now the result below supplementing it by the behaviour of the estimator when conditional independence is violated. It basically says that the distribution of $\widehat{CMI}(X|Y)$ when $X_1,\dots X_d$ are not conditionally independent given Y, the asymptotic distribution is normal whereas in the opposite case of the null hypothesis $(X_1 \perp X_2\dots \perp X_d|Y)$ the distribution is weighted sum of squared independent normally distributed variables. Moreover, in the latter case $\widehat{CMI}$ converges to its theoretical value CMI more quickly: the rate of convergence is n^{-1} instead of $n^{-1/2}$. The result reduces to the known result for Conditional Mutual Information when $d = 2$ for which the weights are equal to ones and the distribution coincides with chi squared distributed random variable with $k = (|\mathcal{X}_1| - 1)(|\mathcal{X}_2| - 1)|\mathcal{Y}|$ degrees of freedom. However, for $d > 2$ this simplification does not hold. It also generalises the result by M. Studený [16] for $|\mathcal{Y}| = 1$ i.e. for the case of unconditional multiinformation.

Theorem 2. *(i) Assume that $CMI(X|Y) \ne 0$. Then we have*

$$n^{1/2}(\widehat{CMI}(X|Y) - CMI(X|Y)) \xrightarrow{d} N(0,\sigma^2_{\widehat{CMI}}), \qquad (6)$$

where $\xrightarrow{d}$ denotes convergence in distribution, $\sigma^2_{\widehat{CMI}}$ equals

$$\sum_{x_1,\dots,x_d,y} p(x_1,\dots,x_d,y)\log^2\frac{p(x_1,\dots,x_d|y)}{p(x_1|y)\cdots p(x_d|y)} - CMI^2(X|Y)$$

$$= \mathrm{Var}\left(\log\frac{p(X_1,\dots X_d|Y)}{p(X_1|Y)\cdots p(X_d|Y)}\right)$$

and $\sigma^2_{\widehat{CMI}} > 0$.

(ii) Assume that $CMI(X|Y) = 0$. Then

$$2n\widehat{CMI}(X|Y) \xrightarrow{d} \sum_{i=1}^{l} \lambda_i(M)Z_i^2, \tag{7}$$

where $l = |\mathcal{X}_1| \cdots |\mathcal{X}_d||\mathcal{Y}|$ and Z_i are independent $N(0,1)$ and $\lambda_i(M)$ are eigenvalues of the matrix M defined as

$$M_{x_1 \ldots x_d y}^{x_1' \ldots x_d' y'} = I(x_1 = x_1', \ldots, x_d = x_d', y = y') - \sum_{i=1}^{d} I(x_i = x_i', y = y') \frac{p(x_1', \ldots, x_d', y')}{p(x_i, y)}$$

$$+ I(y = y') \frac{p(x_1', \ldots, x_d', y')}{p(y)}.$$

with $M_{x_1 \ldots x_d y}^{x_1' \ldots x_d' y'}$ denoting element of M with row index $x_1 \ldots x_d y$ and column index $x_1' \ldots x_d' y'$; $I()$ is an indicator function.

The proof of the above Theorem can be found in on-line supplement. Note that M is a sparse matrix as its elements are non-zero only if one or more row and column indices coincide.

For $d = 2$ as $X_1, \ldots, X_d$ are independent given Y, the above formula reduces to:

$$M_{x_1 x_2 y}^{x_1' x_2' y'} = I(y = y') \left(I(x_1 = x_1') - \frac{p(x_1', y)}{p(y)} \right) \left(I(x_2 = x_2') - \frac{p(x_2', y)}{p(y)} \right).$$

and M can be shown to be indempotent ($M^2 = M$). Thus all its eigenvalues are 0 and 1 and as trace of M equals $(|\mathcal{X}_1| - 1)(|\mathcal{X}_2| - 1)|\mathcal{Y}|$ this yields the known result about asymptotic distribution of conditional mutual information under conditional independence [10]. For general d, matrix M is not necessarily idempotent and asymptotic distribution of $\widehat{CMI}$ deviates from chi-squared distribution. We note that M can be estimated from the sample by its plug-in estimator $\widehat{M}$ and its eigenvalues $\lambda_i(\widehat{M})$ numerically determined. In this way we approximate limiting distribution in (7) and the approximation will serve as a limiting distribution for the proposed test of the conditional independence.

4   Detection of Conditional Dependence: Permutation Versus Asymptotic Method

4.1   Permutation Method

A popular method of checking whether H_0 is violated is the permutation method adapted to the present problem. For a given sample generated from $P_{X,Y}$ and each strata $Y = y$ and $i = 1, \ldots, d$ we randomly permute values of ith coordinate of observations such that the corresponding value of Y equals y (see [19]). Permutations for each i are performed independently. Consequently, performing this operation separately for any value of y occurring in the original sample, we obtain a sample from distribution $P_{X,Y}^{ind}$ which satisfies H_0. We repeat this operation N times drawing N permuted samples in total and calculate corresponding values $\widehat{CMI}_k(X|Y)$ for $k = 1, \ldots, N$. Then empirical p-value

$$\hat{p} = \frac{\#\{k : \widehat{CMI}_k(X|Y) \geq \widehat{CMI}(X|Y)\}}{N},$$

where $\widehat{CMI}(X|Y)$ is empirical CMI, is calculated. Small value of $\hat{p}$ indicates conditional dependence.

4.2   Asymptotic Method

For a given sample pertaining to P_{XY} we calculate $\widehat{CMI}(X|Y)$ and plug-in estimator $\hat{M}$ of matrix M defined in the previous section. We use now the fact that the asymptotic distribution W of $\widehat{CMI}(X|Y)$ is given in (7) and we approximate it by $\widehat{W}$ by plugging in $\lambda_i(\widehat{M})$ for $\lambda_i(M)$, where $\lambda_i(\widehat{M})$ are numerically calculated. Then rejection region for a given significance level α is given by $\{\widehat{CMI}(X|Y) \geq q_{\widehat{W},1-\alpha}\}$, where $q_{\widehat{W},1-\alpha}$ is quantile of order $1 - \alpha$ of distribution $\widehat{W}$. We note that by using asymptotic distribution we avoid the main drawback of permutation method, namely generation of many samples for every value of $Y = y$ which may be very time consuming. R package `eigen` has been used to calculate the eigenvalues and package `COMpQuadForm` for quantiles of $\widehat{W}$.

5   Simulation Study

5.1   Artificial Data Sets

The aim of the simulation experiments was to compare the performance of the tests described in previous section: asymptotic test and permutation test. In addition, we consider semi-parametric test based on the scaled chi squared distribution. It is defined as distribution of $\alpha\chi_d^2 + \beta$, where χ_d^2 is a chi square distribution with $d > 0$; parameters α, d, β are calculated based on permutation samples (see [9]).

To assess the performance of the tests we use ROC-type curves which are generated in the following way. In each simulation, we generate two samples: D_0 and D_1 conforming to the null hypothesis and alternative hypothesis, respectively. So, D_0 is generated from distribution for which $X_1, \ldots, X_d$ are conditionally independent given Y, whereas D_1 is generated from distribution for which $X_1, \ldots, X_d$ are conditionally dependent given Y. Then, we run a considered test for both D_0 and D_1 using significance level $\alpha \in (0,1)$ and report whether the null hypothesis has been rejected. Importantly, the reference distributions of empirical CMI under null hypothesis are different for D_0 and D_1. The above steps are repeated 1000 times which allows to approximate probabilities of rejection in the cases when samples conform and do not conform to H_0, respectively. Each point on the curve corresponds to a different value of α. Observe that the first coordinate of each point is an approximation of type I error for some α obtained using D_0, wheres the second coordinate is an approximation of the power of the test obtained from D_1 for the same value of α. We also report Area Under the Curve (AUC), the larger the value of AUC the better is the performance of the test (we observe larger power for a fixed value of type I error). The above method has two important advantages. First, we control simultaneously the type I error and the corresponding power of the test. Secondly, it is possible to analyze power and type I error for different values of significance levels α. As we found that the actual levels of significance of the asymptotic test may exceed assumed levels of significance in some cases for medium sample sizes we view presented ROC curve analysis to be a more objective way of comparing tests, as they enable comparison of methods at the same level of significance.

We consider the following simulation models.

1. **Simulation model 1.**
   - Sample D_0: Generate $Y \sim B(1, 0.5)$ and $X_1, \ldots, X_d | Y = y \sim N(y, 1)$.
   - Sample D_1: Generate $Y \sim B(1, 0.5)$, $X_1, \ldots, X_{d-1} | Y = y \sim N(y, 1)$ and $X_d | X_{d-1} = x_{d-1} \sim N(\gamma x_{d-1}, 1)$, where γ is a parameter.

Simulation model 1p. Modification of model 1. The only difference is that in D_1, $X_d | X_{d-1} = x_{d-1}, Y = y \sim N(\gamma x_{d-1} + y, 1)$

2. **Simulation model 2.**
   - Sample D_0: Generate $Y \sim B(1, 0.5)$ and $X_1, \ldots, X_d | Y = y \sim N(y, 1)$.
   - Sample D_1: Generate $Y \sim B(1, 0.5)$ and $Z \sim B(1, 0.5)$, where $Y \perp Z$. Next, generate $X_1, \ldots, X_d | Y = y, Z = z \sim N(\gamma(z + y), 1)$, where γ is a parameter.
3. **Simulation model 3.**
   - Sample D_0: Generate $X_1, \ldots, X_d \sim N(0, 1)$ and $Y \sim B(1, 0.5)$.
   - Sample D_1: Generate $X_1, \ldots, X_d \sim N(0, 1)$ and then $Y | X = x \sim B(1, \sigma(\gamma \cdot x^T 1))$ where γ is a parameter and $1 = (1, \ldots, 1)^T$.

The above simulation models are chosen to represent various dependency structures. Figure 1 and 2 show the graphs corresponding to distributions conforming to H_0 and H_1, respectively, for simulation models 1–3. Models 1–2 are generative models, as we first generate Y and then $X_1, \ldots, X_d$, whereas model 3 is a discriminative model, corresponding to scenario of supervised classification. For models 1–2 and sample D_0, variables $X_1, \ldots, X_d$ are conditionally independent given Y, but at the same time they are not unconditionally independent. For model 3 and D_0, all considered variables are independent and also conditionally independent. In all three models, parameter γ controls the difficulty of the problem. For larger γ it is easier to reject the null hypothesis for sample D_1.

Figure 3 shows the ROC-type curves for simulation models 1–3, for $n = 500$ and $d = 7$ (results for other values of d are placed in supplement). As expected,

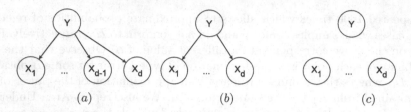

Fig. 1. Dependency structures corresponding to distributions conforming to H_0, for simulation model 1 (a), model 2 (b) and model 3 (c).

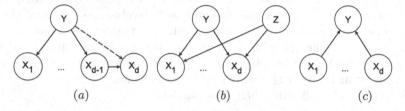

Fig. 2. Dependency structures corresponding to distributions conforming to H_1, for simulation model 1 (a), model 2 (b) and model 3 (c). Dashed line in (a) corresponds to model 1p.

AUC increases with parameter γ, which is due to the fact that for larger γ the conditional dependence is stronger. The proposed asymptotic test works better (in terms of AUC) than the remaining tests for all simulation models except model 1p for which it works on par with permutation test. The advantage of asymptotic test is most pronounced for model 2. The test based on scaled chi squared distribution performs worse than the two competitors, which confirms our theoretical results (see Theorem 2) indicating that the reference distribution of CMI under null hypothesis significantly deviates from chi squared distribution for $d > 2$ and is poorly approximated by the scaled chi squared distribution. We also experimented with smaller $d = 3, 5$, for which all considered methods perform similar (the results are placed in supplement).

5.2   Analysis of Medical Data Set MIMIC-III

We also illustrate the problem of conditional independence testing using real medical dataset MIMIC-III [8] containing information about patients from the intensive care units. We are interested in finding out whether the occurrences of some diseases are conditionally independent given gender. We consider 10 diseases (shortened names and the prevalences estimated from data are given in brackets): hypertension (66%), kidney failure (kidney; 35%), disorders of fluid electrolyte balance (fluid; 37%), hypotension (14%), disorders of lipoid metabolism (lipoid; 37%), liver disease (liver; 7%), diabetes (32%), thyroid disease (thyroid; 14%), chronic obstructive pulmonary disease (copd; 23%) and thrombosis (6%). We analysed all triples, i.e. we tested the null hypothesis of

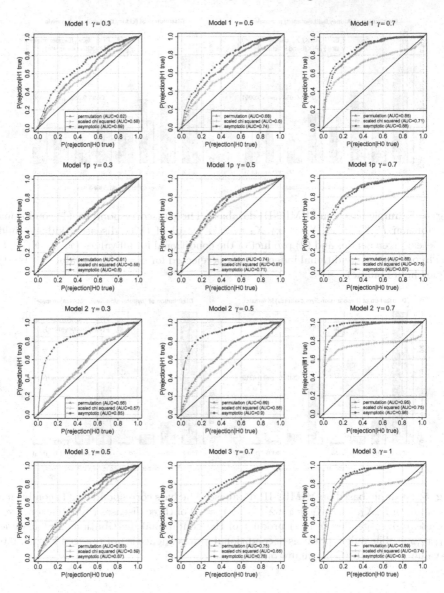

Fig. 3. ROC-type curves for simulation models 1, 1p, 2, 3 and permutation test (red), scaled chi-squared test (green) and asymptotic test (blue). Number of variables $d = 7$ and sample size $n = 500$. (Color figure online)

conditional independence between X_1, X_2, X_3 given Y, where X_1, X_2, X_3 denote occurrences of three out of ten of the above diseases and Y is gender. So, in total, we performed $\binom{10}{3} = 120$ tests. We present the results for triples with the largest (kidney, fluid, diabetes) and the smallest (hypotension, liver, diabetes) value of CMI (Figs. 4 and 5, respectively). Each figure shows the values of joint

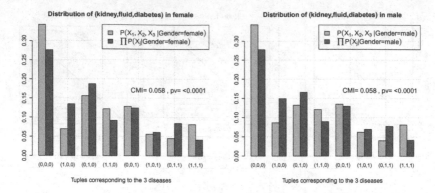

Fig. 4. Example based on MIMIC-III database. The bars correspond to: (1) conditional distribution $P(X_1 = x_1, X_2 = x_2, X_3 = x_3 | Y = y)$ of three diseases (kidney, fluid, diabetes) given gender and (2) product of the conditional for all values (x_1, x_2, x_3). The null hypothesis of conditional independence is rejected for $\alpha = 0.05$ (p-value < 0.0001). (Color figure online)

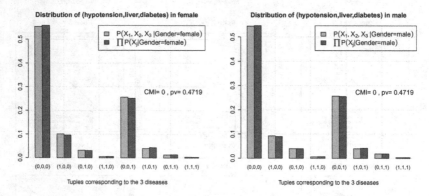

Fig. 5. Example based on MIMIC-III database. The bars correspond to: (1) conditional distribution $P(X_1 = x_1, X_2 = x_2, X_3 = x_3 | Y = y)$ of three diseases (hypotension, liver, diabetes) given gender and (2) product of the conditional distributions for all values (x_1, x_2, x_3). The null hypothesis of conditional independence is not rejected for $\alpha = 0.05$ (p-value = 0.4719). (Color figure online)

conditional probabilities $P(X_1, X_2, X_3 | Y)$ (red bars) and products of marginal conditional probabilities $P(X_1 | Y) P(X_2 | Y) P(X_3 | Y)$ (blue bars). According to the asymptotic test, we reject the null hypothesis in the case of triple: kidney, fluid, diabetes (p-value equal to 0.0004 is smaller than 0.05/120) and we do not reject the null hypothesis in the case of diseases (hypotension, liver, diabetes, p-value equal to 0.4719). We assumed standard significance level $\alpha = 0.05$ and used Bonferroni correction in order to account for multiple tests. As expected, in the latter case, the joint conditional probabilities are very close to the corresponding products of marginal conditional probabilities (see Fig. 5). When the null hypothesis is rejected, the differences between values of the probabilities are significantly larger (Fig. 4). In particular, for kidney, fluid, diabetes, the probability of the occurrence of all diseases at the same time for females is $P(X_1 = 1, X_2 = 1, X_3 = 1 | Y = \text{female}) = 8\%$, whereas the corresponding product is $P(X_1 = 1 | Y = \text{female}) P(X_2 = 1 | Y = \text{female}) P(X_3 = 1 | Y = \text{female}) = 4\%$. Further analysis using Conditional Mutual Information detects pairwise conditional dependencies between kidney and fluid and kidney and diabetes (both p-values of order 10^{-9}) but no dependence is detected between fluid and diabetes (p-value 0.36).

6   Conclusions

In this paper we investigated the properties of conditional multiinformation (CMI), which is a natural measure of a strength of conditional dependence between multiple variables. Our main theoretical contribution is deriving asymptotic distribution of sample CMI (Theorem 2). It is a generalization of well known result for the case of two variables ($d = 2$). Moreover, we constructed a statistical test based on the distribution. Importantly, the asymptotic distribution of sample CMI significantly deviates from chi squared distribution for $d > 2$. This explains why the simple test based on scaled chi squared distribution works poorly when more than two variables are taken into account. The proposed asymptotic test usually outperforms permutation test in terms of power, when the number of variables is moderate. Its advantage over permutation test is that we avoid generating many permutation samples. On the other hand, asymptotic test requires numerical calculation of eigenvalues using matrix whose size significantly increases with the number of variables. Thus, the method may fail when the number of variables d increases. Therefore, the proposed asymptotic test is strongly recommended for moderate number of variables, whereas for larger d we recommend permutation test.

References

1. Bellot, A., van der Schaar, M.: Conditional independence testing using generative adversarial networks. In: Advances in Neural Information Processing Systems, vol. 32, pp. 2199–2208 (2019)
2. Berrett, T.B., Wang, Y., Barber, R.F., Samworth, R.J.: The conditional permutation test for independence while controlling for confounders. J. Roy. Stat. Soc. Ser. B (Stat. Methodol.) **82**(1), 175–197 (2020)
3. Bühlmann, P., van de Geer, S.: Statistics for High-Dimensional Data, 1st edn. Springer, Heidelberg (2015). https://doi.org/10.1007/978-3-642-20192-9
4. Candès, E., Fan, Y., Janson, L., Lv, J.: Panning for gold: model-x knockoffs for high-dimensional controlled variable selection. J. Roy. Stat. Soc. B **80**, 551–577 (2018)
5. Chanda, P., et al.: Ambience: a novel approach and efficient algorithm for identifying informative genetic and environmental associations with complex phenotypes. Genetics **180**, 1191–2010 (2008)

6. Cover, T.M., Thomas, J.A.: Elements of Information Theory. Wiley Series in Telecommunications and Signal Processing. Wiley-Interscience (2006)
7. Dawid, A.P.: Conditional independence in statistical theory. J. Roy. Stat. Soc.: Ser. B (Methodol.) **41**(1), 1–15 (1979)
8. Johnson, A.E.W., et al.: MIMIC-III, a freely accessible critical care database. Sci. Data **3**, 1–9 (2016)
9. Kubkowski, M., Mielniczuk, J.: Asymptotic distributions of interaction information. Methodol. Comput. Appl. Probab. **23**, 291–315 (2020)
10. Kullback, S.: Information Theory and Statistics. Peter Smith (1978)
11. Li, C., Fan, X.: On nonparametric conditional independence tests for continuous variables. WIREs Comput. Stat. **12**, 1–11 (2020)
12. Pearl, J.: Causality. Cambridge University Press, Cambridge (2009)
13. Rowe, T., Troy, D.: The sampling distribution of the total correlation for multivariate gaussian random variables. Entropy **21**, 921 (2019)
14. Runge, J.: Conditional independence testing based on a nearest neighbour estimator of conditional mutual information. In: Proceedings of the 21st International Conference on Artificial Intelligence and Statistics, PMLR, vol. 84, pp. 938–947 (2018)
15. Spirtes, P., Glymour, C., Scheines, R.: Causation, Prediction, and Search, 2nd edn. MIT Press (2000)
16. Studený, M.: Asymptotic behaviour of empirical multiinformation. Kybernetika **23**, 124–135 (1987)
17. Studený, M., Vejnarová, J.: The multiinformation as a tool for measuring stochastic dependence. In: Learning in Graphical Models, pp. 66–82. MIT Press (1999)
18. Tsamardinos, I., Aliferis, C., Statnikov, A.: Algorithms for large scale Markov Blanket discovery. In: FLAIRS Conference, pp. 376–381 (2003)
19. Tsamardinos, I., Borboudakis, G.: Permutation testing improves Bayesian network learning. In: Balcázar, J.L., Bonchi, F., Gionis, A., Sebag, M. (eds.) ECML PKDD 2010. LNCS (LNAI), vol. 6323, pp. 322–337. Springer, Heidelberg (2010). https://doi.org/10.1007/978-3-642-15939-8_21
20. Tsybakov, A.: Introduction to Nonparametric Estimation, 1st edn. Springer, New York (2009). https://doi.org/10.1007/b13794
21. Watanabe, S.: Information theoretical analysis of multivariate correlation. IBM J. Res. Dev. **4**, 66–82 (1960)
22. Zhang, K., Peters, J., Janzing, D., Schölkopf, B.: Kernel-based conditional independence test and application in causal discovery. In: Proceedings of the 27th Conference on Uncertainty in Artificial Intelligence, UAI 2011, pp. 804–813 (2011)

Uncertainty Quantification of Coupled 1D Arterial Blood Flow and 3D Tissue Perfusion Models Using the INSIST Framework

Claire Miller[1]([✉])(iD), Max van der Kolk[1](iD), Raymond Padmos[1](iD),
Tamás Józsa[2](iD), and Alfons Hoekstra[1](iD)

[1] Computational Science Laboratory, Faculty of Science, Informatics Institute,
University of Amsterdam, Science Park 904, 1098 XH Amsterdam, The Netherlands
c.m.miller@uva.nl
[2] Department of Engineering Science, Institute of Biomedical Engineering,
University of Oxford, Parks Road, Oxford OX1 3PJ, UK

Abstract. We perform uncertainty quantification on a one-dimensional arterial blood flow model and investigate the resulting uncertainty in a coupled tissue perfusion model of the brain. The application of interest for this study is acute ischemic stroke. The outcome of interest is infarct volume, estimated using the change in perfusion between the healthy and occluded state (assuming no treatment). Secondary outcomes are the uncertainty in blood flow at the outlets of the network, which provide the boundary conditions to the pial surface of the brain in the tissue perfusion model. Uncertainty in heart stroke volume, heart rate, blood density, and blood viscosity are considered. Results show uncertainty in blood flow at the network outlets is similar to the uncertainty included in the inputs, however the resulting uncertainty in infarct volume is significantly smaller. These results provide evidence when assessing the credibility of the coupled models for use in *in silico* clinical trials.

Keywords: Uncertainty quantification · Blood flow modelling · Tissue perfusion modelling · *In silico* clinical trials · Acute ischemic stroke

1 Introduction

The development of *in silico* clinical trials and physics-based models for personalised medicine is a growing field due to an increased accessibility to compute resources, models, and data. However, such models require a large number of parameters, many of which are expensive, time consuming, or even impossible, to measure within a clinical setting. Consequently, we turn to uncertainty quantification (UQ) to determine how unmeasured parameters impact the results from these models, and which parameters account for any significant variation in the results. A UQ analysis is likely to be critical to prove that these models are

© Springer Nature Switzerland AG 2021
M. Paszynski et al. (Eds.): ICCS 2021, LNCS 12747, pp. 691–697, 2021.
https://doi.org/10.1007/978-3-030-77980-1_52

sufficiently credible to provide evidence of efficacy of new treatments, according to the current standards on simulation of medical devices [1].

Though uncertainty quantification of physics-based models is a well established field, less work has been done on uncertainty quantification of multi-scale and multi-physics models. An approach taken in [14]—a semi-intrusive Monte Carlo method—is to replace the computationally-heavy model in a coupled multi-scale system with a surrogate model. This reduces the number of samples required of the expensive model, hence allowing for a Monte Carlo analysis to be performed in a reasonable computational time.

For the models used in this paper, a non-intrusive uncertainty quantification method is appropriate given the models' complexity. Two well-established non-intrusive methods are quasi-Monte Carlo (QMC) and polynomial chaos [4,13]. In this paper we use a QMC method for its robustness and flexibility.

The intention of the INSIST project is *in silico* clinical trials for acute ischemic stroke [10]. The first elements to these trials are modelling blood flow through the arteries and the resulting perfusion throughout the brain, both with and without an occlusion. We are interested in applying UQ methods to a 1D blood flow model coupled to a 3D model of perfusion [15]. Uncertainty analysis of blood flow through the arterial network has been studied in several models previously [2,3,18]. This study intends to understand how the uncertainty in blood flow impacts the results of the perfusion model in a stroke scenario.

2   Methods

We investigate the uncertainty propagation in a one-way coupled 1D arterial blood flow and 3D tissue perfusion model [15]. We are primarily interested in the uncertainty in the change in perfusion between the healthy state of the system (pre-stroke) and the occluded state (after stroke).

The outcome metric of interest is the volume of tissue that has a 70% decrease in perfusion between the healthy and occluded state, as an estimate of infarct volume without treatment [8]. The secondary outcomes of interest are the variation in blood flow through the occluded vessel in the healthy state; and the uncertainty in the change in flow rates between the healthy and occluded states in the artery outlets, which form the boundary conditions for the tissue perfusion model.

2.1   The INSIST Framework

The blood flow and tissue perfusion models are run as part of the INSIST framework [10]. The framework links a patient-generation model to physics-based models for blood flow, blood perfusion in the brain, thrombolysis, and thrombectomy. The intention of INSIST is to run large cohorts of patients, however for the purposes of this UQ investigation, we instead generate multiple copies of a single patient. The patient has an associated set of parameters generated using a statistical model built on clinical data from the MR CLEAN Registry [5].

The parameters generated by this model are therefore considered to be known parameters, and not investigated in the study.

2.2 The Blood Flow and Tissue Perfusion Models

The two models we are using in this study are an arterial blood flow model, and a tissue perfusion model. The blood flow model is a 1D steady state blood flow model for the artery network from the heart to the pial surface of the brain [15]. The tissue perfusion model uses a 3D finite-element Darcy flow model with three compartments: arteriole, capillary, and venules [8]. The blood flow model provides the boundary conditions for the tissue perfusion model.

2.3 Uncertainty Quantification

In order to efficiently undertake uncertainty quantification (UQ) activities, the INSIST framework is linked to the open-source library EasyVVUQ [16]. We use a quasi-Monte Carlo approach for UQ, and Sobol indices to determine the parameter contributions to output uncertainty, calculated using the Saltelli and Jansen/Saltelli methods for the first and total order indices respectively [6,17]. A total of 3,000 samples were run using Sobol quasi-random sequences to generate the sampling matrices required for the Sobol indices [17] using EasyVVUQ [16].

This paper shows the effect of uncertainty in the arterial blood flow on the estimated infarct volume, calculated using the change in perfusion in the tissue perfusion model. We focus here on aleatoric uncertainty—parameter uncertainty due to inherent variation in the population. The parameters we consider are given in Table 1. Distribution parameters and shapes are based off population studies found in the literature (sources shown in table).

For the purposes of this study, we only consider one occlusion location: the M1 segment in the middle cerebral artery (MCA), as it was the most commonly occluded segment in the MR CLEAN Registry (58% of patients) [5]. We also assume that the clot is impermeable. We intend to use this same method to investigate uncertainty in artery morphology parameters and tissue perfusion model parameters, however these are not included in the results below.

3 Results

3.1 Effect of Uncertainty on Infarct Volume

The primary outcome of interest for the study is the estimate of infarct volume without treatment. This estimate is determined using a threshold on the change in tissue perfusion between the healthy and occluded states (70% reduction) [8]. Figure 1a shows the resulting uncertainty in infarct volume given the uncertainty in the blood flow model parameters given in Table 1. The mean infarct volume is $\mu_{iv} = 268.3 \, \text{mL}$ and the standard deviation is $\sigma_{iv} = 0.5 \, \text{mL}$. This gives a coefficient of variation of $CV_{iv} = 0.002$, which is 2 orders of magnitude lower than

Table 1. The parameters with aleatoric uncertainty in the 1D arterial blood flow model; their distributions and source; and their determined Sobol indices. CV: Coefficient of variation. Heart stroke volume is the volume of blood pumped out the left ventricle per heart beat.

Parameter	Distribution	CV	Source	Sobol first	Sobol total
Blood density (kg.m^{-3})	U(1040, 1055)		[9]	0.00	0.00
Blood viscosity (mPa.s)	N(4.2, 0.9)	0.21	[7]	0.07	0.10
Heart stroke volume (mL)	N(95, 14)	0.15	[11]	0.36	0.40
Heart rate (bpm)	N(73,12.2)	0.17	[12]	0.42	0.50

the variation in the input parameters. We note the right tail of the distribution is associated with low pre-stroke pressures in the occluded vessel, which occur when the product of viscosity, heart rate, and heart stroke volume is very high.

3.2   Effect of Uncertainty on Blood Flow Model Outputs

Though the primary outcome of interest is the infarct volume, it is useful to investigate the uncertainty in the outputs of the blood flow model, which provides the input to the tissue perfusion model. The main artery of interest for this study is the occluded vessel: the right MCA. The resulting distribution of flow rates in this vessel is shown in Fig. 1b. For the 3,000 samples, the mean flow rate through the right MCA vessel is $\mu_{ov} = 2.47 \, \mathrm{mL/s}$, with a standard deviation of $\sigma_{ov} = 0.45 \, \mathrm{mL/s}$, and hence a coefficient of variation of $\mathrm{CV}_{ov} = 0.18$. This is of the same order of magnitude as the uncertainty in the inputs, given in Table 1.

Also relevant are the arteries providing flow to the pial surface of the brain and hence providing the boundary conditions for the coupled perfusion model. On average, the boundary vessels in the brain, excluding the occluded vessel, have a 7% change in their flow rate between the healthy and occluded state. The

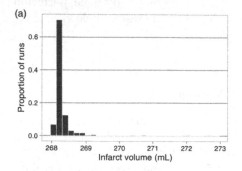
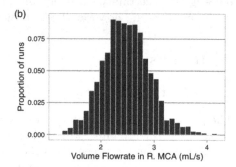

Fig. 1. Effect of uncertainty in arterial blood flow parameters on (a) infarct volume, estimated using a change in perfusion threshold, and (b) healthy state volume flow rate through the occluded vessel: right middle cerebral artery (MCA).

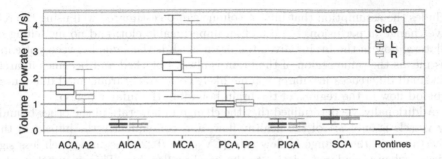

Fig. 2. Quantiles (box) and range (error bars) for blood flow in vessels providing the boundary conditions for the perfusion model. ACA: anterior cerebral artery, AICA: anterior inferior cerebellar artery, MCA: middle cerebral artery, PCA: posterior cerebral artery, PICA: posterior inferior cerebellar artery, SCA: superior cerebellar artery.

uncertainty in blood flow for each of these boundary outlets is shown in Fig. 2, for the healthy state. The mean coefficient of variation for the boundary vessels of the brain in the healthy state is $\overline{CV}_{bv} = 0.19$, similar to the occluded vessel. For the change in flow between healthy and occluded state it is $\overline{CV}_{\delta bv} = 0.22$.

3.3 Sensitivity Analysis

The sensitivity of blood flow in the occluded vessel to each of the parameters considered is given in Table 1. The table shows the first order and total effect for each parameter. It can be seen that almost all the uncertainty in the blood flow is due to uncertainty in the heart rate and heart stroke volume. The product of heart rate and stroke volume determine the flowrate at the input boundary of the artery network (the ascending aorta), explaining their large impact on the output sensitivity.

There is also a small sensitivity to blood viscosity. We would expect flowrate to be inversely proportional to viscosity for laminar flow in a pipe, which explains viscosity's effect. Based on the Hagen-Poiseuille law density and viscosity should have the same weight, so the lack of sensitivity to density is due to the low variation in the parameter. Given this result, when further uncertainties from the blood flow and tissue perfusion models are added, it would be possible to exclude blood density in order to minimise the number of simulations required in an extended UQ.

4   Discussion

The results showed that, although the uncertainty in the arterial blood flow output is on the same order of magnitude as the input uncertainties, the uncertainty in the final determined infarct volume is two orders of magnitude lower. This is likely due to several assumptions and simplifications in the models. This includes the assumption that infarct volume can be estimated well using the relative change in perfusion; the use of an impermeable clot; and no modelling of collateral flow in the brain. Given the brain mesh is the same between patients, this means the same region of the brain is always blocked and becomes infarct. The small differences in infarct volume likely come solely from small differences in blood flow in the regions of the brain bounding the infarct region.

Additionally, we determined that the change in flow rate in the other boundary vessels is only 7% of the healthy flow, and consequently the impact of the uncertainty in the change in flow rate ($\overline{CV}_{\delta bv} = 0.22$) becomes much less significant when considered relative to the baseline flowrate. This would then also significantly decrease the uncertainty in the final infarct volume, given its definition as a reduction in flow of 70%.

Our results show that, though there may be a high level of uncertainty in the outcomes of one model in a framework such as INSIST, this is not indicative of the uncertainty in the whole system. This can help prove increased credibility of the whole workflow for *in silico* clinical trials, as opposed to assessing credibility based off the uncertainty results for each of the models individually.

In future work, we intend to continue to investigate UQ of each model independently and compare this to a UQ of the coupled system. This will involve incorporating uncertainty in further parameters in the arterial blood flow model, as well as parameters in the tissue perfusion model. Additionally, we intend to incorporate a tissue death model, which will likely increase the uncertainty in the infarct volume compared to the current perfusion threshold estimate.

As more parameters are added to the system, computational times will likely become infeasible using the proposed approach. To deal with this increased computational load, we consider two options. Firstly, not including parameters whose uncertainty is not determined to have a significant effect, such as blood density (Table 1). Secondly, exploiting the independence of the models using a semi-intrusive approach. In such an approach, the UQ analysis of tissue perfusion, and any further coupled models, builds on top of the results from the UQ of the blood flow model, as opposed to treating the coupled model as a black box.

Acknowledgments. INSIST has received funding from the European Union's Horizon 2020 research and innovation program under grant agreement No. 777072.

References

1. American Society of Mechanical Engineers: Assessing Credibility of Computational Modeling through Verification and Validation: Application to Medical Devices (2018)
2. Chen, P., Quarteroni, A., Rozza, G.: Simulation-based uncertainty quantification of human arterial network hemodynamics. Int. J. Numer. Meth. Biomed. Eng. **29**(6), 698–721 (2013). https://doi.org/10.1002/cnm.2554
3. Eck, V.G., Sturdy, J., Hellevik, L.R.: Effects of arterial wall models and measurement uncertainties on cardiovascular model predictions. J. Biomech. **50**, 188–194 (2017). https://doi.org/10.1016/j.jbiomech.2016.11.042

4. Eck, V.G., et al.: A guide to uncertainty quantification and sensitivity analysis for cardiovascular applications. Int. J. Numer. Meth. Biomed. Eng. **32**(8), e02755 (2016). https://doi.org/10.1002/cnm.2755

5. Jansen, I.G.H., Mulder, M.J.H.L., Goldhoorn, R.J.B.: Endovascular treatment for acute ischaemic stroke in routine clinical practice: prospective, observational cohort study (MR CLEAN Registry). BMJ **360**, (2018). https://doi.org/10.1136/bmj.k949

6. Jansen, M.J.W.: Analysis of variance designs for model output. Comput. Phys. Commun. **117**(1), 35–43 (1999). https://doi.org/10.1016/S0010-4655(98)00154-4

7. Jeong, S.K., Rosenson, R.S.: Shear rate specific blood viscosity and shear stress of carotid artery duplex ultrasonography in patients with lacunar infarction. BMC Neurol. **13**(1), 36 (2013). https://doi.org/10.1186/1471-2377-13-36

8. Józsa, T.I., Padmos, R.M., Samuels, N., El-Bouri, W.K., Hoekstra, A.G., Payne, S.J.: A porous circulation model of the human brain for in silico clinical trials in ischaemic stroke. Interface Focus **11**(1), 20190127 (2021). https://doi.org/10.1098/rsfs.2019.0127

9. Kenner, T.: The measurement of blood density and its meaning. Basic Res. Cardiol. **84**(2), 111–124 (1989). https://doi.org/10.1007/BF01907921

10. Konduri, P.R., Marquering, H.A., van Bavel, E.E., Hoekstra, A., Majoie, C.B.L.M.: INSIST Investigators: In-silico trials for treatment of acute ischemic stroke. Front. Neurol. **11** (2020). https://doi.org/10.3389/fneur.2020.558125

11. Maceira, A.M., Prasad, S.K., Khan, M., Pennell, D.J.: Normalized left ventricular systolic and diastolic function by steady state free precession cardiovascular magnetic resonance. J. Cardiovasc. Magn. Reson. **8**(3), 417–426 (2006). https://doi.org/10.1080/10976640600572889

12. Mason, J.W., Ramseth, D.J., Chanter, D.O., Moon, T.E., Goodman, D.B., Mendzelevski, B.: Electrocardiographic reference ranges derived from 79,743 ambulatory subjects. J. Electrocardiol. **40**(3), 228–234 (2007). https://doi.org/10.1016/j.jelectrocard.2006.09.003

13. Najm, H.N.: Uncertainty quantification and polynomial chaos techniques in computational fluid dynamics. Annu. Rev. Fluid Mech. **41**(1), 35–52 (2009). https://doi.org/10.1146/annurev.fluid.010908.165248

14. Nikishova, A., Hoekstra, A.G.: Semi-intrusive uncertainty propagation for multiscale models. J. Computat. Sci. **35**, 80–90 (2019). https://doi.org/10.1016/j.jocs.2019.06.007

15. Padmos, R.M., Józsa, T.I., El-Bouri, W.K., Konduri, P.R., Payne, S.J., Hoekstra, A.G.: Coupling one-dimensional arterial blood flow to three-dimensional tissue perfusion models for in silico trials of acute ischaemic stroke. Interface Focus **11**(1), 20190125 (2021). https://doi.org/10.1098/rsfs.2019.0125

16. Richardson, R.A., Wright, D.W., Edeling, W., Jancauskas, V., Lakhlili, J., Coveney, P.V.: EasyVVUQ: a library for verification, validation and uncertainty quantification in high performance computing. J. Open Res. Softw. **8**(1) (2020). https://doi.org/10.5334/jors.303

17. Saltelli, A., Annoni, P., Azzini, I., Campolongo, F., Ratto, M., Tarantola, S.: Variance based sensitivity analysis of model output. Design and estimator for the total sensitivity index. Comput. Phys. Commun. **181**(2), 259–270 (2010). https://doi.org/10.1016/j.cpc.2009.09.018

18. Xiu, D., Sherwin, S.J.: Parametric uncertainty analysis of pulse wave propagation in a model of a human arterial network. J. Comput. Phys. **226**(2), 1385–1407 (2007). https://doi.org/10.1016/j.jcp.2007.05.020

Second Order Moments of Multivariate Hermite Polynomials in Correlated Random Variables

Laura Lyman[✉][iD] and Gianluca Iaccarino

Stanford University, Stanford, CA 94305, USA
{lymanla,jops}@stanford.edu

Abstract. Polynomial chaos methods can be used to estimate solutions of partial differential equations under uncertainty described by random variables. The stochastic solution is represented by a polynomial expansion, whose deterministic coefficient functions are recovered through Galerkin projections. In the presence of multiple uncertainties, the projection step introduces products (second order moments) of the basis polynomials. When the input random variables are correlated Gaussians, calculating the products of the corresponding multivariate basis polynomials is not straightforward and can become computationally expensive. We present a new expression for the products by introducing multiset notation for the polynomial indexing, which allows for simple and efficient evaluation of the second-order moments of correlated multivariate Hermite polynomials.

Keywords: Polynomial chaos · Multivariate Hermite polynomials · Stochastic Galerkin methods

1 Introduction

Uncertainty quantification (UQ) is crucial for developing confidence in predictions resulting from mathematical models of physical phenomena such as those described by partial differential equations (PDEs) [2,12]. A common strategy in computational science is to represent sources of uncertainty by random variables, which causes the solution to the original differential equation(s) to become a function of stochastic parameters [2,5,12]. A polynomial chaos expansion (PCE) expresses the solution as an infinite series of square-integrable orthogonal polynomials of independent random variables [21]. A truncated version of this expansion, as first suggested by Ghanem and Spanos [7], can be used as a solution approximation; further, for the case in which the orthogonal polynomials are Hermite and the random variables centered Gaussians, this truncation is proved to converge in mean-square by Xiu et al. [22] via an application of a theorem by Cameron and Martin [1,5,9].

The most common UQ strategies involve Monte Carlo (MC) sampling, which suffers from a slow convergence rate proportional to the inverse square root of the number of samples [2,14]. If each sample evaluation is expensive—as is often true

© Springer Nature Switzerland AG 2021
M. Paszynski et al. (Eds.): ICCS 2021, LNCS 12747, pp. 698–712, 2021.
https://doi.org/10.1007/978-3-030-77980-1_53

for the solutions of PDEs—this slow convergence rate can make obtaining tens of thousands of samples computationally infeasible [2, 12]. PCE approximations can offer significant computational advantages over Monte Carlo methods in such instances, although there are some exceptions [15].

In [15], Rahman generalizes the classical PCE to account for arbitrary but dependent multi-dimensional Gaussian parameters, proving convergence in mean-square, probability, and distribution. Prior to this work, the multivariate PCE was constrained by the assumption that its input random variables were independent. By introducing correlation into this more general representation, the multi-dimensional analog of Hermite polynomials—referred to as *multivariate Hermite polynomials*—become only *weakly* orthogonal rather than orthogonal [15, 22]. Namely, for two multivariate Hermite polynomials H_α and H_β with multi-indices α and β, which we define formally in Sect. 1.2, weak orthogonality guarantees that

$$\mathbb{E}(H_\alpha(\xi) H_\beta(\xi)) = 0 \quad \text{if } |\alpha| \neq |\beta|, \qquad \xi \sim \mathcal{N}(\mathbf{0}, \Sigma) \tag{1}$$

where $\Sigma \in \mathbb{R}^{n \times n}$ is a real, symmetric positive definite (SPD) covariance matrix. However, $\mathbb{E}(H_\alpha(\xi) H_\beta(\xi))$ can be (and often is) nonzero for *distinct* α, β satisfying $|\alpha| = |\beta|$. The quantities $\mathbb{E}(H_\alpha(\xi) H_\beta(\xi))$ for various α, β are called the *double products* or *second moments* of the multivariate Hermite polynomials. To demonstrate why these double products are important, we will illustrate how a multi-dimensional PCE can be applied in a general setting.

1.1 Application Case and Motivation

For convenience, let $\mathcal{D} \subset \mathbb{R}^n \times \mathbb{R}_{t \geq 0}$ be a compact subset of a spatial and time domain $\mathbb{R}^n_x \times \mathbb{R}_{t \geq 0}$ with initial time $t_0 \geq 0$. Let $u : \mathcal{D} \to \mathbb{R}$ be continuous and differentiable in both its spatial and temporal derivatives; further, let $u \in \mathcal{L}^2(\mathcal{D})$. This u represents the solution a differential equation

$$\mathcal{F}(u, x, t) = 0. \tag{2}$$

Here $\mathcal{F}$ is a general differential operator, often a mix of linear and nonlinear terms. Let $\xi \sim \mathcal{N}(\mathbf{0}, \Sigma)$ be an n-dimensional random variable with known SPD covariance matrix $\Sigma = \mathbb{E}(\xi \xi^T)$. Then ξ has the joint probability density function [13, 22]

$$\phi : \mathbb{R}^n \to \mathbb{R}_{\geq 0} \qquad \phi(x; \Sigma) = \frac{1}{(2\pi)^{n/2} |\det(\Sigma)|^{1/2}} \exp\left(-\frac{1}{2} x^T \Sigma^{-1} x\right). \tag{3}$$

We assume uncertainty is present in the initial condition $u(\cdot, t_0)$ and represent it by setting

$$u(x, t_0; \xi) : \mathbb{R}^n \to \mathbb{R} \qquad u(x, t_0; \xi) = f(x, \xi)$$

where f is a known function of x and ξ. As statistics of ξ, we require that both $u(x, t; \xi)$ and $f(x, \xi)$ have existing second moments.

The multivariate polynomial chaos expansion *separates* the deterministic and random components of u by writing $u(\boldsymbol{x}, t; \boldsymbol{\xi}) = \sum_{\boldsymbol{\alpha} \in \mathbb{N}_0^n} u_{\boldsymbol{\alpha}}(\boldsymbol{x}, t) H_{\boldsymbol{\alpha}}(\boldsymbol{\xi})$ where $\boldsymbol{\alpha} \in \mathbb{N}_0^n$ is a multi-index with ℓ_1 norm $|\boldsymbol{\alpha}|$ (see Sect. 1.2) and $H_{\boldsymbol{\alpha}}$ is a *multivariate Hermite polynomial* (Definition 3). The $u_{\boldsymbol{\alpha}} : \mathcal{D} \to \mathbb{R}$ output deterministic coefficients. The $H_{\boldsymbol{\alpha}}$ are weakly orthogonal (Eq. 1) with respect to the measure $d\boldsymbol{\xi}$ induced by $\boldsymbol{\xi}$. When truncating to some M value, we can define

$$u^{(M)}(\boldsymbol{x}, t; \boldsymbol{\xi}) := \sum_{\substack{\boldsymbol{\alpha} \in \mathbb{N}_0^n \\ 0 \le |\boldsymbol{\alpha}| \le M}} u_{\boldsymbol{\alpha}}^{(M)}(\boldsymbol{x}, t) H_{\boldsymbol{\alpha}}(\boldsymbol{\xi}). \tag{4}$$

As proved in [15], $u^{(M)}(\boldsymbol{x}, t; \boldsymbol{\xi})$ converges to $u(\boldsymbol{x}, t; \boldsymbol{\xi})$ in mean-square, probability, and distribution. The solve Eq. 2, the procedure is to substitute $u^{(M)}$ for u, multiply through by an arbitrary $H_{\boldsymbol{\beta}}$, and integrate with respect to the $d\boldsymbol{\xi}$ measure, repeating for each $H_{\boldsymbol{\beta}}$; this is effectively projecting onto the polynomial basis. From there, orthogonality conditions can be used to eliminate terms. For instance, if Eq. 2 represents the inviscid Burgers' equation,

$$\frac{\partial u}{\partial t} + u \left[\boldsymbol{a}^T \nabla_{\boldsymbol{x}} u \right] = 0 \qquad u(\boldsymbol{x}, 0; \boldsymbol{\xi}) = f(\boldsymbol{x}, \boldsymbol{\xi}) \qquad \boldsymbol{a} \in \mathbb{R}^n \text{ fixed,} \tag{5}$$

then this projection process and weak orthogonality (Eq. 1) gives

$$\sum_{\substack{\boldsymbol{i} \in \mathbb{N}_0^n \\ |\boldsymbol{i}| = |\boldsymbol{k}|}} \frac{\partial u_i^{(M)}}{\partial t} \langle H_i, H_k \rangle + \sum_{\substack{\boldsymbol{i} \in \mathbb{N}_0^n \\ 0 \le |\boldsymbol{i}| \le M}} \sum_{\substack{\boldsymbol{j} \in \mathbb{N}_0^n \\ 0 \le |\boldsymbol{j}| \le M}} u_i^{(M)} (\boldsymbol{a}^T \nabla_{\boldsymbol{x}} u_j^{(M)}) \langle H_i, H_j, H_k \rangle = 0 \tag{6}$$

for *every* $\boldsymbol{k} \in \mathbb{N}_0^n, 1 \le |\boldsymbol{k}| \le M$. Here $\langle H_i, H_k \rangle := \mathbb{E}_{\boldsymbol{\xi}}(H_i(\boldsymbol{\xi}) H_k(\boldsymbol{\xi}))$ denotes the *double product* and $\langle H_i, H_j, H_k \rangle := \mathbb{E}_{\boldsymbol{\xi}}(H_i(\boldsymbol{\xi}) H_j(\boldsymbol{\xi}) H_k(\boldsymbol{\xi}))$ denotes the *triple product*.

Equation 6 is a system of *deterministic* PDEs to solve, where the number of PDEs is equal to the number of $\boldsymbol{k} \in \mathbb{N}_0^n$ such that $1 \le |\boldsymbol{k}| \le M$ for the selected M bound. In particular, a computer can solve such a system via standard numeric techniques if the coefficients $\langle H_i, H_k \rangle$ and $\langle H_i, H_j, H_k \rangle$ are known. Moreover, *any term that is linear in the variable u* in generic $\mathcal{F}(u, \boldsymbol{x}, t)$ of Eq. 2 will generate double product coefficients in the projection process just described; this is not specific to inviscid Burgers' equation.

When the centered $\boldsymbol{\xi}$ is one dimensional, the double and triple products are given by simple expressions [19]. However, when $\boldsymbol{\xi}$ is n-dimensional and correlated, the expressions (first proved in [15]) become cumbersome. *Our contribution provides a new formula (Theorem 2) for the double product of multivariate Hermite polynomials of a centered Gaussian with generic covariance that is both simpler and more computationally efficient to implement than its previous formulation.*

This paper is organized as follows. In Sect. 1.2, we establish the necessary preliminaries for proving our contribution (Theorem 2). Section 2 outlines both the previous formula (Theorem 1) and our contribution for the double product and provides an instructive example (Example 1) in which each formula is applied. In Sect. 3, we compare and discuss computational complexity. Appendices A (Sect. 5) and B (Sect. 6) prove Theorem 2 and report some technical lemmas applied throughout the document.

1.2 Definitions and Notation

The following notation will be utilized throughout this paper:

- $\mathbb{N} :=$ the natural numbers $= \{1, 2, 3, \ldots\}$,
- $\mathbb{N}_0 := \mathbb{N} \cup \{0\} = \{0, 1, 2, \ldots\}$,
- $[n] := \{1, \ldots, n\}$ for any $n \in \mathbb{N}$.

Definition 1 (Multi-index).   *A multi-index α over $[n]$ is an n-tuple $(\alpha_1, \ldots, \alpha_n) \in \mathbb{N}_0^n$ of non-negative integers. Each α_i is referred to as the ith element of the multi-index α.*

For multi-indices $\alpha, \beta \in \mathbb{N}_0^n$, one defines [15,17]

1. *componentwise sum and difference:* $\alpha \pm \beta = (\alpha_1 \pm \beta_1, \ldots, \alpha_n \pm \beta_n)$,
2. *absolute value:* $|\alpha| := \|\alpha\|_1 = \alpha_1 + \cdots + \alpha_n$, which we call the *order* of α,
3. *factorial:* $\alpha! = \alpha_1! \cdots \alpha_n! = \prod_{i=1}^{n} \alpha_i!$, and the
4. *partial derivative:* $D_\alpha^{|\alpha|} = \partial_1^{\alpha_1} \cdots \partial_n^{\alpha_n}$.

Moving forward, any α will denote a multi-index over $[n]$ of order $k \geq 1$ unless otherwise specified.

Sometimes we will need to express α in what we will refer to as its *mutliset notation*. Recall that a multiset is a modification of a set that allows for multiple instances of each of its elements. More formally [8,18], a *multiset M* on a set S is a pair (S, ν), where ν is a function $\nu : S \to \mathbb{N}$ assigning each element $x \in S$ its positive multiplicity i.e. the number of times x is repeated in M. We consider both the multi-index (Definition 1) and the proposed multiset notation (Definition 2) for a label α, because

1. the *multi-index* version is standard in relevant previous literature [13,15,17], and
2. the *multiset* version can be easier to utilize, as later showcased in Theorem 2.

Definition 2 (Multiset notation). *For a multi-index α over $[n]$ of order $|\alpha| = k \geq 1$, let $s(\alpha)$ denote the map*[1]

$$s(\alpha) : [k] \to [n] \qquad s(\alpha)(\ell) := s(\alpha)_\ell = \min\{i \in [n] \mid \ell \leq \sum_{r=1}^{i} \alpha_r\},$$

---

[1] Note that the set $\min\{i \in [n] \mid \ell \leq \sum_{r=1}^{n} \alpha_r\}$ is nonempty, because $\ell \leq k = \sum_{r=1}^{n} \alpha_r$, so $i = n$ always satisfies the condition that $\ell \leq \sum_{r=1}^{i} \alpha_r$.

which we call the multiset notation *for α. It is straightforward (see Lemma 1) to verify that*

$$[s(\alpha)_1, \ldots, s(\alpha)_k] = [\overbrace{1, \ldots, 1}^{\alpha_1 \text{ times}}, \ldots, \overbrace{n \cdots n}^{\alpha_n \text{ times}}] = [1^{\alpha_1}, \ldots, n^{\alpha_n}].$$

Sometimes we use $s(\alpha)$ to refer to the output $[s(\alpha)_1, \ldots, s(\alpha)_k]$ of the map across its whole domain $[k]$ rather than to the map itself, with context making the distinction clear. The elements of $s(\alpha)$ are $s(\alpha)_1, \ldots, s(\alpha)_k$, and the order $|s(\alpha)|$ of $s(\alpha)$ is the total number of such elements k, which is also the order of α. When $k = 0$, so that $\alpha = (0, \ldots, 0)$, we say $s(\alpha)$ is empty.

The name *multiset notation* is chosen, since the outputs of $s(\alpha)$ are reminiscent of a multiset when written as $[s(\alpha)_1 \ldots s(\alpha)_k]$, with each $i \in [n]$ represented with multiplicity α_i in the array (Lemma 1). Note that $\alpha_i = 0$ indicates that an i is not present in the $s(\alpha)$ array.

As an example, if $\alpha = (2, 1, 0, 0, 1)$, then $s(\alpha)$ has order $k = 2 + 1 + 1 = 4$ with $s(\alpha)_1 = s(\alpha)_2 = 1, s(\alpha)_3 = 2$, and $s(\alpha)_4 = 5$. Thus, we represent α via the map $s(\alpha)$ by the multiset notation $s(\alpha) = [1, 1, 2, 5]$.

Finally, the *partial derivative operator* for $s(\alpha)$ is defined as

$$D_{s(\alpha)}^k := \partial_{s(\alpha)_1} \cdots \partial_{s(\alpha)_k} = \partial_1^{\alpha_1} \cdots \partial_n^{\alpha_n}$$

so that $D_{s(\alpha)}^k = D_\alpha^k$ as expected. With this notation in place, we present the multi-dimensional analog of the Hermite polynomial.

Definition 3 (Multivariate Hermite polynomial). *Let $\xi \sim \mathcal{N}(0, \Sigma)$ such that $\Sigma \in \mathbb{R}^{n \times n}$ is symmetric positive definite (SPD) with joint density function $\phi(\xi; \Sigma)$ given by Eq. 3. Then for any multi-index α over $[n]$, the multivariate Hermite polynomial $H_\alpha(\xi; \Sigma)$ indexed by α is a polynomial in ξ of degree $|\alpha|$ defined as[2]*

$$H_\alpha(\xi; \Sigma) = \begin{cases} \frac{(-1)^{|\alpha|}}{\phi(\xi; \Sigma)} D_\alpha^{|\alpha|}(\phi(\xi; \Sigma)) & \text{if } |\alpha| \geq 1 \\ 1 & \text{if } |\alpha| = 0. \end{cases}$$

With multiset notation $s(\alpha)$, note that $H_{s(\alpha)}(\xi; \Sigma) = H_\alpha(\xi; \Sigma)$, since $D_\alpha^{|\alpha|}$ and $D_{s(\alpha)}^{|s(\alpha)|}$ denote identical derivative operators.

To establish notation for the proof of Theorem 2, let

$$T_\alpha = \{(\ell, s(\alpha)_\ell) \mid \ell \in [k]\} \tag{7}$$

be the set of k-tuples, one for each index $\ell \in [k]$. For multi-indices α, β of the same order k, we will consider *bijections between T_α and T_β*, i.e. the ways to pair-off the elements of T_α and T_β. As a heuristic, we can think of this as the

---

[2] Note that these multivariate H_α are not normalized to force $\mathbb{E}(H_\alpha(\xi)^2) = 1$, as is sometimes done in other literature [15].

number of ways to draw lines between the "entries" of $s(\alpha)$ and $s(\beta)$ such that each entry has a unique partner and all entries are covered. For instance, for $s(\alpha) = [1,1,3]$ and $s(\beta) = [2,3,4]$, with $n = 4$, we have the $3! = 6$ options depicted in Fig. 1. Observe how the copies of entries (e.g. 1 in $s(\alpha) = [1,1,3]$) are treated as distinct when drawing these lines; in particular, the number of such pairings for $|\alpha| = |\beta| = k$ will always be $k!$.

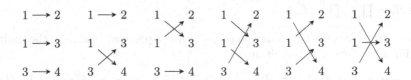

Fig. 1. All possible bijections between T_α and T_β for $\alpha = (2,0,1,0)$ and $\beta = (0,1,1,1)$. Both α, β are multi-indices of order $k = 3$ over $[4]$, with $s(\alpha) = [1,1,3]$ and $s(\beta) = [2,3,4]$.

If we are imagining pairings of the form $(s(\alpha)_\ell, s(\beta)_j)$, why discuss bijections between T_α and T_β rather than bijections between the $s(\alpha)$ and $s(\beta)$ outputs directly? There are several reasons. For one, we would like to have "repeated" mappings counted with multiplicity rather than treated as single entities.[3] For instance, in Fig. 1 the first and third mappings are counted as distinct, even though they choose the same pairings $(1,2), (1,3), (3,4)$.

The bijections between T_α and T_β are the foundation for our new and computationally efficient double-product formula (Theorem 2). As we shall see in Sect. 2, the proposed formula involves multiplying entries of the inverse covariance matrix Σ^{-1}—and the bijections between T_α and T_β determine precisely which entries of Σ^{-1} are selected in this calculation.

2 Double Product Formulations

For both Theorems 1 and 2, let Σ^{-1} denote the known $n \times n$ SPD inverse matrix of the generic covariance Σ given in Definition 3. In this context, the second-order moments of multivariate Hermite polynomials were first proved in the comprehensive work of [15] to equal the expression in Theorem 1 below.

Theorem 1. (Proved in [15]). *Let $\theta \in \mathbb{N}_0^{n \times n}$. Define $r(\theta)$ as the vector of row sums of θ; that is,*

$$r(\theta) = (r_1, \ldots, r_n)^T \text{ with } r_i = \sum_{j=1}^n \theta_{ij} = \|\theta_{i\bullet}\|_1 = \ell_1 \text{ norm of the ith row of } \theta.$$

---

[3] Accordingly, we introduce $s(\alpha)$ as multiset *notation* rather than a *literal* multiset. An underlying philosophy of multisets is that copies of elements cannot be picked out or distinguished by (say) an indexing convention [8,10,16]. For our purposes, however, we want to treat such copies as distinct.

Similarly, let $c(\boldsymbol{\theta})$ be the $n \times 1$ vector such that $c_j = \|\boldsymbol{\theta}_{\bullet j}\|_1$. Then

$$
\langle H_\alpha, H_\beta \rangle = \begin{cases} \alpha!\beta! \displaystyle\sum_{\substack{\boldsymbol{\theta} \in \mathbb{N}_0^{n \times n} \\ r(\boldsymbol{\theta})=\alpha, c(\boldsymbol{\theta})=\beta}} \dfrac{\prod_{p=1}^n \prod_{q=1}^n \left(\Sigma_{pq}^{-1}\right)^{\theta_{pq}}}{\boldsymbol{\theta}!}, & if \ |\alpha| = |\beta| \\[6pt] 0 & else. \end{cases}
$$

where $\boldsymbol{\theta}! = \prod_{i=1}^n \prod_{\ell=1}^n \theta_{i\ell}!$.

We now propose a novel evaluation of the double product based on the multiset notation introduced in Definition 2.

Theorem 2. *Let α, β be multi-indices over $[n]$ with $|\alpha| = |\beta| = k \geq 1$. Then*

$$
\langle H_\alpha, H_\beta \rangle = \sum_{p \in S_k} \prod_{i=1}^k \Sigma^{-1}_{s(\alpha)_i, \, s(\beta)_{p(i)}}
$$

where $\Sigma^{-1}_{s(\alpha)_i, \, s(\beta)_{p(i)}}$ is the $\left(s(\alpha)_i, s(\beta)_{p(i)}\right)$th entry of Σ^{-1} and S_k is the symmetric group on $\{1, \ldots, k\}$, recalling the multiset index notation $s(\alpha)$ given in Definition 2. If $|\alpha| \neq |\beta|$, then $\langle H_\alpha, H_\beta \rangle = 0$.

Note that Theorem 2 only considers when $k > 0$, since the case $k = 0$ is trivial. To explore and illustrate the differences between these two formulas for the double product, we provide the following Example 1.

Example 1. Let $\alpha = (2, 0, 1, 0)$ and $\beta = (0, 1, 1, 1)$ over $[4]$, as they were in Fig. 1. To use Theorem 1, we are searching for $\boldsymbol{\theta} \in \mathbb{N}_0^{4 \times 4}$ such that $r(\boldsymbol{\theta}) = (2, 0, 1, 0)$ and $c(\boldsymbol{\theta}) = (0, 1, 1, 1)$. Let $r^{(i)}$ denote the ith row of $\boldsymbol{\theta}$. For the constraint $r(\boldsymbol{\theta}) = (2, 0, 1, 0)$, noting that rows 2 and 4 must be zeroes, we have $\binom{4+2-1}{2} = 10$ options for $r^{(1)}$ and $\binom{4+1-1}{2} = 3$ options for $r^{(3)}$ [18]. Naively, we could then check all $10 \times 3 = 30$ options for $\boldsymbol{\theta}$ and eliminate those that fail to satisfy the column constraint $c(\boldsymbol{\theta}) = \beta$. To be clever, we can eliminate the $r^{(i)}$ along the way whose entries $r_j^{(i)} \geq \beta_j$, since these rows guarantee that some columns sums in $\boldsymbol{\theta}$ will be too large—which is indicated by the slashes in Fig. 2. Hence, we have $3 \times 3 = 9$ initial matrices $\boldsymbol{\theta}$ to iterate through to find those such that $c(\boldsymbol{\theta}) = \beta$, from which there are 3 final candidates (Fig. 2).

Each of these $\boldsymbol{\theta}$ satisfies $\boldsymbol{\theta}! = 1$. Compute

$$
\langle H_\alpha, H_\beta \rangle = 2[\Sigma_{12}^{-1}\Sigma_{13}^{-1}\Sigma_{34}^{-1} + \Sigma_{12}^{-1}\Sigma_{14}^{-1}\Sigma_{33}^{-1} + \Sigma_{13}^{-1}\Sigma_{14}^{-1}\Sigma_{23}^{-1}],
$$

where we tacitly used that Σ^{-1} is symmetric.

By using the formulation in Theorem 2 instead, we have the $3! = 6$ terms to consider from the start, one for each $p \in S_3$. Once a computer obtains these S_3 entries, which often is elementary and trivially fast to do,[4] the $\Sigma^{-1}_{(s(\alpha)_i, \, s(\beta)_{p(i)})}$

---

[4] For instance, in **Python3**, the **combinatorics** module in **itertools** [20] suffices.

$r^{(1)}$ options: $(2,0,0,0), (0,2,0,0), (0,0,2,0), (0,0,0,2), (1,1,0,0), (1,0,1,0), (1,0,0,1),$

$\qquad\qquad (0,1,1,0), (0,1,0,1), (0,0,1,1)$

$r^{(3)}$ options: $(0,1,0,0), (0,0,1,0), (0,0,0,1)$.

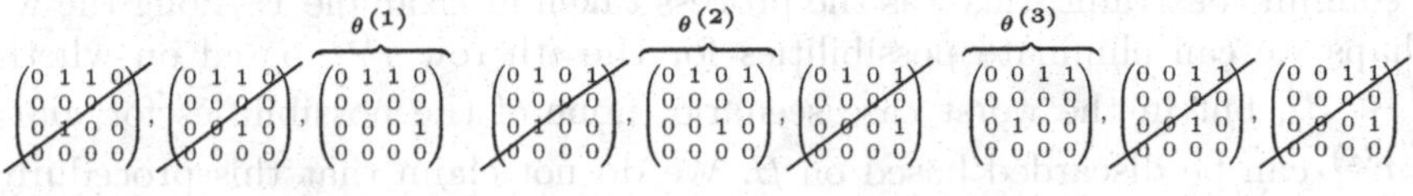

Fig. 2. Determining the possible $\theta \in \mathbb{N}_0^{4\times4}$ such that $r(\theta) = (2,0,1,0)$ and $c(\theta) = (0,1,1,1)$ in Example 1 to use Theorem 1. First we find the options for $r^{(1)}$ and $r^{(3)}$ (rows 1 and 3) of θ such that $r^{(1)}$ sums to 2 and $r^{(3)}$ sums to 1, eliminating the options along the way that have entries $r_j^{(i)} \geq \beta_j$, as indicated by the slashes. The $3 \times 3 = 9$ possible θ are iterated over to see which satisfy the column sums constraint $c(\theta) = (0,1,1,1)$. This leaves the 3 matrices denoted by $\theta^{(1)}, \theta^{(2)}, \theta^{(3)}$ above.

can be evaluated directly. Alternatively, we evaluate Σ^{-1} at the pairs matched by the mappings drawn in Fig. 1. From the final row in Table 1, which sums the entries of the previous rows, we yield the same $\langle H_\alpha, H_\beta \rangle$ as was found with Theorem 1 previously.

What happens to the product in Theorem 2 when the ξ_i are uncorrelated? In this case, every Σ_{ij}^{-1} in which $i \neq j$ equals zero. Similarly, If $\alpha \neq \beta$, then it is easy to show that for every $p \in S_k$ there is at least one $\ell \in [k]$ for which $s(\alpha)_\ell \neq s(\beta)_{p(\ell)}$. As expected, then Theorem 2 gives that $\mathbb{E}(H_\alpha H_\beta) = 0$.

Table 1. Using Theorem 2 to compute $\langle H_\alpha, H_\beta \rangle$ in Example 1. For each permutation p on $\{1, 2, 3\}$, we find the tuples $(s(\alpha)_i, s(\beta)_{p(i)})$ for $s(\alpha) = [1, 1, 3]$ and $s(\beta) = [2, 3, 4]$. Then the product $\prod_{i=1}^3 \Sigma_{s(\alpha)_i, s(\beta)_{p(i)}}^{-1}$ is evaluated at these tuples. Equivalently, the $(s(\alpha)_i, s(\beta)_{p(i)})$ are precisely the pairings shown in the maps of Fig. 1 (left to right), matched by color to show correspondence across the two figures.

$p \in S_3$	$(s(\alpha)_i, s(\beta)_{p(i)}) \; \forall_{i\in[3]}$, i.e. pairings in Fig. 1	$\prod_{i=1}^3 \Sigma_{s(\alpha)_i, s(\beta)_{p(i)}}^{-1}$
(123)	$(1, 2), (1, 3), (3, 4)$	$\Sigma_{12}^{-1} \Sigma_{13}^{-1} \Sigma_{34}^{-1}$
(132)	$(1, 2), (1, 4), (3, 3)$	$\Sigma_{12}^{-1} \Sigma_{14}^{-1} \Sigma_{33}^{-1}$
(213)	$(1, 3), (1, 2), (3, 4)$	$\Sigma_{12}^{-1} \Sigma_{13}^{-1} \Sigma_{34}^{-1}$
(231)	$(1, 3), (1, 4), (3, 2)$	$\Sigma_{13}^{-1} \Sigma_{14}^{-1} \Sigma_{23}^{-1}$
(312)	$(1, 4), (1, 2), (3, 3)$	$\Sigma_{12}^{-1} \Sigma_{14}^{-1} \Sigma_{33}^{-1}$
(321)	$(1, 4), (1, 3), (3, 2)$	$\Sigma_{13}^{-1} \Sigma_{14}^{-1} \Sigma_{23}^{-1}$
$\langle H_\alpha, H_\beta \rangle$	$2[\Sigma_{12}^{-1} \Sigma_{13}^{-1} \Sigma_{34}^{-1} + \Sigma_{12}^{-1} \Sigma_{14}^{-1} \Sigma_{33}^{-1} + \Sigma_{13}^{-1} \Sigma_{14}^{-1} \Sigma_{23}^{-1}]$	

3 Comparing Computational Complexity

We assume that the inverse covariance matrix Σ^{-1} and α, β of order k are given. To use Theorem 1 to compute a single double product, the computation time is dominated by producing all the $\theta \in \mathbb{N}_0^n$ such that $r(\theta) = \alpha, c(\theta) = \beta$.[5] A reasonable, albeit naive, algorithm for finding all such θ is to first generate the possible θ such that $r(\theta) = \alpha$ and then eliminate those which do not satisfy the column constraint; this was the process taken in Example 1. Along the way, perhaps we can eliminate possibilities for the ith row $r^{(i)}$ based on whether $r_j^{(i)} \leq \beta_j$, but in the worst case scenario, none of the possibilities for *any* of the $r^{(i)}$ can be discarded based on β. We do not claim that this procedure is the most efficient computation of $\langle H_\alpha, H_\beta \rangle$ via Theorem 1—but we will use it as the straightforward benchmark for comparison against computing the double product via Theorem 2.

Before considering column constraints,

$$\# \text{ of options for row } i = \# \text{ of } n\text{-tuples whose entries sum to } \alpha_i$$
$$= \binom{n+\alpha_i-1}{n-1}$$

by [8, 10, 15, 16, 18]. Following §4.5.1 in [16], there exist algorithms that output all of the options for $r^{(i)}$ with computational complexity proportional to the number of options, i.e. $\binom{n+\alpha_i-1}{n-1}$. Repeating for all of the rows, there are at least

$$\binom{n+\alpha_1-1}{n-1} \cdots \binom{n+\alpha_n-1}{n-1} = \frac{(n-1)!^n \overbrace{[(n)(n+1)\cdots(n+\alpha_1-1)]}^{\alpha_1 \text{ terms}} \cdots \overbrace{[(n)(n+1)\cdots(n+\alpha_n-1)]}^{\alpha_n \text{ terms}}}{(n-1)!^n \alpha!}$$

$$\geq \frac{n^{\alpha_1}\cdots n^{\alpha_n}}{\alpha!} = \frac{n^{\sum_i \alpha_i}}{\alpha!} = \frac{n^k}{\alpha!} \geq \frac{n^k}{k!} \qquad \left[\text{since } \max_{|\alpha|=k} \alpha! = k!\right]$$

options for $\theta \in \mathbb{N}_0^{n \times n}$ in Eq. 1 such that $r(\theta) = \alpha$. Thus, producing the necessary θ to sum over in Eq. 1 involves iterating over at least $\frac{n^k}{k!}$ matrices in terms of asymptotic complexity. Per θ, computing $\prod_{p=1}^n \prod_{q=1}^n (\Sigma_{pq}^{-1})^{\theta_{pq}}$ involves a total of at least $\sum_{p,q=1}^n \theta_{p,q} = k$ multiplications. Then the computational complexity of implementing Eq. 1 in this direct manner is lower bounded by $\frac{n^k}{(k-1)!}$, which is exponential in k.

When computing the double product via Theorem 2, there are $k!$ terms in the summation, and each summand is the product of k entries of Σ^{-1}. So the cost for computing $\langle H_\alpha, H_\beta \rangle$ in this case is factorial in k, namely $O(kk!) = O(k!)$, for α, β of order k.

4   Conclusion

Polynomial chaos (PC) expansions are effective for incorporating and quantifying uncertainties in problems governed by partial differential equations. In some

---

[5] Counting the number of such index matrices, which are often called *contingency tables with fixed margins* in statistics literature, is well-studied [3,6] and can be done in poly(n) time [4]. This does not mean that the *number* of contingency tables is poly(n) but that algorithms can produce the total count of them in poly(n) time.

contexts, they offer significant computational advantages to classic Monte Carlo sampling methods (for example) [2,15], whose converge rates are especially hindered when each sample evaluation of the PDE is expensive [2,12]. However, when multiple input uncertainties are considered without transformations, PC approaches cannot be generalized in a simple fashion unless the uncertainties are represented in terms of independent variables. Unlike when ξ is one dimensional or uncorrelated, many of the double product coefficients $\langle H_\alpha, H_\beta \rangle := \mathbb{E}_\xi (H_\alpha H_\beta)$ that appear from the Galerkin projections are nonzero and therefore *essential* to compute a priori in order to solve the resulting system numerically.

In this paper, we prove a new formula (Theorem 2) for the double product of two multivariate Hermite polynomials whose n-dimensional input Gaussian random variable has an arbitrary SPD covariance matrix. To do so, we introduce what we call *multiset notation* (Definition 2) for the label indices α. Calculating the double product is computationally more efficient and (arguably) simpler with the proposed approach than doing so with the classical formula in [15] given by Theorem 1. In particular, Sect. 3 analyzes the computational complexity of the two formulations; the implementations considered for each were purposely straightforward and already showcase the reduced cost achieved by the use of the multiset notation.

From the foundational work in this paper, the authors plan to explore the triple product $\langle H_\alpha, H_\beta, H_\gamma \rangle$ calculations in terms of these double product constituents. As demonstrated in inviscid Burgers' equation (Eq. 5), the triple products can arise when the original PDE has quadratic *nonlinear* terms. Establishing these triple product values will be a pivotal building block for handling nonlinear PDEs that incorporate uncertainties in a general setting.

5 Appendix A: Proof of Theorem 2

The proof of Theorem 2 relies on the following Theorems 3 and 4. In the following discussion, assume α is a multi-index over $[n]$ of order $k \geq 1$ unless otherwise specified.

Recall the definition of T_α in Eq. 7 in Sect. 1.2. We assign an ordering to the elements of T_α (or any subset of T_α) based on their first components ascending. That is, $(T_\alpha)_1 = (1, s(\alpha)_1)$, $\ldots, (T_\alpha)_k = (k, s(\alpha)_k)$, and when $A \subseteq T_\alpha$, we label $A_1 = (\ell_1, s(\alpha)_{\ell_1}), \ldots, A_{|A|} = (\ell_{|A|}, s(\alpha)_{\ell_{|A|}})$ such that $\ell_1 < \cdots < \ell_{|A|}$. Before fretting about the specifics, realize that this ordering follows intuition. For example, if $\alpha = (2, 1, 0, 0, 1)$, then $s(\alpha) = [1, 1, 2, 5]$, and $T_\alpha = \{(1,1), (2,1), (3,2), (4,5)\}$. Now, $(T_\alpha)_1 = (1,1), (T_\alpha)_2 = (2,1), (T_\alpha)_3 = (3,2)$, and $(T_\alpha)_4 = (4,5)$. For the subset $A = T_\alpha \setminus \{(2,1)\} = \{(1,1), (3,2), (4,5)\}$ of T_α, we have $A_1 = (1,1), A_2 = (3,2)$, and $A_3 = (4,5)$. In fact, we will use the shorthand

$$(T_\alpha)^{-j} := T_\alpha \setminus \{ (j, s(\alpha)_j) \} \qquad \text{for any } j \in [k], \tag{8}$$

where $(T_\alpha)^{-j}_\ell$ is the ℓth element of $(T_\alpha)^{-j}$ according to this ordering by first components ascending.

For any $i \in [n]$ such that $\alpha_i > 0$, define

$$i^* = \min\left(s(\alpha)^{-1}(\{i\})\right) = \min\{\ell \in [k] \mid s(\alpha)_\ell = i\}. \tag{9}$$

By Lemma 2, we have

$$(T_\alpha)_\ell^{-i^*} = \begin{cases} (\ell, s(\alpha)_\ell) & \text{if } \ell < i^* \\ (\ell+1, s(\alpha)_{\ell+1}) & \text{else} \end{cases} = \begin{cases} (\ell, s(\alpha - e_i)_\ell) & \text{if } \ell < i^* \\ (\ell+1, s(\alpha - e_i)_\ell) & \text{else}. \end{cases}$$

Note that the elements of $T_{\alpha-e_i}$ have the form $(\ell, s(\alpha - e_i)_\ell)$ for $\ell \in [k-1]$. Therefore, *the second coordinate of the ℓth element of $(T_\alpha)_\ell^{-i^*}$ is identical to the second coordinate of the ℓth element of $T_{\alpha-e_i}$.*

Define the projector operator

$$\text{proj} : [k] \times [n] \to [n] \qquad \text{proj}(t_1, t_2) = t_2 \tag{10}$$

that simply ignores the first coordinate of its input. Then for any $\ell \in [k-1]$ and $i \in [n]$ such that $\alpha_i > 0$,

$$\text{proj}((T_{\alpha-e_i})_\ell) = \text{proj}((T_\alpha^{-i^*})_\ell), \qquad i^* = \min s(\alpha)^{-1}(\{i\}). \tag{11}$$

Equation 11, along with the definitions in Eqs. 7, 8, 10, will be utilized in the proof of Theorem 3.

Theorem 3. *Let α, β be two multi-indices over $[n]$ such that $|\alpha| = |\beta| = k \geq 1$. Then*

$$\frac{\partial^{|\beta|} H_\alpha}{\partial \xi_\beta} = \sum_{p \in S_k} \prod_{i=1}^{k} \Sigma_{s(\alpha)_i, s(\beta)_{p(i)}}^{-1}.$$

Proof. We proceed by induction on k. When $k = 1$, $\alpha = e_r$ and $\beta = e_s$ for some $r, s \in [n]$, so the base case is proved by Lemma 4. For the inductive step, let $j = \min\{i \in [n] \mid \beta_i > 0\}$, where we know j exists since $|\beta| = k \geq 1$. Let $g_{\alpha,j,\Sigma}$ be a function on $[n]$ such that $g_{\alpha,j,\Sigma}(i) = \Sigma_{ij}^{-1} H_{\alpha-e_i}$. Then

$$\frac{\partial}{\partial \xi_j}(H_\alpha) = \sum_{i=1}^{n} \alpha_i \Sigma_{ij}^{-1} H_{\alpha-e_i} = \sum_{i=1}^{n} \alpha_i\, g_{\alpha,j,\Sigma}(i) \qquad [\text{Lem. 5 \& def. of } g_{\alpha,j,\Sigma}]$$

$$= \sum_{i=1}^{k} g_{\alpha,j,\Sigma}(s(\alpha)_i) = \sum_{i=1}^{k} \Sigma_{s(\alpha)_i, j}^{-1} H_{\alpha-e_{s(\alpha)_i}} \qquad [\text{Lem. 6}].$$

Substituting,

$$D_\beta^k(H_\alpha) = D_{\beta-e_j}^{k-1} \frac{\partial}{\partial \xi_j}(H_\alpha) = \sum_{i=1}^{k} \Sigma_{s(\alpha)_i, j}^{-1} D_{\beta-e_j}^{k-1}\left(H_{\alpha-e_{s(\alpha)_i}}\right)$$

$$= \sum_{i=1}^{k} \Sigma_{s(\alpha)_i, j}^{-1} \sum_{p \in S_{k-1}} \prod_{\ell=1}^{k-1} \Sigma_{s(\alpha-e_{s(\alpha)_i})_\ell, s(\beta-e_j)_{p(\ell)}}^{-1} \qquad [\text{ind. hypothesis}].$$

Let
$$i^* = \min\{\ell \in [k] \mid s(\alpha)_\ell = s(\alpha)_i\}, \quad j^* = \{\ell \in [k] \mid s(\beta)_\ell = j\}.$$

For all $\ell \in [k-1]$, Eqs. 7, 8, 10, and 11 give
$$\text{proj}\left((T_\alpha^{-i^*})_\ell \right) = \text{proj}\left((T_{\alpha - e_{s(\alpha)_i}})_\ell \right) = s(\alpha - e_{s(\alpha)_i})_\ell$$
$$\text{proj}\left((T_\beta^{-j^*})_\ell \right) = \text{proj}\left((T_{\beta - e_j})_\ell \right) = s(\alpha - e_j)_\ell.$$

Let $h : T_\alpha \times T_\beta \to \mathbb{R}$ such that $h(r, t) = \Sigma^{-1}_{\text{proj}(r), \text{proj}(t)}$, noting that $T_\alpha^{-i^*} \subset T_\alpha$ and $T_\beta^{-j^*} \subset T_\beta$. Then

$$D_\beta^k(H_\alpha) = \sum_{i=1}^k \Sigma^{-1}_{s(\alpha)_i, j} \sum_{p \in S_{k-1}} \prod_{\ell=1}^{k-1} h\left((T_\alpha^{-i^*})_\ell, (T_\beta^{-j^*})_{p(\ell)} \right) \quad \text{[def. of } h\text{]}$$

$$= \sum_{i=1}^k \Sigma^{-1}_{s(\alpha)_i, j} \sum_{b: T_\alpha^{-i^*} \hookrightarrow T_\beta^{-j^*}} \prod_{t \in T_\alpha^{-i^*}} h(t, b(t)) \quad \text{[by Lemma 7, Eq. 12]}$$

where we can match notation from Eq. 12 in Lemma 7 by setting $A = T_\alpha^{-i^*}$ and $B = T_\beta^{-j^*}$. Now,

$$s(\alpha)_i = \text{proj}(\,(i^*, s(\alpha)_i)\,) \quad \text{[def. of proj map in Eq. 10]}$$
$$= \text{proj}(\,(i^*, s(\alpha)_{i^*})\,) \quad \text{[since } s(\alpha)_{i^*} = s(\alpha)_i \text{ by def. of } i^*\text{]}$$
$$= \text{proj}((T_\alpha)_{i^*}) \quad \text{[labeling of } T_\alpha \text{ elements]}.$$

By a similar argument, $j = \text{proj}((T_\beta)_{j^*})$. Therefore,

$$D_\beta^k(H_\alpha) = \sum_{i=1}^k h((T_\alpha)_{i^*}, (T_\beta)_{j^*}) \sum_{b: T_\alpha^{-i^*} \hookrightarrow T_\beta^{-j^*}} \prod_{t \in T_\alpha^{-i^*}} h(t, b(t))$$

$$= \sum_{b: T_\alpha \hookrightarrow T_\beta} \prod_{t \in T_\alpha} h(t, b(t)) \quad \text{[Lem. 7, Eq. 13]}$$

$$= \sum_{p \in S_k} \prod_{\ell=1}^k h((T_\alpha)_\ell, (T_\beta)_{p(\ell)}) \quad \text{[Lem. 7, Eq. 12]}$$

$$= \sum_{p \in S_k} \prod_{\ell=1}^k \Sigma^{-1}_{s(\alpha)_\ell, s(\beta)_{p(\ell)}} \quad \text{[def. of } h\text{]}$$

as desired. $\qquad\square$

Theorem 4. *Let α, β be two multi-indices over $[n]$. Then*

$$\mathbb{E}(H_\alpha H_\beta) = \begin{cases} \dfrac{\partial^{|\beta|} H_\alpha}{\partial \xi_\beta} & \text{if } |\alpha| = |\beta| \\ 0 & \text{else.} \end{cases}$$

Proof. It is proved in [15] that $|\alpha| \neq |\beta|$ implies that $\mathbb{E}(H_\alpha H_\beta) = 0$. So suppose $|\alpha| = |\beta| = k$, and we proceed by induction on k. The $k = 1$ base case is a straightforward consequence of Lemma 3 and the fact that the expected value of any multivariate Hermite polynomial is zero by [15]. For the inductive step, assume $|\alpha| = |\beta| = k$. Let $j = \min\{\ell \in [n] \mid \beta_\ell > 0\}$, where we know such a j exists since $|\beta| > 0$. By Lemma 5 and linearity of expectation,

$$
\begin{aligned}
\mathbb{E}\left(D_\alpha^k(H_\beta)\right) &= \sum_{i=1}^n \alpha_i \Sigma_{ij}^{-1} \mathbb{E}\left(D_{\beta-e_j}^{k-1}(H_{\alpha-e_i})\right) \\
&= \sum_{i=1}^n \Sigma_{ij}^{-1} \mathbb{E}\left(H_{\alpha-e_i} H_{\beta-e_j}\right) && \text{[inductive hypothesis]} \\
&= \mathbb{E}\left(H_{\beta-e_j} \sum_{i=1}^n \alpha_i \Sigma_{ij}^{-1} H_{\alpha-e_i}\right) \\
&= \mathbb{E}\left(H_{\beta-e_j} D_{e_j}(H_\alpha)\right) && \text{[Lem. 5]} \\
&= \mathbb{E}\left(H_{\beta-e_j} H_\alpha H_{e_j}\right) - \mathbb{E}\left(H_{\beta-e_j} H_{\alpha+e_j}\right) && \text{[Lem. 3 for } H_{\alpha+e_j}\text{]}.
\end{aligned}
$$

From $|\beta - e_j| \neq |\alpha + e_j|$, $\mathbb{E}(H_{\beta-e_j} H_{\alpha+e_j}) = 0$. Therefore, $\mathbb{E}(D_\alpha^k(H_\beta)) = \mathbb{E}(H_\alpha H_{\beta-e_j} H_{e_j})$. Applying Lemma 3 to H_β, $\mathbb{E}(H_\alpha H_{\beta-e_j} H_{e_j}) = \mathbb{E}(H_\alpha H_\beta) + \mathbb{E}(H_\alpha D_{e_j}(H_{\beta-e_j}))$. By Lemma 5, we know that $D_{e_j}(H_{\beta-e_j})$ is a linear combination of polynomials of the form $H_{\beta-e_j-e_r}$. Hence, $\mathbb{E}(H_\alpha D_{e_j}(H_{\beta-e_j}))$ is a linear combination of such terms $\mathbb{E}(H_\alpha H_\gamma)$ for $|\alpha| \neq |\gamma|$, each of which is zero. Thus, $\mathbb{E}(D_\beta^k(H_\alpha)) = \mathbb{E}(H_\alpha H_\beta)$. Finally, we know from Theorem 3 that $D_\beta^k(H_\alpha)$ is deterministic (since it is independent of $\boldsymbol{\xi}$), so $\mathbb{E}(H_\alpha H_\beta) = D_\beta^k(H_\alpha) = \frac{\partial^k H_\alpha}{\partial \xi_\beta}$. □

Proof (of Theorem 2). Combining Theorems 3 and 4 immediately gives the desired result.

6   Appendix B

For brevity, several proofs are omitted, but we outline them here. Lemmas 1 and 2 are straightforward. Lemmas 3 and 4 involve differentiating the density ϕ in Definition 3 directly. Lemma 5 is proved by induction and applying Lemma 3. Lemma 6 follows from decomposing $[k]$ into the preimage sets $s(\alpha)^{-1}(\{\ell\})$ for all $\ell \in [n]$. Lemma 7 is a specific application of an elementary combinatorial argument that regards every bijection between two sets as an extension of a bijection on two smaller subsets [11].

Lemma 1. *For multi-index α over $[n]$ of order $k > 0$,*

1. *$s(\alpha)$ is non-decreasing in its indices, i.e. $s(\alpha)_\ell \leq s(\alpha)_{\ell+1}$ for all $\ell \in [k-1]$,*
2. *for fixed $j \in [n]$ such that $\alpha_j > 0$, $\min s(\alpha)^{-1}(\{j\}) = \sum_{r=1}^{j-1} \alpha_r + 1$,*
3. *each $j \in [n]$ appears α_j total times in $[s(\alpha)_1, \ldots, s(\alpha)_k]$.*

Lemma 2. *Let $\boldsymbol{\alpha}$ be a multi-index over $[n]$ such that $|\boldsymbol{\alpha}| = k > 0$. Let $i \in [n]$ such that $\alpha_i > 0$. Define $i^* = \min\{\ell \in [k] \mid s(\boldsymbol{\alpha})_\ell = i\}$. Then for $\ell \in [k-1]$, $s(\boldsymbol{\alpha} - e_i)_\ell = s(\boldsymbol{\alpha})_\ell$ if $\ell < i^*$ and $s(\boldsymbol{\alpha} - e_i)_\ell = s(\boldsymbol{\alpha})_{\ell+1}$ otherwise.*

Lemma 3. *Let $\boldsymbol{\alpha}$ be a multi-index over $[n]$. Then for any $i \in [n]$ such that $\alpha_i > 0$, $H_{\boldsymbol{\alpha}} = H_{\boldsymbol{\alpha} - e_i} H_{e_i} - \frac{\partial}{\partial \xi_i} H_{\boldsymbol{\alpha} - e_i}$.*

Lemma 4. *Let $\boldsymbol{\xi} \sim \mathcal{N}(\mathbf{0}, \Sigma)$ be $\mathbb{R}^n$-valued. Then for any $\ell \in [n]$, $H_{e_\ell}(\boldsymbol{\xi}; \Sigma) = (\Sigma^{-1})_{\ell \bullet} \boldsymbol{\xi}$, where $(\Sigma^{-1})_{\ell \bullet}$ is the ℓth row of the inverse covariance matrix Σ, and e_ℓ is the ℓth standard basis vector written as a multi-index. Thus, $\frac{\partial H_{e_\ell}(\xi; \Sigma)}{\partial \xi_j} = \Sigma_{\ell j}^{-1}$ for any $j \in [n]$.*

Lemma 5. *Let $\boldsymbol{\alpha}$ be a multi-index over $[n]$ such that $|\boldsymbol{\alpha}| = k \geq 1$. Then for any $j \in [n]$, $D_{e_j}^1(H_{\boldsymbol{\alpha}}) = \frac{\partial H_{\boldsymbol{\alpha}}}{\partial \xi_j} = \sum_{i=1}^n \alpha_i \Sigma_{ij}^{-1} H_{\boldsymbol{\alpha} - e_i}$.*

Lemma 6. *Let $\boldsymbol{\alpha}$ be an order-k multi-index for $k \geq 1$ over $[n]$. Let f be a generic function of the indices $[n]$. Then $\sum_{i=1}^n \alpha_i f(i) = \sum_{i=1}^k f(s(\boldsymbol{\alpha})_i)$.*

Lemma 7. *Let A, B be finite sets such that $|A| = |B| = k \geq 1$. Let $M(A, B)$ denote the set of bijections between A and B. Then for fixed $b \in B$ and an arbitrary $h : A \times B \to \mathbb{R}$,*

$$\sum_{p \in S_k} \prod_{\ell=1}^k h(A_\ell, B_{p(\ell)}) = \sum_{f \in M(A,B)} \prod_{a \in A} h(a, f(a)) \tag{12}$$

$$= \sum_{\ell=1}^k h(A_\ell, b) \sum_{g \in M(A \setminus \{A_\ell\}, B \setminus \{b\})} \prod_{a \in A \setminus \{A_\ell\}} h(a, g(a)) \tag{13}$$

where S_k is the symmetric group of permutations on $[k] = \{1, \ldots, k\}$.

References

1. Cameron, R.H., Martin, W.T.: The orthogonal development of non-linear functionals in series of Fourier-Hermite functionals. Ann. Math. **48**(2), 385–392 (1947)
2. Constantine, P.: Spectral methods for parametrized matrix equations. Ph.D. thesis, Stanford University (2009)
3. Diaconis, P., Gangolli, A.: Rectangular arrays with fixed margins. In: Aldous, D., Diaconis, P., Spencer, J., Steele, J.M. (eds.) Discrete Probability and Algorithms, pp. 15–41. Springer, New York (1995). https://doi.org/10.1007/978-1-4612-0801-3_3
4. Dittmer, S.: Counting linear extensions and contingency tables. Ph.D. thesis, University of California, Los Angeles (2019)
5. Ernst, O.G., Mugler, A., Starkloff, H., Ullmann, E.: On the convergence of generalized polynomial chaos expansions. ESAIM: M2AN **46**(2), 317–339 (2012)
6. Gail, M., Mantel, N.: Counting the number of $r \times c$ contingency tables with fixed margins. J. Am. Stat. Assoc. **72**(360), 859–862 (1977)
7. Ghanem, R.G., Spanos, P.D.: Stochastic Finite Elements: A Spectral Approach. Springer, Heidelberg (1991)
8. Hickman, J.: A note on the concept of multiset. Bull. Australian Math. Soc. **22**(2), 211–217 (1980)

9. Janson, S.: Gaussian Hilbert Spaces. Cambridge Tracts in Mathematics, Cambridge University Press (1997). https://doi.org/10.1017/CBO9780511526169
10. Knuth, D.E.: The Art of Computer Programming: A Draft of Section 7.2.1.1, Generating all n-Tuples. Addison-Wesley (06 2004)
11. Lovasz, L., Pelikan, J., Vesztergombi, K.: Discrete Mathematics: Elementary and Beyond. Springer (2003)
12. Lyman, L., Iaccarino, G.: Extending bluff-and-fix estimates for polynomial chaos expansions. J. Comput. Sci. **50**, 101287 (2021)
13. Noreddine, S., Nourdin, I.: On the gaussian approximation of vector-valued multiple integrals. J. Multivariate Anal. **102**(6), 1008–1017 (2011)
14. Owen, A.B.: Monte Carlo theory, methods and examples (2013). https://statweb.stanford.edu/~owen/mc/
15. Rahman, S.: Wiener-Hermite polynomial expansion for multivariate Gaussian probability measures. J. Math. Anal. Appl. **454**(1), 303–334 (2017)
16. Ruskey, F.: Combinatorial Generation. Preliminary working draft. University of Victoria, Victoria, BC, Canada, pp. 71–73, §4.5.1 (2003)
17. Slepian, D.: On the symmetrized kronecker power of a matrix and extensions of Mehler's formula for hermite polynomials. SIAM J. Math. Anal. **3**(4), 606–616 (1972)
18. Stanley, R.P.: Enumerative Combinatorics, vol. 2. Cambridge University Press, Cambridge (2001)
19. Szegő, G.: Orthogonal Polynomials. Colloquium publ, American Mathematical Society, American Math. Soc (1975)
20. Van Rossum, G.: The Python Library Reference, release 3.8.2. Python Software Foundation (2020)
21. Wiener, N.: The homogeneous chaos. Am. J. Math. **60**(4), 897–936 (1938)
22. Xiu, D.: Generalized (Wiener-Askey) Polynomial Chaos. Ph.D. thesis, Brown University (2004)

Author Index

Printed in the United States
by Baker & Taylor Publisher Services